ANNALS OF THE NEW YORK ACADEMY OF SCIENCES

Volume 757

EDITORIAL STAFF
Executive Editor
BILL BOLAND
Managing Editor
JUSTINE CULLINAN
Associate Editor
MARION L. GARRY

The New York Academy of Sciences
2 East 63rd Street
New York, New York 10021

DIVERSITY OF INTERACTING RECEPTORS

ANNALS OF THE NEW YORK ACADEMY OF SCIENCES
Volume 757

DIVERSITY OF INTERACTING RECEPTORS

Edited by Leo G. Abood and Abel Lajtha

The New York Academy of Sciences
New York, New York
1995

⊗ The paper used in this publication meets the minimum requirements of the American National Standard for Information Sciences—Permanence of Paper for Printed Library Materials, ANSI Z39.48-1984.

Cover illustration: The cascade of effects caused by the release of a neurotransmitter.

Library of Congress Cataloging-in-Publication Data

Diversity of interacting receptors/edited by Leo G. Abood and Abel Lajtha.
    p.   cm. — (Annals of the New York Academy of Sciences, ISSN 0077-8923; v. 757)
    Papers presented at a conference entitled Functional Diversity of Interacting Receptors, held in Washington, D.C. on May 25–28, 1994.
    ISBN 0-89766-923-1 (cloth). — ISBN 0-89766-924-X (paper)
    1. Neurotransmitter receptors—Congresses. 2. Cell receptors--Congresses. I. Abood, Leo G. II. Lajtha, Abel. III. Series.
    Q11.N5 vol. 757
    [QP364.7]
    500 s—dc20
    [591.1'88]

          95-11515
          CIP

SP
Printed in the United States of America
ISBN 0-89766-923-1 (cloth)
ISBN 0-89766-924-X (paper)
ISSN 0077-8923

ANNALS OF THE NEW YORK ACADEMY OF SCIENCES

Volume 757
May 10, 1995

DIVERSITY OF INTERACTING RECEPTORS[a]

Editors and Conference Organizers
LEO G. ABOOD AND ABEL LAJTHA

CONTENTS

[a]This volume contains papers presented at a conference entitled Functional Diversity of
Interacting Receptors, which was sponsored by the New York Academy of Sciences and held in
Washington, D.C. on May 25–28, 1994.

Part VIII. Excitatory Amino Acid Receptor Modulation and Pathology

Part IX. GABA Receptor Heterogeneity

Financial assistance was received from:

Major Funder
- THE COUNCIL FOR TOBACCO RESEARCH-U.S.A., INC.

Supporter
- U.S. DEPARTMENT OF THE ARMY

Contributors
- ABBOTT LABORATORIES
- AMERICAN CYANAMID COMPANY
- BRISTOL-MYERS SQUIBB COMPANY
- FISONS PHARMACEUTICALS
- HOFFMANN-LA ROCHE INC.
- PARKE-DAVIS PHARMACEUTICAL RESEARCH
- THE R. W. JOHNSON PHARMACEUTICAL RESEARCH INSTITUTE
- THE UPJOHN COMPANY

Background and Overview
of the Conference

LEO G. ABOOD

Department of Pharmacology
University of Rochester Medical Center
Rochester, New York 14642

BACKGROUND

The discovery that the pharmacologic effects of acetylcholine (ACh) and related agents in the autonomic system were mimicked by muscarine or nicotine,[1] followed by the discovery that ACh was released by the frog vagus nerve,[2] introduced the seminal concepts of chemical transmission and the finding that a single chemical transmitter exhibited multiple functions at distinct effector sites. Since that time the number of neurotransmitters and other chemical mediators along with their multiple receptor subtypes has grown at a phenomenal pace. One of the most challenging problems confronting the biologist is the functional implication of multiple forms of receptors and their interactions. The following are examples of the diversity of multiple receptor subtypes, their contributory factors, and functional implications.

- Differences in the time course of neurally elicited responses.
  Example: Fast excitatory postsynaptic potential of nicotinic receptors and slow excitatory postsynaptic potential of muscarinic cholinergic receptors.[3,4]
- Receptors linked to signal transduction systems require structural alterations for specific molecular associations and reactions.
  Example: β_1-Adrenoreceptor interacts with a stimulatory G_S-protcin to generated cAMP, whereas α_2-adrenoreceptor interacts with an inhibitory G_1-protein to attenuate cAMP production.
- Differences in receptor microenvironment and function at heterologous effector cells.
  Example: Nicotinic receptors at neuromuscular junction and ganglion cells.
- Differences in receptor microenvironment of homologous effector cells.
  Example: nAChR subtypes in various homologous effector cells.
- Differences in subcellular localization associated with functional diversity.
  Example: $GABA_B$ presynaptic autoreceptors and $GABA_A$ postsynaptic receptors.
- Differences in neuronal receptor sensitivity to neurotransmitter.
  Example: Cl^- gated channels of $GABA_A$ receptors containing $\beta_2\gamma_2$ subunits, such as the cerebral cortex, are more responsive to GABA than cells with $\beta_1\gamma_1$ subunits, which are more prevalent in the cerebellum.[5]
- Heterology of regulatory allosteric sites of receptors.
  Example: $GABA_A$ receptors in cerebellar pyramidal and granular cells comprising γ_2 subunits are sensitive to benzodiazepines, whereas cells containing α_2 subunits are insensitive.[6]
- Differences in coupling efficiency of receptors to G-proteins.
  Example: A mutation of the β_2-adrenoreceptor favors its interaction with G_S, resulting in elevated production of cAMP.[7] Spontaneous mutation of G-protein-

1

coupled receptors may result in various pathologies, as exemplified in the rhodopsin system where a point mutation in lys296, which restricts the attachment to retina, leads to increased sensitivity to light and, eventually, retinal degeneration.[8]

- Heteromeric subunit composition imparts differences in ion selectivity and/or conductance.

Example: The AMPA-glutamate receptor subtype exhibits fast conductance, low Ca^{2+} permeability, and Mg^{2+} blockade; whereas the NMDA receptor exhibits slow conductance, high Ca^{2+} permeability, and Mg^{2+} insensitivity.[9] This distinction is attributed to a single amino acid in the M2 membrane spanning segment of the receptor subunit: a glu or arg in the AMPA receptor subunit (GluRB) and an asn in the NMDA subunit (NR1 or NR2).

- Receptor subtypes capable of coupling to different G-proteins.

Example: A chicken m4 gene expressed in CHO cells exhibits forskolin-stimulated cAMP production and agonist-stimulated phosphoinositide turnover, whereas its expression in Y1 cells results in agonist-inhibited cAMP production, but no stimulation of phosphoinositide turnover.[10]

- Changes in receptor subunits during development associated with neuronal maturation.

Example: In the adult nervous system activation of the glutamate-AMPA receptor elicits a current with a fast and steady-state component, whereas in the immature brain the desensitizing component is greatly reduced or absent.[11]

- The distribution of receptor subtypes may be a mechanism for targeting receptors to distinct neurons or neuroanatomic regions.

Example: Type I sodium channels are located in pyramidal and dentate gyrus cell bodies, whereas type II channels are localized in myelinated fibers.[12]

- Variation of coupling efficiency of receptor subtype to G-proteins.

Example: The sevenfold greater D1 receptor-stimulated cAMP production in the corpus striatum as compared to the substantia nigra appears to be related to the greater concentration of G_{olf} present in the striatum.[13] Because low levels of G_S are found in the substantia nigra, it would appear that the difference is due to distinct D1 receptor subtypes.

- Variations in allosteric sites of receptor subtypes regulating excitatory parameters.

Example: The nAChRs exhibit multiple conformational states associated with activation, channel openings, and desensitization, involving sites that are distinct from the ACh recognition site.[14] Included among the physiologic and pharmacologic factors acting on such sites are divalent cations, neuropeptides, hormones, diverse neurotropic drugs, membrane potential, and phosphorylation.

- Phylogenetic changes in receptors.

Example: Neurotransmitters and neuropeptides are present in primitive organisms lacking nervous systems, where they may be linked to signal transduction mechanisms via specific receptors regulating such factors as growth, differentiation, and cell motility.[15] Receptors and their subtypes underwent further development and diversification as neural control of such functions evolved.

- Variations in subunit composition of receptor subtypes associated with differences in conductance characteristics.

Example: Studies with transfected cells expressing $\alpha1\beta2$, $\alpha1\gamma2$, and $\alpha1\beta2\gamma2$ subunits of rat $GABA_A$ receptors showed differences in the time characteristics of chloride channel conductance, depending primarily on the presence or absence of $\beta2$ and $\gamma2$ subunits. GABA-activated currents in cells expressing $\beta2$

subunits exhibited faster desensitization, greater outward rectification, and shorter mean opening time than receptors composed of $\alpha 1 \gamma 2$ subunits.[15]

OVERVIEW

The intent of this conference entitled Functional Diversity of Interacting Receptors was to contribute to a better understanding of the functional significance of receptor diversity, by addressing such topics as the genetic regulation and expression of receptor subtypes; the differential functional characteristics of neurons with distinct subtypes; second messenger and ion channel diversity associated with receptor isoforms; species and regional neuroanatomic differences in receptor subtypes; receptor changes associated with chronic exposure to agonists and antagonists and occurring with age and pathologic states; and the interaction of different receptors in functional regulation. Inasmuch as our knowledge of the structural, genetic, and functional features of the nicotinic cholinergic receptors appears to be more advanced than that of other receptors, the emphasis of the conference was on cholinergic receptors and their interaction with other receptors. Also included were presentations dealing with adrenergic, dopaminergic, histaminergic, gabaminergic, excitatory amino acid, and peptidergic receptors and their interactions and functional implications in health and disease.

The conference began with a presentation reviewing recent findings on the nature of ion channels associated with the various neuronal nAChR subtypes and describing the use of specific ligands for probing ion channels and nicotine's binding site. A discussion of the structure-activity relationships of novel subtype-selective opioid peptides was presented to illustrate the principles involved in the design of agonists and antagonists. With the use of synthetic peptides related to the α subunit of the nAChR, it was shown that α-bungarotoxin binds to amino acids in the 180–200 region (*Torpedo* numbering) of the α subunit, whereas κ-bungarotoxin binds to region 51–70 of the neuronal α_3 subunit. Two-dimensional [^{1}H]NMR studies on complexes formed between α-bungarotoxin and 18-mer peptide, corresponding to 185–196 sequence of the *Torpedo* α subunit, demonstrated the involvement of H186, W187, Y189, and Y190 in the contact zone. Patch clamp studies on cultured rat hippocampal neurons revealed the presence of two nAChR receptors with different properties: an α-bungarotoxin–sensitive one, which was shown by *in situ* hybridization to contain an α_7 subunit, and $\alpha_4\beta_2$.

Presynaptic nAChRs were shown to be involved in the release of norepinephrine (NE) from the CA1, CA3, and DG regions of the hippocampus, in the modulation of NE release from vas deferens and in the release of acetylcholine from the neuromuscular junction. The hippocampal nAChR was believed to comprise the $\alpha_3\beta_2$ subunits. With the use of photoreactive agonists for nicotinic cholinergic and other neurotransmitter receptors, a novel laser-pulse photolysis method with a microsecond time resolution was used to determine the rate constants for ion channel opening and closing, the concentration of the open channel, and the binding constants of inhibitors to sites on both the open and closed channels. The combined use of immunochemical, histochemical, and autoradiographic techniques was described for mapping the muscarinic cholinergic projections from the nucleus basalis to the thalamus and cerebral cortex, from the prepeduncular pontine nucleus to the thalamus, and other projections involving the sensory-limbic and reticular activating systems. By co-transfecting a cAMP response element-driven reporter gene with various muscarinic receptor genes, it was demonstrated that both the m_1 and m_4

increase cAMP production, whereas high levels of expression of m_4 cause agonist-independent inhibition of adenylate cyclase via $G_{i\alpha2}$.

Monoclonal antibodies to the various α and β subunits of nAChR were used to demonstrate that ganglia and retina contain uncertain combinations of α_5, β_2, and β_3 subunits, and brain, mainly the stoichiometry $(\alpha_4)2(\beta_2)3$ as well as two functional α-bungarotoxin-binding subunits, α_7 and α_8. A study of patterns of regulation of nAChR subtypes in various transfected cell lines revealed that ganglionic nAChR mRNA and function are stimulated by nerve growth factor, whereas nicotine exposure increased muscle nAChR numbers but with a loss of function. A presentation on brain nicotinic receptors and cholinergic transmission raised issues concerning (1) the relationship of cholinergic innervation to brain nicotinic cholinergic receptors as determined by such techniques as autoradiography, immunocytochemistry, and lesioning brain pathways; and (2) the functional significance of brain nicotinic receptors.

The afternoon session of the second day was devoted to the functional interaction of various neurotransmitter and neuropeptidergic receptors, describing their neuroanatomic distribution, subtype specificity, and their pathologic-therapeutic implications. Despite the fact that all three β-adrenergic receptors were structurally similar and bound to the same trimeric G_S-protein coupled positively to adenylate cyclase, they exhibited striking differences in their agonist-antagonist profile and regulation, which were attributed to specific point mutations. A number of presentations referred to the role of antipsychotic drugs in elevating brain neurotensin in the neostriatum. On the basis of findings demonstrating that haloperidol both decreased and increased neurotensin mRNA—the former via D_3 and the latter via D_2 receptors in the nucleus accumbens—it was inferred that the antipsychotic drugs alleviated positive symptoms (e.g., hallucinations, anxiety) by acting on D_2 receptors, whereas negative symptoms (e.g., stereopoty, increased self-stimulation) appear to involve D_3 receptors. Neuroanatomic combined with receptor binding studies demonstrated interactions between dopamine receptors and cholecystokinin 8, which produces behavioral effects (sedation, catalepsy, antistereopoty) opposite to those of dopamine. Localization of the peptide binding domain of the substance P receptor was performed by attachment of a photolabile amino acid, p-benzoyl-L-phenylalanine, to positions 4 and 8 of substance P and structural analysis of the photolabeled substance P receptor fragments isolated from receptor-transfected CHO cells.

Renewed interest in the functional role of histamine in the central nervous system has resulted from the cloning and expression of the H_1, H_2, and H_3 receptors. The H_1 receptor increased cAMP, inositol phosphates, and arachidonic acid, whereas the H_2 receptor increased cAMP and decreased arachidonic acid production. 1-Methylhistamine, which has a 1500-fold greater affinity than histamine for the H_3 receptor, may be an endogenous candidate for the neuronal H_3 autoreceptor.

The presentations on opioid receptors dealt with topics ranging from the purification of a μ-receptor, cloning and expression of κ-receptors, structure-activity relationships of subtype-selective opioid peptides, to receptor changes after chronic exposure to opiate agonists and antagonists. With the cloning of the various opioid receptor subtypes the techniques of recombinant expression, immunocytochemistry, and *in situ* hybridization have been used to elucidate the functional as well as subcellular and neuroanatomic localization of subtypes of opioid receptors in brain and spinal cord. Among the novel findings were that δ-receptor is found mainly on axons, the μ-receptor on plasma membranes of axons, cell bodies, and dendrite, and that the enkephalin-containing terminals are proximal to δ- or μ-receptors.

Molecular genetic studies of the melanocortin (MSH) receptor revealed it to be a unique bifunctionally controlled receptor, positively regulated by MSH, resulting in brown-black pigmentation, and negatively regulated by the agouti peptide, resulting in variable pigmentation in coat color.

The final session dealt with the structural, neuroanatomic, functional, and clinical aspects of the various excitatory amino acid (EAA) and GABA receptors. To date three subtypes of EAA receptors are known: a quisqulate- or AMPA-sensitive one existing in four isoforms with 70% homology, a kainate-sensitive one with 75% homology, and a NMDA-sensitive one existing in five isoforms with 25% homology. Studies with selective agonists and antagonists suggest that the NMDA receptor is implicated in brain damage associated with status epilepticus, the quisqulate receptor with general anesthesia, and the NMDA receptor with head injury. The release of GABA from cultured retinal cells was mediated by glutamate receptors and found to be calcium-dependent and blocked by nifendipine. Bilateral lesioning of connections between the rat temporal cortex and the lateral entorhinal cortex resulted in an impairment of retroactive and proactive memory and a reduction in glutamate receptors and loss of terminals in both regions. In addition to its involvement in the release of intracellular Ca^{2+}, EAA receptor-mediated formation of inositol phosphates (IP) stimulates Ca^{2+} entry in synaptosomes via a voltage-gated channel, leading to a further increase in IP.

Following the demonstration that the memory deficit in rats was attenuated by cycloserine, a glycine-receptor agonist, clinical trials were undertaken in Alzheimer patients with encouraging results. NMDA receptors appear to be involved in the negative symptoms associated with schizophrenia that are not responsive to antidopaminergic neuroleptics, but appear, in preliminary clinical trials, to be attenuated by large doses of glycine. The hypothesis that nicotine interacts with the mesolimbic dopaminergic system to improve working memory gained additional support from studies demonstrating that the administration of dopamine agonists along with nicotine improved performance of rats in a radial maze.

Investigations of the subunit composition of the $GABA_A$ receptor by immunocytochemical and *in situ* hybridization techniques revealed the existence of multiple forms in all brain areas with the major subunit composition $\alpha_1 \beta_2 \gamma_2$. Bergmann glia cells of the cerebellum were found to express γ_1 and α_2. A monoclonal antibody for the $GABA_B$ receptor was shown to inhibit the binding of agonists to the receptor and to prevent the GABA-mediated inhibition of adenylate cyclase in cerebral synaptic membrane preparations.

In order to account for the shift in the control of hepatic glycogenolysis from α_{1B} to the inhibitory β_2-adrenergic receptors following hepatic injury or malignancies, a mechanism was proposed involving the translocation of protein kinase C (PKC) from the cytoplasm to the plasma membrane. In addition to PKC, which is involved in the uncoupling of the α_{1B}-receptor and coupling of β_2-receptor, arachidonic acid appears to play a role in the conversion.

The poster sessions included such topics as the modulations of the m_2 muscarinic receptor by nitric oxide; autoregulation of ACh release in rat cerebrum by both a stimulatory m_5 and inhibitory m_1 muscarinic receptor subtype; α_2-adrenergic regulation of body temperature via inhibition of warm-sensitive and inhibition of cold-sensitive hypothalamic neurons and by interaction with serotonergic and dopaminergic systems; interference by peptide YY of cholecystokinin's inhibition of pancreatic secretion; antagonism by polyamines of imipramine-induced immobility in rats; inability to account for the hypotensive action of methionine enkephalin by changes in tyrosine hydroxlase and catecholamine metabolizing enzymes; advantages of combined agonist-antagonist administration in smoking cessation therapy; detection of the Tax gene in human macrophages infected with HTLV-1; expression and functional characterization of an 140–204 amino acid fragment of the α subunit of nAChR; and determination of glutamate receptor subtypes involved in calcium influx and the modulation of dopamine release in hippocampal synaptosomes.

REFERENCES

1. DALE, H. H. 1914. The action of certain esters and ethers of choline and their relation to muscarine. J. Pharmacol. Exp. Ther. **6:** 147–190.
2. LOEWI, O. & E. NAVARITIL. 1926. Uber humorale Ubertragharkeit der Herznervenwirkung. X. Mitteilung, Uber der Schicksal das Vagusstoff. Pfluegers Arch. Gesamte Physiol. **214:** 678–688.
3. ECCLES, R. M. & B. LIBET. 1961. Origin and blockade of the synaptic responses of curarized sympatheti ganglia. J. Physiol. (Lond.) **157:** 484–503.
4. LIBET, B. 1991. Introduction to slow synaptic potentials and their neuromodulation by dopamine. Can. J. Physiol. **70:** S3–S11.
5. DUCIC, I., S. VICINI & E. COSTA. 1993. Efficacy and potency of Gaba action in native recombinant Gaba$_A$ receptors. Neurosci. Abstr. **19:** 851.
6. FRISCHY, J. M., D. BENKE, S. MERTEN, B. GORO & H. MOHLER. 1993. Immunochemical distinction of GabaA receptor subtypes in drug binding profiles and cellular distribution. Neurosci. Abstr. **19:** 476.
7. RATHOUZ, M. M. & D. K. BERG. 1993. Calcium permeability of nicotinic receptors located primarily at synapses on neurons. Neurosci. Abstr. **19:** 464.
8. SAMANA, P., S. COTECCHIA, T. COSTA & R. J. LEFKOWITZ. 1993. A mutation-induced activated state of the beta 2-adrenergic receptor. J. Biol. Chem. **268:** 4625–4636.
9. ROBINSON, P. R., G. B. COHEN, E. A. ZHUKOWSKY & D. D. OPRIAN. 1992. Constitutively active mutants of rhodopsin. Neuron **9:** 719–725.
10. DINGLEDINE, R., R. I. HUME & S. P. HEINEMANN. 1991. Identification of a site in glutamate receptor subunits that control calcium permeability. Science **253:** 1028–1031.
11. TEITJE, K. N., P. S. GOLDMAN & N. M. NATHANSON. 1990. Cloning and functional analysis of a gene encoding a novel muscarinic acetylcholine receptor expressed in chick heart and brain. J. Biol. Chem. **265:** 2828–2834.
12. SOMMER, B., K. KEINÄNEN, T. A. VERDOORN, W. WISDEN, N. BURNASHEV, A. HERB, M. KÖHLER, T. TAKAGI, B. SAKMANN & P. H. SEEBURG. 1990. Flip-flop: A cell-specific functional switch in glutamate operated channels of the CNS. Science **249:** 1580–1585.
13. WESTENBROEK, R. E., D. K. MERRICK & W. A. CATTERALL. 1989. Differential subcellular localization of the RI and RII Na$^+$ channel subtypes in central neurons. Neuron **3:** 695–704.
14. HERVE, D., M. LEVI-STRAUSS, I. MAREY-SEMPER, V. VERNEY, J-P. TESSIN, B. GIOWINSKI & J-A. GIRAULT. 1993. G$_{olf}$ and G$_S$ in rat basal ganglia: Possible involvement of G$_{olf}$ in the coupling of dopamine D1 receptor and adenylyl cyclase. J. Neurosci. **13:** 2237–2248.
15. LENA, C. & J-P. CHANGEUX. 1993. Allosteric modulations of the nicotinic acetylcholine receptor. Trends Neurosci. **16:** 181–186.
16. LANUDER, J. M. 1993. Neurotransmitters as growth regulatory signals: Role of receptors and second messengers. Trends Neurosci. **16:** 233–240.
17. VERDOORN, T. A., A. DRAGUHN, S. YMER, P. H. SEEBURG & B. SAKMANN. 1990. Functional properties of recombinant rat GabaA receptors depend upon subunit composition. Neuron **4:** 919–928.

Molecular Organization of Receptors

Efficacy, Agonists, and Antagonists[a]

VICTOR J. HRUBY,[b] HENRY I. YAMAMURA,[c]
AND FRANK PORRECA[c]

[b]Department of Chemistry
The University of Arizona
Tucson, Arizona 85721

[c]Department of Pharmacology
The University of Arizona
Tucson, Arizona 85724

Intercellular and intracellular communication is of critical importance to the organization, differentiation, and coordination of all complex living systems. The need to maintain homeostasis and adjust to a changing environment is essential to the well-being of any organism. Complex living systems have adapted to their environments and developed highly differentiated cells with specific specialized functions by established communication networks that coordinate the structures and functions necessary for life. In general, the cells utilize a variety of chemical structures that serve as chemical messages to facilitate information transfer. These structures can be quite simple, such as, for example, glutamate, a simple amino acid, or alternatively they can be complex such as the wide variety of hormones and neurotransmitters of which insulin is an example. In general, the message is a signal for the receiving cell to modify or modulate its properties, and most commonly the messenger (hormone, neurotransmitter, growth factor, cytokin, etc.) manifests its effect by interaction with a cell surface receptor or acceptor molecule that generally is a macromolecule such as a protein or glycoprotein. This interaction generally leads to a change in the three-dimensional structure of the receptor. The conformational change in the receptor that accompanies formation of the ligand-receptor complex is the stimulus necessary to trigger a variety of chemical and physical events in the cell such as alternations in enzymatic activity, metabolism, ion channel properties, gene expression, and many other biochemical events. Interestingly, many hormones, neurotransmitters, and other messenger molecules are not highly selective and interact with a variety of receptor types and subtypes (TABLE 1). This promiscuity is also a useful tool for efficiency in the biological system, but presents difficulties in trying to understand the relationship(s) between the structure of a chemical messenger and its biological activity. The situation is further complicated by the fact that often a particular ligand which can interact with a well-defined receptor that has a particular function also can interact with other related receptors; indeed, this is often the case. Various types or subtypes of receptors have been postulated to exist, and in recent years these ideas have been proven by the cloning and expression of multiple receptor types and subtypes (TABLE 2). Although the examples in TABLES 1 and 2 are not comprehensive, they illustrate the complex way in which hormones, neurotrans-

[a]This work was supported by grants from the U.S. Public Health Service (DK 17420 and NS 19972) and the National Institute on Drug Abuse (DA 06284, DA 04248, and DA 08657).

TABLE 1. Examples of Endogenous Ligands That Interact with Multiple Receptor Types and Subtypes

Ligand	Receptor Type or Subtype
β-Endorphin	μ, κ, and δ (and ε) Opioid receptors
Dynorphin	κ, μ, and δ Opioid receptors
α-Melanotropin	MC1, MC3, MC4, MC5 Receptors
Somatostatin	Somatostatin receptors, μ-/ε-opioid receptors
Cholecystokinin	CCK-A and CCK-B Receptors

mitters, and other biological messages interact with a variety of different receptors to modulate and modify biological function.

In fact, the interactions are more comprehensive than implied by these TABLES. Once the interaction of a hormone or neurotransmitter ligand with its receptors has occurred, a variety of further interactions and changes occur. These effects can be viewed from several perspectives. For example, the interaction may lead to coupling of the receptors with various membrane-associated molecules such as ion channels whose properties themselves become modified, leading to changes in cellular functions or properties. Alternatively signal transduction may occur. The classical example is the modulation of the interaction of the receptor with G-proteins to eventually produce molecules which themselves act as signals such as cyclic-adenosine monophosphate (c-AMP), phospholipid metabolites, etc. These in turn activate other enzymes or proteins eventually producing regulation of cell functions. Also, receptor-ligand interactions can lead to dynamic trafficking events in which the receptor-ligand complex is internalized into the cell via endocytosis. In some cases, this process occurs in addition to the signal transduction process or the receptor interaction with membrane-associated molecules. These internalized receptor complexes can elicit or alter various cellular responses in the short term, and in the long term can have dramatic effects because the receptors have been removed from the cell surface, reducing the receptor number and dramatically modifying the possibility for ligand-receptor interactions (down-regulation; the opposite effect also can occur in some cases, that is, up-regulation).

Various time domains are also associated with the ligand-receptor interactions which complicate matters. There are short-term responses on the order of millisecond to minutes due to rapid changes in ion concentrations and other metabolic processes, as well as long-term responses that can take minutes to hours, including those that require protein modification or even protein synthesis. Thus, a variety of

TABLE 2. Selected Neurotransmitter and Hormone Receptors and Their Proposed Receptor Types and Subtypes

Receptor	Types or Subtypes
Opioid	3 Types: μ, δ, κ (5 or more subtypes)
Adrenergic	α_1 (3), α_2 (3)
	β_1, β_2, β_3
Cholecystokinin	CCK_A, CCK_B
Histamine	H_1, H_2, H_3
Dopamine	D1, D2, D3, D4, D5
Adenosine	A_1, A_2(2), A_3, A_4
Tackykinin	NK1, NK2, NK3
Glutamate	$mGluR_1$, $mGluR_2$, $mGluR_3$, $mGluR_4$, $mGluR_5$

regulatory processes in the cell can be modified, and each will depend on complex kinetic and thermodynamic processes. At the moment, the complexity of what previously seemed like a rather "simple" pharmacological or physiological process appears to be increasing daily as new factors, enzymes, protein domains, and substrates increase. On the other hand, it seems clear that the cellular responses to ligand-receptor interactions as understood in terms of chemistry and structure open a new world of opportunity for understanding normal and disease processes and new possibilities for the treatment of disease.

Obviously, this is an exciting and rapidly expanding area of science, and one cannot possibly do justice to it in a short paper. The interested reader is referred to the literature and many excellent books on the subject[1-3] for a more comprehensive treatment. In this paper we briefly discuss the molecular organization of receptors from the standpoint of ligand-receptor interactions in particular, and then concentrate on the design of ligands for obtaining potent and selective agonists and antagonists, with particular emphasis on design considerations that lead to selectivity and high agonist and antagonist potency. Thus, we concentrate primarily on the ligands. Though many transmembrane receptors have been cloned and expressed and are available for functional studies, they still are not available in quantities sufficient for biophysical studies, and it appears that it will be some time before they will be available because of difficulties of purifying membrane-bound proteins and glycoproteins.

GENERAL CONSIDERATIONS

Efforts to understand the relationship between ligand structure and conformation and the dynamics and thermodynamics of the various binding events, as well as the biochemical changes they induce, are critical to future progress. This requires a highly interdisciplinary approach involving aspects of synthetic chemistry including asymmetric synthesis, computer-aided molecular design, biophysical studies of conformation and dynamics in the context of structure-activity relationships, and comprehensive multiple bioassay systems, so that aspects of potency, agonist/antagonist biological activity, and efficacy can be understood in terms of ligand structure, conformation, and topography. FIGURE 1 illustrates the interrelationships of each component of this comprehensive approach. If any component is left out it will be difficult to develop a successful approach.

Synthetic considerations are of primary importance because any effort to obtain highly potent and selective ligands will require a well-developed approach to synthesize designed ligands and then the ability to readily modify them for structure-activity studies, but also for studies of conformation and stability-bioavailability. Often these two latter kinds of studies require highly specialized amino acids, heterocyclics or other structures that contain specific isotopes or radioisotopes that need to be incorporated and that have minimal or no effects on biological properties. Furthermore, the current demands of science and medicine often require that the ligand of interest be prepared in a pure chiral state. This requires development of asymmetric synthesis or chiral resolution methods.

The design of ligands today generally is a combination of the classical medicinal chemistry approach of systematic modification of lead structures, and extensive use of biophysical studies (x-ray crystallography, nuclear magnetic resonance [NMR] spectroscopy, circular dichroism [CD] spectroscopy, fluorescence spectroscopy, and many other biophysical tools) in combination with computer-assisted modeling, molecular mechanics and quantum mechanics calculations, and molecular dynamics studies. Since the major focus of our examples is peptide ligands for protein receptors, we consider the process from two levels, first, that of peptide structure and, second, that of peptide and peptidomimetic design. FIGURE 2 illustrates the principal conformational properties of the backbone of a polypeptide that is composed of ϕ, ψ, and ω torsion angles. The peptide bond is generally *trans* (180°) but also can be *cis* for X-proline and X-*N*-alkylated amino acid bonds. Thus the peptide backbone conformation is primarily a function of the ϕ and ψ angles, and Ramachandran[4,5] and co-workers showed many years ago that the low-energy secondary structures for peptides (α-helices, β-sheets, β-turns, extended conformations, etc.) could be defined by the ϕ and ψ angles and hence ϕ,ψ space often is referred to as Ramachandran space.[6] Given a particular peptide, careful analysis of ϕ,Ψ space can provide insights into the likely secondary structures for that peptide. This in turn can provide a starting point for further molecular design focused on enhancing some

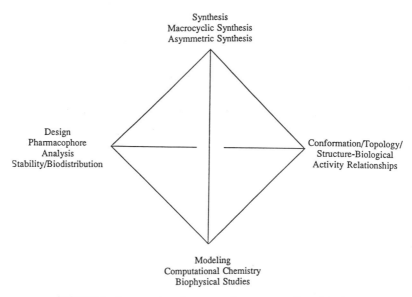

FIGURE 1. A comprehensive approach to receptor ligand design.

secondary structure and conformational property believed to be important for a particular biological effect.

In addition, one must carefully consider the side chain groups on each amino acid residue. Questions regarding their specific requirements for a particular receptor often can be addressed by an alanine scan or a glycine scan. In this approach, each amino acid residue is replaced one at a time either by Ala or by Gly, and the effect of each replacement examined in a binding assay or bioassay. For those compounds that remain highly potent, it is concluded that the side chain group of the particular residue substituted is not important for biological activity, whereas for those compounds that lose all activity, it is assumed to be very important. For those of intermediate biological activity, judgment is reserved. The major caveat for this approach is that if a large conformational change is induced by the substitution, the change in biological activity or potency may not properly reflect the actual situation.

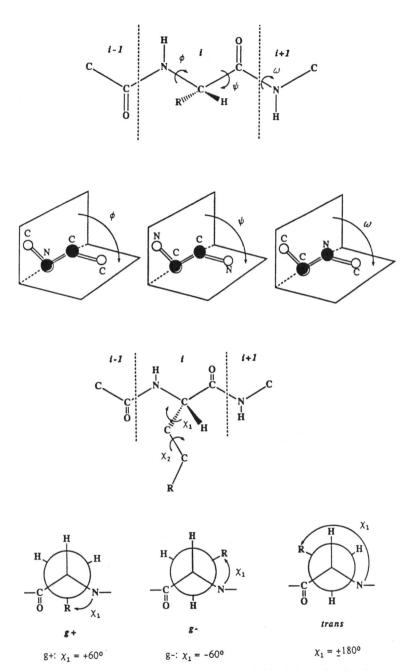

FIGURE 2. Definition of peptide backbone and side chain torsional angles.

Once specific side chains have been recognized as critical, the question arises as to which side chain conformation is important for a particular ligand-receptor interaction. FIGURE 2 defines the three main low-energy side chain conformations for most α-amino acids in a peptide: gauche(−) [g−], gauche(+) [g+], and *trans,* respectively. The chi angles are defined in FIGURE 2. Similar definitions can be given for x_2 and for further removed side chain moieties. Examination of side chain conformational space (often referred to as chi space) is of more recent origin (see, for example, refs. 7 and 8) and can be very illuminating. Recent studies have demonstrated that conformational constraints in chi space can have a profound effect on peptide and peptidomimetic biological potencies and selectivities.[7,9–12] The use of constraints in chi space is expected to play an increasingly important role in peptide and peptidomimetic design.

It is important to emphasize that the developments of the past 15 years have made it possible to develop a systematic approach to peptide and peptidomimetic design (TABLE 3).[12–17] The major goal of this approach is to define a specific

TABLE 3. Steps in Peptidomimetic Design

Define target
  Establish multiple assay systems with positive and negative controls
Obtain peptide lead
  Native ligand
  Peptide libraries
Define key residues for molecular recognition and transduction (often different for agonists and antagonists)
  Consensus sequence
  Discontinuous epitope; address/message
  Alanine scan; D-amino acid scan; etc.
Define pharmacophore
  Local constraints
  Global constraints–Built-in stability
  Topographical constraints–Chi space
  Agonist versus Antagonist
  Selectivity
  Efficacy
Design pseudopeptide, peptoid or nonpeptide scaffold
  Surface and presentation platform
Optimize molecular recognition motif
Fine-tune for selectivity, potency, bioavailability, etc.

pharmacophore of a particular receptor or acceptor molecule, and evaluate its validity by specific design of a ligand with predictable agonist or antagonist activities.

It should be clear that ongoing evaluations of biological activities including potency, selectivity, efficacy, agonist activity, and antagonist activity are critical for the success of any ligand design process. Inasmuch as most native peptide ligands are not selective for one receptor type or subtype, it is critical to develop multiple bioassays and binding assays to be successful. There are now many examples of obtaining highly selective ligands for receptors from a nonselective lead structure, and this only is possible if multiple binding and bioassays are used. As for the assay, it should be remembered that a good assay is a chemical experiment in that valid results can be obtained only if the experiment is done at equilibrium and with full consideration of the laws of thermodynamics. The use of proper controls and the elimination of nonspecific binding effects and nonspecific biological effects are also critical for success.

We now turn to a few specific illustrations of what is possible in the area of design of potent receptor selective ligands. We will emphasize studies from our laboratory with particular emphasis on ligands for the opioid receptors, especially the δ-opioid receptor type and its subtypes. First we will examine agonist activities, and then turn to antagonists.

POTENT AND SELECTIVE AGONISTS

Most naturally occurring hormones, neurotransmitters, growth factors, and other chemical messengers are agonists in interactions with their endogenous receptors, that is, they tend to induce or stimulate some specific activation response in the targeted cell. As previously discussed, most of these compounds do not have high selectivity for specific receptor types and subtypes related to their putative major site of biological activity. Thus, it often is difficult to sort out which biological effects are of primary physiological importance and which are of lesser importance. In view of these problems, it was necessary to develop much more selective ligands for specific receptors. A major hypothesis that we used to develop a systematic and rational approach to the problem is that each receptor type and subtype has specific and different stereostructural and conformational requirements for the ligands. To test this hypothesis requires the development of methods for introducing conformational and topographical constraints.[13–18] Often this approach has led to the development of highly potent and receptor-selective ligands. To illustrate this approach, we use the example of the development of constrained enkephalin and deltorphin/dermenkephalin analogues (TABLE 4).

In our initial approach with enkephalin we used pseudoisosteric cyclization[19] in which the side chain of the Met[5] residue in methionine enkephalin was substituted for a disulfide bridge that was attached to the α-carbon of Gly[2], and then to further constrain the cyclic 13-membered ring by use of geminal dimethyl groups on the β-carbons of both half-cysteine residues. This design led to the ligands of c[D-Pen[2], D-Pen[5]]enkephalin (DPDPE) and c[D-Pen[2], L-Pen[5]]enkephalin (DPLPE).[20] These cyclic peptidomimetic ligands of enkephalin were found to be highly potent ligands and to possess highly δ-opioid receptor selectivity because of their greatly reduced potency at μ-opioid receptors and their virtual inability to bind to κ-opioid receptors.[20,21] Comprehensive biophysical studies using two-dimensional NMR spectroscopy, molecular mechanics calculations, molecular dynamics simulations, and computer modeling[22,23] led to a proposed conformation in which the two aromatic residues were located on one lipophilic surface that was believed to be recognized by the δ-opioid receptor, but not by the μ- or κ-receptors. The question then arose as to what was the optimal topography of the aromatic side chain groups. This was particularly important because structure-activity studies[24] showed that both aromatic residues were important to the biological activity of these analogues. This has taken on added interest since the determination of the x-ray crystal structure of DPDPE,[25] which shows that the cyclic portion of the DPDPE structure in the crystal is very similar to that proposed in solution using NMR, but that significant differences exist in the conformation of the aromatic residues, especially of Tyr.

Stereoelectronic effects of the Phe[4] side chain of DPDPE were explored using a variety of parasubstituted groups.[26] Compounds with much higher potency and selectivity for the δ-opioid receptor were obtained, with the [p-BrPhe[4]]DPDPE (TABLES 5 and 6) being especially selective and potent. We then explored chi space

TABLE 4. Structures of Naturally Occurring Delta Ligands

H-Tyr-Gly-Gly-Phe-Met-OH	Methionine enkephalin
H-Tyr-Gly-Gly-Phe-Leu-OH	Leucine enkephalin
H-Tyr-D-Ala-Phe-Asp-Val-Val-Gly-NH$_2$	Deltorphin I
H-Tyr-D-Ala-Phe-Glu-Val-Val-Gly-NH$_2$	Deltorphin II
H-Tyr-D-Met-Phe-His-Leu-Met-Asp-NH$_2$	Dermenkephalin

using a variety of constrained phenylalanine analogues such as 1,2,3,4-tetrahydroiso-quinoline-3-carboxylic acid (Tic)[27] in which the chi-1 angle takes a preferred conformation of g(+). This compound was found to lose both its potency and selectivity for δ-opioid receptors (TABLES 5 and 6). On the other hand, when all four isomers of the constrained amino acid β-methylphenylalanine (β-MePhe) were placed in position 4 of DPDPE, a large differentiation of binding and biological activities (orders of magnitude in potency)[28] was observed. One of the compounds [the (S,S)-Phe^4-containing compound], which has a preferred g(−) side chain conformation, was found to be the most potent and selective of the four isomers, and though it lost about 10-fold potency at the δ-receptor compared to DPDPE, it was about 10 times more selective in the binding assay (TABLES 5 and 6). In an effort to increase the lipophilicity of DPDPE for crossing the blood–brain barrier (BBB) and to develop a prohormone approach to DPDPE ligands (DPDPE is completely stable to biodegradation in the brain), we have found a remarkable new series of cyclic enkephalins, namely, compounds of the general structure H-Tyr-D-Pen-Gly-Phe(X)-Cys-Phe-OH ([Phe^6]DPLCE) in which X = L, Cl, F, I, etc.[29] These compounds have higher affinity than do DPDPE for δ-opioid receptors, and the p-IPhe^4-analogue is 1300-fold δ-receptor selective (TABLE 5). What is most remarkable is their potency in mouse vas deferens (MVD) bioassays for the δ-receptor where they are found to have potencies in the picomolar range. For example, [Phe^6]DPLCE has an IC$_{50}$ value of 16 picomolar and 5000-fold selectivity for the δ-receptor in the MVD assay (TABLE 6). These findings suggest that there are large differences in stereostructural requirements for the central and peripheral δ-opioid receptors.

TABLE 5. Binding Affinities and Selectivities of δ-Receptor Agonists

	Potencies (nM)		Selectivities
Compound	μ	δ	μ/δ
DPDPE	610	5.2	120
Deltorphin I	2,140	0.60	3,570
Dermenkephalin	1,900	0.47	4,000
H-Tyr-D-Pen-Gly-pBrPhe-D-Pen-OH	418	1.73	242
H-Tyr-D-Pen-Gly-Tic-D-Pen-OH	ND	ND	—
H-Tyr-D-Pen-Gly-(S,S) β-MePhe-D-Pen-OH	14,000	10	1,400
[Phe^6]DPLCE	280	1.6	200
[pIPhe^4,Phe^6]DPLCE	1,600	1.2	1,300
[(2S,3R) β-MePhe^3]Deltorphin	>72,000	2.5	>29,000
[(2S,3R) β-MePhe^3]Dermenkephalin	>70,000	2.4	>29,000
H-Tyr-D-Cys-Phe-Asp-Pen-Val-Gly-NH$_2$	3,760	2.2	1,700
H-Tyr-D-Pen-Phe-Asp-Pen-Nle-Gly-NH$_2$	55,000	4.8	11,500
[(S,S)-TMT^1]DPDPE	720	210	3.5
[(S,S)-TMT^1]Deltorphin	7,500	0.65	4,900

In the meantime, the first truly δ-opioid receptor selective natural ligands were found, the deltorphins and dermenkephalin (TABLE 4).[30,31] As seen in TABLES 5 and 6, these compounds were both more potent and more δ-opioid receptor selective than most of the designed cyclic enkephalin analogues. Given the different structures of these compounds and the cyclic enkephalins, the question arose as to why these compounds were so δ-opioid receptor selective, and what similarities and differences they might have with the cyclic enkephalins. A series of NMR studies in several laboratories showed that the deltorphins and dermenkephalin were highly flexible molecules with several possible low-energy conformations, and that these conformations were different from those of DPDPE. This led to the use of computation methods alone or in combination with NMR studies, to suggest possible topographical similarities that might account for their potent δ-opioid activities.[32-35] In the meantime, it was found that the lipophilic C-terminal tetrapeptide was critical for obtaining a potent and selective δ-ligand, because the N-terminal tripeptide (tetrapeptide) was in fact μ-opioid receptor selective (e.g., ref. 36). Thus, the C-terminal is

TABLE 6. Bioassay Potencies and Selectivities of δ-Receptor Agonists

	Potencies		
Compound	GPI	MVD	Selectivities
DPDPE	7,000	2.2	3,200
Deltorphin I	2,890	0.36	8,000
Dermenkephalin	3,400	0.28	12,000
H-Tyr-D-Pen-Gly-pBrPhe-D-Pen-OH	13,400	1.5	9,000
H-Tyr-D-Pen-Gly-Tic-D-Pen-OH	>300,000	1,500	>200
H-Tyr-D-Pen-Gly-(S,S) β-MePhe-D-Pen-OH	57,400	39	1,500
[Phe⁶]DPLCE	83	0.016	5,200
[pIPhe⁴, Phe⁶]DPLCE	640	0.30	2,100
[(2S,3R) β-MePhe³]Deltorphin	16,000	1.04	25,000
[(2S,3R) β-MePhe³]Dermenkephalin	>100,000	1.75	>57,000
H-Tyr-D-Cys-Phe-Asp-Pen-Val-Gly-NH₂	1100	0.25	4,400
H-Tyr-D-Pen-Phe-Asp-Pen-Nle-Gly-NH₂	140,000	8.8	16,000
[(S,S)-TMT¹]DPDPE	290	170	1.7
[(S,S)-TMT¹]Deltorphin I	3,900	0.70	5,600

modulating the conformational properties of the N-terminal to give high δ receptor potency and selectivity. By use of β-methylphenylalanine in position 3, it was found that the selectivities and potencies in the binding assays and bioassays[37] were greatly dependent on the side chain conformation in ways that were different from those of cyclic enkephalin analogues (TABLES 5 and 6). Furthermore, large differences were seen for potencies at μ-opioid receptors (data not shown). These results led to the design of chimeras of the cyclic enkephalins and the linear deltorphins and dermenkephalin.[38] As expected (TABLE 5 and 6), these compounds were found to be highly potent and selective.

A most intriguing finding is that the efficacy of DPDPE and deltorphin/dermenkephalin analogues vary somewhat in the peripheral bioassays.[39] Generally, less than 10% occupancy of δ-receptors is sufficient for a full biological effect. From these studies, it is clear that much more effort should be placed on studies of those structural features that promote efficacy and more effective ways to measure efficacy. Finally, it should be noted that while these studies were in progress, DPDPE and deltorphin, in functional antinociception assays, were found to be interacting with different δ-opioid receptors (see, e.g., refs. 40–45). By using δ-opioid receptor inhibitors, it was possible to block the antinoceception of one of the ligands while having no effect on the other. This led to the proposal for δ-receptor subtypes, δ_1 for DPDPE-related compounds, and δ_2 for deltorphin-related compounds. Interestingly, although clear-cut differences exist in the *in vivo* assays for δ_1 and δ_2 selective ligands, thus far no suitable radiobinding assay has been developed that can distinguish between them, even using radiolabeled δ_1 and δ_2 ligands. However, we recently have developed methods for the synthesis of tyrosine analogues such as β-methyl-2',6'-dimethyltyrosine (TMT) (all four isomers). We have incorporated the (2S,3S) isomer into DPDPE and deltorphin I and have found the deltorphin I analogue (TABLES 5 and 6) to be highly potent and selective at δ_2-receptors, whereas the corresponding DPDPE analogue lost very significant binding and biological activity potency.[12] It has been demonstrated that these compounds still act at the δ_1- and δ_2-receptors, respectively. It is hoped that a good binding ligand can be developed from the deltorphin I analogue.

POTENT AND SELECTIVE ANTAGONISTS

In general, there is no straightforward way to develop antagonists of hormone and neurotransmitter ligands from knowledge of structure-activity relationships of agonists.[45] First, much evidence exists that agonists and antagonists bind differently to receptors (e.g., ref. 46) and thus, receptor selective antagonists must initially come from leads that are discovered in assays. Because binding per se cannot distinguish between agonists and antagonists, functional assays are required that measure either a second message or a specific bioactivity in tissue in whole animals. The reason agonists and antagonists have different structure-activity relationships is that they serve quite different functions on interacting with the receptor molecule (TABLE 7). In brief, because the conformation of a receptor in its agonist state is different from that in its antagonist or (inactive) state, it follows that the ligand-receptor interaction must be different to lead to a different conformation for the complex. We will discuss a few examples from our laboratory that illustrate a few of the principles of antagonist design. TABLE 8 lists some of the selective opioid receptor antagonists that are available.

An interesting example of antagonist design, based on a lead from a *different* receptor ligand, is the conversion of somatostatin (H-Ala-Gly-Cys-Lys-Asn-Phe-

TABLE 7. Agonist versus Antagonist Biological Activities

Agonist	Antagonist
Binds to a specific site in the receptor	Binds to the agonist site (competitive) or to some other site (noncompetitive) on the receptor
Leads to a change in the receptor	Need not lead to change in receptor conformation, but if it does it must be an inactive conformation
Often leads to phenomena such as patching, desensitization, etc.	Generally does not lead to patching, desensitization, etc.
Residence time on the receptor may be long or short	Generally requires long residence time on the receptor to be effective

TABLE 8. Selected Selective Opioid Receptor Antagonists

Ligand	Receptor Selective
D-Phe-Cys-Tyr-D-Trp-Orn-Thr-Pen-Thr-NH$_2$ (CTOP)	μ
D-Tic-Cys-Tyr-D-Trp-Arg-Thr-Pen-Thr-NH$_2$ (TCTAP)	μ
[D-Ala^2, Leu^5, Cys^6]Enkephalin (DALCE)	δ$_1$
[Cys^4]Deltorphin I	δ$_2$
Naltrindole-5′-isothiocyanate (NTII)	δ$_2$
H-Tyr-Tic-Phe-Phe-NH$_2$ (TIPP)	δ
N,N-diallyl-Tyr-Aib-Aib-Phe-Leu-OH (ICI 174,864)	δ
Naltrindole	δ
Naltrexone	μ
TENA	κ
Nor BNI	κ
β-Funaltrexamine (β-FNA)	μ
Naltriben (NTB)	δ$_2$
BNTX	δ$_1$

Phe-Trp-Lys-Thr-Phe-Thr-Ser-Cys-OH) from a potent somatostatin receptor-specific ligand to a potent highly μ-opioid receptor specific antagonist. The approach came from three considerations.[47] First, it had been observed that at very high concentrations, somatostatin possessed analgesic activity. Second, the Merck group,[48] had determined that a β-turn around the Phe-Trp-Lys-Thr sequence (FIG. 3) could account for most of the potency at the somatostatin receptor. We felt this β-turn would poorly interact with the opioid receptor, and hence could be used as a scaffold (or template) on which to build an opioid ligand. Third, investigators at Sandoz[49] had found that their superpotent cyclic disulfide-containing octapeptide analogue of somatostatin also possessed some inhibiting activity in a neurotransmitter assay. Taking D-Phe-Cys-Phe-D-Trp-Lys-Thr-Cys-Thr-OH as a lead compound, we found that substitution of the Phe^3 residue with Tyr, the Cys^7 residue with Pen and terminating the peptide as a carboxamide group, gave a compound D-Phe-Cys-Tyr-Phe-D-Trp-Lys-Thr-Pen-Thr- NH$_2$ (referred to as CTP) design that had binding properties completely different from somatostatin (TABLE 9). Unlike somatostatin which binds very poorly to μ- and δ-opioid receptors (TABLE 9), CTP binds strongly to the μ-opioid receptor, but only weakly to the δ-opioid and somatostatin receptors in the brain.[47] Subsequent studies showed that it was a potent *in vivo* μ-opioid receptor antagonist.[50] Comprehensive 2D NMR studies showed that CTP maintained the type II′ β-turn and provided considerable insight into the presentation to the receptor of the D-Phe^1, Cys^2, Tyr^3, Pen^7, and Thr-NH$_2$^{8} residues that appear to be involved in binding to the μ-opioid receptor.[51] Interestingly, although the μ-opioid receptor activity of CTP is that of an antagonist, CTP is an agonist at the δ-opioid receptor, but a very weak one.[52] Subsequent studies in which the highly constrained phenylalanine analogue, D-Tic was placed in position 1 gave an analogue, which was more potent and even more selective (TABLE 9),[52,53] and comprehensive 2D NMR and molecular dynamics investigations[54] were able to establish the preferred conformation and topology of [D-Tic^1]CTOP for recognizing the μ-opioid receptor in an antagonist state. Furthermore extensive *in vitro* and *in vivo* studies established that the [D-Tic^1]CTAP analogue showed none of the residual somatostatin-like activities that were found for CTP.[52,55] These compounds are now widely used as μ-opioid receptor antagonists, and are particularly useful for binding and *in vivo* studies

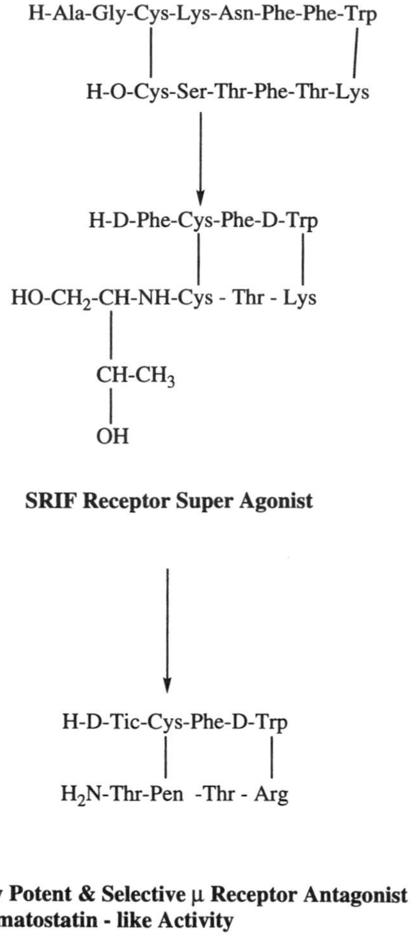

H-Ala-Gly-Cys-Lys-Asn-Phe-Phe-Trp

H-O-Cys-Ser-Thr-Phe-Thr-Lys

H-D-Phe-Cys-Phe-D-Trp

HO-CH$_2$-CH-NH-Cys - Thr - Lys

CH-CH$_3$

OH

SRIF Receptor Super Agonist

H-D-Tic-Cys-Phe-D-Trp

H$_2$N-Thr-Pen -Thr - Arg

**Highly Potent & Selective μ Receptor Antagonist
No Somatostatin - like Activity**

FIGURE 3. Conversion of somatostatin to a μ-opioid receptor antagonist. (Based on results from Bauer et al.[59]; Pelton et al.[47]; Kazmierski and Hruby.[7])

because they are completely stable to proteolytic breakdown and have a very prolonged *in vivo* inhibitory effect at μ-opioid receptors. Hence unlike naloxone, a single dose of compounds such as CTAP blocks *in vivo* μ-agonist activity such as analgesia for several hours.

A very interesting compound recently was discovered in our efforts to further investigate the topographical requirements for CTAP-related compounds. In this investigation, we place a cyclic D-tryptophan analogue D-tetrahydrocarboline (Tca) in position 1 of the cyclic octapeptide series. The analogue obtained, [D-Tca^1]CTAP, had a most unusual activity profile.[56] Although it had only weak binding at μ- and δ-opioid receptors in the rat brain (IC$_{50}$ values of 173 and 220 nM, respectively, TABLE 9), and has corresponding weak agonist activity in the MVD assay, it was a very weak partial agonist at μ-receptor in the GPI assay. However, it was a relatively

TABLE 9. Binding Properties of Selective μ-Opioid Receptor Selective Ligands

Compound	Binding Affinities (nM)			Selectivities	
	μ	δ	Somatostatin	δ/μ	Somatostatin/μ
Somatostatin	27,000	16,000	6	0.00038	.00022
CTP	3.7	8,400	3,690	2,300	1,000
[Orn⁵]CTP (CTOP)	2.8	4,000	23,000	1,400	8,200
[D-Tic¹]CTAP	1.2	1,300	34,300	1,100	29,000
[D-Tic¹]CTOP	1.4	16,000	20,400	11,000	15,000
[D-Tca¹]CTAP	173	211	ND	1.2	—

potent analgesic (similar to DPDPE).[56] Subsequent studies have shown that the compound primarily interacts with the μ- and δ_2-receptors, and may be the first example of a compound that acts at the $\mu\delta_2$-receptor complex by "self potentiation."

A variety of other antagonists for opioid receptors have been developed in several laboratories, especially those of Portoghese[57] and Schiller.[58] The highly potent and δ-opioid receptor selective tetrapeptide of Schiller et al.,[58] H-Tyr-Tic-Phe-Phe-OH (TIPP), is particularly interesting, and demonstrates the power of conformation restriction in ligand design and its use in the discovery of antagonists. The finding that a single conformational change could covert a μ-opioid receptor agonist to a potent δ-opioid receptor antagonist is most intriguing.[58] All the antagonists in TABLES 8 and 9 and others are proving to be very useful in sorting out the complexities of the opioid receptor systems and in identifying the properties of the cloned receptors as they are isolated and expressed.

CONCLUSIONS

The complexity of most hormones and neurotransmitters as manifested by multiple ligands for a particular biological effect, and multiple receptor types and subtypes that have different biological activity profiles provide a profound challenge to chemists, biologists, and physicians who seek to understand these systems and translate that understanding into useful medicine. This complexity suggests that any understanding of these processes must incorporate an array of integrating systems. On the other hand, the unique biological profiles that appear to be associated with specific receptor types and subtypes suggest that highly selective ligands for a particular receptor can provide a highly specific effect and, it is hoped, useful medical treatments. The challenge to sort out and understand the complexities remains undiminished, but the opportunities it provides grow every day. It is hoped that chemists, biologists, biophysists, and medical doctors will cooperate and collaborate more frequently to accelerate this process.

ACKNOWLEDGMENTS

This paper could not have been written without the creativity and work of many outstanding colleagues and collaborators who are listed in the references. A special thanks to H. I. Mosberg, J. Kao, C. A. Gehrig, T. O. Matsanaga, G. Nikiforovich, G. Toth, J. T. Pelton, W. M. Kazmierski, O. Prakash, A. Misicka, A. Lipkowski, R.

Haaseth, H. Bartosz, X. Qian, G. Li, N. Collins, K. Kövér, Professor Tom Davis, and numerous collaborators in biology who made this journey so creative and exciting.

REFERENCES

1. ALBERTS, B., D. BRAY, J. LEWIS, M. ROFF, K. ROBERTS & J. D. WATSON. 1988. Molecular Biology of the Cell. 2nd edit. Garland Publishing Inc., New York.
2. DARNELL, J. E., H. LODISH & D. BALTIMORE. 1990. Molecular Cell Biology. 2nd edit. W. H. Freeman. New York.
3. LAUFFENBURGER, D. A. & J. J. LINDERMANN. 1993. Receptors: Models for Binding Trafficking and Signaling. Oxford University Press. New York.
4. RAMACHANDRAN, G. N., C. RAMAKRISMAN & V. SASISEKHERAN. 1963. J. Mol. Biol. 7: 95–99.
5. RAMACHANDRAN, G. N. & V. SASISEKHERAN. 1968. Adv. Protein Chem. 23: 283–437.
6. HRUBY, V. J. & G. V. NIKIFOROVICH. 1991. In Molecular Conformation and Biological Interactions. P. Balaram & S. Ramaseshan, Eds.: 429–465. Indian Acad. Sci. (Bangalore).
7. KAZMIERSKI, W. M. & V. J. HRUBY. 1988. Tetrahedron 44: 697–710.
8. HRUBY, V. J. 1994. In Peptides: Chemistry, Structure and Biology, Proceedings of the 13th American Peptide Symposium. R. S. Hodges & J. A. Smith, Eds.: 1–17. ESCOM Science Publishers. Leiden.
9. HRUBY, V. J., F. AL-OBEIDI & W. KAZMIERSKI. 1990. Biochem. J. 268: 249–262.
10. HRUBY, V. J., G. TOTH, C. A. GEHRIG, L. F. KAO, R. KNAPP, G. K. LIU, H. I. YAMAMURA, T. H. KRAMER, P. DAVIS & T. F. BURKS. 1991. J. Med. Chem. 34: 1823–1830.
11. FLYNN, G. A., L. L. GIROUX & R. C. DAGE. 1987. J. Am. Chem. Soc. 109: 109: 7914–7916.
12. QIAN, X., K. E. KÖVÉR, M. R. SHENDEROVICH, A. MISICKA, T. ZALEWSKA, R. HORVATH, P. DAVIS, F. PORRECA, H. I. YAMAMURA & V. J. HRUBY. 1994. J. Med. Chem. 34: 1746–1757.
13. KESSLER, H. 1982. Angew. Chem. Int. Ed. Engl. 21: 512–523.
14. HRUBY, V. J. 1982. Life Sci. 31: 189–199.
15. TONIOLO, C. 1990. Int. J. Pept. Protein Res. 35: 287–300.
16. MARSHALL, G. R. 1993. Tetrahedron 49: 3547–3557.
17. RIZO, J. & L. M. GIERASCH. 1992. Annu. Rev. Biochem. 61: 387–418.
18. HRUBY, V. J. 1993. Biopolymers 33: 1073–1082.
19. SAWYER, T. K., P. J. SAN FILIPPO, V. J. HRUBY, M. H. ENGEL, C. B. HEWARD, J. B. BURNETT & M. F. HADLEY. 1980. Proc. Natl. Acad. Sci. USA 77: 5754–5758.
20. MOSBERG, H. I., R. HURST, V. J. HRUBY, K. GEE, H. I. YAMAMURA, J. J. GALLIGAN & T. F. BURKS. 1983. Proc. Natl. Acad. Sci. USA 80: 5871–5874.
21. AKIYAMA, K., K. W. GEE, H. I. MOSBERG, V. J. HRUBY & H. I. YAMAMURA. 1985. Proc. Natl. Acad. Sci. USA 82: 2543–2547.
22. HRUBY, V. J., L. F. KAO, B. M. PETTITT & M. KARPLUS. 1988. J. Am. Chem. Soc. 110: 3351–3359.
23. MOSBERG, H. I., K. SOBCZYK-KOJIRO, P. SUBRAMANIAN, G. M. CRIPPEN, K. RAMALINGA & R. W. WOODARD. 1990. J. Am. Chem. Soc. 112: 822–829.
24. HRUBY, V. J. & C. A. GEHRIG. 1989. Med. Res. Rev. 9: 343–401.
25. FLIPPEN-ANDERSON, J. L., V. J. HRUBY, N. COLLINS, C. GEORGE & B. CUDNAY. 1994. J. Am. Chem. Soc. 116: 7523–7531.
26. TOTH, G., T. K. KRAMER, R. KNAPP, G. LUI, P. DAVIS, T. F. BURKS, H. I. YAMAMURA & V. J. HRUBY. 1990. J. Med. Chem. 33: 249–253.
27. HRUBY, V. J., L. F. KAO, L. D. HIRNING & T. F. BURKS. 1985. In Peptides: Structure and Function, Proceedings of the 9th American Peptide Symposium. C. M. Deber, V. J. Hruby & K. Kopple, Eds.: 691–694. Pierce Chemical Co. Rockford, IL.
28. HRUBY, V. J., G. TOTH, C. A. GEHRIG, L. F. KAO, R. KNAPP, G. K. LUI, H. I. YAMAMURA, T. H. KRAMER, P. DAVIS & T. F. BURKS. 1991. J. Med. Chem. 34: 1823–1830.
29. BARTOSZ-BECHOWSKI, H., P. DAVIS, T. ZALEWSKA, J. SLANINOVÁ, F. PORRECA, H. I. YAMAMURA & V. J. HRUBY. 1990. Biochem. Biophys. Res. Commun. 173: 521–527.

30. ERSPAMER, V., P. MELCHIORRI, G. FALCONIERI-ERSPAMER, L. NAPI, R. CORSI, C. SERERINI, D. BARRA, M. SIAMMACO & G. KREIL. 1989. Proc. Natl. Acad. Sci. USA **86**: 5188–5192.
31. AMICHI, M., S. SAGAN, A. MOR, A. DELFOUR & P. NICOLAS. 1989. Mol. Pharmacol. **35**: 774–779.
32. NIKIFOROVICH, G. & V. J. HRUBY. 1990. Biochem. Biophys. Res. Commun. **173**: 521–527.
33. NIKIFOROVICH, G., V. J. HRUBY, O. PRAKASH & C. A. GEHRIG. 1991. Biopolymers **31**: 941–955.
34. NIKIFOROVICH, G., O. PRAKASH, C. A. GEHRIG & V. J. HRUBY. 1993. J. Am. Chem. Soc. **115**: 3399–3406.
35. BRYANT, S. D., S. SALVADORI, M. ATTLA & L. H. LAZARUS. 1993. J. Am. Chem. Soc. **115**: 8503–8504.
36. CHARPENTIER, S., S. SAGAN, A. DELFOUR & P. NICOLAS. 1991. Biochem. Biophys. Res. Commun. **179**: 1161–1168.
37. MISICKA, A., S. CAVAGNERO, R. HORVATH, P. DAVIS, T. H. KRAMER, H. I. YAMAMURA & V. J. HRUBY. 1993. *In* Peptide Chemistry 1992, Proceedings of the 2nd Japan Symposium on Peptide Chemistry. N. Yanaihara, Ed.: 378–380. ESCOM Science Publishers. Leiden.
38. MISICKA, A., A. W. LIPKOWSKI, R. HORVATH, P. DAVIS, H. I. YAMAMURA, F. PORRECA & V. J. HRUBY. 1994. J. Med. Chem. **37**: 141–145.
39. KRAMER, T. H., P. DAVIS, V. J. HRUBY, T. F. BURKS & F. PORRECA. 1993. J. Pharmacol. Exp. Ther. **226**: 577–584.
40. JIANG, Q., A. E. TAKEMORI, M. SULTANA, P. S. PORTOGHESE, W. D. BOWEN, H. I. MOSBERG & F. PORRECA. 1991. J. Pharmacol. Exp. Ther. **257**: 1069–1075.
41. SOFUOGLU, M., P. S. PORTOGHESE & A. E. TAKEMORI. 1991. J. Pharmacol. Exp. Ther. **257**: 676–680.
42. MATTIA, A., T. VANDERAH, H. I. MOSBERG & F. PORRECA. 1991. J. Pharmacol. Exp. Ther. **258**: 583–587.
43. MATTIA, A., S. C. FARMER, A. E. TAKEMORI, M. SULTANA, P. PORTOGHESE, H. I. MOSBERG, W. D. BOWEN & F. PORRECA. 1992. J. Pharmacol. Exp. Ther. **260**: 518–528.
44. WILD, K. D., V. J. CARLISI, H. I. MOSBERG, W. D. BOWEN, P. S. PORTOGHESE, M. SULTANA, A. E. TAKEMORI, V. J. HRUBY & F. PORRECA. 1992. J. Pharmacol. Exp. Ther. **264**: 831–838.
45. HRUBY, V. J. 1992. *In* Progress in Brain Research. J. Josse, P. M. Bujis & F. J. H. Tilers, Eds. **Vol. 92**: 215–224. Elsevier Science Publishers. Amsterdam.
46. HRUBY, V. J. 1987. Trends Pharmacol. Sci. **8**: 336–339.
47. PELTON, J. T., K. GULYA, V. J. HRUBY, S. P. DUCKLES & H. I. YAMAMURA. 1985. Proc. Natl. Acad. Sci. USA **82**: 236–239.
48. VEBER, D. F., R. M. FREIDINGER, D. SCHWENK-PERLOW, W. J. PALEVEDA, JR., F. W. HOLLY, R. G. STRACHAN, R. F. NUTT & B. H. ARISON. 1981. Nature **232**: 55–57 and references therein.
49. MAURER, R., B. H. GAEHWILER, H. H. BUESCHER, R. C. HILL & D. ROEMER. 1982. Proc. Natl. Acad. Sci. USA **79**: 4815–4818.
50. SHOOK, J. E., J. T. PELTON, W. S. WIRE, L. D. HIRNING, V. J. HRUBY & T. F. BURKS. 1987. J. Pharmacol. Exp. Ther. **240**: 772–777.
51. PELTON, J. T., M. WHALON, W. L. CODY & V. J. HRUBY. 1988. Int. J. Pept. Protein Res. **31**: 109–115.
52. KRAMER, T. H., J. E. SHOOK, W. M. KAZMIERSKI, E. A. AYERS, W. S. WIRE, V. J. HRUBY & T. F. BURKS. 1989. J. Pharmacol. Exp. Ther. **96**: 99–100.
53. KAZMIERSKI, W. M., W. S. WIRE, G. K. LIU, R. J. KNAPP, J. E. SHOOK, T. F. BURKS, H. I. YAMAMURA & V. J. HRUBY. 1988. J. Med. Chem. **31**: 2170–2177.
54. KAZMIERSKI, W. M., H. I. YAMAMURA & V. J. HRUBY. 1991. J. Am. Chem. Soc. **113**: 2275–2283.
55. SHOOK, J. E., J. T. PELTON, P. F. LEMCKE, F. PORRECA, V. J. HRUBY & T. F. BURKS. 1987. J. Pharmacol. Exp. Ther. **242**: 1–7.
56. HORAN, P. J., K. D. WILD, W. M. KAZMIERSKI, R. FERGUSON, V. J. HRUBY, S. J. WEBER,

T. P. Davis, L. Fang, R. J. Knapp, H. I. Yamamura, T. H. Kramer, T. F. Burks, W. D. Bowen, A. E. Takemori & F. Porreca. 1993. Eur. J. Pharmacol. **233:** 53–62.
57. Portoghese, P. S. 1992. J. Med. Chem. **35:** 1927–1937.
58. Schiller, P. W., T. M.-D. Nguyen, G. Weltrowska, B. C. Wilkes, B. J. Marsden, C. Lemieux & N. N. Chung. 1992. Proc. Natl. Acad. Sci. USA **89:** 11871–11875.
59. Bauer, W., U. Briner, W. Doepfner, R. Haller, R. Huguenin, P. Marbach, T. L. Petcher & J. Pless. 1982. Life Sci. **31:** 1133–1140.

Determination of the Chemical Mechanism of Neurotransmitter Receptor-mediated Reactions by Rapid Chemical Kinetic Methods[a]

GEORGE P. HESS,[b] LI NIU, AND RAYMOND WIEBOLDT

Section of Biochemistry, Molecular and Cell Biology
Division of Biological Sciences
216 Biotechnology Building
Cornell University
Ithaca, New York 14853-2703

To understand the brain, we must know how nerve cells behave and how they interact.

Francis Crick
The Astonishing Hypothesis[1]

How do nerve cells receive information from within and without the organism? How can they store information for short and long periods of time? What role do chemical reactions play in the perception and storage of information? Signal initiation between cells of the nervous system is regulated to a large extent by membrane-bound proteins, the neurotransmitter receptors. The proteins in the membrane of one cell bind chemical signals, neurotransmitters, released by a neighboring cell. The reactions involve ligand-binding steps and conformational changes of the protein. Upon binding a specific neurotransmitter, the receptors form transiently open transmembrane channels. Receptors for excitatory neurotransmitters allow sodium and potassium ions to pass through the open channels. Receptors for inhibitory neurotransmitters allow the passage of chloride ions. The properties of the receptor-mediated reactions are fundamental to the ability of a neuron to function. They determine the changes in transmembrane voltage that trigger signal transmission between neurons.

Many of the receptor-mediated reactions occur on a sub-millisecond time scale. Until recently, chemical kinetic techniques suitable for use with membrane-bound receptor proteins that must be studied in intact cells and vesicles in the sub-millisecond time region were not available. When it was observed that desensitization (the transient and reversible inactivation) of a neurotransmitter receptor can occur in the millisecond time region,[2,3] such techniques were developed.[3-7] They allow one to determine (1) the ligand-binding properties of the active form both before and after the receptor has desensitized,[5] (2) the rate constants for receptor desensitization,[5] and (3) the rate constants for the formation of the open transmembrane channel.[8,9]

Here we describe three aspects of our work. First, we outline why we use chemical kinetic approaches to study receptor-mediated reactions on the sub-millisecond time scale. Then we describe a newly developed chemical kinetic method, laser-pulse photolysis of caged neurotransmitters, with a time resolution on

[a] This research was supported by grants from the National Institutes of Health (GM04842 and NS 08527) and the National Science Foundation (922061).

[b] Corresponding author.

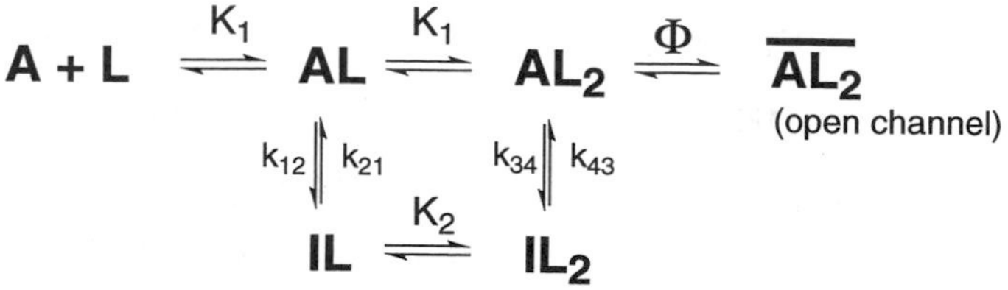

$$A + L \; \overset{K_1}{\rightleftharpoons} \; AL \; \overset{K_1}{\rightleftharpoons} \; AL_2 \; \overset{\Phi}{\rightleftharpoons} \; \overline{AL_2}$$

(open channel)

$$k_{12} \updownarrow k_{21} \qquad k_{34} \updownarrow k_{43}$$

$$IL \; \overset{K_2}{\rightleftharpoons} \; IL_2$$

SCHEME 1. Chemical mechanism for the acetylcholine receptor from *E. electricus* electroplax. A and I represent the active and inactive (desensitized) receptor forms, respectively, and K_1 and K_2 the receptor:neurotransmitter dissociation constants for the A and I forms, respectively. The subscript indicates the number of neurotransmitter molecules (L) bound. Φ is the equilibrium constant. $AL_2/\overline{AL_2} = k_{cl}/k_{op}$ where k_{op} and k_{cl} are the rate constants for channel opening and closing, respectively. Rate constants $k_{12} k_{21}$, k_{34}, and k_{43} are for the interconversion between A and I receptor forms.

the millisecond scale. Finally, we describe the type of information that can be obtained.

The minimum mechanism in SCHEME 1 is based on a mechanism originally proposed for the nicotinic acetylcholine receptor in frog muscle cells by Katz and Thesleff[10] to describe the results of classical electrophysiological experiments. Once new techniques were developed the constants for the mechanism could be determined by studying the reaction before the receptor desensitized to a form with altered ligand-binding properties and biological activity. The minimum mechanism shown accounts for results obtained with (1) the muscle type nicotinic acetylcholine receptor in the electric organ of *Electrophorus electricus* (reviewed in ref. 5) and *Torpedo* sp. (reviewed in ref. 11) using quench- and stopped-flow techniques with a 5-ms time resolution; (2) the acetylcholine receptor in BC3H1 muscle cells[4,9,12] using cell-flow[4] and laser-pulse photolysis[13] techniques with 5-ms and 100-μs time resolutions respectively; (3) the neuronal acetylcholine receptor in PC12 cells[8] using the cell-flow technique; (4) the inhibitory γ-aminobutyric type A (GABA$_A$) receptor in rat brain membrane vesicles[14,15] and in cerebral cortical cells of embryonic mice,[16] using quench-flow and cell-flow techniques with ~5- to 10-ms time resolution respectively, and (5) the inhibitory glycine receptor in embryonic mouse spinal cord cells.[17]

In the minimum mechanism in Scheme 1, A represents the receptor in its active form, and I the inactive, desensitized receptor form. The activating ligand is represented by L, and the subscript the number of ligand molecules bound to the receptor protein. $\overline{AL_2}$ represents the open-channel form of the receptor through which inorganic ions exchange across the cell membrane. K_1 and K_2 are the intrinsic dissociation constants of the ligand for the active and desensitized receptor forms respectively. We assume that each site on the active receptor form has the same affinity for ligand K_1, and that each site on the inactive form has the same affinity for ligand K_2. The channel-opening equilibrium constant Φ^{-1} was introduced[18] to account for the cooperativeness observed in the binding of neurotransmitter to the receptor. The rate constants k_{12}, k_{21}, k_{34}, k_{43} are for the interconversion between active and inactive receptor forms. The mechanism and the constants allow one to determine the concentration of the open receptor-channel, $\overline{AL_2}$ in the mechanism, as a function of neurotransmitter concentration and time.

Why is this of interest? Determination of the rate coefficient for the movement of inorganic ions across the membrane is the first step toward the calculation of the receptor-controlled change in transmembrane voltage. It is useful to look at the underlying physical theory that relates chemical mechanisms to changes in transmembrane voltage. About 100 years ago, Max Planck[19] derived the relationship between the rate of movement of inorganic cations and anions across a porous barrier and the resulting electric field.

$$d[M^+]_1/dt = k_{obs}f_1[M^+]_2 - k_{obs}f_2[M^+]_1$$

$[M^+]$ represents the concentration of a monovalent cation, and the subscript refers to the sides of a porous barrier; and k_{obs} the voltage-independent rate coefficients for the transfer of inorganic ions across the barrier; f_1 and f_2 are the factors that account for the acceleration or retardation of the ions as they move in an electric field where $f_2/f_1 = \exp(V_mF)/(RT)$. V_m represents the transmembrane voltage, and F, R, and T the Faraday constant, the molar gas constant, and the absolute temperature, respectively. k_{obs} is given by a specific reaction rate constant, J, multiplied by the concentration of the open receptor-channel, $\overline{AL}_n$, where n represents the number of neurotransmitter molecules bound to the receptor.[20]

$$k_{obs} = J[\overline{AL}_n]_{t,L}$$

Providing we can determine J and $(\overline{AL})_n$, we can determine k_{obs}, and then calculate the transmembrane voltage.[20]

$$J = \gamma RT/F^2[M^+]NA$$

J is proportional to the conductance, γ, of the open channel. NA represents Avogadro's number and all the other terms have been defined. We can determine γ conveniently by using the single-channel technique developed by Neher and Sakmann.[21] The second task is to determine $\overline{AL}_n$ as a function of neurotransmitter concentration and time. Achieving this objective depends on knowing the chemical mechanism of the receptor-controlled reaction and the constants pertaining to it.

The interconversion between active and inactive receptor forms can occur in the 100-ms time region in the cases of the acetylcholine, $GABA_A$, and glycine receptors[3,11,14-17] and in the 10-ms time region in the case of the glutamate receptor.[22] The concentration of the open acetylcholine receptor-channel changes in the microsecond-to-millisecond time region.[23] The results in TABLE 1 illustrate why it is important to make kinetic investigations using appropriate time scales. The values were obtained with a variety of techniques in which the intermediates of the reaction are allowed to come to a quasi-equilibrium before the first measurement is made; the values differ by almost two orders of magnitude.

We decided to try a different approach. Our aim was to develop rapid chemical kinetic techniques that would allow us to investigate the sequential (pre-steady-

TABLE 1. Acetylcholine Receptor—Electrophysiological Determinations[a]

Tissue	Year	Temperature (°C)	k_{op} (s^{-1})
Frog end plate	1981	~10	2,300
	1983	~10	40,000
	1988	~10	20,000
BC_3H1 cells	1984	11	7,700
	1986	11	320
	1987	11	450

[a]Values were taken from reference 23.

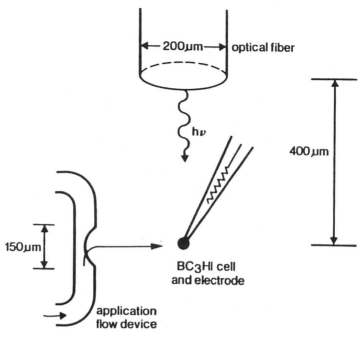

FIGURE 1A. Laser-pulse photolysis apparatus. A BC$_3$H1 cell, approximately 15 μm in diameter, attached to an electrode for recording whole-cell currents, was equilibrated with caged carbamoylcholine. The beam from a Candela SLL500 dye laser with a 600-ns pulse length was introduced from an optical fiber. The cell-flow method was used before and after each laser pulse to determine the concentration of liberated carbamoylcholine and to detect cell damage.

state) steps of a reaction before it reaches an equilibrium and that could be used with membrane-bound proteins in intact single cells and vesicles. What are the advantages of such an approach?

The major aim of rapid reaction techniques is to identify the sequence of reaction steps. This is a well-known and notoriously difficult problem in the kinetic analysis of quasi-equilibrium mixtures of complex reactions. However, with rapid reaction techniques it is possible to separate sequential steps of a complex reaction along the time axis. It is also possible to determine which step in a reaction occurs first, and which occurs later. Once the individual steps of a complex reaction are separated along the time axis, the kinetics for each step often follow simple rate laws. This allows one to evaluate the pertinent rate constants. Another advantage is the ability to use a wide range of reactants. A good time resolution makes it possible to use high concentrations of both receptor and neurotransmitter, thus allowing one to identify reaction intermediates that exist in such low concentrations that they may not otherwise be observed.

One of the methods we developed is a laser-pulse photolysis technique. With this it is possible to equilibrate receptors with caged neurotransmitter and then to photolyze the caged compound. This leads to the release of neurotransmitter in the μs time domain. A diagram of the apparatus is given in FIGURE 1A. A receptor-containing cell is attached to an electrode, and the solution surrounding the cell contains a photolabile precursor of a neurotransmitter. The precursor is inactive towards the receptors so one can equilibrate it with them on the cell surface, and then release the neurotransmitter by the application of a single pulse of laser light, causing the removal of the protecting group within microseconds. FIGURE 1B is an example of the photolysis of caged carbamoylcholine which produces free carbamoylcholine plus a 2-nitrosophenyl-α-keto acid. Thus, the time resolution is improved because diffusional barriers are overcome before the activating ligand is released. The neurotransmitter binds to its specific receptors, which form open channels through which current flows. We record the resulting whole-cell current, using a technique developed by Hamill *et al.*[24]

To produce photolabile precursors of neurotransmitters, we took advantage of research on the photochemistry of *ortho*-nitrobenzyl derivatives pioneered by De Mayo in 1960,[25] Barltrop in 1966,[26,27] and Patchornik and Woodward in 1970.[28] The first applications of *ortho*-nitrobenzyl derivatives to biological problems were made by Kaplan (reviewed in ref. 29) and Trentham (reviewed in ref. 30), who synthesized photolabile precursors of biologically interesting phosphates ("caged phosphates"), for example, ATP. The compounds can be photolyzed at a rate of 10^3 s^{-1}, which is about three orders of magnitude faster than the original compounds of De Mayo[25] and Barltrop.[26,27]

We found that by attaching a carboxyl group to the benzylic carbon of the nitrobenzyl protecting group we greatly increased the photolysis rate (TABLE 2),[31] and used this approach to make photolabile precursors of carbamoylcholine. Caged carbamoylcholine is photolyzed to carbamoylcholine at a rate of 17,000 s^{-1} and with a product quantum yield of 0.8.

Unlike acetylcholine, other neurotransmitters such as γ-aminobutyric acid, glycine, and glutamate contain carboxyl groups. We could use the α-carboxy-*o*-

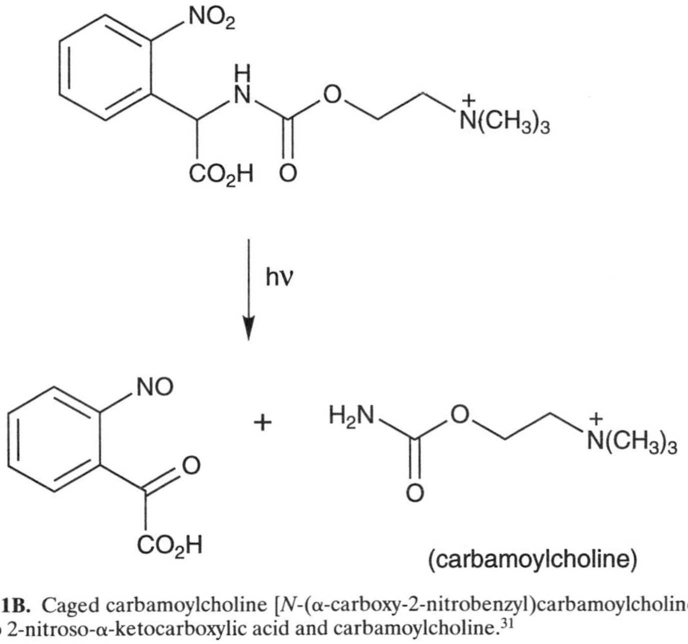

FIGURE 1B. Caged carbamoylcholine [*N*-(α-carboxy-2-nitrobenzyl)carbamoylcholine] is photolyzed to 2-nitroso-α-ketocarboxylic acid and carbamoylcholine.[31]

TABLE 2. Caged Neurotransmitters

Caging Group	Compound Caged	Photolysis Rate^a (s^{-1})	Product Quantum Yield	Reference
	$-NH-\overset{\displaystyle O}{\overset{\|}{C}}-(CH_2)_2-\overset{+}{N}(CH_3)_3$ carbamoylcholine	1.7×10^4	0.8	31
α-carboxy-*o*-nitrobenzyl	$-O-\overset{\displaystyle O}{\overset{\|}{C}}-(CH_2)_2$ $HO\overset{\displaystyle O}{\overset{\|}{C}}-CH-\overset{+}{NH_3}$ glutamic acid	3.3×10^4	0.14	33
	$-O-\overset{\displaystyle O}{\overset{\|}{C}}$ $(CH_2)_2$ $CH_2-\overset{+}{NH_3}$ γ-amino-butyric acid	2.3×10^4	0.15	45
2-methoxy-5-nitrophenyl	$-\overset{\displaystyle O}{\overset{\|}{C}}-CH_2-\overset{+}{NH_3}$ glycine	$\sim 70 \times 10^4$	0.2	35

^{a} 22 °C, pH 7.4.

nitrobenzyl group to protect the carboxyl group of the neurotransmitters glutamate[32,33] and γ-aminobutyric acid.[34] These derivatives allow us to investigate reaction steps with rate constants as high as 20,000 s^{-1}. The derivatives are stable in aqueous solutions and are biologically inert. We have used a new photolabile group to protect the neurotransmitter glycine. The 2-methoxy 5-nitrophenyl group (TABLE 2) is photolyzed 30 times faster than the α-carboxy-*o*-nitrobenzyl group and is biologically inert.[35] It is, however, less convenient to use because it is slowly hydrolyzed in aqueous solutions at neutral pH.

What type of information can one obtain by using the laser-pulse photolysis technique? We shall discuss its use first with the nicotinic acetylcholine receptor in

BC$_3$H1 muscle cells and then with the glutamate receptor in hippocampal neurons. A current trace obtained in an early whole-cell experiment in which we used caged carbamoylcholine is shown in FIGURE 2A. Caged carbamoylcholine (400 μM) was equilibrated with acetylcholine receptors on the surface of a single BC$_3$H1 muscle cell before photolysis was induced. In this experiment, the caged derivative was photolyzed within 100 μs, and the current reached its maximum value within 2 ms. The falling phase of the current, which reflects the desensitization reaction, is shown on a different time scale.

From experiments such as the one shown in FIGURE 2A one may obtain a considerable amount of information about the mechanism. The rising phase of the current contains two types of information. At low concentrations of neurotransmitter it reflects the rate constants for the neurotransmitter-binding steps. At high concentrations of neurotransmitter it reflects the rate constants for the channel-opening process. The maximum amplitude of the current is a measure of the concentration of receptors in the open-channel form. From the effect of neurotransmitter concentration on the maximum amplitude, one can determine the equilibrium constants for neurotransmitter binding to the receptor and for the channel-opening process. Thus, the laser-pulse photolysis technique accomplishes the aim of chemical kinetic measurements. One can spread out sequential steps of a complex reaction along the time axis, so that the kinetics of individual steps can be measured separately.

The observed rate constant for the rise of the current trace in FIGURE 2A is shown as a function of carbamoylcholine concentration in FIGURE 2B. Over 90% of the rise is governed by a single exponential, suggesting that a single rate process is observed. The effect of carbamoylcholine concentration on the observed rate constant indicates that it reflects the rate constants for both channel opening and closing. The slope of the line gives a channel-opening rate constant of about 10,000 s^{-1}, and the intercept a channel-closing rate constant of 500 s^{-1}.

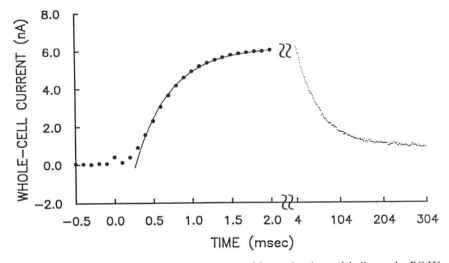

FIGURE 2A. A laser-pulse photolysis experiment with caged carbamoylcholine and a BC$_3$H1 cell, pH 7.4, 22–23 °C, and −60 mV. The whole-cell current was generated by photolysis of 400 μM caged carbamoylcholine. The laser excitation wavelength was 328 nm. The solid line through the points represents the rise of the current fitted to a single exponential (k_{obs} = 2140 s^{-1}).

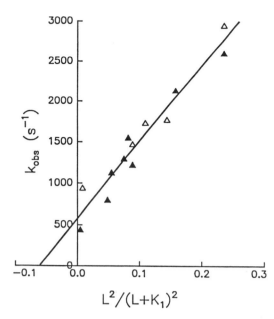

FIGURE 2B. Evaluation of the kinetic parameters for channel opening. Rate constants k_{op}, k_{cl}, and K_1 (listed in TABLE 2) were evaluated using a nonlinear least-squares fitting procedure, and the values were used to construct the solid line. The different symbols represent data from different experiments using different laser dyes with outputs at 318 or 328 nm.[12]

Can we obtain independent evidence for our measurements? Results obtained with different methods are compared in TABLE 3. All the constants obtained by laser-pulse photolysis, with the exception of k_{op}, can be evaluated by independent measurements. The rate constant for channel closing is expected to agree with the lifetime of the open channel. This lifetime was measured independently using single-channel current recordings.[12] The fraction of receptors in the open-channel form when the receptor is saturated with neurotransmitter was determined by both laser-pulse photolysis and a cell-flow technique. The value for K_1, the equilibrium constant for neurotransmitter binding to the receptor site controlling channel opening, was obtained by both photolysis and cell-flow methods. All the values obtained are in good agreement.

What other information can we obtain using the new method? An intriguing question concerned inhibition of the acetylcholine receptor by such compounds as the local anesthetic procaine,[36] the abused drug cocaine,[37] or by acetylcholine itself at high concentrations.[38] Most reports on this subject agree on a mechanism in which

TABLE 3. Comparison of the Value of Constants Obtained with BC$_3$H1 Cells Using Various Methods[a]

Constant	Method	Value of Constant
k_{cl}	Laser-pulse photolysis	500 ± 100 s^{-1}
	Single-channel current	400 ± 130 s^{-1}
$k_{op} (k_{op} + k_{cl})^{-1}$	Laser-pulse photolysis	0.94
	Cell-flow	0.84
K_1	Laser-pulse photolysis	210 ± 90 μM
	Cell-flow	240 μM

[a]pH 7.4, 22–23 °C, −60 mV.

the receptor-channel has to open first before the inhibitor blocks the open channel.[36,38] Allosteric mechanisms in which an inhibitor binds to one form of a protein, preventing formation of an open channel, are not often invoked.

The simplest suggestion for a channel-blocking mechanism is shown in SCHEME 2. The active, non-desensitized receptor is represented by A and the neurotransmitter by L. After binding the neurotransmitter, the receptor forms an open channel, indicated by $\overline{AL_2}$. Procaine, represented by I, then binds in the open channel and blocks it. This mechanism predicts that the rate constant for channel closing will decrease as the inhibitor concentration is increased. An allosteric mechanism also accounts for all the results obtained in single-channel recordings. In the noncompetitive mechanism for allosteric inhibition, the inhibitor binds to a regulatory site on the receptor before and after the channel opens. The rate constant for channel closing is again predicted to decrease as the inhibitor concentration is increased. However, the regulatory mechanism in which the inhibitor binds to the receptor before the channel opens makes a clear-cut prediction that is not made by the channel-blocking mechanism: the channel-opening rate is expected to decrease as the inhibitor

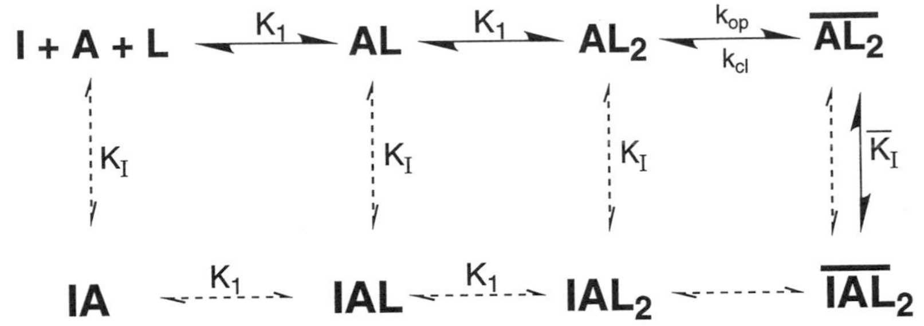

SCHEME 2. Alternative mechanisms for inhibition of the acetylcholine receptor. L represents the channel-activating ligand. A represents the active nondesensitized receptor forms, K_1 the dissociation constant for the receptor:ligand complex, and $\overline{AL_2}$ the open-channel form of the receptor. I represents the inhibitor, and K_I and $\overline{K_I}$ the dissociation constants of the receptor: inhibitor complexes of the closed- and open-channel forms of the receptor.

concentration is increased. This is not the case with the channel-blocking mechanism. The two mechanisms can, therefore, be distinguished. However, the effect of inhibitors on the channel-opening rate constant was not measured because the time resolution of existing methods was not sufficient.

In order to distinguish between the two mechanisms we determined the effects of procaine, QX222, and cocaine on the channel-opening rate constant using the new technique. The results of two experiments done with procaine are shown in FIGURE 3. The questions we asked were: Does procaine inhibit only the channel-closing rate constant, as is predicted by the channel-blocking mechanism, or does it inhibit both the opening and closing rate constants, as is predicted by an allosteric mechanism in which the receptor has a regulatory site? The effect of procaine on k_{obs} for the current rise, measured at low concentrations of released carbamoylcholine, is illustrated in FIGURE 3; under these conditions the effect of procaine on the channel-closing rate constant, k_{cl} is determined.[9] The rate constant decreased as the procaine concentration was increased, as is predicted by both mechanisms.

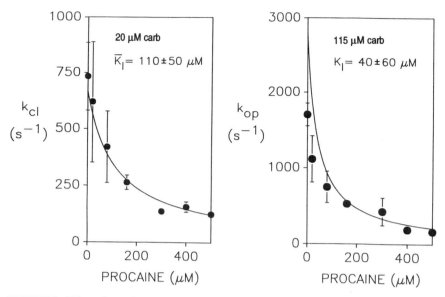

FIGURE 3. Effect of procaine on k_{cl} and k_{op} determined by the laser-pulse photolysis technique at pH 7.4, 22 °C, and −60 mV.[9]

The effect of procaine on k_{obs} for the current rise, measured at high concentrations of released carbamoylcholine, is also illustrated in FIGURE 3.[9] Under these conditions, the rate constant for channel opening[12] is much larger than that for channel closing, and k_{obs} is a measure of the channel-opening rate constant. The channel-opening rate constant decreases as the procaine concentration is increased. This is consistent with the inhibitor binding to an inhibitory site before the channel opens. Thus, the acetylcholine receptor can be inhibited before the channel opens. The new laser-pulse photolysis technique by which the constants for both channel opening and closing can be measured conveniently in the same experiment made it possible to obtain these results.

Does it make a difference which mechanism is correct? Channel opening is very fast; a receptor-channel can open in a few milliseconds, for example, the nicotinic acetylcholine receptor at a neuromuscular junction.[12,23,38,39] If the channel-blocking mechanism operates, the channel opens and a signal is initiated before the channel is blocked. Channel blocking can occur by any compound that can enter the channel but cannot pass through. The physiological meaning of the channel-blocking mechanism is, therefore, not obvious. A regulatory mechanism, in which an inhibitor binds to the receptor before the channel opens, may have physiological significance. It indicates that the receptor has a regulatory site to which specific compounds can bind and inhibit the receptor before the channel opens, thus preventing signal transmission. Acetylcholine at high concentrations inhibits the acetylcholine receptor, perhaps also by binding to a regulatory site.[40]

What other information can we obtain using a rapid chemical kinetic approach? A current trace recorded from a rat hippocampal cell and induced by photolysis of caged glutamate within ∼ 60 μs is shown on the left of FIGURE 4A. It gives the same type of information that we obtained with caged carbamoylcholine and the acetylcholine receptor in BC₃H1 muscle cells. But we also learned something else. Caged glutamate (500 μM) was used in this experiment. The maximum current obtained was 3000 pA, and the $t_{1/2}$ of the falling phase of the current, indicative of receptor desensitization, was about 15 ms. On the right-hand side (FIG. 4A inset) is an experiment in which 300 μM glutamate was applied to the same cell using a cell-flow device. The maximum current obtained was only 150 pA, which is 5% of the current obtained in the photolysis experiment. Furthermore, the $t_{1/2}$ of the falling phase of the current was 100 ms, which is 10 times longer than we observed in the photolysis experiment. Thus, when we used the cell-flow technique we lost all the information about the receptor form that desensitizes rapidly, because this form desensitizes during the time it takes for the receptors to equilibrate with glutamate in the flowing solution. One can improve the time resolution by using membrane patches rather than whole cells. An experiment done with a sealed membrane vesicle, with a diameter of about 7 μm, is shown in FIGURE 4B. The vesicle was suspended from a recording electrode and neurotransmitter flowed over the cell. The current induced is a measure of the concentration of open receptor-channels, and the graph shows the current produced as a function of time. The trace first rises to a maximum, within 10 ms, and then decreases due to desensitization. In this case, the rate coefficient for desensitization is about 50 s^{-1}. This means that about half the current disappears within about 15 ms; half the receptors present have desensitized within 15 ms.

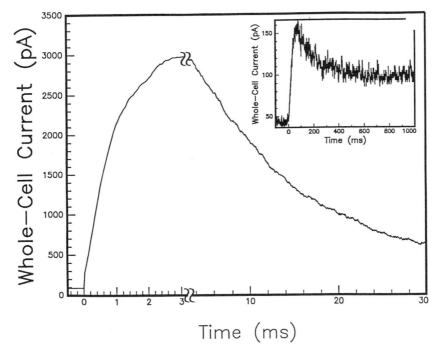

FIGURE 4A. Comparison of whole-cell current obtained with a rat hippocampal neuron by activation by laser pulse photolysis of caged glutamate and rapid flow application of a glutamate solution (pH 7.4, 22–23 °C, transmembrane voltage −60 mV). A 3-nA current is obtained from a neuron when 500 μM γ-(*o*-nitrophenyl)-α-carboxy glutamic acid is photolyzed at the cell surface by a flash of 343 nm laser light. The inset shows the response of the same neuron when it was exposed to a rapid flow of a 300 μM glutamate solution.

Even in this experiment, in which the current reaches a maximum value within 10 ms, a problem remains which has not yet been widely recognized. The receptors on the cell surface facing the flow device are in contact with the neurotransmitter and desensitize before the neurotransmitter reaches the far side of the cell. To deal with this problem, we[4] have taken into account the available hydrodynamic theory[41] to correct the observed current for the desensitization that occurs while the receptors equilibrate with the neurotransmitter. The dotted line at the top, parallel to the abscissa of the graph, gives the current after it has been corrected for desensitization. So even when a neurotransmitter equilibrates with receptors within 10 ms, as it did in this experiment, the concentration of active receptor forms can be determined only if one corrects the observed current for the desensitization that occurs during the period of equilibration. For the same reasons, the equilibrium constants that

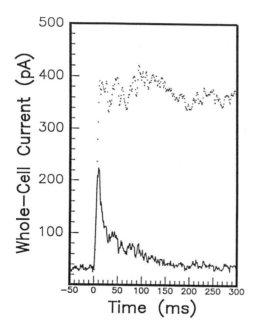

FIGURE 4B. Whole-cell recording and rapid flow of a 600-μM solution of free glutamate over a 7-μM vesicle pulled from a neuron similar to the one used for FIGURE 4A produces a reduced amplitude current response.

determine the concentration of open receptor-channels—which is what we want to determine—can be measured only if the observed current is corrected for desensitization. If we do not make this correction, we obtain apparent equilibrium constants, which also depend on the rate of receptor desensitization and the time resolution of the method used. The determined constants then show the same variability as they did before the use of rapid reaction techniques.

The slowly desensitizing phase of the receptors seen in the inset of FIGURE 4A is no longer seen in FIGURE 4B, because with a membrane vesicle we observe a smaller population of glutamate receptors than in a whole-cell experiments. Therefore, although we can improve the time resolution by using small membrane patches, we lose information about a minor component of the reaction. Why not use higher flow rates to improve the time resolution? First, because by using lower flow rates and

FIGURE 4C. Activation of different receptor types on mouse central nervous system neurons by photolytic release of amino acid neurotransmitters. The three caged compounds used were *N*-(α-carboxy-2-nitrophenyl)glycine, *N*-α-carboxy-2-nitrophenyl)-γ-aminobutyric acid and *N*-α-carboxy-2-nitrophenyl) glutamic acid. The experiment in the upper left corner with caged carbamoylcholine and BC₃H1 cells is the same as that in FIGURE 2A and is shown for comparison. pH 7.4, 22–23 °C, and −60 mV.

Acetylcholine receptor in BC₃H1 muscle cell
100 μM caged carbamoylchline

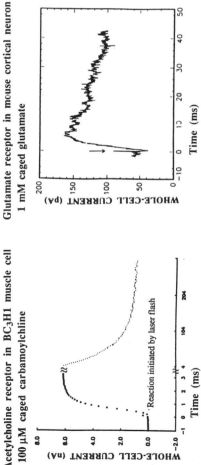

Glutamate receptor in mouse cortical neuron
1 mM caged glutamate

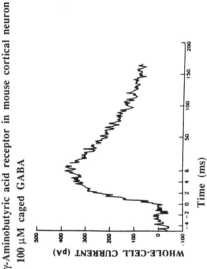

Glycine receptor in mouse spinal cord neuron
620 μM caged glycine

γ-Aminobutyric acid receptor in mouse cortical neuron
100 μM caged GABA

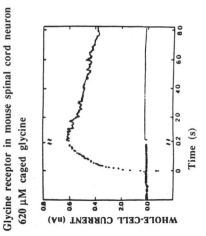

correcting the current, the seal between a cell and an electrode remains stable for much longer. This increases the number of measurements that can be made with each cell and, therefore, decreases the statistical error of the measurements. There is, however, a more serious problem. By increasing the flow rate we can shorten the apparent time it takes for a ligand to equilibrate with the receptors, but we pay a price. In the cell-flow experiments a cell is surrounded by a diffusion layer of physiological salt solution. It is important that the flowing solution containing the neurotransmitter replace the layer of buffer solution immediately surrounding the cell without mixing with it, that is to say, laminar flow is required. If the rate at which fluid flows over a cell is increased, we obtain turbulent flow. In that case neurotransmitter in the flowing solution mixes with the buffer solution surrounding the cell, and we no longer know the concentration of the neurotransmitter equilibrating with the receptors.

All the experiments we have done with the acetylcholine receptor can also be done with excitatory glutamate receptors and inhibitory GABA and glycine receptors (FIG. 4C). Very little is known about these receptor mechanisms when compared to what we know about the muscle acetylcholine receptor mechanism. All these receptors are important in controlling signal transmission. All are involved in various diseases. All are targets of clinically relevant compounds. Chemical kinetic techniques can now be used to study these processes on the molecular level.

TABLE 3 summarizes some of the results we have obtained with a cell from the central nervous system using rapid reaction techniques. We know from molecular biological experiments carried out in many laboratories that different forms of a receptor, activated by the same neurotransmitter, can coexist in one cell. Using chemical kinetic methods, we determined that different receptor forms exist in different concentrations in the same cell, that each receptor form has a different affinity for its neurotransmitter and a different rate constant for desensitization. We also have indications that the channel-opening rate constants will be different. At least eight different receptor forms may be present in a single neuron.[14-17] Some of the receptors are excitatory and some are inhibitory. All this has a significant bearing on the rate coefficients for transmembrane ion flux and, therefore, on transmembrane voltage changes and signal transmission.

Using chemical kinetic techniques, we arrived at the minimum reaction scheme shown in SCHEME 1. This scheme is based on intensive investigations of the muscle type of nicotinic acetylcholine receptor in the electric organ of two fish, *E. electricus* (reviewed in ref. 5) and *T. californica* (reviewed in ref. 11), and in BC$_3$H1 cells,[4,9,12] in both membrane vesicles and single cells, using a combination of rapid mixing techniques, laser-pulse photolysis, and single-channel current measurements. The same mechanism also accounts for the results of chemical kinetic investigations of a neuronal type of acetylcholine receptor in PC12 cells,[8] the inhibitory GABA$_A$ receptor in rat brain vesicles[14,15] and mouse cerebral cortical cells,[16] and the inhibitory glycine receptor in mouse spinal cord cells.[17] The overall mechanism is typical of regulatory proteins, including those that regulate metabolism and DNA biosynthesis.[42] The upper line shows the sequential neurotransmitter-binding steps leading to an open channel, which can form on the sub-millisecond time scale; an important outcome of the use of rapid reaction techniques is the determination of the constants associated with this upper line. The inactive receptor species, which can form within milliseconds, are shown on the lower line. We have determined all the rate and equilibrium constants for this mechanism, and we can predict the concentration of the open acetylcholine receptor-channel in BC$_3$H1 cells over a 100-fold range of neurotransmitter concentration and as a function of time.[9,12]

In this paper we illustrate the use of one of the rapid chemical kinetic techniques we developed, and show how we evaluate some of the constants that determine the concentration of open receptor-channels over a wide range of neurotransmitter concentration and time. We give preliminary results that the diversity of receptors in central nervous system neurons can be expressed in terms of the diversity of values of the constants that determine the concentration of open receptor-channels, and indicate how this affects the transmembrane voltage changes that trigger signal initiation between neurons. We discuss a minimum reaction scheme containing the minimum number of constants, and we present one example in which chemical kinetic techniques gave additional insight into the mechanism of inhibition of the receptor-mediated reaction. We can thus learn the conditions under which a signal is transmitted by a cell in the central nervous system, and how these conditions change in response to external stimuli, diseases of the nervous system, and clinically relevant compounds. Elucidation of receptor mechanism has a bearing on many current problems in neurobiology. The results obtained with the new techniques presented previously[5-7] and here suggest that rapid chemical kinetic techniques, which are essential in elucidating mechanisms of reactions mediated by soluble proteins,[42-44] can also be applied to proteins that must be studied in a membrane-bound form in a cell or vesicle. Several methods that can be used in such studies are now available.[4,7,12] Accounting for the initiation, inhibition, and alteration of signal transmission in basic units (cells) from different areas of the nervous system in terms of well-characterized chemical reactions is, therefore, becoming an attainable goal. The new knowledge is expected to increase our understanding of the coordinated response of cells responsible for perception of and reaction to external stimuli, and integration and storage of information.

REFERENCES

1. CRICK, F. H. C. 1994. The Astonishing Hypothesis. Scribner & Sons. New York.
2. HESS, G. P., S. LIPKOWITZ & G. E. STRUVE. 1978. Acetylcholine-receptor-mediated ion flux in electroplax membrane microsacs (vesicles): Change in mechanism produced by asymmetrical distribution of sodium and potassium ions. Proc. Natl. Acad. Sci. USA **75:** 1703–1707.
3. HESS, G. P., D. J. CASH & H. AOSHIMA. 1979. Acetylcholine receptor-controlled ion fluxes in membrane vesicles investigated by fast reaction techniques. Nature **282:** 329–331.
4. UDGAONKAR, J. B. & G. P. HESS. 1987. Chemical kinetic measurements of a mammalian acetylcholine receptor using a fast reaction technique. Proc. Natl. Acad. Sci. USA **84:** 8758–8762.
5. HESS, G. P., D. J. CASH & H. AOSHIMA. 1983. Acetylcholine receptor-controlled ion translocation: Chemical kinetic investigations of the mechanism. Annu. Rev. Biophys. Bioeng. **12:** 443–473.
6. HESS, G. P., J. B. UDGAONKAR & W. L. OLBRICHT. 1987. Chemical kinetic measurements of transmembrane processes using rapid reaction techniques: Acetylcholine receptor. Annu. Rev. Biophys. Biophys. Chem. **16:** 507–534.
7. HESS, G. P. 1993. Determination of the chemical mechanism of neurotransmitter receptor-mediated reactions by rapid chemical kinetic techniques. Biochemistry **32:** 989–1000.
8. MATSUBARA, N. & G. P. HESS. 1992. On the mechanism of a mammalian neuronal-type nicotinic acetylcholine receptor in PC12 cells: What can one learn from chemical kinetic measurements with a 20-millisecond time resolution? Biochemistry **31:** 5477–5487.
9. NIU, L. & G. P. HESS. 1993. An acetylcholine receptor regulatory site in BC$_3$H1 cells: Characterized by laser-pulse photolysis in the microsecond-to-millisecond time domain. Biochemistry **32:** 3831–3835.
10. KATZ, B. & S. THESLEFF. 1957. A study of the "desensitization" produced by acetylcholine at the motor end-plate. J. Physiol. (Lond.) **138:** 63–80.

11. OCHOA, E. L. M., A. CHATTOPADHYAY & M. G. MCNAMEE. 1989. Desensitization of the nicotinic acetylcholine receptor: Modular mechanisms and effect of modulators. Cell. Mol. Neurobiol. **9:** 141–177.
12. MATSUBARA, N., A. P. BILLINGTON & G. P. HESS. 1992. Laser pulse photolysis of caged carbamoylcholine in investigations of a mammalian nicotinic acetylcholine receptor in BC_3H1 cells: What can one learn from chemical kinetic measurements in the microsecond time region? Biochemistry **31:** 5507–5514.
13. BILLINGTON, A. P., N. MATSUBARA, W. W. WEBB & G. P. HESS. 1992. Protein conformational changes in the μs time region investigated with a laser pulse photolysis technique. Adv. Protein Chem. **3:** 417–427.
14. CASH, D. J. & R. SUBBARAO. 1987a. Desensitization of γ-aminobutyric acid receptor from rat brain: Two distinguishable receptors on the same membrane. Biochemistry **26:** 7556–7562.
15. CASH, D. J. & R. SUBBARAO. 1987b. Two desensitization processes of GABA receptor from rat brain-rapid measurements of chloride-ion flux using quench-flow techniques. FEBS Lett. **217:** 129–133.
16. GEETHA, N. & G. P. HESS. 1992. On the mechanism of a mammalian γ-aminobutyric acid receptor in primary brain cells. Chemical kinetic measurements with a 10-millisecond time resolution. Biochemistry **31:** 5488–5499.
17. WALSTROM, K. M. & G. P. HESS. 1994. Mechanism for the channel-opening reaction of strychnine-sensitive glycine receptors on cultured embryonic mouse spinal cord cells. Biochemistry **33:** 7718–7730.
18. CASH, D. J. & G. P. HESS. 1980. Molecular mechanism of acetylcholine receptor-controlled ion translocation across cell membranes. Proc. Natl. Acad. Sci. USA **77:** 842–846.
19. PLANCK, M. 1890. Uber die Potentialdiferenz zwischen verduennten Losungen binärer Electrolytes. Annalen Physik u. Chemie **40:** 561–576.
20. HESS, G. P., H.-A. KOLB, P. LÄUGER, E. SCHOFFENIELS & W. SCHWARZE. 1984. Acetylcholine receptor (from *Electrophorus electricus*): A comparison of single-channel current recordings and chemical kinetic measurements. Proc. Natl. Acad. Sci. USA **81:** 5281–5285.
21. NEHER, E. & B. SAKMANN. 1976. Single-channel currents recorded from membrane of denervated frog muscle fibres. Nature **260:** 861–863.
22. TRUSSELL, L. O., L. L. THIO, C. F. ZORUMSKI & G. D. FISCHBACH. 1988. Rapid desensitization of glutamate receptors in vertebrate central neurons. Proc. Natl. Acad. Sci. USA **85:** 4562–4566.
23. MADESON, B. W. & R. O. EDESON. 1988. Nicotinic receptors and the elusive β. Trends Pharmacol. Sci. **9:** 315–316.
24. HAMILL, O. P., A. MARTY, E. NEHER, B. SAKMANN & F. J. SIGWORTH. 1981. Improved patch clamp techniques for high-resolution current recording from cells and cell-free membrane patches. Pfluegers Arch. **391:** 85–100.
25. DE MAYO, P. 1960. Ultraviolet photochemistry of simple unsaturated systems Adv. Org. Chem. **2:** 367–425.
26. BARLTROP, J. A., P. J. PLANT & P. SCHOFIELD. 1966. Photosensitive protective groups. Chem. Commun. 822–823.
27. BARLTROP, J. A. & N. J. BUNCE. 1968. Organic photochemistry. VIII. The photochemical reduction of nitro compounds. J. Chem. Soc. C. no. **12:** 1467–1474.
28. PATCHORNIK, A., B. AMIT & R. B. WOODWARD. 1970. Photosensitive protecting groups. J. Am. Chem. Soc. **92:** 6333–6335.
29. KAPLAN, J. H. 1990. Photochemical manipulation of divalent cation levels. Annu. Rev. Physiol. **52:** 887–914.
30. CORRIE, J. E. T. & D. R. TRENTHAM. 1993. Caged nucleotides and neurotransmitters. Biological applications of photochemical switches. *In* Bioorganic Photochemistry. H. Morrison, Ed. Vol. **2:** 243–305. Wiley. New York.
31. MILBURN, T., N. MATSUBARA, A. P. BILLINGTON, J. B. UDGAONKAR, J. W. WALKER, B. K. CARPENTER, W. W. WEBB, J. MARQUE, W. DENK, J. A. MCCRAY & G. P. HESS. 1989.

Synthesis, photochemistry, and biological activity of a caged photolabile acetylcholine receptor ligand. Biochemistry **29:** 49–55.

32. WILCOX, M., R. W. VIOLA, K. W. JOHNSON, A. P. BILLINGTON, B. K. CARPENTER, J. A. MCCRAY, A. GUZIKOWSKI & G. P. HESS. 1990. Synthesis of photolabile "precursors" of amino-acid neurotransmitters. J. Org. Chem. **55:** 1585–1589.

33. WIEBOLDT, R., K. GEE, D. RAMESH, B. K. CARPENTER & G. P. HESS. 1994. Photolabile precursors of glutamate: Synthesis, photochemical properties, and activation of glutamate receptors on a microsecond time scale. Proc. Natl. Acad. Sci. USA **91:** 8752–8756.

34. WIEBOLDT, R., D. RAMESH, B. K. CARPENTER & G. P. HESS. 1993. Synthesis and photochemistry of photolabile derivatives of γ-aminobutyric acid for chemical kinetic investigations of the GABA receptor in the millisecond time region. Biochemistry **33:** 1526–1533.

35. RAMESH, D., R. WIEBOLDT, L. NIU, B. K. CARPENTER & G. P. HESS. 1993. Photolysis of a protecting group for the carboxyl function of neurotransmitters within 3 μs and with product quantum yield of 0.2. Proc. Natl. Acad. Sci. USA **90:** 11074–11078.

36. ADAMS, P. R. 1976. Drug blockade of open end-plate channels. J. Physiol. (Lond.) **260:** 531–532.

37. SWANSON, K. L. & E. X. ALBUQUERQUE. 1987. Nicotinic acetylcholine receptor ion channel blockade by cocaine. The mechanism of synaptic action. J. Pharmacol. Exp. Ther. **243:** 1202–1210.

38. OGDEN, D. C. & D. COLQUHOUN. 1985. Ion channel block by acetylcholine, carbachol and suberyldicholine at the frog neuromuscular junction. Proc. Roy. Soc. Lond. B. **225:** 329–355.

39. MAGLEBY, K. L. & C. F. STEVENS. 1972. A quantitative description of end-plate currents. J. Physiol. **223:** 173–179.

40. SHIONO, S., K. TAKEYASU, J. B. UDGAONKAR, A. H. DELCOUR, N. FUJITA & G. P. HESS. 1984. Regulatory properties of acetylcholine receptor: Evidence for two different inhibitory sites, one for acetylcholine and the other for a noncompetitive inhibitor of receptor function (procaine). Biochemistry **23:** 6889–6893.

41. LANDAU, V. G. & E. M. LIFSHITZ. 1959. Fluid Mechanics.: 219. Pergamon. Oxford.

42. JOHNSON, K. A. 1992. Transient state kinetic analysis of enzyme reaction pathways. *In* The Enzymes **20:** 1–61. Academic Press. New York.

43. HAMMES, G. G. 1982. Enzyme Catalysis and Regulation. Academic Press. New York.

44. FERSHT, A. 1985. Enzyme Structure and Mechanism. W. H. Freeman. New York.

45. GEE, K. R., R. WIEBOLDT & G. P. HESS. 1994. Synthesis and photochemistry of a new photolabile derivative of GABA. Neurotransmitter release and receptor activation in the microsecond time region. J. Am. Chem. Soc. **116:** 8366–8367.

Ligands, Receptor Models, and Evolution

VIC COCKCROFT,[a] MARCELO ORTELLS,
AND GEORGE LUNT

Biochemistry Department
Bath University
Bath BA2 7AY, United Kingdom

With the coming of age of molecular neurobiology, many components of signaling systems in the brain are now characterized to some extent at the molecular level, and a reductionism approach towards a more complete synthesis of neurofunction now seems possible. In the case of neurotransmitter receptors the pace at which their gene sequences and derived primary structures have been acquired has been astounding. Several surprises have emerged from this work, the most notable of which is that only a few structural classes of neurotransmitter receptors exist. Two major superfamilies have been identified, the ligand-gated ion-channel (LGIC) receptors,[1] and the G-protein coupled receptors (GPCR).[2] They mediate the fast "all or nothing" and the slow-graded modes of chemical communication, respectively. Importantly, they encompass the majority of the receptor neuropharmacology that has accumulated to date.

In the case of the LGIC receptors, the structural and conformational requirements for agonist and antagonist binding to the receptors of each of them separately have been extensively covered in the literature. In this paper a more unified pharmacophore is proposed for the LGIC receptors.

LIGAND-GATED ION-CHANNEL AGONISTS

FIGURE 1 shows a collection of agonist structures for the different types of LGIC receptors. In rows from top to bottom are the neurotransmitters, semirigid analogues with restricted rotatable dihedral bonds, and the almost totally rigid agonist analogues. Several comparisons are possible in the context of the superfamily. The common features for binding to the LGIC receptors can be identified as (1) an amine group, (2) a π-bonded system, (3) an sp^2 electronegative atom within the π-system, and (4) small size (i.e. < 15 nonhydrogen atoms). The unified receptor pharmacophore model incorporates these conserved features. The broad similarity of agonists is strikingly demonstrated by comparison with the almost totally rigid analogues cytisine and THIP (4,5,6,7-tetrahydroisoxazolo[5,4-c]pyridin-3-ol), which are agonists of the evolutionary distantly related nicotinic acetylcholine and GABA$_A$ receptors, respectively.

For the amine group, this can vary from a primary through to a quaternary group, even for a given receptor type (see nAChR column in FIG. 1). Acetylcholine is an informative molecule because it establishes that the role for the conserved nitrogen center is to provide a positively charged group for interaction with the receptor site.

---

[a] Present address: Orion-Farmos, BioCity, P.O. Box 425, 20101 Turku, Finland.

The π-bonded system part of the molecule can also vary, both in shape and electronic properties, even for a given receptor (e.g., the pyridyl ring in nicotine instead of the acetyl group in acetylcholine). Nonetheless, it displays several unusual and interesting features that are conserved, which may be important for recognition. It is rigid-coplanar and contains an electronegative atom center that is capable of forming a hydrogen bonding interaction. Of particular note, the electronegative center forms a delocalizable local dipole over the π-system.[3]

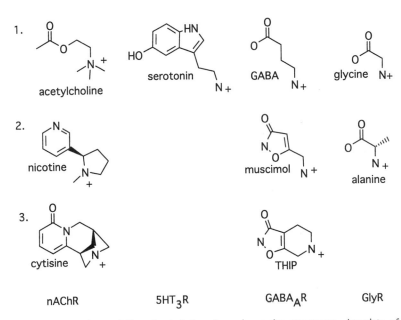

FIGURE 1. Comparison of ligand-gated ion-channel agonist structures. Agonists of the nicotinic acetylcholine-R (nAChR), 5HT$_3$-R (5H$_3$R), γ-amino butyric acid$_A$-R (GABA$_A$ R), and glycine-R (GlyR) are shown. Row 1, endogenous neurotransmitters; row 2, semirigid analogues; row 3, almost totally rigid analogues. The symbol (+) highlights the positively charged amine group. Note that none of the endogenous neurotransmitter molecules contains a chiral center.

SPECIFICITY OF RECEPTOR RECOGNITION

Features leading to specific recognition can be deduced by looking for consistent differences in the ligands for the various types of receptor. However, this is a less sure exercise than identification of the common features. Differences proposed as having a role in specific recognition are:

Position and types of polar atoms within the π-system region. For 5HT$_3$-R as compared to the nicotinic acetylcholine-R there is an apparent switch from a requirement for a hydrogen bond donor to a hydrogen bond acceptor.

Distance between the electronegative center of the π-system and the positive pole. This is apparent for glycine-R (3.5 Å) as compared to the GABA$_A$-R ($\sim$5.5 Å).

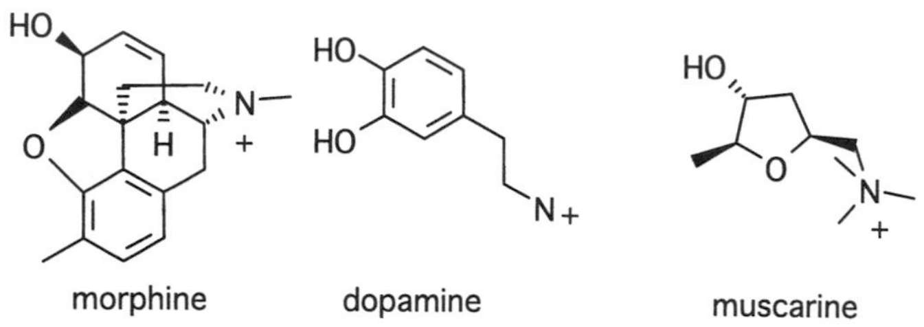

FIGURE 2. Examples of monoamine G-protein coupled receptor agonist structures. Common features are (1) the positively charged amine group and (2) the occurrence of a ring hydroxyl at position that is six bond lengths away from the amine group.

Hydrophobic and polar-charged nature of the π-system region. This places the cationic channel and anionic channel receptors into two separate groupings.

Size of the π-system. This might be of secondary importance to (1) above in achieving specific recognition at the $5HT_3$-R and nicotinic acetylcholine-R.

Of course differences in the ligands must also reflect differences in the receptors, but in the context of a conserved binding site for the superfamily.

OTHER RECEPTOR SUPERFAMILIES

A comparison of LGIC receptor agonists with those of the G-protein coupled receptors is useful to assess the significance of the proposed common features of the LGIC agonists. This is possible because some of the endogenous neurotransmitters act on receptors of both classes. This is the case at least for acetylcholine and serotonin. Of the structures shown, muscarine (FIG. 2) is an informative molecule because it establishes that a π-bonded system is not essential for agonist interaction at G-protein coupled receptors.

Initially it was expected that glutamate receptor subunits would also fall into the nicotinic-containing LGIC superfamily. Instead, based on analysis of multiple aligned sequences it can be seen that no apparent homology exists between the classes. Indeed, the extracellular domain of the glutamate receptors is more similar to a family of bacterial amino acid binding proteins than it is to any LGIC receptor.[4,5] This raises the possibility that in the LGIC superfamily the amine has to be an isolated group and cannot be functionalized such as in the form of an amino acid moiety.

LGIC ANTAGONISTS AND MODULATORS

In some antagonist structures the skeleton of their corresponding agonists can be identified (FIG. 3). For the nicotinic acetylcholine-R, the natural product methylly-caconitine (MLA) is a clear example.[6] This is of interest because the parent structure aconitine is known to act on voltage-gated sodium channels, but has no reported activity at nACh receptors. With the ester linkage in MLA a framework is introduced that can be fitted to acetylcholine. For GABA, it has already been proposed that bicucullin contains a region with the electronic properties resembling the carboxylate group of the endogenous transmitter.[7] This correspondence can also be extended to the GABA$_A$-R modulators flumazenil and ethyl-β-carboline-3-carboxylate. When such compounds are overlayed using the three site-points present in the agonist unified pharmacophore model, an impression of structural conservation outside the agonist binding region is obtained (see FIG. 4).

UNIFIED LGIC PHARMACOPHORE MODEL

FIGURE 5 shows a unified LGIC receptor pharmacophore model. The agonist site comprises the positive pole and the oriented local dipole with the electronegative

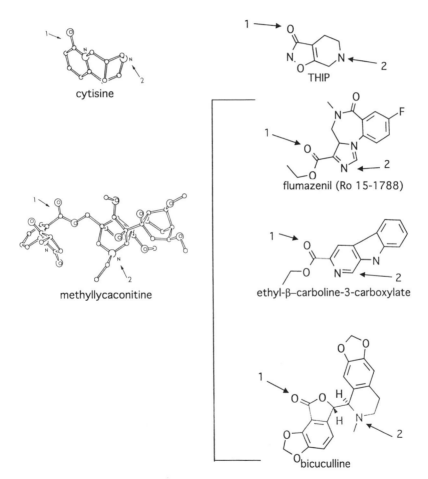

FIGURE 3. Comparison of ligand-gated ion-channel antagonists, modulators, and agonists. Methyllycaconitine is a competitive antagonist of neuronal nicotinic acetylcholine-R;[6] cytisine is a potent agonist of nicotinic acetylcholine-R; bicuculline is a competitive antagonist of GABA$_A$-R; ethyl-*b*-carboline-3-carboxylate is an inverse agonist of GABA$_A$-R;[7] flumazenil is a positive modulator of GABA$_A$-R; THIP is a potent agonist of GABA$_A$-R.

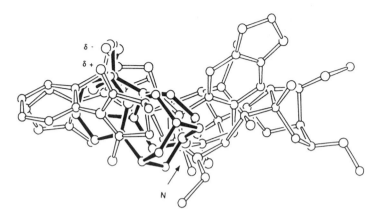

FIGURE 4. Overlay of dissimilar ligand-gated ion-channel antagonist structures. The rigid agonists cytisine and THIP (*solid bonds*) and the antagonists methyllycaconitine and bicuculline are shown. Superpositioning was done using three points: the nitrogen atom of the positive pole and the electronegative and electropositive sp^2 atom centers of the oriented dipole. The agonist-antagonist pairs THIP-bicuculline and cytisine-methyllycaconitine were first superimposed. These pairs were then overlaid by superpositioning of the agonists THIP and cytisine. THIP, 4,5,6,7-tetrahydroisoxazolo[5,4-c] pyridin-3-ol.

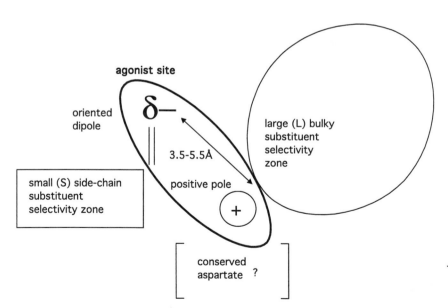

FIGURE 5. Unified ligand-gated ion-channel pharmacophore model. *Symbols*: ⊕ = the positive pole; δ− = electronegative atom of the π-system. The large (L) and small (S) regions outside the agonist core are regions of the binding site that can be occupied by antagonists.

atom center available for hydrogen bonding. Outside the agonist core two regions are proposed as being occupied by antagonist molecules. The large region (L) corresponds to the aconitine portion of methyllycaconitine, whereas the smaller region (S) is occupied by its *N*-phenylsuccinimide moiety. In the case of the benzodiazepine modulators only the large region is occupied.

Inasmuch as the *N*-phenylsuccinimide moiety of methyllycaconitine is topologically equivalent to the reactive group in ligands that label the α-subunit cysteines 192–193 of the *Torpedo* nicotinic acetylcholine-R, it can be expected that this moiety contacts the surface area of the binding cavity. In contrast, the oxygen-rich large bulk

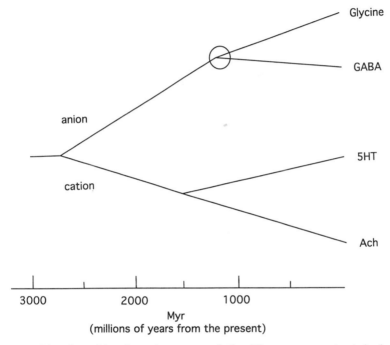

FIGURE 6. Ligand-gated ion-channel receptor evolution. The tree was constructed using 106 nucleic acid sequences.[8] The tree presented shows only branch points for early events in receptor evolution. The circling of the bifurcation point for the GABA and glycine branches signifies that they are not sister branches.

of the aconitine portion of methyllycaconitine could be facing out towards the solvent. Interaction with parts of neighboring subunits could then be possible, as is observed with *d*-tubocurarine, a competitive antagonist of nicotinic acetylcholine-R.[8]

From the unified pharmacophore model a hypothetical recognition pathway model is proposed. Starting from when an agonist molecule is within 12 Å of the assumed conserved aspartate of the binding site, long-range electrostatic interaction between the negative charge of this residue and the positive pole of the agonist is sufficient to cause the agonist to be attracted towards it. On closer approach, at 6 Å, the local dipole of the agonist could become oriented in the electrostatic field of the aspartate residue. The steering of the ligand at this stage may assist the entry of agonist into the binding cavity. On formation of the binary complex, the overwhelming electrostatic field of the aspartate residue increases the size of the local dipole of the π-system, causing a shift in electron density over the electronegative atom. Synergism for the interaction of agonist with the receptor could be obtained by this induction step, because the electronegative atom becomes a more effective hydrogen bonding group.

RECEPTOR EVOLUTION

The evolutionary tree in FIGURE 6 shows branch points representing the early separation of different types of LGIC receptor types.[9] Under the assumption of the analysis, the initial branch point would have been at least 2500 million years ago. This would roughly exceed current estimates for the time of origin of eukaryotes. Although this is a surprisingly early start for the LGIC receptors, in the case of G-protein coupled receptors it is now reasonably well founded that bacteriorhodopsin is structurally homologous to the vertebrate rhodopsins. It is also apparent from the tree that the origins of the distinct receptor types are also ancient, occurring at some time when multicellular organisms were starting to evolve.

Given the hierarchical nature of divergent evolution, the question can be asked, How might receptor evolution have had an impact on current day pharmacological probing of receptor superfamilies? Not too surprisingly, examples have appeared in the literature where the selectivity of a ligand acting at a subset of receptors reflects evolutionary relatedness. The adrenergic pharmacological classification scheme appears to be a good example of this.[10] However, there are other examples where a particular ligand interacts with a receptor distantly related to its classical pharmacologically defined site.[11] Examination of these unusual instances, termed as leakage specificity, will no doubt allow understanding of how an isolated pairwise interaction between a group in the ligand and an amino acid side-chain in receptors can influence pharmacological specificity.

In summary, evaluating classical pharmacology in the context of a superfamily can involve comparison of ligands and construction of receptor homology models, and should be viewed in an evolutionary setting.

REFERENCES

1. SCHOFIELD, P. R., M. G. DARLISON, N. FUJITA, D. BURT, F. A. STEPHENSON, H. RODRIGUEZ, L. M. RHEE, J. RAMACHANDRAN, V. REALE, T. A. GLENCORSE, P. H. SEEBURG & E. A. BARNARD. 1987. Sequence and functional expression of the GABA$_A$ receptor shows a ligand gated receptor superfamily. Nature **328:** 221–227.
2. DIXON, R. A. F., B. K. KOBILKA, D. J. STRADER, J. L. BENOVIC, H. G. DOHLMAN, T. FRIELLE, M. A. BOLANOWSKI, C. D. BENNETT, E. RANDS, R. E. DIEHL, R. A. MUMFORD, E. E. SLATER, I. S. SIGAL, M. G. CARON, R. J. LEFKOWITZ & C. D. STRADER. 1986. Cloning of the gene and cDNA for mammalian β-adrenergic receptor and homology with rhodopsin. Nature **321:** 650–656.
3. COCKCROFT, V. B., D. J. OSGUTHORPE, E. A. BARNARD & G. G. LUNT. 1990. Modelling of agonist binding to the ligand-gated ion-channel superfamily of receptors. Proteins Struct. Funct. Genet. **8:** 386–397.
4. NAKANISHI, N., N. A. SHNEIDER & R. AXEL. 1990. A family of glutamate receptor genes: Evidence for the formation of heteromeric receptors with distinct channel properties. Neuron **5:** 569–581.

5. COCKCROFT, V. B., M. ORTELLS & G. G. LUNT. 1993. Homologies and disparities of glutamate receptors: A critical analysis. Neurochem. Int. **23**: 583–594.
6. WARD, J. M., V. B. COCKCROFT, G. G. LUNT, F. S. SMILLIE & S. WONNACOTT. 1990. Methyllycaconitine: A selective probe for neuronal alpha-bungarotoxin binding sites. FEBS Lett. **270**: 45–48.
7. APRISON, M. H. & K. B. LIPKOWITZ. 1989. On the $GABA_A$ receptor: A molecular modelling approach. J. Neurosci. **23**: 129–135.
8. PEDERSON, S. E. & J. B. COHEN. 1990. *d*-Tubocurarine binding sites are located at α–γ and α–δ subunit interfaces of the nicotinic acetylcholine receptor. Proc. Natl. Acad. Sci. USA **87**: 2785–2789.
9. ORTELLS, M. O. & G. G. LUNT. Evolutionary history of the ligand gated superfamily of receptors. In preparation.
10. KOBILKA, B. K., T. S. KOBILKA, K. W. DANIEL, J. W. REAGAN, M. G. CARON & R. J. LEFKOWITZ. 1988. Chimeric $\alpha2$–$\beta2$ adrenergic receptors: Delineation of domains involved in effector coupling and ligand binding specificity. Science **240**: 1310–1316.
11. HOYER, D., G. ENGEL & H. O. KALKMAN. 1985. Molecular pharmacology of $5HT_1$ serotonin receptors in human brain. I. Characterization and autoradiographic localization of $5\text{-}HT_{1A}$ recognition sites in rat and pig brain membranes: Radioligand binding studies with 5-HT, 8-OH-DPAT, (−) iodocyanopindolol, and mesulergine. Eur. J. Pharmacol. **118**: 13–23.

Nicotinic Receptor Function in the Mammalian Central Nervous System[a]

EDSON X. ALBUQUERQUE,[b,c] EDNA F. R. PEREIRA,[b,c]
NEWTON G. CASTRO,[b,c] MANICKAVASAGOM ALKONDON,[b]
SIGRID REINHARDT,[d] HANNSJÖRG SCHRÖDER,[e]
AND ALFRED MAELICKE[d]

[b]Department of Pharmacology and Experimental Therapeutics
University of Maryland School of Medicine
Baltimore, Maryland 21201

[c]Laboratory of Molecular Pharmacology II
Institute of Biophysics "Carlos Chagas Filho"
Federal University of Rio de Janeiro
Rio de Janeiro, RJ 21944, Brazil

[d]Institute of Physiological Chemistry and Pathobiochemistry
Johannes-Gutenberg University Medical School
Duesbergweg 6
Mainz, Germany

[e]Department of Anatomy
University of Köln
Köln, Germany

Earlier studies on the characterization of nicotinic acetylcholine receptors (nAChRs) in the central nervous system (CNS) relied on assays of the binding of nicotinic radioligands to various brain regions by the use of radioligand binding assays.[1–5] However, it was not until very recently that the functional properties of some of these receptors could be electrophysiologically addressed.[6–20]

Initially, on the basis of the binding of [3H]nicotine and [125I]α-bungarotoxin (α-BGT) to various brain regions, it was demonstrated that two distinct populations of presumed nAChRs existed in the CNS.[2,3] Neuronal nAChRs that could bind [3H]nicotine with high affinity were identified in the interpeduncular nucleus, superior colliculus, medial habenula, substantia nigra pars compacta and ventral tegmental area, molecular layer of the dentate gyrus, presubiculum, cerebral cortex, and most of the thalamic nuclei.[3] Studies carried out in brain synaptosomes led to the suggestion that these neuronal nAChRs were localized mostly presynaptically and were responsible for the control of transmitter release in various neurotransmitter

[a]This work was supported in the United States by National Institutes of Health grants NS25296 and ES05730, National Institute of Mental Health Center for Neuroscience and Schizophrenia grant P-50MH44211, and a Conselho Nacional de Pesquisa e Desenvolvimento graduate student fellowship from Brazil. In Germany, this work was supported by DFG grant Ma 599/17-1 and a grant from Fonds der Chemischen Industrie.
[b]Address to which reprint requests should be sent.

systems.[21-23] On the other hand, [^{125}I]α-BGT was found to label several brain areas, namely, the cerebral cortex, hippocampus, hypothalamus, inferior and superior colliculus, and some brain-stem nuclei,[3-5] whereas no α-BGT-sensitive nicotinic responses could be found in these areas.[24] For this reason, α-BGT-binding proteins at one time were believed to mediate functions unrelated to nAChR.[25] Further, it was not completely clear whether nAChRs binding [^{3}H]nicotine with high affinity consisted of a homogenous population of a single nAChR subtype, or a population of various subtypes of nAChRs that in spite of being distinct from one another could have similar affinities for [^{3}H]nicotine. Several complementary experimental approaches, including ligand-binding assays, molecular biological procedures, and electrophysiological techniques were necessary to unveil the functional, pharmacological, and structural diversity of CNS nAChRs.[26,27]

Like their counterparts in the muscle, the nAChRs in the CNS are ligand-gated cation channels. Inasmuch as neuronal and muscle nAChRs are encoded by homologous genes, cDNAs encoding muscle nAChR subunits have been used to screen libraries of mRNAs isolated from neurons from various CNS areas of chick, rat, and humans. To date, at least eight α- (α2 through α9) and three β- (β2 through β4) subunits have been cloned from chick, rat, and human neuronal tissues.[26,28] A CNS nAChR subunit is classified as α if it has the characteristic ACh-binding domain found in the muscle nAChR α-subunit, that is, the vicinal Cys residues in the N-terminal region. The CNS nAChR β-subunits comprise a group of proteins that, in addition to playing an important role in defining the structure of the receptor, may also contribute to the sensitivity of the nAChR to a number of pharmacological agents. Whereas the CNS nAChR α-subunits are considerably homologous with their muscle counterparts, the CNS nAChR β-subunits show very low homology to the muscle nAChR β-subunit.[29] It is most likely that the diverse properties of CNS nAChRs are largely due to the fact that these receptors can be made up of different associations of α- and β-subunits. Indeed, transient expression of a variety of combinations of α- and β-subunits can give rise to a number of functionally distinct neuronal nAChRs in *Xenopus* oocytes.[30] Molecular biological studies have also proven that α7, α8, and α9 nAChR subunits can form homomeric, functional nAChR channels that are remarkably and uniquely sensitive to blockade by α-BGT.[28,31,32]

Although several studies have dealt with the characterization of recombinant neuronal nAChRs,[26] it was not until recently that studies were directed at characterizing functionally and pharmacologically native CNS nAChRs *in situ*.[27] Research in our laboratory has been aimed at identifying the various functional nAChRs expressed throughout the brain, at understanding the mechanisms by which nAChR activity can be modulated, and at defining the ion selectivity of α-BGT-sensitive CNS nAChRs. In the present paper, the following topics are discussed: (1) how CNS nAChR subtypes can be identified on the basis of the pharmacological and kinetic properties of nicotinic currents activated in various CNS neuronal preparations, (2) the modulation of nAChR activity via an ACh-insensitive binding site and via intracellular mechanisms, and (3) the ion permeability of α-BGT-sensitive CNS nAChRs. An understanding of the pharmacological and functional characteristics of CNS nAChRs may serve as the cornerstone for the development of useful therapeutic strategies to treat and/or prevent neuropathological conditions such as Alzheimer's[33] and Parkinson's[34] diseases and nicotine addiction, in which the function of CNS nicotinic systems is known to be severely compromised, and will help to unveil the physiological roles of nicotinic synaptic transmission in the brain.

CHARACTERIZATION OF FUNCTIONAL nAChRs IN MAMMALIAN CNS NEURONS

By the end of the 1980s, molecular biological studies revealed that mRNAs coding for several nAChR α- and β-subunits can be found in various regions of the CNS,[26] and it was not until the early 1990s that reports regarding the identification and characterization of functional nAChRs in the mammalian CNS by electrophysiological techniques started to appear.[8–10,12,13] Previously, only two studies had reported the existence of functional nAChRs in CNS neurons by direct electrophysiological techniques.[6,7] Our studies on the characterization of hippocampal nAChRs began in 1987 when we demonstrated, for the first time, that nicotinic agonists could activate single-channel currents in cultured hippocampal neurons patch-clamped under the cell-attached condition.[6] The relative difficulty in recording nicotinic currents from CNS neuronal preparations may have contributed to there being until recently practically no electrophysiological studies of nicotinic responses in CNS neurons. Some of the difficulties underlying studies of neuronal nicotinic currents were overcome by the development of devices that could guarantee that nicotinic agonists could be rapidly applied to and immediately removed from the vicinity of the cells.[12,13] The device we initially used consisted of a U-shaped tube fashioned from a thin capillary glass. At the apex of the U tube a pore of 250–400 μm in diameter was made, and through this pore the test solutions were delivered to and removed from the vicinity of the neurons.[13,17] Using this device and others similar to it, it was possible to measure reliably nicotinic responses in biological preparations of CNS neurons.[12–19]

In our initial studies on nicotinic responses in CNS neurons, we found that α-cobratoxin[10] or α-BGT[13] could block nicotinic whole-cell currents activated in cultured hippocampal neurons, demonstrating unequivocally for the first time the existence of α-BGT-sensitive nicotinic currents in CNS neurons. In these neurons, which were cultured from the hippocampi of rat fetuses (embryonic day 17–18), the peak-current amplitude increased with the number of days the neurons were maintained in culture and leveled off by 4–5 weeks after the plating of the neurons.[13] For instance, if a neuron was considered to display nicotinic sensitivity when it responded to anatoxin-a (AnTX, 10 μM) with a peak whole-cell current of at least 10 pA at −50 mV, about 33% of the 6–14-day-old cultured neurons, 88% of the 21–22-day-old cultured neurons, and 98% of the 25–50-day-old cultured neurons showed nicotinic sensitivity.[13] When the hippocampi were removed from the brain of postnatal rats at various ages and the neurons were acutely dissociated by mechanical means, a similar pattern of increase in the peak amplitude of the nicotinic currents was observed.[14]

The pharmacological identification of functional hippocampal nAChRs sensitive to α-BGT relied on the α-BGT-induced blockade of whole-cell currents activated by nicotinic agonists. However, the slowness of onset and pseudoirreversible nature of the blockade induced by this toxin limited its utility. Under the experimental conditions used, the peak amplitude of α-BGT-sensitive nicotinic currents had a tendency to run down with time. This rundown, whose magnitude was variable from cell to cell, was apparently independent of agonist identity and concentration.[13,15] It was then questioned whether the substantial decrease of the peak amplitude of hippocampal nicotinic currents observed after incubation of the neurons with α-BGT reflected the effect of the toxin or rundown of the nicotinic responses. Therefore, we had to search for a new antagonist that could specifically and reversibly block α-BGT-sensitive CNS nAChRs.

Methyllycaconitine (MLA), an alkaloid isolated from the seeds of *Delphinium brownii*, was initially shown to inhibit specifically and potently the binding of [^{125}I]α-BGT to brain membrane preparations.[35] Only at very high concentrations could MLA antagonize the binding of [^{3}H]nicotine to the brain membrane preparations or the binding of α-BGT to muscle nAChRs.[35] Although MLA could reliably be used to identify α-BGT-sensitive CNS nAChRs in binding studies, its mechanisms of action on CNS nAChRs remained unknown until its effects on nicotinic currents activated in cultured hippocampal neurons were investigated. When applied to the neurons via the bath perfusion, MLA could specifically and reversibly decrease the peak amplitude of whole-cell currents elicited by nicotinic agonists.[15] The onset of the MLA effect was very rapid. At concentrations of MLA that could completely inhibit nicotinic currents, the responses of the neurons to *N*-methyl-D-aspartate (NMDA), quisqualate, kainate, or GABA remained unchanged. The IC_{50} for MLA in inhibiting nicotinic currents in hippocampal neurons was found to be about 150

ACh (3 mM)

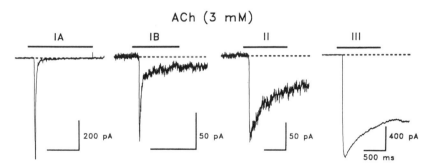

FIGURE 1. Family of ACh-activated whole-cell currents in cultured hippocampal neurons. Sample recordings of whole-cell currents evoked in four hippocampal neurons cultured for 20 days. Recordings were made 4–18 min after obtaining the whole-cell patch. Holding potential = −56 mV. ACh (3 mM) pulses were applied for 0.5 to 1 s as indicated by the solid bars on the top of the traces. The external bath solution (pH, 7.3; osmolarity, 330 mosm) consisted of NaCl 165 mM, KCl 5 mM, CaCl$_2$ 2 mM, glucose 10 mM, 4-(2-hydroethyl)-1-piperazineethanesulfonic acid (HEPES) 5 mM, and tetrodotoxin (TTX) 0.3 μM. The internal pipette solution (pH, 7.3; osmolarity, 340 mosm) consisted of CsCl 80 mM, CsF 80 mM, ethyleneglycoltetraacetic acid (EGTA) 10 mM, and HEPES 10 mM.

pM, and the inhibition caused by MLA was competitive with that caused by α-BGT. Therefore, MLA is a potent, reversible, and competitive nicotinic antagonist specific for α-BGT-sensitive CNS nAChRs.[15,17]

The knowledge that accumulated throughout the years on the characteristics of recombinant and native CNS nAChRs prompted us to address in detail the diversity of CNS nAChRs expressed in various brain regions.[17,19,36] Using a number of nicotinic agonists and antagonists, we were able to demonstrate that at least three distinct types of nicotinic currents can be elicited in hippocampal neurons. According to their pharmacological and kinetic properties, these currents were classified into types IA, II, and III[17,19] (see FIG. 1). It was suggested that each of these currents is carried by a distinct type of CNS nAChR channel.

Type IA currents, which are the predominant nicotinic responses that can be elicited in hippocampal neurons, desensitize fast, have low affinity for ACh, and are highly sensitive to blockade by the neurotoxins α-BGT and MLA.[17] Based on EC_{50}s in eliciting type IA currents, AnTX was the most potent agonist, followed by dimethylphenylpiperazinium (DMPP), (−)nicotine, cytisine, ACh, carbachol, and (+)nicotine. At saturating concentrations of any given nicotinic agonist tested, the decay phase of type IA currents at a holding potential of −50 mV could be fit by a double-exponential function, with the fast decay-time constant being as short as 6 ms and averaging 26.7 ± 1.9 ms. In outside-out patches excised from the soma of hippocampal neurons that showed at least 50 pA of acetylcholine (ACh, 1 mM)-elicited type IA currents, ACh activated a type of single channel whose open time is about 110 μs at −60 mV and that has a conductance of 73 pS.[18] This type of ACh-activated channel is sensitive to blockade by MLA, and is consistently found in outside-out patches excised from neurons that display type IA whole-cell currents. In addition, these single-channel currents show the fast kinetics of activation and inactivation characteristic of type IA currents. Thus, single channels that inactivate rather fast, have a high conductance, and a brief lifetime account for the α-BGT-sensitive IA currents.[18] Because the activation and desensitization rates of these channels are so similar, the response is always transient, and its magnitude depends upon the rate of agonist application. This may explain the failure of previous studies, which used slow agonist perfusion, to detect the brief-lifetime, fast-inactivating, α-BGT-sensitive single-channel currents.

Only about 10% of the cultured hippocampal neurons studied to date have exhibited type II and III nicotinic currents.[16,18] These currents desensitize relatively slowly, are relatively insensitive to blockade by α-BGT or MLA, and can be selectively blocked by either dihydro-β-erythroidine (DHβE, type II currents) or mecamylamine (type III currents). A distinct rank order of potency of agonists was found for each of these two types of nicotinic currents. Whereas the most potent agonist in eliciting type II currents was ACh, followed by AnTX, (−) nicotine, DMPP, carbamylcholine, cytisine, and (+) nicotine, the most potent agonist in eliciting type III currents was AnTX, followed by cytisine, (−)nicotine, DMPP, ACh, carbamylcholine, and (+)nicotine.[17] In some neurons, whole-cell currents with characteristics of both type IA and type II currents have been observed (FIG. 1). These composite currents, which are referred to as IB, display a rapidly decaying component that can be specifically blocked by α-BGT or MLA, and a slowly decaying component that can be specifically blocked by DHβE.[17] In addition, fast application of ACh (1 mM) to a few outside-out patches excised from hippocampal neurons can activate simultaneously two distinct types of single channels: a fast-desensitizing, brief-lifetime, and high-conductance channel that accounts for the type IA current, and a slowly desensitizing, long-lifetime, and low-conductance channel that accounts for the type II currents.[18] Although these findings indicate that some neurons can express two types of nAChRs and that in some cases both nAChR subtypes can be expressed in the same region of the neuron, it is still unclear whether topological segregation of the two nAChR subtypes in the neuronal soma, dendrites, or axon also takes place.

The kinetic and pharmacological properties of the three "pure" types of whole-cell currents activated by nicotinic agonists in hippocampal neurons led us to suggest that structurally distinct nAChRs subserved each of these currents.[16] Comparison of the properties of currents evoked by activation of recombinant nAChRs expressed in oocytes with those of the nicotinic whole-cell currents activated in hippocampal neurons led us to hypothesize that an α7-bearing nAChR accounted for the type IA response, an α4β2 nAChR for the type II response, and an α3β4 nAChR for the type III response.[17] These inferences were supported by the results obtained in in situ hybridization studies using digoxigenin-labeled cDNA probes specific for nAChR α7, α4, and β2 subunits. The mRNAs coding for CNS nAChR α7, α4, and β2 subunits were found to be expressed in cultured hippocampal neurons[19] (FIG. 2), the α7- and α4-subunit mRNAs being detected mostly in the perinuclear cytosol of the cultured cells. In addition, the α7-subunit mRNA was found to be expressed in about 70–80% of all cells, whereas the α4-subunit mRNA was found to be expressed in only 20–30% of the cells, indicating that the proportion of neurons that express the α7- or α4-subunit mRNAs is strongly correlated with the probability of eliciting type IA or type II currents. Nevertheless, it remains unclear whether type IA currents are subserved by a homomeric α7 nAChR or a heteromeric nAChR that bears in its structure the α7 subunit, because whereas the kinetic and pharmacological properties of IA responses are the same as those of currents evoked by activation of homomeric α7 nAChRs transiently expressed in oocytes, some of the properties of the ion channels that account for IA responses are quite different from those of

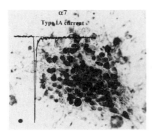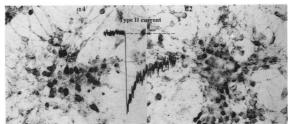

FIGURE 2. Identification of nAChR subunit mRNAs expressed in rat cultured hippocampal neurons. **Left panel:** Expression of α7-subunit mRNA in cultured hippocampal neurons as revealed by *in situ* hybridization using a specific digoxigenin-labeled cDNA probe for 350 bps (1617–1968) of the 3′ end. A sample recording of a type IA current recorded from a hippocampal neuron cultured for 20 days and held at −56 mV is superimposed. As stated in the text, the characteristics of this nicotinic current resembles those of currents triggered by activation of homomeric α7 nAChRs in *Xenopus* oocytes. **Middle and right panels:** Expression of α4-subunit mRNA (*middle*) and β4-subunit mRNA (*right*) in cultured hippocampal neurons as revealed by *in situ* hybridization using specific digoxigenin-labeled cDNA probes for 202 bps (1903–2105) of the 3′ end of the α4-subunit mRNA and for 205 bps (1–205) of the 5′ end of the β2-subunit mRNA. A sample recording of a type II current recorded from a hippocampal neuron cultured for 20 days and held at −56 mV is superimposed to the middle and right panels. As stated in the text, the characteristics of this nicotinic current resembles those of currents triggered by activation of α4β2 nAChRs in various expression systems. Methodology used has been described elsewhere.

homomeric α7-nAChR channels expressed in oocytes.[18,37] For instance, whereas the single channels that account for type IA currents appear as isolated, short-lived events, homomeric α7 nAChRs expressed in oocytes are typically activated in bursts of 10-ms duration.[37] These functional discrepancies indicate that posttranslational modifications and/or actual subunit compositions differ between the native α-BGT-sensitive hippocampal nAChR and the homomeric α7 nAChR expressed in oocytes. It is also conceivable that these discrepancies could be accounted for by the membrane compositions of *Xenopus* oocytes and hippocampal neurons.

We also demonstrated that ACh and other nicotinic agonists can activate fast-desensitizing, MLA-/α-BGT-sensitive whole-cell currents in cultured neurons from the rat olfactory bulb[36] (FIG. 3). The pharmacological and kinetic profile of these nicotinic currents suggested that olfactory bulb neurons express a neuronal

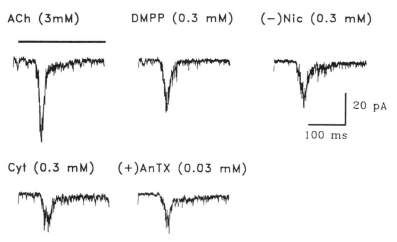

FIGURE 3. Nicotinic currents activated in rat cultured olfactory bulb neurons. Sample recordings of whole-cell currents evoked in one olfactory bulb neuron cultured for 29 days. The nicotinic agonists were applied to the cell as 250-ms pulses separated by 2-min intervals. Holding potential = −56 mV. Compositions of the external and internal solutions were the same as those described in FIGURE 1.

nAChR with properties similar to those reported above for α-BGT-sensitive nAChRs expressed in hippocampal neurons. To date, no other type of nicotinic response has been found in cultured olfactory bulb neurons, although mRNAs encoding for α4 and β2 have been found in these neurons by *in situ* hybridization (unpublished observations).

The discovery of substances that can act specifically on one receptor type in the CNS is of immense therapeutic implications given that such substances would have more selective neurological effects and as a result may be less toxic. Very recently, the alkaloid epibatidine, which was originally isolated from frog skin, was shown to have a potent analgesic activity that is insensitive to naloxone, but can be completely antagonized by the nicotinic antagonist mecamylamine.[38] The idea that CNS nAChRs could mediate the analgesic activity of epibatidine was supported by the finding that this alkaloid can induce mecamylamine-sensitive ion fluxes into cultured cells that express different neuronal nAChR subtypes.[38] The (−) and (+) stereoisomers of epibatidine were approximately equipotent as nicotinic agonists in the tested cell systems.[38] To characterize the specificity of epibatidine on CNS nAChRs, we tested the effects of both isomers in cultured hippocampal neurons under whole-cell patch-clamp conditions. Both (+) and (−) epibatidine were able to act as full agonists in eliciting type IA currents (FIG. 4), and practically no stereoselectivity was observed, because the EC_{50}s for (−) and (+) epibatidines were of about 1 and 2 μM, respectively (FIG. 5). In contrast, low nanomolar concentrations of (+) or (−) epibatidine were able to activate type II currents (FIG. 4), and the (+) isomer was much more effective than the (−) isomer in eliciting type II currents (FIG. 5). Based on our results, epibatidine would seem to be a more potent and selective agonist for the hippocampal α4β2 nAChR than for the α7-bearing hippocampal nAChR. It remains to be determined whether epibatidine would be even more potent in activating other types of neuronal nAChRs.[39] Epibatidine, in addition to being an extremely potent agonist on neuronal nAChRs, has also been proven to be a highly specific nicotinic agonist, because it does not interact with other neuronal neurotransmitter receptors, such as GABA receptors, NMDA receptors, serotoninergic receptors, and dopaminergic receptors.[38] Thus, the characterization of the interactions of this alkaloid with CNS nAChRs may be of major importance to the understanding of how these receptors are involved in the analgesia induced by nicotinic agonists.[40]

MODULATION OF nAChR ACTIVITY IN THE MAMMALIAN CNS

Modulation of nAChR Activity by Extracellular Ligands

The coexistence of neuromodulators and neurotransmitters in many synapses is associated with the presence of specific receptor-binding sites that are distinct from the agonist sites and are presumed to recognize the modulator. The NMDA type of glutamatergic receptor, for example, which is involved in learning, memory, and neuronal function, has a number of regulatory sites that are targets for endogenous

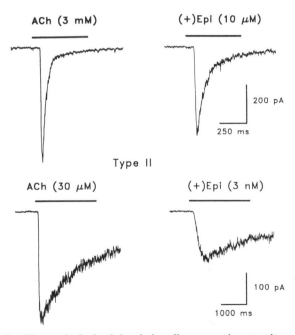

FIGURE 4. Epibatidine-evoked nicotinic whole-cell currents in rat cultured hippocampal neurons. **Top, left:** Sample recording of ACh-activated type IA current in a hippocampal neuron cultured for 21 days. **Top, right:** Sample recording of epibatidine-activated current in the same neuron as in the left panel. **Bottom, left:** Sample recording of ACh-activated type II current in a hippocampal neuron cultured for 11 days. **Bottom, right:** Sample recording of epibatidine-activated current in the same neuron as in the left panel. The nicotinic agonists were applied to the cell every 2 min. Duration of the agonist pulses is indicated by the solid bars on the top of the traces. Holding potential = −56 mV. Currents were classified as type I or II on the basis of the kinetics of their decay phase and their sensitivity to MLA (1 nM) or DHβE (100 nM). Compositions of the external and internal solutions were the same as those described in FIGURE 1.

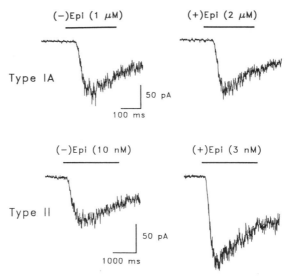

FIGURE 5. Enantiomers of epibatidine distinguish the subtypes of nAChRs present on hippocampal neurons. **Top:** Sample recordings of type IA current activated by (−) (*left*) or (+) (*right*) epibatidine. The recordings were obtained from a hippocampal neuron cultured for 18 days. **Bottom:** Sample recordings of type II current activated by (−) (*left*) or (+) (*right*) epibatidine. The recordings were obtained from a hippocampal neuron cultured for 11 days. The nicotinic agonists were applied to the cell every 2 min. Duration of the agonist pulse is indicated on the top of the traces. Holding potential = −56 mV. Currents were classified as type I or II on the basis of their sensitivity to MLA (1 nM) or DHβE (100 nM). Compositions of the external and internal solutions were the same as those indicated in FIGURE 1.

and exogenous compounds, such as (1) glycine, which is known to decrease NMDA receptor desensitization; (2) Mg^{2+}, MK-801, and phencyclidines, each of which can cause voltage-dependent blockade of NMDA-induced responses via different mechanisms; (3) Zn^{2+}, which allosterically inhibits NMDA receptor activity; and (4) polyamines, which either potentiate or inhibit the activation of NMDA receptors.[41]

In contrast to the vast knowledge on neuromodulators that can bind to the NMDA receptor and control its activation, little is known with respect to neuromodulators that can control nAChR activity in the CNS or at the neuromuscular junction. For example, about 15 years ago, perhydrohistrionicotoxin ($H_{12}HTX$) was shown to have differential effects on end-plate currents (EPCs) depending upon whether the EPCs were activated by nerve stimulation or by iontophoretic application of ACh.[42] It was hypothesized then that an endogenous neuromodulator and ACh could be simultaneously released from the nerve terminal, and that such a modulator could act upon the muscle nAChR to protect it from the inhibitory effect of $H_{12}HTX$ on nerve-evoked EPCs.[42] However, the nature of this putative neuromodulator and its site on action on the nAChR remain obscure to date.

Very recently, a new binding site was identified on neuronal and muscle nAChRs through which the ion-channel activity can be modulated. This site, which is insensitive to ACh, recognizes as agonists the anticholinesterases physostigmine (PHY) and galanthamine (GAL), as well as the muscle relaxant benzoquinonium (BZQ), and the morphine derivative codeine (FIG. 6).[43–51] It was initially shown that at therapeutically relevant concentrations, PHY can interact directly with the muscle nAChR as an agonist and an open-channel blocker.[43,44] Subsequently, the agonist effect of PHY was observed in several other preparations, each of which expressed different types of nAChRs. Indeed, PHY, as well as BZQ and GAL, can activate the nAChRs expressed on frog muscle fibers, electric organs of *Torpedo,* cultured hippocampal neurons, M10 fibroblasts, and clonal pheochromocytoma (PC12) cells.[43–52] The important conclusion, however, was that PHY, GAL, or BZQ could activate the nAChR channels present in these preparations by binding to a site distinct from that for ACh. Such a conclusion was inferred from the findings that (1) the nAChR-specific monoclonal antibody FK1 can block the agonist actions of these three compounds, without affecting those of ACh, and (2) competitive nicotinic antagonists have no effect on the agonist actions of PHY, GAL, and BZQ, but block those of ACh.[45–53] Supporting the notion that the effects of PHY are mediated via an ACh-insensitive nAChR site, PHY, but not ACh, displaces the binding of FK1 to hippocampal neurons,[48] and PHY can activate ion flux into *Torpedo* vesicles even after the nAChR has been desensitized by high concentrations of ACh.[46]

In an attempt to identify the nAChR binding site for PHY, a study using photoaffinity-labeling techniques was carried out in membrane preparations from *Torpedo* electroplax.[50] In that study, [^{3}H]PHY was activated by ultraviolet light to react covalently with its binding site on the membrane-bound *Torpedo* nAChR, and predominantly the Lys-125 residue of the nAChR α-subunit was found to be radiolabeled. This residue is not situated in the sequence regions presently suggested to contain elements of the ACh-binding site, but is located within the extracellular, amino-terminal region of all the nAChR α subunits cloned to date, a domain suitable for the binding of extracellular ligands[48,54] (FIG. 7). Basically two main domains can be defined in the amino-terminal region of nAChR subunits. One of these domains

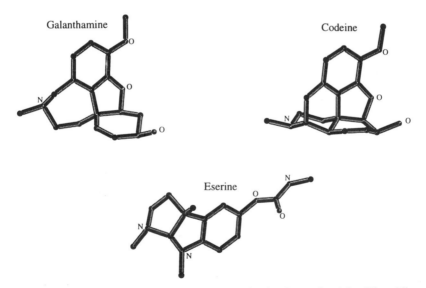

FIGURE 6. Chemical structures of physostigmine, galanthamine, and codeine. The tridimensional structures of physostigmine (eserine), galanthamine, and codeine are illustrated. In the structures, **O** and **N** represent the oxygen and nitrogen atoms, respectively. (Adapted from Storch *et al.*[52])

contains the invariant Cys 192 and 193 residues to which ACh and nicotinic antagonists (such as the monoclonal antibody WF6) can bind, and the other includes the invariant Lys-125 to which PHY and related ligands (such as the monoclonal antibody FK1) can bind. Using synthetic peptides that represented the amino acid sequences 181–200 and 118–137 of the *Torpedo* nAChR α subunit, it was demonstrated that whereas FK1 binds with high affinity to the peptide α118–137 and with low affinity to the peptide α181–200, WF6 binds with low affinity to the peptide α118–137 and with high affinity to the peptide α181–200.[54] Taken together with the results from functional and ligand competition studies, these findings indicate that there is practically no overlap between the binding sites for PHY and ACh, and partial overlap between the binding sites for FK1 and WF6. The sequence region of nAChR α subunits in which the binding site for PHY and related compounds is

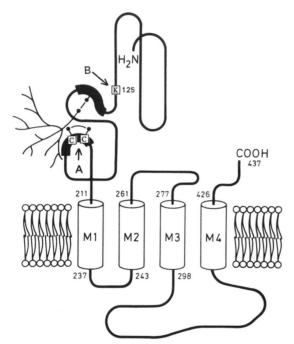

FIGURE 7. Location of the PHY-binding site on the nAChR α subunit. In this model of the nAChR α subunit, the amino acid residues are numbered according to the primary structure of the *Torpedo* nAChR. Arrow **A** points to the ACh-binding site, and **B** points to the PHY-binding site (From Pereira *et al.*[48] Reproduced, with permission, from the *Journal of Pharmacology and Experimental Therapeutics.*)

located (α118–137) is amphipathic.[54] The hydrophobicity of the nAChR region that contains the PHY-binding site may account, at least in part, for the ligand selectivity of this site for PHY and related compounds. In fact, in contrast to compounds that bind to the ACh site, which are essentially hydrophilic, all the compounds found to bind to the PHY site are extremely hydrophobic. Inasmuch as the region surrounding and including Lys-125 is highly conserved in all the nAChR α subunits sequenced to date,[48] and PHY has been shown to interact with a variety of nAChR subtypes, it is most likely that the novel identified ligand site on the neuronal nAChR plays a key role in the control of the activation of most, if not all, nAChR subtypes. The identification of an endogenous ligand that could bind to this site and control nAChR activity would ensure the physiological relevance of this newly identified site in the process of nicotinic synaptic transmission.

During the process of synaptic transmission in the neuromuscular junction, depolarization of the presynaptic nerve terminal causes Ca^{2+} influx, mobilization of the ACh-containing vesicles, and increase of ACh release into the synaptic cleft. ACh, then, diffuses across the synaptic cleft from the presynaptic terminal to the muscle, binds to the postsynaptic nAChRs, and increases with high efficacy the probability of opening of the nAChR channel.[55,56] In contrast, the efficacy of PHY and PHY-like compounds as nicotinic agonists, either on muscle or CNS nAChRs, is apparently so low that although these compounds can activate nicotinic single-channel currents they are unable to evoke macroscopic currents.[47,48,51,52] At this stage, one cannot rule out the possibility that these compounds can also exert other effects on the nAChR, which oppose and outweigh their agonist effects, and impair their ability to activate whole-cell responses. Indeed, all the compounds that have been shown to bind to the newly described nAChR site also have open-channel blocking properties or desensitizing effects on the nAChRs, and the concentrations at which such compounds can block or inactivate the nAChR channels overlap those at which they act as agonists.[48,51] Nevertheless, it is tempting to speculate that ligands that would bind to this newly described nAChR site could act as co-agonists rather than agonists, thereby potentiating channel activation by the natural transmitter (unpublished observations). Such a co-agonist action has been observed for glycine on the NMDA receptor.[58,59] Considering that there might be an endogenous ligand that binds to the PHY/GAL site and modulates the CNS nAChR activity, such a ligand could be part of a "chemical network." In such a network, endogenous neurohormones and/or neurotransmitters, in addition to serving their primary receptors, could modulate the activation of the nAChR by ACh.

An attempt has been made to identify endogenous ligands that could bind specifically to the PHY/GAL-binding site.[52] Based on molecular modeling, phenanthrene-type opium alkaloids were found to be structurally related to PHY and GAL (FIG. 8). In outside-out patches excised from PC12 cells, the morphine derivative codeine could activate single-channel currents via the same mechanism as PHY and GAL.[52] Therefore, endogenous opioid-type compounds, for example, endorphin and/or enkephalin, may serve as endogenous ligands for this newly described nAChR-binding site.

Modulation of Activity of CNS nAChRs by Phosphorylation

Receptor phosphorylation is a well-known mechanism by which muscle nAChR activity can be modulated. In the neuromuscular junction, the motoneurons release a peptide named calcitonin-gene-related peptide (CGRP), which can bind to G_s-coupled CGRP receptors located postsynaptically in muscles.[59] Binding of CGRP to its receptors can stimulate cAMP-dependent phosphorylation of muscle nAChRs, which can then be desensitized more rapidly by ACh.[59] The phosphorylation state of proteins closely associated with ion channels has also been associated with the rundown of currents triggered by activation of a number of voltage- and neurotransmitter-gated ion channels. For instance, the rundown of NMDA-activated whole-cell currents and voltage-activated Ca^{2+} currents can be prevented by the use of an ATP-regenerating internal solution.[60,61] The ATP-regenerating internal solution consisted of ATP, phosphocreatine, and creatine phosphokinase, and its main function was to keep the intracellular levels of ATP constant in spite of the dialysis of the intracellular contents by the pipette solution during the whole-cell experiments. Apparently, the mechanism by which ATP prevents the rundown of NMDA-activated currents and voltage-gated Ca^{2+} currents is unrelated to a direct phosphor-

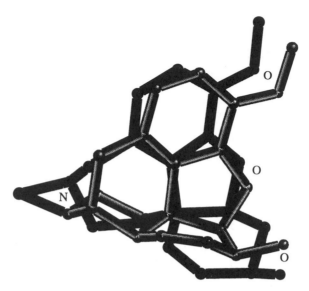

FIGURE 8. Structural relationship between the structures of galanthamine and codeine. The tridimensional chemical structures of galanthamine and codeine are superimposed to depict the common structural features between these two compounds. (Adapted from Storch *et al.*[52])

ylation of the ion-channel protein. Instead, it seems that ATP can influence the activity of these ion channels by altering the state of actin polymerization.[60] A model has been proposed in which Ca^{2+} binds to and modulates the function of a regulatory protein whose interactions with the NMDA receptor are dependent upon the integrity of underlying cytoskeletal elements, particularly actin.[60] This model accounts for both the Ca^{2+} and ATP dependence of rundown of NMDA-activated and voltage-gated Ca^{2+} currents. We have demonstrated that addition of ATP-regenerating compounds to the internal solution can prevent rundown of type IA currents without changing their kinetic properties in cultured hippocampal neurons[17,19] from fetal rats (FIG. 9). Remarkably, phosphocreatine per se is able to prevent the rundown of type IA currents to the same extent as the entire ATP-regenerating solution, in this way suggesting that phosphocreatine may be readily removed during the dialysis of the intracellular contents.[19] It is conceivable that the removal of intracellular phosphocreatine may affect substantially the phosphorylation state of the $\alpha7$-bearing nAChR, that is, the nAChR that subserves type IA currents, or of proteins closely associated with this nAChR. The phosphorylation/dephosphorylation state of such proteins may not affect the fast nAChR inactivation that accounts for the fast decay of IA currents in the presence of ACh, but may be essential for the slow nAChR inactivation that accounts for the rundown of the currents with time. Under our experimental conditions, the mechanism by which phosphorylation would play a role in preventing rundown of type IA currents seems to be unrelated to the microtubule or microfilament components of the cytoskeleton, because the presence of taxol, a microtubule stabilizer, or phalloidin, a microfilament stabilizer, in the internal solution did not alter the rate of rundown of type IA currents.[19] Of interest, rundown is not observed in type II currents, which are supposedly subserved by an $\alpha4\beta2$ neuronal nAChR.

Modulation of Activity of CNS nAChRs by Intracellular Mg²⁺

Rectification is another property of voltage- and neurotransmitter-gated ion channels that can be modulated. Rectification can occur because of lower probability of channel openings in one range of membrane potentials versus another, or because of a voltage-dependent blockade of the ion channels by a given element. For instance, it is well known that at physiological concentrations extracellular Mg^{2+} can cause a voltage-dependent blockade of NMDA receptors, thereby causing NMDA-activated whole-cell currents to rectify at membrane potentials more negative than -50 mV.[62] Also, ACh-activated currents in PC12 cells show an inward rectification that can be accounted for both by a blockade of the nAChR channels by physiological concentrations of intracellular Mg^{2+} and by a low probability of neuronal nAChR channel opening at positive potentials.[63]

Although most of the recombinant neuronal nAChRs have been known to give rise to currents that rectify inwardly, in our initial studies type IA currents in hippocampal neurons displayed a mild inward rectification at the beginning of the whole-cell current recordings, and this rectification tended to disappear with time.[17,19] These results indicated that a component of the intracellular contents, which could account for the inward rectification of the currents, was probably being removed during the recording by dialysis. Because our internal solution had a high concentra-

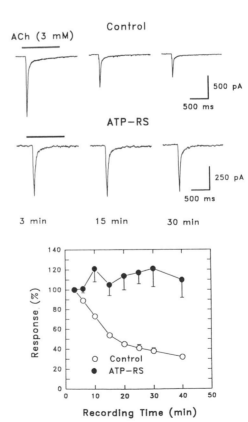

FIGURE 9. ATP-regenerating compounds can prevent the rundown of type IA currents. Top traces represent sample recordings of ACh-evoked whole-cell currents using an internal solution devoid of ATP-regenerating compounds. ACh was applied to a 40-day-old cultured hippocampal neuron in 1-s pulses separated by 2-min intervals. Bottom traces represent sample recordings of ACh-evoked whole-cell currents using an ATP-regenerating internal solution (ATP-RS). ACh was applied to a 20-day-old cultured hippocampal neuron as 1-s pulses separated by 2-min intervals. Holding potential = -56 mV. Bottom graph depicts the effect of the ATP-regenerating solution on the progressive decrease of the peak-current amplitude with time. The peak-current amplitude activated by the first ACh pulse was considered as 100%.

tion of F^-, an anion whose solubility product with Mg^{2+} is extremely low and leads to precipitation of Mg^{2+} in the form of MgF_2, it was hypothesized that intracellular Mg^{2+} could account, at least in part, for the inward rectification of type IA currents, and that its removal from the intracellular medium would result in attenuation of the inward rectification. To test this hypothesis, Mg^{2+}-containing, F^--free internal solutions were used in experiments directed at studying IA current-voltage (I-V) relationships. At the negative range of membrane potentials, the I-V relationship for type IA responses was the same regardless of the presence or absence of Mg^{2+} (2 or 5 mM) in the internal solution. However, at the positive range of membrane potentials, Mg^{2+} in the internal solution inhibited the outward currents. In the range of 0 to 50 mV, the normalized peak-current amplitudes were smaller than those obtained in the absence of added Mg^{2+}.[19] Therefore, under physiological conditions, intracellular Mg^{2+} may play a key role in the control of α-BGT-sensitive nAChR activity in the CNS. In contrast to type IA currents, type II currents rectified both in the presence and in the absence of Mg^{2+} (2 or 5 mM).[19] Thus, whereas intracellular Mg^{2+} may modulate the activity of native α7-bearing CNS nAChRs in vivo, it may not be the major factor that affects the activity of native α4β2 CNS nAChRs.

Activation of NMDA receptors in neurons can result in a substantial increase in the intracellular concentrations of free Mg^{2+}.[64] Therefore, although the normal physiological levels of intracellular Mg^{2+} may be enough to prevent to a great extent the activation of α-BGT-sensitive nAChRs at depolarized potentials, it is tempting to speculate that if NMDA receptors and α-BGT-sensitive nAChRs are expressed in the same neuron, when the NMDA receptor is activated at depolarized membrane potentials and the intracellular levels of free Mg^{2+} are increased, the nAChR activation would be completely prevented. Such a cross-talk between the two neurotransmitter systems, that is, the glutamatergic and the cholinergic systems, would be of pivotal relevance to cell function, given that both NMDA receptors and α-BGT-sensitive nAChRs are highly permeable to Ca^{2+}. This cross-talk may represent a means by which rapid rise in intracellular Ca^{2+} concentrations via activation of NMDA receptors and nAChRs could be tightly controlled, so that intracellular Ca^{2+} overloading could be avoided.

ION PERMEABILITY OF α-BGT-SENSITIVE CNS nAChRs

Intracellular Ca^{2+} is important in a variety of physiological processes ranging from modulation of neurotransmitter release to control of cell survival. Several pathways, each of which has distinct temporal characteristics and total capacities, can lead to a rise in intracellular Ca^{2+} concentration.[65] The fastest ones rely upon extracellular Ca^{2+} entry through selective ion channels, such as the voltage-gated Ca^{2+} channels and some neurotransmitter-gated channels. Among the latter, the glutamate-activated NMDA channels are the most widely distributed and best characterized in the CNS. Recent studies have shown that neuronal nAChRs also have a significant Ca^{2+} permeability,[66–71] presenting the interesting possibility that central cholinergic transmission might be involved in fast intracellular Ca^{2+} signals. It is now clear that the Ca^{2+} permeability differs among nAChR subtypes, and that a cellular specificity of the signal may occur depending upon the nAChRs expressed. In the case of chick ciliary ganglion neurons, nAChRs sensitive to α-BGT and MLA induce a marked rise in intracellular Ca^{2+}, whereas α-BGT-insensitive nAChRs in the same cells generate a much smaller Ca^{2+} signal.[71] Also, among the recombinant nAChR channels, the α7-based homomers are unique not only in being sensitive to α-BGT but also in having the highest relative Ca^{2+} permeability.[67–69] Thus, the Ca^{2+}

permeability of the native, α-BGT-sensitive, hippocampal nAChR channel (which probably bears α7-subunits in its structure[17,19]) could be high, and this could be the clue to its physiological function. Recently, our research has been aimed at determining relative ion permeabilities of this nAChR channel in cultured rat hippocampal neurons, according to classical Goldman-Hodgkin-Katz (GHK) modeling.

The reversal potential (V_R) of ACh-induced IA currents was measured using physiological solutions of various compositions.[72–74] In order to ascertain that only the α-BGT-sensitive currents were being tested, it was essential to perform the experiments in the presence of DHβE (0.1 μM) to inhibit the activation of type II (α4β2) nicotinic channels. Permeability ratios relative to Cs^+ were calculated independently of the intracellular medium, using a GHK equation for V_R shifts in the presence of Ca^{2+}. To ascertain the accuracy of the calculations, V_Rs were corrected for liquid-junction potentials, and ion activities were used instead of concentrations. Also, experiments with NMDA-gated currents were carried out in parallel, providing an internal standard for our permeability measurements. Replacing extracellular Cl^- with methanesulfonate caused a negative shift in the ACh current V_R, ruling out a significant contribution of Cl^- to type IA currents. Upon switching from 150 mM Cs^+- to 150 mM Na^+-containing external solutions, the ACh-gated current V_R showed a small positive shift that could be accounted for by the higher Na^+ activity, so that P_{Na}/P_{Cs} was close to the unity. Then, the Ca^{2+} permeability was investigated using Cs^+-based external solutions containing various Ca^{2+} concentrations. The V_R of the ACh-activated currents became more positive when the extracellular Ca^{2+} concentration was raised from 1 to 10 mM (FIG. 10), with the shifts yielding a P_{Ca}/P_{Cs} of about 6. In similar experiments, P_{Ca}/P_{Cs} for the NMDA currents was about 10. Thus, the native α7-bearing nAChR in the rat hippocampus is a cation channel considerably permeable to Ca^{2+}, nearly as much as the NMDA channel.[74] Although both these ACh- and glutamate-gated channels can mediate Ca^{2+} influx, their I-V relationships are quite different, suggesting nonoverlapping roles in the regulation of the neuronal Ca^{2+} concentration.

PHYSIOLOGICAL FUNCTIONS OF CNS nAChRs

The current knowledge of the physiological roles of neuronal nAChRs is still very poor, which is most regrettable in view of the increasing evidence implicating alterations in nAChR function and/or expression in physiopathological processes as in Alzheimer's and Parkinson's diseases.[33,34] Our approach to this problem has been to advance in the characterization of the functional, structural, and pharmacological properties of the brain nAChRs at the molecular and cellular levels.

As alluded to previously, for a number of years the α-BGT-binding proteins in the brain were believed to be unrelated to functional nAChRs.[26] It was not until the cloning of the chick nAChR α7 subunit and the demonstration that this subunit can form a functional ACh-gated ion channel that the α-BGT-binding proteins in the CNS were seen as putative functional nAChRs.[31] Another indirect evidence that CNS α-BGT-binding proteins could be functional nAChRs appeared when affinity-purified α-BGT-binding proteins from the chick optic lobe were shown to yield functional nAChR channels when reconstituted in planar lipid bilayer.[75] The direct demonstration that α-BGT-sensitive currents can be activated in a neuron came with the recordings made in our laboratory,[12–15] soon followed by others.[16,20] This "novel," α-BGT-sensitive receptor channel, which is likely to be distributed in all the brain areas where α-BGT binding and α7-subunit mRNA have been detected, can now be functionally probed. However, the intrinsic properties of this receptor channel, including fast kinetics and a tendency to rundown, still challenge our technical capabilities. To date, no one has been able to identify the endogenous functional activity of these receptors, like, for instance, an α-BGT-sensitive postsynaptic potential. Although the whole-cell and single-channel currents activated by fast

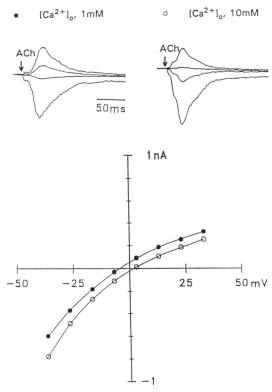

FIGURE 10. Ca^{2+} permeability of α-BGT-sensitive nAChR channels in cultured hippocampal neurons. **Top traces:** Sample recordings of ACh (1 mM)-activated type IA currents in a cultured hippocampal neuron held at the following holding potentials (from top to bottom traces): +23, +3, −7, and −27 mV. The compositions of the external solutions were (in mM): $CsCH_3SO_3$ 100, CsCl 50, HEPES 10, D-glucose 10, $CaCl_2$ 1 (●) or 10 (○), and N-methyl-D-glucamine·HCl 35 or 20 (pH, 7.3; osmolarity, 330 mosm). DHβE (0.1 μM), TTX (0.3 μM), and atropine (1 μM) were also added to the external solutions. The composition of the internal solution was (in mM): CsCl 60, CsF 60, CsOH 38.5, $MgCl_2$ 5, EGTA 10, HEPES 10, ATP 5, phosphocreatine 20, Tris·HCl 52.5, and creatine phosphate 50 U/mL (pH, 7.3; osmolarity, 340 mosm). Notice that upon increasing the concentration of extracellular Ca^{2+} from 1 to 10 mM, the currents decayed faster. **Bottom graph:** ACh current-voltage relationship obtained under the two different experimental conditions. Notice that upon increasing the extracellular Ca^{2+} concentration, the reversal potential of ACh current was shifted to the right. ACh (1 mM) was applied to the neuron in a 1-s pulse every 30 s.

application of nicotinic agonists to cultured CNS neurons demonstrate that the α-BGT-sensitive receptors are not presynaptic, it is not clear whether they are located synaptically or extrasynaptically. Electron-microscopic autoradiography studies have shown that α-BGT-binding sites can be located in dendritic and somatic membranes and that in some cases the labeling is confined to postsynaptic regions.[76] In contrast, in chick ciliary ganglion neurons, α-BGT binding sites occur mostly extrasynaptically, on the somatic membrane.[77] The magnitude and the time course of the changes in agonist concentration in these two circumstances are likely to be quite different, and so must be the dynamics of activation of the α-BGT-sensitive receptor channels.

The data discussed in previous sections showed that α-BGT-sensitive nAChR channels are inward-rectifying cation channels highly permeable to Ca^{2+}. Therefore, they can simultaneously depolarize the neuronal membrane and produce Ca^{2+} influx into the neuronal cells. This influx, in contrast to that mediated by voltage-gated Ca^{2+} channels and NMDA channels, increases with hyperpolarization. The nAChR-mediated rise in intracellular Ca^{2+} concentration can occur even if the current through the channel is not sufficient to depolarize the neuronal membrane. For a given receptor density, the amplitude of that current depends strongly upon the rate of agonist application, because of the fast kinetics of receptor activation/inactivation. Thus, for the nAChRs located in a cholinergic synapse, where the transmitter concentration is supposed to rise and fall sharply, the endogenous response could be an excitatory synaptic potential, associated with an intracellular "Ca^{2+} spike." Alternatively, if the nAChRs are extrasynaptic receptors, which the transmitter presumably reaches in lower concentration after diffusing from a distant source, the response can be a long-lasting Ca^{2+} influx, independent of neuronal excitation. Considering the two extremes, the receptor is capable of modulating both fast and slow Ca^{2+}-dependent processes, ranging from the on-switching of postsynaptic Ca^{2+}-calmodulin-dependent protein kinase II[78] to the activation of immediate early genes.[79] Another process dependent upon Ca^{2+} influx is the assembly of cytoskeletal elements, which may underlie the α-BGT-sensitive nicotinic effects on neurite outgrowth[80] and on retraction of growth cone phylopodia.[81]

The different α-BGT-insensitive nAChRs studied to date are also inward-rectifying cationic channels, but are less permeable to Ca^{2+}, desensitize much more slowly, and have much higher affinities for ACh than the α-BGT-sensitive nAChRs. Several lines of evidence have indicated the presence of these neuronal nAChRs in presynaptic terminals of the peripheral and central nervous systems.[22] A population of nAChRs believed to be composed of a combination of $\alpha 4$ and $\beta 2$ subunits[82] is located in presynaptic terminals of GABAergic and dopaminergic systems, and can control the release of GABA and dopamine into the synaptic cleft.[21-23] It has been proposed that related nAChRs in the rat interpeduncular nucleus are actually located proximal to the axon terminal in GABAergic neurons, thus being able to trigger fast, tetrodotoxin-sensitive GABA release.[23] In autonomic ganglia, α-BGT-insensitive nAChRs mediate synaptic transmission.[26] However, in the CNS, the physiological role of α-BGT-insensitive postsynaptic nAChRs, which have been detected by whole-cell current recording in neurons from the rat hippocampus (type II and III currents)[17] and habenula,[9] is still unclear.

More studies are required to determine what nAChR subtypes are located at spines on the axodendritic region of CNS neurons, and the role of these receptors in synaptic transmission. It is likely that such nAChRs would be involved in the regulation of synaptic transmission by modulating the release of neurotransmitters from the presynaptic terminals and/or by controlling the activation of second messenger systems at the postsynaptic terminals. Moreover, although acetylcholinesterase is found in high levels throughout the hippocampal region of the brain (FIG. 11), its role in terminating nicotinic responses induced by activation of hippocampal nAChRs remains to be determined.

Another important issue with respect to the physiological roles of CNS nAChRs *in vivo* is related to the existence of a newly described nAChR site through which nAChR activity may be modulated.[43-54] Although we are still seeking an endogenous neuromodulator that could bind to this site, thereby controlling nAChR activity, the idea of the existence of a neurochemical network that would allow for a single neurotransmitter or neurohormone to act in different receptors by binding to specific recognition sites is exciting. Such a network seems to exist with respect to the

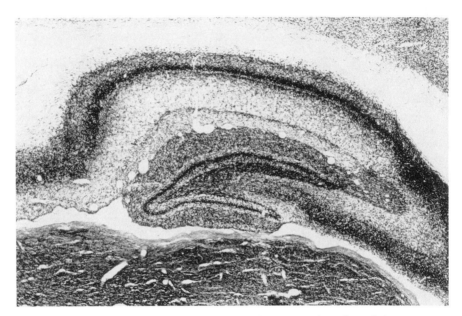

FIGURE 11. Distribution of acetylcholinesterase in a coronal section of the rat dorsal hippocampus. A dense differential labeling of the various hippocampal regions is shown in this micrograph of a cholinesterase-stained hippocampal section. Paraformaldehyde-fixed cryostat sections of rat hippocampi were treated according to a method adapted from Koelle and Fridenwald.[83] Briefly, the sections were air-dried, incubated in a solution containing ethopropazine, glycine, copper sulfate, acetylthiocholine iodide, and sodium acetate, and developed in sodium sulfide·HCl. The labeling was intensified using a silver nitrate solution. In the dendate gyrus, intensely labeled layers (zones) of fibers can be seen in the molecular layer, mainly in the supragranular sublaminae of both the supra- and infralaminar blade of the dendate gyrus, followed by a lighter stained zone and a dense region which extends to the pial surface. Another intensely labeled zone lines the inner border of the granular cell layer towards the hilus of the dendate gyrus, which also displays an intense labeling. In the corpus amonis (CA), differential acetylcholinesterase-staining is also observed. In the CA1 region, the pyramidal cell layer is ensheathed by a broader zone that extends towards the label stratum oriens and a smaller band at the border between the pyramidal layer and the stratum radiatum. The latter and the stratum lacunosum-moleculare—divided by a darker zone—are rather lightly stained. (From Schröder *et al.,* in preparation).

modulation of the NMDA receptor by glycine.[57,58] Glycine, in addition to acting as an inhibitory neurotransmitter by binding to the agonist binding site of glycine-gated Cl⁻ channels, can also decrease the desensitization of NMDA-induced currents by binding to an allosteric site on the NMDA receptor. Such a mode of action would make centrally acting co-agonists interesting drugs in the treatment of diseases in which an enhanced sensitivity of a neuroreceptor to its natural transmitter would be advantageous.

In summary, the diversity of CNS nAChRs may represent a means by which differential modulation of receptor activity, desensitization, down- or up-regulation, and cross-talk with other neurotransmitter systems can be mediated by a single neurotransmitter, ACh. It is possible that segregation of different subtypes of CNS nAChRs may occur not only in different areas of the brain, but even more important, in different regions of the neuronal surface. Therefore, in order to unveil the physiological functions of neuronal nAChRs in the CNS, the next step will be to determine the subcellular localization of the diverse nAChR subtypes in various regions of the brain.

SUMMARY

The diversity of neuronal nicotinic receptors (nAChRs) in addition to their possible involvement in such pathological conditions as Alzheimer's disease have directed our research towards the characterization of these receptors in various mammalian brain areas. Our studies have relied on electrophysiological, biochemical, and immunofluorescent techniques applied to cultured and acutely dissociated hippocampal neurons, and have been aimed at identifying the various subtypes of nAChRs expressed in the mammalian central nervous system (CNS), at defining the mechanisms by which CNS nAChR activity is modulated, and at determining the ion permeability of CNS nAChR channels. Our findings can be summarized as follows: (1) hippocampal neurons express at least three subtypes of CNS nAChRs—an $\alpha 7$-subunit-bearing nAChR that subserves fast-inactivating, α-BGT-sensitive currents, which are referred to as type IA, an $\alpha 4 \beta 2$ nAChR that subserves slowly inactivating, dihydro-β-erythroidine-sensitive currents, which are referred to as type II, and an $\alpha 3 \beta 4$ nAChR that subserves slowly inactivating, mecamylamine-sensitive currents, which are referred to as type III; (2) nicotinic agonists can activate a single type of nicotinic current in olfactory bulb neurons, that is, type IA currents; (3) $\alpha 7$-subunit-bearing nAChR channels in the hippocampus have a brief lifetime, a high conductance, and a high Ca^{2+} permeability; (4) the peak amplitude of type IA currents tends to rundown with time, and this rundown can be prevented by the presence of ATP-regenerating compounds (particularly phosphocreatine) in the internal solution; (5) rectification of type IA currents is dependent on the presence of Mg^{2+} in the internal solution; and (6) there is an ACh-insensitive site on neuronal and nonneuronal nAChRs through which the receptor channel can be activated. These findings lay the groundwork for a better understanding of the physiological role of these receptors in synaptic transmission in the CNS.

ACKNOWLEDGMENTS

The technical assistance of M. A. Zelle, B. Marrow, V. Pondeljak, and H. Taschner is gratefully acknowledged. The authors would like to thank Dr. J. G. Montes for his helpful comments on the manuscript. The authors are also grateful to Drs. Ulrich Bolbach and Uwe Metzinger, who participated in the studies of acetylcholinesterase labeling of rat hippocampal sections.

REFERENCES

1. LUKAS, R. J. & E. L. BENNETT. 1980. Interaction of nicotinic receptor affinity reagents with central nervous system α-bungarotoxin binding entities. Mol. Pharmacol. **17:** 149–155.
2. LUKAS, R. J. 1984. Detection of low affinity α-bungarotoxin binding sites in the rat central nervous system. Biochemistry **23:** 1160–1164.
3. CLARKE, P. B. S., R. D. SCHWARTS, S. M. PAUL, C. B. PERT & A. PERT. 1985. Nicotinic binding in rat brain: Autoradiographic comparison of [³H]acetylcholine, [³H]nicotine, and [¹²⁵I]alpha-bungarotoxin. J. Neurosci. **5:** 1307–1315.
4. SWANSON, L. W., D. M. SIMMONS, P. J. WHITING & J. LINDSTROM. 1987. Immunohistochemical localization of neuronal nicotinic receptors in the rodent central nervous system. J. Neurosci. **7:** 3334–3342.
5. SORENSON, E. M. & V. A. CHIAPPINELLI. 1992. Localization of [³H]nicotine, [¹²⁵I]kappa-bungarotoxin, and [¹²⁵I]alpha-bungarotoxin binding to nicotinic sites in the chicken forebrain and midbrain. J. Comp. Neurol. **323:** 1–12.
6. ARACAVA, Y., S. S. DESHPANDE, K. L. SWANSON, H. RAPOPORT, S. WONNACOTT, G. LUNT & E. X. ALBUQUERQUE. 1987. Nicotinic acetylcholine receptors in cultured neurons from the hippocampus and brain stem of the rat characterized by single channel recording. FEBS Lett. **222:** 63–70.
7. LIPTON, S. A., E. AIZENMAN & R. H. LORING. 1987. Neuronal nicotinic acetylcholine responses in solitary mammalian retinal ganglion cells. Pflügers Arch. **410:** 37–43.
8. ZHANG, Z. W. & P. FELTZ. 1990. Nicotinic acetylcholine receptors in porcine hypophyseal intermediate lobe cells. J. Physiol. (Lond.). **422:** 83–101.
9. MULLE, C. & J.-P. CHANGEUX. 1990. A novel type of nicotinic receptor in the rat central nervous system characterized by patch-clamp techniques. J. Neurosci. **10:** 169–175.
10. ALKONDON, M. & E. X. ALBUQUERQUE. 1990. α-Cobratoxin blocks the nicotinic acetylcholine receptor in rat hippocampal neurons. Eur. J. Pharmacol. **191:** 505–506.
11. RAMOA, A. S., M. ALKONDON, Y. ARACAVA, J. IRONS, G. G. LUNT, S. S. DESHPANDE, S. WONNACOTT, R. S. ARONSTAM & E. X. ALBUQUERQUE. 1990. The anticonvulsant MK-801 interacts with peripheral and central nicotinic acetylcholine receptor ion channels. J. Pharmacol. Exp. Ther. **254:** 71–82.
12. ALBUQUERQUE, E. X., A. C. S. COSTA, M. ALKONDON, K.-P. SHAW, A. S. RAMOA & Y. ARACAVA. 1991. Functional properties of the nicotinic and glutamatergic receptors. J. Recept. Res. **11:** 603–625.
13. ALKONDON, M. & E. X. ALBUQUERQUE. 1991. Initial characterization of the nicotinic acetylcholine receptors in rat hippocampal neurons. J. Recept. Res. **11:** 1001–1021.
14. ISHIHARA, K., M. ALKONDON, J. G. MONTES & E. X. ALBUQUERQUE. 1995. Nicotinic responses in acutely dissociated rat hippocampal neurons and the selective blockade of fast-desensitizing nicotinic currents by lead. J. Pharmacol. Exp. Ther. In press.
15. ALKONDON, M., E. F. R. PEREIRA, S. WONNACOTT & E. X. ALBUQUERQUE. 1992. Blockade of nicotinic currents in hippocampal neurons defines methyllycaconitine as a potent and specific receptor antagonist. Mol. Pharmacol. **41:** 802–808.
16. ZORUMSKI, C. F., L. L. THIO, K. E. ISENBERG & D. B. CLIFFORD. 1992. Nicotinic acetylcholine currents in cultured postnatal rat hippocampal neurons. Mol. Pharmacol. **41:** 931–936.
17. ALKONDON, M. & E. X. ALBUQUERQUE. 1993. Diversity of nicotinic acetylcholine receptors in rat hippocampal neurons: I. Pharmacological and functional evidence for distinct structural subtypes. J. Pharmacol. Exp. Ther. **265:** 1455–1473.
18. CASTRO, N. G. & E. X. ALBUQUERQUE. 1993. Brief-lifetime, fast-inactivating ion channels account for the α-bungarotoxin-sensitive nicotinic response in hippocampal neurons. Neurosci. Lett. **164:** 137–140.
19. ALKONDON, M., S. REINHARDT, C. LOBRON, B. HERMSEN, A. MAELICKE & E. X. ALBUQUERQUE. 1994. Diversity of nicotinic acetylcholine receptors in rat hippocampal neurons: II. Rundown and inward rectification of agonist-elicited whole-cell currents and identification of receptor subunits by *in situ* hybridization. J. Pharmacol. Exp. Ther. **271:** 494–506.
20. ZHANG, Z. W., S. VIJAYARAGHAVAN & D. K. BERG. 1994. Neuronal acetylcholine

receptors that bind α-bungarotoxin with high affinity function as ligand-gated ion channels. Neuron **12:** 167–177.

21. WONNACOTT, S., J. IRONS, G. G. LUNT, C. M. RAPIER & E. X. ALBUQUERQUE. 1988. αBungarotoxin and presynaptic nicotinic receptors: Functional studies. *In* Nicotinic Acetylcholine Receptors in the Nervous System. F. Clementi, C. Gotti & E. Sher, Eds.: 41–60. Springer-Verlag. Berlin.

22. WONNACOTT, S., A. DRASDO, E. SANDERSON & P. POWELL. 1990. Presynaptic nicotinic receptors and the modulation of transmitter release. *In* Ciba Foundation Symposium. The Biology of Nicotine Dependence. 87–105. Wiley. Chichester.

23. LÉNA, C., J.-P. CHANGEUX & C. MULLE. 1993. Evidence for "preterminal" nicotinic receptors on GABAergic axons in the rat interpeduncular nucleus. J. Neurosci. **13:** 2680–2688.

24. CLARKE, P. B. 1993. The fall and rise of neuronal alpha-bungarotoxin binding proteins. Trends Pharmacol. Sci. **13:** 407–413.

25. CHIAPPINELLI, V. A. 1991. Kappa-neurotoxins: Effects on neuronal nicotinic acetylcholine receptors. *In* Snake Toxins. A. L. Harvey, Eds.: 223–258. New York.

26. SARGENT, P. B. 1993. The diversity of neuronal acetylcholine receptors. Annu. Rev. Neurosci. **16:** 403–433.

27. MONTES, J. G., M. ALKONDON, E. F. R. PEREIRA & E. X. ALBUQUERQUE. 1994. Nicotinic acetylcholine receptor of the mammalian central nervous system. *In* Handbook of Membrane Channels: Molecular and Cellular Physiology. C. Peracchia, Ed.: 269–286. Academic Press. San Diego, CA.

28. ELGOYHEN, A. B., D. JOHNSON, J. BOULTER, D. VETTER & S. HEINEMANN. 1994. α9: An acetylcholine receptor with novel pharmacological properties expressed in rat cochlear hair cells. Cell **79:** 705–715.

29. DENERIS, E. S., J. CONNOLLY, S. W. ROGERS & R. DUVOISIN. 1991. Pharmacological and functional diversity of neuronal nicotinic acetylcholine receptors. Trends Pharmacol. Sci. **12:** 34–40.

30. ROLE, L. W. 1992. Diversity in primary structure and function of neuronal nicotinic acetylcholine receptor channels. Curr. Opin. Neurosci. **2:** 254–262.

31. COUTURIER, S., D. BERTRAND, J. M. MATTER, M. C. HERNANDEZ, S. BERTRAND, N. MILLAR, S. VALERA, T. BARKAS & M. BALLIVET. 1990. A neuronal nicotinic acetylcholine receptor subunit (α7) is developmentally regulated and forms a homoligomeric channel blocked by αBTX. Neuron **5:** 847–856.

32. GERZANICH, V., R. ANAND & J. LINDSTROM. 1994. Homomers of α8 and α7 subunits of nicotinic receptor exhibit similar channel but contrasting binding site properties. Mol. Pharmacol. **45:** 212–220.

33. SCHRÖDER, H., E. GIACOBINI, R. G. STRUBLE, K. ZILLES & A. MAELICKE. 1991. Nicotinic cholinoceptive neurons of the frontal cortex are reduced in Alzheimer's disease. Neurobiol. Aging **12:** 259–262.

34. LANGE, K. W., F. R. WELLS, P. SENNER & C. D. MARSDEN. 1993. Altered muscarinic and nicotinic receptor densities in cortical and subcortical brain regions in Parkinson's disease. J. Neurochem. **60:** 197–203.

35. MACALLAN, D. R. E., G. G. LUNT, S. WONNACOTT, K. L. SWANSON, H. RAPOPORT & E. X. ALBUQUERQUE. 1988. Methyllycaconitine and (+)-anatoxin-a differentiate between nicotinic receptors in vertebrate and invertebrate nervous system. FEBS Lett. **226:** 357–363.

36. ALKONDON, M. & E. X. ALBUQUERQUE. 1994. Presence of alpha-bungarotoxin-sensitive nicotinic acetylcholine receptors in rat olfactory bulb neurons. Neurosci. Lett. **176:** 152–156.

37. REVAH, F., D. BERTRAND, J.-L. GALZI, A. DEVILLERIS-THIÉRY, C. MULLE, N. HUSSY, S. BERTRAND, M. BALLIVET & J.-P. CHANGEUX. 1991. Mutations in the channel domain alter desensitization of a neuronal nicotinic receptor. Nature (Lond.) **353:** 846–849.

38. BADIO, B. & J. DALY. 1994. Epibatidine, a potent analgesic and nicotinic agonist. Mol. Pharmacol. **45:** 563–569.

39. ALKONDON, M. & E. X. ALBUQUERQUE. 1994. Enantiomers of epibatidine as potent

      nicotinic agonists at two identified subtypes of nicotinic acetylcholine receptors (nAChRs) in rat hippocampal neurons. Soc. Neurosci. Abstr. **20:** 1135.

40. SALLY, T. L. & G. G. BERNSTON. 1979. Antinociceptive effects of central and systemic administration of nicotine in rat. Psychopharmacology **65:** 279–283.

41. SCATTON, B. 1993. The NMDA receptor complex. Fundam. Clin. Pharmacol. **7:** 389–400.

42. ALBUQUERQUE, E. X., P. W. GAGE & A. C. OLIVEIRA. 1979. Differential effect of perhydrohistrionicotoxin on "intrinsic" and "extrinsic" endplate responses. J. Physiol. (Lond.) **297:** 423–442.

43. SHAW, K.-P., Y. ARACAVA, A. AKAIKE, J. W. DALY, D. L. RICKETT & E. X. ALBUQUERQUE. 1985. The reversible cholinesterase inhibitor physostigmine has channel-blocking and agonist effects on the acetylcholine receptor-ion channel complex. Mol. Pharmacol. **28:** 527–538.

44. ALBUQUERQUE, E. X., Y. ARACAVA, W. M. CINTRA, A. BROSSI, B. SCHÖNENBERGER & S. S. DESHPANDE. 1988. Structure-activity relationship of reversible cholinesterase inhibitors: Activation, channel blockade and stereospecificity of the nicotinic acetylcholine receptor-ion channel complex. Braz. J. Med. Biol. Res. **21:** 1173–1196.

45. OKONJO, K. O., J. KUHLMANN & A. MAELICKE. 1991. A second pathway for the activation of the *Torpedo* acetylcholine receptor. Eur. J. Biochem. **200:** 671–677.

46. KUHLMANN, J., K. O. OKONJO & A. MAELICKE. 1991. Desensitization is a property of the cholinergic binding region of the nicotinic acetylcholine receptor, not of the receptor-integral ion channel. FEBS Lett. **279:** 216–218.

47. PEREIRA, E. F. R., M. ALKONDON, T. TANO, N. G. CASTRO, M. M. FRÓES-FERRÃO, R. ROZENTAL, R. S. ARONSTAM & E. X. ALBUQUERQUE. 1993. A novel agonist binding site on nicotinic acetylcholine receptors. J. Recept. Res. **13:** 413–436.

48. PEREIRA, E. F. R., S. REINHARDT-MAELICKE, A. SCHRATTENHOLZ, A. MAELICKE & E. X. ALBUQUERQUE. 1993. Identification and characterization of a new agonist site on nicotinic acetylcholine receptors of cultured hippocampal neurons. J. Pharmacol. Exp. Ther. **265:** 1474–1491.

49. SCHRATTENHOLZ, A., T. COBAN, B. SCHRÖDER, K. O. OKONJO, J. KUHLMANN, E. F. R. PEREIRA, E. X. ALBUQUERQUE & A. MAELICKE. 1993. Biochemical characterization of a novel channel-activating site on nicotinic acetylcholine receptors. J. Recept. Res. **13:** 393–412.

50. SCHRATTENHOLZ, A., J. GODOVAC-ZIMMERMAN, H.-J. SCHÄFER, E. X. ALBUQUERQUE & A. MAELICKE. 1993. Photoaffinity labeling of *Torpedo* acetylcholine receptor by the reversible cholinesterase inhibitor physostigmine. Eur. J. Biochem. **216:** 671–677.

51. PEREIRA, E. F. R., M. ALKONDON, S. REINHARDT, A. MAELICKE, X. PENG, J. LINDSTROM, P. WHITING & E. X. ALBUQUERQUE. 1994. Physostigmine and galanthamine characterize the presence of the novel binding site on the α4β2 subtype of neuronal nicotinic acetylcholine receptors stably expressed in fibroblast cells. J. Pharmacol. Exp. Ther. **270:** 768–778.

52. STORCH, A., J. C. COOPER, O. GUTBROD, K.-H. WEBER, S. REINHARDT, C. LOBRON, B. HERMSEN, V. SOSKIC, A. SCHRATTENHOLZ, E. X. ALBUQUERQUE, C. METHFESSEL & A. MAELICKE. 1994. Physostigmine and galanthamine act as non-competitive agonists on clonal rat pheochromocytoma cells. Eur. J. Pharmacol. Submitted.

53. DUNN, S. M. J. & M. A. RAFTERY. 1993. Cholinergic binding sites on the pentameric acetylcholine receptor of *Torpedo californica*. Biochemistry **32:** 8608–8615.

54. SCHRÖDER, B., S. REINHARDT-MAELICKE, A. SCHRATTENHOLZ, K. E. MCLANE, B. M. CONTI-TRONCONI & A. MAELICKE. 1994. Monoclonal antibodies FK1 and WF6 define two neighboring ligand binding sites on *Torpedo* acetylcholine receptor α-polypeptide. J. Biol. Chem. **269:** 10407–10416.

55. TRAUTMANN, A. 1983. A comparative study of the activation of the cholinergic receptor by various agonists. Proc. R. Soc. Lond. Biol. Sci. **218:** 241–251.

56. KARLIN, A. 1993. Structure of nicotinic acetylcholine receptors. Curr. Opin. Neurobiol. **3:** 299–309.

57. BENVENISTE, M., J. CLEMENTES, L. VYCKLICKY & M. L. MAYER. 1990. A kinetic analysis of the modulation of *N*-methyl-D-aspartate receptors by glycine in mouse cultured hippocampal neurons. J. Physiol. (Lond.) **428:** 337–357.

58. JOHNSON, J. & P. ASCHER. 1992. Equilibrium and kinetic study of glycine action on the NMDA receptor in cultured mouse brain neurons. J. Physiol. (Lond.) **455:** 339–365.
59. MILES, K., P. GREENGARD & R. L. HUGANIR. 1989. Calcitonin gene-related peptide regulates phosphorylation of the nicotinic acetylcholine receptor in rat myotubes. Neuron **2:** 1517–1524.
60. ROSEMUND, C. & G. L. WESTBROOK. 1993. Calcium-induced actin depolymerization reduces NMDA channel activity. Neuron **10:** 805–814.
61. JOHNSON, B. D. & L. BYERLY. 1993. A cytoskeletal mechanism for calcium channel metabolic dependence and inactivation by intracellular calcium. Neuron **10:** 797–804.
62. NELSON, M. & E. X. ALBUQUERQUE. 1994. 9-Aminoacridines act at a site different from that for Mg^{2+} in blockade of the N-methyl-D-aspartate receptor channel. Mol. Pharmacol. **46:** 151–160.
63. IFUNE, C. K. & J. H. STEINBACH. 1992. Inward rectification of acetylcholine-elicited currents in rat pheochromocytoma cells. J. Physiol. (Lond.) **457:** 143–165.
64. BROCARD, J. B., S. RAJDEV & I. J. REYNOLDS. 1993. Glutamate-induced increases in intracellular free Mg^{2+} in cultured cortical neurons. Neuron **11:** 751–757.
65. BLAUSTEIN, M. P. 1988. Calcium transport and buffering in neurons. Trends Neurosci. **11:** 438–443.
66. MULLE, C., D. COQUET, H. KORN & J.-P. CHANGEUX. 1992. Calcium influx through nicotinic receptors in rat central neurons: Its relevance to cellular regulation. Neuron **8:** 135–143.
67. SÉGUÉLA, P., J. WADICHE, K. DINELY-MILLER, J. A. DANI & J. W. PATRICK. 1993. Molecular cloning, functional properties, and distribution of rat brain α7: A nicotinic cation channel highly permeable to Ca^{2+}. J. Neurosci. **13:** 596–604.
68. SANDS, S. B., A. C. S. COSTA & J. W. PATRICK. 1993. Barium permeability of neuronal nicotinic receptor α7 expressed in *Xenopus* oocytes. Biophys. J. **65:** 2614–2621.
69. BERTRAND, D., N. L. GALZI, A. DEVILLERS-THIERY, S. BERTRAND & J.-P. CHANGEUX. 1993. Mutations at two distinct sites within the channel domain M2 alter calcium permeability of neuronal α7 nicotinic receptor. 1993. Proc. Natl. Acad. Sci. USA **90:** 6971–6975.
70. VERNINO, S., M. AMADOR, C. W. LUETJE, J. PATRICK & J. A. DANI. 1992. Calcium modulation and high calcium permeability of neuronal nicotinic acetylcholine receptors. Neuron **8:** 127–134.
71. VIJAYARAGHAVAN, S., P. C. PUGH, Z.-W. ZANG, M. M. RAFTERY & D. K. BERG. 1992. Nicotinic receptors that bind α-bungarotoxin on neurons raise intracellular free calcium. Neuron **8:** 353–362.
72. CASTRO, N. G., A. T. ELDEFRAWI & E. X. ALBUQUERQUE. 1993. Fast kinetics and calcium permeability of α-bungarotoxin-sensitive hippocampal nicotinic receptor channels. Soc. Neurosci. Abstr. **19:** 464.
73. CASTRO, N. G. & E. X. ALBUQUERQUE. 1994. Calcium permeability of α-bungarotoxin-sensitive nicotinic acetylcholine receptors in rat hippocampal neurons. Biophys. J. **66:** A214.
74. CASTRO, N. G. & E. X. ALBUQUERQUE. 1995. The α-bungarotoxin-sensitive hippocampal nicotinic acetylcholine receptor channel has a high calcium permeability. Biophys. J. In press.
75. GOTTI, C., A. ESPARIS OGANDO, W. HANKE, R. SCHLUE, M. MORETTI & F. CLEMENTI. 1991. Purification and characterization of an α-bungarotoxin receptor that forms a functional nicotinic channel. Proc. Natl. Acad. Sci. USA **88:** 3258–3262.
76. HUNT, S. P. & J. SCHMIDT. 1978. The electron microscopy autoradiographic localization of α-bungarotoxin binding sites within the central nervous system of the rat. Brain Res. **142:** 152–159.
77. JACOB, M. H. & D. K. BERG. 1983. The ultrastructural localization of alpha-bungarotoxin binding sites in relation to synapses on chick ciliary ganglion neurons. J. Neurosci. **3:** 260–271.
78. BRONSTEIN, J. M., D. B. FARBER & C. G. WASTERLAIN. 1993. Regulation of type-II calmodulin kinase: Functional implications. Brain Res. Rev. **18:** 135–147.

79. BADING, H., D. D. GINTY & M. E. GREENBERG. 1993. Regulation of gene expression in hippocampal neurons by distinct calcium signaling pathways. Science **260:** 181–186.
80. PUGH, P. C. & D. K. BERG. 1994. Neuronal acetylcholine receptors that bind α-bungarotoxin mediate neurite retraction in a calcium-dependent manner. J. Neurosci. **14:** 889–896.
81. CHAN, J. & M. QUIK. 1993. A role for the nicotinic α-bungarotoxin receptor in neurite outgrowth in PC12 cells. Neuroscience **56:** 441–451.
82. WHITING, P., R. SCHOEPFER, J. LINDSTROM & T. PRIESTLEY. 1991. Structural and pharmacological characterization of the major brain nicotinic acetylcholine receptor subtype stably transfected in mouse fibroblasts. Mol. Pharmacol. **40:** 463–472.
83. KOELLE, G. B. & J. S. FRIEDENWALD. 1949. A histochemical method for localizing cholinesterase activity. Proc. Soc. Exp. Biol. Med. **70:** 617–622.

Nicotinic Receptors and Cholinergic Neurotransmission in the Central Nervous System

PAUL B. S. CLARKE

Department of Pharmacology and Therapeutics
McGill University
Montreal H3G 1Y6, Canada

Classic studies of the neuromuscular junction and autonomic ganglia have provided us with a clear picture of nicotinic cholinergic transmission in these tissues. Within the central nervous system (CNS), the elegant experiments of Curtis, Eccles, and associates revealed that nicotinic cholinergic transmission occurs in the ventral horn of the spinal cord, between motoneurons and Renshaw cells. However, as discussed below, evidence for nicotinic cholinergic transmission at other sites in mammalian spinal cord and brain is remarkably fragmentary. We may have been too influenced by earlier studies, assuming too readily that all nAChRs are cholinergic (activated by physiologically released ACh) as well as being cholinoceptive.

The existence and prevalence of CNS nicotinic cholinergic transmission is germane to several important issues. First, we should be aware that studying the actions of nicotine in isolation may not tell us as much as we would like about sites of nicotinic cholinergic transmission. Second, with increasing interest in the use of nicotinic antagonists to combat tobacco smoking, it becomes important to identify possible consequences of blocking endogenous nicotinic cholinergic tone. Third, there is a growing expectation that nicotinic cholinergic transmission may be impaired in Alzheimer's disease. This view depends greatly on recent reports that nicotine can alleviate certain cognitive deficits in afflicted individuals[1,2] and that the nicotinic antagonist mecamylamine can impair cognitive performance in normal human subjects.[3] Before central cholinergic systems can be implicated definitively in these drug effects, it is important to know more about possible sites of nicotinic cholinergic transmission and the extent to which such systems are tonically active. This paper serves as a brief summary of current knowledge. Throughout, the abbreviation "nAChRs" refers to nicotinic cholinoceptors, whether or not they are activated by endogenous ACh.

NEUROANATOMICAL MAPPING OF CHOLINERGIC SYSTEMS AND NICOTINIC RECEPTORS

The major cholinergic cell groups have been extensively mapped at the light microscopic level in rats, principally by choline acetyltransferase immunocytochemistry combined with retrograde tracing.[4] Mapping in higher mammals has provided much less detailed hodological information, but the available evidence suggests that the main cholinergic pathways are conserved during evolution, even in humans.[5,6] Light microscopic studies have revealed the location of cholinergic cell bodies and terminal fields (the latter shown by varicosities), and synaptic formations have been described in a number of regions examined by electron microscopy.

73

Brain nicotinic receptors have been mapped at low resolution using autoradiography. Several radioligands have been used, revealing two prominent populations. One population is preferentially labeled with high affinity (nanomolar k_d) by tritiated agonists such as [³H]nicotine, [³H]ACh, and [³H]cytisine, whereas the other is selectively labeled by the elapid snake toxin [¹²⁵]-α-bungarotoxin.[7–9] These two receptor populations are quite differently distributed in brain, and can also be distinguished by affinity isolation, immunological characterization, and molecular genetic approaches.[10,11] Two other populations of putative nAChR have been identified, characterized by high affinity for the antagonists [³H]dihydro-β-erythroidine[12] and [¹²⁵I]neuronal bungarotoxin,[13] respectively. The latter population appears to represent a distinct receptor subtype, but is present only in low density, at least in rat brain.

Much less is known about the ultrastructural location of nAChRs. The first studies addressing this issue used labeled α-bungarotoxin[14–16] and provided evidence for a (post)synaptic location. However, these pioneering studies should probably be viewed with caution, particularly because statistical analysis was not performed that might have more clearly demonstrated a preferential synaptic location. Another reason for caution is that on mammalian skeletal muscle fibers, α-bungarotoxin binding sites are overwhelmingly *extrasynaptic,* and at autonomic ganglia, predominantly synaptic or *extrasynaptic* locations have both been reported, possibly depending on the species and/or ganglion in question.[17–20] Statistical analysis of [¹²⁵I]-α-bungarotoxin binding in an area of rat hypothalamus nevertheless appeared to favor a synaptic location.[21]

More recently, several groups have raised antibodies to nAChRs or related peptide fragments and reported ultrastructural observations in rat brain. Schröder and colleagues have localized putative nAChRs using a monoclonal antibody (WF6) originally raised against the *Torpedo* electroplaque electric organ nAChR.[22] These authors described a distribution of immunoreactivity that was frequently postsynaptic in rat cerebral cortex. The receptor subunit selectivity of this antibody has yet to be determined. Another antibody, mAb 270, which is known to recognize β2 nAChR subunits, has provided low resolution film autoradiographs[23] and light microscopic images[24] in rat brain. However, thus far it has not proved compatible with tissue preparation procedures required for electron microscopic examination. Another group has recently reported immunocytochemical mapping of rat brain using polyclonal antibodies raised against fusion proteins corresponding to the β2 subunit; little evidence was obtained for synaptically located nAChRs.[25] In contrast, junctionally located immunoreactivity has been reported in material stained using mAb 299, which is selective for α4 nAChR subunits.[26]

At the light microscopic level, cholinergic projections and nAChRs are both widely dispersed throughout the neuroaxis. The pervasiveness of cholinergic systems has become particularly apparent with the development of sensitive anti-ChAT antibodies and immunocytochemical visualization methods,[27] so that previous reports of brain areas that possess nAChRs but lack cholinergic innervation must now be treated with caution.[28] In this respect, detailed ultrastructural information would be invaluable. However, very little is known at the ultrastructural level about the location of nAChRs with respect to cholinergic innervation. The early studies of Hunt, Arimatsu, Lentz, and their associates predated the development of a specific marker for cholinergic terminals, and provide no indication of the neurotransmitter phenotype(s) involved. Double-labeling studies in the CNS should by now be possible, but appear not to have been attempted.

Thus far, I have accepted the common assumption that in order to play a role in nicotinic cholinergic transmission, nAChRs must be preferentially located at synapses. However, even where cholinergic transmission is synaptic in nature rather than paracrine, the critical question is really whether a *sufficient* number of nAChRs are present at the synapse to mediate the actions of released transmitter. How much is sufficient? In general, we do not know. This issue takes on more than academic interest when it is realized that some nAChRs may not be accurately located over the cell membrane. Although the evidence on this point is only suggestive, there are a number of neuronal systems where nAChRs appear to exist both at the somatodendritic level and on or near terminals. Examples include certain retinofugal pathways,[23] mesolimbic and nigrostriatal dopamine systems,[29] a number of thalamocortical projections,[30,31] and the habenulointerpeduncular pathway.[23,32] The prevalence of this phenomenon has led certain authors to refer to the possibility of "routing accidents."[23] In some neuronal systems, the receptors may indeed mediate cholinergic transmission at both levels of the neuron, but this is far from proven. For example, in the neostriatum, ultrastructural visualization has revealed few signs of cholinergic (or other) axoaxonic contacts.[33-35] These findings indicate that presynaptic cholinergic modulation of dopamine release is unlikely to be important physiologically, unless it occurs at some distance to ACh release sites, despite the existence of presynaptic nAChRs in this brain region.

FUNCTIONAL STUDIES RELEVANT TO NICOTINIC CHOLINERGIC TRANSMISSION

Anatomical receptor mapping studies using radioligands and antibodies have indicated that cholinergic projections and nAChRs are widely encountered in the brain. However, little attempt has as yet been made to marry the two in order to assess the prevalence of nicotinic cholinergic transmission. Similarly, functional studies have concentrated on the actions of exogenous nicotine and acetylcholine, revealing that many neurons are sensitive to nicotinic agonists. However, few efforts have been directed at identifying possible sites in the CNS where *endogenous* ACh might be released onto nAChRs. Indeed, reasonably strong evidence for nicotinic cholinergic transmission in the brain has been obtained only in thalamus, substantia nigra pars compacta, and nucleus ambiguus. This work is now briefly reviewed.

Most thalamic nuclei exhibit high levels of nAChRs and related mRNA,[10,36] and within several thalamic nuclei, direct application of nicotinic agonists has been shown to increase neuronal firing rates.[37,38] The thalamus receives a major cholinergic input from the pedunculopontine tegmental nucleus and adjacent laterodorsal tegmental nucleus of the brain stem,[4] and electron microscopic evidence also exists for synaptic innervation by ChAT-positive terminals.[39] Electrical stimulation in the vicinity of the pedunculopontine tegmental nucleus has been shown to excite thalamic relay neurons in the dorsal lateral geniculate nucleus with short latency, and this excitation was blocked by direct application of the nicotinic antagonist hexamethonium, providing evidence for nicotinic cholinergic transmission.[40]

In the rat substantia nigra pars compacta, dopamine (DA) cells express nAChRs, as indicated by *in situ* hybridization histochemistry,[36] receptor autoradiography,[29] and electrophysiological recording.[41] Electron microscopic visualization of ChAT-like immunoreactivity indicates that the DA neurons receive a cholinergic innervation,[42] and retrograde tracing studies have identified the ipsilateral pedunculopontine tegmental nucleus as a major source of cholinergic fibers.[43,44] Infusion of the excitant kainic acid in the vicinity of this ACh cell body group resulted in a dose-related and prompt excitation of identified nigral DA cells.[44] This excitation was shown to be induced by ipsilateral but not by contralateral kainate infusion (consistent with the hodological findings), and was prevented by systemic administration of the nicotinic antagonist mecamylamine. Although a polysynaptic input cannot be ruled out, it appears likely that a direct cholinergic link exists between cell bodies in the pedunculopontine tegmental nucleus and the dopamine cells of the nigra.

Recently, evidence has also been provided for a nicotinic cholinergic input to motoneurons of the rat nucleus ambiguus. This nucleus possesses high levels of nAChR-related mRNA and protein.[23,36] In anesthetized rats, responses to local administration of ACh and to the nicotinic agonist DMPP were shown to be blocked by nicotinic antagonists.[45] In subsequent intracellular recordings, application of ACh resulted in inward currents. These drug responses were most likely direct, because they persisted in the presence of tetrodotoxin and high extracellular manganese which inhibit sodium-dependent spikes and calcium-dependent stimulus-secretion coupling, respectively. Zhang et al. subsequently used retrograde tracing to identify a probable cholinergic input from cell bodies located in the rostral medulla, and the existence of nicotinic cholinergic transmission then received strong support in experiments in which electrical stimulation combined with direct antagonist application was performed in vitro.[46]

IN VIVO REGULATION OF CNS nAChRs BY TREATMENT WITH ACETYLCHOLINESTERASE INHIBITORS

Another approach to assessing whether nAChRs are cholinergically innervated is to examine the effects of chronic in vivo administration of acetylcholinesterase inhibitors (AChEIs) on nAChR density in the brain. This kind of pharmacological treatment tends to increase cholinergic tone, and has been shown to decrease the density of brain muscarinic receptors.[47] In the first such experiments, high-affinity nicotinic agonist binding sites were also found to be depleted,[47,48] suggesting that the corresponding nAChRs were cholinergically innervated. Subsequently, however, chronic treatment with AChEIs was found to up-regulate these sites.[49,50] In another study, $[^{125}I]$-α-bungarotoxin binding sites were unaltered by chronic AChEI treatment despite concomitant decreases in $[^3H]$nicotine binding.[51]

The use of AChEIs for defining possible sites of nicotinic cholinergic transmission suffers from a number of problems. Not only are the results discordant, but there are additional interpretational issues that have only recently surfaced. The first relates to possible neurotoxic effects of esterase inhibition[52] that could conceivably lead to an irreversible loss of cells expressing nAChRs. In this context, it should be noted that the reversibility of AChEI-induced nAChR changes has yet to be examined.

The second issue relates to direct actions that certain AChEIs exert on CNS nAChRs. Albuquerque and colleagues have amassed considerable evidence that in rat brain the AChEI physostigmine can produce activation via a direct action on nAChRs which is mediated by a site on α-subunits that is distinct from the ACh binding domain;[53,54] at higher concentrations, an antagonist action was seen.[55] It should be noted that the receptors under study in these experiments were of a subtype in which responses to classical nicotinic agonists were blocked by α-bungarotoxin.

In our own experiments, we have studied the effects of several AChEIs on a nicotinic response that is mediated by receptors insensitive to α-bungarotoxin. The response in question is $[^3H]$dopamine release from superfused rat striatal synaptosomes, which can be evoked by nicotine and other agonists in a concentration- and calcium-dependent manner.[56] Physostigmine, neostigmine, tacrine, and diisopropylfluorophosphate (DFP) all reduced nicotine-induced DA release in a concentration-dependent manner (FIG. 1), but physostigmine and tacrine were clearly more potent than DFP.[57] Tests of pharmacological selectivity involved comparisons between nicotine and other secretagogues (other nicotinic agonists, amphetamine, and high K^+) (FIG. 2). These tests revealed that physostigmine blocked nicotinic responses in a selective fashion, indicating a probable action at nAChRs. Tacrine, in contrast, acted nonselectively, and may have exerted its actions entirely independently of

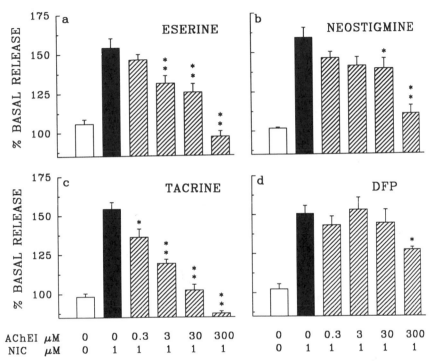

| AChEI μM | 0 | 0 | 0.3 | 3 | 30 | 300 | | 0 | 0 | 0.3 | 3 | 30 | 300 |
| NIC μM | 0 | 1 | 1 | 1 | 1 | 1 | | 0 | 1 | 1 | 1 | 1 | 1 |

FIGURE 1. Effects of eserine (**a**), neostigmine (**b**), tacrine (**c**), and DFP (**d**) on [³H]dopamine release induced by nicotine from striatal synaptosomes. Synaptosomes were superfused with buffer in the presence or absence of AChE inhibitor (0.3–300 μM) for 35 min prior to challenge with nicotine 1 μM or buffer. The vertical axis shows mean ± SE mean peak release expressed as a percentage of basal release. Superfusion channels per condition: n = 10–20 (**a**), 8–11 (**b**), 7–10 (**c**), 7–14 (**d**). *p < 0.05, **p < 0.01 vs. nicotine alone (Dunnett's test). (Clarke et al.[57] Reproduced, with permission, from the *British Journal of Pharmacology*.)

nAChRs. Antagonism by physostigmine was not preceded by a nicotine-like stimulation, suggesting that it was not the result of agonist-induced desensitization. We also considered the possibility that the observed blockade resulted from inhibition of AChE, which might conceivably have produced sufficiently elevated ACh levels in the perfusate to lead to nAChR desensitization. However, a clear dissociation was found between nicotinic block and esterase inhibition, supporting the conclusion that physostigmine directly blocks the nAChRs under study. The same concentrations of physostigmine that produced nicotinic antagonism are commonly used *in vitro* in order to inhibit AChE; whether *in vivo* physostigmine treatment would significantly affect nAChR function is not clear.

The third issue that complicates the interpretation of nAChR regulation experiments relates to uncertainties concerning the mechanisms by which receptor changes

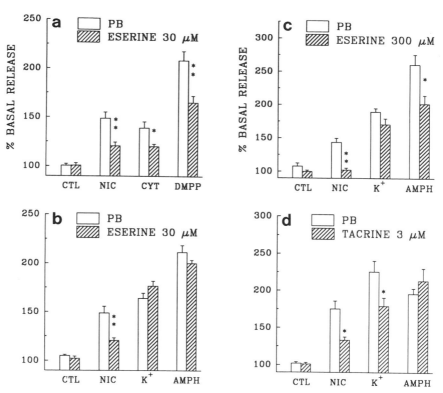

FIGURE 2. Pharmacological selectivity of blockade by eserine or tacrine. In all assays, synaptosomes were superfused for 35 min with or without AChE inhibitor (*cross-hatched* and *open bars*, respectively), prior to acute challenge with [³H]dopamine secretagogue. (**a**) Challenge with nicotine 1 μM, cytisine 10 μM, DMPP 10 μM or buffer alone, with or without eserine (30 μM). (**b**) Challenge with nicotine 1 μM, K⁺ 12 mM, (+)-amphetamine 0.3 μM or buffer alone, with or without eserine 30 μM. (**c**) As for (**b**), except eserine 300 μM was used. (**d**) As for (**b**), except tacrine 3 μM was used. The vertical axis shows mean ± SE mean peak release expressed as a percentage of basal release. Superfusion channels per condition: $n = 8$–12 (**a**), 13–17 (**b**), 5–8 (**c**), 6–13 (**d**). *$p < 0.05$, **$p < 0.01$ vs. AChEI-free condition at same agonist concentration (Student's t test with Bonferroni adjustment). (Clarke *et al.*[57] Reproduced, with permission, from the *British Journal of Pharmacology.*)

are triggered. Paradoxically, chronic *in vivo* treatment with nicotine or other centrally acting agonists typically *up-regulates* high-affinity [³H]agonist binding site density.[58] It has been suggested that this occurs because in the doses administered, nicotine may act as a "time averaged" antagonist.[59,60] However, this plausible notion seems to be put in some doubt by our recent observations that treatment with the quasi-irreversible CNS nicotinic antagonist chlorisondamine did not significantly alter [³H]nicotine binding site density, and did not prevent the up-regulation resulting from chronic nicotine treatment, despite demonstrated CNS nicotinic blockade.[61]

CONSEQUENCES OF CNS NICOTINIC BLOCK

If nicotinic cholinergic transmission is important to CNS functioning, the consequences of CNS nAChR blockade should be serious. Many studies have examined the effects of centrally active nicotinic antagonists on neurochemical and behavioral effects in animals, but the interpretation of these studies is complicated by several issues. First, almost all investigators have relied on the antagonist mecamylamine. It is not clear how selective this drug is for nAChRs over NMDA-type glutamate receptors.[62] In addition, because mecamylamine has been found to act in an insurmountable fashion in the CNS,[62] doses of this drug that completely block the effects of administered nicotine should also be sufficient to block the actions of endogenous ACh. However, where mecamylamine has been found to be active in behavioral tests, this has often only occurred at high doses.

A second problem lies in the frequent use of hexamethonium as a control for the peripheral effects of mecamylamine. Although both nicotinic antagonists are ganglion blockers, there appears to be little information available on relative potency and duration of action in the rat. Thus, the commonly performed comparison of mecamylamine 1 mg/kg s.c. versus hexamethonium 5 mg/kg s.c. is based more upon tradition than on hard data.

The possible existence of nicotinic autoreceptors poses further problems of interpretation. Thus, in certain isolated tissues, ACh release is enhanced by a direct action of nicotinic agonists.[63,64] It is conceivable, therefore, that nicotinic antagonists, by blocking autoreceptors, may inhibit cholinergic transmission *in vivo* at any synapses where *postsynaptic* cholinoceptors are muscarinic rather than nicotinic.

Although most investigators have used mecamylamine as the nicotinic antagonist of choice, the bisquaternary compound chlorisondamine provides an alternative method with which to investigate the effects of central nAChR blockade. Central blockade can be achieved by administering chlorisondamine either in a high subcutaneous dose or in a much lower dose given intracerebroventricularly.[65] In contrast to the persistent central block, which lasts for many weeks after a single administration, ganglionic blockade is only transient.[62] Where studied, chlorisondamine has been found to antagonize the effects of nicotine in an insurmountable fashion.[56,66] Thus far, it does not appear that the chronic nicotinic blockade that follows chlorisondamine administration results in any major functional impairment.[65,67,68] This might imply that nicotinic cholinergic transmission in the CNS is not critical to important psychobiological processes, but may also reflect the capacity of the nervous system to adapt.

CONCLUSION

Studying the actions of nicotine should in principle tell us something about cholinergic neurotransmission in the brain. However, the relationship appears to be complicated, and it is not at all clear how many CNS nAChRs are really innervated by ACh. Although cholinergic fibers and nAChRs have been mapped in some detail, few attempts have been made to demonstrate transmission. Indeed, some observations suggest that nAChR localization may *not* be precisely controlled. Another way to identify possible sites of cholinergic transmission is to examine the *in vivo* regulation of nAChRs by cholinesterase inhibitors. However, recent evidence suggests that functional status may *not* be an important factor in the regulation of CNS nAChRs.

REFERENCES

1. JONES, G. M. M., B. J. SAHAKIAN, R. LEVY, D. M. WARBURTON & J. A. GRAY. 1992. Effects of acute subcutaneous nicotine on attention, information processing and short-term memory in Alzheimer's disease. Psychopharmacology (Berl.) **108:** 485–494.
2. NEWHOUSE, P. A., T. SUNDERLAND, P. N. TARIOT, C. L. BLUMHARDT, H. WEINGARTNER, A. MELLOW & D. L. MURPHY. 1988. Intravenous nicotine in Alzheimer's disease: A pilot study. Psychopharmacology (Berl.) **95:** 171–175.
3. NEWHOUSE, P. A., A. POTTER, J. CORWIN & R. LENOX. 1992. Acute nicotinic blockade produces cognitive impairment in normal humans. Psychopharmacology (Berl.) **108:** 480–484.
4. WOOLF, N. J. 1991. Cholinergic systems in mammalian brain and spinal cord. Prog. Neurobiol. **37:** 475–524.
5. MESULAM, M.-M. & C. GEULA. 1992. Overlap between acetylcholinesterase-rich and choline acetyltransferase-positive (cholinergic) axons in human cerebral cortex. Brain Res. **577:** 112–120.
6. MESULAM, M.-M., D. MASH, L. HERSH, M. BOTHWELL & C. GEULA. 1992. Cholinergic innervation of the human striatum, globus pallidus, subthalamic nucleus, substantia nigra, and red nucleus. J. Comp. Neurol. **323:** 252–268.
7. MARKS, M. J., J. A. STITZEL, E. ROMM, J. M. WEHNER & A. C. COLLINS. 1986. Nicotinic binding sites in rat and mouse brain: Comparison of acetylcholine, nicotine, and alpha-bungarotoxin. Mol. Pharmacol. **30:** 427–436.
8. MARTINO-BARROWS, A. M. & K. J. KELLAR. 1987. ^{3}H-acetylcholine and ^{3}H-nicotine label the same recognition site in rat brain. Mol. Pharmacol. **31:** 169–174.
9. PABREZA, L. A., S. DHAWAN & K. J. KELLAR. 1991. [^{3}H]Cytisine binding to nicotinic cholinergic receptors in brain. Mol. Pharmacol. **39:** 9–12.
10. CLARKE, P. B. S., R. D. SCHWARTZ, S. M. PAUL, C. B. PERT & A. PERT. 1985. Nicotinic binding in rat brain: Autoradiographic comparison of ^{3}H-acetylcholine, ^{3}H-nicotine, and ^{125}I-alpha-bungarotoxin. J. Neurosci. **5:** 1307–1315.
11. LINDSTROM, J., R. SCHOEPFER & P. WHITING. 1987. Molecular studies of the neuronal nicotinic acetylcholine receptor family. Mol. Neurobiol. **1:** 281–337.
12. WILLIAMS, M. & J. L. ROBINSON. 1984. Binding of the nicotinic cholinergic antagonist, dihydro-beta-erythroidine, to rat brain tissue. J. Neurosci. **4:** 2906–2911.
13. SCHULZ, D. W., R. H. LORING, E. AIZENMAN & R. E. ZIGMOND. 1991. Autoradiographic localization of putative nicotinic receptors in the rat brain using ^{125}I-neuronal bungarotoxin. J. Neurosci. **11:** 287–297.
14. HUNT, S. P. & J. SCHMIDT. 1978. The electron microscopic autoradiographic localization of alpha-bungarotoxin binding sites within the central nervous system of the rat. Brain Res. **142:** 152–159.
15. ARIMATSU, Y., A. SETO & T. AMANO. 1978. Localization of alpha-bungarotoxin binding sites in mouse brain by light and electron microscopic autoradiography. Brain Res. **147:** 165–169.
16. LENTZ, T. L. & J. CHESTER. 1977. Localization of acetylcholine receptors in central synapses. J. Cell Biol. **75:** 258–267.
17. MARSHALL, L. M. 1981. Synaptic localization of alpha-bungarotoxin binding which blocks nicotinic transmission at frog sympathetic neurons. Proc. Natl. Acad. Sci. USA **78:** 1948–1952.
18. JACOB, M. H. & D. K. BERG. 1983. The ultrastructural localization of alpha-bungarotoxin binding sites in relation to synapses on chick ciliary ganglion neurons. J. Neurosci. **3:** 260–271.

19. SMOLEN, A. J. 1983. Specific binding of alpha-bungarotoxin to synaptic membranes in rat sympathetic ganglion: Computer best-fit analysis of electron microscope radioautographs. Brain Res. **289:** 177–188.
20. LORING, R. H., L. M. DAHM & R. E. ZIGMOND. 1985. Localization of alpha-bungarotoxin binding sites in the ciliary ganglion of the embryonic chick: An autoradiographic study at the light and electron microscopic level. Neuroscience **14:** 645–660.
21. MILLER, M. M. & E. ANTECKA. 1987. Internalization of ^{125}I-alpha-bungarotoxin into rat suprachiasmatic nucleus neurons and dendrites. Brain Res. Bull. **19:** 429–437.
22. SCHRODER, H., K. ZILLES, A. MAELICKE & F. HAJOS. 1989. Immunohisto- and cytochemical localization of cortical nicotinic cholinoceptors in rat and man. Brain Res. **502:** 287–295.
23. SWANSON, L. W., D. M. SIMMONS, P. J. WHITING & J. LINDSTROM. 1987. Immunohistochemical localization of neuronal nicotinic receptors in the rodent central nervous system. J. Neurosci. **7:** 3334–3342.
24. BRAVO, H. & H. J. KARTEN. 1992. Pyramidal neurons of the rat cerebral cortex, immunoreactive to nicotinic acetylcholine receptors, project mainly to subcortical targets. J. Comp. Neurol. **320:** 62–68.
25. HILL, J. A., JR., M. ZOLI, J.-P. BOURGEOIS & J.-P. CHANGEUX. 1993. Immunocytochemical localization of a neuronal nicotinic receptor: The β2-subunit. J. Neurosci. **13:** 1551–1568.
26. OKUDA, H., S. SHIODA, Y. NAKAI, H. NAKAYAMA, M. OKAMOTO & T. NAKASHIMA. 1993. Immunocytochemical localization of nicotinic acetylcholine receptor in rat hypothalamus. Brain Res. **625:** 145–151.
27. ICHIKAWA, T. & Y. HIRATA. 1986. Organization of choline acetyltransferase-containing structures in the forebrain of the rat. J. Neurosci. **6:** 281–292.
28. HUNT, S. & J. SCHMIDT. 1979. The relationship of alpha-bungarotoxin binding activity and cholinergic termination within the rat hippocampus. Neuroscience **4:** 585–592.
29. CLARKE, P. B. S. & A. PERT. 1985. Autoradiographic evidence for nicotine receptors on nigrostriatal and mesolimbic dopaminergic neurons. Brain Res. **348:** 355–358.
30. PRUSKY, G. T., C. SHAW & M. S. CYNADER. 1987. Nicotine receptors are located on lateral geniculate nucleus terminals in cat visual cortex. Brain Res. **412:** 131–138.
31. CLARKE, P. B. S. 1991. Nicotinic receptors in rat cerebral cortex are associated with thalamocortical afferents. Soc. Neurosci. Abstr. **17:** 384.18.
32. CLARKE, P. B. S., G. S. HAMILL, N. S. NADI, D. M. JACOBOWITZ & A. PERT. 1986. ^{3}H-nicotine- and ^{125}I-alpha-bungarotoxin–labeled nicotinic receptors in the interpeduncular nucleus of rats. II. Effects of habenular deafferentation. J. Comp. Neurol. **251:** 407–413.
33. BOUYER, J. J., D. H. PARK, T. H. JOH & V. M. PICKEL. 1984. Chemical and structural analysis of the relation between cortical inputs and tyrosine hydroxylase-containing terminals in rat neostriatum. Brain Res. **302:** 267–275.
34. WAINER, B. H., J. P. BOLAM, T. F. FREUND, Z. HENDERSON, S. TOTTERDELL & A. D. SMITH. 1984. Cholinergic synapses in the rat brain: A correlated light and electron microscopic immunohistochemical study employing a monoclonal antibody against choline acetyltransferase. Brain Res. **308:** 69–76.
35. PHELPS, P. E., C. R. HOUSER & J. E. VAUGHN. 1985. Immunocytochemical localization of choline acetyltransferase within the rat neostriatum: A correlated light and electron microscopic study of cholinergic neurons and synapses. J. Comp. Neurol. **238:** 286–307.
36. WADA, E., K. WADA, J. BOULTER, E. DENERIS, S. HEINEMANN, J. PATRICK & L. W. SWANSON. 1989. Distribution of alpha 2, alpha 3, alpha 4, and beta 2 neuronal nicotinic receptor subunit mRNAs in the central nervous system: A hybridization histochemical study in the rat. J. Comp. Neurol. **284:** 314–335.
37. MCCORMICK, D. A. & D. A. PRINCE. 1987. Actions of acetylcholine in the guinea-pig and cat medial and lateral geniculate nuclei, in vitro. J. Physiol. (Lond.) **392:** 147–165.
38. CURRO DOSSI, R., D. PARÉ & M. STERIADE. 1991. Short-lasting nicotinic and long-lasting muscarinic depolarizing responses of thalamocortical neurons to stimulation of mesopontine cholinergic nuclei. J. Neurophysiol. **65:** 393–405.

39. DE LIMA, A. D., V. M. MONTERO & W. SINGER. 1985. The cholinergic innervation of the visual thalamus: An EM immunocytochemical study. Exp. Brain Res. **59:** 206–212.

40. HU, B., M. STERIADE & M. DESCHENES. 1989. The effects of brainstem peribrachial stimulation on neurons of the lateral geniculate nucleus. Neuroscience **31:** 13–24.

41. LICHTENSTEIGER, W., F. HEFTI, D. FELIX, T. HUWYLER, E. MELAMED & M. SCHLUMPF. 1982. Stimulation of nigrostriatal dopamine neurones by nicotine. Neuropharmacology **21:** 963–968.

42. BOLAM, J. P., C. M. FRANCIS & Z. HENDERSON. 1991. Cholinergic input to dopaminergic neurons in the substantia nigra: A double immunocytochemical study. Neuroscience **41:** 483–494.

43. BENINATO, M. & R. F. SPENCER. 1987. A cholinergic projection to the rat substantia nigra from the pedunculopontine tegmental nucleus. Brain Res. **412:** 169–174.

44. CLARKE, P. B. S., D. W. HOMMER, A. PERT & L. R. SKIRBOLL. 1987. Innervation of substantia nigra neurons by cholinergic afferents from the pedunculopontine nucleus in rats: Neuroanatomical and electrophysiological evidence. Neuroscience **23:** 1011–1020.

45. WANG, Y. T., R. S. NEUMAN & D. BIEGER. 1991. Nicotinic cholinoceptor-mediated excitation in ambigual motoneurons of the rat. Neuroscience **40:** 759–767.

46. ZHANG, M., Y. T. WANG, D. M. VYAS, R. S. NEUMAN & D. BIEGER. 1993. Nicotinic cholinoceptor-mediated excitatory postsynaptic potentials in rat nucleus ambiguus. Exp. Brain Res. **96:** 83–88.

47. COSTA, L. G. & S. D. MURPHY. 1985. Antinociceptive effect of diisopropylphosphofluoridate: Development of tolerance and lack of cross-tolerance to morphine. Neurobehav. Toxicol. Teratol. **7:** 251–256.

48. SCHWARTZ, R. D. & K. J. KELLAR. 1983. Nicotinic cholinergic receptor binding sites in the brain: Regulation in vivo. Science **220:** 214–216.

49. DE SARNO, P. & E. GIACOBINI. 1989. Modulation of acetylcholine release by nicotinic receptors in the rat brain. J. Neurosci. Res. **22:** 194–200.

50. BHAT, R. V., S. L. TURNER, M. J. MARKS & A. C. COLLINS. 1990. Selective changes in sensitivity to cholinergic agonists and receptor changes elicited by continuous physostigmine infusion. J. Pharmacol. Exp. Ther. **255:** 187–196.

51. VAN DE KAMP, J. L. & A. C. COLLINS. 1992. Species differences in diisopropylfluorophosphate-induced decreases in the number of brain nicotinic receptors. Pharmacol. Biochem. Behav. **42:** 131–141.

52. TANAKA, D., JR., S. J. BURSIAN & E. LEHNING. 1990. Selective axonal and terminal degeneration in the chicken brainstem and cerebellum following exposure to *bis*(1-methylethyl)phosphorofluor idate (DFP). Brain Res. **519:** 200–208.

53. PEREIRA, E. F. R., S. REINHARDT-MAELICKE, A. SCHRATTENHOLZ, A. MAELICKE & E. X. ALBURQUERQUE. 1993. Identification and functional characterization of a new agonist site on nicotinic acetylcholine receptors of cultured hippocampal neurons. J. Pharmacol. Exp. **265:** 1474–1491.

54. SCHRATTENHOLZ, A., T. COBAN, B. SCHRÖDER, K. O. OKONJO, J. KUHLMANN, E. F. PEREIRA, E. X. ALBUQUERQUE & A. MAELICKE. 1993. Biochemical characterization of a novel channel-activating site on nicotinic acetylcholine receptors. J. Recept. Res. **13:** 393–412.

55. SHAW, K.-P., Y. ARACAVA, A. AWAIKE, J. W. DALY, D. L. RICKETT & E. X. ALBUQUERQUE. 1985. The reversible cholinesterase inhibitor physostigmine has channel-blocking and agonist effects on the acetylcholine receptor-ion channel complex. Mol. Pharmacol. **28:** 527–538.

56. EL-BIZRI, H. & P. B. S. CLARKE. 1994. Blockade of nicotinic receptor-mediated release of dopamine from striatal synaptosomes by chlorisondamine and other nicotinic antagonists administered *in vitro*. Br. J. Pharmacol. **111:** 406–413.

57. CLARKE, P. B. S., M. REUBEN & H. EL-BIZRI. 1994. Blockade of nicotinic responses by physostigmine, tacrine, and other cholinesterase inhibitors in rat striatum. Br. J. Pharmacol. **111:** 695–702.

58. WONNACOTT, S. 1990. The paradox of nicotinic acetylcholine receptor upregulation by nicotine. Trends Pharmacol. Sci. **11:** 216–219.

59. SCHWARTZ, R. D. & K. J. KELLAR. 1985. In vivo regulation of ^{3}H-acetylcholine recognition sites in brain by nicotinic cholinergic drugs. J. Neurochem. **45:** 427–433.
60. MARKS, M. J., J. B. BURCH & A. C. COLLINS. 1983. Effects of chronic nicotine infusion on tolerance development and nicotinic receptors. J. Pharmacol. Exp. Ther. **226:** 817–825.
61. EL-BIZRI, H. & P. B. S. CLARKE. 1994. Regulation of nicotinic receptors in rat brain following quasi-irreversible nicotinic blockade by chlorisondamine and chronic treatment with nicotine. Br. J. Pharmacol. In press.
62. CLARKE, P. B. S., I. CHAUDIEU, H. EL-BIZRI, P. BOKSA, M. QUIK, B. A. ESPLIN & R. CAPEK. 1994. The pharmacology of the nicotinic antagonist, chlorisondamine, investigated in rat brain and autonomic ganglion. Br. J. Pharmacol. **111:** 397–405.
63. ROWELL, P. P. & D. L. WINKLER. 1984. Nicotinic stimulation of [3H]acetylcholine release from mouse cerebral cortical synaptosomes. J. Neurochem. **43:** 1593–1598.
64. ARAUJO, D. M., P. A. LAPCHAK, B. COLLIER & R. QUIRION. 1988. Characterization of N-[3H]methylcarbamylcholine binding sites and effect of N-methylcarbamylcholine on acetylcholine release in rat brain. J. Neurochem. **51:** 292–299.
65. CLARKE, P. B. S. 1984. Chronic central nicotinic blockade after a single administration of the bisquaternary ganglion-blocking drug chlorisondamine. Br. J. Pharmacol. **83:** 527–535.
66. EL-BIZRI, H. & P. B. S. CLARKE. 1994. Blockade of nicotinic receptor-mediated release of dopamine from striatal synaptosomes by chlorisondamine administered *in vivo*. Br. J. Pharmacol. **111:** 414–418.
67. CLARKE, P. B. S. & H. C. FIBIGER. 1990. Reinforced alternation performance is impaired by muscarinic but not by nicotinic receptor blockade in rats. Behav. Brain Res. **36:** 203–207.
68. CORRIGALL, W. A., K. B. J. FRANKLIN, K. M. COEN & P. B. S. CLARKE. 1992. The mesolimbic dopaminergic system is implicated in the reinforcing effects of nicotine. Psychopharmacology (Berl.) **107:** 285–289.

Neurochemical Evidence of Heterogeneity of Presynaptic and Somatodendritic Nicotinic Acetylcholine Receptors[a]

E. S. VIZI,[b,c] H. SERSHEN,[d] A. BALLA,[b] Á. MIKE,[b]
K. WINDISCH,[b] ZS. JURÁNYI,[b] AND A. LAJTHA[d]

[b]Department of Pharmacology
Institute of Experimental Medicine
Hungarian Academy of Sciences
P.O. Box 67
H-1450 Budapest, Hungary

[d]The Nathan Kline Institute for Psychiatric Research
Center for Neurochemistry
140 Old Orangeburg Road
Orangeburg, New York 10962

Since Langley[1] first reported the existence of a "receptive substance" for nicotine in skeletal muscle that could mediate the effects of nicotine, evidence was obtained that acetylcholine receptors (AChRs) could be involved in synaptic transmission in many parts of the central nervous system (CNS), and also could be responsible for signal transmission through autonomic ganglia. Langley[1] wrote, "The stimuli passing the nerve can only affect the contractile molecule by the radical which combines with nicotine and curare. And this seems in its turn to require that the nervous impulse should not pass from nerve to muscle by an electrical discharge, but by the secretion of a special substance at the end of the nerve." In the paper in which he discussed the data obtained in striated muscle, he was talking about a "contractile molecule" and "radical" that combines with nicotine and curare; he rejected electrical transmission and suggested chemical transmission by a "special substance" secreted from the nerve terminals. He was a visionary.[1,2] Since then enormous progress has been made, and the endogenous special substance was identified as acetylcholine (ACh), and its receptors were shown to be located post- and presynaptically. Sir Henry Dale[2] was the first to suggest in a paper published in 1914 that ACh exerts its action via two different receptors, via muscarinic and nicotinic receptors. At first it was believed that the nicotinic action of ACh is limited to the neuromuscular junction and that its muscarinic action is manifested only at intestinal parasympathetic effector sites. Later it became evident that ACh also stimulates preganglionic nicotinic receptors of the autonomic nervous system, presynaptic nicotinic AChRs (nAChRs) located on the motor nerve terminal, cholinergic receptors located on axon terminals in the CNS and the autonomic nervous system, and cholinergic effector cells in various organs (e.g., head, intestines). ACh, like a master key, fits and stimulates all such muscarinic and nicotinic receptors. Therefore, the effect of ACh released from cholinergic axon terminals depends on the localization of these ACh-sensitive

[a]This study was supported in part by grants from the Council for Tobacco Research-USA, Inc. (No. 3377) and the Hungarian Research Fund (OTKA).
[c]Corresponding author.

receptors. It was recently shown that nicotinic receptors on neurons represent a family of receptors distinct from the well-characterized AChRs of skeletal muscle.[3,4] This area of research has received significant attention because of studies with receptor-ligand binding and molecular biological cloning techniques showing that there are subtypes of nAChRs: receptors located on the muscle are different from those present in the brain. The disadvantage of the receptor binding technique is that it cannot be used for precise localization of the receptors because the ligands used are not selective. It does give an overall characterization of nicotinic receptors of different morphological localization; however, neither of these two methods could differentiate between the receptors in their different neuronal localization. Nevertheless, functional diversity of neuronal nAChRs was shown.[5]

AChR has become one of the best-characterized ligand-gated ion channels with respect of structure, kinetics, pharmacology, and molecular composition. Nicotinic receptors have been postulated to be involved mostly in postsynaptic signaling responses. At present we know much more about the binding properties and

TABLE 1. Neurochemical Evidence That Stimulation of nAChRs Results in Transmitter Release in the Central Nervous System

Transmitter	nAChR Agonist	Brain Region	Reference
NE	DMPP	Human neuroblastoma	63
	Nic	Hippocampus	8
DA	Nic	Striatal synaptosome	11, 12
	Nic	Nucleus accumbens	64
	Nic	Mouse striatal slice	16, 17
	Nic	Rat striatal slice	65
	Nic	Rat striatum	13, 66
ACh	Nic	Cerebral synaptosome	7
	MCC	Rat brain	67
	MCC	Cerebellar synaptosome	33, 68
	Nic	Cerebral cortex slice	69
GABA	Nic	Hippocampal synaptosome	70
	Nic	Cerebral cortex slice	69

Abbreviations: NE, norepinephrine; DMPP, dimethylphenyl piperazinium iodide; Nic, nicotinic; DA, dopamine; ACh, acetylcholine; MCC, methylcarbamylcholine; GABA, γ-aminobutyric acid.

structure of neuronal nAChRs than other aspects. However, a full understanding of the process of chemical neurotransmission in both the peripheral and central nervous system must take into account the role of modulatory presynaptic/prejunctional nicotinic receptors. Neurochemical evidence has been provided that stimulation of nAChRs located on the axon terminals results in release of different transmitters. In the CNS it was found (TABLE 1) that nAChR stimulation enhances the release of ACh from the cortex[6,7] and that of norepinephrine (NE)[8] and serotonin[9] from the hippocampus, and that it increases the resting release of dopamine (DA) from both striatal synaptosomal[10–12] and slice preparations.[13–17] In the peripheral nervous system, nicotinic receptor stimulation facilitates the electrical stimulation-evoked release of ACh from the phrenic nerve terminals in the diaphragm[18] and the release of NE from the pulmonary artery[19] and vas deferens.[20] Several subtypes of nAChRs have been distinguished by pharmacological methods[4,20–24] and molecular biological techniques.[25–27]

In this report, we describe studies of the effects of different nAChR agonists on the release of different transmitters and the modulatory role of presynaptic nicotinic receptors in chemical signal transmission. In addition, using different nAChR agonists and antagonists we attempted to characterize the presynaptic nAChRs located on the noradrenergic axon terminals in the hippocampus and vas deferens and the somatodendritic nAChRs located on the myenteric plexus. The advantage of release studies is that one can be certain where the transmitter comes from, and thus can locate the site of action of different nAChR agonists and antagonists, making the characterization of the receptors rather specific.

RESULTS AND DISCUSSION

Functional Role of Presynaptic nAChRs in Chemical Transmission

The functional role of nicotinic postsynaptic/postjunctional cholinoceptors in chemical neurotransmission is clearly established: in fact, these are the most thoroughly studied of all receptors. However, to understand the effect of stimulation by ACh, the endogenous ligand of nAChRs, on the process of chemical neurotransmission and the mechanism of action of nicotine in both the peripheral and central nervous system, one must take into account its effects on modulatory presynaptic/prejunctional nicotinic receptors, which are located on different axon terminals containing transmitters. Several subtypes of nAChRs have been distinguished by ligand binding studies, (cf. refs. 23 and 28) protein chemistry,[29] and molecular biological techniques (cf. refs. 3 and 25). The disadvantage of these techniques is that the ligands used until now lacked sufficient selectivity and could not be used for precise anatomical localization of the nAChRs, in particular, to distinguish between pre- and postsynaptic sites and the types of neurons on which these sites are localized. Because *in situ* hybridization studies suggest that more than one variant of α and β subunit combinations exist in the CNS, the study of a broad concentration range of different agonists and antagonists for their ability to release NE and inhibit NE release, respectively, from the hippocampus enabled us to classify the subtype of presynaptic nAChRs.

Presynaptic nAChRs

We examined NE release in the hippocampus. Several studies provided evidence that a high concentration of $(-)$nicotine (> 100 μM) stimulates transmitter release in a $[Ca^{2+}]_o$-independent manner and that the release cannot be blocked by classical nAChR antagonists.[30,31] Many other studies, in contrast, have shown that $(-)$nicotine is able to release transmitter in a $[Ca^{2+}]_o$-dependent manner.[12,20,31-34] It has been reported[23] that at least three different subtypes of nAChR may exist in the hippocampus, possibly with unique functional characteristics. Although the $\alpha7$ subunit appears to be the predominant subtype on the basis of *in situ* hybridization studies in the hippocampus[35] and electrophysiological responses,[36] $\alpha2$–5 and $\beta2$ subunits have also been demonstrated. To classify the subtype of nAChR involved in the $[Ca^{2+}]_o$-dependent release of NE from the hippocampus, the effect of different nAChR agonists and antagonists was studied.

After the tissues had been loaded with [^{3}H]NE, the spontaneous and stimulation-evoked release of [^{3}H] NE was measured (FIG. 1), and the effect of different agonists and antagonists on resting and stimulation-evoked release was studied. The release of $[^3H]NE$ at rest is $[Ca^{2+}]_o$-independent and is not subject to presynaptic modulation. In contrast, the release from noradrenergic axon terminals in the hippocampus in response to field stimulation is tetrodotoxin (TTX)-sensitive and $[Ca^{2+}]_o$-dependent, and is subject to modulation via presynaptic receptors (e.g., α_2-

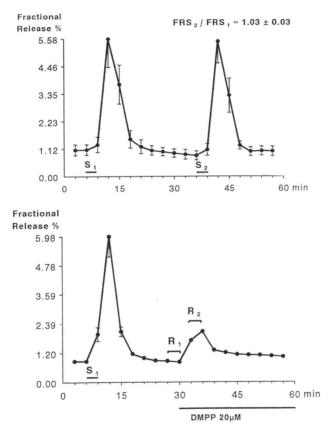

FIGURE 1. Release of $[^3H]$norepinephrine at rest and in response to field stimulation (2 Hz, 360 shocks, 1-ms impulse duration, supramaximal stimulation) and to DMPP administration. Rat hippocampal slice preparation. The release was expressed as percentage of radioactivity (fractional release) in the tissue at the time the sample was collected. Microvolume perfusion system. Means ± SE of the mean. Krebs solution was used and the temperature was 37 °C. DMPP, dimethylphenyl piperazinium iodide.

adrenoceptors). With the tissue loaded with $[^3H]NE$, the fractional release of radioactivity at rest, which is independent of $[Ca^{2+}]_o$, was 0.70 ± 0.02, 0.54 ± 0.01, and 0.55 ± 0.02% of the total radioactivity content measured in the CA1, CA3, and dentate gyrus subregions of the hippocampus, respectively ($n = 3$). DMPP (1–100 μM) enhanced the release of NE (TABLE 2) in a concentration-dependent manner.

TABLE 2. Norepinephrine Releasing Effect of Dimethylphenyl Piperazinium Iodide (DMPP) in Isolated CA1, CA3, and Dentate Gyrus Subregion of Hippocampus

| | Fractional Release at $Rest_2$/Fractional Release at $Rest_1$ | | | |
	CA1	CA3	DG	Hippocampus (CA1 + CA3 + DG)
DMPP				
1 μM	0.91 (1)[a]	0.91 (1)	1.12 (1)	1.14 ± 0.03[b] (4)
3 μM	1.26 ± 0.07[b] (3)	1.02 ± 0.07 (3)	1.42 ± 0.13[b] (3)	
10 μM	1.67 ± 0.12[b] (5)	1.30 ± 0.09[b] (5)	1.73 ± 0.12[b] (5)	1.85 ± 0.04[b] (4)
20 μM	1.66 ± 0.12[b] (10)	1.60 ± 0.15[b] (8)	2.02 ± 0.14[b] (8)	2.23 ± 0.09[b] (4)
50 μM	1.66 ± 0.09[b] (4)	1.58 ± 0.09[b] (4)	2.11 ± 0.26[b] (4)	
100 μM	1.66 ± 0.14[b] (3)	1.59 ± 0.09[b] (4)	2.33 ± 0.31[b] (3)	4.41 ± 0.10[b] (4)
Control	0.94 ± 0.04 (5)	0.90 ± 0.02 (5)	0.92 ± 0.04 (6)	0.91 ± 0.02 (4)

NOTE: Male Wistar rats, weighing 150–160 g, were killed by a blow to the head under light ether anesthesia. The brain was quickly removed and the hippocampus dissected on ice and sliced into 0.4 mm thick coronal sections with a McIlwain chopper. Slices were dissected into tissue parts containing mainly CA1, CA3, and dentate gyrus. The tissues were incubated at 37 °C for 40 min in Krebs solution (in mM: NaCl, 113; KCl, 4.7; $CaCl_2$, 2.5; KH_2PO_4, 1.2; Mg_2SO_4, 1.2; $NaHCO_3$, 25.0; and glucose, 11.5) containing 1-[7,8-^{3}H]norepinephrine (490 KBq/mL, 555 GBq/mmol spec. act., Amersham). Krebs solution contained ascorbic acid (3×10^{-3} M), Na_2EDTA (10^{-4} M). Release of tritium was expressed in Bq/g and as percentage of the radioactivity (fractional release, FR) in the tissue at the time the sample was collected. DMPP was added to the Krebs solution in the 8th fraction and kept throughout the experiments. FRR_2/FRR_1 was calculated.
[a]Number of experiments is in parentheses.
[b]Significance at the level of $p < 0.05$. Mean ± SEM.

No significant difference was found in the effect of DMPP in the three subregions of the hippocampus. In addition various nAChR agonists were tested for their ability to release NE from hippocampal slice preparations. (+)Anatoxin-a is an important probe for characterizing nAChRs[37] and has much greater affinity for high-affinity (−)nicotine sites (K_i = 0.34–3.5 nM)[34] than for muscarinic binding sites. It inhibits muscle twitch in rat[38] and at 1 μM concentration blocks the indirect twitch of the frog sciatic nerve-sartorius preparation in a short time.[39] This effect is associated with its AChR ion channel-blocking action exerted at higher concentration.[40] At nanomolar concentrations it is a selective agonist.[41] In our experiments it was also effective in releasing NE from hippocampal slices. The rank order of potency of the nAChR agonist for releasing NE from the hippocampal slice preparation was (+)anatoxin-a > DMPP > (−)nicotine ≫ cytisine, with the equipotent (ED_{20} of DMPP) values 1.2, 2.3, 10.2, and 22.0 μM, respectively. DMPP, unlike other agonists, enhanced the stimulation-evoked release, but in contrast, this effect was not antagonizable with nAChR antagonists. To rule out the possibility that the effect of nAChR agonists is indirect, mediated through glutamic acid release,[40] experiments were performed in the presence of MK-801 (3 μM). Under these conditions DMPP enhanced the release of NE to the same extent. Thus, it is unlikely that glutamic acid released by nAChR stimulation is responsible for the release of NE.

Various nAChR antagonists were tested for their ability to prevent the release of NE induced by 20 μM DMPP, a concentration of DMPP that gives near half-maximal release (see TABLE 2). Mecamylamine and other antagonists [(+)tubocurarine and hexamethonium] prevented the increase by DMPP of NE release. α-Bungarotoxin (αBGT), a potent nAChR blocker at the neuromuscular junction, applied at a concentration of 3 μM for 10 min, and dihydro-β-erythroidine (DHβE) (10 μM) did not affect the response to DMPP (TABLE 3). The rank order of potency was mecamylamine > (+)tubocurarine > hexamethonium ≫ αBGT = DHβE.

[Ca²⁺]ₒ-Dependence of nAChR-mediated Release of Norepinephrine

When $[Ca^{2+}]_o$ was omitted and EGTA (1 mM) added to the Krebs solution, DMPP (20 μM) had no effect on [³H]NE release from the CA1 region: FRR2/FRR1 = 0.97 ± 0.11 (*n* = 4) measured in the presence of DMPP was not different from control (0.94 ± 0.04, *p* > 0.1). The nAChR is a ligand-operated ion channel whose activation results in membrane depolarization and Ca^{2+} influx and subsequent release of different transmitters. It has been suggested that Ca^{2+} influx through nAChRs accounts for some effects of nAChR agonists in the nervous system[33,42,43] and that this is a distinctive property of neuronal receptors in contrast to muscle nicotinic receptors.[43] The findings that the effect of DMPP was $[Ca^{2+}]_o$-dependent, that Cd^{2+} (100 μM) partly prevented the effect of DMPP (data not shown), and that nifedipine, even at a high concentration, had no effect indicate that the stimulation of nAChRs opens N-type Ca^{2+} channels.

Atropine, a muscarinic antagonist (1 μM) that totally blocks muscarinic responses in other systems, and oxotremorine (1 μM), a muscarinic agonist, were without effect on the release of NE.[44] This makes the noradrenergic pathway different from others, where noradrenergic axon terminals express inhibitory muscarinic receptors. The presence of nAChRs means that a possibility exists for excitatory input by cholinergic neurons.

The release of [³H]NE from rat vas deferens[20] was also measured. nAChR agonists released NE with the following rank order of potency: (+)anatoxin-a > DMPP > (−)nicotine > cytisine; ED_{20} = 0.8 > 5.1 > 10.2 > 52.3 μM (ED_{20} indicates that the release was enhanced by 20%).

TABLE 3. Apparent Dissociation Constants (K_d) of Different nAChR Antagonists at Presynaptic nAChRs of Noradrenergic Axon Terminals in the Hippocampus of the Rat

Antagonist	Dissociation Constant (μM)
Mecamylamine	1.05 ± 0.05 (6)
(+)Tubocurarine	17.65 ± 1.40 (5)
Hexamethonium	35.20 ± 2.15 (4)
Pancuronium	21.21 ± 2.00 (5)
α-Bungarotoxin	≫ 30 (5)
Dihydro-β-erythroidine	≫ 30 (3)

NOTE: In hippocampal slice preparations the apparent equilibrium dissociation constant (K_d) for nAChR antagonists was determined by the dose-ratio method. DMPP was used as agonist. The following equation was used to relate the dissociation constant (K_d) to the dose-ratio and the antagonist concentration K_d = a/DR-1, where DR is the concentration-ratio, i.e., the EC_{50} value for agonist in the presence of the antagonist divided by the ED_{50} value in the absence of antagonist and *a* is the concentration of antagonist. EC_{50} indicates the concentration of agonist needed to produce a 50% increase of R_2/R_1 value. Three to six different concentrations of agonists were used to establish a concentration-response curve. Mean ± SEM. Number of experiments is in parentheses.

Modulation by α_2-Adrenoceptors of nAChR-mediated Release of Norepinephrine

It is known that the action potential-evoked release of NE is subject to α_2-adrenoceptor-mediated inhibition; therefore, the question arises whether the nAChR-mediated release, which is $[Ca^{2+}]_o$-dependent, but not associated with axonal firing, can be modulated. Neurochemical evidence was obtained that xylazine, an α_2-adrenoceptor agonist, prevented DMPP from releasing NE. 7,8(Methylenedioxy)-14α-hydroxyalloberbane (CH-38083), a selective α_2-adrenoceptor antagonist,[45] antagonized the effect of xylazine, indicating that this action was mediated via α_2-adrenoceptors (TABLE 4). The effect of α_2-adrenoceptor stimulation might have been mediated by decrease of presynaptic Ca^{2+} channels[46] and/or by increase of K^+ conductance of the presynaptic axon terminal. Evidence has been obtained that the

TABLE 4. Inhibition by α_2-Adrenoceptor Stimulation of [^{3}H]Norepinephrine Release Induced by DMPP in Rat Hippocampal Slices

Treatment	FRR_2/FRR_1	Significance p
1. Control	0.95 ± 0.03	
2. Xylazine, 3 μM	0.98 ± 0.10	
3. CH-38083, 1 μM	0.94 ± 0.06	
4. DMPP, 20 μM	2.23 ± 0.09	4:1 < 0.001
		5:1 > 0.05
5. Xylazine, 3 μM + DMPP, 20 μM	1.19 ± 0.06	
6. CH-38083, 1 μM + DMPP, 20 μM	2.21 ± 0.17	6:1 < 0.001
		6:4 > 0.05
		6:5 < 0.01
7. CH-38083, 1 μM + xylazine, 3 μM	1.92 ± 0.20	7:1 < 0.001
+ DMPP, 20 μM		7:4 > 0.05
		7:5 < 0.01
		7:6 > 0.05

NOTE: CH-38083, 7,8(methylendioxy)-14α-hydroxyalloberbane, a selective α_2-adrenoceptor antagonist,[45] was added to the perfusion fluid 6 min prior to administration of xylazine, an α_2-adrenoceptor agonist. Fractional release at rest, FRR_1, was estimated at the third and fourth fractions, and FRR_2 at the tenth and eleventh fractions. Drugs were added between the two, 10 min prior to R_2 and kept throughout the experiment. Krebs solution, 37 °C. Significance, one-way analysis of variance followed by a Tukey-Kramer multiple comparisons test. DMPP, dimethylphenyl piperazinium iodide.

stimulation of α_2-adrenoceptors reduces the activity of N-type Ca^{2+} channels,[46] the dominant calcium entry pathway triggering sympathetic transmitter release.[47] It is conceivable that the effect of nAChR is mediated via stimulation of N-type Ca^{2+} channel located on axon terminals.

In contrast to the nAChR located on the striated muscle, which is assembled from four different subunits arranged as the pentamer $\alpha2\beta\gamma\delta$, several findings suggest that the neuronal receptor is composed of α and β subunits only.[29]

The finding that the rank order of antagonists was mecamylamine > (+)tubocurarine > hexamethonium and αBGT or DHβE were completely ineffective indicates that the presynaptic nAChRs involved in NE release do not contain the α7 subunit, one of the predominant subtypes of nAChRs in the hippocampus and one involved in long-term potentiation.[36] Binding studies suggest that cytisine is a more potent

TABLE 5. Rank Order of Potency of Different nAChR Agonists and Antagonists and the Possible Subunit Composition of nAChRs

Subunits		Rank Order of Potency of	
		Agonist	Antagonist
Neuronal nAChRs			
Presynaptic			
NE release			
α3β2	Hippocampus (1)	(+)ATX-a > DMPP > (−)Nic ≫ Cyt	Mec > dTC > Hex ≫ α-BGT = DHβE
	Vas deferens (2)	(+)ATX-a > DMPP > (−)Nic > Cyt	
DA release			
α3β2	Striatum (3)	ACh > (−)Nic = Cyt > DMPP > Anab > Lob	n-BGT > Mec > DHβE > Hex > dTC ≫ α-BGT
	(4)	Cyt = (−)Nic = DMPP	Neosurtx > DHβE > Mec ≫ α-BGT
ACh release			
	Neuromuscular junction (5)		
	(6)	(−)Nic = Cyt	dTC > Panc > Hex ≫ α-BGT
Postsynaptic			
Somatodendritic			
α4β2	Myenteric plexus (7)	(+)ATX-a > DMPP = (−)Nic > Cyt	DHβE = Mec > dTC > Hex ≫ α-BGT
	Substantia nigra (8)	(−)Nic	DHβE = Mec
Muscle nAChRs			
α2β1εδ	Striated muscle (9)		Panc > dTC > α-BGT ≫ Hex = Mec

Abbreviation: NE, norepinephrine; DA, dopamine; ACh, acetylcholine; (+)ATX-a, (+)anatoxin-a; DMPP, 1,1-dimethyl-4-phenylpiperazinium iodide; (−)Nic, (−)Nicotine; Anab, (±)Anabasine; Lob, (−)lobeline HCl; Hex, hexamethonium HBr; dTC, (+)tubocurarine; α-BGT, α-bungarotoxin; Mec, mecamylamine; DHβE, dihydro-β-erythroidine, Cyt, cytisine; Panc, pancuronium; Neosurtx, neosurugatoxin.

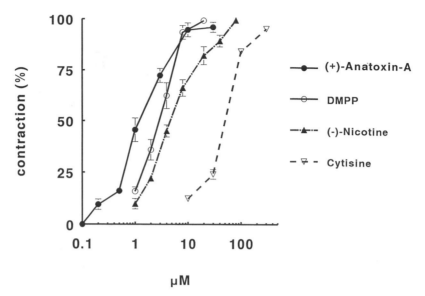

FIGURE 2. Effect of nAChR agonists on isolated longitudinal muscle strip with myenteric plexus attached (guinea pig ileum). For method see reference 49.

agonist than (−)nicotine.[31,48] However, its potency depends on the subunit composition of the nAChR. *In situ* hybridization has revealed the existence of α- and β-subunits in different regions of the brain, including the hippocampus.[26,27] Cytisine can distinguish between receptors containing β4, which are highly sensitive to cytisine, and receptors containing β2, which are completely insensitive. Therefore it is suggested that the presynaptic nAChRs, whose stimulation by nAChR agonists [(+)anatoxin-a > DMPP > (−)nicotine ≫ cytisine] resulted in an increase of NE in the three different regions of the hippocampus, contain the β2 subunit (TABLE 5). αBGT-insensitive nAChRs, which represent a large population of nAChRs present throughout the hippocampus, are most likely composed of a combination of α3 and β2 subunits, and are involved in the release of NE by nAChR stimulation.

In the striatum the dopaminergic axon terminals are equipped with similar nAChRs.[31] It is very likely that these nAChRs are composed of α3 and β2 subunits. Although n-bungarotoxin and neurosuguratoxin were very potent in inhibiting DA release evoked by nAChR stimulation, αBGT had no effect at all.[12,31] Because the potency of nAChR agonists depends on the α-β combination,[24] the rank order of agonist potency for DA release obtained by Grady *et al.*[31] suggests an α3β4 combination (relatively high cytisine activity), although it is also not a perfect match. However, inasmuch as nBTX is not active on nAChRs composed of α3β4 subunits, the suggestion of Grady *et al.*[31] seems very unlikely, and therefore the nAChRs responsible for DA release are most likely composed of α3β2 subunits.

There is convincing neurochemical evidence that at the mouse neuromuscular junction (phrenic nerve–hemidiaphragm preparation) there are nAChRs presynaptically involved in positive feedback modulation of ACh release.[18] Presynaptically the rank order of potency of nAChR antagonists was (+)tubocurarine > pancuronium > hexamethonium, and αBGT had no effect. In contrast, postsynaptically αBGT was very active (TABLE 5), and hexamethonium had no effect. This, in fact, is the first evidence that in a cholinergic synapse the pre- and postsynaptically located nAChRs are heterogeneous and their subunit composition is different.

Somatodendritic nAChRs

The nAChR agonists produced increasing contractions of the longitudinal muscle of guinea-pig ileum[49,50] (FIG. 2), an effect certainly due to the release of ACh from cholinergic interneurons of the Auerbach plexus,[50–53] because the smooth muscle is not equipped with nicotinic receptors, and the somatodendritic part of cholinergic interneurons is equipped with nAChRs. In fact, nAChR agonists stimulate nAChRs located on the somatodendritic part of the neurons[50,51] subsequently producing firing

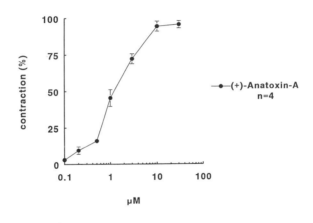

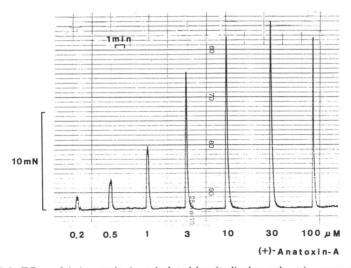

FIGURE 3. Effect of (+)anatoxin-A on isolated longitudinal muscle strip preparation. For method see reference 49.

of the neurons and resulting in release of ACh.[51,53] First the concentration-response curve for the specific nicotinic agonist DMPP was determined. At 100 μM concentration, DMPP produced the maximum contraction ($ED_{50} = 2.96 \pm 0.34$ μM), then decreased responses at higher concentrations. (−)Epibatidine, (+)anatoxin-a, (−)nicotine, and cytisine also produced contractions in a concentration-dependent manner. (−)Epibatidine is an alkaloid isolated from the skin of the poison frog, *Epipedobates tricolor*.[54] A nAChR agonist,[55] it inhibits [³H]cytisine binding with an

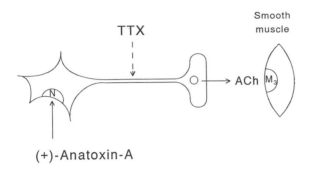

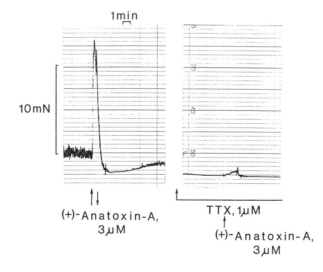

FIGURE 4. Evidence that the site of effect of (+)anatoxin-A is on the somatodendritic part of the cholinergic interneurons. Note that tetrodotoxin (TTX) inhibited the effect of (+)anatoxin-A to produce contraction. TTX blocks nerve conduction; therefore, the stimulation of somatodendritic nAChRs does not produce firing and subsequent ACh release and M_3 muscarinic subtype-mediated contraction of the smooth muscle.

IC_{50} of 70 pM and with a K_i of 43 pM. In our experiments the ED_{50} was 0.03 ± 0.01 μM. Its effect was readily antagonized by mecamylamine ($K_d = 0.8$ μM). The contraction was very fast and showed no tendency to build up desensitization. Atropine and TTX blocked its effect to produce contraction, indicating that its action is mediated via ACh released in response to somatodendritic stimulation of cholinergic interneurons. Maximum responses occurred at a concentration of 10 μM of (+)anatoxin-a (FIG. 3). Peak contraction developed quickly at concentrations above

TABLE 6. Effect of Different nAChR Agonists on Somatodendritic Nicotinic Receptors Located on Cholinergic Interneurons

Agonists	EC_{50} (μM)	N
(−)Epibatidine[a]	0.03 ± 0.01	6
(+)Anatoxin-a	1.38 ± 0.23	4
DMPP[b]	2.96 ± 0.34	11
(−)Nicotine	5.87 ± 0.24	14
Cytisine	50.00 ± 2.00	3

NOTE: To study the effect of nAChR agonists and antagonists on somatodendritic nAChRs, 4–5 cm long, longitudinal muscle strips of guinea-pig ileum[49] were suspended in Krebs solution in a 2-mL organ bath. A resting tension of 10 mN was applied to the muscle. The dose-response of nAChR agonists was determined by adding successively higher concentrations to the bath. From the computer-derived, best-fit, log dose-response regression line, the concentration of nAChR agonist that increased the isometric force of muscle contraction to 50% of the maximum (EC_{50}) was calculated.
[a]Exo-2(6-chloro-3-pyridyl)-7-azabicyclo heptane.[54,55]
[b]1,1-Dimethyl-4-phenylpiperazinium iodide.

5 μM but decreased before washout, that is, the contraction was not maintained. This decrease represented either receptor desensitization or a block induced by the agonist applied. When the preparation was exposed to ACh (0.05–3 μM) the contraction was maintained. Atropine completely blocked the responses, indicating that the response of the smooth muscle is mediated by ACh released from varicosities via stimulation of muscarinic receptors located on the smooth muscle cells. In addition, TTX, a sodium channel blocker, completely abolished smooth muscle contractions (FIG. 4), and the release of ACh evoked by nAChR agonists such as DMPP, (−)nicotine,[51] or (+)anatoxin-a,[52] and rather selective nAChR antagonists antagonized the effects of these agonists. Therefore, it is suggested that this method is useful for studying the potency of different nAChR agonists and antagonists,

TABLE 7. Apparent Dissociation Constant (K_d) of Different nAChR Antagonists Estimated on Somatodendritic Nicotinic Receptors

Antagonists/Agonists	K_d (μM)	
	DMPP	(−)Nicotine
Dihydro-β-erythroidine	0.63 ± 0.08 (4)	
Mecamylamine	0.95 ± 0.28 (7)	1.26 ± 0.26 (4)
(+)Tubocurarine	2.87 ± 0.61 (6)	1.36 ± 0.41 (8)
Hexamethonium	10.37 ± 2.79 (10)	14.41 ± 3.46 (6)
α-Bungarotoxin	> 20	

NOTE: To estimate the apparent dissociation constant of different antagonists versus different agonists the following equation was used: a/DR-1. The apparent dissociation constant (K_d) of antagonists, a concentration required to double the ED_{50} of agonists for the increase of the force of muscle contraction by nAChR agonists, was calculated. The negative logarithm of this concentration, pA_2 is commonly accepted as a measure of antagonist affinity. The maximal contractile response to nAChR agonists was designated as 100%, and the ED_{50} of agonists were calculated. The effect of oxotremorine, a selective muscarine receptor agonist that induced contraction of the muscles, was also investigated in the presence of different concentrations of nAChR antagonists studied. The nAChR antagonists did not affect oxotremorine on the smooth muscle, indicating that they do not exert antimuscarinic activity. Number of experiments is in parentheses.

provided they do not possess antimuscarinic activity at the concentration applied. When the equipotent concentration (EC_{50}) of different agonists was calculated (TABLE 6), (−)epibatidine, a nAChR agonist with strong analgesic activity, proved to the most potent agonist and cytisine the least potent. However, when the apparent dissociation constants (K_d) of antagonists were estimated (TABLE 7) DHβE was seen to be the most potent, and the rank order of potency was DHβE > mecamylamine > (+)-tubocurarine > hexamethonium ≫ αBGT. The rank order of potency of agonists for somatodendritic receptors [(+)anatoxin-a > DMPP = (−)nicotine > cytisine] differs from that found in the hippocampus. The finding that cytisine is less potent than DMPP and (−)nicotine (EC_{50} = 47.5 ± 2.0 μM) indicates that the β2 subunit is involved in forming somatodendritic nAChRs. In addition, that DHβE was very active and αBGT had no effect suggests that the α4-subunit is also present. Flores *et al.*[56] suggested that α4 and β2 subunits are associated in forming the predominant, and possibly the only, subtype of neuronal nicotinic receptor with high affinity for agonists. They found that all α4-subunits that were labeled by [³H]cytisine were coupled to β2 subunits. In addition it was shown[57] that neuronal α4β2 nAChRs are insensitive to αBGT, but they are sensitive to blockade by DHβE.[48] Taking into account our data, it is suggested that on the postsynaptic site in the neuron-neuron synapse the nAChRs are composed of α4β2 subunits.

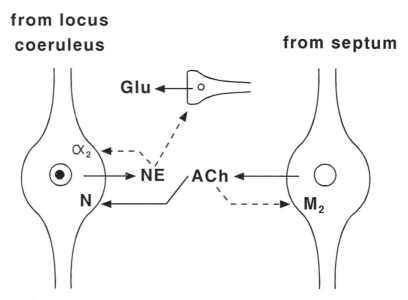

FIGURE 5. Interaction between cholinergic input and noradrenergic axon terminals in the hippocampus. ACh released from the cholinergic terminals may be able to stimulate nAChRs located on the noradrenergic terminals and release norepinephrine (NE). Note that the noradrenergic axon terminals are not equipped with inhibitory muscarinic receptors;[44] thus the effect of ACh on nAChRs is unopposed.

TABLE 8. Characteristics of nAChR Stimulation-mediated Release of Norepinephrine. Different Types of Transmitter Release

	Associated with AP[a]	$[Ca^{2+}]_o$-Dependent	Subject to Presynaptic Modulation	Carrier-mediated
Resting	Not	Not	Not	Yes (partly)
Electrical stimulation	Yes	Yes	Yes	Not
nAChR-mediated	Not	Yes	Yes	Not
Reversal of the carrier	Not	Not	Not	Yes

[a]AP, action potential.

Possible Role of Presynaptic nAChR Located on the Noradrenergic Axon Terminals in the Hippocampus

Inasmuch as the hippocampus receives noradrenergic innervation from the locus coeruleus and NE inhibits excitatory glutamate-mediated synaptic transmission in area CA3,[37,58,59] the nAChR-mediated NE-releasing action of cholinergic innervation and the α_2-adrenoceptor-mediated inhibition of this interaction described in this paper may have a functional role. The effectiveness of this interaction is multiplied by the fact that the noradrenergic axon terminals are not subjected to muscarinic receptor-mediated inhibition (FIG. 5). It means that the firing of cholinergic afferents and the subsequent release of ACh may produce an increase of NE release that is not associated with axonal firing. The release of NE in response to activation of cholinergic input in the hippocampus may result in a decrease of the evoked release of transmitter from excitatory terminals of both mossy fibers and CA3 pyramidal cell recurrent collaterals;[59] thus the stimulation of nAChRs located on the noradrenergic axon terminals would result in reduction of the activity of pyramidal cells.

It is generally accepted that both the cholinergic system and the hippocampus play a significant role in memory and learning. Recent data, however, suggest that the interaction of cholinergic and noradrenergic pathways may be even more important for memory processes.[60] Several studies concerning presynaptic inhibition of transmitter release from the hippocampus have shown, using electrophysiological[cf.61] and neurochemical[42,62] methods, that the release of different transmitters can be modulated via stimulation of presynaptic receptors. The recent finding of excitatory interaction between cholinergic input and noradrenergic neurons through nAChRs located on noradrenergic axon terminals, and of the absence of inhibitory muscarinic receptors on the noradrenergic axon terminals[44]—in contrast to the axon terminals with opposite muscarinic and nicotinic actions on the same population of axon terminals (e.g., the nigrostriatal pathway)—make this interaction very effective.

In summary, the stimulation of nAChRs located on the axon terminals results in a release of NE and other transmitters, but unlike that observed at rest, it is $[Ca^{2+}]_o$-dependent and is subject to presynaptic modulation (TABLE 8).

REFERENCES

1. LANGLEY, J. N. 1906. Proc. R. Soc. Lond. B. **78:** 83–196.
2. DALE, H. 1914. J. Pharmacol. Exp. Ther. **6:** 147–190.
3. BERG, D. K., R. T. BOYD, S. W. HALVERSEN, L. S. HIGGINS, M. H. JACOB & J. F. MARGIOTTA. 1989. Trends Neurosci. **12:** 16–21.
4. SARGENT, P. B. 1993. Annu. Rev. Neurosci. **16:** 403–443.

5. PATRICK, J., P. SÉQUÉLA, S. VERNINO, M. AMADOR, C. LUETJE & J. A. DANI. 1993. Prog. Brain Res. **98:** 113–120.
6. CHIOU, C. Y., J. P. LONG, R. POTREPKA & J. L. SPRATT. 1970. Arch. Int. Pharmacodyn. Ther. **187:** 88–96.
7. ROWELL, P. P. & D. L. WINKLER. 1984. J. Neurochem. **43:** 1593–1598.
8. HALL, G. H. & D. M. TURNER. 1972. Biochem. Pharmacol. **21:** 1829–1838.
9. HERY, F., S. BOURGOIN, N. HAMON, J. P. TERNAUX & J. GLOWINSKY. 1977. Arch. Int. Pharmacodyn. Ther. **296:** 91–97.
10. DE BELLEROCHE, J. & H. F. BRADFORD. 1978. Brain. Res. **142:** 53–68.
11. RAPIER, C., G. G. LUNT & S. WONNACOTT. 1988. J. Neurochem. **50:** 1123–1130.
12. RAPIER, C., G. G. LUNT & S. WONNACOTT. 1990. J. Neurochem. **54:** 937–945.
13. WESTFALL, T. C. 1974. Neuropharmacology **13:** 693–700.
14. GIORGUIEFF, M. F., M. L. LE FLOCH, T. C. WESTFALL, J. GLOWINSKI & M. J. BESSON. 1976. Brain Res. **106:** 117–131.
15. SCHULZ, D. W. & R. E. ZIGMOND. 1989. Neurosci. Lett. **98:** 310–316.
16. HÁRSING, L. G., JR., H. SERSHEN & A. LAJTHA. 1992. J. Neurochem. **59:** 48–54.
17. HÁRSING, L. G., JR., H. SERSHEN, E. S. VIZI & A. LAJTHA. 1992. Neurochem. Res. **17:** 729–734.
18. VIZI, E. S., G. T. SOMOGYI, H. NAGASHIMA, D. DUNCALF, I. A. CHAUDHRY, O. KOBAYASHI, P. L. GOLDINER & F. F. FOLDES. 1987. Br. J. Anaesth. **59:** 226–231.
19. SU, C. & J. A. BEVAN. 1970. J. Pharmacol. Exp. Ther. **175:** 533–540.
20. TODOROV, L., K. WINDISCH, H. SERSHEN, A. LAJTHA, M. PAPASOVA & E. S. VIZI. 1990. Br. J. Pharmacol. **102:** 186–190.
21. SERSHEN, H., M. E. A. REITH, A. LAJTHA & J. GENNARO. 1981. J. Recept. Res. **2:** 1–15.
22. ALKONKON, M. & E. X. ALBUQUERQUE. 1991. J. Recept. Res. **11:** 505–506.
23. ALKONKON, M. & E. X. ALBUQUERQUE. 1993. J. Pharmacol. Exp. Ther. **256:** 1455–1473.
24. LUETJE, C. W. & J. PATRICK. 1991. J. Neurosci. **11:** 837–845.
25. LUKAS, R. J. & M. BENCHERIF. 1992. Int. Rev. Neurobiol. **34:** 25–131.
26. WADA, E., K. WADA, E. BOULTER, E. S. DENERIS, S. HEINEMANN, J. PATRICK & L. SWANSON. 1989. J. Comp. Neurol. **284:** 314–335.
27. SÉQUÉLA, P., J. WADICHE, K. MILLER, J. DANI & J. PATRICK. 1992. J. Neurosci. **13:** 569–604.
28. WONNACOTT, S. 1987. Hum. Toxicol. **6:** 343–353.
29. WHITING, P. & J. LINDSTROM. 1987. Proc. Natl. Acad. Sci. USA **84:** 595–599.
30. WESTFALL, T. C., H. GRANT & H. PERRY. 1983. Gen. Pharmacol. **14:** 321–325.
31. GRADY, S., M. J. MARKS, S. WONNACOTT & A. D. COLLINS. 1992. J. Neurochem. **59:** 848–856.
32. WONNACOTT, S., J. IRONS, C. RAPIER, B. THORNE & G. G. LUNT. 1989. Prog. Brain Res. **79:** 157–163.
33. WONNACOTT, S., J. IRONS, C. RAPIER, B. THORNE & G. G. LUNT. 1989. Prog. Brain Res. **79:** 157–163.
34. WONNACOTT, S., S. JACKMAN, K. L. SWANSON, H. RAPOPORT & E. X. ALBUQUERQUE. 1991. J. Pharmacol. Exp. Ther. **259:** 387–391.
35. SÉQUÉLA, P., J. WADICHE, K. MILLER, J. A. DANI & J. W. PATRICK. 1993. J. Neurosci. **13:** 596–604.
36. HUNTER, B. E., C. M. DE FIBRE, R. L. PAPKE, W. R. KEM & E. M. MEYER. 1994. Neurosci. Lett. **168:** 130–134.
37. THOMAS, P., M. STEPHENS, G. WILKIE, M. AMAR, G. G. LUNT, P. WHITING, T. GALLAGHER, E. PEREIRA, M. ALKONDON, E. X. ALBUQUERQUE & S. WONNACOTT. 1993. J. Neurochem. **60:** 2308–2311.
38. CARMICHAEL, W. M., D. F. BIGGS & P. R. GORHAM. 1975. Science **187:** 542–544.
39. SWANSON, K. L., R. S. ARONSTAM, S. WONNACOTT, H. RAPOPORT & E. X. ALBUQUERQUE. 1991. J. Pharmacol. Exp. Ther. **259:** 377–386.
40. KOFUJI, P., Y. ARACAVA, K. L. SWANSON, R. S. ARONSTAM & H. RAPOPORT. 1990. J. PET **252:** 517–525.
41. SWANSON, K. L., C. N. ALLEN, R. S. ARONSTAM, H. RAPOPORT & E. X. ALBUQUERQUE. 1986. Mol. Pharmacol. **29:** 250–257.

42. MULLE, C., D. CHOQUET, H. KORN & J. P. CHANGEUX. 1992. Neuron **8**: 135–143.
43. VERNINO, S., M. AMADOR, C. W. LUETJE, J. PATRICK & J. A. DANI. 1992. Neuron **8**: 127–134.
44. MILUSHEVA, E., M. BARANYI, T. ZELLES, Á. MIKE & E. S. VIZI. 1994. Eur. J. Neurosci. **6**: 187–192.
45. VIZI, E. S., L. G. HÁRSING, JR., J. GAAL, J. KAPOCSI, B. BERNATH & G. T. SOMOGYI. 1986. J. Pharmacol. Exp. Ther. **238**: 701–706.
46. LIPSCOMBE, D., S. KONGSAMUT & R. N. TSIEN. 1989. Nature **340**: 639–642.
47. HIRNING, L. D., A. P. FOX, E. W. MCCLESKEY, B. M. OLIVERA, S. A. THAYER, R. J. MILLER & R. W. TISIEN. 1988. Science **239**: 57–61.
48. WHITING, P., R. SCHOEPHER, J. LINDSTROM & T. PRIESTLEY. 1991. Mol. Pharmacol. **40**: 463–472.
49. PATON, W. D. M. & E. S. VIZI. 1969. Br. J. Pharmacol. **35**: 10–25.
50. PATON, W. D. M. & E. S. VIZI. 1971. J. Physiol. (Lond.) **215**: 819–848.
51. VIZI, E. S. 1973. Br. J. Pharmacol. **47**: 765–777.
52. GORDON, R. K., R. R. GRAY, C. B. REAVES, D. L. BUTLER & P. K. CHIANG. 1992. J. Pharmacol. Exp. Ther. **263**: 997–1002.
53. TOROCSIK, A., F. OBERFRANK, H. SERSHEN, A. LAJTHA, K. NEMESSY & E. S. VIZI. 1991. Eur. J. Pharmacol. **202**: 297–302.
54. SPANDE, T. F., H. M. GARRAFFO, M. W. EDWARDS, H. J. C. YEH, L. PANNELL & J. W. DALY. 1992. J. Am. Chem. Soc. **114**: 3475–3478.
55. QIAN, G., T. LI, T. V. SHEN, L. LIBERTINE-GARAHAN, J. ECKMAN, T. BIFTA & S. IP. 1993. Eur. J. Pharmacol. **250**: R13–R14.
56. FLORES, C. M., S. W. ROGERS, L. A. PABREZA, B. B. WOLFE & K. J. KELLAR. 1992. Mol. Pharmacol. **41**: 31–37.
57. PEREIRA, E. F. R., S. REINHARDT-MAELICKE, A. SCHRATTENHOLZ, A. MAELICKE & E. X. ALBUQUERQUE. 1993. J. Pharmacol. Exp. Ther. **265**: 1474–1491.
58. SCANZIANI, M., M. CAPOGNA, B. H. GAHWILER & S. M. THOMPSON. 1992. Neuron **9**: 919–927.
59. SCANZIANI, M., B. H. GAHWILER & S. M. THOMPSON. 1993. J. Neurosci. **13**: 5393–5401.
60. AYYAGARI, V., L. E. HARRELL & D. S. PARSONS. 1991. J. Neurosci. **11**: 2848–2854.
61. THOMPSON, S. M., M. CAPOGNA & M. SCANZIANI. 1993. Trends Neurosci. **16**: 222–227.
62. TARCILUS, E., J. SCHOCK & H. BREER. 1994. Neurochem. Int. **24**: 349–361.
63. VAUGHAN, P. F. T., D. F. KAYE, H. L. REEVE, S. G. BALL & C. PEERS. 1993. J. Neurochem. **60**: 2159–2166.
64. ROWELL, P. P., L. A. CARR & A. C. GARNER. 1987. J. Neurochem. **49**: 1149–1454.
65. GIORGUIEFF-CHESSELET, M. F., M. L. KEMEL, D. WANDSCHEER & J. GLOWINSKI. 1979. Life Sci. **25**: 1257–1262.
66. SCHULZ, D. W. & R. E. ZIGMOND. 1989. Neurosci. Lett. **98**: 310–316.
67. ARAUJO, D. M., P. A. LAPCHAK, B. COLLIER & R. QUIRION. 1988. J. Neurochem. **51**: 292–299.
68. LASPCHAK, P. A., D. M. ARAUJO, R. QUIRION & B. COLLIER. 1989. J. Neurochem. **53**: 1843–1851.
69. BEANI, L., C. BIANCHI, L. FERRARO, L. NILSSON, A. NORDBERG, L. ROMANELLI, P. SPALLUTO, A. SUNDWALL & S. TANGANELLI. 1989. Prog. Brain Res. **79**: 149–155.
70. WONNACOTT, S., L. FRYER & G. G. LUNT. 1987. J. Neurochem. **48**: Suppl. 72B.
71. SERSHEN, H., E. S. VIZI, A. BALLA & Å. LAJTHA. In press.
72. TODOROV, L., K. WINDISCH, H. SERSHEN, A. LAJTHA, M. PAPASOVA & E. S. VIZI. 1991. Br. J. Pharmacol. **102**: 186–190.
73. WESSLER, H. I. & H. KILBINGER. 1987. Naunyn-Schmiedebergs **355**: R77.
74. LICHTENSTEIGER, W., F. HEFTI, D. FELIX, S. HUWYLER, E. MELAMED & M. SCHLUMPF. 1982. Neuropharmacology **21**: 963–968.
75. CLARKE, P. B., D. W. HOMMER, A. PERT, L. R. SKIRBOLL. 1985. Br. J. Pharmacol. **85**: 827–835.
76. TOROCSIK, A., I. A. CHAUDRY, K. BIRO, H. NAGASHIMA, M. KINJO, D. DUNCALF, R. NAGASHIMA, F. F. FOLDES, P. L. GOLDINER & E. S. VIZI. 1989. Arch. Int. Pharmacodyn. Ther. **299**: 247–253.

Neuronal Nicotinic Receptor Subtypes

JON LINDSTROM, RENÉ ANAND, XIAO PENG,

VOLODYMYR GERZANICH, FAN WANG,

AND YUEBING LI

Department of Neuroscience
Medical School of the University of Pennsylvania
Philadelphia, Pennsylvania 19104-6074

Nicotinic acetylcholine receptors (AChRs) are members of a gene superfamily of ligand-gated ion channels that includes $GABA_A$ receptors, glycine receptors, and 5HT-3 serotonin receptors,[1,2] but does not appear to include glutamate receptors.[3] AChRs are the predominant excitatory receptors in the mammalian peripheral nervous system, whereas glutamate receptors are the predominant excitatory receptors in the central nervous system. In the peripheral nervous system AChRs are critical for controlling skeletal muscles and consequently are the target of toxins such as snake venom toxins like α-bungarotoxin (αBgt), and muscle AChRs are the target of an antibody-mediated autoimmune response in myasthenia gravis.[4] Although outnumbered by glutamate receptors in the vertebrate central nervous system, neuronal AChRs are widespread. Their functional roles are not yet well known. They have been implicated in learning and memory, and their number is decreased in Parkinson's syndrome and Alzheimer's disease.[5] One measure of their significance is that through nicotine they mediate the addiction to tobacco, which is predicted to cause a quarter of a billion premature deaths by the turn of the century.[6]

The three branches of the AChR gene family[7] are (1) muscle AChRs, (2) neuronal AChRs that, unlike those of muscle, do not bind αBgt, and (3) neuronal AChRs that do bind αBgt. All are ACh-gated cation channels formed probably from five subunits of usually two to four homologous types.[8,9] These subunits are thought to be organized like barrel staves around a central cation channel. The synthesis, structure, and function of muscle-type AChRs are known in relatively great detail, whereas the more diverse structures, functional properties, and functional roles of neuronal AChRs are much less well characterized.[7–9] The functional properties of various combinations of neuronal AChR subunits expressed in *Xenopus* oocytes have been better characterized than have the functional properties of diverse neuronal AChR subtypes *in vivo,* and the actual functional roles in the nervous system of many of the real and potential subtypes of neuronal AChRs remain to be determined. It is becoming apparent that many neuronal AChRs are likely to have functional roles that differ from the straightforward postsynaptic type of critical link in neurotransmission exemplified by muscle AChRs.

We attempt to review briefly the diversity of neuronal AChRs primarily from the perspective of what has been done in our laboratory. Muscle AChRs, which serve as an archetype for the gene family, are considered only briefly as a model for comparison with neuronal AChRs. Most emphasis is given to neuronal AChRs that bind αBgt, because they have been the most recent focus of research attention in this field. Data are presented that compare and contrast some of the properties of representatives of each of the three branches of the AChR gene family as revealed by expressing their cRNAs in *Xenopus* oocytes.

MUSCLE AChRs

Schematic representations of the known AChR subunit sequences are depicted in FIGURE 1. All AChR subunits consist of homologous sequences[7-9] that start with a signal sequence cleaved during translation. The N-terminal ≈ 200 amino acids of the mature subunits are thought to form a large domain on the extracellular surface. This is followed by ≈ 90 amino acids comprising three closely spaced hydrophobic sequences that are thought to form transmembrane domains (M1—M3). M2 is thought to line the ion channel. Following M3 is a large cytoplasmic domain of ≈ 100–200 amino acids. The large cytoplasmic domain is the most variable part of the sequence between subunits. In many AChR subunits this region contains the most immunogenic epitopes that are recognized by antibodies in both native and denatured subunits.[10] Thus, antibodies raised against bacterially expressed peptides from this region have been very effective at recognizing both native and denatured AChRs.[11,12] The large cytoplasmic domain is followed by a fourth hydrophobic sequence (M4) leading to a small ≈ 10–30 amino acid extracellular domain at the C-terminus. Among the methods that have been applied to determining the transmembrane orientation of the AChR subunit polypeptide chains is the "reporter epitope" method.[13] In this method, epitope sequences are inserted by *in vitro* mutagenesis, and then the transmembrane orientation of the tagged part of the sequence is determined using a monoclonal antibody (mAb) to the epitope to label the extracellular or cytoplasmic surface of the AChR expressed from cRNA in *Xenopus* oocytes. The reporter epitope method has the technical virtue that it avoids the slow and expensive process of making new antibodies. It should be able to probe the transmembrane orientation of any part of a subunit sequence that is on the protein surface and that is not part of an active site. The basic method should be applicable to any cloned membrane protein.

AChR subunit genes are scattered over several human chromosomes: 1(β2), 2(α1,γ,δ), 8(α2), 15(α3, α5, α7, β4), 17(β1), and 20(α4).[14-19] Genes for subunits that are components of a native AChR are not necessarily located together. Thus, the genes for α1, β1, γ, and δ subunits of muscle AChR are on two chromosomes, as is the case for the genes for α4 and β2 subunits of the brain AChR subtype with high affinity for nicotine. However, the α3, β4, and α5 genes that encode subunits of ganglionic AChRs are contiguous on a single chromosome.

Muscle α1 subunits mature in their conformation after synthesis and before assembly with other subunits.[20] This conformational maturation is marked by their acquisition of the ability to bind αBgt to what will become the ACh binding site (and is in part composed of amino acids in the region 180–200)[8] and mAbs to the main immunogenic region (which is in part composed of amino acids in the region 66–76).[21,22] This conformational maturation is thought to be associated with formation of a disulfide bond between cysteines α128 and α142[23] to form a loop which is conserved in all AChR subunits, and which in most contains an N-glycosylation site at 141.[7-9] Similar maturation events may occur with all AChR subunits, but have not been studied. Ability to bind ACh and small cholinergic ligands is not acquired until α1 associates with γ, δ, or ϵ subunits; the binding site is thought to be formed at the interface between subunits.[24-26] The extracellular domain is thought to be largely responsible for the specific associations between muscle AChR subunits.[25] Sequence homologies suggest that similar considerations apply to the assembly of neuronal AChR subunits and the formation of ACh binding sites at some subunit interfaces.

At mature neuromuscular junctions the AChRs are composed of two α1 subunits in combination with one each of β1, δ, and ϵ subunits.[8] Extrajunctional AChRs found

SUBUNITS OF MUSCLE NICOTINIC AChRs

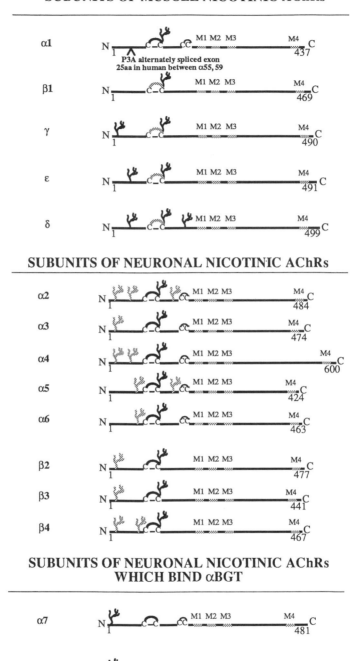

SUBUNITS OF NEURONAL NICOTINIC AChRs

SUBUNITS OF NEURONAL NICOTINIC AChRs WHICH BIND αBGT

FIGURE 1. Comparison of AChR subunit sequences. The sequences of known AChR subunits are schematically represented to indicate the homologies in their structures. The complete aligned amino acid sequences of all rat AChR subunits are listed in reference 7.

before innervation or after denervation differ from mature AChRs by substituting γ for ε subunits, resulting in increased duration of channel opening and an increased rate of turnover, among other changes. The subunits are thought to be arranged around a central cation channel in the order α1, δ, α1, ε, β in such a way that two acetylcholine binding sites are formed, one at the interface between α and δ subunits and the other between α1 and ε subunits. Evidence suggests that the channel of both muscle and neuronal AChRs is lined by amino acids from the second transmembrane domain (M2).[7–9]

The structure of muscle type AChRs from *Torpedo* electric organ has been determined to a resolution of 9 Å by Nigel Unwin's studies of two dimensional crystalline arrays of AChRs in membrane fragments.[27] This analysis reveals in side view an 80-Å-wide 120-Å-long protein extending 60 Å on the extracellular surface and 20 Å on the cytoplasmic surface. Viewed from the top, a pentagonal array of subunits is seen to surround a channel of 25 Å in diameter at its mouth which narrows abruptly at the level of the lipid bilayer and then flares open again on the cytoplasmic surface. Only a single transmembrane α helix was observed in each subunit. It was suggested that this α helix might correspond to the channel lining domain M2, but other evidence indicates that at least part of M2 is in a β conformation.[28] In any case, it seems that most of the transmembrane sequences of AChR subunits must be in β rather than α helical conformations. Because of the basic homologies in sequence between the subunits of all AChR subunits, it seems likely that they all have a basically similar size and pentagonal shape.

NEURONAL AChRs THAT DO NOT BIND α-BUNGAROTOXIN

Putative neuronal AChR subunit cDNAs have been identified by low stringency hybridization starting with muscle AChR probes.[9,29,30] These have been termed[29] α2–α6 if they contained a cysteine pair homologous to the pair at 192 and 193 in α1 subunits which can be affinity-labeled by ACh analogues.[8] Homologous cDNAs that did not contain this cysteine pair were termed β2–β4. Pairwise combinations of α2, α3, or α4 with β2 or β4 subunits form ACh-gated cation channels when coexpressed in *Xenopus* oocytes.[9,29,30] This allows for many potential AChR subtypes to be expressed in various regions or at different times during development. Although the localizations, developmental, and functional roles of these subtypes have not been worked out in detail, some simplifying generalizations have emerged. In the mature mammalian brain the predominant AChR subtype that does not bind αBgt is composed of α4 and β2 subunits.[31,32] These α4β2 AChRs comprise at least 90% of the high-affinity nicotine binding sites. In retina[33] and autonomic ganglia,[11] the predominant AChR subtype which does not bind αBgt includes α3 subunits.

AChRs with high affinity for nicotine have been immunoaffinity-purified from brains of chickens,[34,35] rats,[31] cattle, and humans[32] using antibodies. Such AChRs have subsequently been affinity-purified from rat brains by affinity chromatography using bromoacetylcholine-agarose.[36,37] In rats and chickens N-terminal amino acid sequencing identified α4 and β2 subunits as the components of this AChR subtype.[33,38,39] This conclusion has subsequently been confirmed by immunological analysis as well, using subunit-specific mAbs[32] and antipeptide sera.[33,40] Whereas in mammals[32] α4β2 AChRs comprise ≥ 90% of the brain AChRs with high affinity for nicotine, in chickens nearly half of such AChRs are formed by β2 in combination with an α subunit yet to be identified.[35] Evidence shows that a small fraction of α4β2 AChRs may have α5 subunits associated.[41] Mouse fibroblasts permanently transfected with α4 and β2 subunits exhibit ACh-gated cation channels and the same pharmacological properties as native brain AChRs for binding of [³H]nicotine and competing ligands.[42,43] The subunit stoichiometry was shown to be $(\alpha4)_2(\beta2)_3$ using a method that involved purifying metabolically labeled $\alpha4\beta2$ AChRs.[44] This method involved expressing these cDNAs in [³⁵S]methionine-labeled *Xenopus* oocytes, isolating fully assembled AChRs of the same ~ 10S size as native AChRs by sucrose gradient sedimentation, affinity-purifying these AChRs, purifying their subunits by PAGE, determining the ratio of [³⁵S]methionine label in the subunits, and then correcting for the methionine composition of these subunits.[44] This general method for stoichiometry determination was also shown to work for muscle AChR, and should work for any cloned multisubunit receptor similarly expressed. The same $(\alpha4)_2(\beta2)_3$ stoichiometry was also inferred from an electrophysiological method.[45] Thus, in the one case where the stoichiometry of a neuronal AChR subtype is known, the pentagonal symmetry of $(\alpha1)_2\beta1\gamma\delta$ AChRs of muscle is conserved, as presumably is the presence of two ACh binding sites; however, only two kinds of subunits compose that neuronal AChR subtype rather than the four kinds of subunits which compose muscle type AChRs.

Immunohistological localization of β2 subunits throughout rat brain using a subunit-specific mAb revealed a wide distribution[46] which closely paralleled that of high-affinity sites for binding [³H]ACh or [³H]nicotine, and which overlapped but was distinct from the pattern of [¹²⁵I]αBgt binding.[47] These results have subsequently been confirmed by others using an antipeptide serum to β2 antibodies.[48] Monoclonal antibodies have also been used to localize β2 subunits in chick brain and retina.[49-52] Transport of AChRs down the axons of retinal ganglion cells to their termination in the superior colliculus and other nuclei was demonstrated by showing that removal of an eye eliminated all labeling by a β2-specific mAb of the contralateral superior colliculus.[46] β2 was also located in dorsal root ganglion cells.[46] These results are consistent with the idea that many α4β2 AChRs are located presynaptically where they may function to modulate the release of ACh or another transmitter. This is consistent with evidence for transport of AChRs in the habenulointerpeduncular tract,[51] and with evidence that AChRs can modulate the release of dopamine and other transmitters from synaptosomes.[53,54]

The α4β2 AChRs account for > 90% of the high-affinity nicotine binding sites in mammalian brains,[32] and the amount of α4β2 AChRs in brain is increased by chronic exposure to nicotine.[40] This effect does not appear to result from an increase in transcription of α4 or β2 subunits.[55] Instead, this effect appears to result from a decrease in the rate of destruction of α4β2 AChRs.[56] This was demonstrated by showing that chronic exposure to 1 μM nicotine caused a twofold increase in the amount of α4β2 AChRs in transfected fibroblasts, and that after prevention of synthesis of new proteins by cyclohexamide the AChRs already in the membrane were destroyed much less rapidly in the presence of nicotine.

The α3 AChRs are much more abundant in chick retina than are α4β2 AChRs.[33] Chick ciliary ganglia have been shown, using subunit-specific antibodies, to contain AChRs which include the α3 subunit.[11] The α3 AChRs in ciliary ganglia appear to also contain β4 subunits and some may contain β2 subunits.[57] These AChRs also contain α5 subunits,[41] and are usually identified using an mAb to the main immunogenic region on α1 subunits,[58] which also crossreacts with a similar sequence on α5 subunits.[41] Human α5 subunits will assemble with human α3 and β2[59] or β4 subunits and be expressed on the surface of *Xenopus* oocytes, but α5 is not expressed on the surface when expressed alone or in combination with β2 subunits.[60] The exact subunit composition or stoichiometry of native α3 AChRs is not known, but it is clear that they do not include α7 subunits which are expressed in the same neurons.[57] The α3 AChRs have been immunohistologically located to the postsynaptic regions of the cholinergic synapses on these neurons,[61,62] whereas the α7 AChRs have been localized away from these synapses on pseudodendrites[63] where the source of ACh to stimulate them is not obvious. However, it is not clear whether α3 AChRs will always be found at postsynaptic localizations or whether α7 AChRs will always be found at extrasynaptic locations.

NEURONAL AChRs THAT BIND α-BUNGAROTOXIN

α-Bungarotoxin binding proteins affinity-purified from chick brain yielded a partial amino acid sequence that[64] was used to design an oligonucleotide probe that was used to identify a cDNA for α7 subunits.[12] This cDNA was used to identify an α8 cDNA from chick brain which exhibited a very similar sequence.[12] α7 has subsequently also been cloned from rats[65] and humans.[15,16,66] α7 and α8 are homologous in sequence to, but about equally distinct from, muscle α1 and neuronal α2–α6.[12]

Demonstration that α7 and α8 are components of αBgt-binding AChRs depended on using mAbs to bacterially expressed large cytoplasmic domain fragments of α7 and α8 to immunoprecipitate [^{125}I]αBgt-labeled AChRs.[12] The epitopes for these mAbs have been mapped more precisely using synthetic α7 and α8 peptides.[67] Synthetic peptides have also been used to show that αBgt binds to amino acids within the sequence 180–200 of α7 and α8 subunits.[68] Most brain AChRs that could bind αBgt (75%) were found to contain α7 subunits, whereas a minority were found to contain both α7 and α8 subunits.[12] By contrast, in retina the majority of AChRs that can bind αBgt (69%) contain α8 subunits, and both α7α8 AChRs (17%) and α7 AChRs (14%) comprise minority populations.[69] The complete subunit composition or stoichiometry of these α7, α8 or α7α8 AChRs is not known, but it is clear from immunoprecipitation experiments that they do not contain other known AChR subunits. All preparations of purified αBgt binding neuronal AChRs that have been reported contain several peptide components,[31,70,71] and it is likely that some of these peptides correspond to AChR subunits that have yet to be identified.

Ballivet and co-workers first showed that α7 had the remarkably useful property of efficiently forming homo-oligomeric ACh-gated cation channels when expressed in *Xenopus* oocytes.[72] FIGURE 2 shows that α7 homomers are expressed as efficiently on the surface of *Xenopus* oocytes as are the native AChR subunit combinations α4β2 and α1β1γδ. FIGURE 3 shows that α7 homomers form ACh-gated cation channels, although the currents detected per α subunit are lower than for α4β2 AChRs or α1β1γδ AChRs. As will be discussed below, this is due in part to rapid desensitization of the α7 homomers. These three AChR subtypes differ in pharmacological properties, as illustrated in FIGURE 4. The α7 homomers have higher affinity for nicotine than ACh, whereas for α1β1γδ AChRs the opposite is true. Nicotine and ACh are most potent at activating the α4β2 AChR subtype, and they also have, by far, the highest equilibrium binding affinities for the desensitized conformation of this subtype.[32,42,75] Nicotine is a full agonist on α7 homomers and α4β2 AChRs, but only a partial agonist on α1β1γδ AChRs. All three subtypes exhibited Hill coefficients between 1.5 and 1.8. This suggests that not only for (α1)$_2$β1γδ AChRs (as has long been known),[76] but also for (α4)$_2$(β2)$_3$ AChRs and even (α7)$_5$ homomers, ACh must bind at two sites to provide sufficient energy to activate opening of the channel. This presumably reflects the basic homologies in the structures of these subtypes. Presumably only two binding sites are present in the subtypes with two α subunits, whereas activation of only two of five potential subunits is sufficient to activate α7 homomers.

The existence of efficiently expressed functional homomers of α7 greatly simplifies mutagenesis and expression studies, and has led to a series of very instructive

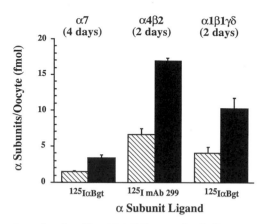

FIGURE 2. Comparison of surface ▨ and total ■ expression in *Xenopus* oocytes of chicken α7 homomers, chicken α4β2 AChRs, and *Torpedo* α1β1γδ AChRs. Note that the fraction of AChR on the surface is very similar for both α7 homomers (40%), α4β2 AChRs (35%), and α1β1γδ AChRs (38%). Oocytes were injected with 15 ng of cRNA for the indicated subunits and assayed 3 days later in this and all subsequent figures. Surface α7 and α1β1γδ AChRs on intact oocytes were measured using [^{125}I]αBgt. Surface α4β2 AChRs were measured using [^{125}I]mAb 299 to α4 subunits.[32] Total AChR concentrations were measured with the same [^{125}I]ligands using immunoisolated, detergent solubilized AChRs. Triton X-100 extracts of the oocytes were applied to Immulon-4 microwells coated with mAb 318 to α7 subunits,[73] or mAb 270 to β2 subunits,[35] or mAb 210 to α1 subunits.[74]

experiments by Changeux and co-workers. These include the demonstration that ion channel selectivity could be changed from cation to anion by mutagenesis of as few as two amino acids in the M2 region of α7 to correspond to amino acids found in homologous positions of glycine and GABA receptors,[77] and the demonstration that replacing the N-terminal domain of a 5HT-3 receptor with that of α7 produced ACh-gated channels characteristic of 5HT-3 receptors.[78] These experiments strikingly illustrate the fundamental homologies in structure between the different members of the ligand-gated ion channel superfamily which includes AChRs.

The α7 AChRs and α8 AChRs differ in their pharmacological properties.[69,73,79] The α7 AChRs have lower affinity for αBgt (K_d = 2 nM) than does muscle AChR, which binds αBgt nearly irreversibly, but α7 AChRs have higher affinity for αBgt than do α8 AChRs (K_d = 20 nM).[69,73] Because of their low affinity for αBgt, α8 AChRs may not be detected by the 1 nM concentrations of αBgt frequently used in binding studies. The α7 AChRs have much lower affinity for small cholinergic ligands than do α8 AChRs.[73] Also, α8 AChRs exhibit heterogeneity in ligand binding not seen with α7 AChRs. For example, the IC$_{50}$ value for nicotine is 1.3 μM for α7 AChRs, 0.012 μM for 78% of the α8 AChR [^{125}I]αBgt binding sites, and 11 μM for 22% of the α8 AChR binding sites.[73] The heterogeneity in α8 AChR binding affinity may suggest that α8 associates with more than one kind of structural subunit. The heterogeneity could occur between two different ACh binding sites in one α8 AChR protein. For example, in a muscle AChR the affinity for curare of the ACh binding site formed by α1 and γ differs from that formed by α1 and δ.[24] Alternatively, two α8 AChR subtypes may occur which differ in subunit composition. Another interesting feature of the pharmacology of α7 AChRs and α8 AChRs is their relatively high affinity for the classic glycinergic antagonist strychnine and the classic muscarinic antagonist atropine. The classic nicotinic antagonist curare has an $IC_{50} = 7$ μM for α7 AChRs and IC_{50} values of 0.79 μM and 65 μM for the two α8 AChR sites. Strychnine is nearly as potent, with an IC_{50} of 10 μM for α7 AChRs and IC_{50} values of 2 μM and 18 μM for the two α8 AChR sites. Atropine has an IC_{50} for α7 AChRs of 160 μM, but 0.031 μM and 390 μM for the two α8 AChR sites.

Pharmacological properties of α7 and α8 AChRs are generally, but not precisely, reflected by the properties of α7 and α8 homomers expressed in *Xenopus* oocytes.[79,80] The differences may be due to differences in posttranslational modifications in the expression system, but are more likely to be due to the lack of structural subunits normally associated with α7 and α8 subunits *in vivo*. The α7 homomers are expressed on the surface of oocytes at about the same 40% of total rate observed with α1β1γδ AChRs or α4β2 AChRs expressed in oocytes (FIG. 2), whereas <5% of α8 homomers are expressed on the surface at similar cRNA doses (15 ng).[79] At very high cRNA doses (100 ng), α8 homomer expression on the surface can increase to nearly 15%.[79] Separation of homomers on a sucrose gradient reveals that only α7 homomers of the ~ 10S size of native α7 AChRs can bind [¹²⁵I]αBgt,[80] whereas α8 homomers in various states of aggregation can bind [¹²⁵I]αBgt.[79] The α7 homomers immunoisolated on mAb-coated microwells had monotonic ligand binding properties,[80] but α8 homomers under these conditions displayed very broad binding curves.[79] This may reflect the observation in muscle AChRs that the ACh binding sites are formed at the interfaces between subunits.[8] Thus, ligands may only bind when adjacent α7 subunits are properly positioned in a fully assembled pentamer, whereas various associations

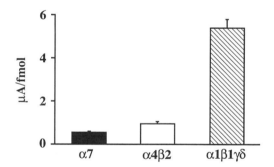

FIGURE 3. Comparison of current per surface α subunit for expression in *Xenopus* oocytes of chicken α7 homomers, chicken α4β2 AChRs, and *Torpedo* α1β1γδ AChRs, using saturating ACh at −90 mV. The experiments shown here and in subsequent figures used inhibitors of the Cl⁻ channel to prevent enhanced currents due to activation of the Cl⁻ channel by Ca^{2+} entering through the AChR channels. Note that the current/α subunit is much lower for α7 homomers than for the other AChR subtypes. As may become evident from subsequent figures, this was probably due to a combination of several factors including (1) rapid desensitization of α7, (2) possible side effects of the inhibitors of Cl⁻ channels used, and (3) the presence of five α subunits in an α7 homomer rather than two in an $(α4)_2(β2)_3$ AChR or an $(α1)_2β1γδ$ AChR. Surface α subunits were measured as shown in FIGURE 2. In this and all subsequent figures a two-electrode voltage clamp (Oocyte Clamp OC-725, Warner Instrument Corporation) was used as described below. The chamber was continuously perfused with 96 mM NaCl, 2 mM KCl, 1.8 mM CaCl₂, and 5 mM HEPES buffer pH 7.5 at 10 mL/min. The perfusing solution contained 0.5 μM atropine to block muscarinic receptors. The perfusing solution for α7 homomers and α4β2 AChRs also contained 100 μM each of niflumic and flufenamic acids to block Ca^{2+}-activated Cl⁻ channels. These were not added to α1β1γδ AChRs because their Ca^{2+} conductance was quite low.

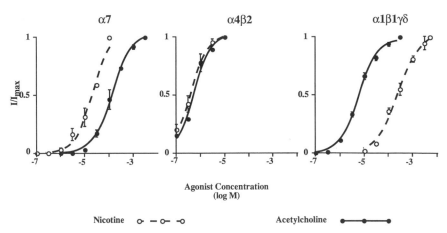

FIGURE 4. Comparison of ACh and nicotine dose/response curves for chicken α7 homomers, chicken α4β2 AChRs, and *Torpedo* α1β1γδ AChRs expressed in *Xenopus* oocytes. Note the contrasting pharmacological properties of the three AChR subtypes: (1) α7 homomers, like native α7 AChRs,[73] have higher affinity for nicotine (EC_{50} = 9 μM) than ACh (EC_{50} = 110 μM); (2) nicotine (EC_{50} = 0.42 μM) and ACh (EC_{50} = 0.45 μM) have equally high affinity for α4β2 AChRs; and (3) muscle-type AChRs have higher affinity for ACh (EC_{50} = 5.9 μM) than nicotine (EC_{50} = 245 μM). Saturating concentrations of both ACh and nicotine induced similar maximum currents with both α7 homomers and α4β2 AChRs. However, nicotine behaved as a partial agonist on α1β1γδ AChRs and, at saturating concentrations, produced currents 2–3-fold less than produced by saturating concentrations of ACh. Hill coefficients calculated at low concentrations of agonists (to avoid a reduction due to rapid desensitization at high concentrations) ranged between 1.5 and 1.8 in all cases. This suggests that not only for $(α1)_2β1γδ$ AChRs[76] and $(α4)_2(β2)_3$ AChRs, but also for $(α7)_5$ homomers, ACh acting at two sites is required to provide sufficient binding energy to activate opening of the channel.

of α8 subunits bind αBgt, but with differing affinities. Surface α8 homomers assayed for pharmacological effects on function exhibit monotonic dose/response curves,[79] suggesting that only properly assembled pentamers may be expressed on the surface. The limited sequence differences between the N-terminal extracellular domains of α7 and α8 subunits[12] presumably account for their pharmacological differences. Expression of a series of mosaics between α7 and α8 subunits reveal that virtually all of the pharmacological differences between α7 and α8 can be accounted for by amino acids between 179 and 208.[81]

The ion channel properties of α7 and α8 homomers are identical, reflecting their virtually identical sequences in the M1—M3 region.[79] As initially noted by Patrick and co-workers,[65] the most striking feature of these channels is that they are at least as selective for Ca^{2+} as are NMDA receptor channels. The increased Ca^{2+} selectivity of neuronal AChRs and α7 homomers, in particular, as compared to muscle type AChRs is illustrated in FIGURE 5. The Ca^{2+} that enters the channels can act as a second messenger and in *Xenopus* oocytes activates a Cl^- channel.[66,79] *In vivo,* this property could permit these AChRs to regulate many channels and processes, for example, neurite extension.[83–85] In *Xenopus* oocytes, activation of the Cl^- channel can be used to amplify the weak signal from α8 homomers. Another striking feature of α7 and α8 homomer channels is the rapidity with which the response desensitizes. Native α7 AChRs have been reported to desensitize so rapidly that especially fast agonist application and fast electronics are required to measure channel opening accurately.[86–88] This no doubt has been responsible for the failure to detect the activity of these AChRs until quite recently. In oocytes, the relative slowness with which the large cells can be perfused may prolong the response and be used to experimental advantage. The rapid rate of α7 homomer desensitization as compared to α4β2 AChR or α1β1γδ AChR desensitization is shown in FIGURE 6. A third feature of their channels is strong inward rectification. Thus, as the cell depolarizes, α7 and α8 homomer channels close. This is the opposite of the rectification exhibited by NMDA receptors. Neuronal α4β2 AChRs, like α7 homomers, exhibit rectification, by contrast with muscle type AChRs, as shown in FIGURE 7. Curiously, both the rapid desensitization and inward rectification combine to minimize sustained ion flux through these channels. These self-limiting responses seem especially curious when considering the case of α7 AChRs at extrasynaptic locations removed from obvious sources of ACh.

The α7 AChRs have been histologically localized by binding of [[125I]αBgt[47] and mAbs,[50,52,69,89] and by *in situ* hybridization.[65] In rat brain α7 is prominent, for example

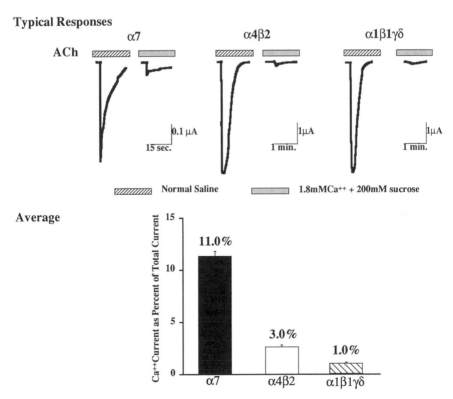

FIGURE 5. Comparison of the fraction of the current carried by Ca^{2+} for AChRs expressed in *Xenopus* oocytes by chicken α7 homomers, chicken α4β2 AChRs, and *Torpedo* α1β1γδ AChRs. Note that a much larger fraction of the total current through α7 homomers is carried by Ca^{2+} than is the case with the other AChR subtypes. This is consistent with other observations in oocytes and neurons.[65,82] All extracellular cations were replaced by 200 mM sucrose leaving only 1.8 mM Ca Cl₂ in the bathing solution.

in the hippocampus.[65,89] This is a region not rich in α4β2 AChRs,[46] but rich in NMDA receptors. Albuquerque and co-workers have found α7-like AChR function to be present on nearly all hippocampal neurons in tissue culture.[86–88]

The α8 subunits have been immunohistologically localized in chick brain and retina.[50,52,69] α3, α7, α8, and β2 have also been immunohistologically localized in chick brain and retina.[33,49–52,69,90–92] Double-labeling studies have been employed to co-localize different subunits to particular cells.[91] Developmental studies in retina have traced the early development of neurons expressing α8 and α3 subunits starting from embryonic day 4.5.[92] Various types of amacrine and ganglion cells contain α3, β2, α7, and α8 subunits, whereas bipolar cells have only α8 subunits.[91] High-resolution localization of these AChRs in combination with electrophysiological studies will be necessary to understand their physiological roles.

Chick cochlear hair cells were shown by Fuchs and co-workers to exhibit a response to ACh from efferent endings of brain stem neurons which could be blocked by αBgt.[93,94] This response exhibits several interesting properties which might exemplify the sorts of synaptic mechanisms in which α7 and α8 AChRs may participate. It was found that ACh invoked Ca^{2+} influx through these AChRs which resulted in a long-lasting inhibitory hyperpolarizing response due to activation of Ca^{2+}-dependent K^+ channels. These AChRs could be blocked by αBgt, curare, atropine, and strychnine. We found that cochlear sensory epithelium contains α7 mRNA and have immunoisolated α7 but not α8 or α1 AChRs from cochlear sensory epithelium.[95] These immunoisolated cochlear AChRs display pharmacological properties identical to those of brain α7 AChRs. Newly identified α9 AChRs have also been detected in rat cochlear hair cells, and their pharmacology closely resembles that observed in chick cochlea.[96] Thus, hair cells express several AChR subunits.

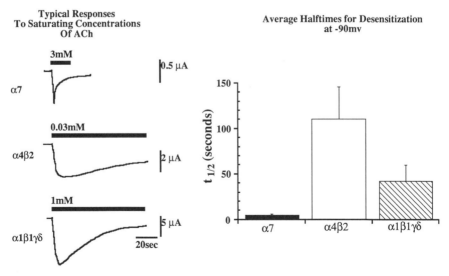

FIGURE 6. Comparison of rates of desensitization of chicken α7 homomers, chicken α4β2 AChRs, and *Torpedo* α1β1γδ AChRs expressed in *Xenopus* oocytes. Note that α7 homomers desensitize much more rapidly than the other AChR subtypes. The rate of desensitization may be underestimated due to the time necessary to perfuse the oocytes.

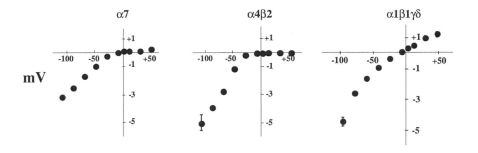

Normalized Current ($^I/^I\text{-50mV}$)

FIGURE 7. Comparison of rectification properties of chicken α7 homomers, chicken α4β2 AChRs, and *Torpedo* α1β1γδ AChRs expressed in *Xenopus* oocytes. Note that both neuronal α7 homomers and α4β2 AChRs exhibit strong inward rectification and thus close their channels at depolarizing potentials, whereas muscle-type AChRs exhibit a nonrectifying, relatively linear I/V curve.

CONCLUDING REMARKS

AChR subtypes from muscles and nerves exhibit structural and functional homologies dictated by the common evolutionary origin of their subunits, but they also differ significantly in important subtleties of their subunit compositions, pharmacological properties, ion channel properties, and functional roles. Muscle-type AChR is the best characterized ligand-gated ion channel, and is frequently used as a model for understanding both homologous and unrelated ligand-gated ion channels; however, as its structure is determined with increasing precision, surprises continue to be revealed. Much less is known about the structures and functions of neuronal AChRs, but it is likely that in addition to serving as the direct postsynaptic link in neurotransmission exemplified by muscle-type AChRs, some will also serve in different functional roles to which their structures have adapted.

REFERENCES

1. BETZ, H. 1990. Ligand-gated ion channels in the brain: The amino acid receptor superfamily. Neuron **5:** 383–392.
2. BARNARD, E. 1992. Receptor classes and the transmitter-gated ion channels. TIBS **17:** 368–374.
3. SEEBURG, P. 1993. The molecular biology of mammalian glutamate receptor channels. Trends Neurosci. **16:** 359–364.
4. LINDSTROM, J., G. SHELTON & Y. FUJII. 1988. Myasthenia gravis. Adv. Immunol. **42:** 233–284.
5. WONACOTT, S., M. RUSSELL & I. STOLERMAN. 1990. Nicotine Psychopharmacology: Molecular, Cellular, and Behavioral Aspects. Oxford Science Publications, Oxford, UK.
6. PETO, R., A. LOPEZ, J. BOREHAM, M. THUN & C. HEATH. 1992. Mortality from tobacco in developed countries: Indirect estimation from national vital statistics. Lancet **339:** 1268–1278.
7. LINDSTROM, J. 1994. Nicotine acetylcholine receptors. *In* CRC Handbook of Receptors. Alan North, Ed.: 153–176. CRC Press. Boca Raton, FL.

8. KARLIN, A. 1993. Structure of nicotinic acetylcholine receptors. Curr. Opin. Neurobiol. **3:** 299–309.
9. SARGENT, P. 1993. The diversity of neuronal nicotinic acetylcholine receptors. Annu. Rev. Neurosci. **16:** 403–443.
10. DAS, M. & J. LINDSTROM. 1991. Epitope mapping of antibodies to acetylcholine receptors. Biochemistry **30:** 2470–2477.
11. SCHOEPFER, R., S. HALVORSEN, W. CONROY, P. WHITING & J. LINDSTROM. 1989. Antisera against an α3 fusion protein bind to ganglion but not to brain nicotinic acetylcholine receptors. FEBS Lett. **257:** 393–399.
12. SCHOEPFER, R., W. CONROY, P. WHITING, M. GORE & J. LINDSTROM. 1990. Brain α-bungarotoxin-binding protein cDNAs and mAbs reveal subtypes of this branch of the ligand-gated ion channel superfamily. Neuron **5:** 35–48.
13. ANAND, R., L. BASON, M. SAEDI, V. GERZANICH, X. PENG & J. LINDSTROM. 1993. Reporter epitopes: A novel approach to examine transmembrane topology of integral membrane proteins applied to the α1 subunit of the nicotinic acetylcholine receptor. Biochemistry **32:** 9975–9984.
14. ANAND, R. & J. LINDSTROM. 1992. Chromosomal localization of seven neuronal nicotinic receptor subunit genes in humans. Genomics **13:** 962–967.
15. CHINI, B., E. RAIMOND, A. ELGOYHEN, D. MORALLI, M. BALZARETTI & S. HEINEMANN. 1994. Molecular cloning and chromosomal localization of the human α7 nicotinic receptor subunit gene (CHRNA7). Genomics **19:** 379–381.
16. DONCETTESTAMM, L., L. MONTEGGIA, D. DONNELLY ROBERTS, M. WONG, J. LEE, J. TIAN & T. GIORDANO. 1993. Cloning and sequence of the human α7 nicotinic acetylcholine receptor. Drug Dev. Res. **30:** 252–256.
17. LOBOS, E., C. RUDNICK, M. WATSON & K. ISENBERG. 1989. Linkage disequilibrium study of RFLPs detected at the human muscle nicotinic acetylcholine receptor subunit genes. Am. J. Hum. Genet. **44:** 522–533.
18. BEESON, D., S. JEREMIAH, L. WEST, S. POVEY & J. NEWSOME-DAVIS. 1990. Assignment of the human nicotinic acetylcholine receptor genes: The α and δ subunit genes to chromosome 2 and the β subunit gene to chromosome 17. Ann. Hum. Genet. **54:** 199–208.
19. ENG, C., C. KOZAK, A. BEADET & H. ZOGHBI. 1991. Mapping of multiple subunits of the neuronal nicotinic acetylcholine receptor to chromosome 15 in man and chromosome 9 in mouse. Genomics **9:** 278–282.
20. MERLIE, J., R. SEBBANE, S. GARDNER, E. OLSON & J. LINDSTROM. 1983. The regulation of acetylcholine receptor expression in mammalian muscle. Cold Spring Harbor Symposia on Quantitative Biology **158:** 135–146.
21. TZARTOS, S., M. CUNG, P. DEMANGE, H. LOUTRARI, A. MAMALAKI, M. MARRAUD, I. PAPADOULI, C. SAKARELLOS & V. TSIKARIS. 1991. The main immunogenic region (MIR) of the nicotinic acetylcholine receptor and the anti-MIR antibodies. Mol. Neurobiol. **5:** 1–29.
22. SAEDI, M., R. ANAND, W. CONROY & J. LINDSTROM. 1990. Determination of amino acids critical to the main immunogenic region of intact acetylcholine receptors by in vitro mutagenesis. FEBS Lett. **267:** 55–59.
23. BLOUNT, P. & J. MERLIE. 1990. Mutational analysis of muscle nicotinic acetylcholine receptor subunit assembly. J. Cell Biol. **111:** 2612–2622.
24. BLOUNT, P. & J. MERLIE. 1989. Molecular basis of the two nonequivalent ligand binding sites of the muscle nicotinic acetylcholine receptor. Neuron **3:** 349–357.
25. VERRALL, S. & Z. HALL. 1992. The N-terminal domains of acetylcholine receptor subunits contain recognition signals for the initial steps of receptor assembly. Cell **68:** 23–31.
26. SAEDI, M., W. CONROY & J. LINDSTROM. 1991. Assembly of *Torpedo* acetylcholine receptor in *Xenopus* oocytes. J. Cell Biol. **112:** 1007–1015.
27. UNWIN, N. 1993. Nicotinic acetylcholine receptor at 9Å resolution. J. Mol. Biol. **229:** 1101–1124.
28. AKABAS, M., D. STAUFFER, M. XU & A. KARLIN. 1992. Acetylcholine receptor channel structure probed in cysteine-substitution mutants. Science **258:** 307–310.
29. DENERIS, E., J. CONNOLLY, S. ROGERS & R. DUVOISIN. 1991. Pharmacological and

functional diversity of neuronal nicotinic acetylcholine receptors. Trends Pharmacol. Sci. **12**: 34–40.

30. HEINEMANN, S., J. BOULTER, J. CONNOLLY, E. DENERIS, R. DUVOISIN, M. HARTLEY, I. HERMANS-BORGMEYER, M. HOLLMANN, A. O'SHEA-GREENFIELD, A. PAPKE, S. ROGERS & J. PATRICK. 1991. The nicotinic receptor genes. Clin. Neuropharmacol. **14**: S45–S61.

31. WHITING, P. & J. LINDSTROM. 1987. Purification and characterization of a nicotinic acetylcholine receptor from rat brain. Proc. Natl. Acad. Sci. USA **84**: 595–599.

32. WHITING, P. & J. LINDSTROM. 1988. Characterization of bovine and human neuronal nicotinic acetylcholine receptors using monoclonal antibodies. J. Neurosci. **8**: 3395–3404.

33. WHITING, P., R. SCHOEPFER, W. CONROY, M. GORE, K. KEYSER, S. SHIMASAKI, F. ESCH & J. LINDSTROM. 1991. Differential expression of nicotinic acetylcholine receptor subtypes in brain and retina. Mol. Brain Res. **10**: 61–70.

34. WHITING, P. & J. LINDSTROM. 1986. Purification and characterization of a nicotinic acetylcholine receptor from chick brain. Biochemistry **25**: 2082–2093.

35. WHITING, P., R. LIU, B. MORLEY & J. LINDSTROM. 1987. Structurally different neuronal nicotinic acetylcholine receptor subtypes purified and characterized using monoclonal antibodies. J. Neurosci. **7**: 4005–4016.

36. NAKAYAMA, H., M. SHIRASE, T. NAKASHIMA, Y. KUROGOCHI & J. LINDSTROM. 1990. Affinity purification of nicotinic acetylcholine receptor from rat brain. Mol. Brain Res. **7**: 221–226.

37. NAKAYAMA, H., H. OKUDA & T. NAKASHIMA. 1993. Phosphorylation of rat brain nicotinic acetylcholine receptor by cAMP-dependent protein kinase in vitro. Mol. Brain Res. **20**: 171–177.

38. WHITING, P., F. ESCH, S. SHIASAKI & J. LINDSTROM. 1987. Neuronal nicotinic acetylcholine receptor β subunit is coded for by the cDNA clone α4. FEBS Lett. **219**: 459–463.

39. SCHOEPFER, R., P. WHITING, F. ESCH, R. BLACHER, S. SHIMASAKI & J. LINDSTROM. 1988. cDNA clones coding for the structural subunit of a chicken brain nicotinic acetylcholine receptor. Neuron **1**: 241–248.

40. FLORES, C., S. ROGERS, L. PABREZA, B. WOLFE & K. KELLAR. 1992. A subtype of nicotinic cholinergic receptor in rat brain is composed of α4 and β2 subunits and is upregulated by chronic nicotine treatment. Mol. Pharmacol. **41**: 31–37.

41. CONROY, W., A. VERNALLIS & D. BERG. 1992. The α5 gene product assembles with multiple acetylcholine receptor subunits to form distinctive receptor subtypes in brain. Neuron **9**: 1–20.

42. WHITING, P., R. SCHOEPFER, J. LINDSTROM & T. PRIESTLY. 1991. Structural and pharmacological characterization of the major brain nicotinic acetylcholine receptor subtype stably expressed in mouse fibroblasts. Mol. Pharmacol. **40**: 463–472.

43. PEREIRA, E., M. ALKONDON, S. REINHARDT-MAELICKE, X. PENG, J. LINDSTROM, P. WHITING & E. ALBUQUERQUE. 1994. Physostigmine and galanthamine reveal the presence of the novel binding site on the α4β2 subtype of neuronal nicotinic acetylcholine receptors stably expressed in fibroblast cells. J. Pharmacol. Exp. Ther. **270**: 768–778.

44. ANAND, R., W. CONROY, R. SCHOEPFER, P. WHITING & J. LINDSTROM. 1991. Chicken neuronal nicotinic acetylcholine receptors expressed in *Xenopus* oocytes have a pentameric quaternary structure. J. Biol. Chem. **266**: 11192–11198.

45. COOPER, E., S. COUTURIER & M. BALLIVET. 1991. Pentameric structure and subunit stoichiometry of a neuronal nicotinic acetylcholine receptor. Nature **350**: 235–238.

46. SWANSON, L., D. SIMMONS, P. WHITING & J. LINDSTROM. 1987. Immunohistochemical localization of neuronal nicotinic receptors in the rodent central nervous system. J. Neurosci. **7**: 3334–3342.

47. CLARKE, P., R. SCHWARTZ, S. PAUL, C. PERT & A. PERT. 1985. Nicotinic binding in rat brain: Autoradiographic comparison of [³H] nicotine, and [¹²⁵I] α-bungarotoxin. J. Neurosci. **5**: 1307–1315.

48. HILL, J., M. ZOLI, J. BOURGEOIS & J. CHANGEUX. 1993. Immunocytochemical localization of a neuronal nicotinic receptor: The β2 subunit. J. Neurosci. **13**: 1551–1568.

49. KEYSER, K., T. HUGHES, P. WHITING, J. LINDSTROM & H. KARTEN. 1988. Cholinoceptive

neurons in the retina of the chick: An immunohistochemical study of the nicotinic acetylcholine receptors. Visual Neurosci. **1:** 349–366.

50. BRITTO, L., D. HAMASSAKI-BRITTO, E. FERRO, K. KEYSER, H. KARTEN & J. LINDSTROM. 1992. Neurons of the chick brain and retina expressing both α-bungarotoxin-sensitive and α-bungarotoxin-insensitive nicotinic acetylcholine receptors: An immunohistochemical analysis. Brain Res. **590:** 193–200.

51. CLARKE, P., G. HAMILL, N. NADI, D. JACOBOWITZ & A. PERT. 1986. ^{3}H-Nicotine and ^{125}I-α-bungarotoxin-labeled nicotinic receptors in the interpeduncular nucleus of rats. II. Effects of habenular deafferentation. J. Comp. Neurol. **251:** 407–413.

52. BRITTO, L., K. KEYSER, J. LINDSTROM & H. KARTEN. 1992. Immunohistochemical localization of nicotinic acetylcholine receptor subunits in the mesencephalon and diencephalon of the chick (*Gallus gallus*). J. Comp. Neurol. **317:** 325–340.

53. RAPIER, C., G. LUNT & S. WONNACOTT. 1988. Stereoselective nicotine-induced release of dopamine from striatal synaptosomes: Concentration dependence and repetitive stimulation. J. Neurochem. **50:** 1123–1130.

54. GRADY, S., M. MARKS & A. COLLINS. 1994. Desensitization of nicotine-stimulated ^{3}H dopamine release from mouse striatal synaptosomes. J. Neurochem. **62:** 1390–1398.

55. MARKS, M., J. PAULY, D. GROSS, E. DENERIS, I. HERMANS-BORGMEYER, S. HEINMANN & A. COLLINS. 1992. Nicotine binding and nicotinic receptor subunit RNA after chronic nicotine treatment. J. Neurosci. **12:** 2765–2784.

56. PENG, X., V. GERZANICH, R. ANAND, P. WHITING & J. LINDSTROM. 1994. Nicotine-induced increase in neuronal nicotinic receptors results from a decrease in the rate of receptor turnover. Mol. Pharmacol. **46:** 523–530.

57. VERNALIS, A., W. CONROY & D. BERG. 1993. Neurons assemble acetylcholine receptors with as many as three kinds of subunits while maintaining subunit segregation among receptor subtypes. Neuron **10:** 451–464.

58. TZARTOS, S., D. RAND, B. EINARSON & J. LINDSTROM. 1981. Mapping of surface structures on *Electrophorus* acetylcholine receptor using monoclonal antibodies. J. Biol. Chem. **256:** 8635–8645.

59. ANAND, R. & J. LINDSTROM. 1990. Nucleotide sequence of the human nicotinic acetylcholine receptor β2 subunit gene. Nucleic Acids Res. **18:** 4272.

60. WANG, F. & J. LINDSTROM. Unpublished results.

61. JACOB, M., D. BERG & J. LINDSTROM. 1984. A shared antigenic determinant between the *Electrophorus* acetylcholine receptor and a synaptic component on chick ciliary ganglion neurons. Proc. Natl. Acad. Sci. USA **81:** 3223–3227.

62. JACOB, M., J. LINDSTROM & D. BERG. 1986. Surface and intracellular distribution of a putative neuronal nicotinic acetylcholine receptor. J. Cell Biol. **103:** 205–214.

63. JACOB, M. & D. BERG. 1983. The ultrastructural localization of α-bungarotoxin binding sites in relation to synapses on chick ciliary ganglion neurons. J. Neurosci. **3:** 260–271.

64. CONTI-TRONCONI, B., S. DUNN, E. BARNARD, J. DOLLY, F. LAI, N. RAY & M. RAFTERY. 1985. Brain and muscle nicotinic acetylcholine receptors are different but homologous proteins. Proc. Natl. Acad. Sci. USA **82:** 5208–5212.

65. SÉGUÉLA, P., J. WADICHE, K. DINELLY-MILLER, J. DANI & J. PATRICK. 1993. Molecular cloning, functional properties, and distribution of rat brain α7: A nicotinic cation channel highly permeable to calcium. J. Neurosci. **13:** 596–604.

66. PENG, X., M. KATZ, V. GERZANICH, R. ANAND & J. LINDSTROM. 1994. Human α7 acetylcholine receptor: Cloning of the α7 subunit from the SH-SY5Y cell line and determination of pharmacological properties of native receptors and functional α7 homomers expressed in *Xenopus* oocytes. Mol. Pharmacol. **45:** 546–554.

67. MCLANE, K., X. WU, J. LINDSTROM & B. CONTI-TRONCONI. 1992. Epitope mapping of polyclonal and monoclonal antibodies against two α-bungarotoxin binding subunits from neuronal nicotinic receptors. J. Neuroimmunol. **38:** 115–128.

68. MCLANE, K., X. WU, R. SCHOEPFER, J. LINDSTROM & B. CONTI-TRONCONI. 1991. Identification of sequence segments forming the α-bungarotoxin binding sites on two nicotinic acetylcholine receptor α subunits from the avian brain. J. Biol. Chem. **266:** 15230–15239.

69. KEYSER, K., L. BRITTO, R. SCHOEPFER, P. WHITING, J. COOPER, W. CONROY, A.

BROZOZOWSKA-PRECHTL, H. KARTEN & J. LINDSTROM. 1993. Three subtypes of α-bungarotoxin sensitive nicotinic acetylcholine receptors are expressed in chick retina. J. Neurosci. **13**: 442–454.

70. GOTTI, C., A. OGANDO, W. HANKE, R. SCHULE, M. MORETTI & F. CLEMENTI. 1991. Purification and characterization of an α-bungarotoxin receptor that forms a functional nicotinic channel. Proc. Natl. Acad. Sci. USA **88**: 3258–3262.

71. GOTTI, C., M. MORETTI, R. LONGHI, L. BRISCINI, E. MANURA & F. CLEMENTI. 1993. Antipeptide specific antibodies for the characterization of different α subunits of α-bungarotoxin binding acetylcholine receptors present in chick optic lobe. J. Recept. Res. **13**: 453–465.

72. COURTURIER, S., D. BERTRAND, J. MATTER, M. HERNANDEZ, S. BERTRAND, N. MILLAR, S. VALERA, T. BARKAS & M. BALLIVET. 1990. A neuronal nicotinic acetylcholine receptor subunit (α7) is developmentally regulated and forms a homomeric channel blocked by α-bungarotoxin. Neuron **5**: 847–856.

73. ANAND, R., X. PENG, J. BALLESTA & J. LINDSTROM. 1993. Pharmacological characterization of α-bungarotoxin sensitive AChRs immunoisolated from chick retina: Contrasting properties of α7 and α8 subunit-containing subtypes. Mol. Pharmacol. **44**: 1046–1050.

74. CONROY, W., M. SAEDI & J. LINDSTROM. 1990. TE671 cells express an abundance of a partially mature acetylcholine receptor α subunit which has characteristics of an assembly intermediate. J. Biol. Chem. **265**: 21642–21651.

75. WHITING, P. & J. LINDSTROM. 1986. Pharmacological properties of immunoisolated neuronal nicotinic receptors. J. Neurosci. **6**: 3061–3069.

76. JACKSON, M. 1989. Perfection of a synaptic receptor: Kinetics and energetics of the acetylcholine receptor. Proc. Natl. Acad. Sci. USA **86**: 2199–2203.

77. GALZI, J., A. DEVILLERS-THIERY, N. HUSSY, S. BERTRAND, J. CHANGEUX & D. BERTRAND. 1992. Mutations in the channel domain of a neuronal nicotinic receptor convert ion selectivity from cationic to anionic. Nature **359**: 500–505.

78. EISILE, J., S. BERTRAND, J. GALZI, A. DEVILLERS-THIERY, J. CHANGEUX & D. BERTRAND. 1993. Chimaeric nicotinic-serotonergic receptor combines distinct ligand binding and channel specificities. Nature **366**: 479–409.

79. GERZANICH, V., R. ANAND & J. LINDSTROM. 1994. Homomers of α8 subunits of nicotinic receptors functionally expressed in *Xenopus* oocytes exhibit similar channel but contrasting binding site properties compared to α7 homomers. Mol. Pharmacol. **45**: 212–220.

80. ANAND, R., X. PENG & J. LINDSTROM. 1993. Homomeric and native α7 acetylcholine receptors exhibit remarkably similar but nonidentical pharmacological properties suggesting that the native receptor is a heteromeric protein complex. FEBS Lett. **327**: 241–246.

81. ANAND, R. & J. LINDSTROM. Unpublished results.

82. VERNINO, S., M. AMADOR, C. LUETJE, J. PATRICK & J. DANI. 1992. Calcium modulation and high calcium permeability of neuronal nicotinic acetylcholine receptors. Neuron **8**: 127–134.

83. LIPTON, S., M. FROSH, M. PHILLIPS, D. TAUCK & E. AIZENMAN. 1988. Nicotinic antagonists enhance process outgrowth by rat retinal ganglion cells in culture. Science **239**: 1293–1296.

84. ZHANG, Z., S. VIJAYAROGHAVEN & D. BERG. 1994. Neuronal acetylcholine receptors that bind α-bungarotoxin with high affinity function as ligand-gated ion channels. Neuron **12**: 167–177.

85. PUGH, P. & D. BERG. 1994. Neuronal acetylcholine receptors that bind α-bungarotoxin mediate neurite retraction in a calcium-dependent manner. J. Neurosci. **14**: 889–896.

86. ALKONDON, M. & E. ALBUQUERQUE. 1991. Initial characterization of the nicotinic acetylcholine receptors in rat hippocampal neurons. J. Recept. Res. **11**: 1101–1201.

87. ALKONDON, M. & E. ALBUQUERQUE. 1993. Diversity of nicotinic acetylcholine receptors in rat hippocampal neurons. I. Pharmacological and functional evidence for distinct structural subtypes. J. Pharmacol. Exp. Ther. **265**: 1455–1473.

88. CASTRO, N. & E. ALBUQUERQUE. 1993. Brief-lifetime, fast-inactivating ion channels account for the α-bungarotoxin-sensitive nicotinic response in hippocampal neurons. Neurosci. Lett. **164**: 137–140.

89. DEL TORO, J. JUIZ, X. PENG, J. LINDSTROM & M. CRIADO. 1994. Immunocytochemical localization of the α7 subunit of the nicotinic acetylcholine receptor in the rat central nervous system. J. Comp. Neurol. **349:** 325–342.

90. HAMASSAKI-BRITTO, D., A. BRZOZOWSKA-PRECHTL, H. KARTEN, J. LINDSTROM & K. KEYSER. 1991. GABA-like immunoreactive cells containing nicotinic acetylcholine receptors in the chick retina. J. Comp. Neurol. **313:** 394–408.

91. HAMASSAKI-BRITTO, D., A. BRZOZOWSKA-PRECHTHL, H. KARTEN & J. LINDSTROM. 1994. Bipolar cells of the chick retina containing α-bungarotoxin-sensitive nicotinic acetylcholine receptors. Visual Neurosci. **11:** 63–70.

92. HAMASSAKI-BRITTO, D., P. GARDINO, J. HOKOC, K. KEYSER, H. KARTEN, J. LINDSTROM & L. BRITTO. 1994. Differential development of α-bungarotoxin-sensitive and α-bungarotoxin-insensitive nicotinic acetylcholine receptors in the chick retina. J. Comp. Neurol. **347:** 161–170.

93. FUCHS, P. & B. MURROW. 1992. Cholinergic inhibition of short (outer) hair cells of the chick's cochlea. J. Neurosci. **12:** 800–809.

94. FUCHS, P. & B. MURROW. 1992. A novel cholinergic receptor mediates inhibition of chick cochlear hair cells. Proc. R. Soc. Lond. B. **248:** 35–40.

95. ANAND, R., V. GERZANICH, J. COOPER, X. PENG, P. FUCHS & J. LINDSTROM. Unpublished results.

96. ELGOYHEN, A., D. JOHNSON, J. BOULTER, D. VETTER & S. HEINEMANN. 1994. α9: An acetylcholine receptor with novel pharmacological properties expressed in rat cochlear hair cells. Cell **79:** 705–715.

Methods for Increasing the Expression Level of a Soluble Fusion Protein Encoding a 62 Amino Acid Fragment of the α-Subunit of the Nicotinic Acetylcholine Receptor[a]

LISA N. GENTILE AND EDWARD HAWROT

Section of Molecular and Biochemical Pharmacology
Brown University
Providence, Rhode Island 02912

We are interested in expressing a large soluble fragment of the α-bungarotoxin (BGTX) binding domain of the nicotinic acetylcholine receptor (nAChR) to facilitate structural studies by high-resolution NMR spectroscopy.[1] As such, we need to express the receptor fragment (with and without the possibility for metabolic ^{15}N and/or ^{13}C labeling) in multimilligram quantities. Toward this end we have constructed a synthetic gene from overlapping oligonucleotides of a T62mer (*Torpedo californica* α-subunit residues 143–204: TMKLGIWTYDGTKVSISPESDRPDLST-FMESGEWVMKDYRGWKHWVYYTCCPDTPYLDITYH). The T62mer receptor gene was appended to the C-terminus of gene 9 (in plasmid pSR9), a highly soluble bacteriophage *T7* coat protein, separated by an acid cleavage (DP) site. The expression of the fusion protein is cytoplasmic in *Escherichia coli,* and under the inducible control of the *lacUV5* promoter of *T7* RNA polymerase (see FIG. 1; a similar approach was used to construct and express a synthetic gene for α-BGTX).[2] Gene 9 was chosen as a fusion partner for our receptor fragment for a number of reasons: its solubility, its lack of Cys residues that could interfere with the oxidation of Cys 192–193 of the receptor, and its high levels of expression (up to 50% of the soluble protein) in other systems.[3] At present, we are able to obtain gram quantities of the fusion protein in rich media (LB) and 500–600 mg in minimal media (M9) per 2-L fermenter preparation.

MATERIALS AND METHODS

A typical protocol for harvesting the total soluble cytoplasmic proteins after induction includes centrifugation to pellet the cells, washing of the cell pellet in PBS followed by resuspension of the cells in lysis buffer, french press lysis at 2100 psi, high-speed centrifugation to collect soluble fractions (note, the protein is not found in inclusion bodies in the pellet), DEAE column to remove nucleic acids, and

[a]This work, supported by grants from the National Science Foundation (BNS 90-21227), the National Institutes of Health (GM32629), and a NIH predoctoral training grant (GM07601), was done in partial fulfillment of the requirements for a Ph.D. degree (LNG) from Brown University.

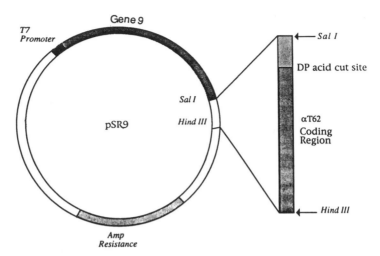

FIGURE 1. The synthetic gene 9/αT62 fusion protein. The DNA sequence for αT62 was derived from back-translation of the primary amino acid sequence, and optimized for codon usage in *Escherichia coli*. A DP acid cut site was engineered in at the 5′ end of the T62mer coding sequence to allow later liberation of the 62mer from gene 9.

ammonium sulfate precipitation. Protease inhibitors are present throughout this procedure.

RESULTS AND DISCUSSION

To ensure that we were expressing as much fusion protein as possible for our system, we performed protein expression checks of the cultures derived from a number of different BL21(DE3) isolates containing fusion protein insert. We found, but cannot explain, the presence of both "high" and "low" expressing colonies from a single transformation of our fusion protein expressing plasmid into our host strain. Growth experiments were performed under various conditions to maximize the final OD_{600} of the culture. The results of these studies are shown in TABLE 1 below. The greatest yields of fusion protein were obtained when the cultures were grown in a 2-L bench-top fermenter, in LB media supplemented with glucose.[4] Under these condi-

TABLE 1. Comparison of Cell Growth in Various Media

Prep Conditions (37 °C)	Medium	OD_{600} at Induction	Final Uninduced OD_{600}	Induced Protein
2L/Fernbach	LB	0.7	1.4	350 mg
2L/Fermenter	LB	2.4	4.3	470 mg
2L/Fermenter	LB + 4 g glucose	4.7	8.5	1080 mg
2L/Fermenter	M9	1.6	3.7	N/A
2L/Fermenter	M9 + 7.2 g glucose spike at $OD_{600} \sim 1$	2.3	4.8	540 mg

tions, and with the IPTG induction at mid-log phase, we harvested >1 g of total protein at the ammonium sulfate precipitation step. We estimate that approximately one-third of this is our induced fusion protein as judged by SDS-PAGE (further purification can be achieved by FPLC on a Mono-Q column).

As we are interested in ^{15}N and/or ^{13}C metabolic labeling of our receptor fragment for NMR studies, it is important that we can express high levels of the fusion protein in a defined medium. From 2-L cultures grown in M9 medium, where NH_4Cl is the only source of nitrogen and either glucose or NaOAC are the only sources of carbon, we obtained >0.5 g of our fusion protein. Although it may be possible to increase the levels of expression even further by monitoring the glucose level in the fermenter and supplying the bacteria with increased O_2, the inexpensive modifications made above should enable us to produce the materials needed for structural studies.

REFERENCES

1. BASUS, V. J., G. SONG & E. HAWROT. 1993. Biochemistry **32:** 12290.
2. ROSENTHAL, J. A., S. H. HSU, N. J. MESSIER, D. SCHNEIDER, C. A. VASLET, L. N. GENTILE & E. HAWROT. 1994. J. Biol. Chem. **269:** 11178.
3. HOWELL, M. L. & K. M. BLUMENTHAL. 1989. J. Biol. Chem. **264:** 15268.
4. PILLET, L., O. TREMEAU, F. DUCANIEL, P. DREVET, S. JUSTIN-ZINN, S. PINKASFELD, J-C. BOULAIN & A. MENEZ. 1993. J. Biol. Chem. **268:** 909.

Probing Ion Channels and Recognition Sites of Neuronal Nicotinic Cholinergic Receptors with Novel Nicotine Affinity and Other Ligands[a]

NICOLE LERNER-MARMAROSH,[b] ANDREW S. KENDE,[c]
DAVID X. WANG,[b] AND LEO G. ABOOD[b,d]

Departments of Pharmacology[b] and Chemistry[c]
University of Rochester
Rochester, New York 14642

One of the fundamental problems of receptor biology concerns the interrelationships of the receptor agonist recognition site to voltage-dependent ion channels associated with excitability and other bioelectric phenomena. There are two known classes of receptors within the nervous system: those in which the ion channels are an integral component of the receptor complex comprising multiple subunits and those in which receptors are linked to ion channels via G-proteins. The former includes the nicotinic cholinergic, gabaminergic, glycinergic, and glutaminergic receptors, and comprises a superfamily with remarkable sequence homology in the β subunits and the seven membrane-spanning regions.[1] The G-coupled receptors, which are a far more extensive group for endogenous mediators, include muscarinic cholinergic, biogenic amines (catecholamines and serotonin), peptides, hormones, and growth factors. In the *Torpedo* membrane the ionic pore results from the pentameric array of the nAChR subunits.[2] Although neuronal nAChR receptors are composed of only α and β subunits, lacking the δ and γ subunits of the *Torpedo* nAChR, the high degree of sequence homology in uncharged segments of the membrane-spanning regions of all four subunits[3] is suggestive of ion-channel functional homology in all nAChRs. The issue of whether the nAChR ion channel comprises a single or multiple subunits is still unresolved.

Subtypes of nAChR, comprising α_7 or α_8 subunits and exhibiting a high affinity for α-bungarotoxin and a low affinity for nicotine, have been shown to contain ligand-gated Ca^{2+} ion channels.[4] It remains to be determined whether the nAChR receptors in chick ciliary ganglion, which contain voltage-gated Ca^{2+} channels that may be associated with signaling function as well as synaptic transmission, comprise α_7 or α_8 subunits.[5]

REGULATION OF ION CHANNELS BY G-PROTEINS

Ion conductance within nicotinic cholinergic (nAChR) and excitatory amino acid receptors results directly from conformational changes in the receptor protein

[a]This research was supported by National Institutes of Health grant DA 00464. N.L.-M. was a fellow on National Institute on Drug Abuse training grant DA 07232.
[d]Corresponding author.

complex, whereas K^+ conductance initiated by muscarinic cholinergic receptors, such as the M2 subtype in heart muscle, is mediated via a pertussin toxin-sensitive G-protein coupled to the ion channel.[6] Although some controversy remains regarding a direct G-protein gating of ion channels, the most definitive evidence for the hypothesis derives from inside-outside membrane patch studies in muscarinic-sensitive heart muscle demonstrating that the application of either the GTPgS subunit or a purified pertussis toxin G_i-protein purified from erythrocytes results in activation of K^+ channels.[7,8] In contrast to atrium, where a G_i-protein is involved, the muscarinic sensitive K^+ channel in hippocampal neurons is coupled to G_O[9].

When expressed in *Xenopus* oocytes, the α_7 subunit, unlike other nAChR subtypes, exhibits significant agonist-mediated Ca^{2+} conductance (TABLE 1). A major portion of the current through the α_7 channels is carried by Ca^{2+}, and the incoming Ca^{2+} in turn activates Ca^{2+}-dependent Cl^- conductance. Agonist activation results in a biphasic current consisting of an initial inward current through α_7 channels followed by an outward current through Ca^{2+}-dependent Cl^- channels.[10] It has been speculated that the α_7 homoligomer, which is present in a number of limbic

TABLE 1. Ion Permeability Associated with Various nAChR Subtypes

Subtype	Ions	Source
α_4	Ca^{2+}	Ganglia
α_7	Voltage-gated Ca^{2+} Ca^{2+}-dependent Cl^-	Hippocampus, cochlea
α_8	Same as α_7	Chick retina
β_2	Inhibitory	Rat dorsolateral septal nucleus
$\alpha_2-\alpha_4$	Excitatory	Rat medial vestibular nucleus
	Cation selective	Cardiac ganglion
α_2	Cs^+, Na^+, Ca^{2+} $P_{Cl}/P_{Na} = 0.05$	

system areas, may be involved in the activation of Ca^{2+}-dependent mechanisms; however, it is difficult to reconcile this notion with the observation that α-bungarotoxin was pharmacologically inactive when administered intraventricularly in high concentrations to rats.[11] Both an α-bungarotoxin sensitive and α-bungarotoxin-insensitive ($\alpha_3\beta_4$) nAChR have been found in chick ciliary ganglion; the former is synaptic in origin and exhibits agonist-mediated voltage sensitive Ca^{2+} permeability, whereas the latter is monosynaptic and exhibits voltage-insensitive Ca^{2+} permeability.[12] Acetylcholine or nicotine-evoked currents in cultured neurons dissociated from rat parasympathetic cardiac ganglia exhibited a strong inward rectification and cation selectivity; the permeabilities of Cs^+, Ca^{2+}, and Na^+ were comparable, and the Cl^- permeability was one-twentieth of that of Na^+.[13] Because the response was inhibited by both mecamylamine and hexamethonium in a dose-dependent manner, it was concluded that the ACh-activated ion channels of the postganglionic neurons are mediated by nAChRs. A summary of the voltage-gated ion channels of the nAChR is presented in TABLE 1.

[³H]MECAMYLAMINE AS A LIGAND FOR ION CHANNELS
OF THE nAChR

Mecamylamine, which had been originally developed as a ganglionic blocking agent,[14] is an antagonist to the peripheral and central actions of nicotine. It has been shown to block acetylcholine-induced currents in crustacean muscle in a concentration- and voltage-dependent manner; recovery of the blockade requires the presence of an agonist.[15] Because mecamylamine has a very low affinity ($K_i > 1 \times 10^{-4}$ M) for the nAChR recognition site, it appears to be a noncompetitive inhibitor. Mecamylamine and pempidine have been shown to noncompetitively inhibit ^{86}Rb flux in mouse brain synaptosomes.[16] On the basis of pharmacological studies with various mecamylamine and pempidine derivatives exploring the structural requirements for nicotine agonists and agonits, it was inferred that mecamylamine exhibited both competitive and noncompetitive properties in antagonizing the central effects of nicotine.[17]

[³H]Mecamylamine binding studies have recently been employed to investigate the characteristics of the nAChR ion channels.[18,19] Although [³H]mecamylamine binding is displaceable by mecamylamine in the submicromolar range and correlates well with the pharmacological efficacy of agents structurally related to mecamylamine, the method has some limitations.[18] Binding is sensitive to very low concentrations of monovalent and divalent inorganic cations, Ca^{2+} and Mg^{2+}; and although the sensitivity to inorganic cations is to be expected of a ligand acting at voltage-gated ion channels, it is difficult to control for variations in ionic strength contributed by the test ligands (FIG. 1). The method is suitable, however, for ligands structurally related to mecamylamine. Inasmuch as [³H]mecamylamine binding is equally sensitive to Na^+, K^+, and Rb^+, there appears to be no selectivity for monovalent ions.

CORRELATION OF [³H]MECAMYLAMINE BINDING WITH ANTAGONISM
OF NICOTINE'S PHARMACOLOGICAL ACTION

With a series of mecamylamine, pempidine, and camphene derivatives a reasonably good correlation was found between the K_i values for [³H]mecamylamine

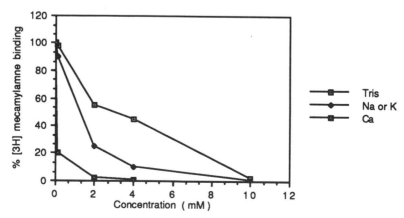

FIGURE 1. Effect of concentration of Tris and inorganic cations on [³H]mecamylamine binding to calf brain membranes.

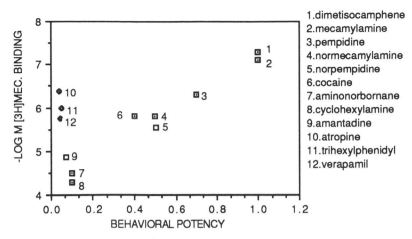

FIGURE 2A. Plot of IC_{50} for [^{3}H]mecamylamine binding versus antagonism of nicotine behavioral effects of various antagonists. Behavioral potency is expressed relative to mecamylamine as 1. Mice were given 10 μmoles/kg nicotine intraperitoneally followed by 50 μmoles/kg of test agent after occurrence of behavioral effects, which included prostration, tremors, seizures, straub tail, respiratory depression, and decreased motor activity. Five mice were used for each agent. Procedures for [^{3}H]mecamylamine binding and behavioral measurements are described elsewhere.[27]

binding and the ability of the agents to block nicotine-induced prostration and seizures in rats; however, a number of structurally and pharmacologically unrelated agents, including various nicotine analogues, had K_i values in the μM range (FIGS. 2A and 2B). Cocaine, which appears to block nicotine-induced prostration, but not seizures, has a K_i value over a magnitude greater than that for mecamylamine, whereas the muscarinic antagonists, atropine and trihexylphenidyl, are ineffective against nicotine's action.

COMPARISON OF [^{3}H]MECAMYLAMINE WITH [^{3}H]METHYLCARBAMYLCHOLINE AND [^{3}H](R,S)-3-PYRIDYL-1-[^{3}H]METHYL-2-AZETIDINE BINDING IN VARIOUS REGIONS OF CALF BRAIN

Because the density of nAChRs showed marked variations in the various brain regions, a comparison of the extent of [^{3}H]mecamylamine with [^{3}H]methylcarbamylcholine ([^{3}H]MCC) binding may help resolve the issue concerning the significance of the [^{3}H]mecamylamine binding in relation to nAChR function. The aim of the study was to compare [^{3}H]mecamylamine binding to membranes from various calf brain regions with the binding of two different receptor ligands, one chemically related to acetylcholine and the other to nicotine. A comparison of the [^{3}H]MCC and [^{3}H](R,S)-3-pyridyl-1-[^{3}H]methyl-2-azetidine ([^{3}H]MPA) (a nicotine analogue[20]) binding in membranes from various calf brain regions revealed that the density[20] of nAChR receptors was threefold greater in striatum and substantia nigra compared to frontal cortex, hippocampus, and locus coeruleus, whereas the density in the cerebellum was one-tenth that of the frontal cortex (TABLE 2). With [^{3}H]MPA as the ligand, total

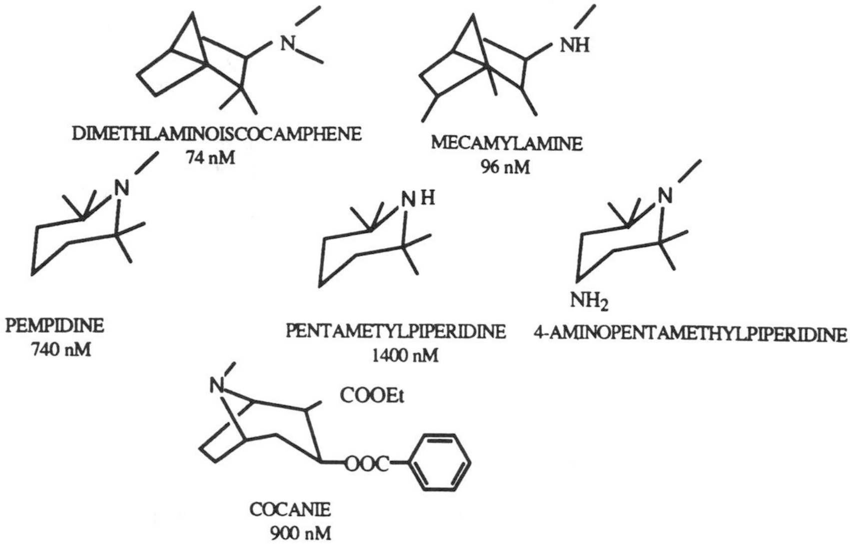

FIGURE 2B. Ion channel blockers for nicotinic receptors. Chemical structures of mecamylamine analogues and related agents.

binding in the various brain regions was comparable to that observed with [³H]MCC, with some notable exceptions. The extent of [³H]MPA binding in the cerebellum was 33% that of the frontal cortex as compared to 10% seen with [³H]MCC, whereas the density of [³H]MPA binding in the locus coeruleus was 40% that in the frontal cortex. [³H]MCC binding in the substantia nigra was greater than that seen in striatum, whereas [³H]MPA binding was slightly less in the substantia nigra than in the striatum. The difference in [³H]MCC and [³H]MPA binding in the various brain regions may be attributable to differences in the nAChR subtypes.

The saturation plots of [³H]mecamylamine binding for the various brain regions show the greatest binding in the substantia nigra, intermediate binding in the

TABLE 2. Comparison of [³H]MCC and [³H]MPA Binding in Membranes of Various Calf Brain Regions[a]

| | [³H]MCC | | [³H]MPA | |
	fmole/mg	% Frontal Cortex	fmole/mg	% Frontal Cortex
Frontal cortex	13	—	15	—
Hippocampus	14	108	14	93
Striatum	35	270	36	242
Substantia nigra	43	330	33	219
Cerebellum	1.3	10	5	33
Locus coeruleus	11.5	90	6	40

Abbreviations: [³H]MCC: [³H]methylcarbamylcholine; [³H]MPA: (R,S)-3-pyridyl-1-[³H]methyl-2-azetidine, an analogue of nicotine.[20]

[a]The results are an average of three separate experiments run in triplicate and agreeing within 8%.

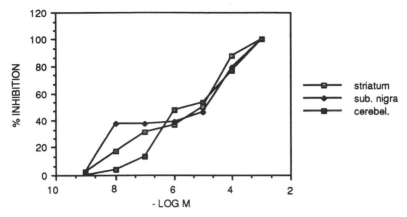

FIGURE 3A. Saturation analysis of [³H]mecamylamine binding in various calf brain regions.

striatum, and little or no binding in the cerebellum within the 1–10 nM range (FIGS. 3A and 3B). At concentrations of 100 nM and greater no difference was found among the various brain regions. [³H]MCC binding differed markedly among various brain regions, with the highest density in the substantia nigra and striatum and the lowest in the cerebellum (TABLE 2). The rank order of the densities of [³H]MCC binding for the various tissues is similar to the rank order of densities for [³H]mecamylamine binding in the lower but not upper concentration range. Because a variety of alkaloids, including nicotine analogues, have K_i values in the micromolar range (FIG. 2A) but do not block the behavioral effects of nicotine, one might infer that only [³H]mecamylamine binding occurring near the submicromolar range is reflective of nAChRs. It also appears that the [³H]mecamylamine binding curve is biphasic, suggestive of lower and higher affinity sites. It remains to be seen whether the lower affinity binding is of any functional significance.

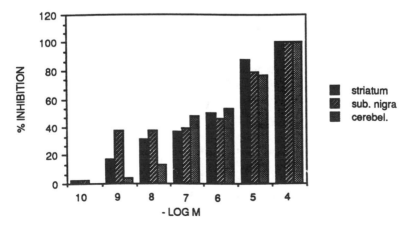

FIGURE 3B. [³H]Mecamylamine binding in various calf brain regions. Bar graph representation of FIGURE 3A.

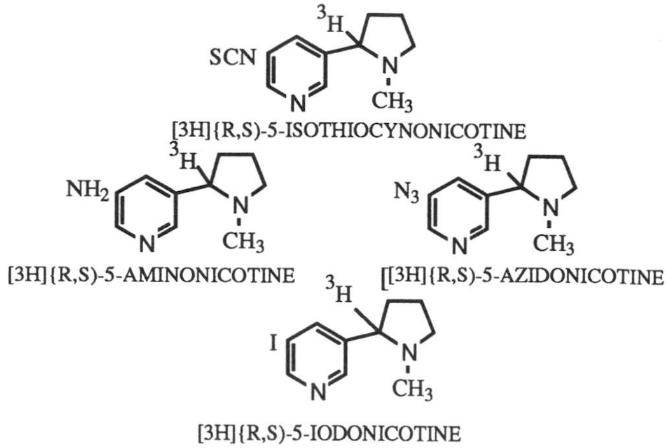

[3H]{R,S)-5-ISOTHIOCYNONICOTINE

[3H]{R,S)-5-AMINONICOTINE

[[3H]{R,S)-5-AZIDONICOTINE

[3H]{R,S)-5-IODONICOTINE

FIGURE 4. Chemical structures of [³H]-labeled nicotine affinity and photoaffinity ligands.

CHEMICAL NATURE OF THE nAChR AGONIST RECOGNITION SITE

The techniques of affinity labeling and construction of chimeric receptor subunits have been used to map the ligand recognition site of muscle and neuronal nAChRs. The nAChR is known to contain a readily reducible sulfhydryl site adjacent to the negative site for acetylcholine binding;[21] alkylating agents, such as [³H]-4-(N-maleimido)benzyltrimethylammomium, have been used to identify and purify the α-subunit of the *Torpedo* nAChR.[22] The site of interaction of the sulfhydryl agent has been shown to be Cys192 and Cys193, which normally form a disulfide bond; this region in the vicinity of the disulfide bond is conserved in both muscle and neuronal nAChRs.[23] In addition to the Cys192 and Cys193 of the α-subunit, a number of other amino acids appear to play a role in agonist binding and function, including Tyr89,[24] Trp149,[25] and Tyr198.[2] Recently, site-directed mutagenesis of the *Torpedo* δ subunit has demonstrated the involvement of the acidic residues, Asp180 and Glu189, for acetylcholine binding.[25] A schematic model for the nAChR has been proposed in which acetylcholine is bound in the Asp180–Glu189 region and surrounded by an array of other amino acids (Cys192, Cys193, Tyr93, Trp149, and Tyr198) that contribute to its binding.[26] The structural perturbation resulting from the interaction of acetylcholine with this array of amino acids is then transmitted to the ion channel to somehow regulate ion conductance.

(R,S)-5-ISOTHIOCYANONICOTINE AND [³H]NICOTINOID PHOTOAFFINITY LIGANDS AS TOOLS FOR PROBING nAChRs

As part of an effort to examine the specificity of nicotine for nAChR subtypes and determine its sites of interaction a series of unlabeled and [³H]-labeled nicotine affinity ligands were prepared[26,27] (FIG. 4). The most useful ligand was (R,S)-5-isothiocyanonicotine (SCN-nic), which was found to irreversibly inhibit the binding of [³H]MCC to brain membranes in the nanomolar range. SCN-nic also inhibited mouse brain nicotinic receptors *in vivo* in a dose-dependent manner, the inhibition being 50% at a dose of 20 mmoles/kg. Behavioral studies in mice revealed that SCN-nic had less than one-fifth the potency of nicotine in producing muscle weakness and seizures with a considerably greater duration of action. Other nicotinoids included three photoaffinity agents: [³H-1′](R,S)-5-azidonicotine, [³H-1′](R,S)-5-iodononicotine, and [³H-1′](R,S)-5-aminononicotine.

Photolysis of iodoaryl compounds results in cleavage of iodide to form an aryl radical,[28] which is capable of extracting H from SH, NH₂, and OH groups of amino acids, whereas photolysis of aryl amines is believed to result in the extraction of an amino H to form a aryl imino cation.[29] Photolysis of membrane preparations in the presence of 10 μM [³H](R,S)-5-iodononicotine, [³H](R,S)-5-azidonicotine, and [³H](R,S)-5-aminonicotine resulted in 46, 4, and 23% inhibition of [³H]MCC binding, respectively, as compared with 13, 42, and 2% inhibition of [³H]3-quinuclidinyl-benzilate ([³H]QNB).

EFFECT OF DITHIOTHREITOL ON INHIBITION OF [³H]MCC BINDING BY NICOTINE AFFINITY LIGANDS

The IC_{50} for irreversible inhibition of [³H]MCC SCN-nic was recently shown to be 1×10^{-7} M in the absence of dithiothreitol (DTT) and 2×10^{-9} M in its presence.[27] It was also observed that alkylation of SH groups by N-ethylmaleimide did not alter the inhibition of [³H]MCC binding by SCN-nic (FIG. 5). Pretreatment of calf brain cortical membranes with DTT prior to photolysis also resulted in a marked increase in the inhibition of [³H]MCC binding by (R,S)-5-azidonicotine with no effect on the inhibition of [³H]QNB binding. The findings are consistent with the notion that a vicinal disulfide bond is involved in agonist binding to brain nAChRs, and although the presence of free SH groups enhances the covalent interaction of SCN-nic, they are not required for the covalent interaction of SCN-nic. The inability

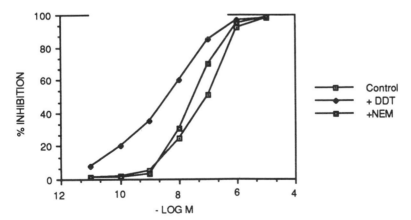

FIGURE 5. Effect of dithiothreitol (DTT) and N-ethylmaleimide (NEM) on the irreversible inhibition of [³H]MCC binding by SCN-nic. Membranes of calf frontal cortex were incubated with either 1 mM DTT or 3 mM NEM for 30 min at room temperature, centrifuged and washed once with 0.04 mM sodium phosphate, pH 7.0, exposed to varying concentrations of SCN-nic, and washed three times before determining [³H]MCC binding. The results are an average of three separate experiments in triplicate, agreeing within 7%. Experimental details are described elsewhere.[18]

of DTT to alter the photolytic inhibition of (R,S)-5-iodononicotine and (R,S)-5-aminonicotine also supports the notion that SH groups are not involved in the reactivity of the photoreactive nicotinoids with nAChRs. Photolabeling with [³H]nicotine has been shown to primarily involve Tyr-198.[30] Both the pyridine[31] and pyrrolidine[32] rings of nicotine appear to participate in radical mediated photoreactions; however, information on the nature of the reactions involved is lacking. The present finding that the presence of either an iodo or amino group in position 5 significantly increases the photoreactivity of the nicotine molecule supports the notion that the pyridine ring is involved.

EFFECT OF DITHIOTHREITOL TREATMENT ON IRREVERSIBLE INHIBITION OF [³H]MCC AND [³H]MPA BINDING BY SCN-NIC

To further explore the involvement of SH groups in the ligand binding region of nAChRs, a study was undertaken to determine the effect of DTT treatment on

TABLE 3. Effect of Dithiotreitol on SCN-nic Inhibition of [³H]MCC and [³H]MPA Binding to Membranes from Various Calf Brain Areas[a]

Brain Area	[³H]MCC IC_{50} nM			[³H]MPA IC_{50} nM		
	Control	DTT	Change	Control	DTT	Change
Frontal cortex	100	2	50-fold	10	5	2-fold
Striatum	98	10	10-fold	100	30	3.3-fold
Substantia nigra	90	12	7.5-fold	80	60	1.3-fold
Locus coeruleus	75	10	7.5-fold	80	80	0

[a]Membranes were exposed to 2 mM DTT and washed once with 0.04 mM sodium phosphate, pH 7.5. They were then exposed to various concentrations of SCN-nic, washed thrice, and measured for [³H]MCC and [³H]MPA binding. The results are an average of three separate experiments run in triplicate and agreeing within 8%. DTT, dithiotreitol; SCN-nic, (R,S)-5-isothiocyanonicotine.

irreversible inhibition of [³H]MCC and [³H]MPA binding by SCN-nic for various calf brain regions. The enhancement of SCN-nic inhibition by DTT was significantly greater for [³H]MCC than for [³H]MPA binding; the frontal cortex showed a 50-fold enhancement and other brain regions from 7.5–10-fold (TABLE 3). Only a slight enhancement in the irreversible inhibition was noted with [³H]MPA as the ligand. The findings are consistent with the notion that the two radioligands differ in their affinity to nAChR subtypes.

NEURONAL nAChR SUBTYPES THAT ARE RESISTANT TO SULFHYDRYL REAGENTS

Although the disulfide bond at positions 192 and 193 of the α subunit is required for activation of the skeletal muscle nAChR, the requirement for the vicinal cystine residue appears to be different for the various neuronal nAChR subtypes. It has been shown that agonist activation of an inhibitory nAChR in the rat dorsolateral septal nucleus is unaffected by DTT followed by alkylation of disulfide bonds by bromoace-

TABLE 4. Effect of Dithiotreitol on [³H]MCC and [³H]MPA Binding to Membranes from Various Calf Brain Regions[a]

Brain Area	[³H]MCC (pmole/mg)			[³H]MPA (pmole/mg)		
	Control	DTT	% Change	Control	DTT	% Change
Frontal cortex	12.2	6.2	−49	11.4	5.2	−54
Striatum	32.5	15.0	−54	32.0	33.0	+3
Substantia nigra	42.0	26.5	−40	40.2	36.5	−10
Locus coeruleus	11.2	11.3	0	6	6.2	0

[a]Membranes were exposed to 2 mM dithiothreitol (DTT) for 30 min and washed twice with 0.04 mM sodium phosphate, pH 7.5, prior to determining receptor binding. The results are an average of three separate experiments run in triplicate and agreeing within 7%.

tylcholine; however, an excitatory nAChR in the rat medial vestibular nucleus is inhibited upon treatment with DTT.[33] Because the septal nucleus does not appear to contain α_2, α_3, or α_4 subunits, but does contain β_2 subunits,[34] it was suggested that a novel inhibitory receptor contains a β_2 subunit.

To explore further the possibility that [³H]MCC and [³H]MPA differ in their affinities for nAChR subtypes we examined the effect of DTT on [³H]MCC and [³H]MPA binding in various regions of calf brain (TABLE 4). With [³H]MCC as the ligand, binding was inhibited 40–50% in all brain regions except the locus coeruleus. With [³H]MPA as the ligand, the only brain region inhibited was the frontal cortex. The findings are suggestive of differences in the nAChR subtypes in the various brain regions and in the affinity of the two ligands for the subtypes.

EFFECT OF SCN-NIC ON [³H]MECAMYLAMINE BINDING

A study was undertaken to determine if SCN-nic affected [³H]mecamylamine binding to calf brain membranes. Calf brain membranes were treated with 1×10^{-5} M SCN-nic to completely inhibited [³H]MCC binding, and then assayed for [³H]mecamylamine binding. A plot of the % inhibition of binding versus concentration of unlabeled mecamylamine showed only 25% inhibition at 1×10^{-5} M and none at 1×10^{-6} M (data not shown). It was also found that [³H]mecamylamine binding was unchanged in membranes photolytically exposed to 5-azidonicotine and the other nicotine photoaffinity ligands. This finding indicates that the site of interaction of

TABLE 5. Reaction of [³H]SCN-nic Labeled Proteins from Rat Brain Cortex with Antibodies of nAChR Subunits[a]

nAChR Subunit	Reaction
α_3	+
α_4	+
α_7	+
α_8	−
β_2	−

[a]Membranes from whole rat brain were labeled with [³H]SCN-nic, separated by SDS-gel electrophoresis, transferred to nitrocellulose membranes, and exposed to 1:5000 dilution of antibodies. Immunoblots were compared with [³H]SCN-nic labeled bands from radiograms of duplicate gels. The chicken monoclonal antibodies were a gift of Jon Lindstrom.

mecamylamine with neuronal nAChRs is distinct from that of the nicotinic recognition site. Furthermore, occupancy of the recognition site by a variety of covalent nicotine ligands does not affect the interaction of mecamylamine with the nAChR receptor. Although there are limitations to the use of [³H]mecamylamine to determine the allosteric sites of nAChRs, mecamylamine and related nicotine antagonists with K_i values of 10 μM or lower appear to correlate reasonably well with their ability to antagonize the pharmacological effects of nicotine.

REACTION OF ANTIBODIES OF nAChR SUBUNITS TO [³H]SCN-NIC LABELED PROTEINS

A study was performed to determine which nAChR subunits were labeled by [³H]SCN-nic. Membranes from whole rat brain were labeled with [³H]SCN-nic, separated by acrylamide SDS-gel electrophoresis, transferred to nitrocellulose membranes, and exposed to 1:5000 dilution of the antibodies. Immunoblots obtained with various monoclonal antibodies derived from chicken nAChR subunits (gift of Jon Lindstrom) were compared with specifically [³H]SCN-nic labeled bands on radiograms of duplicate gels. A positive reaction was obtained with α_3, α_4 and α_7 subunits, whereas α_8, and β_2 subunits were negative (TABLE 5). The findings support the notion that [³H]SCN-nic has selectivity for some of the neuronal nAChR subunits.

In summary, some novel affinity and photoaffinity nicotine analogues have been utilized to examine the receptor binding characteristics of membranes prepared from various calf brain regions. A re-examination of the use of [³H]mecamylamine as a probe for the ion channel of nAChRs suggests that its usefulness is limited to analogues of mecamylamine and pempidine with K_i values in the micromolar range. Immunoblot studies with [³H]SCN-nic and the differences observed in the reactivity of the nicotine affinity ligands and the binding affinities of [³H]MCC and [³H]MPA in various calf brain regions suggest that various ligands may prove useful in characterizing nAChR subtypes.

REFERENCES

1. SCHOFIELD, P. R., M. G. DARLISON, N. FUJITA, D. R. BURT, F. A. STEPHENSON, L. M. RODRIGUEZ, J. RAMACHANDRAN, V. REALE, GLENCOURSE, P. H. SEEBURG & E. A. BARNARD. 1987. Sequence and functional expression of the GABAa receptor shows a ligand-gated receptor superfamily. Nature 228: 221–227.
2. BRISSON, A. & P. N. T. UNWIN. 1989. Quaternary structure of the acetylcholine receptor. Nature 315: 474–477.
3. NODA, M., T. TAKAHASHI, M. TANABE, S. TOYOSATO, T. KIKYOTANI, T. FURUTANI & S. NUMA. 1988. Structural homology of *Torpedo* acetylcholine receptor subunits. Nature 302: 528–532.
4. GERZANICH, V., R. ANAND & J. LINDSTROM. 1993. Ca⁺⁺ permeable a₈ AChR functionally expressed in *Xenopus* oocytes exhibits significantly different pharmacology from a7 AChR. Neurosci. Abstr. 19: 465.
5. RATHOUZ, M. M. & D. K. BERG. 1993. Calcium permeability of nicotinic receptors located primarily at synapses on neurons. Neurosci. Abstr. 19: 464.
6. TEITJE, K. N., P. S. GOLDMAN & N. M. NATHANSON. 1990. Cloning and functional analysis of a gene encoding a novel muscarinic acetycholine receptor expressed in chick heart and brain. J. Biol. Chem. 265: 2828–2834.
7. YATANI, A., J. CONDINA, A. M. BROWN & L. BIRNBAUMER. 1975. Direct activation of mammalian atrial muscarinic K channels by a human erythrocyte pertussis toxin-sensitive G-protein, G_k. Science 735: 207–211.

8. CERBAI, E., U. LOECKNER & G. ISENBERG. 1988. The α subunit of the GTP binding protein activates muscarinic potassium channels of the atrium. Science **240:** 1782–1784.

9. VAN DONGEN, A., J. CODINA, J. OLATE, R. MATTERA, R. JOHO, L. BIRNBAUMER & A. M. BROWN. 1988. Newly identified brain potassium channels gated by the guanine nucleotide binding (G) protein, Go. Science **242:** 1433–1437.

10. WADICHE, J., K. DINELY-MILLER, J. A. DANI & J. W. PATRICK. 1993. Molecular cloning, functional properties, and distribution of rat brain alpha-7: A nicotinic cation channel highly permeable to calcium. J. Neurosci. **13:** 596–604.

11. ABOOD, L. G., D. T. REYNOLDS, H. BOOTH & J. M. BIDLACK. 1981. Sites and mechanisms for nicotine's action in the brain. Neurosci. Biobehav. Rev. **5:** 479–486.

12. RATHOUZ, M. M. & D. K. BERG. 1993. Calcium permeability of nicotinic receptors located primarily at synapses on neurons. Neurosci. Abstr. **19:** 464.

13. FIEBER, L. A. & A. J. ADAMS. 1991. Acetylcholine-evoked currents in cultured neurones dissociated from rat parasympathetic cardiac ganglia. J. Physiol. (Lond.) **434:** 215–217.

14. STONE, C. A., M. L. TORCHIANA, K. L. MECKELNBERG & J. STAVORSKI. 1962. Chemistry and structure-activity relationships of mecamylamine and derivatives. J. Med. Chem. **5:** 665–686.

15. LINGLE, C. Blockade of cholinergic channels by chlorisondamine on a crustacean muscle. J. Physiol. (Lond.). **339:** 395–417.

16. CAO, W., M. J. MARKS & A. C. COLLINS. 1993. The nicotinic antagonists, mecamylamine and pempidine are noncompetitive inhibitors of brain nicotinic receptor function. Both mec and pemp noncompetitively inhibited ^{86}Rb flux in mouse brain synaptosomes. Neurosci. Abstr. **19:** 1533.

17. MARTIN, T. J., J. SUCHOCKI, E. L. MAY & B. R. MARTIN. 1990. Pharmacological evaluation of the antagonism of nicotine's central effects by mecamylamine and pempidine. J. Pharmacol. Exp. Ther. **254:** 45–51.

18. BANERJEE, S., J. S. PUNZI, K. KREILICK & L. G. ABOOD. 1990. [3]Mecamylamine binding to rat brain membranes. Biochem. Pharmacol. **40:** 205–210.

19. LONDON, E. D. & M. D. MAJEWSKI. 1989. Binding of [3]mecamylamine to the nicotinic cholinergic complex in the rat brain: Modulation by ATP. Neurosci. Abstr. **14:** 64.

20. ABOOD, L. G., X. LU & S. BANERJEEE. 1987. Receptor binding characteristics of a ^{3}H-labeled azetidine analogue of nicotine. Biochem. Pharmacol. **34:** 2337–2341.

21. KARLIN, A. 1969. Chemical modification of the active site of the acetylcholine receptor. J. Gen. Physiol. **54:** 245s.

22. KAO, P. N., A. J. DWORK, R. J. KALADNY, M. L. SILVER, J. WIDEMAN, S. STEIN & A. KARLIN. 1984. Identification of the α subunit half-cystine specifically labelled by an affinity reagent for the acetylcholine receptor binding site. J. Biol. Chem. **261:** 8085–8088.

23. GALZI, J. L., F. REVAH, D. BLACK, M. GOELDNER, C. HIRTH & J. P. CHANGEUX. 1990. Biochemistry **265:** 10430–10437.

24. DENNIS, M., J. GIRAUDAT, F. KOTZYBA-HIBERT, M. GOELDNER, C. HIRTH, J. Y. CHANG, C. LAZURE, M. CHRETIEN & J. P. CHANGEUX. 1988. Biochemistry **27:** 2346–2357.

25. CZAJKOWSKI, C., C. KAUFMANN & A. KARLIN. 1993. Negatively charged amino acid residues in the nicotinic receptor δ subunit that contribute to the binding of acetylcholine. Proc. Natl. Acad. Sci. USA **90:** 6285–6289.

26. KIM, K., N. LERNER-MARMAROSH, M. SARASWATI, A. S. KENDE & L. G. ABOOD. [3]Labeled affinity and photoaffinity nicotine analogues for probing brain nicotinic cholinergic receptors. Biochem. Pharmacol. Submitted.

27. KIM, K., N. LERNER-MARMAROSH, M. SARAWATI, A. S. KENDE & L. G. ABOOD. 1994. (R,S)-5-isothiocyanonicotine: A high affinity irreversible ligand for brain nicotinic cholinergic receptors. Biochem. Pharmacol. In press.

28. RAHN, R. F. 1992. Photochemistry of halogen pyrimidines. Photochem. Photobiol. **56:** 9–15.

29. SHAW, A. A., L. A. WAINSCHEL & M. D. SHETLAR. 1992. The photochemistry of *p*-aminobenzoic acid. Photochem. Photobiol. **55:** 647–656.

30. MIDDLETON, R. E. & J. C. COHEN. 1991. Mapping of the acetylcholine binding site of the nicotinic acetylcholine receptor: [3]nicotine as agonist photoaffinity ligand. Biochemistry **31:** 6987–6997.

31. CAPLAIN, S., A. CASTELLANO, J. P. CATTEAU & A. LABLACHE-COMBIER. 1971. Etudes photochimiques-VI. Mechanisme de la photosubstitution de la pyridine en solution. Tetrahedron **27**: 3541–3553.
32. HUBERT-BIERRE, Y., D. HERLEM & F. KHUONG-HUU. 1975. Oxydation photochimique d'amines tertiaires et d'alcaloides-VI. Oxydation photosensibilisée d'alcaloides comportant un heterocycle N-methyl (nicotine, N-methyl anabasine, aimaline). Tetrahedron **31**: 3049–3054.
33. SORENSON, E. M. & J. P. GALLAGHER. 1993. The reducing agent dithiothreitol (DTT) does not abolish the inhibitory nicotinic response recorded from rat dorsolateral septal neurons. Neurosci. Lett. **152**: 137–140.
34. WADA, E., K. WADA, J. BOULTER, E. S. DENERIS, S. HEINEMANN, J. PATRICK & L. W. SWANSON. 1989. Distribution of alpha2, alpha3, alpha4 and beta 2 neuronal nicotinic receptor subunit mRNA in the central nervous system: A hybridization histochemical study in the rat. J. Comp. Neurol. **284**: 330–334.

Binding Sites for Neurotoxins and Cholinergic Ligands in Peripheral and Neuronal Nicotinic Receptors

Studies with Synthetic Receptor Sequences[a]

BIANCA M. CONTI-FINE,[b-d] ALFRED MAELICKE,[e]
SIGRID REINHARDT-MAELICKE,[e]
VINCENT CHIAPPINELLI,[f] AND KATYA E. McLANE[g]

[b]Department of Biochemistry
University of Minnesota
1479 Gortner Avenue
St. Paul, Minnesota 55108

[c]Department of Pharmacology
University of Minnesota
435 Delaware Street
Minneapolis, Minnesota 55455

[e]Institute of Physiological Chemistry and Pathobiochemistry
Johannes-Gutenberg University Medical School
55099 Mainz, Germany

[f]Department of Pharmacology and Physiology
St. Louis University School of Medicine
St. Louis, Missouri 63104

[g]Department of Chemistry
University of Minnesota-Duluth
Duluth, Minnesota 55812

The nicotinic acetylcholine receptors (AChRs) are formed by different combinations of homologous subunits: this is the structural basis for their pharmacological heterogeneity. The AChR has two binding sites for agonists and competitive antagonists that reside primarily on the two AChR α subunits. Inasmuch as the sites are at the interface between an α subunit and a neighboring one, non-α subunits also contribute to their formation and influence the pharmacological properties of the resulting AChR (for review see ref. 1).

The α subunit of *Torpedo* AChR was first identified as contributing structural elements of cholinergic sites recognized by snake α-neurotoxins, such as α-bungarotoxin (α-BTX), because *Torpedo* AChR expressed in *Xenopus* oocytes does not bind α-BTX unless the α subunit is expressed,[2] and the denatured isolated α subunits binds [^{125}I]α-BTX,[3-5] thus indicating that one or more continuous segment(s) of its sequence contributes to the α-BTX binding site.

In this paper we focus on the studies we carried out, using a synthetic peptide

[a]The studies reported here were supported in part by the U.S. National Institute on Drug Abuse program project grants 5P01-DA05695 and 1P01-DA08131 (to B.M.C-T.).
[d]Corresponding author (previously known as Bianca M. Conti-Tronconi).

approach, on the structural elements contributed by the α subunits of different AChR isotypes to the binding sites for for α- and κ-neurotoxins from snake venoms, and for monoclonal antibodies (mAbs) able to compete with cholinergic ligands.

Using libraries of overlapping synthetic peptides corresponding to the deduced amino acid sequences of different AChR α subunits, we first identified sequence regions of different peripheral and neuronal α subunits able to bind cholinergic probes, such as α- or κ-neurotoxins or cholinergic competitive mAbs, in the absence of other surrounding structural elements (prototopes[6,7]), which are therefore likely to contribute to the formation of cholinergic binding sites. We then used single-residue substituted peptide analogues of these segments to identify structural requirements for binding of different cholinergic probes.[8–19]

These studies were carried out on different α subunit isotypes contributing to AChRs with different pharmacological properties. The results allow construction of possible models of the cholinergic sites, and shed light on the structural features underlying the pharmacological heterogeneity of peripheral and neuronal AChRs.

SUBUNIT COMPOSITION OF AChRs FROM PERIPHERAL AND NEURONAL TISSUES: DIFFERENT SUBTYPES OF α AND β SUBUNITS

AChR from peripheral tissues such as fish electric organ and vertebrate striated muscle comprise four subunits in a stoichiometry $\alpha_2\beta\gamma(\text{or }\epsilon)\delta$ (reviewed in ref. 1). Different α and β subunit isoforms have been described in $Xenopus$[20] and/or mammalian muscle,[21,22] but their functional significance is not known.

Several AChR subunits have been identified and sequenced from neurons (reviewed in ref. 1). Neuronal AChR subunits can be classified, on the basis of the sequence homology among themselves and with the subunits of peripheral tissue AChRs, as α subunits, which contain within their N-terminal extracellular segment a vicinal pair of cysteine residues, the hallmark of all AChR α subunits, and β (or non-α) subunits. Conventionally, muscle AChR subunits are indicated with the postscript 1 (α_1, β_1), neuronal subunits with postscript numbers that indicate the order in which a particular subunit was identified and sequenced. Neuronal subunits from α_2 to α_9 and from β_2 to β_5 have been described thus far (reviewed in ref. 1).

Subunits corresponding to the muscle γ, ϵ, and δ subunits have not yet been described in neuronal systems. Expression studies demonstrated that, whereas muscle AChRs contain four different subunits, neuronal AChRs may contain only two subunits, α and β, or even one subunit only (α_7) (reviewed in ref. 1).

FUNCTIONAL HETEROGENEITY IN ACh SUBTYPES INDUCED BY DIFFERENT SUBUNIT COMBINATIONS

Functional diversity conferred by different combinations of subunits was first described for the AChRs of embryonic and adult mammalian muscle (reviewed in ref. 1). Coexpression of different neuronal α and β subunits in $Xenopus$ oocytes also results in AChR having different conductance, open times and burst kinetics, and different pharmacology (reviewed in ref. 1).

The α subunit subtype is most important in determining the differential sensitivity of the resulting AChR complex to neurotoxins from invertebrates and snake venoms. Two classes of snake neurotoxins from the venoms of *Bungarus multicinctus, Bungarus flavus,* and different *Naja* species distinguish between AChR subtypes, that is, α-neurotoxins, such as α-BTX and α-najatoxin (α-NTX), and κ-neurotoxins, such as κ-bungarotoxin (κ-BTX) and κ-flavitoxin (κ-FTX) (reviewed in ref. 23) κ-BTX and α-BTX were initially regarded as specific antagonists of neuronal AChRs and muscle AChRs, respectively (reviewed in ref. 23). Later studies, however, revealed that this simple dichotomy does not hold true (reviewed in refs. 1 and 23).

α-BTX irreversibly blocks [$t_{1/2} > 200$ h] AChRs formed by the subunit combinations $\alpha_1\beta_1\gamma\delta$ and $\alpha_1\beta_2\gamma\delta$ (reviewed in ref. 1). Several neuronal AChRs comprising different α/β subunit combinations—$\alpha_2\beta_2$, $\alpha_3\beta_2$, $\alpha_4\beta_2$, and $\alpha_3\beta_4$—are insensitive to α-BTX (reviewed in refs. 1 and 23). Neuronal AChRs formed by $\alpha_3\beta_2$ and $\alpha_4\beta_2$ subunit combinations are sensitive to κ-BTX.[24] The sensitivity of the $\alpha_3\beta_2$ AChR to κ-BTX is 10-fold greater than the $\alpha_4\beta_2$ complex.[25,26] Interestingly, the $\alpha_3\beta_4$ complex is insensitive to κ-BTX,[27] indicating that the β subunit affects the pharmacological properties of the resulting AChR. The $\alpha_2\beta_2$ neuronal AChR is insensitive to both α-BTX and κ-BTX.[28]

The chicken α_7 subunit can form homomeric AChRs sensitive to α-BTX (reviewed in ref. 1). However, the α_7 subunit may also contribute to formation of α-BTX insensitive AChR in the chick sympathetic ganglia, and therefore might also form AChR complexes, perhaps involving other α subunit subtypes, with different α-BTX binding properties than the complexes formed by α_7 subunit alone.[29]

Insect neuronal AChRs are sensitive to α-BTX (reviewed in ref. 1). Locust AChRs seem to be homomeric complexes of one type of subunit, although in *Drosophila* they may comprise both α and non-α (also known as β) subunits (reviewed in ref. 1).

The α subunit present in an AChR influences the sensitivity of the resulting complex to other neurotoxins. For example, neosugarotoxin, from the Japanese ivory shell, blocks AChRs formed by the β_2 subunit in combination with the α_2, α_3 or α_4 subunit, whereas the $\alpha_1\beta_1\gamma\delta$ AChR is relatively insensitive.[25,26] In contrast, α-conotoxins from the venom of marine snails block only the $\alpha_1\beta_1\gamma\delta$ AChR.[30] Lophotoxin, a cyclic diterpene from gorgonian corals, covalently labels Tyr$_{190}$ of the *Torpedo* α subunit[31]—a residue conserved in all AChR α subunits but the neuronal α_5 subunit. As predicted from the presence of this tyrosine, AChR formed by combinations of $\alpha_1\beta_1\gamma\delta$, $\alpha_2\beta_2$, $\alpha_3\beta_3$, and $\alpha_4\beta_2$ subunits are sensitive to lophotoxin, although the $\alpha_2\beta_2$ AChR is less sensitive for reasons that remain unclear.[25] TABLE 1 summarizes the toxin sensitivity of AChRs resulting from combinations of different neuronal subunits.

USE OF THE SYNTHETIC SEQUENCE OF THE AChR α SUBUNIT FOR STRUCTURAL STUDIES OF CHOLINERGIC SITES: ADVANTAGES AND CAVEATS

We extensively used synthetic AChR sequences to investigate the structure of AChR domains which interact with the snake α- and κ-neurotoxins, and with mAbs against the cholinergic site.

Using peptides for those purposes required careful assessment of the legitimacy of those approaches, because most surface domains of proteins are formed by residues from discontinuous sequence regions, whereas synthetic peptides can be successfully used to map tridimensional protein domains only if a substantial portion of the surface domain to be studied is formed by residues within the same short sequence segment. If folded properly, the corresponding synthetic sequence may then be representative of the structure of the domain, or of part of it.

That this can commonly occur in large proteins has been proven by a study that systematically investigated the ability of mAbs against different segments of the *Torpedo* AChR α-subunit sequence to cross-react with native AChR.[32] Monoclonal antibodies raised against three particular sequence segments of the α subunit fully cross-reacted with native AChR, indicating that those sequence regions are largely exposed on the AChR surface. By analogy, one would expect that the same will apply to most large proteins.

Protein ligands recognizing surface regions that include largely exposed sequence segments should be able to interact with the corresponding synthetic sequence. Therefore, synthetic peptides may be more confidently used to study the structural

TABLE 1. Contribution of Different Subunits to the Toxin Sensitivity of Neuronal AChRs

AChR	Neurotoxin Sensitivity
α1β1γδ α1β2γδ α7	α-BTX sensitive[24,71,72]
α2β2 α3β4	α-BTX insensitive, κ-BTX insensitive[24,27,28]
α3β2 α4β2	α-BTX low sensitivity, κ-BTX sensitive[24–26]
α2β2 α3β2 α4β2	Neosugarotoxin sensitive, α-conotoxin insensitive[25,26]
α1β1γδ	Neosugarotoxin insensitive, α-conotoxin sensitive[25,26]
α1β1γδ α2β2 α3β3 α4β2	Lophotoxin sensitive[25,26]

domain interacting with protein ligands than with small ligands, because high affinity protein/protein binding generally involves large surface interactions.

Experimental strategies employing a sequence region excised from the structural context of the cognate native protein have other important caveats. First, a sequence segment containing several residues involved in formation of a binding site may fold, in the absence of the structural constraints of the native protein, in a manner incompatible for ligand interaction, leading to false negative conclusions. On the other hand, one might obtain a positive result with a peptide corresponding to a sequence which in the native protein is inaccessible to the ligand because of obstruction by surrounding residues.

The latter possibility is exemplified by the findings that although coexpression in *Xenopus* oocytes of the α_3 subunit—which contains important constituent elements of cholinergic binding sites—together with the β_2 subunit results in an AChR capable of irreversibly binding the neuronal AChR-specific antagonist κ-BTX (reviewed in refs. 1 and 23), expression of the α_3/β_4 subunit pair yields an AChR unable to bind κ-BTX,[27] or able to bind if only reversibly,[33] demonstrating that the β subunit contributes elements to the cholinergic site formation which interfere with the binding of κ-BTX to the α_3 domain. Similarly, the α_7 subunit contains a prototope for α-BTX binding,[14] and may form homomeric AChR complexes sensitive to the α-BTX (reviewed in ref. 1). However, the α_7 subunit may also be part of heteromeric AChRs that are insensitive to α-BTX.[34] Also, a prototope for α-BTX has been identified on the rat α_5 subunit,[9] which may not actually bind cholinergic ligands when part of a native neuronal AChR.[35]

Despite their potential pitfalls, studies with synthetic AChR sequences have proven to have a reliable predictive value (reviewed in ref. 36). Synthetic peptide studies of the main immunogenic region (MIR)—the set of largely overlapping epitopes on the extracellular surface of the AChR α subunit recognized by the majority of autoimmune antibodies (Abs) in the human disease myasthenia gravis,[37] correctly located the MIR to the sequence α67–76 on human muscle and *Torpedo* α subunits, as verified by comparisons of naturally occurring mutations in AChRs which do not bind anti-MIR Abs, and by the results of expression studies of mutant AChRs (reviewed in ref. 36). The sequence region that surrounds two vicinal cysteine residues, at positions approximately 192 and 193—the hallmark of AChR α subunits—has been identified as contributing to the formation of the α-BTX binding site by several studies employing synthetic and biosynthetic peptides (reviewed in ref. 1). This identification is supported by the results of studies using mutated AChRs, or cholinergic affinity ligands (reviewed in refs. 1 and 36). The results of studies using single-residue substituted peptides to identify residues within that sequence region important for AChR/α-BTX interactions,[11,18] agree well with those of *Xenopus* oocyte expression studies of mutated AChRs.[38]

Consistently, the affinity of the cholinergic probes for synthetic AChR sequences is several orders of magnitude lower than that for the native AChR. This is to be expected, because even when a short sequence region contributes several residues to a surface domain, other residues are contributed by discontinuous sequence regions (reviewed in ref. 39). That this is the case for the α- and κ-neurotoxin binding site is supported by the large toxin surface area involved in interaction with the AChR (reviewed in refs. 1 and 36) and the extremely high affinity for the AChRs of these toxins,[23] whose structural basis must be the intimate complementarity of large surface regions.

This reduced affinity of peptides for synthetic sequences excised from the context of the native protein may be beneficial when investigating individual residues involved in formation of a binding site by the use of single-residue substituted analogues: the relatively low affinity of the ligand/peptide interaction makes this a sensitive system to small changes in affinity. The same changes might be difficult to detect for substitutions of the same residues in the native AChR, given the very high binding affinity, and the stabilizing influence on the binding of the many more residues in the native AChR than in the peptide.

Solid phase assays, due to immobilization of the peptide sequence in a conformation noncompatible for ligand binding, might give falsely negative results. Therefore, it is advisable to use more than one type of solid phase assay and, whenever possible, a competition assay using peptides in solution. The latter assay, although more labor intensive than direct solid-phase binding assays, has the advantage that the peptide in solution may fold into any low energy conformation, which may include one corresponding to that in the native protein.

IDENTIFICATION OF SEQUENCE SEGMENTS CONTRIBUTING TO CHOLINERGIC SITES ON THE α SUBUNIT OF *TORPEDO* AND MAMMALIAN MUSCLE AChR

Several studies using synthetic or biosynthetic AChR sequences have indicated that the sequence region of the *Torpedo* α subunit flanking the vicinal Cys residues at position 192 and 193 forms a prototope for α-BTX (reviewed in ref. 1). We further defined structural elements of α-BTX binding site(s) on *Torpedo* AChR, by using a panel of overlapping synthetic peptides corresponding to the complete α subunit sequence as representative structural elements of the AChR.[8] We investigated whether, in addition to the sequence flanking the cysteinyl residues at positions 192 and 193, other sequence regions of the AChR α subunit could bind α-BTX, and/or several mAbs able to compete with α-BTX and with other cholinergic ligands for AChR binding. We also used overlapping peptides corresponding to the sequence segments of each *Torpedo* AChR subunit homologous to α166–203.

The mAbs used (WF6, WF5, and W2) were raised against native *Torpedo* AChR and specifically recognize the α subunit.[40,41] The binding of WF5 and W2 to *Torpedo* AChR is inhibited by all cholinergic ligands, that of WF6 by agonists, but not by low molecular weight antagonist.[40,41] The differential competition between the mAbs and cholinergic ligands, and the incomplete mutual inhibition by these mAbs for AChR binding, suggest that they bind to distinct overlapping parts of the area recognized by cholinergic ligands, and that within this area subsites may exist, recognized either by all small cholinergic ligands or by cholinergic antagonists alone. Binding subsites for different cholinergic ligands were suggested long ago, based on the results of pharmacological studies.[42]

α-BTX and WF6 bound to the synthetic sequence α181–200 and also, albeit to a lesser extent, to α55–74. The two other mAbs predominantly bound to α55–74, and to a lesser extent to α181–200. Peptides α181–200 and α55–74 both inhibited binding of [^{125}I]α-BTX to native *Torpedo* AChR. None of the peptides corresponding to sequence segments from other subunits bound α-BTX, WF6 or the other mAbs, or interfered with their binding.

Interestingly, results of studies from our and other laboratories[43,44] on the binding of α-NTX to synthetic sequences of *Torpedo* electric organ and mammalian muscle AChRs indicated that α-NTX, in spite of its high homology with α-BTX, with which it fully competes for binding to native AChR, does not recognize synthetic peptides corresponding to the sequence region flanking the cysteine residues 192–193. On the other hand, α-NTX binds effectively to the synthetic sequence α55–74 of *Torpedo* and muscle AChRs.[44] Therefore, as with the different mAbs against the cholinergic site described above, also highly homologous, mutually competitive α-neurotoxins may recognize different overlapping regions of the same binding area.

These results indicate that the cholinergic binding site is not a single narrow sequence region, but rather that two or more discontinuous sequence segments within the N-terminal extracellular region of the AChR α subunit, folded together in the native structure of the receptor, contribute to form a cholinergic binding region. Such a structural arrangement is similar to the "discontinuous epitopes" observed by X-ray diffraction studies of Ab-antigen complexes.[39] The multipoint attachments of α-BTX to the α subunit gives a structural basis for the high affinity of the α-neurotoxin/AChR interaction.

The structural characteristics of the *Torpedo* peptides α55–74 and α181–200 have been studied by circular dichroism (CD) and fluorescence spectroscopy (reviewed in refs. 1 and 11). Both peptides have a high content of β-sheet and β-turn. Differential CD-spectroscopy, in the presence and absence of α-BTX, indicates that peptides

α55–74 and α181–200 undergo structural changes upon α-BTX binding, with a net increase in the β-structure component (reviewed in refs. 1 and 11). These structural changes may reflect a mechanistic basis for the essentially irreversible inactivation of the AChR by α-BTX.

Similar studies carried out with panels of synthetic peptides corresponding to the complete α subunit sequence of mouse and human AChR demonstrated that only the sequence region flanking the vicinal cystines 192–193 forms a prototope for α-BTX.[13] Lack of detectable binding of α-BTX to peptide sequences of human and mouse AChRs homologous to the sequence *Torpedo* α55–74, which binds α-BTX, may be due to different reasons, including (1) improper folding of the peptide, different from that of the same sequence in the native AChR; (2) mutation in the mammalian muscle sequence region α55–74 of amino acid residues directly involved in α-BTX binding, and reduction of the affinity of this prototope for α-BTX to levels incompatible with detection of α-BTX binding in the assays used; and (3) lesser contribution to the structure of the α-BTX binding site by the sequence region α55–74 in mammalian muscle AChR than in *Torpedo* AChR.

IDENTIFICATION OF SEQUENCE SEGMENTS CONTRIBUTING TO CHOLINERGIC SITES ON THE α SUBUNIT OF κ-NEUROTOXIN SENSITIVE NEURONAL AChRs

Neuronal AChRs containing the α_3 subunit bind κ-neurotoxins, either irreversibly or reversibly (reviewed in refs. 1 and 33). Using overlapping peptides corresponding to the α_3 subunit sequence, we mapped a potential constituent segment of the binding sites to the sequence region $\alpha_3$51–70.[10,16] κ-BTX and κ-FTX bind to this sequence specifically; α-BTX does not bind to any α_3 peptides. The sequence $\alpha_3$51–70—which largely overlaps the homologous sequence regions of *Torpedo* AChR, α55–74, which forms a prototope for α-BTX (see above)—contains several negatively charged residues that may interact with the K and R residues present in the sequence loops of κ-BTX believed to interact with the AChR. It also contains several aromatic amino acids, which are a consistent structural feature of the α- and κ-neurotoxin binding prototopes (see below).

Two other largely overlapping peptide sequences that bind κ-BTX, $\alpha_3$180–199 and $\alpha_3$183–201, were identified using a competition assay with native neuronal AChR on PC-12 cells (most likely the $\alpha_3\beta_2$ subtype).[10] Both peptides contain the vicinal Cys pair, and are homologous to, although relatively divergent from, the muscle-type α-BTX binding sequence $\alpha_1$181–200 of different species. Therefore, in the α_3 neuronal AChRs, the sequence region surrounding the vicinal cysteines is also likely to contribute to the cholinergic binding site. An involvement of this region of the α_3 subunit in κ-BTX binding is supported by *Xenopus* oocyte expression studies using α_2/α_3 chimeras.[45]

Thus, like α-BTX, κ-neurotoxins appear to have multipoint attachments to the α_3 subunit, and the segments of the α_3 subunit contributing to these binding sites are homologous to those contributing to the α-BTX site in the *Torpedo* α subunits.

IDENTIFICATION OF SEQUENCE SEGMENTS CONTRIBUTING TO CHOLINERGIC SITES ON THE α SUBUNITS OF NEURONAL AChRs SENSITIVE TO α-BTX

We used similar experimental strategies to identify prototopes for α-BTX in two neuronal α subunits, the α_7 and α_8.[14] The cDNAs for these neuronal α subunits were isolated using oligonucleotides corresponding to the aminoterminal sequence region of an α subunit from an AChR protein(s) isolated from chick brain using α-BTX affinity chromatography.[46] The α_7 subunit forms homo-oligomeric AChRs able to bind to α-BTX (reviewed in ref. 1); the α_8 subunit is highly homologous to the α_7 subunit.[47] We synthesized a panel of overlapping synthetic peptides and tested them for ability to bind [^{125}I]α-BTX. The peptides were 20 residues long, and corresponded to the complete chick brain α_7 subunit and to residues 166–215 of the chick brain α_8 subunit.[14]

We found that the synthetic sequences $\alpha_7$181–200 and $\alpha_8$181–200 consistently and specifically bound [^{125}I]α-BTX, although with different affinities—$\alpha_7$181–200 with an apparent affinity 10 times lower than that of $\alpha_8$181–200. The ability of these peptides to bind α-BTX was significantly decreased by reduction and alkylation of the Cys residues at positions 190/191, whereas oxidation had little effect on α-BTX binding activity.

Therefore, α_7 and α_8 are ligand binding subunits, able to bind α-BTX at sites homologous with the AChR α subunits from muscle and electric tissue, and the small sequence differences between these two highly homologous subunits may confer differential ligand binding properties to the AChR subtypes of which they are components.

We have used synthetic peptides to obtain clues about the neurotoxin sensitivity of neuronal AChRs containing an α subunits not successfully expressed in *Xenopus* oocytes as functional complexes, that is, the α_5 subunit.[9] Expression of α_5 mRNA correlates with the presence of neuronal α-BTX binding AChRs in several cell lines.[48] On the other hand, the α_5 subunits may contribute, together with the α_3 and β_4 subunit, to neuronal AChRs expressed in sympathetic ganglia that are sensitive to κ-BTX and insensitive to α-BTX.[29]

Overlapping peptides corresponding to the sequence region 171–205 of the α_5 subunit, of mouse muscle α_1 and rat neuronal α_2, α_3, and α_4 subunits, which all contain the vicinal Cys residues, were compared for ability to bind α-BTX.[9] In a solid phase assay testing the direct binding of [^{125}I]α-BTX to synthetic peptides, as well as in two different competition assays in which peptides were tested for their ability to sequester [^{125}I]α-BTX from binding to native AChR on postsynaptic membrane fragments of *Torpedo* electric organ or PC-12 cells, only peptides corresponding to the mouse muscle α_1 and rat neuronal α_5 subunits bound α-BTX.[9] These results are consistent with the known pharmacology of the α_1, α_2, α_3, and α_4 AChR subtypes (reviewed in ref. 1), and suggest that the α_5 subunit could bind α-BTX, and that the sequence $\alpha_5$180–199 may contribute to an α-BTX binding site. It is unclear in view of the results summarized above, whether the α-BTX binding prototope $\alpha_5$180–199 contributes in the cognate native neuronal AChRs to a cholinergic site able to bind α-BTX, or if in such native AChRs binding of α-BTX is impeded by surrounding structural elements.

IDENTIFICATION OF INDIVIDUAL RESIDUES INTERACTING WITH α- AND κ-NEUROTOXINS BY THE USE OF SYNTHETIC SEQUENCES: GENERAL CONSIDERATIONS

Following identification of AChR α subunit prototopes for binding of α- and κ-neurotoxins and of competitive mAbs, synthetic peptide approaches have also proven useful in identifying individual amino acid residues involved in the binding of those cholinergic probes.

We used two different approaches for these purposes. In the first, binding of α-BTX to synthetic peptides corresponding to homologous prototopes of different α-BTX binding peripheral and neuronal AChR was studied to obtain clues about the effect of naturally occurring amino acid substitutions on α-BTX binding. As described below, those studies demonstrated a lack of sequence motifs identifying α-BTX binding proteins. Therefore a second approach was used in attempts to determine the structural requirements for α-BTX and κ-neurotoxin binding, that is, investigation of the effect of individual nonconservative or conservative substitutions on the binding of α-BTX or competing mAbs, or κ-neurotoxins, to the sequence regions of peripheral or neuronal AChRs found to bind these cholinergic probes. The results of these studies will be compared and, if necessary, contrasted with those of investigations employing mutation analysis of AChRs expressed in oocytes.

COMPARISON OF HOMOLOGOUS SEQUENCES α181–200 FROM MUSCLE AChR OF DIFFERENT SPECIES

Muscle AChR α subunits are highly conserved between different species. Because the sequence α181–200 in *Torpedo* electric organ, mouse and human muscle AChRs forms a prototope for α-BTX, we systematically investigated the ability to bind α-BTX and the affinity of such binding of synthetic peptides that correspond to the homologous sequence regions from the muscle of other vertebrate species, whose AChR α subunit sequence is known.[13] Included in the species studied were the sequence regions α181–200 of *Naja* and mongoose muscle AChRs. These animals are notable exceptions to the rule that α-BTX blocks muscle AChR function: several snake species, including *Naja*, are resistant to the blocking action of α-BTX,[49] and the mongoose, a mammal well known for its ability to kill and eat poisonous snakes (e.g., see ref. 50), is also resistant to snake neurotoxins.[51]

All the synthetic sequences α181–200 tested contained an α-BTX binding prototope, with the notable exceptions of the *Naja* and the mongoose sequences.[13] As illustrated in FIGURE 1, the sequence region α181–200 is very conserved in *Torpedo* and vertebrate muscle AChRs. However, this does not apply to the *Naja* and mongoose sequences. Six amino acid residues of this sequence region in the snake α subunit differ from the other α subunits, and may be important for α-BTX binding. Nonconservative substitutions in the snake α181–200 sequence include replacements of K_{185}, W_{187}, Y_{189}, and P_{194} by W, S, N, and L, respectively. Single residue mutations of the *Torpedo* α sequence to each of the six substitutions of the *Naja* α sequence demonstrated that conversion of Y_{189} to N or P_{194} to L in the *Torpedo* sequence suffices to eliminate α-BTX binding.[52] In the mongoose sequence, W_{187}, Y_{189}, and P_{194} are nonconservatively substituted to N, T, and L respectively.[49] In addition, P_{197}, conserved in all known peripheral AChR α subunits, is nonconservatively substituted to H, and at position 195 the negatively charged residue present on most α-BTX binding prototypes and proposed to be involved in interaction with α-BTX,[53] is conservatively substituted by T.[49]

Comparison of the vertebrate muscle α sequences (FIG. 1) shows that the sequence VVY at positions 188–190 is common to the peptides that bind α-BTX with high affinity (*Torpedo,* frog, chick), and that substitution of the amino acid at position 189 to F (as in the calf and mouse sequences), N (as in the *Naja* sequence) or T (as in the human and mongoose sequences) correlates with a reduced affinity of this prototope for α-BTX.[15] The convergence of these results strongly indicates Y_{189} as a critical residue in the interaction of α subunits from different muscle AChRs with α-BTX.

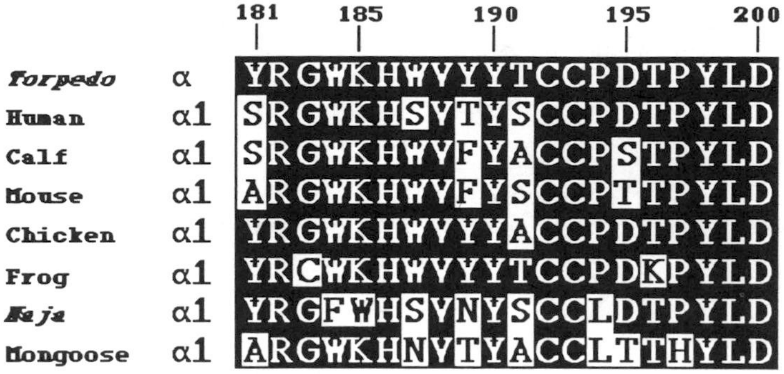

```
                  181    185     190     195     200
                   |      |       |       |       |
Torpedo    α      YR GWK HWV YY T CCP DTP YLD
Human      α1     SR GWK HSV TY S CCP DTP YLD
Calf       α1     SR GWK HWV FYA CCP STP YLD
Mouse      α1     AR GWK HWV FYS CCP TTP YLD
Chicken    α1     YR GWK HWV YYA CCP DTP YLD
Frog       α1     YR CWK HWV YYT CCP DKP YLD
Raja       α1     YR GFW HSV NYS CCL DTP YLD
Mongoose   α1     AR GWK HNV TYA CCL TTH YLD
```

FIGURE 1. Alignment of the sequence region 181–200 from the α subunit of different vertebrate muscle AChRs. This sequence region is highly conserved in most muscle α subunits. Identical residues (as compared to the *Torpedo* sequence, which is aligned at the top of the sequences) are enclosed in black boxes. The amino acid residues which are substituted, as compared to the *Torpedo* α subunit, are in white boxes. Several nonconservative substitutions within this region are present in the α subunits of cobra and mongoose muscle AChRs. This may be related to the resistance of muscle from these species to α-neurotoxin block, and the inability of synthetic peptides corresponding to the cobra and mongoose sequence region α181–200 to bind α-BTX. See text for further details. (From Conti-Tronconi *et al.*[1] Reproduced with permission of the *CRC Critical Reviews in Biochemistry and Molecular Biology.*)

COMPARISON OF HOMOLOGOUS SEQUENCES α181–200 FROM NEURONAL AChRs OF DIFFERENT SPECIES AND SUBTYPES

The sequence region surrounding the vicinal cysteines 192/193 of the neuronal α-BTX binding α subunits is highly diverged with respect to the *Torpedo* and muscle α1 subunits (FIG. 2). The low predictive value of sequence homology to infer neurotoxin sensitivity can be best appreciated by comparison of the sequence regions of AChR α subunits that bind α-BTX (*Torpedo* α, vertebrate muscle α_1, rat neuronal α_5, chick brain α_7 and α_8, and *Drosophila* ALS and SAD subunits), with the homologous sequence regions of AChRs that do not bind α-BTX (the *Naja* and mongoose muscle α_1, and the neuronal α_2, α_3, and α_4 subunits). Seven amino acid residues are characteristic of all α subunits regardless of their α-BTX binding ability: that is, G_{183} (or the conservative substitution A), Y_{190}, C_{192}, C_{193}, D_{195} (or the conservative substitution E), Y_{198}, and D_{200}. All of the α subunits in FIGURE 2 that bind α-BTX have Y_{189} (or the conservative substitution F) and P_{197}, whereas K_{189} and I_{197} are characteristic of α subunits that do not bind α-BTX. This general rule, however, does not hold true for the α_5 subunit, whose sequence is highly divergent from other AChR α subunits.

The inability to correlate critical structural features required for α-BTX binding with a particular amino acid sequence indicates in a broader sense a serious limitation to the use of sequence homology to define families of functionally and structurally related proteins. It is obvious from comparison of the α-BTX binding sequences that different primary sequences must fold into three-dimensional structures with comparable hydrophobic, hydrogen-bonding, and charge interactions. Compensatory, multiple nonconservative substitutions that occurred during the evolution of α-BTX binding proteins have obscured a "universal" α-BTX binding motif. This fact is also illustrated by the lack of sequence homology between any of the nicotinic AChR α subunits and the ACh binding sites of the muscarinic ACh receptor and acetylcholinesterase.[54,55] The failure to find a common α-BTX binding motif is similar to the search for targeting sequence signals involved in sorting proteins into different cellular compartments. In those cases, instead of primary

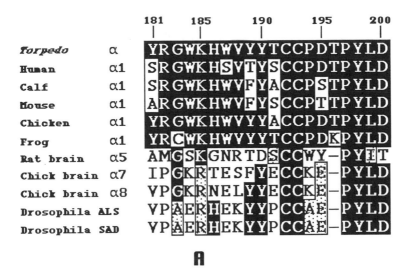

FIGURE 2. Alignment of the sequence region 181–200 from the α subunit of different muscle and neuronal AChRs. The residues are numbered with reference to the *Torpedo* α subunit sequence. Identical residues (as compared to the *Torpedo* sequence, which, for sake of comparison, is also aligned at the top of the sequences that cannot bind α-BTX) are enclosed in black boxes, conservative substitutions on a dotted background. (**A**) AChR α subunit sequences that bind α-BGT; (**B**) AChR α subunit sequences that do not bind α-BGT. See text for further details. (From Conti-Tronconi *et al.*[1] Reproduced with permission of the *CRC Critical Reviews in Biochemistry and Molecular Biology.*)

sequence conservation, compositional motifs are found, in which certain amino acids or residues with similar physical characteristics are common between proteins destined to the same cellular organelle or membrane compartment (see ref. 56).

INDIVIDUAL RESIDUES OF THE *TORPEDO* α SUBUNIT SEQUENCE INTERACTING WITH α-NEUROTOXINS: STUDIES WITH SINGLE RESIDUE SUBSTITUTED PEPTIDE ANALOGUES

We attempted to elucidate the structural requirement for ligand binding to the cholinergic subsites formed by the sequences α181–200 and α55–74 of *Torpedo* AChR by investigating the effect on the binding of α-BTX of nonconservative and conservative substitutions of individual residues within these sequence regions. For the sequence region α181–200, we also investigated the binding of mAb WF6 and for α55–74, of α-NTX.

We first used a panel of substituted peptide analogues of the *Torpedo* sequence α181–200, carrying single amino acid substitutions of glycine or alanine for each native residue.[11] Circular dichroism spectral analysis indicated that the substituted analogues had comparable structures in solution, and they could therefore be used to analyze the influence of single amino acid substitutions on ligand binding. Several peptide analogues clearly differed from the unsubstituted parental sequence in their ability to bind α-BTX or mAb WF6, or both.

Distinct clusters of amino acid residues, discontinuously positioned along the sequence α181–200, seem to serve as attachment points for the two ligands studied. The residues necessary for binding of α-BTX are different from those crucial for binding of WF6. In particular, residues at positions 188–190 (VYY) and 192–194 (CCP) were necessary for binding of α-BTX, whereas residues W_{187}, T_{191}, and Y_{198} and the three residues at positions 193–195 (CPD) were necessary for binding of WF6. Several other residues flanking the two clusters VYY and CCP also seemed to be involved in α-BTX binding (W_{184}, K_{185}, W_{187}, D_{195}, T_{196}, P_{197}, and Y_{198}).

Comparison of the CD spectra of the toxin/peptide complexes, and those obtained for the same peptides and α-BTX in solution, indicates that structural changes of the ligand(s) occur upon binding, with a net increase of the β-structure component. The increase in the order of the structure may reflect a structural rearrangement of the peptide upon binding to the high affinity α-BTX "matrix."

These results further demonstrate that within this relatively large structure, cholinergic ligands bind with multiple points of attachment, and ligand-specific patterns of the attachment points exist. This may be the molecular basis of the wide spectra of binding affinities, kinetic parameters, and pharmacological properties observed for the different cholinergic ligands.

The results described above were verified by other studies. A study using single amino acid substitutions (to G or A) of the sequence segment *Torpedo* α188–197, indicated that Y_{189}, Y_{190}, and D_{195} are important for α-BTX binding.[53] Another study, which tested the effect of multiple amino acid substitutions of the *Torpedo* sequence α166–211 expressed as a bacterial fusion protein, confirmed that Y_{189} and P_{194} are critical for α-BTX binding.[52]

Other studies investigated the binding of α-BTX and ACh to mouse muscle α subunit mutants expressed as AChR complexes in *Xenopus* oocytes.[57] Although the affinity for ACh was markedly reduced when Y_{190} was substituted to F, none of the substitution administered affected the α-BTX binding detectably. The different results of this study with respect to the ability to bind α-BTX and of studies using synthetic "mutated" sequences of *Torpedo* and mouse α subunits could be due to (1) species differences, or (2) differences in the amino acid substitutions made, or—most likely—(3) the fact that, because binding of α-BTX to native AChR occurs via large interacting surfaces, mutation of one of the several residues involved in such interaction may not suffice to change the binding affinity detectably.

To further define the structural requirements for α-BTX binding to the prototope α181–200 of *Torpedo* AChR, we investigated the effect on α-BTX binding to this synthetic sequence of conservative substitutions of residues that previous studies had indicated as important for binding of α-BTX.[18] Amino acid substitutions for this mutational analysis were chosen in order to determine (1) physico-chemical attributes of the amino acid side chains that mediate α-BTX binding (i.e., steric,

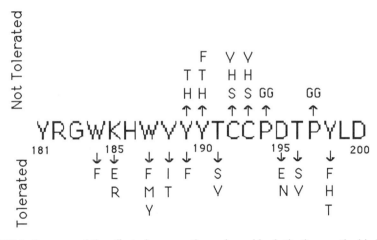

FIGURE 3. Summary of the effect of conservative amino acid substitutions on the binding of [^{125}I]α-BTX to *Torpedo* AChR α181–200. Consistent results obtained for both the solution and the solid phase assays are summarized. The native sequence of *Torpedo* α181–200 is indicated at the center of the figure, and the residue position numbers indicated under amino acids are relative to the *Torpedo* α subunit. The amino acid substitutions that are tolerated and result in equal or slightly lower α-BTX binding activity are indicated by arrows from below the positions of the native sequence, which are replaced. Similarly, those substitutions that are not tolerated and abolish α-BTX binding activity of the peptides are indicated by arrows above the native sequence. (From McLane *et al.*[18] Reproduced with permission of *Biochemistry*.)

hydrophobic, hydrogen-bonding, and/or electrostatic interactions) and (2) how the secondary structural constraints imposed by the presence of prolyl residues or a disulfide bond confer α-BTX binding activity to this prototope.

The conclusions of those studies,[18] based on consensus of solid and solution phase assays, are summarized in FIGURE 3. Amino acid substitutions that had a profound effect on α-BTX binding are indicated above the native *Torpedo* α181–200 sequence (Not Tolerated), and those that slightly or moderately affected α-BTX binding are indicated below the native sequence (Tolerated). Conservative substitutions of Y_{189}, Y_{190}, C_{192}, and C_{193} abolished or strongly affected α-BTX binding, whereas V_{188} could be conservatively replaced by I or T with minor effects on α-BTX binding. We previously suggested that these residues form important contacts with α-BTX, based on the results obtained when these residues were substituted with G.[11] Structural changes of the peptide α181–200 induced by substitution of P_{194} or P_{197} with two adjacent G residues, or insertion of a G between C_{192} and C_{193}, were also incompatible with α-BTX binding. Conservative substitutions of other aliphatic and aromatic residues, and of residues K_{185} and D_{195}, had minor effects on α-BTX binding.

Therefore, binding of α-BTX to the prototope α181–200 involves important interactions with Y_{189}, Y_{190}, C_{192}, and C_{193} that are highly specific to the amino acid residues at these positions. Residues Y_{190}, C_{192}, and C_{193} are highly conserved, because the two vicinal C residues are the hallmark of the α subunits, and Y_{190} is found in all but the $α_5$ α subunits (reviewed in ref. 1). However, these three residues *per se* are not correlated with α-BTX binding activity, because they are conserved in α subunits that do not bind α-BTX (reviewed in ref. 1). The presence of Y_{189}, on the other hand, correlates with high-affinity α-BTX binding (reviewed in refs. 1 and 23). P_{184} and P_{187} play important structural roles in maintaining the correct conformation of the peptide to display the α-BTX binding motif.

α-BTX is believed to contact the AChR through a region of extensive β-sheet that forms a carapace, excluding water from the AChR/α-BTX interface.[58–60] It can be expected that the α-BTX binding site on the AChR itself will also be formed by a β-sheet. Both protopes α181–200 and α55–74 spontaneously fold in solution to form a β-sheet.[1,11] The results of our study using conservatively substituted analogues of the prototope α180–200 further support the possibility that a β-sheet folding of this sequence is necessary for α-BTX binding, because substitution of either C_{192} or C_{193} by residues with side chains of similar size (S and V) is not tolerated, nor is substitution by H, which shares polar and hydrophobic properties. C_{192} and C_{193} form a highly reactive vicinal disulfide in the native AChR (reviewed in ref. 1) which, however, is not required for α-BTX binding, either to native *Torpedo* AChR or to the synthetic sequence α181–200 (reviewed in ref. 1). Therefore, the important attributes of cysteinyl residues at these positions do not involve a propensity to form a vicinal disulfide, but rather are a function of the size and hydrophobic nature of the cysteinyl side chains.

Structurally, the importance of the C/C pair for α-BTX binding is also demonstrated by the inability to bind α-BTX of a peptide analogue of α180–200, where a glycine was inserted between C_{192} and C_{193}. The introduction of any conformational flexibility within this region is deleterious for α-BTX binding, as shown by the inability to bind α-BTX of peptide analogues of α180–200, where two glycines were substituted for either P_{184} or P_{187}.

Both vicinal disulfides and proline imides have a propensity to form nonplanar, *cis* peptide bonds.[61,62] Two adjacent *cis* peptide bonds could act together to cause a turn in the peptide backbone, providing an important element of secondary structure for α-BTX activity of this prototope. Molecular modeling of this region indicates that the vicinal disulfide, P_{194} and P_{197}, are likely to induce a β-turn.[63,64] This would allow the β-strand formed by this peptide to fold-back on itself and stabilize this conformation by β-sheet formation.[62] Such structure would complement the β-sheet folding of the binding area of the α-BTX molecule.

The guanidium group of R_{37} on α-BTX, common to all α-neurotoxins, may be analogous to the quaternary ammonium group of ACh and other nicotinic agonists and antagonists.[65] A complementary negative subsite on the α subunit of the AChR was proposed to involve D_{195},[66] but affinity labeling studies have indicated that this residue is not involved in cation stabilization of acetylcholine.[67] Our studies also indicated that D_{195} can be substituted by N with little effect on α-BTX binding.

An alternative model for a complementary binding subsite on the AChR α subunit for the quaternary ammonium of ACh is suggested by two other systems that indicate that cation stabilization can be provided by π-electrons from aromatic rings: (1) a synthetic ACh "receptor" composed entirely of aromatic rings[68] and (2) the structure of acetylcholinesterase, whose active site is at the bottom of a gorge lined with aromatic amino acids.[69] Thus, the AChR subsite required for cation stabilization might be provided by the π-electrons present in the numerous aromatic residues of both the prototopes $\alpha 181$–200 and $\alpha 55$–74 (see below). Many of those aromatic residues are crucial or important for α-BTX binding. Also, in neuronal AChR α subunits, the prototopes for α- and κ-neurotoxins are rich in aromatic residues, whose substitution is frequently poorly or not tolerated.

We used a similar approach to investigate the residues within the sequence *Torpedo* $\alpha 55$–74 involved in the binding of α-BTX and α-NTX.[17] This sequence region overlaps the sequence segment $\alpha 67$–76, which has been shown to contribute important structural elements of the MIR (reviewed in ref. 1). However, α-BTX and anti-MIR mAbs bind to the AChR in a nonmutually exclusive way, suggesting that the two sides do not overlap. We used a panel of single residue substituted analogues of the sequence T$\alpha 55$–74 to identify the residues involved in α-neurotoxin binding and those involved in the binding of anti-MIR mAbs. Binding of α-BTX and α-NTX was similarly affected by substitutions within this sequence. The overlap between the residues important for α-neurotoxin binding and those involved in anti-MIR Ab binding was minimal. Substitution of several positively charged or aromatic residues strongly inhibited α-BTX binding; of those, only one (W_{60}) significantly reduced the binding of two anti-MIR mAbs. Only substitution of residue N_{68} strongly reduced the binding of both α-BTX and anti-MIR mAbs. These results are consistent with a model in which the MIR and the α-neurotoxin binding site, although within the same large surface area of the native AChR and very close to each other, have minimal overlap, and in which the α-neurotoxin binding site is rich in aromatic residues (see above).

RESIDUES OF THE NEURONAL α_5 SUBUNIT SEQUENCE INTERACTING WITH α-BTX: STUDIES WITH SINGLE RESIDUE SUBSTITUTED PEPTIDE ANALOGUES

The α-BTX binding sequence region 180–200 of the rat α_5 subunit is relatively divergent as compared with the homologous sequence regions of *Torpedo* and muscle AChRs. Amino acid residues critical for α-BTX binding were identified by testing the effects of single amino acid substitutions to G or A for each residue of the rat $\alpha_5 180$–199 sequence on binding of α-BTX to the substituted peptide analogues.[15] Substitutions of four residues (K_{184}, R_{187}, C_{191}, and P_{195}) abolished α-BTX binding; other substitutions (G_{185}, N_{186}, D_{189}, W_{193}, Y_{194}, and Y_{196}) lowered its affinity. On the other hand, substitutions of C_{192} (homologous to C_{193} of *Torpedo* AChR α subunit) did not affect the ability of the α_5 prototope to bind α-BTX.

The lack of effect of substitution of one of the vicinal cysteines underlines that, as discussed above, a disulphide bridge between them is not crucial for α-BTX binding. Presumably, in the α_5 prototope the several nonconservative changes, as compared to the homologous *Torpedo* and muscle prototopes, stabilize its β-sheet conformation, even in the absence of the stabilizing action of a planar *cis*-disulphide bridge.

The importance of several aromatic amino acids for α-BTX binding to the α_5 peptide is analogous to the findings reported above for the *Torpedo* α180–200 sequence. Thus, despite the apparent divergence of the α_5 sequence from other α-BTX binding α subunits, some structural features, such as an abundance of aromatic residues and amino acids able to contribute electrostatic and/or hydrogen bond interactions, have been conserved.

RESIDUES OF THE NEURONAL α_3 SUBUNIT SEQUENCE INTERACTING WITH κ-NEUROTOXINS: STUDIES WITH SINGLE RESIDUE SUBSTITUTED PEPTIDE ANALOGUES

Synthetic peptide analogues of the sequence $\alpha_3$50–71, in which each amino acid was sequentially replaced by a glycine, were used to identify the amino acid side chains involved in the interaction of this prototope with κ-neurotoxins by testing their ability to bind $[^{125}I]$κ-BTX and $[^{125}I]$κ-FTX.[12,16] No single substitution obliterated κ-BTX binding, but several substitutions lowered the affinity of this peptide for κ-BTX—two negatively charged residues (E_{51} and D_{62}) and several aliphatic and aromatic residues (L_{54}, L_{56}, and Y_{63}).[12]

$[^{125}I]$κ-FTX binding was more sensitive to amino acid substitutions than that of $[^{125}I]$κ-BTX.[16] Similar to κ-BTX, aliphatic and aromatic amino acid residues were important for κ-FTX binding (L_{54}, L_{56}, and Y_{63}, also involved in κ-BTX binding, and additional W residues at positions 55, 60 and 67). In contrast to κ-BTX, however, positively rather than negatively charged amino acids appeared to mediate electrostatic interactions with κ-FTX—K residues at positions 57, 64, 66, and 68.

These differences in amino acid specificity can be correlated with sequence differences of κ-BTX and κ-FTX, and provide clues as to the reason for these different charge requirements and the residue interactions at the κ-neurotoxin subunit interface.

κ-FTX and κ-BTX are both highly basic (pI 8.8 and 9.1, respectively) and share 82% sequence identity: the different residues (12 out of 66) are clustered in two sequence regions believed to interact with the AChR, region I (positions 23–33) and region II (positions 43–54).[70] Region I of the homologous α-neurotoxins interacts with the peripheral AChR α subunit.[58–61] Within region I of the κ-neurotoxins, the amino acid substitutions are primarily conservative and do not result in differences in charge distribution. This region might determine the overall binding specificity of κ-FTX and κ-BTX for the sequence segment $\alpha_3$51–70. The amino acid substitutions of region II result in a different overall charge and differences in the spatial arrangement of charged groups. E48 of κ-FTX, which is Q in κ-BTX, may determine the relative importance for positively charged groups of the $\alpha_3$51–70 sequence for binding. The different spatial arrangements of an arginine residue (positions 50 and 52 of κ-BTX and κ-FTX, respectively) may account for the unique sensitivity of κ-BTX binding to substitution of negatively charged residues in the peptide $\alpha_3$51–70.

A common requirement for the binding of both κ-neurotoxins is the presence of several aliphatic and aromatic residues. These are features similar to those identified as structural requirements for the binding of α-neurotoxins to their AChR prototopes, supporting the notion that similar mechanisms may apply to the interaction of the different AChR isotypes with α- or κ-neurotoxins, and, therefore, that the cholinergic binding site may have similar structure in all AChRs.

SUMMARY OF THE RESULTS OF STUDIES ON THE STRUCTURE OF CHOLINERGIC BINDING SITES BY USE OF SYNTHETIC OR BIOSYNTHETIC PEPTIDES

The studies summarized in the previous sections allow the following conclusions:

1. At least two sequence segments of the α subunits, which always include the segment containing the vicinal cysteines, contribute to form the cholinergic binding sites recognized by α- and κ-neurotoxins. These sites therefore are complex surface areas, formed by clusters of amino acid residues from different sequence regions.
2. The sequence segments contributing to the cholinergic site are in similar positions along the α subunit sequences, suggesting that the extracellular domain of all α subunits folds in a similar manner.
3. The sequence region α180–200 is very conserved, and well defined clusters of residues surrounding and including the residues Cys_{192} and Cys_{193} are involved in interaction with α-BTX. The homologous sequence region is not well conserved in neuronal AChRs that bind α-BTX, and the residues identified as crucial for interaction with α-BTX are at positions different from those of peripheral AChRs. Therefore, there is no universal sequence motif with predictive value for α-BTX binding, and multiple, nonconservative substitutions in these sequence regions during evolution of the AChR proteins have both obscured the original ancestral sequence and reestablished, as a result of new mutual interactions, a structure compatible with α-BTX binding.
4. Although Cys_{192}/Cys_{193} are involved in forming the toxin/α-subunit interface, a vicinal disulfide bond is not required for α-BTX binding.
5. Within the relatively large area of the cholinergic site, cholinergic ligands bind with multiple points of attachment, and ligand-specific patterns of attachment points exist. This may be the molecular basis of the broad spectra of binding affinities, kinetic parameters, and pharmacological properties observed for the different cholinergic ligands.
6. The sequence regions α181–200 and α50–75 are unusually rich in aromatic residues, whose substitution frequently affects α- and κ-neurotoxin binding. These findings suggest that the anionic cholinergic binding site of the AChR is formed not by a single negatively charged residue, but rather by interaction of the π-electrons of aromatic rings, as demonstrated for the cholinergic site of acetylcholinesterase.

CONCLUSIONS

The structural requirements for α- and κ-neurotoxin binding to peripheral AChRs, as predicted from the x-ray structure of α-neurotoxins, and the deduced structure of the homologous κ-neurotoxin, include that the toxin/receptor interface is a large area composed of residues that are hydrophobic or aromatic. These structural predictions have been born out by studies employing synthetic peptides. Qualitatively similar results have been obtained by studying the binding of α- and κ-neurotoxin prototopes, supporting the notion that the cholinergic binding site may have similar structure in all AChRs.

We have summarized previously the several caveats of experimental approaches which employ synthetic sequences excised from the structural context of the cognate protein, as representative structural elements of the native protein. Great caution must be exercised in extrapolating the results of low-affinity binding to synthetic sequences to the very high-affinity binding sites of the native, heterooligomeric complexes of the intact AChR. The relative contribution to binding of α- or κ-neurotoxins of an individual sequence segment observed in our experiments using peptides may not be representative of the accessibility or conformation of that sequence in the corresponding native AChR, or of its actual importance for high-affinity toxin binding to native AChRs. On the other hand, identification of potential sequence regions that might contribute to the cholinergic binding site on different AChR subtypes will make it possible to use targeted mutagenesis and expression of functional AChRs, to test the actual importance of these sequence regions in formation of the high-affinity sites in the intact receptor. This sort of information could ultimately be used for the rational design by protein engineering of peptide toxins specific for a given AChR subtype, which are not naturally occurring.

[NOTE: For recent reviews published by our group on these matters, see references 1 and 36; also, McLane, K. E., S. J. M. Dunn, A. A. Manfredi, B. M. Conti-Tronconi & M. A. Raftery, 1994, "The Nicotinic Acetylcholine Receptor as a Model of a Superfamily of Ligand-Gated Ion Channel Proteins," in Handbook for Protein and Peptide Design, P. R. Carey, Ed., Academic Press, Orlando, FL.

Portions of this article have been reprinted, by permission of CRC Press, Boca Raton, Florida, from "The Nicotinic Acetylcholine Receptor: Structure and Autoimmune Pathology" by B. M. Conti-Tronconi, K. E. McLane, M. A. Raftery, S. A. Grando, and M. P. Protti, published in the CRC Critical Reviews in Biochemistry and Molecular Biology, 1994, Vol. 29: 69–123; and by permission of Academic Press, Inc. from "Use of Synthetic Sequences of the Nicotinic Acetylcholine Receptor to Identify Structural Determinants of Binding Sites for Neurotoxins and Antibodies to the Main Immunogenic Region" by K. E. McLane, J. L. Wahlsten, and B. M. Conti-Tronconi, published in Methods: A Companion to Methods Enzymology, 1993, Vol. 5: 201–211.]

REFERENCES

1. CONTI-TRONCONI, B. M., K. E. MCLANE, M. A. RAFTERY, S. GRANDO & M. P. PROTTI. 1994. CRC Crit. Rev. Biochem. Mol. Biol. **29:** 69–123.
2. MISHINA, M., T. KUROSAKI, T. TOBIMATSU, Y. MORIMOTO, M. NODA, T. YAMAMOTO, M. TERAO, J. LINDSTROM, T. TAKAHASHI, M. KUNO & S. NUMA. 1984. Nature **307:** 604–608.
3. HAGGERTY, J. G. & S. C. FROEHNER. 1981. J. Biol. Chem. **256:** 8294–8297.
4. TZARTOS, S. J. & J.-P. CHANGEUX. 1983. EMBO J. **2:** 381–387.
5. GERSHONI, J. M., E. HAWROT & T. L. LENTZ. 1983. Proc. Natl. Acad. Sci. USA **80:** 4973–4977.
6. HOUSE, C. & B. E. KEMP. 1987. Science **238:** 1726–1728.
7. WILSON, P. T., E. HAWROT & T. L. LENTZ. 1988. Mol. Pharmacol. **34:** 643–651.
8. CONTI-TRONCONI, B. M., F. TANG, B. M. DIETHELM, S. R. SPENCER, S. REINHARDT-MAELICKE & A. MAELICKE. 1990. Biochemistry **29:** 6221–6230.
9. MCLANE, K. E., X. WU & B. M. CONTI-TRONCONI. 1990. J. Biol. Chem. **265:** 9816–9824.
10. MCLANE, K. E., F. TANG & B. M. CONTI-TRONCONI. 1990. J. Biol. Chem. **265:** 1537–1544.
11. CONTI-TRONCONI, B. M., B. M. DIETHELM, X. WU, F. TANG, A. BERTAZZON & A. MAELICKE. 1991. Biochemistry **30:** 2575–2584.
12. MCLANE, K. E., X. WU & B. M. CONTI-TRONCONI. 1991. Biochem. Biophys. Res. Commun. **176:** 11–18.
13. MCLANE, K. E., X. WU, B. M. DIETHELM & B. M. CONTI-TRONCONI. 1991. Biochemistry **30:** 4925–4934.

14. McLane, K. E., X. Wu, R. Schoepfer, J. M. Lindstrom & B. M. Conti-Tronconi. 1991. J. Biol. Chem. 266: 15230–15239.
15. McLane, K. E., X. Wu & B. M. Conti-Tronconi. 1991d. Biochemistry 30: 10730–10738.
16. McLane, K. E., W. R. Weaver, S. Lei, V. A. Chiappinelli & B. M. Conti-Tronconi. 1993. Biochemistry 32: 6988–6994.
17. Wahlsten, J. L., J. M. Lindstrom & B. M. Conti-Tronconi. 1993. J. Recept. Res. 13: 989–1008.
18. McLane, K. E., X. Wu & B. M. Conti-Tronconi. 1994. Biochemistry 33: 2576–2585.
19. Schröder, B., S. Reinhardt-Maelicke, A. Schrattenholz, K. E. McLane, A. Kretschmer, B. M. Conti-Tronconi & A. Maelicke. 1994. J. Biol. Chem. 269: 10407–10416.
20. Hartman, D. S. & T. Claudio. 1990. Nature 343: 372–373.
21. Beeson, D., A. Morris, A. Vincent & J. Newsom-Davis. 1990. EMBO J. 9: 2101–2106.
22. Goldman, D. & K. Tanai. 1989. Nucleic Acids Res. 25: 3049–3056.
23. Chiappinelli, V. A. 1993. *In* Natural and Synthetic Neurotoxins. A. L. Harvey, Ed.: 65–128. Academic Press. London.
24. Deneris, E. S., J. Connolly, J. Boulter, E. Wada, K. Wada, L. W. Swanson, J. Patrick & S. Heinemann. 1988. Neuron 1: 45–54.
25. Luetje, C. W., K. Wada, S. Rogers, S. N. Abramson, K. Tsuji, S. Heinemann & J. Patrick. 1990. J. Neurochem. 55: 632–640.
26. Luetje, C. W., J. Patrick & P. Seguela. 1990. FASEB J. 4: 2753–2760.
27. Duvoisin, R. M., E. S. Deneris, J. Patrick & S. Heinemann. 1989. Neuron 3: 487–496.
28. Wada, K., M. Ballivet, J. Boulter, J. Connolly, E. Wada, E. S. Deneris, L. W. Swanson, S. Heinemann & J. Patrick. 1988. Science 240: 330–334.
29. Vernallis, A. B., W. G. Conroy & D. K. Berg. 1993. Neuron 10: 451–464.
30. Gray, W. R., B. M. Olivera & L. J. Cruz. 1988. Annu. Rev. Biochem. 57: 665–700.
31. Abramson, S. N., Y. Li, P. Culver & P. Taylor. 1989. J. Biol. Chem. 264: 12666–12672.
32. Lei, S., M. A. Raftery & B. M. Conti-Tronconi. 1993. Biochemistry 32: 91–100.
33. Papke, R. L., R. M. Duvoisin & S. F. Heinemann. 1993. Proc. R. Soc. Lond. 252: 141–148.
34. Listerud, M., A. B. Brussaard, P. Devay, D. R. Colman & L. W. Role. 1991. Science 254: 1518–1521.
35. Conroy, W. G., A. B. Vernallis & D. K. Berg. 1992. Neuron 9: 679–691.
36. McLane, K. E., J. L. Wahlsten & B. M. Conti-Tronconi. 1993. Methods: A Companion to Methods Enzymol. 5: 201–211.
37. Tzartos, S. J., H. V. Loutrari, F. Tang, A. Kokla, S. L. Walgrave, R. P. Milius & B. M. Conti-Tronconi. 1990. J. Neurochem. 54: 51–61.
38. Tomaselli, G. F., J. T. McLaughlin, M. E. Jurman, E. Hawrot & G. Yellen. 1991. Biophys. J. 60: 721–727.
39. Davies, D. R., S. Sheriff & E. A. Padlan. 1988. J. Biol. Chem. 263: 10541–10544.
40. Watters, D. & A. Maelicke. 1983. Biochemistry 22: 1811–1819.
41. Fels, G., R. Plumer-Wilk, H. Schreiber & A. Maelicke. 1986. J. Biol. Chem. 261: 15746–15754.
42. Beers & Reich. 1970. Nature 228: 917–922.
43. Gotti, C., G. Mazzola, R. Longhi, D. Fornasari & F. Clementi. 1987. Neurosci. Lett. 82: 113–119.
44. Wahlsten, J., M. Raftery & B. M. Conti-Fine. Unpublished data.
45. Luetje, C. W., M. Plattoni & J. Patrick. 1992. Mol. Pharmacol. 44: 657–666.
46. Conti-Tronconi, B. M., S. M. J. Dunn, E. A. Barnard, J. O. Dolly, F. A. Lai, N. Ray & M. A. Raftery. 1985. Proc. Natl. Acad. Sci. USA 82: 5208–5212.
47. Schoepfer, R., W. G. Conroy, P. Whiting, M. Gore & J. Lindstrom. 1990. Neuron 4: 35–48.
48. Chini, B., F. Clementi, N. Hukovic & E. Sher. 1992. Proc. Natl. Acad. Sci. USA 89: 1572–1576.
49. Fuchs, S., D. Barchan, S. Kachalsky, D. Neumann, M. Aladjem, Z. Vogel, M. Ovadia & E. Kochva. 1993. Ann. N.Y. Acad. Sci. 681: 126–139.
50. Kipling, R. 1895. The Jungle Book. Harper and Brothers. New York.

51. OVADIA, M. & E. KOCHVA. 1977. Toxicon **15:** 541–548.
52. CHATURVEDI, V., D. L. DONNELLY-ROBERTS & T. L. LENTZ. 1992. Biochemistry **31:** 1370–1375.
53. TZARTOS, S. J. & M. S. REMOUNDOS. 1990. J. Biol. Chem. **265:** 21462–21467.
54. SCHUMACHER, M., S. CAMP, Y. MAULET, M. NEWTON, K. MACHPHEE-QUIGLEY, S. S. TAYLOR, T. FRIEDMAN & P. TAYLOR. 1986. Nature (Lond.) **319:** 407–409.
55. HULME, E. C., N. J. M. BIRDSALL & N. J. BUCKLEY. 1990. Annu. Rev. Pharmacol. Toxicol. **30:** 633–673.
56. DICE, J. F. 1990. Trends Biochem. Sci. **15:** 305–309.
57. TOMMASELLI, G. F., J. T. MCLAUGHLIN, M. E. JURMAN, E. HAWROT & G. YELLEN. 1991. Biophys. J. **60:** 721–724.
58. LOVE, R. A. & R. M. STROUD. 1986. Protein Eng. **1:** 37–46.
59. BASUS, V. J. & R. M. SCHEEK. 1988. Biochemistry **27:** 2772–2775.
60. BASUS, V. J., M. BILLETER, R. A. LOVE, R. M. STROUD & I. D. KUNTZ. 1988. Biochemistry **27:** 2763–2771.
61. STEWART, D. E., A. ARKAR & J. E. WAMPLER. 1990. J. Mol. Biol. **214:** 253–260.
62. CREIGHTON, T. E. 1993. Proteins: Structure and Molecular Properties, 2nd edit. Freeman. New York.
63. CHATUVERDI, V., D. L. DONNELLY-ROBERTS & T. L. LENTZ. 1993. Biochemistry **32:** 9570–9576.
64. WILMOT, C. M. & J. M. THORNTON. 1988. J. Mol. Biol. **203:** 221–232.
65. KARLSSON, E. 1979. Handb. Exp. Pharmacol. **52:** 159–212.
66. OHANA, B. & J. M. GERSHONI. 1990. Biochemistry **29:** 6409–6415.
67. COHEN, J. B., S. D. SHARP & W. S. LIU. 1991. J. Biol. Chem. **266:** 23364–23365.
68. DOUGHERTY, D. A. & D. A. STAUFFER. 1990. Science **250:** 1558–1560.
69. SUSSMAN, J. L., M. HAREL, F. FROLOW, C. OEFNER, A. GOLDMAN, L. TOKER & I. SILMAN. 1991. Science **253:** 872–879.
70. GRANT, G. A., M. W. FRAZIER & V. A. CHIAPPINELLI. 1988. Biochemistry **27:** 3794–3798.
71. COUTURIER, S., D. BERTRAND, J.-M. MATTER, M.-C. HERNANDEZ, S. BERTRAND, N. MILLAR, S. VALER, T. BARKAS & M. BALLIVET. 1990. Neuron **5:** 847–856.
72. BERTRAND, D., A. DEVILLERS-THIERY, F. REVAH, J.-L. GALZI, N. HUSSY, C. MULLE, S. BERTRAND, M. BALLIVET & J.-P CHANGEUX. 1992. Proc. Natl. Acad. Sci. USA **89:** 1261–1265.

Diversity and Patterns of Regulation of Nicotinic Receptor Subtypes[a]

RONALD J. LUKAS

Division of Neurobiology
Barrow Neurological Institute
350 West Thomas Road
Phoenix, Arizona 85013

EVIDENCE FOR nAChR SUBTYPE AND SUBUNIT DIVERSITY

Nicotinic acetylcholine receptors (nAChR) are now known to exist as a heterogenous group of macromolecules.[1-4] In retrospect, it is surprising that the concept of nAChR diversity was not more rapidly and widely accepted. Comparative pharmacological studies dating from the times of Dale, Paton, and their contemporaries clearly indicated differences in nicotinic drug actions at the vertebrate neuromuscular junction and at autonomic ganglia.[5-7] Other pharmacological studies also suggested that nAChR subtypes distinct from those found in muscle and ganglia exist in the central nervous system (CNS) as well, based on differing sensitivity to small nicotinic agonists or antagonists.[8-9] More recent studies extended those observations to different classes of neurotoxins, such as the bungarotoxins and neosurugatoxin, and to more reduced preparations.[2,4,10-12] This evidence for nAChR functional diversity in muscle, ganglia, and brain can only be explained—given that possible complications of pharmacokinetics are excluded—if there is structural diversity of nAChR.

Following the success in elucidation of the structure of muscle-type nAChR from the electric organ and the demonstration that muscle-type nAChR are composed of homologous, but distinct, subunits,[13-14] studies employing protein chemical, radioligand binding, and immunochemical techniques also demonstrated the existence of structurally distinct nAChR subtypes.[2-4] About the time that pharmacological features were being assigned to these distinct entities on the basis of functional and radioligand binding studies, the application of recombinant DNA techniques revealed the existence of a family of homologous, but genetically distinct, nAChR subunits that were expressed not in muscle, but in the nervous system.[2-4,15] Currently, at least 15 different nAChR subunit genes, including 5 expressed in muscle and 10 "neuronal" nAChR subunits, have been identified. Those neuronal nAChR subunits that have tandem cysteine residues near the putative nicotinic ligand-binding active site (i.e., in which features of the ligand-binding domain of the $\alpha 1$ subunit are preserved) are defined as α subunits, whereas those neuronal nAChR subunits that lack the tandem cysteine residues but that retain other features of muscle nAChR subunits are defined as non-α or, more popularly, β subunits. Tissue- and/or brain region-specific expression of some of these genes restricts possible assignments to

[a]This work was supported, at different stages, by grants from the National Institute on Drug Abuse (DA07319), the National Institute of Neurological Disorders and Stroke (NS16821), the Smokeless Tobacco Research Council (0277-01), the Arizona Disease Control Research Commission (82-1-098), the American Parkinson Disease Association, and the Council for Tobacco Research-U.S.A. (1683A), and by faculty endowment and laboratory capitalization funds from the Men's and Women's Boards of the Barrow Neurological Foundation.

identified nAChR subtypes, but work continues to elucidate the rules that define which subunits combine to form unique nAChR subtypes. Nevertheless, it is clear that diversity in nAChR subunit genes provides at least a partial basis for nAChR subtype diversity.

RELATIONSHIPS BETWEEN DIVERSE nAChR SUBUNITS

Alignment and analysis of some of the deduced amino acid sequences of rat nAChR subunits provide interesting perspectives on possible patterns of nAChR subunit evolution that are presently consistent with known properties and distributions of the products of nAChR subunit genes (FIG. 1, TABLE 1). "Structural" subunits of rat muscle nAChR, which include β1, δ, fetal γ, and adult ε subunits,[14] form a branch in the nAChR subunit phylogenetic tree that appears to have diverged early in evolution from a separate branch that contains the "ligand-binding" α subunits. The fetal γ/adult ε subunit pair shares (a perhaps surprisingly low) 52% amino acid sequence identity, and the δ subunit is 42/47% identical to the γ/ε subunit. However, no more than 40% sequence identity exists between these subunits and the β1 subunit or between β1, γ, δ, or ε subunits and any of the other subunits in the ligand-binding branch.

The α7 subunit, which is a constituent of at least some of the neuronal/nicotinic α-bungarotoxin binding sites of the CNS and probably also of the autonomic nervous system (ANS),[10,16–18] is the most distant of any of the members of the nAChR subunit

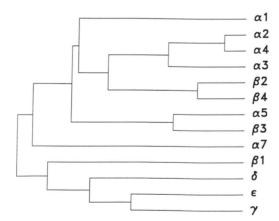

FIGURE 1. Dendrogram or phylogenetic tree diagram showing relationships between rat nAChR subunit amino acid sequences. Amino acid sequences for rat nAChR subunits were aligned and analyzed for sequence similarity according to the PCGENE program CLUSTAL using a k-tuple value of 1, a gap penalty of 5, a window size of 10, a filtering level of 2.5, and open gap and unit gap costs of 10. The program fit the indicated sequences to a reference sequence of 677 amino acids. Branch points farthest to the right indicate subunits with highest degrees of identity or identity plus similarity (65 and 73%, respectively, for α2 and α4), whereas branch points to the left on the diagram indicate lower degrees of identity or identity plus similarity (e.g., 32 and 47% for β1, γ, δ, and ε globally versus the other subunits). Although isolation of a rat nAChR α6 subunit gene has been reported, we did not have access to the amino acid sequence, and whereas a chick nAChR α8 subunit gene has been identified, no rat homolog has been described.

TABLE 1. Percentage Identity/Identity Plus Similarity for Rat nAChR Subunits[a]

Subunit	α1	α2	α3	α4	α5	α7	β1	β2	β3	β4	γ	δ	ε
α1		46/59	51/65	51/64	42/67	38/52	35/49	43/59	42/58	40/55	32/45	36/53	31/47
α2			54/64	65/73	50/65	39/51	37/51	44/57	53/66	46/59	32/43	36/50	30/43
α3				58/68	48/65	36/50	34/48	45/58	47/63	44/56	32/47	35/50	32/49
α4					52/68	36/48	37/52	50/62	53/67	48/61	33/44	34/49	30/43
α5						32/48	36/54	41/60	64/76	43/59	32/47	35/53	30/48
α7							28/41	31/46	33/46	38/52	25/35	23/33	22/34
β1								40/55	36/53	38/53	40/54	35/49	35/50
β2									40/58	61/71	35/47	39/54	34/47
β3										41/56	29/44	36/54	27/43
β4											37/49	39/54	33/46
γ												42/53	52/65
δ													47/63
ε													

[a]Table matrix shows (percentage identity)/(percentage identity plus similarity) for indicated rat nAChR subunits. Amino acid sequences for rat nAChR subunits were aligned and analyzed pairwise for identity and similarity according to the PCGENE program ALIGN using the structure-genetic comparison matrix, an open gap cost of 7 and a unit gap cost of 1 where similar amino acid groups are (single letter code) A,S,T; D,E; N,Q; R,K; I,L,M,V; and F,Y,W.

family. It shares less than 40% identity with other members of the ligand-binding branch of nAChR subunits and less than 25% amino acid sequence identity with members of the muscle nAChR structural subunit branch. Perhaps the α7 subunit is just the first member in a third branch or subfamily of rat nAChR subunits that will be shown to also include other, currently unidentified members (the chick α7 subunit shares 72% identity with the chick α8 subunit,[19] but the rat homolog of α8 has not yet been identified).

The α5 (found in both the CNS and ANS) and β3 (found, to date, only in the CNS) pair of subunits share 64% identity and constitute a subunit group of the ligand-binding subunit branch or subfamily.[20-21] It is interesting that no functional role for either of these subunits has yet been found,[2-4] but that they are more closely related to the five nAChR subunits that are known to have functional roles in the ANS and CNS and to the muscle ligand-binding α1 subunit (41–52% and 40–53% identity to α1–4, β2, and β4 for α5 and β3, respectively) than is the α7 subunit.

The muscle nAChR α1 subunit is more like the α2–4 subunits (46–51% identity) than it is like any of the others (31–43% identity to α5, α7, β1–4, γ, δ or ε). It seems to have diverged more recently than the α7 subunit (and at about the same time as the α5/β3 and β2/β4 pairs) in evolution from the other truly ligand-binding subunits (α2–4).

The β2/β4 pair of structural subunits, both of which are found in the ANS and CNS, share 61% sequence identity, are 40–50% identical to α1–4 subunits, and are more closely related to the α-bungarotoxin-insensitive, neuronal nAChR α2–4 subunits than to β1, β3, or other muscle nAChR structural subunits. β2 and β4 subunits are distinctive in that they can substitute for the β1 subunit as a functional partner with muscle α1, γ, and δ subunits, and they can combine to form functional nAChR with either α2, α3, or α4 subunits in the Xenopus oocyte nAChR expression system,[2-4,15] thereby truly fulfilling part of the definition of structural subunits. However, because they can also influence ligand interactions with neuronal α2–4 subunits, they represent a group of functionally relevant subunits as might be expected for structural subunits "displaced" in the ligand-binding branch of the nAChR subunit phylogenetic tree.

Diverging next from the main branch of the ligand-binding subunit subfamily is the α3 subunit, which shares 54–58% identity with α2 and α4 subunits and seems to function as the predominant ligand-binding subunit in autonomic ganglia.[22–25] The α3 subunit is also expressed in a narrowly defined range of brain structures, perhaps as a component of some CNS nAChR that may help to control neurotransmitter release.[2–4,15,26–28] The remaining α2 and α4 subunits share 65% identity and represent a minor and the dominant ligand-binding (high-affinity agonist binding) subunit, respectively, in the rat CNS.

In the phylogenetic tree, α3, α5, and β4 subunits, which are found in a tight gene cluster in the rat or chicken,[20–21] in the human on chromosome 15,[29] and probably in the mouse on chromosome 9,[30] are more closely related to other subunits than to each other, suggesting that translocations to other chromosomes occurred more recently than the tandem duplication of a common ancestor that may have given rise to this cluster of genes. Along with α7 and β2 genes, clustered α3, α5, and β4 subunit genes are expressed in autonomic ganglia and related clonal cells[21–25,31,32] and are coordinately regulated in response to a number of stimuli (see below). Hence, it is perhaps not surprising that they have been found, in chick, to combine in formation of a unique ganglionic nAChR subtype. What is not evident from the data presented in FIGURE 1 and TABLE 1 is the basis for much of the divergence in amino acid sequences between nAChR subunits. All nAChR subunits have similar patterns of amino acid sequence suggesting that they contain an extended, N-terminal, extracellular domain, four transmembrane domains, an extended cytoplasmic loop between the third and fourth transmembrane domains, and an extracellular C-terminal tail. The putative, extended cytoplasmic loops are almost totally unique from one subunit to another not only in amino acid sequence, but also sometimes in length. For example, the rat α4 subunit is over 120 amino acids longer than the α3 subunit in the putative cytoplasmic domain, and the α2/α3 subunit pair shares 54% amino acid identity across the full sequence, 71% identity when the signal sequence and extended cytoplasmic loop are excluded, and only 20% identity through the signal sequence and cytoplasmic loop. Thus, the most unique feature of any nAChR subunit is its cytoplasmic domain.

FUNCTIONAL RELEVANCE OF nAChR SUBUNIT AND SUBTYPE DIVERSITY

A question that arises from these discoveries is, What is the functional relevance of nAChR subtype and subunit diversity?[2] Part of the answer must derive from the difficulties presented in defining functional roles of nAChR in the ANS and CNS and from recent insights gleaned from those studies.[2–4,33] Advances in understanding neuronal nAChR have come slowly, perhaps because of technical limitations, but also perhaps because some aspects of the functional relevance of nAChR diversity are already evident and run counter to dogma based on our classical understanding of muscle and ganglionic nAChR. For example, some neuronal nAChR desensitize so quickly in the presence of agonists that it now is not surprising that their functions (and perhaps those of other, related nAChR subtypes) escaped detection for so long.[11,12,16,34,35] Some neuronal nAChR may be located only on nerve terminals where their activation may not produce easily detectable currents but where they could have profound effects on neurotransmitter release.[27,28] Some neuronal nAChR may function as ligand-gated Ca^{2+} channels as well as or rather than ligand-gated Na^+

channels.[10–12,36,37] Established and more contemporary evidence indicates that sensitivity to drugs, including exogenous agents such as nicotine and endogenous acetylcholine, differs across tissues, cell types, and brain regions (i.e., across nAChR subtypes) and that diverse nAChR subunits play significant roles in the process of drug selectivity.[2,4,15,16,22,23,38–41] From this evidence and these lines of reasoning, it is clear that part of the functional relevance of nAChR diversity is that it allows closely related members of this single class of receptors to have diverse and physiologically relevant functions, ligand sensitivities, and tissue, regional, and cellular/subcellular distributions.

There is at least one example of how nAChR functional diversity can have physiological ramifications and teleological implications. The switch from a fetal to an adult phenotype of muscle nAChR that occurs about the time of motor neuronal innervation of recently differentiated myotubes simply reflects the switch from γ to ε subunit expression.[2,42] Nevertheless, fetal, γ-containing nAChR have lower conductance and longer mean open channel times than do adult, ε-containing nAChR, and fetal nAChR are expressed all over the noninnervated myotube surface, whereas adult nAChR expression is restricted to synaptic junctions. We can speculate that this subunit switch became stabilized through evolution because it conferred a functional advantage to developing muscle, in which fetal nAChR initially function as low-gain, poorly time-resolved, innervation sensors distributed widely over the muscle cell, whereas adult nAChR expressed only subsynaptically in innervated muscle function as high-gain elements best designed to respond to rapid chemical transients as would occur at the active neuromotor synapse. Hence, even a modest switch in expression of two closely related subunits contributes to a dramatic remodeling of a critical synapse as well as changes in nAChR function, distribution, and perhaps drug sensitivity.

Another perspective on the functional relevance of nAChR diversity comes from an understanding of the critical roles that nAChR play in nervous system function throughout the organism.[2] They are the exclusive mediators of excitatory neurotransmission at the neuromuscular junction, thereby regulating movement, and at preganglionic-postganglionic synapses, thereby regulating all autonomic activity. Given their widespread localization throughout the brain and just the beginnings of their functional characterization at those sites, neuronal nAChR must share with ionotropic glutamate receptors and serotonin (5-HT₃) receptors the mediation of excitatory neurotransmission that completes brain neuronal circuitry and contributes to higher-order brain processes including perception, cognition, and emotion. Collectively, the functional importance of nAChR centrally and in the periphery suggests that they are ideal targets for the regulation of nervous system function.

Does nAChR diversity allow for diversity in regulation of nAChR expression as well as for diversity in function, ligand selectivity, and distribution?[2] Does diversity in regulation of nAChR expression dovetail with diversity in nAChR and their functions to enhance the potential for plasticity in what we now are beginning to realize is a plastic and dynamic nervous system? If so, then we are only beginning to appreciate the functional relevance of nAChR diversity. We can speculate that the ability of a neuronal cell to express a different nAChR subtype during development or in response to changes in its environment could confer a functional benefit to that cell, just as it seems to be the case for muscle cells expressing fetal versus adult nAChR and subunit genes as a function of innervation state. For example, early in development, a nicotinic cholinoceptive cell may be functioning unconditionally as a circuit element whenever its dendritic nAChR sense release of acetylcholine. If the dendritic tree grows and encounters axons whose transmitters cause dendritic levels of some second messenger to increase, and if the function of the expressed, dendritic nAChR is inhibited in the presence of those second messengers, then that cholinoceptive cell may have altered patterns of firing, responding now as a conditional circuit element active only in response to nicotinic stimulation, but not in response to simultaneous nicotinic stimulation and activation of second messenger production. Could such a cell change its pattern of nAChR subunit expression to alleviate second messenger sensitivity? As that cell innervates its targets, or as the cellular dendritic tree and its contacts with retrograde circuit elements are altered, would there be changes in retrograde, anterograde, or extracellular matrix-bound growth factor influences on nAChR expression and, hence, cell function? How would such a change in cell environment be sensed, and how could a message to alter expression of nAChR subunits be transmitted?

MECHANISMS OF REGULATION OF DIVERSE nAChR SUBTYPES

Here we report a sampling of an ongoing series of studies in this laboratory concerning patterns of nAChR regulation. These studies are designed in part to define some mechanisms of regulation of nAChR expression and function and the roles of nAChR diversity in those processes. We considered the potential roles of diversity (1) in nAChR subunit genes in differential transcriptional control of nAChR expression and (2) in nAChR subunit amino acid sequences in differential posttranslational control of nAChR expression and modification of nAChR function as we described and discussed in detail previously.[2] We also took note of the uniqueness of amino acid sequences in putative cytoplasmic domains across nAChR subunits and for a given subunit across species. Even though sequences for a given subunit across mammalian species typically share 80–90% identity and about 95% identity plus similarity, cross-species differences in sequence most often appear in the cytoplasmic domains and at sites that may have functional relevance. For example, three putative protein kinase C phosphorylation sites are present in the cytoplasmic domain of the rat $\alpha 3$ subunit, but only one of those sites is preserved in the human $\alpha 3$ sequence. The human $\alpha 5$ subunit contains four putative protein kinase C phosphorylation sites in its cytoplasmic domain, but only one of these sites is expressed in the rat $\alpha 5$ subunit sequence. Therefore, we considered the possibility that homologous subunits in different species may be regulated differently. Aside from the possible roles of divergent cytoplasmic domains in conferring differential sensitivity of nAChR subtypes and subunits to actions of interacting receptors altering second messenger activity, protein kinase activity, and nAChR phosphorylation state, we also considered roles that cytoplasmic domains might play in stabilizing nAChR at the cell surface, in targeting nAChR to different locales, and in modulating nAChR function through their interactions with cytoskeletal elements. We also have examined potential roles for nicotinic ligands in the regulation of expression of their own receptors, and of growth factor-mediated effects on nAChR expression.

We have been advocates of the use of clonal cell lines as models for the identification and characterization of diverse nAChR subtypes, and some of our previous (and ongoing) work with clonal cells contributes to continuing development of the concept of nAChR diversity. We now further advocate the use of clonal cell lines as models for studies of mechanisms of nAChR regulation. We argue that clonal cells are well suited for such an application, in that they are homogenous, can be generated in quantities suitable for virtually any type of biochemical application, and can be subjected to electrophysiological or chemical measures of nAChR function. Clonal cells naturally express the same nAChR subtypes as are found in analogous non-neoplastic tissues from which the tumor cells are derived; that such expression is under the control of natural and not artificial promoters is particularly relevant to studies of nAChR regulation, which may have a transcriptional component. We found—and will continually be challenged to show—that clonal cells also possess similar, if not identical, intracellular signaling mechanisms as are found in their non-neoplastic analogs. At a minimum, this allows us not only to define mechanisms of nAChR regulation, but also to narrow the realm of possibilities for regulation of a given nAChR subtype by an interacting neurotransmitter, growth factor, or matrix receptor system in a given cell type. Obviously, a future challenge will be to extend our observations to non-neoplastic tissues. However, at present, particularly if the phenomenology of a drug treatment on nAChR in a clonal cell is similar to that for the same drug treatment on a non-neoplastic cellular analog, the intensive work needed to establish mechanisms involved is more expeditiously done using the clonal cell line models.

The three cell lines that we used as models for our studies are (1) the human TE671/RD clone, which expresses $\alpha1$, $\beta1$, γ, and δ subunits in a fetal muscle-type nAChR that mediates α-bungarotoxin (Bgt)-sensitive, nicotinic agonist-gated monovalent cation channel activity and binds [³H]-labeled acetylcholine ([³H]ACh) or [¹²⁵I]-labeled α-bungarotoxin (I-Bgt) with high affinity;[2] (2) the BC₃H-1 mouse muscle cell line, which also expresses fetal muscle nAChR subunits as a Bgt-sensitive, fetal, muscle-type nAChR; and (3) and the SH-SY5Y human neuroblastoma, which expresses nAChR $\alpha3$, $\alpha5$, $\alpha7$, $\beta2$, and $\beta4$ subunits and two types of neuronal nAChR.[22] We use the term neuronal/nicotinic Bgt binding sites (nBgtS) in reference to high-affinity I-Bgt binding sites found in these cells that seem to contain $\alpha7$ subunits but for which we have found (to date) no evidence for ion channel function, and we use the term ganglia-type nAChR in reference to other neuronal nAChR found in these cells that appear to contain $\alpha3$ and $\beta4$ subunits that mediate both monovalent cation and Ca^{2+} conductances in response to nicotinic agonists in a Bgt-insensitive manner and that bind [³H]ACh with high affinity, also in a Bgt-insensitive manner.[22,43]

EFFECTS OF CHRONIC NICOTINE TREATMENT

Previous studies from this laboratory have established that chronic nicotine treatment (1 mM) leads to an increase in numbers of muscle-type nAChR (I-Bgt binding sites) in crude membrane preparations of TE671/RD cells that is evident within 6–12 h of drug treatment and reaches and maintains maximal values (2–3-fold increase) after one day and for as long as five days of drug treatment.[44] Other previous work established that nAChR function is lost, however, even during this phase of nicotine-induced up-regulation of nAChR numbers.[45] Phenomenologically, these results are similar to those observed for effects of chronic nicotine treatment on cell ganglia-type nAChR or for chronic nicotine treatment on mouse or rat brain nAChR.[26,44,45] New results shown here indicate that these effects on nAChR numbers do not involve either transient or sustained increases in levels of messenger RNA coding for TE671 cell $\alpha1$, $\beta1$, γ or δ subunits (FIG. 2). We made similar observations using nicotine treatment of primary rat cortical neurons in culture as a model—in this case, in reference to high-affinity [³H]nicotine binding sites and mRNA encoding the $\alpha4$ and $\beta2$ subunits that make up those nAChR, using a lower (1 μM) concentration of chronic nicotine.[46] Other previous work using an intact mouse model showed that 10 days of chronic nicotine infusion, which is adequate to affect an increase in [³H]nicotine binding sites, failed to induce a sustained increase in nAChR $\alpha4$ or $\beta2$ subunit mRNA.[26] Our results add new information that excludes a

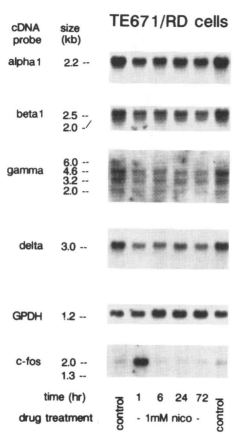

cDNA probe	size (kb)	TE671/RD cells
alpha1	2.2 --	
beta1	2.5 -- 2.0 -/	
gamma	6.0 -- 4.6 -- 3.2 -- 2.0 --	
delta	3.0 --	
GPDH	1.2 --	
c-fos	2.0 -- 1.3 --	

time (hr)   control   1 6 24 72   control

drug treatment   - 1mM nico -

FIGURE 2. Temporal pattern of the effects of nicotine treatment on muscle-type nAChR subunit mRNA levels in TE671/RD cells. Northern blot analyses were conducted using established techniques and cDNA probes previously described[22,53] for TE671/RD cells that had been treated for the indicated periods with 1 mM nicotine (nico) or with no added drug (control). Approximate sizes (kb) of nAChR subunit transcripts corresponding to $\alpha 1$, $\beta 1$, γ, and δ genes are shown as is mRNA for glyceraldehyde-3-phosphate dehydrogenase (GPDH), which was used as a control for mRNA loading. Quantitative densitometric analysis of this and other experiments indicates that there are no significant increases in nAChR subunit mRNA levels in nicotine-treated cells relative to drug-free controls when data are normalized to levels of GPDH or total (measured by densitometric analysis of photographic negatives of ethidium-stained mRNA in the smear between 18S and 28S ribosomal RNAs) mRNA. Note that nicotine treatment also transiently induces expression of c-fos mRNA.

transient effect of nicotine treatment on nAChR mRNA levels and probably on nAChR gene transcription. Our findings also reveal for the first time, in the absence of possible pharmacokinetic effects, that whereas chronic nicotine treatment can induce comparable increases in muscle-type nAChR, ganglia-type nAChR, or CNS $\alpha 4/\beta 2$-nAChR, the dose of nicotine required to affect those changes differs from one nAChR subtype to another. We hypothesize that chronic nicotine exposure first induces a loss of nAChR function that somehow is sensed by the cell and/or triggers an increase in nAChR expression measurable by radioligand binding, but that this increase does not involve activation of nAChR subunit gene transcription. Continued studies are necessary to reveal the possible posttranslational mechanisms (enhanced assembly of nAChR subunits into detectable nAChR or enhanced recruitment of nAChR to the cell surface from a microsomal pool) involved in these effects. However, the current work illustrates how mechanisms of nAChR regulation can sometimes be blind to nAChR diversity. Moreover, these new findings indicate that nAChR diversity can contribute to differences in ligand sensitivity, not only in regard to concentrations of drug needed to acutely trigger nAChR function, but also to those needed to chronically alter nAChR expression and/or function. That the predominant brain nAChR subtype (but not muscle- or ganglia-type nAChR) is affected by concentrations of nicotine known to be present in the blood of smokers suggests that the current findings are relevant to the process of nicotine addiction as an event of the CNS and helps to explain why chronic nicotine exposure at levels seen in smokers is comparatively ineffective in chronically altering neuromuscular or ganglionic function.

EFFECTS OF NERVE GROWTH FACTOR

Previous reports in the literature indicate that treatment of cells of the PC12 rat pheochromocytoma with nerve growth factor (NGF) causes an increase in nAChR function[47-49] and numbers,[50,51] and a recent study indicated that those effects are accompanied by an increase in levels of nAChR α3, α5, α7, β2, and β4 subunit mRNA levels.[52] Cells of the SH-SY5Y human neuroblastoma express the same range of nAChR subunits and subtypes as are expressed in PC12 cells or non-neoplastic autonomic neurons of neural crest origin.[22] As shown in FIGURE 3, new results indicate that levels of nAChR α3, α5, and β4 subunit mRNA increase in SH-SY5Y cells treated for as little as six hours or for as long as five days with human

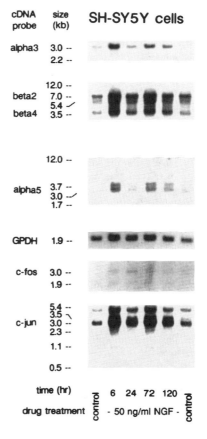

FIGURE 3. Temporal pattern of the effects of nerve growth factor (NGF) treatment on nAChR subunit mRNA levels in SH-SY5Y cells. Northern blot analyses were conducted using established techniques and cDNA probes previously described[22,53] for SH-SY5Y cells that had been treated for the indicated periods with 50 ng/mL NGF or with no added growth factor (control). Approximate sizes (kb) of nAChR subunit transcripts corresponding to α3, β2, β4, and α5 genes are shown as is mRNA for glyceraldehyde-3-phosphate dehydrogenase, which was used as a control for mRNA loading. Note also that there is sustained elevation of both c-fos and c-jun mRNAs on NGF treatment.

recombinant NGF (50 ng/mL). Quantitative densitometric analysis indicates that about a threefold increase in mRNA levels occurs for the indicated subunits, but not for β2 subunit mRNA, levels of which matched those of the loading control (glyceraldehyde-3-phosphate dehydrogenase; GPDH). These results indicate that NGF treatment coordinately increases the expression in human SH-SY5Y cells of the three neuronal nAChR subunit genes that are found as a gene cluster in chickens and rats, as occurs upon treatment with phorbol ester (see below). This observation supports the hypothesis that the gene cluster acts as a functional unit.[20-21] The finding that human β2 gene expression in SH-SY5Y cells is not altered by NGF treatments that affect rat β2 gene expression in PC12 cells[52] (see also ref. 31) indicates that this gene can be uncoupled from the others. It also suggests that differences must exist in sequences of human and rat β2 genes in a region(s) that confers responsiveness to nuclear transacting factors sensitive to actions of NGF and perhaps other agents. A prediction of the current results is that ganglia-type nAChR numbers and function will be increased in SH-SY5Y cells subjected to NGF treatment. Given that NGF effects on nAChR subunit mRNA levels occur in wild-type and protein kinase A-deficient PC12 cells[52] and that both NGF and phorbol ester treatments have the same effects on nAChR subunit mRNA levels in SH-SY5Y cells (see below), another prediction is that the protein kinase C signaling pathway may mediate some of the effects of NGF or that the NGF and C-kinase signaling pathways converge at some point proximal to the activation of nAChR subunit gene expression. The current results illustrate how nAChR subunit diversity across species may influence responses of nAChR to regulatory influences, indicate that nAChR levels can be altered by pretranslational mechanisms, and demonstrate that growth factors can profoundly influence nAChR expression.

POSSIBLE ROLE FOR THE CYTOSKELETON IN nAChR REGULATION

One question posed above was, How might a cell sense a change in its environment, and how could it transmit signals to alter expression or function of nAChR? The classical picture of the role of the cytoskeleton in receptor biology is one of a relatively passive, structural element that maintains cell shape and possibly participates in transport and local organization of receptors, perhaps in coordination with outgrowth of dendritic or axonal processes. However, some of our recent work,[53] particularly from the perspective of other reports in the literature, suggests that the cytoskeleton may play a more dynamic and functional role in the regulation of nAChR expression and function. Treatment of TE671/RD cells with any one of a variety of cytochalasins (done to exclude possible effects of a narrower subset of cytochalasins on cell functions or processes other than stability of actin microfilaments) produces changes in cell size and shape consistent with disruption of actin microfilament networks, that is, a rapid loss (within an hour) of submembrane, phalloidin staining of F-actin fibers coordinate with condensation of the cell into a rounded shape, followed by enlargement of the cell soma and bipolar process outgrowth apparently driven by extensive polymerization and parallel organization of tubulin filaments, clearly evident within two days of the start of cytochalasin treatment (R.J. Lukas, unpublished results). Cytochalasin treatment also produces a steady increase over four days of drug treatment in muscle-type nAChR numbers (i.e., I-Bgt binding sites) in total membrane fractions, but not on the cell surface.[53] nAChR function also declines (detectably so within hours), but 3–5-fold increases occur in levels of nAChR α1, β1, γ, and δ subunit mRNA levels at two days of cytochalasin treatment. All of these effects appear to be specific, in that levels of other mRNAs and numbers of other receptors (e.g., m3-type muscarinic acetylcholine receptors) are not affected. These results suggest that actin microfilament integrity is necessary for maintenance of nAChR function. They also suggest that actin microfilament integrity may play a role in tonic inhibition of nAChR subunit gene expression, perhaps providing a mechanism for transmitting signals (possibly to the nucleus) concerning the status of the cell membrane and its receptors.[53] Given that process outgrowth/retraction in neurons can be influenced by neurotransmitters including those that act at nAChR[54,55] and that those processes must involve reorganization of the cytoskeleton, it is clear that cytoskeletal structure and nAChR activity and expression can influence and be influenced by each other.[53] These studies suggest, for example, that synaptic remodeling may be a constant in the nervous system, influenced by interacting receptor activity and perhaps mediated by the cytoskeleton as well as by more mobile and compact signaling molecules. Another area ripe for investigation is the role that specific actin- or tubulin-associated proteins play as chaperons targeting nAChR via their unique cytoplasmic domains to different cellular destinations and as mediators of receptor-cytoskeletal signaling.

ACTIONS ON nAChR EXPRESSION OF AGENTS TARGETING THE NUCLEUS

As a model for studies of extra- or intracellular signals that might be involved in mediating the presumably neuronal influences on nAChR gene expression in developing muscle, we initially investigated the effects of dibutyryl cyclic AMP (dbcAMP) on muscle-type nAChR and nAChR subunit gene expression in BC_3H-1 cells.[56] We found that dbcAMP treatment induced a dramatic loss in numbers of muscle-type nAChR (i.e., I-Bgt binding sites) and nAChR function. Northern blot analysis indicated that substantial (60–80%) declines occurred in levels of nAChR $\alpha1$, $\beta1$, and δ subunit mRNA, just as occurs in normal muscle about the time of neuronal innervation, but also that there was a virtually complete loss (quantifiable as a 250-fold decline, also as occurs in normal muscle about the time of neuronal innervation) in γ subunit mRNA levels. The loss in γ subunit gene expression clearly accounted for the loss of nAChR numbers and function, suggesting that gamma-less nAChR expression was very inefficient, at best, and that dbcAMP treatment may access a signaling pathway that is involved in the innervation-induced changes in nAChR subunit and phenotype expression *in vivo*. Studies of the time course for these effects and pharmacological dissection indicated, however, that the effects of dbcAMP in BC_3H-1 cells on nAChR subunit gene expression were attributable to the actions of butyrate, which must be generated in BC_3H-1 cells treated with dbcAMP by hydrolysis of dbcAMP to butyrate and monobutyryl cyclic AMP. Butyrate is known to specifically alter the expression of several genes, but the mechanism for those effects is presently being debated.[57,58] Among the possibilities are the effects of butyrate on histone acetylation, thereby affecting nucleosome structure, or on transacting or transcriptional factors; putative butyrate response elements have been identified in promoter/enhancer regions of viral genes, for example. The BC_3H-1 butyrate model provides an opportunity to investigate mechanisms of action of butyrate as well as possible nuclear mechanisms involved in the nAChR γ/ϵ subunit switch. These studies illustrate how nAChR subunit gene expression can be regulated differentially and how expression of nAChR can be influenced by agents that perhaps are specifically targeted to the nucleus.

SECOND MESSENGER REGULATION OF nAChR EXPRESSION

An obvious potential mechanism for the control of nAChR expression and function, particularly in the light of reports in the literature about the roles of nAChR phosphorylation in functional desensitization and in posttranslational processing of nAChR, is via second messengers that alter protein kinase activity.[2,59] Some of our previous studies have shown, for example, that muscle-type nAChR expression in human TE671/RD cells is subject to second messenger-sensitive

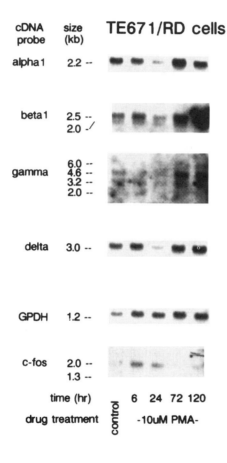

FIGURE 4. Temporal pattern of the effects of phorbol-12-myristate-13-acetate (PMA) treatment on muscle-type nAChR subunit mRNA levels in TE671/RD cells. Northern blot analyses were conducted using established techniques and cDNA probes previously described[22,53] for TE671/RD cells that had been treated for the indicated periods with 10 µM PMA or with no added drug (control). Approximate sizes (kb) of nAChR subunit transcripts corresponding to α1, β1, γ, and δ genes are shown, as is mRNA for glyceraldehyde-3-phosphate dehydrogenase, which was used as a control for mRNA loading. Note that PMA treatment also transiently elevates c-fos mRNA levels.

regulation.[60–61] Here, we describe and present Northern analysis data indicating that some, but not all, of these effects have a potential transcriptional basis. Treatment of TE671/RD cells with phorbol-12-myristate-13-acetate (PMA) induces a temporally biphasic effect on nAChR expression (i.e., I-Bgt binding sites in total membrane preparations) characterized by a transient loss of nAChR followed by an increase to 200% of control levels after 4–5 days of drug treatment. By contrast, treatment with dbcAMP (which does not seem to lead to production of butyrate in these cells), induces little-to-no change in nAChR numbers. Whereas Northern analysis reveals no change in muscle-type nAChR subunit mRNA levels in dbcAMP-treated TE671/RD cells, in cells treated with PMA, levels of mRNA corresponding to $\alpha1$, $\beta1$, γ, and δ subunits follow the same temporal pattern of an early decline followed by an up-regulation as is seen for I-Bgt binding sites (FIG. 4). Hence, in this case, it is likely that a protein kinase C-sensitive pathway produces changes in nAChR subunit gene transcription to influence nAChR numbers. Interestingly, treatment of SH-SY5Y cells with PMA also produces a longer-term increase (of about 4–5-fold but without a transient decline as seen in TE671/RD cells) in transcript levels corresponding to nAChR $\alpha3$, $\alpha5$, and $\beta4$ subunits (but not $\beta2$ subunits; data not shown), suggesting that a variety of nAChR genes may be sensitive to protein kinase C-mediated signals. Collectively, these studies are consistent with posttranslational effects of protein kinase A-mediated signaling on nAChR and with both transcriptional and posttranslational effects (not discussed here) of protein kinase C-mediated signaling.

SUMMARY

In our studies we explored the functional relevance of nAChR diversity, in part from the perspective of nAChR as ideal targets for regulatory influences, including those mediated via actions of ligands at other "interacting" receptors. We explored possible mechanisms for nAChR regulation and roles played by nAChR subtype and subunit diversity in those processes. We showed that regulatory factors can influence nAChR numbers at transcriptional and posttranscriptional levels and can affect nAChR function and subcellular distribution. We also demonstrated that nAChR expression can be influenced (1) by nicotinic ligands, (2) by second messengers, (3) by growth factors, (4) by agents targeting the nucleus, and (5) by agents targeting the cytoskeleton. We found common effects of some regulatory influences on more than one nAChR subtype, and we found instances where regulatory influences differ for different cell and nAChR types. Even from the very limited number of these initial studies, it is evident that nAChR subunit and subtype diversity, which alone can provide diversity in nAChR functions, localization, and ligand sensitivity, dovetails with diversity in cellular signaling mechanisms that can affect nAChR expression to amplify the potential functional plasticity of cholinoceptive cells. As examples, we discussed potential roles for nAChR diversity and regulatory plasticity in synapse remodeling and in changes in neuronal circuit conditions. These examples illustrate how nAChR diversity could play important roles in the regulation of nervous system function.

ACKNOWLEDGMENTS

The author gratefully acknowledges contributions to this work by colleagues Merouane Bencherif, Renaldo Drisdel, Cynthia M. Eisenhour, Anna M. Joy, Lei Ke, Linda Lucero, Sylvia A. Norman, Elzbieta Puchacz, and Hal N. Siegel. The author also thanks Drs. Jim Patrick, Steven Heinemann, and Jim Boulter for kind gifts of nAChR subunit cDNA probes.

REFERENCES

1. CHANGEUX, J.-P., J.-L. GALZI, A. DEVILLERS-THIERY & D. BERTRAND. 1992. The functional architecture of the acetylcholine receptor explored by affinity labeling and site directed mutagenesis. Q. Rev. Biophys. **25:** 395–432.
2. LUKAS, R. J. & M. BENCHERIF. 1992. Heterogeneity and regulation of nicotinic acetylcholine receptors. Int. Rev. Neurobiol. **34:** 25–131.
3. ROLE, L. W. 1992. Diversity in primary structure and function of neuronal nicotinic acetylcholine receptor channels. Curr. Opin. Neurobiol. **2:** 254–262.
4. SARGENT, P. B. 1993. The diversity of neuronal nicotinic acetylcholine receptors. Annu. Rev. Neurosci. **16:** 403–443.
5. PATON, W. D. M. & E. J. ZAIMIS. 1952. The methonium compounds. Pharmacol. Rev. **4:** 219–253.
6. DALE, H. H. 1954. The beginnings and the prospects of neuro-humoral transmission. Pharmacol. Rev. **6:** 7–13.
7. KHARKEVICH, D. A., ED. 1980. Pharmacology of Ganglionic Transmission. Springer-Verlag. Berlin.
8. KRENJEVIC, K. 1975. Acetylcholine receptors in the vertebrate CNS. In Handbook of Psychopharmacology. L. L. Iversen, S. D. Iversen & S. H. Snyder, Eds. Vol. **6:** 97–125. Plenum Press. New York.
9. MARTIN, B. R. 1986. Nicotine receptors in the central nervous system. In The Receptors. P. M. Conn, Ed. Vol. **3:** 393–415. Academic Press. New York.
10. VIJAYARAGHAVAN, S., P. C. PUGH, Z.-W. ZHANG, M. M. RATHOUS & D. K. BERG. 1992. Nicotinic receptors that bind α-bungarotoxin on neurons raise intracellular free Ca. Neuron **8:** 353–362.
11. CASTRO, N. & E. X. ALBUQUERQUE. 1993. Brief-lifetime, fast-inactivating ion channels account for the α-bungarotoxin-sensitive nicotinic response in hippocampal neurons. Neurosci. Lett. **164:** 137–140.
12. ZHANG, Z.-W., S. VIJAYARAGHAVAN & D. K. BERG. 1994. Neuronal acetylcholine receptors that bind α-bungarotoxin with high affinity function as ligand-gated ion channels. Neuron **12:** 167–177.
13. GALZI, J.-L., F. REVAH, A. BESSIS & J.-P. CHANGEUX. 1991. Functional architecture of the nicotinic acetylcholine receptor: From electric organ to brain. Annu. Rev. Pharmacol. **31:** 37–72.
14. KARLIN, A. 1993. Structure of nicotinic acetylcholine receptors. Curr. Opin. Neurobiol. **3:** 299–309.
15. DENERIS, E. S., J. CONNOLLY, S. W. ROGERS & R. DUVOISIN. 1991. Pharmacological and functional diversity of neuronal nicotinic acetylcholine receptors. Trends Pharmacol. **12:** 34–40.
16. BERTRAND, D., A. DEVILLERS-THIERY, F. REVAH, J.-L. GALZI, N. HUSSY, C. MULLE, S. BERTRAND, M. BALLIVET & J.-P. CHANGEUX. 1992. Unconventional pharmacology of a neuronal nicotinic receptor mutated in the channel domain. Proc. Natl. Acad. Sci. USA **89:** 1261–1265.
17. ANAND, R., X. PENG & J. LINDSTROM. 1993. Homomeric and native α7 acetylcholine receptors exhibit remarkably similar but non-identical pharmacological properties, suggesting that the native receptor is a heteromeric protein complex. FEBS Lett. **327:** 241–246.
18. SEGUELA, P., J. WADICHE, K. DINELEY-MILLER, J. A. DANI & J. W. PATRICK. 1993. Molecular cloning, functional properties, and distribution of rat brain α7: A nicotinic cation channel highly permeable to calcium. J. Neurosci. **13:** 596–604.
19. SCHOEPFER, R., W. G. CONROY, P. WHITING, M. GORE & J. LINDSTROM. Brain α-bungarotoxin binding protein cDNAs and mAbs reveal subtypes of this branch of the ligand-gated ion channel gene family. Neuron **5:** 35–48.
20. BOULTER, J., A. O'SHEA-GREENFIELD, R. M. DUVOISIN, J. G. CONNOLLY, E. WADA, A. JENSEN, P. D. GARDNER, M. BALLIVET, E. S. DENERIS, D. MCKINNON, S. HEINEMANN & J. PATRICK. 1990. α3, α5, and β4: Three members of the rat neuronal nicotinic receptor-related gene family form a gene cluster. J. Biol. Chem. **265:** 4472–4482.

21. COUTURIER, S., L. ERKMAN, S. VALERA, D. RUNGGER, S. BERTRAND, J. BOULTER, M. BALLIVET & D. BERTRAND. 1990. α5, α3, and non-α3: Three clustered avian genes encoding neuronal nicotinic acetylcholine receptor-related subunits. J. Biol. Chem. **265:** 17560–17567.

22. LUKAS, R. J., S. A. NORMAN & L. LUCERO. 1993. Characterization of nicotinic acetylcholine receptors expressed by cells of the SH-SY5Y human neuroblastoma clonal line. Mol. Cell. Neurosci. **4:** 1–12.

23. LUKAS, R. J. 1993. Expression of ganglia-type nicotinic acetylcholine receptors and nicotinic ligand binding sites by cells of the IMR-32 human neuroblastoma clonal line. J. Pharmacol. Exp. Ther. **265:** 294–302.

24. VERNALLIS, A. B., W. G. CONROY & D. K. BERG. 1993. Neurons assemble acetylcholine receptors with as many as three kinds of subunits while maintaining subunit segregation among receptor subtypes. Neuron **10:** 451–464.

25. LISTERUD, M., A. B. BRUSSAARD, P. DEVAY, D. R. COLMAN & L. W. ROLE. 1991. Functional contribution of neuronal AChR subunits revealed by antisense oligonucleotides. Science **254:** 1518–1521.

26. MARKS, M. J., J. R. PAULY, S. D. GROSS, E. S. DENERIS, I. HORMANS-BORGMEYER, S. F. HEINEMANN & A. C. COLLINS. 1992. Nicotine binding and nicotinic receptor subunit mRNA after chronic nicotine treatment. J. Neurosci. **12:** 2765–2784.

27. GRADY, S., M. J. MARKS, S. WONNACOTT & A. C. COLLINS. 1992. Characterization of nicotinic receptor-mediated [3H]dopamine release from synaptosomes prepared from mouse striatum. J. Neurochem. **59:** 848–856.

28. MARKS, M. J., D. A. FARNHAM, S. R. GRADY & A. C. COLLINS. 1993. Nicotinic receptor function determined by stimulation of rubidium efflux from mouse brain synaptosomes. J. Pharmacol. Exp. Ther. **264:** 542–552.

29. ANAND, R. & J. LINDSTROM. 1992. Chromosomal localization of seven neuronal nicotinic acetylcholine receptor subunit genes in humans. Genomics **13:** 962–967.

30. BESSIS, A., D. SIMON-CHAZOTTES, A. DEVILLERS-THIERY, J.-L. GUENET & J.-P. CHANGEUX. 1990. Chromosomal localization of the mouse genes coding for α2, α3, α4 and β2 subunits of neuronal nicotinic acetylcholine receptor. FEBS Lett. **264:** 48–52.

31. ROGERS, S. W., A. MANDELZYS, E. S. DENERIS, E. COOPER & S. HEINEMANN. 1992. The expression of nicotinic acetylcholine receptors by PC12 cells treated with NGF. J. Neurosci. **12:** 4611–4623.25.

32. CORRIVEAU, R. A. & D. K. BERG. 1993. Coexpression of multiple acetylcholine receptor genes in neurons: Quantification of transcripts during development. J. Neurosci. **13:** 2662–2671.

33. CLARKE, P. B. S. 1992. The fall and rise of neuronal α-bungarotoxin binding proteins. Trends Pharmacol. Sci. **13:** 407–413.

34. REVAH, F., D. BERTRAND, J.-L. GALZI, A. DEVILLERS-THIERY, C. MULLE, N. HUSSY, S. BERTRAND, M. BALLIVET & J.-P. CHANGEUX. 1991. Mutations in the channel domain alter desensitization of a neuronal nicotinic receptor. Nature **353:** 846–849.

35. ALKONDON, M. & E. X. ALBUQUERQUE. 1993. Diversity of nicotinic acetylcholine receptors in rat hippocampal neurons. I. Pharmacological and functional evidence for distinct structural subtypes. J. Pharmacol. Exp. Ther. **265:** 1455–1473.

36. BERTRAND, D., J.-L. GALZI, A. DEVILLERS-THIERY, S. BERTRAND & J.-P. CHANGEUX. 1993. Mutations at two distinct sites within the channel domain M2 alter calcium permeability of neuronal α7 nicotinic receptor. Proc. Natl. Acad. Sci. USA **90:** 6971–6975.

37. VERNINO, S., M. AMADOR, C. W. LUETJE, J. PATRICK & J. A. DANI. 1992. Calcium modulation and high calcium permeability of neuronal nicotinic acetylcholine receptors. Neuron **8:** 127–134.

38. LUETJE, C. & J. PATRICK. 1991. Both α- and β-subunits contribute to the agonist sensitivity of neuronal nicotinic acetylcholine receptors. J. Neurosci. **11:** 837–845.

39. PAPKE, R. L. & S. HEINEMANN. 1991. The role of the β4 subunit in determining the kinetic properties of rat neuronal nicotinic acetylcholine α3 receptors. J. Physiol. (Lond.) **440:** 95–112.

40. ANAND, R., X. PENG, J. J. BALLESTA & J. LINDSTROM. 1993. Pharmacological characteriza-

tion of α-bungarotoxin-sensitive acetylcholine receptors immunoisolated from chick retina: Contrasting properties of α7 and α8 subunit-containing subtypes. Mol. Pharmacol. **44:** 1046–1050.

41. THOMAS, P., M. STEPHENS, G. WILKIE, M. AMAR, G. G. LUNT, P. WHITING, T. GALLAGHER, E. PEREIRA, M. ALKONDON, E. X. ALBUQUERQUE & S. WONNACOTT. 1993. (+)-Anatoxin-a is a potent agonist at neuronal nicotinic acetylcholine receptors. J. Neurochem. **60:** 2308–2311.

42. WITZEMANN, V., H.-R. BRENNER & B. SAKMANN. 1991. Neural factors regulate AChR subunit mRNAs at rat neuromuscular synapses. J. Cell Biol. **114:** 125–141.

43. EISENHOUR, C. M., E. PUCHACZ & R. J. LUKAS. 1993. Calcium ion influx in clonal cell lines expressing nicotinic acetylcholine receptors. Soc. Neurosci. Abstr. **19:** 1533, 1993.

44. SIEGEL, H. N. & R. J. LUKAS. 1988. Nicotinic agonists regulate α-bungarotoxin binding sites of TE671 human medulloblastoma cells. J. Neurochem. **50:** 1272–1278.

45. LUKAS, R. J. 1991. Effects of chronic nicotinic ligand exposure on functional activity of nicotinic acetylcholine receptors expressed by cells of the PC12 rat pheochromocytoma or the TE671/RD human clonal line. J. Neurochem. **56:** 1134–1145.

46. BENCHERIF, M., K. FOWLER, R. J. LUKAS & P. LIPPIELLO. 1994. Nicotine up-regulates high affinity α4β2 nicotinic acetylcholine receptors through posttranslational mechanisms. International Symposium on Nicotine.: 43. Birkhäuser Verlag. Basel.

47. DICHTER, M. A., A. S. TISCHLER & L. A. GREENE. 1977. Nerve growth factor-induced increase in electrical excitability and acetylcholine sensitivity of a rat pheochromocytoma cell line. Nature **268:** 501–504.

48. AMY, C. M. & E. L. BENNETT. 1983. Increased sodium ion conductance through nicotinic acetylcholine receptor channels in PC12 cells exposed to nerve growth factors. J. Neurosci. **3:** 1547–1553.

49. IFUNE, C. K. & J. H. STEINBACH. 1990. Regulation of sodium currents and acetylcholine responses in PC12 cells. Brain Res. **506:** 243–248.

50. WHITING, P. J., R. SCHOEPFER, L. W. SWANSON, D. M. SIMMONS & J. M. LINDSTROM. 1987. Functional acetylcholine receptor in PC12 cells reacts with a monoclonal antibody to brain nicotinic receptors. Nature **327:** 515–518.

51. MADHOK, T. C. & B. M. SHARP. 1992. Nerve growth factor enhances [³H]nicotine binding to a nicotinic cholinergic receptor on PC12 cells. Endocrinology **67:** 825–830.

52. HENDERSON, L. P., M. J. GDOVIN, C.-L. LIU, P. D. GARDNER & R. A. MAUE. 1994. Nerve growth factor increases nicotinic ACh receptor gene expression and current density in wild-type and protein kinase A-deficient PC12 cells. J. Neurosci. **14:** 1153–1163.

53. BENCHERIF, M. & R. J. LUKAS. 1993. Cytochalasin modulation of nicotinic cholinergic receptor expression and muscarinic receptor function in human TE671/RD cells: A possible functional role of the cytoskeleton. J. Neurochem. **61:** 852–864.

54. MATTSON, M. P. 1988. Neurotransmitters in the regulation of neuronal architecture. Brain Res. Rev. **13:** 179–212.

55. LIPTON, S. A. & S. B. KATER. 1989. Neurotransmitter regulation of neuronal outgrowth, plasticity and survival. Trends Neurosci. **12:** 265–270.

56. LUCERO, L. & R. J. LUKAS. 1992. On the transcriptional basis of butyrate- and cyclic AMP-induced downregulation of nicotinic acetylcholine receptor expression by BC₃H-1 cells. J. Cell. Biochem. **16E:** T313.

57. KRUH, J., N. DEFER & L. TICHONICKY. 1991. Molecular and cellular action of butyrate. C. R. Soc. Biol. **186:** 12–25.

58. FREGEAU, C. J., C. D. HELGASON & R. C. BLEACKLEY. 1992. Two cytotoxic proteinase genes are differentially sensitive to sodium butyrate. Nucleic Acids Res. **20:** 3113–3119.

59. SWOPE, S. L., S. J. MOSS, C. D. BLACKSTONE & R. L. HUGANIR. 1992. Phosphorylation of ligand-gated ion channels: A possible mode of synaptic plasticity. FASEB J. **6:** 2514–2523.

60. SIEGEL, H. N. & R. J. LUKAS. 1988. Morphological and biochemical differentiation of the human medulloblastoma cell line TE671. Dev. Brain Res. **44:** 269–280.

61. BENCHERIF, M. & R. J. LUKAS. 1991. Differential regulation of nicotinic acetylcholine receptor expression by human TE671/RD cells following second messenger modulation and sodium butyrate treatments. Mol. Cell. Neurosci. **2:** 52–65.

Cholinergic Pathways and the Ascending Reticular Activating System of the Human Brain[a]

M-MARSEL MESULAM

*Center for Behavioral and Cognitive Neurology
and the Alzheimer Program
Departments of Neurology and Psychiatry
Northwestern University Medical School
Chicago, Illinois 60611*

The primate brain contains eight major cholinergic cell groups that project to other central nervous system structures. Many of these cholinergic cell groups do not respect traditional nuclear boundaries and their constituent cells are intermixed with other noncholinergic neurons. We therefore introduced the Ch1–Ch8 nomenclature in order to designate the cholinergic (i.e., choline acetyltransferase-containing) neurons within these eight cell groups.[1]

According to this nomenclature, Ch1 designates the cholinergic cells associated with the medial septal nucleus, Ch2 those associated with the vertical nucleus of the diagonal band, Ch3 those associated with the horizontal limb of the diagonal band nucleus, Ch4 those associated with the nucleus basalis of Meynert, Ch5 those associated with the pedunculopontine nucleus of the rostral brain stem, Ch6 those associated with the laterodorsal tegmental nucleus also in the rostral brain stem, Ch7 those in the medial habenula, and Ch8 those in the parabigeminal nucleus.

Tracer experiments in a number of animal species have shown that Ch1 and Ch2 provide the major cholinergic innervation for the hippocampal complex, Ch3 for the olfactory bulb, Ch4 for the cerebral cortex and amygdala, Ch5 and Ch6 for the thalamus, Ch7 for interpeduncular nucleus, and Ch8 for the superior colliculus. There are also lesser connections from Ch1–Ch4 and Ch8 to the thalamus and from Ch5–Ch6 to the cerebral cortex.[1,2]

In the rodent brain, intrinsic cholinergic interneurons may provide up to 30% of the cholinergic innervation in the cerebral cortex. No such cholinergic interneurons have been reported in the adult primate cerebral cortex or in the thalamus of any species studied thus far. The cholinergic innervation of the adult primate cerebral cortex and thalamus is therefore almost exclusively extrinsic.

CHOLINERGIC NEURONS OF THE BASAL FOREBRAIN

The basal forebrain of the primate brain contains four overlapping constellations of cholinergic projection neurons. Studies in the monkey brain show that approximately 10% of perikarya within the boundaries of the medial septal nucleus are cholinergic and belong to the Ch1 cell group; approximately 70% of neurons in the

[a]This work was supported in part by a Javits Neuroscience Investigator Award (NS20285).

[b]Address correspondence to M-Marsel Mesulam, M.D., Center for Behavioral and Cognitive Neurology and the Alzheimer Program, 320 East Superior Street, Chicago, IL 60611.

vertical limbic nucleus of the diagonal band are cholinergic and belong to the Ch2 cell group; less than 5% of neurons in the horizontal nucleus of the diagonal band are cholinergic and belong to the Ch3 cell group; approximately 90% of the large neurons in the nucleus basalis of the substantia innominata are cholinergic and belong to the Ch4 cell group. Of these four cholinergic cell groups, the Ch4 group is by far the largest and the one that has been most extensively studied in the human brain.[3,4]

Because nearly 90% of the nucleus basalis (NB) neurons in the human brain express choline acetyltransferase (and therefore belong to Ch4), this cell group can also be designated as the NB-Ch4 complex. The more general term "NB" can be used to designate all of the components in this nucleus (large and small cells, cholinergic and noncholinergic), whereas the more restrictive Ch4 designation is reserved for the contingent of cholinergic NB neurons as revealed by ChAT immunohistochemistry.

The human NB-Ch4 extends from the level of the olfactory tubercle to that of the anterior hippocampus, spanning a distance of 13–14 mm in the sagittal plane. It attains its greatest mediolateral width of 18 mm within the substantia innominata (subcommissural gray). Arendt et al.[5] estimated that the human NB-Ch4 complex contains 200,000 neurons in each hemisphere. Thus, the NB-Ch4 contains at least 10 times as many neurons as the nucleus locus coeruleus, which has approximately 15,000 neurons in the adult human brain.[6] On topographical grounds, the constituent neurons of the human NB-Ch4 complex can be subdivided into six sectors that occupy its anteromedial (Ch4am), anterolateral (NB-Ch4al), anterointermediate (NB-Ch4ai), intermediodorsal (NB-Ch4id), intermedioventral (NB-Ch4iv), and posterior (NB-Ch4p) regions.

Gorry[7] has shown that the NB displays a progressive evolutionary trend, becoming more and more extensive and differentiated in more highly evolved species, especially in primates and cetacea. Our observations are consistent with this general view and show that the human NB-Ch4 is a highly differentiated and relatively large structure. Although many morphological features of the human NB-Ch4 are similar to those described for the rhesus monkey, there is also a sense of increased complexity and differentiation. For example, a prominent Ch4ai sector is easily identified in the human brain but not in the rhesus monkey. In addition to these "compact" neuronal sectors, the Ch4 complex also contains "interstitial" elements that are embedded within the internal capsule, the diagonal band of Broca, the anterior commissure, the ansa peduncularis, the inferior thalamic peduncle, and the ansa lenticularis (FIG. 1). The physiological implications of this intimate association with fiber bundles are unknown. Conceivably, the NB-Ch4 complex, and especially its interstitial components, could monitor and perhaps influence the physiological activity along these fiber tracts. The presence of these interstitial components outside the traditional boundaries of the nucleus basalis is another reason why Ch4 and NB are not synonymous terms.

No strict delineation exists between the boundaries of NB-Ch4 and adjacent cell groups such as those of the olfactory tubercle, preoptic area, hypothalamic nuclei, striatal structures, nuclei of the diagonal band, amygdaloid nuclei, and globus pallidus. In addition to this "open" nuclear structure, the neurons of NB-Ch4 are heteromorphic in shape and have an isodendritic morphology with overlapping dendritic fields, many of which extend into fiber tracts traversing the basal forebrain. These characteristics are also present in the nuclei of the brain-stem reticular formation and have led to the suggestion that the NB-Ch4 complex could be conceptualized as a telencephalic extension of the brain-stem reticular core.[8]

All neurons of the Ch1–Ch4 cell groups contain AChE and ChAT in the perikarya, dendrites, and axons. Approximately 90% of Ch1–Ch4 neurons express the p75 low affinity nerve growth factor receptor (NGFr).[9,10] Nearly all Ch1–Ch4 cholinergic neurons of the human brain also express calbindin D28K.[11] Considerable interspecies differences are present in the cytochemical signature of basal forebrain cholinergic neurons.[11] For example, 20–30% of cholinergic neurons in the basal forebrain of the rat contain reduced nicotinamide-adenine-dinucleotide-phosphate-diaphorase (NADPHd) activity (which is now known to overlap with nitric oxide synthase activity[12]), whereas none of the basal forebrain cholinergic neurons in the monkey or human brain do so. Furthermore, the basal forebrain cholinergic neurons of the rat do not express calbindin D28K, whereas almost all Ch1–Ch4 neurons of the monkey and human do. Differences exist among primates as well. For example,

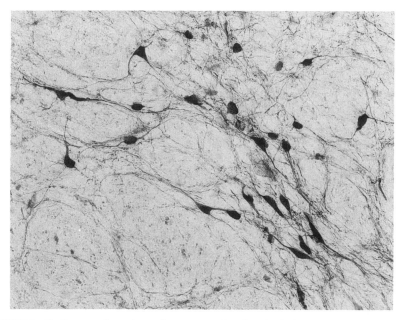

FIGURE 1. Choline acetyltransferase immunocytochemistry in the human brain shows interstitial cholinergic neurons of Ch4 embedded within the internal capsule. Dorsal is to the top and medial to the left. (Magnification, 150×; reduced to 75%.)

Ch1–Ch4 neurons of the monkey express galanin whereas this does not occur in the human brain.[13] Such cytochemical differences need to be taken into account when developing animal models for human diseases that affect the basal forebrain cholinergic cell groups.

Experimental neuroanatomical methods in the monkey brain have shown that different cortical areas receive their major cholinergic input from individual sectors of the NB-Ch4 complex. Thus, Ch4am provides the major source of cholinergic input to medial cortical areas including the cingulate gyrus; Ch4al to frontoparietal and opercular regions and the amygdaloid nuclei; Ch4id-Ch4iv to laterodorsal frontoparietal, peristriate, and midtemporal regions; and Ch4p to the superior temporal and temporopolar areas.[4] The experimental methods that are needed to reveal this topographical arrangement cannot be used in the human brain. However, indirect evidence for the existence of a similar topographical arrangement can be gathered from patients with Alzheimer's disease. We described two patients in whom extensive loss of cholinergic fibers in temporopolar but not frontal opercular cortex was associated with marked cell loss in the posterior (Ch4p) but not the anterior (Ch4am + Ch4al) sectors of Ch4.[3] This relationship is consistent with the topography of the projections in the monkey brain.

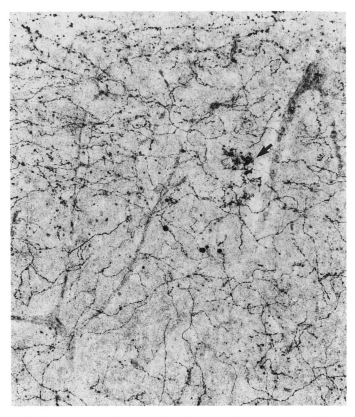

FIGURE 2. Choline acetyltransferase-positive (cholinergic) axons in area 18 of the human brain. Layer I is at the top, layer III at the bottom. The arrow points to a complex preterminal profile. (Magnification, 266×; reduced to 85%.)

The distribution of cholinergic innervation in the human cerebral cortex has been studied in detail with the help of AChE histochemistry, ChAT immunocytochemistry, and NGFr immunocytochemistry.[14–16] All cytoarchitectonic regions and layers of the cerebral cortex display a dense cholinergic innervation (FIG. 2). These fibers have numerous varicosities and, on occasion, complex preterminal profiles arranged in the form of dense clusters. The density of cholinergic axons is higher in the more superficial layers (layers I, II, and the upper parts of layer III) of the cerebral cortex.

Major and statistically significant differences also are found in the overall density of cholinergic axons among the various cytoarchitectonic areas. The cholinergic innervation of primary sensory, unimodal, and heteromodal association areas is significantly lighter than that of paralimbic and limbic areas. Within unimodal association areas, the density of cholinergic axons and varicosities is lower in the upstream (parasensory) sectors than in the downstream sectors. Within paralimbic regions, the nonisocortical sectors have a higher density of cholinergic innervation than the isocortical sectors. The highest density of cholinergic axons occurs in core limbic structures such as the hippocampus and the amygdala.

Within the hippocampal complex, the highest density of AChE-rich cholinergic fibers is seen in a thin band along the inner edge of the molecular layer of the dentate gyrus and within parts of the CA2, CA3, and CA4 sectors. The subiculum has a cholinergic innervation that is lighter than that of the other hippocampal sectors.[17] In the amygdala, each nucleus has a slightly different profile of cholinergic innervation.[18] The density is highest in the central and basal lateral nuclei and lightest in the lateral nucleus. The medial nucleus is the only region of the amygdala that has virtually no cholinergic innervation.

In all cortical and hippocampal fields, NGFr axonal staining is of approximately equivalent density to that of axonal ChAT, providing further evidence that the majority of cholinergic innervation to these regions arises from the Ch1–Ch4 cell groups.[19] The one exception occurs in the amygdala, especially in the basolateral nucleus, which contains very light NGFr staining, raising the possibility that the cholinergic innervation to this nucleus and perhaps to other parts of the amygdala arises from NGFr-negative Ch1–Ch4 neurons or from cholinergic neurons in the brain stem.

POSTSYNAPTIC COMPONENTS OF CORTICAL CHOLINERGIC PATHWAYS

Electronmicroscopic studies in rodents indicate that most cortical cholinergic axons are unmyelinated and that they make symmetrical and asymmetrical synaptic contacts with large numbers of cortical neurons.[20,21] It is also thought that some acetylcholine may be released outside of traditional synaptic contacts and that it may exert its effect by diffusion into receptor-containing sites.[22]

The acetylcholine released from presynaptic cholinergic axons of the cerebral cortex exert their neurotransmitter effects through the mediation of nicotinic and muscarinic receptors. Muscarinic receptors predominate in the mammalian cerebral cortex. Five subtypes of muscarinic cholinergic receptors (m1–m5) have been recognized, each the product of a different gene.[23,24] Three muscarinic receptor subtypes have been characterized pharmacologically (M1–M3), and of these the M1 and M2 subtypes have received the greatest attention. Autoradiographic experiments in the rhesus monkey showed that the pirenzepine-sensitive M1 receptors were far more numerous than M2 receptors. The M1 receptor density reaches the highest levels in components of limbic and association cortex. In contrast, the M2 receptors reach their highest densities in primary sensory and motor areas of the cortex.[25]

Immunocytochemical studies in the human brain have identified cortical neurons which express nicotinic and muscarinic receptors. Such neurons are localized predominantly in the pyramidal neurons of layers III and V. Approximately 30% of immunopositive pyramidal neurons were found to display immunoreactivity for both muscarinic and nicotinic receptors.[26]

It is thought that all cholinoceptive neurons express AChE in order to hydrolyze acetylcholine. However, only a subset of cholinoceptive neurons give an AChE-rich histochemical reaction.[27,28] Some of these AChE-rich neurons are polymorphic in shape and are distributed preferentially in the deeper cortical layers and the subjacent white matter. Others are pyramidal in shape and are located in layers III and V. Cholinergic axons are thought to express presynaptic autoreceptors which may be involved in the autoregulation of acetylcholine release.

PHYSIOLOGICAL AND BEHAVIORAL IMPLICATIONS

The physiological effect of acetylcholine on cholinoceptive cortical neurons is exceedingly complex. The major effect of acetylcholine is to cause a relatively prolonged reduction of potassium conductance so as to make cortical cholinoceptive neurons more susceptible to other excitatory inputs.[29,30] However, the effect of acetylcholine on cortical neurons can also be inhibitory, either directly or through the mediation of GABAergic interneurons.

Because all regions of the cerebral cortex receive intense cholinergic innervation, it is not surprising that all aspects of cortical function are influenced by cholinergic neurotransmission. In primary visual cortex, for example, cholinergic stimulation does not alter the orientation specificity of a given neuron but increases the likelihood that the neuron will fire in response to its preferred stimulus.[26] An analogous effect has been described in somatosensory cortex.[31]

The Ch1–Ch4 cell groups of the basal forebrain can be considered as a telencephalic extension of the brain-stem reticular formation and also as a direct extension of basomedial limbic cortex. This dual identity helps to explain why arousal and memory are the two major behavioral affiliations of the Ch1–Ch4 cell groups. Experiments in rats have shown that the cortical cholinergic projections from the basal forebrain plays a major role in sustaining at least one component of the hippocampal theta rhythm and also the arousal-related low voltage fast activity of the cortical EEG.[32,33] In a number of animal species, lesions of the Ch4 cell group can cause severe impairments of memory that can be reversed by the systemic administration of agonists.[34,35]

Single-unit studies in monkeys have shown that the neurons of the NB (Ch4) are particularly sensitive to stimulus novelty and to the motivational relevance of sensory cues.[36,37] The novelty and behavioral significance of a sensory event can therefore influence the cortical release of acetylcholine which, in turn, modulates the cortical response to the sensory event. Cortical cholinergic pathways are thus in a position to alter the neural impact of sensory experiences according to their behavioral significance. It is easy to see how such a circuitry would have a major influence on cortical arousal. In keeping with this interpretation, the muscarinic blocking agent scopolamine attenuates the cortical P-300 arousal response that is normally elicited by novel or surprising stimuli.[38]

The relationship of the Ch1–Ch4 cell groups and of cortical cholinergic innervation to memory function is quite complex. Limbic and paralimbic regions of the cerebral cortex are known to play a critical role in memory and learning. The preferential concentration of cholinergic innervation in these parts of cortex may explain why cholinergic antagonists and cholinoactive drugs seem to have a preferential effect on memory, learning, and other limbic functions such as mood, motivation, and aggression.[39–41] The role of acetylcholine in hippocampal long-term potentiation[42] may provide another mechanism that underlies the relationship of cholinergic pathways to memory.

Recent brain slice experiments in piriform cortex of the rat have shown that acetylcholine can selectively suppress intrinsic synaptic transmission through a presynaptic mechanism, while leaving extrinsic afferent input unaffected. This selective suppression could prevent interference from previously stored patterns during the learning of new patterns. Hasselmo and colleagues[43] argued that this could provide a novel mechanism through which cortical cholinergic innervation could participate in new learning. Buzsaki[44] proposed a model according to which the cholinergic innervation, especially of the hippocampal complex, plays a major role in switching from on-line attentive processing, characterized by the hippocampal theta rhythm, to an off-line period of consolidation, characterized by sharp wave activity (see ref. 45 for review).

Another mechanism that links cholinergic axons to memory and learning may be related to the differential regional density of cortical cholinergic innervation. Experimental evidence leads to the conclusion that sensory-limbic pathways play pivotal roles in a wide range of behaviors related to emotion, motivation, and especially memory.[46,47] The process starts within the primary sensory areas of the cerebral cortex which provide a portal for the entry of sensory information into cortical circuitry. These primary areas project predominantly to upstream (parasensory) unimodal sensory association areas, which, in turn, project to downstream unimodal areas and heteromodal cortex. The heteromodal and downstream unimodal areas collectively provide the major source of sensory information into paralimbic and limbic areas of the brain. Our observations show that the density of cholinergic innervation is lower within unimodal and heteromodal association areas than in paralimbic areas of the brain. In the unimodal areas, moreover, the downstream sectors have a higher density of cholinergic innervation than the upstream sectors. Core limbic areas such as the amygdala and hippocampus contain the highest densities of cholinergic innervation. This pattern of differential distribution led us to suggest that sensory information is likely to come under progressively greater cholinergic influence as it is conveyed along the multisynaptic pathways leading to the limbic system. As a consequence of this arrangement, cortical cholinergic innervation may help to channel (or gate) sensory information into and out of the limbic system in a way that is sensitive to the behavioral relevance of the associated experience. The memory disturbances that arise after damage to the Ch1–Ch4 cell groups or after the systemic administration of cholinergic antagonists may therefore reflect a disruption of sensory-limbic interactions which are crucial for effective memory and learning.

TOWARDS AN EXPANDED ASCENDING RETICULAR ACTIVATING SYSTEM

In addition to Ch1–Ch4, two cholinergic cell groups in the upper brain stem, the Ch5 neurons of the pedunculopontine nucleus and the Ch6 cell group of the laterodorsal tegmental nucleus, are also intimately involved in the modulation of arousal. The Ch5 and Ch6 cell groups provide the major cholinergic innervation of the thalamus. Moruzzi and Magoun[48] had introduced the concept of a brain-stem ascending reticular activating system (ARAS) that acted to desynchronize the cortical electroencephalogram via a relay in the thalamus. Subsequent work revealed that a most important component in this system consists of a cholinergic reticulothalamic pathway that facilitates the activation of corticopetal relay neurons in the thalamus.[49–53]

The physiological relevance of this pathway to the reticular activating system was demonstrated by Kayama and colleagues.[54] They identified the Ch5–Ch6 neurons with NDPHd histochemistry and showed that electrical stimulation of these neurons causes a scopolamine-sensitive activation of lateral geniculate neurons and even an occasional enhancement of their response to photic stimulation. Electrical stimulation of Ch5 also causes a hyperpolarization of GABAergic neurons in the reticular nucleus of the thalamus. Because the neurons of the reticular nucleus have an inhibitory effect on thalamic relay neurons, the net effect of Ch5 stimulation is to disinhibit thalamic relay nuclei.[30] Thus, the Ch5–Ch6 neurons can facilitate the transthalamic (and ultimately corticopetal) processing of sensory information in ways that could further modulate arousal and attention.

These observations show that the original concept of the ARAS needs to be expanded to include at least two sources of ascending cholinergic projections, a traditional one in the upper brain stem (Ch5–Ch6) and a second one in the basal forebrain (Ch1–Ch4). Noncholinergic regulatory pathways that arise from the hypothalamus (histaminergic), ventral tegmental area (dopaminergic), nucleus locus coeruleus (noradrenergic), and brain-stem raphe (serotonergic) and that send widespread projections to the cerebral cortex and thalamus are also part of this expanded ARAS (see ref. 55 for review). Each of these cholinergic and noncholinergic projections can exert a powerful influence on the information processing state of the thalamus and cerebral cortex in ways that influence attentional, emotional, motivational, and arousal states. The collective activity of these ascending regulatory pathways provides the physiological matrix (or state) within which the discrete, point-to-point projections that interconnect cortex, thalamus, and the basal ganglia can set the vectors of complex behaviors related to cognition and comportment. The rapidly accumulating information on the ascending cholinergic projections provides a blueprint for investigating the characteristics of the other components of this extremely important neural system in the human brain.

ACKNOWLEDGMENTS

I want to thank Leah Christie for expert secretarial assistance.

REFERENCES

1. MESULAM, M-M. 1988. Central cholinergic pathways: Neuroanatomy and some behavioral implications. *In* Neurotransmitters and Cortical Function. M. Avoli, T. A. Reader, R. W. Dykes & P. Gloor, Eds.: 237–260. Plenum Publishing. New York.
2. WAINER, B. H. & M-M. MESULAM. 1990. Ascending cholinergic pathways in the rat brain. *In* Brain Cholinergic Systems. M. Steriade & D. Biesold, Eds. **2**: 65–119. Oxford University Press.
3. MESULAM, M-M. & C. GEULA. 1988. Nucleus basalis (Ch4) and cortical cholinergic innervation of the human brain: Observations based on the distribution of acetylcholinesterase and choline acetyltransferase. J. Comp. Neurol. **275**: 216–240.
4. MESULAM, M-M., E. J. MUFSON, A. I. LEVEY & B. H. WAINER. 1983. Cholinergic innervation of cortex by the basal forebrain: Cytochemistry and cortical connections of the septal area, diagonal band nuclei, nucleus basalis (substantia innominata) and hypothalamus in the rhesus monkey. J. Comp. Neurol. **214**: 170–197.
5. ARENDT, T., V. BIGL, A. TENNSTEDT & A. ARENDT. 1985. Neuronal loss in different parts of the nucleus basalis is related to neuritic plaque formation in cortical target areas in Alzheimer's disease. J. Neurosci. **14**: 1–14.
6. VIJAYASHANKAR, N. & H. BRODY. 1979. A quantitative study of the pigmented neurons in the nuclei locus coeruleus and sub coeruleus in man as related to aging. J. Neuropathol. Exp. Neurol. **38**: 490–497.

7. GORRY, J. D. 1963. Studies on the comparative anatomy of the ganglion basale of Meynert. Acta Anat. **55:** 51–104.
8. RAMON-MOLINER, E. & W. J. H. NAUTA. 1966. The isodendritic core of the brain. J. Comp. Neurol. **126:** 311–336.
9. MESULAM, M-M., C. GEULA, M. A. BOTHWELL & L. B. HERSH. 1989. Human reticular formation: Cholinergic neurons of the pedunculopontine and laterodorsal tegmental nuclei and some cytochemical comparisons to the forebrain cholinergic neurons. J. Comp. Neurol. **281:** 611–633.
10. MUFSON, E. J., M. BOTHWELL, L. B. HERSH & J. H. KORDOWER. 1989. Nerve growth factor receptor immunoreactive profiles in the normal, aged human basal forebrain: Colocalization with cholinergic neurons. J. Comp. Neurol. **285:** 196–217.
11. GEULA, C., C. R. SCHATZ & M-M. MESULAM. 1993. Differential localization of NADPH-diaphorase and calbindin-D 28K within the cholinergic neurons of the basal forebrain, striatum and brainstem in the rat, monkey, baboon and human. Neuroscience **54:** 461–476.
12. VINCENT, S. R. & B. T. HOPE. 1992. Neurons that say no. TINS **15:** 108–113.
13. KORDOWER, J. H. & E. J. MUFSON. 1990. Galanin-like immunoreactivity within the primate basal forebrain: Differential staining patterns between humans and monkeys. J. Comp. Neurol. **294:** 281–292.
14. GEULA, C. & M-M. MESULAM. 1989. Cortical cholinergic fibers in aging and Alzheimer's disease: A morphometric study. Neuroscience **33:** 469–481.
15. MESULAM, M-M. & C. GEULA. 1992. Overlap between acetylcholinesterase-rich and choline acetyltransferase-positive (cholinergic) axons in human cerebral cortex. Brain Res. **577:** 112–120.
16. MESULAM, M-M., L. B. HERSH, D. C. MASH & C. GEULA. 1992. Differential cholinergic innervation within functional subdivisions of the human cerebral cortex: A choline acetyltransferase study. J. Comp. Neurol. **318:** 316–328.
17. GREEN, R. C. & M-M. MESULAM. 1988. Acetylcholinesterase fiber staining in the human hippocampus and parahippocampal gyrus. J. Comp. Neurol. **273:** 488–499.
18. EMRE, M., S. HECKERS, D. C. MASH, C. GEULA & M-M. MESULAM. 1993. Cholinergic innervation of the amygdaloid complex in the human brain and its alterations in old age and Alzheimer's disease. J. Comp. Neurol. **336:** 117–134.
19. MESULAM, M-M., D. MASH, L. HERSH, M. BOTHWELL & C. GEULA. 1992. Cholinergic innervation of the human striatum, globus pallidus, subthalamic nucleus, substantia nigra and red nucleus. J. Comp. Neurol. **323:** 252–268.
20. FROTSCHER, M. & C. LERANTH. 1985. Cholinergic innervation of the rat hippocampus as revealed by choline acetyltransferase immunocyto-chemistry: A combined light and electron microscopic study. J. Comp. Neurol. **239:** 237–246.
21. WAINER, B. H., J. P. BOLAM, T. F. FREUND, Z. HENDERSON, S. TOTTERDELL & A. D. SMITH. 1984. Cholinergic synapses in the rat brain: A correlated light and electron micro-scopic immunohistochemical study employing a monoclonal antibody against choline acetyltransferase. Brain Res. **308:** 69–76.
22. UMBRIACO, D., K. F. WATKINS, L. DESCARRIES, C. COZZARI & B. HARTMAN. 1990. Ultrastructural features of acetylcholine axon terminals in adult rat cerebral cortex. Soc. Neurosci. Abstr. **16:** 1057.
23. BONNER, T. I. 1989. The molecular basis of muscarinic receptor diversity. TINS **12:** 148–151.
24. HOSEY, M. M. 1992. Diversity of structure, signaling and regulation within the family of muscarinic cholinergic receptors. FASEB J. **6:** 845–852.
25. MASH, D. C., W. F. WHITE & M-M. MESULAM. 1988. Distribution of muscarinic receptor subtypes within architectonic subregions of the primate cerebral cortex. J. Comp. Neurol. **278:** 265–274.
26. SATO, H., V. HATA, K. HAGIHARA & T. TSUMOTO. 1987. Effects of cholinergic depletion on neuron activities in the cat visual cortex. J. Neurophysiol. **58:** 781–794.
27. MESULAM, M-M. & C. GEULA. 1988. Acetylcholinesterase-rich pyramidal neurons in the human neocortex and hippocampus: Absence at birth, development during the life span, and dissolution in Alzheimer's disease. Ann. Neurol. **24:** 765–773.

28. MESULAM, M-M. & C. GEULA. 1991. Acetylcholinesterase-rich neurons of human cerebral cortex: Cytoarchitectonic and ontogenetic patterns of distribution. J. Comp. Neurol. **306:** 193–220.

29. McCORMICK, D. A. 1990. Cellular mechanisms of cholinergic control of neocortical and thalamic neuronal excitability. *In* Brain Cholinergic Systems. M. Steriade & D. Biesold, Eds.: 236–264. Oxford University Press.

30. STERIADE, M., P. GLOOR, R. R. LLINAS, F. H. LOPES DA SILVA & M-M. MESULAM. 1990. Basic mechanisms of cerebral rhythmic activities. Electroencephalogr. Clin. Neurophysiol. **76:** 481–508.

31. METHERATE, R., N. TREMBLAY & R. W. DYKES. 1988. The effects of acetylcholine on response properties of cat somatosensory cortical neurons. J. Neurophysiol. **59:** 1231–1252.

32. BUZSAKI, G., R. G. BICKFORD, G. PONOMAREFF, L. J. THAL, R. MANDEL & F. H. GAGE. 1988. Nucleus basalis and thalamic control of neocortical activity in the freely moving rat. J. Neurosci. **8:** 4007–4026.

33. STEWARD, D. F., D. F. MACFABE & C. H. VANDERWOLF. 1984. Cholinergic activation of the electrocorticogram: Role of the substantia innominata and effects of atropine and quinuclidinyl benzilate. Brain Res. **322:** 219–232.

34. IRLE, E. & H. J. MARKOWITSCH. 1987. Basal forebrain-lesioned monkeys are severely impaired in tasks of association and recognition memory. Ann. Neurol. **22:** 735–743.

35. RIDLEY, R. M., T. K. MURRAY, J. A. JOHNSON & H. F. BAKER. 1986. Learning impairment following lesion of the basal nucleus of Meynert in the marmoset: Modification by cholinergic drugs. Brain Res. **376:** 108–116.

36. WILSON, F. A. W. & E. T. ROLLS. 1990. Neuronal responses related to reinforcement in the primate basal forebrain. Brain Res. **509:** 213–231.

37. WILSON, F. A. W. & E. T. ROLLS. 1990. Neuronal responses related to novelty and familiarity of visual stimuli in the substantia innominata, diagonal band of Broca and periventricular region of the primate basal forebrain. Exp. Brain Res. **80:** 104–120.

38. HAMMOND, E. J., K. J. MEADOR, R. AUNQ-DIN & B. J. WILDER. 1987. Cholinergic modulation of human P3 event-related potentials. Neurology **37:** 346–350.

39. GILLIN, J. C., L. SUTTON, C. RUIZ, J. KELSOE, R. M. DUPONT, D. DOVKO, S. C. RISCH, S. GOLSHAN & D. JANOWSKY. 1991. The cholinergic rapid eye movement induction test with arecholine in depression. Arch. Gen. Psychiatry **48:** 264–270.

40. YEOMANS, J. S., O. KOFMAN & V. McFARLANE. 1984. Cholinergic involvement in lateral hypothalamic rewarding brain stimulation. Brain Res. **329:** 19–26.

41. YOSHIMURA, H. & S. UEKI. 1977. Biochemical correlates in mouse-killing behavior of the rat: Prolonged isolation and brain cholinergic function. Pharmacol. Biochem. Behav. **6:** 193–196.

42. TANAKA, Y., M. SAKURAI & S. HAYASHI. 1989. Effect of scopolamine and HP029, a cholinesterase inhibitor, on long-term potentiation in hippocampal slices of guinea pig. Neurosci. Lett. **98:** 179–183.

43. HASSELMO, M. E., B. P. ANDERSON & J. M. BOWER. 1992. Cholinergic modulation of cortical associative memory function. J. Neurophysiol. **67:** 1230–1246.

44. BUZSAKI, G. 1989. Commentary. Two-stage model of memory trace formation: A role for "noisy" brain states. Neuroscience **31:** 551–570.

45. CHURCHLAND, P. S. & T. J. SEJNOWSKI. 1992. The computational brain. Plasticity: Cells, circuits, brains and behavior.: 239–329. MIT Press. Cambridge, MA.

46. MESULAM, M-M. 1985. Patterns in behavioral neuroanatomy. *In* Principles of Behavioral Neurology, Contemporary Neurology Series. M-M. Mesulam, Ed.: 1–70. F. A. Davis Co. Philadelphia, PA.

47. MISHKIN, M. 1982. A memory system in the monkey. Philos. Trans. R. Soc. Lond. B. **298:** 85–92.

48. MORUZZI, G. & H. W. MAGOUN. 1949. Brain stem reticular formation and activation of the EEG. Electroencephalogr. Clin. Neurophysiol. **1:** 459–473.

49. DINGLEDINE, R. & J. S. KELLY. 1977. The brainstem stimulation and the acetylcholine-invoked inhibition of neurons in the feline nucleus reticularis thalami. J. Physiol. **271:** 135–154.

50. HOOVER, D. B. & R. H. BAISDEN. 1980. Localization of putative cholinergic neurons innervation the anteroventral thalamus. Brain Res. Bull. **5:** 519–524.
51. HOOVER, D. B. & D. M. JACOBOWITZ. 1979. Neurochemical and histochemical studies of the effect of a lesion of the nucleus cuneiformis on the cholinergic innervation of discrete areas of the rat brain. Brain Res. **70:** 113–122.
52. MCCANCE, I., J. W. PHILLIS & R. A. WESTERMAN. 1986. Acetylcholine-sensitivity of thalamic neurons: Its relationship to synaptic transmission. Br. J. Pharmacol. **32:** 635–651.
53. PHILLIS, J. W., A. K. TEBECIS & D. H. YORK. 1967. A study of cholinoceptive cells in the lateral geniculate nucleus. J. Physiol. **192:** 695–713.
54. KAYAMA, J., M. TAKAGI & T. OGAWA. 1986. Cholinergic influence of the laterodorsal tegmental nucleus on neuronal activity in the rat lateral geniculate nucleus. J. Neurophysiol. **56:** 1297–1309.
55. MESULAM, M-M. 1990. Large scale neurocognitive networks and distributed processing for attention, language and memory. Ann. Neurol. **28:** 597–613.

Regulation of Muscarinic Acetylcholine Receptor Expression and Function[a]

JACQUES C. MIGEON, PHYLLIS S. GOLDMAN,
BETH A. HABECKER, AND NEIL M. NATHANSON[b]

Department of Pharmacology, SJ-30
University of Washington
Seattle, Washington 98195

Muscarinic acetylcholine receptors (mAChR) are members of the superfamily of G-protein-coupled cell surface receptors that are characterized by the presence of seven putative transmembrane domains. Other members of this family include alpha- and beta-adrenergic receptors, the opsins, olfactory receptors, the dopamine receptors, and the receptors for many neuropeptide and glycoprotein hormones (see Dohlman *et al.*[1] for review). The mAChR themselves represent a small gene family: the genes for five subtypes of mAChR have been identified by molecular cloning. Muscarinic receptors are present in both the central and peripheral nervous systems, cardiac and smooth muscle, and various exocrine glands (see Nathanson[2] and Bonner[3] for review). Muscarinic receptors produce functional responses either by regulating the activity of enzymes such as adenylyl cyclase (AC) or phospholipase C (PLC) which produce intracellular second messengers, or by regulation of ion channels such as potassium and calcium channels. The m1, m3, and m5 receptors couple preferentially to stimulation of PLC, and the m2 and m4 receptors couple preferentially to inhibition of AC. However, the specificity of mAChR functional coupling is dependent both on levels of receptor expression and on the cell type in which a given receptor or subtype is expressed (see Tietje & Nathanson[4] and references therein). Muscarinic receptor number can be altered in response to sustained agonist exposure. Short-term agonist exposure (s to min) causes a rapid removal of mAChR from the cell surface (sequestration) whereas agonist exposure for longer periods of time (h) causes a decrease in total receptor number (down-regulation). In this report we describe studies on the functional analysis of mAChR coupling mechanisms using a reporter gene system for the detection of mAChR-mediated changes in intracellular cAMP levels. We also demonstrate that a tyrosine residue in the intracellular carboxyl tail of the m2 receptor is important in agonist-mediated down-regulation but is not required for agonist-mediated functional responsiveness or receptor sequestration. Finally, we also demonstrate that long-term agonist activation of mAChR in chick heart cells leads to decreases in the level of mAChR mRNA levels.

[a]This research was supported by grants from the National Institutes of Health, and a grant-in-aid from the American Heart Association.
[b]Corresponding author.

REGULATION OF cAMP-MEDIATED GENE EXPRESSION
BY MUSCARINIC RECEPTORS AND G-PROTEINS

We used a luciferase reporter gene under the transcriptional control of a cAMP response element (CRE) to measure the regulation of intracellular cAMP levels and cAMP-regulated gene expression in response to mAChR activation.[5] Treatment of transiently transfected JEG-3 human choriocarcinoma cells with forskolin, which increases intracellular cAMP levels due to the activation of adenylyl cyclase, results in a greater than 10-fold increase in luciferase activity. Expression of the CRE-luciferase reporter gene in JEG-3 cells is not significantly regulated by changes in intracellular calcium levels. Treatment of untransfected cells with the muscarinic agonist carbachol does not result in an increase in luciferase expression, even though the cells express a low level of endogenous mAChR which mediate a 3–5-fold carbachol-stimulation of phosphatidylinositol turnover. Furthermore, treatment with the calcium ionophore A23187 or with the phorbol 12-myristate,13-acetate, or both together only results in small increases in luciferase expression. Treatment of JEG-3

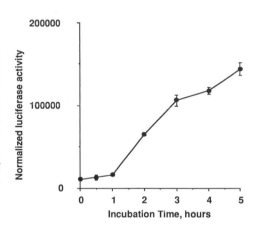

FIGURE 1. Time course of m1-mediated increase in luciferase activity. Two hours are required to detect agonist stimulation of CRE-mediated luciferase expression in m1-transfected JEG-3 cells. Cells transfected with the m1 receptor and the CRE-luciferase reporter gene were treated with carbachol (1 mM) for varying amounts of time prior to harvesting and determination of normalized luciferase activity. Data are the mean of triplicate determinations ± SD. CRE, cAMP response element.

cells transiently transfected with mouse m1 and chick m4 mAChR results in 8–10-fold and 3–5-fold increases, respectively, in luciferase activity. In order to determine the amount of time necessary to detect changes in luciferase expression, m1 transfected cells were incubated with carbachol for varying amounts of time and harvested immediately thereafter (FIG. 1). Increases in luciferase activity are first detectable at the 2-h time point and level off between 3 and 5 h. In order to determine the minimal duration of stimulation required to observe a subsequent increase in luciferase gene activity, JEG-3 cells were treated with either carbachol or forskolin for varying amounts of time, after which the drug was removed and the cells were washed and reincubated with fresh media. All cultures were harvested 5 h from the beginning of the drug treatment period (FIG. 2). Stimulation of luciferase activity mediated by m1 mAChR was easily detectable when the carbachol was removed after only a 5-min incubation and peaked after 3–5 h of incubation. The carbachol-mediated responses are as sensitive to inhibition by cotransfection with a cAMP-dependent protein kinase (PKA) inhibitor peptide or a dominant negative mutant PKA regulatory subunit as is the forskolin-stimulated response. Thus, m1- and m4-mediated responses are dependent upon a functional cAMP-dependent protein kinase.

The m1- and m4-mediated increases in luciferase expression are not blocked by pretreatment with pertussis toxin, and the m4 response was potentiated by it. Thus, the G-proteins G_i and G_o are not involved. The most reasonable hypothesis to explain the apparent increase in intracellular cAMP is that the m1 and m4 receptors can activate adenylyl cyclase by interacting with the stimulatory G-protein G_s. Consistent with this hypothesis, both the m1[6] and m4[7] receptors can activate adenylyl cyclase activity in membranes prepared from stably transfected cells.

Atropine treatment of JEG-3 cells transiently transfected with high levels of m4 mAChR, but not m1, causes an elevation in basal levels of CRE-mediated luciferase expression in the absence of agonist.[5] Furthermore, treatment of m4- but not m1-transfected cells with pertussis toxin caused a significant increase in the level of luciferase expression in the absence of agonist. These results suggest that the m4 receptor is spontaneously active and can cause constitutive inhibition of adenylyl

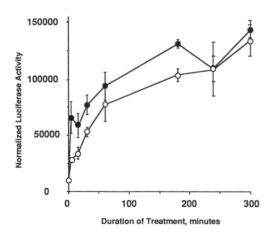

FIGURE 2. Induction of luciferase expression after carbachol or forskolin treatment. Five minutes of carbachol or forskolin treatment are sufficient to detect stimulation of CRE-mediated luciferase expression in m1-transfected JEG-3 cells. Cells transfected with the m1 receptor and the CRE-luciferase reporter gene were treated with carbachol (1 mM; *open circles*) or forskolin (1 μM; *closed circles*) for varying amounts of time (0–5 h), after which the cells were washed and allowed to continue incubation until 5 h after the initial administration of drug. Data are plotted as normalized luciferase activity, and values are the mean of triplicate determinations ± SD. CRE, cAMP response element.

cyclase that is relieved by treatment with atropine. This suggests that the unbound m4 receptor exhibits spontaneous activity and that binding of antagonist induces a change in the conformation of the receptor that prevents it from interacting with G-proteins. This hypothesis is supported by the observation that atropine dissociates mAChR–G-protein complexes in rat heart membranes.[8] Furthermore, electrophysiological evidence indicates that mAChR in the heart may exhibit spontaneous activity that can be blocked by antagonist treatment.[9,10] Our results demonstrate that an antagonist can produce a subtype-specific functional response in the absence of agonist.

Surprisingly, the m4 receptor exhibits little if any agonist-induced inhibition of either basal or forskolin-stimulated luciferase expression at either low or high levels of receptor expression.[5] JEG-3 cells express $G_i\alpha$-1 and $G_i\alpha$-3, but not $G_i\alpha$-2. Cotransfection of $G_i\alpha$-1 or $G_i\alpha$-3 does not alter the m4-mediated response, whereas cotransfection with $G_i\alpha$-2 greatly increases the ability of the m4 receptor to inhibit forskolin-stimulated luciferase expression. Thus, the m4 receptor requires $G_i\alpha$-2 for optimal agonist-mediated inhibition of adenylyl cyclase activity.

DIFFERENTIAL ROLE OF CARBOXYL-TERMINAL TYROSINE IN AGONIST-INDUCED DOWN-REGULATION OF THE m2 RECEPTOR

Tyrosine residues located in the cytoplasmic tails of many membrane receptors are necessary for agonist-induced internalization of these receptors via clathrin-coated pits (see Goldman & Nathanson[11] for references). In addition, either of the two tyrosine residues located in the cytoplasmic tail of the β_2-adrenergic (β_2AR) is required for the down-regulation but not the sequestration of receptors in response to agonist.[12] The mammalian m2 receptor has a single tyrosine residue (Tyr-459) in the carboxyl terminal cytoplasmic domain. We have used site-directed mutagenesis to determine the role of this tyrosine residue in the regulation and function of the m2 receptor.[11] We found that substitution of Tyr-459 with Phe, Trp, Ala, or Ile affected neither agonist nor antagonist binding, and did not alter the ability of the m2 receptor to inhibit the accumulation of intracellular cAMP. Although elimination of Tyr-459 did not affect agonist-induced sequestration of the m2 receptor, the sensitivity as well as the rate and extent of agonist-induced down-regulation was attenuated in the Ala, Phe, and Trp mutant m2 receptors. Mutation of the Tyr to Ile resulted in a slight decrease in the sensitivity of the m2 receptor to agonist-induced down-regulation, although the maximal decrease in receptor number was not significantly different from the wild-type receptor.

The attenuation of down-regulation observed with the Phe and Trp mutations indicates that the aromatic side chain of the Tyr residue is not sufficient for down-regulation of the m2 receptor. The similar time courses for agonist-induced down-regulation between the wild-type and Ile mutant suggest that the hydroxyl group of Tyr-459 is not required for agonist-induced down-regulation of the m2 receptor. The results suggest that Tyr-459 is important as part of a structural motif in which subtle changes in amino acid structure or size might perturb the overall structure of this domain. These results are the first to identify a site in the mAChR involved in agonist-induced down-regulation and the first to make mutations that affect the long-term down-regulation, but not the rapid sequestration of the mAChR.

REGULATION OF mAChR mRNA BY mAChR ACTIVATION

Cloning studies have demonstrated that at least two subtypes of mAChR are expressed in the embryonic chick heart, the cm2 and cm4 receptors.[4,13] We found that persistent activation of the mAChR in cultured chick heart cells with the cholinergic agonist carbachol causes significant decreases in the levels of both cm2 and cm4 mRNA. The level of mAChR mRNA was measured by solution hybridization using subtype-specific riboprobes. Treatment with carbachol did not alter the half-lives of either the cm2 or the cm4 mRNAs, indicating that the agonist most likely regulates mRNA levels by regulating the rate of gene transcription.[14] The regulation of mAChR mRNA by the agonist has important functional consequences, as it regulates the rate of reappearance of muscarinic receptor number during the recovery from agonist-induced down-regulation after agonist withdrawal.[15]

The regulation of mAChR mRNA levels most likely results from a change in the level of one or more intracellular second messenger. In chick heart, activation of mAChR causes both inhibition of adenylyl cyclase activity and stimulation of PLC activity. Several different approaches were used to test the role of second messenger systems in the regulation of mAChR mRNAs. Embryonic chick heart expresses A1 adenosine receptors, which inhibit AC but do not activate PLC, and angiotensin II receptors, which stimulate PLC but do not inhibit AC. Although activation of each receptor separately had relatively little effect on mAChR mRNA levels, concomitant activation of both of these heterologous receptor pathways resulted in significantly greater decreased mAChR mRNA levels. These results suggest that regulation of both adenylyl cyclase and PLC activities are involved in the regulation of mAChR gene expression by mAChR activation.[14]

We also used pharmacological and biochemical experiments to test if both second messenger pathways were involved in regulation of mAChR mRNA levels following homologous receptor activation.[15] Treatment of cells with pertussis toxin, which blocks coupling of muscarinic receptors to AC but not PLC, blocked the muscarinic receptor-mediated decrease of both cm2 and cm4 mRNA levels. Thus, inhibition of AC activity was required for homologous regulation of mAChR mRNA. Incubation of chick heart cells with the partial agonist pilocarpine, which causes inhibition of AC activity but not stimulation of PLC, resulted in significantly smaller decreases of receptor mRNA levels than agonists such as carbachol that regulate both second messenger systems. Thus, coupling to both the AC and PLC pathways is required for maximal regulation of cm2 and cm4 mRNA levels in response to mAChR activation.

CONCLUSIONS

We showed that the CRE-luciferase assay is a convenient system for the determination of mAChR-mediated changes in intracellular cAMP levels. This system should be useful for the determination of the G-protein coupling specificity of other mAChR subtypes as well as structure-function analyses of G-protein subunits. The identification of a potential down-regulation motif in the m2 mAChR will allow the determination of whether similar sequences have similar roles in other mAChR subtypes. Finally, the identification of regulatory sequences in the mAChR genes will allow determination of the molecular and cellular basis for the regulation of muscarinic receptor gene expression following receptor activation by acetylcholine.

REFERENCES

1. DOHLMAN, H. G., M. G. CARON & R. J. LEFKOWITZ. 1987. A family of receptors coupled to guanine nucleotide regulatory proteins. Biochemistry **26:** 2657–2664.
2. NATHANSON, N. M. 1987. Molecular properties of the muscarinic acetylcholine receptor. Annu. Rev. Neurosci. **10:** 195–236.
3. BONNER, T. I. 1989. The molecular basis of muscarinic receptor diversity. Trends Neurosci. **12:** 148–151.
4. TIETJE, K. M. & N. M. NATHANSON. 1991. Embryonic chick heart expresses multiple muscarinic acetylcholine receptor subtypes: Isolation and characterization of a gene encoding a novel m2 muscarinic acetylcholine receptor with a high affinity for pirenzipine. J. Biol. Chem. **266:** 17382–17387.
5. MIGEON, J. C. & N. M. NATHANSON. 1994. Differential regulation of cAMP-mediated gene expression by m1 and m4 muscarinic acetylcholine receptors: Preferential coupling of m4 receptors to Giα-2. J. Biol. Chem. **269:** 9767–9773.
6. GURWITZ, D., R. HARING, E. HELDMAN, C. M. FRASER, D. MANOR & A. FISHER. 1994.

Discrete activation of transduction pathways associated with acetylcholine m1 receptor by several muscarinic ligands. Eur. J. Pharmacol. **267:** 21–31.

7. DITTMAN, A. H., J. P. WEBER, T. J. HINDS, E.-J. CHOI, J. C. MIGEON, N. M. NATHANSON & D. R. STORM. 1994. A novel mechanism for coupling of m4 muscarinic acetylcholine receptors to calmodulin-sensitive adenylyl cyclases: Crossover from G-protein coupled inhibition to stimulation. Biochemistry **33:** 943–951.

8. MATESIC, D. V. & G. R. LUTHIN. 1991. Atropine dissociates complexes of muscarinic acetylcholine receptor and guanine nucleotide-binding protein in heart membranes. FEBS Lett. **284:** 184–186.

9. SOEJIMA, M. & A. NOMA. 1984. Mode of regulation of the ACh-sensitive K-channel by the muscarinic receptor in rabbit atrial cells. Pfluegers Arch. **400:** 424–431.

10. HANF, R., Y. LI, G. SZABO & R. FISCHMEISTER. 1993. Agonist-independent effects of muscarinic antagonists on Ca^{2+} and K^+ currents in frog and rat cardiac cells. J. Physiol. **461:** 743–765.

11. GOLDMAN, P. S. & N. M. NATHANSON. 1994. Differential role of the carboxy-terminal tyrosine in down-regulation and sequestration of the m2 muscarinic acetylcholine receptor. J. Biol. Chem. **269:** in press.

12. VALIQUETTE, M., H. BONIN & M. BOUVIER. 1993. Mutation of tyrosine-350 impairs the coupling of the beta 2-adrenergic receptor to the stimulatory guanine nucleotide binding protein without interfering with receptor down-regulation. Biochemistry **32:** P4979–P4985.

13. TIETJE, K. M., P. S. GOLDMAN & N. M. NATHANSON. 1990. Cloning and functional analysis of a gene encoding a novel muscarinic acetylcholine receptor expressed in chick heart and brain. J. Biol. Chem. **265:** 2828–2834.

14. HABECKER, B. A. & N. M. NATHANSON. 1992. Regulation of muscarinic acetylcholine receptor mRNA expression by activation of homologous and heterologous receptors. Proc. Natl. Acad. Sci. USA **89:** 5035–5038.

15. HABECKER, B. A., H. WANG & N. M. NATHANSON. 1993. Multiple second messenger pathways mediate agonist regulation of muscarinic receptor mRNA expression. Biochemistry **32:** 4986–4990.

Development of Selective Antisera for Muscarinic Cholinergic Receptor Subtypes

BARRY B. WOLFE AND ROBERT P. YASUDA

Department of Pharmacology
Georgetown University
School of Medicine
3900 Reservoir Road N.W.
Washington, D.C. 20007

Muscarinic cholinergic receptors belong to the family of G-protein–coupled receptors that modulate several second messenger systems including those of cyclic AMP and inositol trisphosphate. The actions of drugs on these receptors were initially described by Dale,[1] and subsequently large numbers of compounds have been shown to be agonists or antagonists at these receptors. In the 1970s and 1980s several investigators suggested on the basis of pharmacological data that subtypes of muscarinic receptors must exist; Goyal and Rattan[2] coined the terms M1 and M2 muscarinic receptors to refer to receptors having high and low affinity, respectively, for the drug pirenzepine. Later workers further subdivided the M2 class of receptors on the basis of selective affinity for the compound AF-DX-116 and defined the M3 class of receptors having low affinity for both AF-DX-116 and pirenzepine.[3]

During the late 1980s several laboratories reported cloning the cDNA and/or genes encoding muscarinic receptors.[4-10] In all, five distinct genes were identified and these molecularly defined subtypes were given the nomenclature m1–m5, using lowercase m to distinguish them from the pharmacologically defined M1–M3 receptors described above. The m1, and to some extent the m4, receptor corresponded with the M1 receptor; the m2, and to some extent the m4, receptor corresponded with the M2 receptor; the m3 receptor corresponded with the M3 receptor; it was unclear where the m5 receptor would be classified pharmacologically. Thus, the pharmacological classification methodology available does not have the tools to define unequivocally a given molecularly defined subtype of muscarinic receptor. In response to this problem, we embarked on a program to develop a set of antisera that selectively recognizes each of the five molecularly defined subtypes.

The general approach to obtaining selective antisera was to determine which parts of the primary sequence of the proteins, as deduced from the DNA sequence, were unique to each subtype and to utilize proteins or peptides based on these sequences as antigens. The muscarinic receptors, as for all known G-protein–coupled receptors, are monomers that have seven membrane-spanning regions with the N-terminus on the outside and the C-terminus on the inside of the cell. In general, the regions that are most unique between receptor subtypes include the N-terminal region, the third intracellular loop (I3), and the C-terminal region. Among these the I3 loop is large (approximately 170–230 amino acids) and, with the exception of the first 10 to 15 amino acids nearest the membrane, is quite different for all five subtypes. In addition, the I3 loop is fairly hydrophilic, a property that often increases the antigenicity of a protein. This region was thus chosen as a good immunogen to raise antisera against.

The cDNA encoding the I3 loops of each of the five muscarinic receptor subtypes was subcloned into expression vectors. Different vectors were used for the proteins because the experiments were performed over a two-year period in which new and better vectors became available. Thus, the m1, m2, and m4 cDNA was cloned into pRIT, an expression vector that is driven by a strong, constitutively-active promoter and expresses the IgG-binding domains of protein A.[11-13] Thus, the resultant fusion proteins were easily purified using an IgG-Sepharose column. For the m5 receptor the cDNA was cloned into pET3, which has an inducible promoter that drives expression levels to extremely high levels.[13] The protein produced from this vector does not, however, have an "affinity handle," and it was thus purified by size-exclusion chromatography. The m3 receptor cDNA initially proved difficult in that the proteins expressed using pRIT and pET3 were both badly degraded. Thus, initial antisera were raised to a unique small peptide corresponding to a region in the C-terminus.[14] In later experiments, however, the cDNA encoding the I3 loop of the m3 receptor was cloned into the vector pGEX3, which has an inducible promoter, and the resultant protein is fused to the enzyme glutathione-S-transferase (Li and Wolfe, unpublished data). The fusion protein is thus easily purified over a glutathione-agarose column.

Using these fusion proteins, or peptide in the case of m3, we injected rabbits; antisera were tested to determine whether antibodies were present that selectively recognized a specific subtype of muscarinic receptor. The assay that proved to be the most fruitful was immunoprecipitation of the radiolabeled receptor. As shown in FIG. 1, in this assay, receptors in membrane preparations are first labeled at 32 °C with the high-affinity muscarinic antagonist [^{3}H]QNB. This ligand is very specific for muscarinic receptors in that it does not seem to bind to other sites and has very low levels of "nonspecific" binding.[15] This ligand seems to have similar affinities for all five subtypes of muscarinic receptor with a K_d value near 10 pM (Wolfe, unpublished data). Receptors are labeled with a concentration (0.5–2 nM) of [^{3}H]QNB that occupies nearly all receptors. Samples are then cooled to 4 °C at which temperature the ligand has negligible dissociation over a two-day period.[11] The labeled receptors are then solubilized in digitonin and a high speed supernatant is produced. This supernatant containing the labeled receptors is incubated with the putative antiserum for 40–55 h at 4 °C. The IgG molecules in the antiserum are then complexed to Protein A on the surface of fixed bacteria to allow for the protein A/IgG/receptor/[^{3}H]QNB complex to be easily sedimented in a microcentrifuge.[11]

For these immunoprecipitation experiments, cells transfected with the cDNA encoding a single, defined subtype were used as a source of muscarinic receptors. It was found that a given antiserum would immunoprecipitate approximately 90% of the labeled receptors from cells expressing the appropriate receptor and less than 2% of labeled receptors from cells expressing the inappropriate receptor.[11-14] Thus, the antisera were selective and quantitative in their ability to recognize each subtype of muscarinic receptor and represented tools with which to study the distribution and density of the muscarinic receptor subtypes.

A number of studies have been carried out using these antisera. Initial experiments determined the general distribution of muscarinic receptor subtypes in rat brain and certain peripheral organs. Thus, as shown in FIGURE 2, all subtypes except m5 are expressed at reasonably high (> 200 fmol/mg protein) levels compared to a number of other receptors in the rat brain. The m1 receptor, for example, is expressed at high levels (> 800 fmol/mg) in the rostral portions of the rat brain including the cortex, hippocampus, striatum, and olfactory tubercule. In the midbrain, and pons-medulla it was expressed at much lower levels, and in cerebellum m1 receptors are only barely detectable. The profile for the m2 receptor, on the other

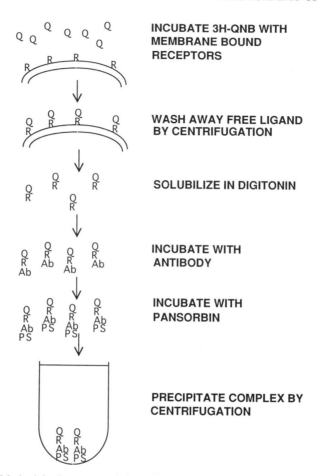

FIGURE 1. Method for immunoprecipitation of radiolabeled muscarinic receptor subtypes. Membrane-bound muscarinic receptors (R) are incubated at 32 °C with [³H]QNB (Q) at a concentration (0.5–2 nM) that will occupy essentially all of each subtype. Samples are cooled to 4 °C, centrifuged, and the membranes washed to remove free ligand. Labeled receptors are solubilized in 1% digitonin and the high speed (80,000 × g) supernatant is incubated with a subtype-selective antiserum (Ab). Following 40 to 55 h of incubation at 4 °C, Pansorbin (PS; fixed bacteria expressing protein A on their cell wall) is added, and the [³H]QNB/receptor/ antibody/Pansorbin complex is sedimented to separate labeled subtypes.

hand, shows that this receptor subtype is distributed throughout the brain in a relatively homogeneous manner with levels in most areas being around 250–400 pmol/mg. In the cerebellum m2 receptor density is only about 150 pmol/mg, but this still represents the majority of the muscarinic receptors expressed in this brain area. The m3 subtype is expressed in a pattern somewhat similar to that seen for m1 albeit at lower levels. Thus, there is a clear rostral-to-caudal gradient of m3 receptor expression. The m4 receptor is most highly expressed in the striatum where levels are nearly 1300 fmol/mg, the highest level of any single subtype in any area of the brain examined. This subtype is also poorly expressed in the caudal portions of the brain. The m5 muscarinic receptor was found, on the other hand, to be in very low abundance anywhere in the brain with reliably detectable values (15–25 fmol/mg) found only in the hippocampus, striatum, and midbrain.

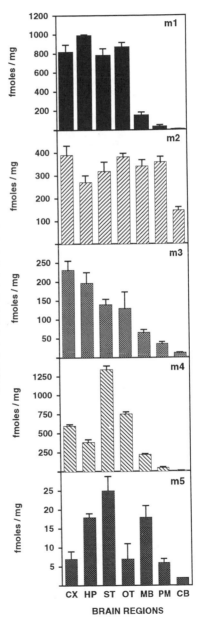

FIGURE 2. Regional distribution of muscarinic receptor subtypes in rat brain. Rat brain was dissected into seven regions: cortex (CX), hippocampus (HP), striatum (ST), olfactory tubercule (OT), midbrain (MB), pons-medulla (PM), and cerebellum (CB). Midbrain contained structures such as the thalamus, hypothalamus, and substantia nigra. Tissues were processed as described in FIGURE 1 to obtain the percent of each subtype in a given tissue. Total density of muscarinic receptors was determined by [³H]QNB binding. The percentage of a specific receptor subtype in a given tissue was multiplied by the total density of muscarinic receptors in that tissue to obtain the density (in fmol/mg protein) of the subtype. Data are adapted from Wall et al.,[11,14] Li et al.,[12] and Yasuda et al.[13]

It is interesting to compare the distribution of the protein determined by the immunoprecipitation studies to the distribution of the mRNA described using *in situ* hybridization techniques. Thus, the m1 mRNA distribution is not dissimilar to that found for protein distribution.[8,11] The distribution of m2 mRNA and protein are, however, quite different with the mRNA being concentrated in areas of cholinergic cell bodies whereas the protein is found to be widely distributed.[8,12] Additionally, the levels of m2 mRNA have been found to be very low relative to m1, m3, and m4.[8] Interestingly, the levels of m2 protein are higher than those of the m3 protein in many areas of the brain including, for example, the hippocampus in which the mRNA levels are quite the opposite.[8,12,14] Therefore, mRNA distribution and density do not necessarily predict protein distribution and density. On the other hand, the mRNA for m4 receptors has been shown to be very highly expressed in the striatum,[8] an area found to be very rich in m4 protein. Similarly, m5 mRNA was initially reported not to be found in rat brain but a subsequent study demonstrated the expression of this mRNA in the substantia nigra, pars compacta,[16] localized, presumably, in the dopaminergic neurons projecting to the striatum. Immunoprecipitation studies found the m5 receptor at low levels in the midbrain (which would include the substantia nigra) and the striatum, suggesting a predictive relationship between mRNA and protein for the m5 receptor.

Other studies utilizing the selective antisera have focused on the regulation of receptor subtypes in rat brain following loss of cholinergic input. A guiding hypothesis for these studies is that if neurotransmitter access to a receptor is blocked, either by blocking the receptor or destroying the neurons containing the neurotransmitter, the receptor will often respond by up-regulating its density. If this hypothesis is correct and uniform for all receptors (probably a stretch), then the level of up-regulation gives some estimate of the level of neurotransmitter "tone" in the normal situation. On the other hand, if a receptor is expressed on a neuron that is destroyed by a lesion, one would expect the density of that receptor to decrease. Two paradigms have been examined. The first involves chronic administration of the muscarinic receptor antagonist atropine.[17] This drug is nonselective and binds with high affinity and specificity to all five subtypes of muscarinic receptor. When rats were treated for 14 days with high levels of atropine and the effects on receptor densities were examined using immunoprecipitation, it was found that only some subtypes of muscarinic receptors were affected (FIG. 3). Levels of m2 receptors were unaffected whereas levels of m1 and m4 receptors increased modestly by about 10 to 20%. On the other hand, levels of m3 receptors were up-regulated by nearly 70%, indicating a high degree of "plasticity" and a high degree of tone for this receptor subtype. In a second, related paradigm, cholinergic tone was stopped by lesioning the cholinergic neurons innervating the dorsal hippocampus.[18] Either 10 or 24 days after the cholinergic lesion the densities of muscarinic receptor subtypes were determined. At both times the results were identical and data were combined (FIG. 3). In general, results were similar to those obtained by chronic blockade with atropine. Thus, m1 and m4 receptor densities were increased modestly by 10 to 30% whereas m3 receptor density was more strongly up-regulated by 77 percent. On the other hand, levels of m2 receptor *decreased* by more than 20% in contrast to the observed result from chronic atropine blockade. These data could be interpreted to indicate that, as concluded for the atropine experiment, inasmuch as the m1 and m4 receptors are not as "plastic" as m3 receptors they may not receive as strong a tone as the m3 receptors under normal conditions. The decrease in m2 receptors can be interpreted to indicate that some m2 receptors reside on the cholinergic nerve terminals that were destroyed by the lesion. This is consistent with much of the data demonstrating a functional presynaptic autoreceptor on these terminals that regulates acetylcholine release, which may be an m2 receptor.[19]

Lastly, experiments have been performed examining the normal developmental profile of the subtypes of muscarinic receptor in whole rat brain.[20] In general, each of the five subtypes developed along approximately the same time course with levels being very low at 3 to 4 days after birth, but rising rapidly over the next several days to reach adult levels by about 14 to 20 days after birth. This developmental profile is

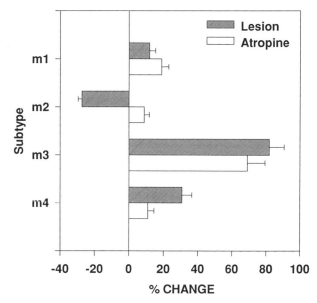

FIGURE 3. Changes in receptor subtype densities following either chronic blockade with atropine or chronic denervation by fimbria-fornix lesion. In one set of experiments rats were unilaterally lesioned by cutting the fimbria-fornix and the rostral supracallosal stria/cingulum bundle, pathways by which cholinergic neurons innervate the dorsal hippocampus. Either 10 or 24 days post lesion, dorsal hippocampus tissue was taken, and the densities of muscarinic receptor subtypes were determined as described in FIGURE 1. Data shown in the hatched bars (adapted from Wall *et al.*[18]) are presented as percent change from the control (unlesioned) side. In other experiments, rats were treated with atropine to block all five subtypes of muscarinic receptors for 14 days. Cortex and dorsal hippocampus were assayed for subtypes of muscarinic receptors as described above. Data shown in open bars (adapted from Wall *et al.*[17]) are presented as percent change from the control (sham-treated) rats. All values were statistically ($p < 0.05$) different from control except m2 receptors in atropine-treated tissues.

similar to that of carbachol-stimulated phosphoinositide hydrolysis,[21] suggesting that the appearance of receptors is responsible for the appearance of responsiveness. On the other hand, Lee *et al.*[21] reported that carbachol-mediated inhibition of adenylyl cyclase does not appear before age 14 days. Thus, for this signal transduction system and the receptors (m2 and m4) that mediate it, the appearance of receptors does not mandate the appearance of functional response and that the latter is more likely tied to the regulation of receptor/G-protein coupling.[21]

In conclusion, the development of a set of antisera that can localize and quantify each of the five subtypes of muscarinic cholinergic receptor has led to a number of novel observations. Future experiments examining, for example, the relationship of these receptors and the loss of cognitive abilities in aged rats or the decline in cognition in humans with Alzheimer's disease may provide new data with which to formulate hypotheses regarding therapeutic intervention in various disease states.

REFERENCES

1. DALE, H. H. 1914. The action of certain esters and ethers of choline and their relation to muscarine. J. Pharmacol. Exp. Ther. **6:** 147–190.
2. GOYAL, R. K. & S. RATTAN. 1978. Neurohormonal, hormonal, and drug receptors for the lower esophageal sphincter. Prog. Gastroenterol. **74:** 598–619.
3. GIACHETTI, A., R. MICHELETTI & E. MONTAGNA. 1986. Cardioselective profile of AF-DX 116, a muscarinic M2-receptor antagonist. Life Sci. **38:** 1663–1672.
4. KUBO, T., K. FUKUDA, A. MIKAMI, A. MAEDA, H. TAKAHASHI, M. MISHINA, T. HAGA, K. HAGA, A. ICHIYAMA, K. KANGAWA, M. KOJIMA, H. MATSUO, T. HIROSE & S. NUMA. 1986. Cloning, sequencing and expression of complementary DNA encoding the muscarinic acetylcholine receptor. Nature **323:** 411–416.
5. KUBO, T., A. MAEDA, K. SUGIMOTO, I. AKIBA, A. MIKAMI, H. TAKAHASHI, T. HAGA, K. HAGA, A. ICHIYAMA, M. KANGAWA, H. MATSUO, T. HIROSE & S. NUMA. 1986. Primary structure of porcine cardiac muscarinic acetylcholine receptor deduced from the cDNA sequence. FEBS Lett. **209:** 367–372.
6. PERALTA, E. G., J. W. WINSLOW, G. L. PETERSON, D. H. SMITH, A. ASHKENAZI, J. RAMACHANDRAN, M. I. SCHIMERLIK & D. J. CAPON. 1987. Primary structure and biochemical properties of an M2 muscarinic receptor. Science **236:** 600–605.
7. PERALTA, E. G., A. ASHKENAZI, J. W. WINSLOW, D. H. SMITH, J. RAMACHANDRAN & D. J. CAPON. 1987. Distinct primary structures, ligand-binding properties and tissue-specific expression of four human muscarinic acetylcholine receptors. EMBO J. **6:** 3923–3929.
8. BONNER, T. I., N. J. BUCKLEY, A. C. YOUNG & M. R. BRANN. 1987. Identification of a family of muscarinic acetylcholine receptor genes. Science **237:** 527–532.
9. BRAUN, T., P. R. SCHOFIELD, B. D. SHIVERS, D. B. PRITCHETT & P. H. SEEBURG. 1987. A novel subtype of muscarinic receptor identified by homology screening. Biochem. Biophys. Res. Commun. **149:** 125–132.
10. BONNER, T. I., A. C. YOUNG, M. R. BRANN & N. J. BUCKLEY. 1988. Cloning and expression of the human and rat m5 muscarinic acetylcholine receptor genes. Neuron **1:** 403–410.
11. WALL, S. J., R. P. YASUDA, F. HORY, S. FLAGG, B. M. MARTIN, E. I. GINNS & B. B. WOLFE. 1991. Production of antisera selective for m1 muscarinic receptors using fusion proteins: Distribution of m1 receptors in rat brain. Mol. Pharmacol. **39:** 643–649.
12. LI, M., R. P. YASUDA, S. J. WALL, A. WELLSTEIN & B. B. WOLFE. 1991. Distribution of m2 muscarinic receptors in rat brain using antisera selective for m2 receptors. Mol. Pharmacol. **40:** 28–35.
13. YASUDA, R. P., W. CIESLA, L. R. FLORES, S. J. WALL, M. LI, S. A. SATKUS, J. S. WEISSTEIN, B. V. SPAGNOLA & B. B. WOLFE. 1993. Development of antisera selective for m4 and m5 muscarinic cholinergic receptors: Distribution of m4 and m5 receptors in rat brain. Mol. Pharmacol. **43:** 149–157.
14. WALL, S. J., R. P. YASUDA, M. LI & B. B. WOLFE. 1991. Development of an antiserum against m3 muscarinic receptors: Distribution of m3 receptors in rat tissues and clonal cell lines. Mol. Pharmacol. **40:** 783–789.
15. YAMAMURA, H. I. & S. H. SNYDER. 1974. Muscarinic cholinergic binding in rat brain. Proc. Natl. Acad. Sci. USA **71:** 1725–1729.
16. WEINER, D. M., A. I. LEVEY & M. R. BRANN. 1990. Expression of muscarinic acetylcholine and dopamine receptor mRNAs in rat basal ganglia. Proc. Natl. Acad. Sci. USA **87:** 7050–7054.

17. WALL, S. J., R. P. YASUDA, M. LI, W. CIESLA & B. B. WOLFE. 1992. Differential regulation of subtypes (m1–m5) of muscarinic receptors in forebrain by chronic atropine administration. J. Pharmacol. Exp. Ther. **262:** 584–588.
18. WALL, S. J., B. B. WOLFE & L. F. KROMER. 1994. Cholinergic deafferentation of dorsal hippocampus by fimbria-fornix lesioning differentially regulates subtypes (m1–m5) of muscarinic receptors. J. Neurochem. **62:** 1345–1351.
19. MARCHI, M. & M. RAITERI. 1989. Interaction of acetylcholine-glutamate in rat hippocampus: Involvement of two subtypes of M-2 muscarinic receptors. J. Pharmacol. Exp. Ther. **248:** 1255–1260.
20. WALL, S. J., R. P. YASUDA, M. LI, W. CIESLA & B. B. WOLFE. 1992. The ontogeny of m1–m5 muscarinic receptor subtypes in rat brain. Dev. Brain Res. **66:** 181–185.
21. LEE, W. L., K. J. NICKLAUS, D. R. MANNING & B. B. WOLFE. 1990. Ontogeny of cortical muscarinic receptor subtypes and muscarinic receptor mediated responses in rat. J. Pharmacol. Exp. Ther. **252:** 482–490.

Studies on the Diversity of Muscarinic Receptors in the Autoregulation of Acetylcholine Release in the Rodent Cerebrum Using Furan Analogs of Muscarine[a]

B. V. RAMA SASTRY,[b,c] O. S. TAYEB,[d,e]
AND N. JAISWAL[d,f]

Departments of Anesthesiology[b] and Pharmacology[d]
Vanderbilt University Medical Center
Nashville, Tennessee 37232-2125

Previous studies have indicated that two feedback mechanisms, one positive and the other negative, regulate the rate of acetylcholine (ACh) release.[1–3] The positive feedback system has at least three components: a muscarinic receptor (Ms), release of substance P (SP), and influx of extracellular Ca^{2+}. If the amount of ACh released from the presynaptic nerve terminal is low, the released ACh activates Ms, which stimulates the release of SP. Substance P increases Ca^{2+} influx, resulting in the release of further quantities of ACh for effective cholinergic transmission. Similarly, three components are present in the negative feedback system: a presynaptic muscarinic receptor (Mi), release of methionine enkephalin (MEK), and inhibition of Ca^{2+} influx. If the ACh in the synaptic gap is high, it activates Mi, resulting in the release of MEK. Methionine enkephalin decreases Ca^{2+} influx which decreases the rate of release of ACh from the cholinergic nerve terminal. The critical first step in both of these feedback systems is activation of presynaptic muscarinic receptors, Ms and Mi. No selective agonists have been described for presynaptic muscarinic receptors. Therefore, we studied the effects of 5-methylfurfuryltrimethylammonium (5-MFT) and 5-hydroxyfurfurltrimethylammonium (5-HMFT) on the simultaneous release of ACh, SP, and MEK from the superfused mouse cerebral slices in our search for selective agonists for Ms and Mi receptors. 5-MFT and 5-HMFT were selected for this purpose because they have diverse effects in the peripheral nervous system.[4,5]

[a]This work was supported by grants from The Council for Tobacco Research-U.S.A., Inc., United States Department of Health and Human Services–National Institute on Drug Abuse (DA 06207), and The Study Center for Anesthesia Toxicology.
[c]Corresponding author.
[e]Present address: School of Medicine and Allied Sciences, King Abdulaziz University, Jeddah 21483, Saudi Arabia.
[f]Present address: Department of Brain and Vascular Research, Cleveland Clinic Foundation, Cleveland, OH 44195.

METHODS AND RESULTS

Mouse cerebral slices were incubated in a modified Kreb's ringer buffer containing methyl-[³H]choline (0.1 mM; 0.25 μCi/mL) for 60 min. They were filtered, washed, and transferred to a microbath set up for superfusion with the above buffer containing hemicholinium-3 (10 μM). The release of [³H]ACh into the superfusate was measured as a function of time.[1-3] The effects of 5-MFT and 5-HMFT on the simultaneous release of labeled [³H]ACh, SP, and MEK from the superfused mouse cerebral slices were determined. The [³H]ACh was measured by a radiometric method. SP and MEK were measured by selective radioimmunoassays.[6] Both spontaneous and electrically evoked release of [³H]ACh, SP, and MEK were

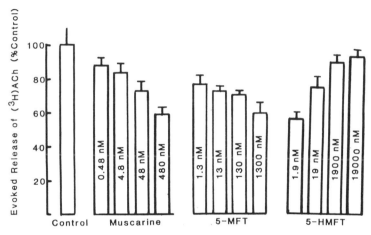

FIGURE 1. Effects of muscarine, 5-MFT, and 5-HMFT on [³H]ACh-evoked release as a function of concentration. Detailed conditions for the preparation of slices, their incubation with [³H]choline to form [³H]ACh, and the superfusion of the slices and collection of superfusion samples for 1 h were described elsewhere.[1-3] During 30–60 min, the slices were subjected to field electrical stimulation to measure evoked release of [³H]ACh in the presence and absence of 5-MFT or 5-HMFT. All values expressed as percentage of control values. Each bar is a mean ± SE from six values. Similar results are obtained for spontaneous release during 0–30 min of superfusion. The effect of muscarine and 5-MFT increased with increasing concentration, whereas the effect of 5-HMFT decreased with increasing concentration.

measured. 5-MFT (13–1300 nM) decreased spontaneous ACh release (60% of control) in a concentration-dependent manner (FIG. 1). 5-HMFT (1.9 nM) decreased the spontaneous release of ACh to 40% of control. This effect is inversely related to increasing concentrations of 5-HMFT. At 190 μM of 5-HMFT, the effect was only 15% of control. The effect of 5-MFT was antagonized by atropine (1 μM) but not naloxone (55 nM). The effect of 5-HMFT was antagonized by scopolamine (10 nM) and naloxone. 5-MFT was considerably more potent than 5-HMFT in stimulating muscarinic receptors (M) in smooth muscle. Both 5-MFT and 5-HMFT inhibited MEK release whereas 5-HMFT was more potent in causing SP release (FIG. 2). These observations indicate that autoregulation of ACh release operates through two subtypes of M, one stimulatory (Ms) and the other inhibitory (Mi). The

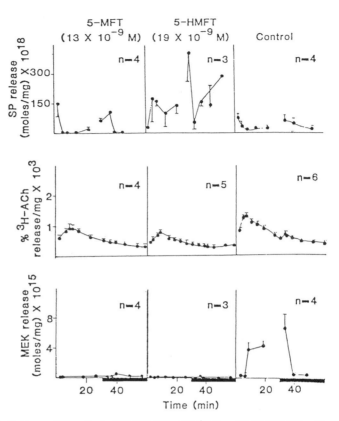

FIGURE 2. Patterns of the effect of 5-MFT (13×10^{-9} M) and 5-HMFT (19×10^{-9} M) on the simultaneous release of substance P (SP), acetylcholine (ACh), and methionine enkephalin (MEK) from mouse cerebral slices. Each point represents a mean; the vertical lines represent the standard error. Inhibition of the release of MEK predominates both with 5-MFT and 5-HMFT. Depressions in ACh release seem to trigger the enhanced SP release and depressed MEK release. Thick horizontal bars (*bottom panel*) indicate periods of field electrical stimulation.

identity of Ms and Mi receptors with M1, M2, or M3 muscarine receptor subtypes is not yet determined.[7,8]

REFERENCES

1. SASTRY, B. V. R. & O. S. TAYEB. 1982. Adv. Biosci. **38:** 165–172.
2. SASTRY, B. V. R., V. E. JANSON, N. JAISWAL & O. S. TAYEB. 1983. Pharmacology **26:** 61–72.
3. SASTRY, B. V. R. & O. S. TAYEB. 1988. Regul. Pept. **22(1–2):** 168.
4. SASTRY, B. V. R., L. K. OWENS & R. F. OCHILLO. 1990. Ann. N.Y. Acad. Sci. **604:** 566–568.
5. HORST, M. A., B. V. R. SASTRY & E. J. LANDON. 1987. Arch. Int. Pharmacodyn. Ther. **288:** 87–99.
6. SASTRY, B. V. R., O. S. TAYEB, V. E. JANSON & L. K. OWENS. 1981. Placenta (Suppl.) **3:** 327–337.
7. STEPHAN, C. C. & B. V. R. SASTRY. 1982. Cell. Mol. Biol. **38(7):** 701–712.
8. SASTRY, B. V. R. 1993. Anesth. Pharmacol. Rev. **1:** 6–19.

Phosphorylation of the Nicotinic Acetylcholine Receptor by Protein Tyrosine Kinases[a]

SHERIDAN L. SWOPE,[b] ZHICAN QU,
AND RICHARD L. HUGANIR

Department of Neuroscience
Howard Hughes Medical Institute
The Johns Hopkins University School of Medicine
Baltimore, Maryland 21205

Protein phosphorylation is a major posttranslational modification widely recognized to regulate almost all cellular processes.[1,2] Based on their amino acid specificities, protein kinases may be classified as either protein serine/threonine kinases or protein tyrosine kinases. Protein tyrosine kinases were originally identified as retroviral oncogene products and subsequently shown to have normal cellular homologs.[3] Initially, protein tyrosine kinases were believed to be important primarily in the regulation of growth and mitogenesis. However, more recent studies implicate protein tyrosine kinases in the regulation of differentiated cell function. For example, many protein tyrosine kinases are most highly expressed in the neurons of the central nervous system.[4] Furthermore, protein tyrosine kinases can mediate neuronal differentiation; activation of the nerve growth factor receptor[5] or expression of v-*src*[6] in PC12 chromaffin cells induces a sympathetic neuron-like phenotype. These studies suggest an involvement of protein tyrosine kinases in neuronal differentiation as well as in synaptic transmission.

The transfer of information from a neuron to its target is the process of synaptic transmission. Chemical synapses, at which the postsynaptic response is mediated by ligand-gated ion channels, provide many opportunities for synaptic plasticity. Our understanding of the molecular mechanisms underlying synapse formation, function, and modulation derives primarily from studies examining the neuromuscular junction. At the neuromuscular junction, the acetylcholine receptor (AChR) is the ligand-gated ion channel that mediates the rapid excitatory postsynaptic response of the muscle. Because of its enrichment in the electric organs of *Torpedo californica*, the AChR has served as a model for the study of the structure, function, and regulation of ligand-gated ion channels. The AChR is a 250-kDa pentameric complex comprised of four homologous subunits in a stoichiometry of $\alpha_2\beta\gamma\delta$ (FIG. 1 and ref. 7). At least three protein kinase activities phosphorylate the AChR (FIG. 1 and ref. 8). Protein kinase A phosphorylates the γ and δ subunits, protein kinase C phosphorylates the δ subunit, and a protein tyrosine kinase activity phosphorylates the β, γ, and δ subunits. Phosphorylation of the AChR by all three protein kinase

[a]This research was supported by the Muscular Dystrophy Association and the National Institutes of Health.

[b]Address correspondence to Dr. Sheridan L. Swope, Department of Neuroscience, The Johns Hopkins University School of Medicine, 725 N. Wolfe Street, 900 PCTB, Baltimore, MD 21205-2185.

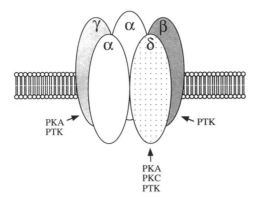

FIGURE 1. Model of the structure and subunit specificity for phosphorylation of the AChR. The α, β, γ, and δ subunits are as indicated. The specificity for phosphorylation of the AChR subunits by protein kinase A (PKA), protein kinase C (PKC), and protein tyrosine kinase(s) (PTK) are as indicated.

activities results in an alteration of channel desensitization kinetics.[8] In addition, activation of protein tyrosine kinase activity, including tyrosine phosphorylation of the AChR, is associated with a clustering of the AChR during the formation of postsynaptic specializations. In this report, we review some of our recent studies to identify the extracellular factors that stimulate AChR tyrosine phosphorylation as well as the protein tyrosine kinase(s) that may be involved in the regulation of synaptic function at the neuromuscular junction by directly phosphorylating the AChR.

Methodologies used are as described in references 9–12.

RESULTS

An Extracellular Factor That Stimulates AChR Tyrosine Phosphorylation

The AChR of *Torpedo californica* electric organ and innervated rat skeletal muscle is phosphorylated on tyrosine residues.[9,13,14] In contrast, the AChR of cultured myotubes[15] and BC3H1 myocytes[16] contains very little phosphotyrosine. These studies suggest that innervation may regulate the tyrosine phosphorylation of the AChR. Another effect of the neuron on the AChR is to induce receptor clustering at the site of nerve and muscle contact.[17] In addition, some reports have suggested that phosphorylation of the AChR may be involved in regulating receptor distribution.[10,18-20] It is now known that the neuronally derived extracellular maxtrix protein agrin mediates AChR clustering induced by the neuron.[21] Therefore, we have examined whether the effect of agrin to stimulate AChR aggregation is associated with phosphorylation of the receptor by a protein tyrosine kinase.

Cultured chick myotubes were labeled with anti-phosphotyrosine antibodies and rhodamine-conjugated α-bungarotoxin. The small number of aggregates of AChRs found on the surface of control myotubes show only slight labeling with the anti-phosphotyrosine antibody (FIG. 2A and B). After 4 h of agrin treatment, AChR aggregates increase in size and the clusters stain intensely for phosphotyrosine (FIG. 2C and D). There is a striking correspondence in the localization of the expanding AChR clusters and phosphotyrosine during the time course of agrin treatment (FIG. 2C–H). Thus, tyrosine-phosphorylated proteins aggregate at agrin-induced AChR clusters.

To determine whether the AChR itself is phosphorylated in response to agrin, chick myotube cultures were treated with or without agrin and, after affinity purification, the AChR phosphotyrosine content was examined by Western blotting using the α-phosphotyrosine antibody. Agrin induces a dramatic increase in the phosphorylation of the chicken AChR β subunit on tyrosine residue(s) (FIG. 3). Thus, agrin induces both AChR clustering and tyrosine phosphorylation.

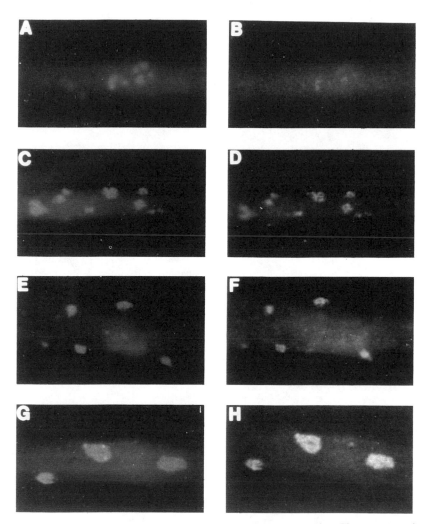

FIGURE 2. Effect of agrin on AChR and phosphotyrosine aggregation. Fluorescence micrographs of segments of cultured chick myotubes treated with agrin for 0 h (**A** and **B**), 4 h (**C** and **D**), 8 h (**E** and **F**), and 18 h (**G** and **H**) were fixed, permeabilized, and labeled with rhodamine-conjugated α-bungarotoxin to reveal the distribution of the AChR (**A, C, E,** and **G**) and with antiphosphotyrosine monoclonal antibody PY20 and a fluorescein-conjugated secondary antibody to visualize phosphotyrosine residues (**B, D, F,** and **H**) as described.[10] (From Wallace *et al.*[10] Reproduced, with permission, from *Neuron.*)

Identification of Synaptic Protein Tyrosine Kinases That Associate with the AChR

Because of the enrichment of synaptic components, including the AChR, in *Torpedo californica* electric organ, we predicted that the kinase that phosphorylates the AChR would also be enriched in that tissue. Therefore, our initial goal was to identify protein tyrosine kinases of *Torpedo californica* electric organ by molecular cloning. Protein tyrosine kinases contain several highly conserved subdomains within the catalytic region[22] making them particularly amenable to analysis by the polymerase chain reaction (PCR).[23] Using cDNA derived from *Torpedo* electric organ as template, we used oligonucleotides based on the sequences of subdomains VII and VIII of the *src* class of protein tyrosine kinases to generate PCR products encoding protein tyrosine kinase fragments. *Src*-like protein tyrosine kinase cDNA clones were identified by screening a *Torpedo* electric organ γgt10 cDNA library with the

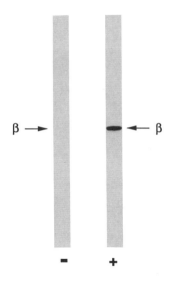

β →   ← β

−   +

AGRIN

FIGURE 3. Effect of agrin on the phosphotyrosine content of the AChR. Cultured chicken myotubes were treated without (−) or with (+) agrin for 12 h after which the AChR was isolated, resolved by SDS-PAGE, transferred to nitrocellulose, and analyzed for phosphotyrosine content using a monoclonal antibody as described.[10] The position of the β subunit is as indicated. (Modified from Wallace *et al.*[10] and used with permission from *Neuron.*)

cloned PCR products. Positive clones could be divided into two classes based on Southern analysis using the PCR products.

Within the coding regions, the DNA sequences of the two clones have 76% dispersed identity (with three gaps) demonstrating that they are derived from homologous but unique genes (data not shown). Each deduced translation product encodes an Src-like protein tyrosine kinase that contains all the hallmark subdomains including a myristylation site, Src homology 3 (SH3) and Src homology 2 (SH2) domains, and all the recognized subdomains of the catalytic region (FIG. 4). The amino acid sequences of the two clones show 82% identity to each other. Homology searches demonstrated that both clones are most similar to *fyn*, a member of the *src* class of protein tyrosine kinases (FIG. 4). Two *fyn* protein tyrosine kinases, the neuronal and thymic variants, have been identified; these forms arise from alternative splicing.[24,25,28] As shown in FIGURE 4, both *Torpedo* clones contain the neuronal alternative splice insert, suggesting one of the clones is the *Torpedo* neuronal *fyn* and

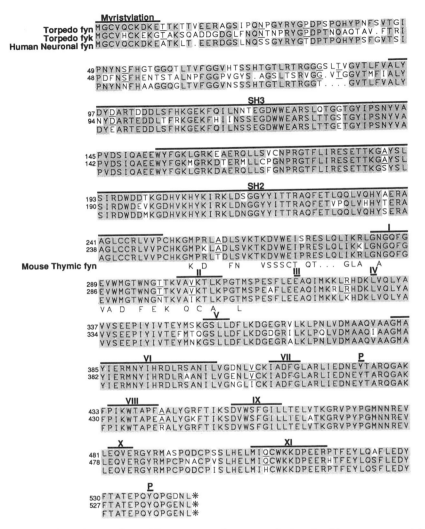

FIGURE 4. Complete amino acid sequences of two protein tyrosine kinases expressed in *Torpedo* electric organ. The homologies of the deduced amino acid sequences of the two *Torpedo* electric organ protein tyrosine kinases to known proteins were determined as described.[11] For each clone, amino acid identities with the neuronal form of human Fyn[24] are indicated by shading. Amino acids of nonidentity with human Fyn but identity between Fyn and Fyk are indicated by underlining. The divergence between the neuronal and the alternatively spliced thymic form of murine Fyn[25] is shown. Conserved protein tyrosine kinase subdomains are indicated by horizontal bars. SH3 and SH2, Src homology domains 3 and 2;[26,27] P, putative phosphorylation sites; I–XI, conserved subdomains.[22] The amino acid positions in the sequence of each clone are indicated on the left. (From Swope and Huganir.[11] Reproduced, with permission, from the *Journal of Biological Chemistry*.)

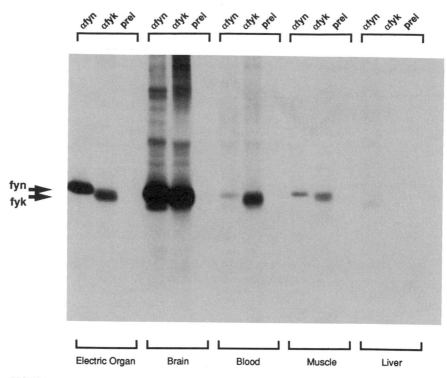

FIGURE 5. Expression of Fyn and Fyk protein tyrosine kinase activities in various *Torpedo* tissues. Membrane proteins (500 μg) prepared from the tissues indicated were solubilized at 1 mg/mL, immunoprecipitated with anti-*fyn* (αfyn), anti-*fyk* (αfyk), or preimmune (prei) serum as indicated and incubated under phosphorylating conditions at 30 °C for 30 min in the presence of [γ-32P]ATP followed by analysis by SDS-PAGE and autoradiography for 1 h as described.[11] (From Swope and Huganir.[11] Reproduced, with permission, from the *Journal of Biological Chemistry.*)

the other clone a novel gene product. The upper clone in FIGURE 4 shows 78% identity with human neuronal *fyn* at the nucleotide level and 90% identity at the amino acid level. The second clone shows 75% identity with human neuronal *fyn* at the nucleotide level and 82% at the amino acid level. Between members of the *src* family of protein tyrosine kinases, the amino acid sequence between the myristylation motif and the SH3 domain is the region that is most unique.[29] Therefore, to clarify the identity of our two clones, we compared the unique domain of each clone to known protein tyrosine kinases. The upper clone in FIGURE 4 is most homologous to *fyn,* showing 69% identity. In contrast, the second clone has only 39% identity with *fyn* in this N-terminal region. In addition, the second clone is almost equally homologous, 32% identity, with the *yes* protein tyrosine kinase.[30] These data support the upper clone in FIGURE 4 as the *Torpedo* neuronal *fyn* protein tyrosine kinase. The second clone is a novel protein tyrosine kinase that we have named *fyk* for *fyn-yes*-kinase.

To examine the expression of these two tyrosine kinases at the protein level, we generated antibodies to bacterial fusion proteins containing the N-terminal unique domains of *fyn* and *fyk*. These antibodies were used in an immunoprecipitation/ kinase assay in which the immunoprecipitated proteins were incubated under phosphorylating conditions in the presence of [γ-^{32}P]-ATP. Two protein tyrosine kinase activities of *Torpedo* electric organ are distinguished with antibodies to *fyn* and *fyk* (FIG. 5). The antibody to *fyn* immunoprecipitated a 55-kDa protein that contained only phosphotyrosine (data not shown), presumably due to autophosphorylation. In contrast, anti-*fyk* antibodies immunoprecipitated a doublet of 56- and 53-kDa (FIG. 5). The appearance as a doublet may be due to differing phosphorylation states because the immunoprecipitated phosphorylated Fyk contained both phosphotyrosine and phosphoserine (data not shown). The distributions of Fyn and Fyk protein tyrosine kinase activities in various *Torpedo* tissues were examined using the immunoprecipitation/kinase assay. *Torpedo* electric organ and brain are highly enriched in both kinases (FIG. 5). In addition, Fyn and Fyk activities are detected in muscle, a tissue that also expresses AChR and from which electric organ is embryonically derived. It is interesting to note that the relative abundance of the two kinases in electric organ and muscle is proportional to the expression of the AChR in these tissues.

Protein tyrosine kinases of the Src class are typically associated with the membrane. *Torpedo* electroplax membranes can be fractionated on discontinuous sucrose gradients to enrich for postsynaptic versus nonpostsynaptic membranes.[9,31] We determined the contribution of Fyn and Fyk to the total protein tyrosine kinase activity present in postsynaptic membranes of *Torpedo* electric organ. Detergent extracts of postsynaptic membranes were depleted of Fyn and Fyk by immunoprecipitation after which the kinase activity in the supernatant was measured using the synthetic polypeptide substrate POLY (Glu-Na,Tyr)4:1. Phosphorylation of POLY (Glu-Na,Tyr)4:1 by solubilized membrane proteins after immunoprecipitation with preimmune serum was defined as 100% of the protein tyrosine kinase activity (FIG. 6). When the membrane proteins are immunoprecipitated with the anti-*fyn* antibody, the kinase activity in the supernatant is reduced to 66% of control (FIG. 6). Thus, Fyn

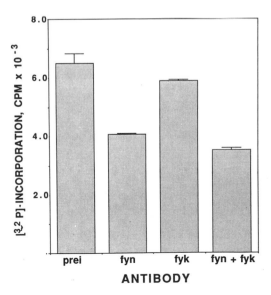

FIGURE 6. Percentage of protein tyrosine kinase activity of postsynaptic membranes represented by Fyn and Fyk. Postsynaptic membranes proteins (25 μg) of *Torpedo* electric organ were solubilized at 1 mg/mL and immunoprecipitated with preimmune (prei), anti-*fyn* (fyn), anti-*fyk* (fyk), or anti-*fyn* plus anti-*fyk* (fyn + fyk) serum as indicated. The supernatants were removed and incubated with poly (Glu-Na,Tyr), 4:1 at 1 μg/μL under phosphorylating conditions in the presence of [γ-^{32}P]ATP. After 30 min at 30 °C the reactions were stopped, and the supernatants were analyzed for ^{32}P incorporation as described.[11] The data represent the mean value ± SE, *n* = 5. (From Swope and Huganir.[11] Reproduced, with permission, from the *Journal of Biological Chemistry*.)

represents 34% of the total protein tyrosine kinase activity. In the experiment shown in FIGURE 6, anti-*fyk* antibody depleted 4% of the kinase activity whereas anti-*fyn* plus anti-*fyk* decreased the activity to 57% of control. Thus, Fyn plus Fyk constituted 43% of the kinase activity. In three experiments, Fyn and Fyk represent 36 ± 2% and 8 ± 3%, respectively of the total activity (mean ± SEM). Thus, together these two protein tyrosine kinases comprise a major fraction (44 ± 4%) of the protein tyrosine kinase activity in the postsynaptic membrane.

A "kinase trapping" strategy has been used by others to show that Fyn as well as the Lyn and Yes protein tyrosine kinases are associated with p21^{ras}GAP, a tyrosine-phosphorylated substrate.[32] We used the same strategy to examine whether Fyn or Fyk are in a complex with the AChR. To examine whether tyrosine kinase activity is associated with the receptor, the AChR was isolated from detergent extracts of postsynaptic membranes by immunoprecipitation with a monoclonal antibody (88b) and then incubated under phosphorylating conditions. As shown in FIGURE 7A, the β, γ, and δ subunits of the AChR are phosphorylated in the immunoprecipitate. Phosphoamino acid analysis demonstrated that all three subunits contain phosphotyrosine (data not shown). These data suggest that a protein tyrosine kinase is coimmunoprecipitated with the AChR. To examine whether Fyn or Fyk is complexed with the AChR, we did the reciprocal experiment shown in FIGURE 7B. Postsynaptic membranes solubilized under stringent conditions were incubated with either preimmune, anti-*fyn*, or anti-*fyk* antibodies and then the immunoprecipitates were analyzed for the presence of the AChR using a pool of monoclonal antibodies to the individual subunits. Immunoprecipitation with anti-*fyn* and anti-*fyk* antibodies

---

FIGURE 7. Coimmunoprecipitation of the AChR with Fyn and Fyk. **(A)** Postsynaptic membrane proteins (150 μg) prepared from *Torpedo* electric organ were solubilized at 1 mg/mL, immunoprecipitated in the presence (+88b) or absence (−) of monoclonal antibody 88b to the AChR, and incubated at 30 °C for 60 min in the presence of [γ-^{32}P]ATP with inhibitors of both protein kinase A and protein kinase C followed by analysis by SDS-PAGE and autoradiography for 6 hr as described.[11] The positions of the AChR β, γ, and δ subunits are as indicated. **(B)** Postsynaptic membrane proteins (250 μg) were solubilized at 1 mg/mL with immunoprecipitation buffer containing 1% Triton X-100, 0.5% deoxycholate, and 0.1% SDS and then immunoprecipitated with preimmune (prei), anti-*fyn* (αfyn), or anti-*fyk* (αfyk) serum in the absence or presence of kinase-specific fusion protein (+FP) or backbone fusion protein (+BB). The immunoprecipitates were resolved by SDS-PAGE, blotted onto nitrocellulose, and analyzed with a pool of monoclonal antibodies to the AChR as described.[11] The positions of the AChR α, β, γ, and δ subunits are as indicated. **(C)** Postsynaptic membrane proteins (250 μg) were solubilized at 1 mg/mL with immunoprecipitation buffer containing 1% Triton X-100, 0.5% deoxycholate, and 0.1% SDS and then immunoprecipitated with preimmune (prei), anti-*fyn* (αfyn), or anti-*fyk* (αfyk) serum. The immunoprecipitates and 10 μg of *Torpedo* electric organ postsynaptic membrane proteins (PSM) were resolved by SDS-PAGE, blotted onto nitrocellulose, and analyzed with a monoclonal antibody to the 43-kDa protein as described.[11] The position of 43 kDa is as indicated. **(D)** Partially purified AChR (250 μg), prepared as described,[11] was immunoprecipitated with preimmune (prei), anti-*fyn* (αfyn), or anti-*fyk* (αfyk) antiserum. The immunoprecipitates were incubated at 30 °C for 30 min in the presence of [γ-^{32}P]ATP followed by analysis by SDS-PAGE and autoradiography for 4 h as described.[11] The positions of the 50-kDa (β), 58-kDa (γ), and 65-kDa (δ) proteins are as indicated. **(E)** Partially purified AChR (25 μg), prepared as described,[11] was immunoprecipitated with preimmune (prei), anti-*fyn* (αfyn), or anti-*fyk* (αfyk) serum. The immunoprecipitates were resolved by SDS-PAGE, blotted onto nitrocellulose, and analyzed with a pool of monoclonal antibodies to the AChR as described.[11] The positions of the AChR α, β, γ, and δ subunits are as indicated. (From Swope and Huganir.[11] Reproduced, with permission from the *Journal of Biological Chemistry.*)

results in specific coimmunoprecipitation of the AChR, with the anti-*fyk* immunopre-cipitates being more enriched in AChR than the anti-*fyn* immunoprecipitates. The coimmunoprecipitation of the AChR with Fyn and Fyk is blocked when the antibod-ies are preincubated with the kinase-specific fusion proteins used as immunogens, but not with the backbone fusion protein (FIG. 7B). These data indicate that the complex formation is due to an association of the AChR with the kinases. In addition, coimmunoprecipitation of the receptor with Fyn and Fyk is apparently not due to a nonspecific association of the abundant AChR with the kinases because the 43-kDa protein, another synaptic component that is stoichiometrically expressed with the receptor,[33] is not coimmunoprecipitated with Fyn and Fyk (FIG. 7C). Taken together, these data indicate that the AChR forms a specific and tight complex with

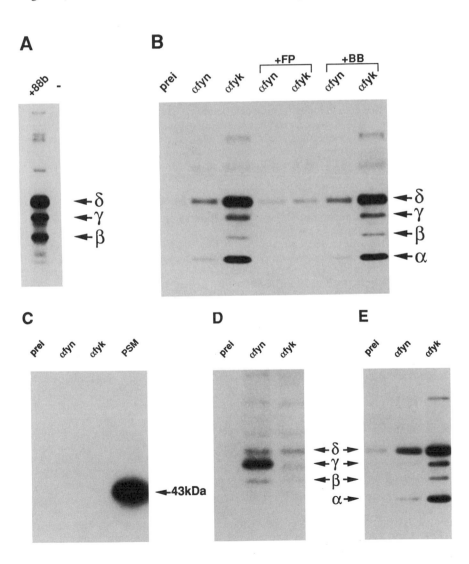

Fyn and Fyk, suggesting that these kinases may catalyze the tyrosine phosphorylation of the receptor.

Association of the AChR with Fyn and Fyk is also supported by the experiment shown in FIGURE 7D and E. Purified AChR was analyzed for the presence of the protein tyrosine kinases by immunoprecipitating with anti-*fyn* or anti-*fyk* antibodies followed by incubation of the precipitates under phosphorylating conditions. Phosphorylated bands of 50, 58, and 65 kDa are observed in the antiprotein tyrosine kinase but not preimmune serum precipitates (FIG. 7D). Analysis of the precipitates by Western blotting demonstrates that the phosphorylated bands comigrate with the β, γ, and δ AChR subunits (FIG. 7D and E). These data indicate that Fyn and Fyk, in a complex with the AChR, copurify with the receptor on acetylcholine-affinity resin, that the receptor is subsequently coimmunoprecipitated with the kinases, and that both Fyn and Fyk are capable of phosphorylating the AChR in the precipitate.

Molecular Basis for the Association of the AChR with Fyn and Fyk

Like other members of the Src class of protein tyrosine kinases,[34] Fyn and Fyk each contain a unique region, an SH3 domain, an SH2 domain, and a catalytic domain. SH2 domains mediate the association of Src-like tyrosine kinases as well as a variety of other SH2 domain containing signaling components with phosphotyrosine containing proteins, including autophosphorylated growth factor receptors and tyrosine kinase substrates.[27] Therefore, we investigated the involvement of the SH2 domains of Fyn and Fyk in the association of the AChR with these two protein tyrosine kinases.

Using glutathione-S transferase fusion proteins as affinity reagents, we examined whether the coimmunoprecipitation of the AChR with Fyn and Fyk is mediated by binding of the receptor to the SH2 domains. Upon incubation of solubilized *Torpedo* electric organ postsynaptic membrane proteins representing between 30 and 1 μg of protein with the affinity reagents, the AChR binds to the Fyn and Fyk SH2 domain fusion proteins in a concentration-dependent manner (FIG. 8). By comparing the relative intensity of the AChR detected in the flow through to that in the bound, we estimate that between 5% (30 μg) and 20% (1 μg) of the receptor binds to the Fyn and Fyk SH2 affinity resin under these conditions. The binding of the AChR to the SH2 domain fusion proteins is specific because little if any AChR binds to the glutathione-S transferase backbone upon incubation with the highest concentration of protein (30 μg).

The binding of the AChR to the Fyn and Fyk SH2 domain fusion proteins could be mediated by an ancillary protein or may be due to a direct interaction of one or more receptor subunits with the kinase fragment. To address this question, postsynaptic membrane proteins were treated under denaturing conditions, 2% SDS, to dissociate the AChR from any bound proteins as well as to dissociate the receptor into individual subunits. After dilution of the membrane proteins into Triton X-100, we examined the ability of the individual receptor subunits to bind to the SH2 fusion proteins. After solubilization under native conditions, incubation of postsynaptic membrane proteins with the SH2 domain affinity resins results in AChR binding, and all four subunits are detected (FIG. 9) as demonstrated above. However, after denaturation only the AChR δ subunit is observed to bind to the Fyn and Fyk SH2 domain fusion proteins (FIG. 9). These data support a direct interaction between the AChR δ subunit and the Fyn and Fyk SH2 domain fusion proteins. In addition, these data indicate that not all tyrosine phosphorylated postsynaptic membrane proteins, including the AChR β and γ subunits, bind to Fyn and Fyk SH2 domain fusion proteins.

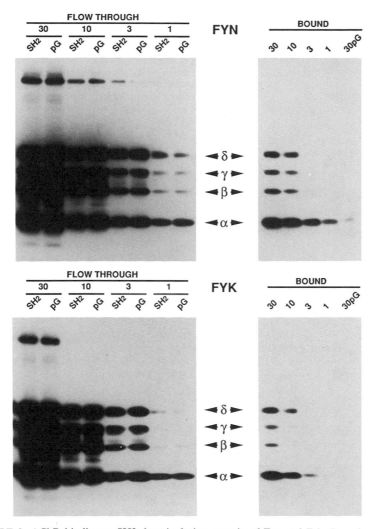

FIGURE 8. AChR binding to SH2 domain fusion protein of Fyn and Fyk. *Torpedo* electric organ postsynaptic membranes, representing between 1 and 30 μg of membrane protein, were incubated under phosphorylating conditions, solubilized, and incubated with glutathione agarose containing 5 μg of either Fyn or Fyk SH2 domain fusion protein (SH2) or glutathione-S-transferase backbone protein (pG) as described.[12] The proteins in both the supernatants (Flow Through) and those bound to the agarose (Bound) were resolved by SDS-PAGE, blotted onto nitrocellulose, and analyzed with antiserum to the AChR as described.[12] The positions of the α, β, γ, and δ subunits are as indicated. (From Swope and Huganir.[12] Reproduced, with permission, from the *Journal of Biological Chemistry*.)

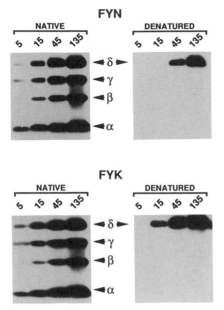

FIGURE 9. Subunit specificity for binding of the AChR to Fyn and Fyk SH2 domain fusion proteins. *Torpedo* electric organ postsynaptic membranes (100 μg) were incubated under phosphorylating conditions, treated without (Native) or with (Denatured) 2% SDS, and diluted to a final concentration of 0.4% SDS/2% Triton X-100. The solubilized proteins were incubated with glutathione agarose containing the indicated micrograms of Fyn or Fyk SH2 domain fusion protein as described.[12] The bound proteins were resolved by SDS-PAGE, blotted onto nitrocellulose, and analyzed with antiserum to the AChR as described.[12] The positions of the α, β, γ, and δ subunits are as indicated. (From Swope and Huganir.[12] Reproduced, with permission, from the *Journal of Biological Chemistry.*)

To examine whether the association of the AChR with Fyn and Fyk during immunoprecipitiation is mediated by binding of the receptor to the SH2 domain of the kinases, the effect of postsynaptic membrane protein denaturation was determined. Under native conditions, the AChR is enriched in the anti-Fyn and anti-Fyk immunoprecipitates (FIG. 10) as shown above (FIG. 7). All four AChR subunits are detected in the anti-Fyk (FIG. 10) and anti-Fyn immunoprecipitates (longer expo-

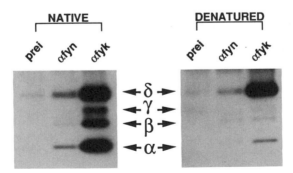

FIGURE 10. Subunit specificity for coimmunoprecipitation of the AChR with Fyn and Fyk. *Torpedo* electric organ postsynaptic membranes (250 μg) were incubated under phosphorylating conditions, treated without (Native) or with (Denatured) 2% SDS, and diluted to a final concentration of 0.4% SDS/2% Trition X-100. The solubilized proteins were immunoprecipitated with preimmune (prei), anti-*fyn* (αfyn), or anti-*fyk* (αfyk) serum, resolved by SDS-PAGE, and analyzed by Western blot using a pool of monoclonal antibodies to the AChR as described.[12] The positions of the α, β, γ, and δ subunits are as indicated. (From Swope and Huganir.[12] Reproduced, with permission, from the *Journal of Biological Chemistry.*)

sures of film; data not shown). Upon denaturation, the anti-Fyk immunoprecipitate is specifically enriched in the AChR δ subunit (FIG. 10). In addition, as shown in FIGURE 10 and by scanning densitometry (data not shown), after denaturation the anti-Fyn immunoprecipitate is also enriched in the AChR δ subunit as well as the β subunit. These data indicate that coimmunoprecipitation of the AChR with Fyn and Fyk is mediated by a direct association of the receptor with the protein tyrosine kinases. In addition, the fact that the dissociated AChR δ subunit coimmunoprecipitates with Fyn and Fyk is consistent with the results obtained using the fusion proteins and supports an association of the receptor with the kinases via the SH2 domains of Fyn and Fyk.

DISCUSSION

The results presented here demonstrate that agrin induces the clustering of AChR and phosphotyrosine-containing proteins. One of these phosphoproteins appears to be the AChR itself, because agrin induces the tyrosine phosphorylation of the receptor. Whether agrin-induced AChR phosphorylation mediates receptor clustering or is a consequence of clustering is not presently known. One attractive hypothesis is that agrin-induced tyrosine phosphorylation of the AChR β subunit of the AChR might alter its interaction with the cytoskeletal elements in such a way as to cause receptor immobilization. At developing neuromuscular junctions, phosphorylated receptors would accumulate at synaptic sites as a result of the local activation of a protein tyrosine kinase by agrin released from the axon terminal. However, it is possible that agrin-induced phosphorylation of the AChR is a consequence of receptor aggregation. For example, protein tyrosine kinases might accumulate together with the AChR in agrin-induced specializations, phosphorylating receptors as they aggregate.

We used the electric organ of *Torpedo californica* as a system for identifying protein tyrosine kinases involved in the regulation of synaptic function. Two kinases of the Src class were identified, cloned, sequenced, and functionally characterized. As demonstrated by examining the homology of the conserved and unique regions of our two clones to known protein tyrosine kinases, one of the clones is the *Torpedo* homolog of neuronal Fyn. The other is a novel protein tyrosine kinase we have named Fyk because of its homology to the Fyn and Yes protein tyrosine kinases.

As for many other types of protein tyrosine kinases,[25,35,36] the activities of Fyn and Fyk are highest in the brain, suggesting an involvement in neuronal function. In neurons, protein tyrosine kinases are believed to be involved in differentiation[5,6] as well as synaptic function.[4] Our results suggest that Fyn and Fyk are important for postsynaptic function at the neuromuscular junction and may be involved in the tyrosine phosphorylation of the AChR. For example, the enrichment of Fyn and Fyk activities in the electric organ as compared to muscle suggests that the two kinases are not simply "housekeeping" proteins of muscle and electroplax, which is derived from muscle, but are coenriched in electric organ with other synaptic proteins. In addition, both kinases are detected in the postsynaptic membrane fraction that is enriched in the AChR; Fyn and Fyk comprise 36% and 8%, respectively, of the total protein tyrosine kinase activity. Furthermore, our data demonstrate that anti-AChR immunoprecipitates contain a protein tyrosine kinase activity that phosphorylates the receptor. Conversely, the AChR but not the 43-kDa protein is enriched in immunoprecipitates of both Fyn and Fyk. Finally, phosphorylation of the AChR β, γ, and δ subunits occurs when affinity-purified AChR is immunoprecipitated with anti-*fyn* and anti-*fyk* antibodies followed by incubation under phosphorylating conditions. These data indicate that Fyn and Fyk form a tight complex with the AChR and suggest that the coimmunoprecipitated AChR is phosphorylated by these protein tyrosine kinases. In addition, these data suggest that Fyn and Fyk may catalyze the endogenous tyrosine phosphorylation of the AChR.

The N-terminal half of Src-like protein tyrosine kinases contains four discrete functional domains including the SH2 domain.[34] The SH2 domain was originally identified as a noncatalytic region that is conserved among cytoplasmic protein tyrosine kinases and modifies kinase function.[37] Subsequently, SH2 domains were shown to be contained in a variety of signaling proteins, playing a crucial role in their association with activated growth factor receptors.[27,38] SH2 domains bind tyrosine phosphorylated proteins including autophosphorylated protein tyrosine kinases and tyrosine kinase substrates[27] via the recognition of specific phosphopeptide sequences.[39,40] Inasmuch as the AChR is a substrate for a protein tyrosine kinase(s), we postulated that the coimmunoprecipitation of the receptor with Fyn and Fyk may be mediated by an association between the SH2 domain of each kinase with the tyrosine-phosphorylated receptor. To address this possibility, we employed affinity chromatography using fusion proteins derived from the SH2 domain of Fyn and Fyk.

We found that the AChR bound to the SH2 domain fusion proteins in a manner dependent on protein concentration. As demonstrated by two independent methods, the binding of the AChR to the SH2 domains requires tyrosine phosphorylation. First, the effect of several phosphoamino acids on AChR binding was determined; both phosphotyrosine and phenylphosphate block binding whereas phosphoserine and phosphothreonine do not.[12] In addition, the extent of AChR binding to the Fyn and Fyk affinity resin correlates with the extent of receptor tyrosine phosphorylation.[12]

The AChR is a 250-kDa pentameric complex comprising four homologous transmembrane subunits in a stoichiometry of $\alpha_2\beta\gamma\delta$.[7] The β, γ, and δ subunits are each phosphorylated on a single tyrosine residue that is contained on the major intracellular loop between transmembrane domains three and four.[41] We investigated the subunit specificity for the binding of the AChR to the Fyn and Fyk SH2 domain fusion proteins. Subsequent to dissociation of the AChR by SDS, the binding of the individual receptor subunits to the affinity resin was examined. A dramatically specific binding of the δ subunit to the Fyn and Fyk SH2 fusion protein was observed. These data suggest that only the δ subunit binds directly to the SH2 domain affinity resin. Alternatively, the binding sites on the β and γ subunits may have been inactivated by the SDS treatment. Therefore, the subunit specificity for binding of the intact native AChR was examined using phosphotyrosine-containing peptides derived from the known phosphorylation sites of the receptor.[41] The δ subunit phosphopeptide is at least 10-fold more potent than the β or γ subunit phosphopeptides in blocking the binding of the native AChR to the Fyn and Fyk SH2 domain fusion proteins.[12] Examination of the peptide sequence for the β, γ, and δ subunits reveals that the subunit specificity observed here is in agreement with previous reports. Phosphopeptides containing asparagine at position +2 to the phosphotyrosine and especially isoleucine at postion +3 are preferred for binding by Src and Fyn SH2 domains.[40] The sequences of the tyrosine phosphorylation sites of the β, γ, and δ subunits are consistent with these data; the δ subunit sequence is YFNI, whereas the β and γ subunit sequences are YFIR and YILK, respectively.[41] Thus, the association of the AChR with the SH2 domain of Fyn and Fyk appears to be mediated by the tyrosine phosphorylation site contained within the major intracellular loop between transmembrane domains three and four of the receptor δ subunit.

The AChR specifically coimmunoprecipitates with Fyn and Fyk. To examine whether the association of the AChR with Fyn and Fyk is mediated by binding of the receptor to the SH2 domains of the kinases, the effect of postsynaptic membrane protein denaturation was determined. In agreement with the results using SH2 domain fusion protein affinity resin, the coimmunoprecipitation of the AChR with Fyn and Fyk appears to be mediated by the δ subunit: after denaturation by SDS, the anti-*fyn* and anti-*fyk* immunoprecipitates are also enriched in the AChR δ subunit. Thus, binding of the tyrosine-phosphorylated intracellular loop of the AChR δ subunit with the SH2 domain of Fyn and Fyk appears to mediate the association of the receptor with these two protein tyrosine kinases.

The functional consequence of association between the AChR and Fyn and Fyk remains to be determined. It has been postulated that binding of tyrosine-phosphorylated proteins to the SH2 domain of Src-like kinases and adapter proteins stabilizes the bound substrates by preventing rapid dephosphorylation.[27] In fact, binding of Src SH2 domain fusion protein to pp[125] focal adhesion kinase, protects the major site of pp[125] focal adhesion kinase autophosphorylation from dephosphorylation upon treatment with a tyrosine phosphatase *in vitro.*[42] Alternatively, association between tyrosine-phosphorylated substrates and SH2 domain containing proteins may provide a means for forming complexes of proteins containing a variety of functions. Thus, association between Fyn and Fyk and the AChR may protect the receptor from dephosphorylation. However, the association may promote the recruitment of enzymatic and structural elements involved in synaptic transmission at the neuromuscular junction.

Synaptogenesis is a complex developmental process for which the neuromuscular junction has served as a model. The clustering of postsynaptic components at the nerve-muscle contact appears to be regulated by factors contained in the extracellular matrix[43] including agrin[44,45] and basic fibroblast growth factor.[46,47] The action of these components may be mediated by tyrosine phosphorylation. In fact, a role for protein tyrosine kinase activation has been implicated in the clustering of synaptic components induced by a variety of neuronal and nonneuronal stimuli including nerve,[9] polymer beads,[48,49] electric field,[50] expression of exogenous 43-kDa protein,[51] and *in vitro* treatment with basic fibroblast growth factor[47] or agrin.[10] Formation of the postsynaptic specializations is associated with a rearrangement of cytoskeletal elements[52] and an alteration in gene expression.[53] The ubiquitous expression of Src-like protein tyrosine kinases as well as their substrate promiscuity suggests that these kinases regulate a wide range of cellular functions.[29,54] Although the data presented here indicate that Fyn and Fyk function to phosphorylate the AChR, it will be interesting to examine additional roles of these two protein tyrosine kinases during synaptogenesis.

The molecular mechanisms involved in the regulation of Fyn and Fyk in the postsynaptic membrane of *Torpedo* electric organ and in skeletal muscle are not known. Tyrosine phosphorylation of the AChR is regulated by innervation in rat skeletal muscle,[9] a process that appears to be mediated by agrin.[10,51] Agrin interacts with a specific receptor in the muscle plasma membrane that is responsible for the nerve and agrin-induced clustering of the AChR at the synapse.[55,56] An agrin receptor has recently been identified as the extracellular peripheral membrane subunit of α-dystroglycan.[57–59] It will be interesting to determine whether this agrin binding protein functions to transduce the activation of an intracellular protein tyrosine kinase such as Fyn and/or Fyk.

SUMMARY

Most neurotransmitter receptors examined to date are either regulated by phosphorylation or contain consensus sequences for phosphorylation by protein kinases. The nicotinic acetylcholine receptor (AChR), which mediates depolarization at the neuromuscular junction, has served as a model for the study of the structure, function, and regulation of ligand-gated ion channels. The AChR is phosphorylated by protein kinase A, protein kinase C, and an unidentified protein tyrosine kinase. Tyrosine phosphorylation of the AChR is correlated with a modulation of the rate of receptor desensitization and is associated with AChR clustering. We showed that agrin, a neuronally derived extracellular matrix protein, induces AChR clustering and tyrosine phosphorylation. In addition, we identified two protein tyrosine kinases, Fyn and Fyk, that appear to be involved in the regulation of synaptic transmission at the neuromuscular junction by phosphorylating the AChR. The two kinases are highly expressed in *Torpedo* electric organ, a tissue enriched in synaptic components including the AChR. As demonstrated by coimmunoprecipitation, Fyn and Fyk associate with the AChR. Furthermore, the AChR is phosphorylated in Fyn and Fyk immunoprecipitates. We investigated the molecular basis for the association of the AChR with Fyn and Fyk using fusion proteins derived from the kinases. The AChR bound specifically to the SH2 domain fusion proteins of Fyn and Fyk. The association of the AChR with the SH2 domains is dependent on the state of AChR tyrosine phosphorylation and is mediated by the δ subunit of the receptor. These data provide evidence that the protein tyrosine kinases Fyn and Fyk may act to phosphorylate the AChR *in vivo*.

ACKNOWLEDGMENTS

We thank Drs. Susan K. H. Gillespie, Michael Ehlers, and Eric Fung for critical review of the manuscript, and Cindy Finch for secretarial support.

REFERENCES

1. EDELMAN, A. M., D. K. BLUMENTHAL & E. G. KREBS. 1987. Annu. Rev. Biochem. **56:** 567–613.
2. COHEN, P. 1989. Annu. Rev. Biochem. **58:** 453–508.
3. HUNTER, T. & J. A. COOPER. 1985. Annu. Rev. Biochem. **54:** 897–930.
4. WAGNER, K. R., L. MEI & R. L. HUGANIR. 1991. Curr. Opin. Neuro. **1:** 65–73.
5. DICHTER, M. A., A. S. TISCHLER & L. A. GREENE. 1977. Nature **268:** 501–504.
6. ALEMÀ, S., P. CASALBORE, E. AGOSTINI & F. TATÒ. 1985. Nature **316:** 557–559.
7. GALZI, J.-L., F. REVAH, A. BESSIS & J.-P. CHANGEUX. 1991. Annu. Rev. Pharmacol. **31:** 37–72.
8. SWOPE, S. L., S. J. MOSS, C. D. BLACKSTONE & R. L. HUGANIR. 1992. FASEB J. **6:** 2514–2523.
9. QU, Z., E. MORITZ & R. L. HUGANIR. 1990. Neuron **2:** 367–378.
10. WALLACE, B. G., Z. QU & R. L. HUGANIR. 1991. Neuron **6:** 869–878.
11. SWOPE, S. L. & R. L. HUGANIR. 1993. J. Biol. Chem. **268:** 25152–25161.
12. SWOPE, S. L. & R. L. HUGANIR. 1994. J. Biol. Chem. **269:** 29817–29824.
13. HUGANIR, R. L., K. MILES & P. GREENGARD. 1984. Proc. Natl. Acad. Sci. USA **81:** 6968–6972.

14. HOPFIELD, J. F., D. W. TANK, P. GREENGARD & R. L. HUGANIR. 1988. Nature **336:** 677–680.
15. MILES, K., P. GREENGARD & R. L. HUGANIR. 1989. Neuron **2:** 1517–1524.
16. SMITH, M. M., J. P. MERLIE & J. C. LAWRENCE, JR. 1987. Proc. Natl. Acad. Sci. USA **84:** 6601–6605.
17. HALL, Z. W. & J. R. SANES. 1993. Cell 72/Neuron **10:** 99–121.
18. ANTHONY, D. T., S. M. SCHUETZE & L. L. RUBIN. 1984. Proc. Natl. Acad. Sci. USA **81:** 2265–2269.
19. WALLACE, B. G. 1988. J. Cell Biol. **107:** 267–278.
20. ROSS, A., M. RAPUANO & J. PRIVES. 1988. J. Cell Biol. **107:** 1139–1145.
21. MCMAHAN, U. J. 1990. Cold Spring Harbor Symp. Quant. Biol. **55:** 407–418.
22. HANKS, S. K., A. M. QUINN & T. HUNTER. 1988. Science **241:** 42–52.
23. WILKS, A. F. 1991. Methods Enzymol. **200:** 533–546.
24. KAWAKAMI, T., C. Y. PENNINGTON & K. C. ROBBINS. 1986. Mol. Cell Biol. **6:** 4195–4201.
25. COOKE, M. P. & R. M. PERLMUTTER. 1989. New Biol. **1:** 66–74.
26. STAHL, M. L., C. R. FERENZ, K. L. KELLEHER, R. W. KRIZ & J. L. KNOPF. 1988. Nature **332:** 269–272.
27. KOCH, C. A., D. ANDERSON, M. F. MORAN, C. ELLIS & T. PAWSON. 1991. Science **252:** 668–674.
28. SEMBA, K., M. NISHIZAWA, N. MIYAJIMA, M. C. YOSHIDA, J. SUKEGAWA, Y. YAMANASHI, M. SASAKI, T. YAMAMOTO & K. TOYOSHIMA. 1986. Proc. Natl. Acad. Sci. USA **83:** 5459–5463.
29. COOPER, J. A. 1990. *In* Peptides & Protein Phosphorylation. B. E. Kemp, Ed.: 85–113. CRC Press Inc., Boca Raton, FL.
30. STEELE, R. E., M. Y. IRWIN, C. L. KNUDSEN, J. W. COLLETT & J. B. FERO. 1989. Oncogene Res. **4:** 223 233.
31. SOBEL, A., M. WEBER & J.-P. CHANGEUX. 1977. Eur. J. Biochem. **80:** 215–224.
32. CICHOWSKI, K., F. MCCORMICK & J. S. BRUGGE. 1992. J. Biol. Chem. **267:** 5025–5028.
33. LAROCHELLE, W. J. & S. C. FROEHNER. 1986. J. Biol. Chem. **261:** 5270–5274.
34. SUDOL, M. 1993. *In* The Molecular Basis of Human Cancer. B. G. Neel & R. Kumar, Eds.: 203–223. Futura Publishing. Mount Kisco, NY.
35. COTTON, P. C. & J. S. BRUGGE. 1983. Mol. Cell. Biol. **3:** 1157–1162.
36. SUDOL, M. & H. HANAFUSA. 1986. Mol. Cell. Biol. **6:** 2839–2846.
37. SADOWSKI, I., J. C. STONE & T. PAWSON. 1986. Mol. Cell. Biol. **6:** 4396–4408.
38. PAWSON, T. & G. D. GISH. 1992. Cell **71:** 359–362.
39. FANTL, W. J., J. A. ESCOBEDO, G. A. MARTIN, C. W. TURCK, M. DEL ROSARIO, F. MCCORMICK & L. T. WILLIAMS. 1992. Cell **69:** 413–423.
40. SONGYANG, Z., S. E. SHOELSON, M. CHAUDHURL, G. GISH, T. PAWSON, W. G. HASER, F. KING, T. ROBERTS, S. RANOFSKY, R. J. LECHIEDER, B. G. NEEL, R. B. BIRGE, J. E. FAJARDO, M. M. CHOU, H. HANAFUSA, B. SCHAFTHAUSEN & L. C. CANTLEY. 1993. Cell **72:** 767–778.
41. WAGNER, K., K. EDSON, L. HEGINBOTHAM, M. POST, R. L. HUGANIR & A. J. CZERNICK. 1991. J. Biol. Chem. **266:** 23784–23789.
42. COBB, B. S., M. D. SCHALLER, T.-H. LEU & J. T. PARSONS. 1994. Mol. Cell. Biol. **14:** 147–155.
43. BURDEN, S. J., P. B. SARGENT & U. J. MCMAHAN. 1979. J. Cell Biol. **82:** 412–425.
44. MAGILL-SOLC, C. & U. J. MCMAHAN. 1990. J. Exp. Biol. **153:** 1–10.
45. WALLACE, B. G. 1989. J. Neurosci. **9:** 1294–1302.
46. RAPRAEGER, A. C., A. KRUFKA & B. B. OLWIN. 1991. Science **252:** 1705–1708.
47. PENG, H. B., L. P. BAKER & Q. CHEN. 1991. Neuron **6:** 237–246.
48. BAKER, L. P. & H. B. PENG. 1992. J. Cell Biol. **120:** 185–195.
49. BAKER, L. P., Q. CHEN & H. B. PENG. 1992. J. Cell Sci. **102:** 543–555.
50. PENG, H. B., L. P. BAKER & Z. DAI. 1993. J. Cell Biol. **120:** 197–204.
51. QU, Z. & R. L. HUGANIR. Unpublished results.
52. BLOCH, R. J. & D. W. PUMPLIN. 1988. Am. J. Physiol. **254:** C345–C364.

53. MARTINOU, J. C., D. L. FALLS, G. D. FISCHBACH & J. P. MERLIE. 1991. Proc. Natl. Acad. Sci. USA **88:** 7669–7673.
54. COOPER, J. A. & T. HUNTER. 1983. Curr. Top. Microbiol. Immunol. **107:** 125–161.
55. NASTUK, M. A., E. LIETH, J. MA, C. A. CARDASIS, E. B. MOYNIHAN, B. A. MCKECHNIE & J. R. FALLON. 1991. Neuron **7:** 807–818.
56. MA, J., M. A. NASTUK, B. M. MCKECHNIE & J. R. FALLON. 1993. J. Biol. Chem. **268:** 25108–25117.
57. BOWE, M. A., K. A. DEYST, J. D. LESZYK & J. R. FALLON. 1994. Neuron **12:** 1173–1180.
58. CAMPANELLI, J. T., S. L. ROBERDS, K. P. CAMPBELL & R. H. SCHELLER. 1994. Cell **77:** 663–674.
59. GEE, S. H., F. MONTANARO, M. H. LINDENBAUM & S. CARBONETTO. 1994. Cell **77:** 675–686.

S-Nitrosylation of m2 Muscarinic Receptor Thiols Disrupts Receptor–G-Protein Coupling[a]

ROBERT S. ARONSTAM,[b,c] DAN C. MARTIN,[d]
ROBERT L. DENNISON,[d] AND HEATHER G. COOLEY[b]

[b]Guthrie Research Institute
One Guthrie Square
Sayre, Pennsylvania 18840

[d]Department of Anesthesiology
Medical College of Georgia
Augusta, Georgia 30912

Muscarinic acetylcholine receptors possess nine cysteine residues which are conserved among all five receptor subtypes: four are in membrane-spanning helices; one is in the cytoplasmic NH_2-terminal tail and is probably palmitoylated; two are in the third extracellular loop; and two cysteines located in the first and second extracellular loops participate in disulfide bond formation.[1,2] In addition, each receptor has a variable number (1–6) of cysteine residues in the large third intracellular loop. Although biochemical manipulation of these groups has little effect on [³H]antagonist binding, it has been known for at least 10 years that the redox state of these sulfhydryl moieties influences the functional coupling of receptors to transducer G-proteins.[3,4] However, natural mediator(s) that may act on these groups to regulate receptor activity have not been identified. The work of Lipton and co-workers[5] raises the possibility that S-nitrosylation of reactive sulfhydryl centers may be a common pathway for molecular control of protein function. Nitric oxide (NO) group transfer (of NO^+) from nitric oxide depends on the redox chemistry intrinsic to the NO molecule.[5] Sodium nitroprusside (SNP) has a strong NO^+ character and nitrosylates protein thiols. In the present study, we examined the possibility that nitrosylation of sulfhydryl groups by NO^+ affects receptor–G-protein coupling.

Rat atrial membranes contain m2 muscarinic receptors whose binding of [³H]N-methylscopolamine increases dramatically in the presence of Gpp(NH)p and after physical treatments (heat, low pH) that engender m2 dissociation from G-proteins.[4,6] Exposure of atrial m2 receptors to SNP (1–5 mM for 30 minutes at room temperature) produced a moderate (18–30%) increase in [³H]MS binding (FIG. 1A). Gpp(NH)p (10 μM) increased the apparent density of muscarinic binding sites by 60% (FIG. 1B). However, marked stimulation of [³H]MS binding by Gpp(NH)p was moderated by SNP (FIG. 1B). Fully two-thirds of the increase in binding engendered by Gpp(NH)p was eliminated by 5 mM SNP, so that [³H]MS binding was reduced to a level lower than that measured after exposure to 5 mM SNP alone. Under redox conditions (i.e., exposure to ascorbate) which promoted $NO^·$, as opposed to NO^+, formation from SNP, m2 binding and activity were severely depressed or eliminated

[a]This work was supported by grants GM46408, NS25296, Department of the Army contract DAMD17-94-J-4011, and the Research Service of the Veterans Administration.
[c]Corresponding author.

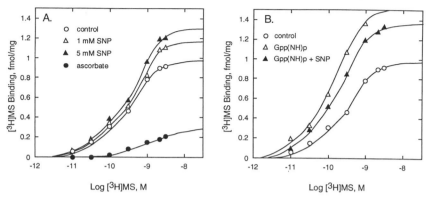

FIGURE 1. Influence of sodium nitroprusside (SNP) on the binding of [³H]MS to atrial m2 receptors. Each point represents the mean from 3–6 experiments. (A) The binding of the indicated concentrations of [³H]MS was measured in the absence of SNP (○) or after exposure to 1 (△) or 5 (▲) mM SNP. In one series of experiments, atrial membranes were treated with 1 mM ascorbate for 10 min at room temperature before being exposed to 1 mM SNP (●). Nonlinear regression analyses using a single receptor population model indicated no change in [³H]MS affinity but increases in apparent receptor density from 0.99 ± 0.07 to 1.18 ± 0.12 and 1.30 ± 0.16 fmol/mg protein after exposure to 1 and 5 mM SNP, respectively. (B) The binding of [³H]MS was measured in the absence of modulating agents (○) or in the presence of 10 μM Gpp(NH)p (△) or 10 μM Gpp(NH)p after exposure to 1 mM SNP (▲). Nonlinear regression analyses using a single receptor population model indicated an increase in receptor density in the presence of Gpp(NH)p (from 0.99 ± 0.07 to 1.57 ± 0.05 fmol/mg protein) with no change in [³H]MS binding affinity. The apparent receptor density in the presence of Gpp(NH)p was reduced to 1.39 ± 0.11 and 1.19 ± 0.08 after exposure to 1 (▲) and 5 (not shown) mM SNP, respectively.

(FIG. 1A). Exposure to ascorbate alone did not alter [³H]MS binding. Sodium nitroprusside (5 mM) decreased surface and total thiol content by 27% with an IC_{10} of 1.0 ± 0.2 mM ($n = 3$) (FIG. 2). These effects of SNP were not observed under conditions (i.e., heat and low pH) in which receptor–G-protein coupling was eliminated. Thus, S-nitrosylation appears to uncouple m2 receptors from transducer

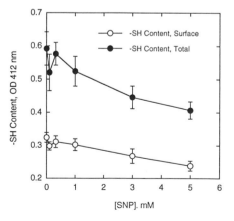

FIGURE 2. Influence of sodium nitroprusside (SNP) on the sulfhydryl content of atrial membranes. Sulfhydryl content was measured after a 10-min exposure to SNP in undisrupted membranes (○) or in SDS-solubilized (●) membranes. Measurements were performed spectroscopically (412 nm) using 5,5'-dithio-*bis*-2-nitrobenzoate as a specific probe.

G-proteins. It is interesting that SNP effects were not observed in cultured human neuroblastoma cells (Sk-N-SH), which predominantly express m3 muscarinic receptors. Preliminary experiments also indicate that treatment with 5 mM SNP disrupts m2 receptor control of G-protein GTPase activity, although the nature of this effect (i.e., a receptor, G-protein, or coupling effect) is not yet clear.

These findings suggest a mechanism for modulation of muscarinic receptor function that depends on both the redox milieu and nitric oxide production. Although the identity of the relevant reactive sulfhydryl center awaits confirmation by mutational analysis, the fact that m2 but not m3 receptors are selectively affected might indicate involvement of the cysteine moiety on the second intracellular loop, which is only present in m2 and m4 receptors. This work supports the suggestion by Lipton and co-workers[5] that nitrosylation of reactive sulfhydryl groups is a common biochemical mechanism for the control of protein function.

REFERENCES

1. HULME, E. C., N. J. M. BIRDSALL & N. J. BUCKLEY. 1990. Muscarinic receptor subtypes. Annu. Rev. Pharmacol. Toxicol. **30:** 633–673.
2. WESS, J. 1993. Mutational analysis of muscarinic acetylcholine receptors: Structural basis of ligand/receptor/G protein interactions. Life Sci. **53:** 1447–1463.
3. ARONSTAM, R. S., L. G. ABOOD & W. HOSS. 1978. Influence of sulfhydryl reagents and heavy metals on the functional state of the muscarinic acetylcholine receptor in rat brain. Mol. Pharmacol. **14:** 575–586.
4. ARONSTAM, R. S., M. L. KIRBY & M. SMITH. 1985. Muscarinic acetylcholine receptors in chick heart: Influence of heat and *N*-ethylmaleimide on receptor conformations and interactions with guanine nucleotide-dependent regulatory proteins. Neurosci. Lett. **54:** 289–294.
5. LIPTON, S. A., Y.-B. CHOI, Z.-H. PAN, S. Z. LEI, H.-S. V. CHEN, N. J. SUCHER, J. LOSCALZO, D. J. SINGEL & J. S. STAMLER. 1993. A redox-based mechanism for the neuroprotective and neurodestructive effects of nitric oxide and related nitroso-compounds. Nature **364:** 626–632.
6. ANTHONY, B. L. & R. S. ARONSTAM. 1986. Effect of pH on muscarinic receptors from rat brainstem. J. Neurochem. **46:** 556–561.

Combined Administration of Agonist-Antagonist as a Method of Regulating Receptor Activation

JED E. ROSE,[a,b] EDWARD D. LEVIN,[b]
FREDERIQUE M. BEHM,[b] ERIC C. WESTMAN,[a,c]
ROY M. STEIN,[a,b] JAMES D. LANE,[b] AND GAIL V. RIPKA[b]

[a]VA Medical Center
Durham, North Carolina 27705

Departments of Psychiatry[b] and Medicine[c]
Duke University
Durham, North Carolina

Co-administration of an agonist with an antagonist may occupy a greater number of receptors than either drug alone, thereby attenuating further response to a phasic stimulus. At the same time, it is hypothesized that the tonic actions of the agonist and antagonist offset one another, regulating receptor activation and avoiding over- or understimulation that would be caused by administration of either agonist or antagonist alone. This hypothesis was tested in the context of drug reward. Twelve smokers rated the rewarding effects of cigarette smoke after separate and combined administration of nicotine and the nicotinic antagonist mecamylamine. Subjects rated test cigarettes after administration of mecamylamine (10 mg) versus placebo capsules, and a nicotine (1.1 mg) versus non-nicotine smoke preload. Smoking withdrawal symptoms, task performance, and cardiovascular activity were also measured. As predicted, mecamylamine significantly attenuated smoking satisfaction, liking, and airway sensations (FIG. 1). The nicotine preload similarly reduced the enjoyable aspects of subsequent test cigarettes, and this action of the preload was *not* prevented by mecamylamine. In contrast, mecamylamine did block nicotine-related increases in heart rate and systolic blood pressure (FIG. 2). Conversely, nicotine counteracted the sedative effects of mecamylamine on tapping speed (FIG. 3) and orthostatic blood pressure response (FIG. 2). Although each drug offset potential side effects of the other, they acted in unison to attenuate smoking satisfaction. Thus, combined agonist-antagonist administration may attenuate drug reward while preventing the withdrawal symptoms or other side effects associated with presentation of agonist alone or antagonist alone. Unlike partial agonists, the current approach of co-administering agonist and antagonist allows for flexible titration of the ratio of agonist-to-antagonist doses.

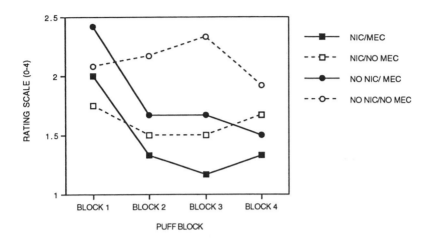

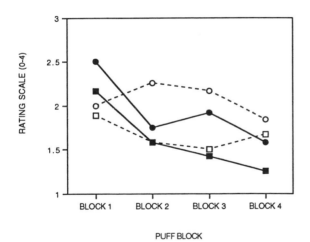

FIGURE 1. Effects of nicotine and mecamylamine on smoking satisfaction (**A**) and on liking (**B**) test cigarettes.

[*FIGURES 2 and 3 are on following pages.*]

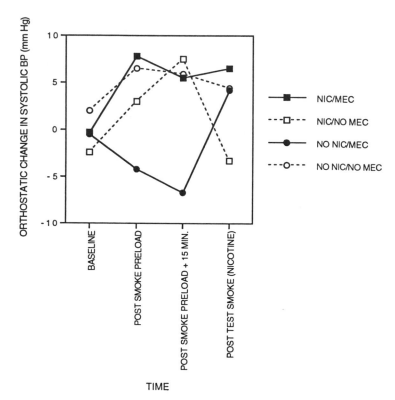

FIGURE 2. Effects of nicotine and mecamylamine on orthostatic response of systolic blood pressure.

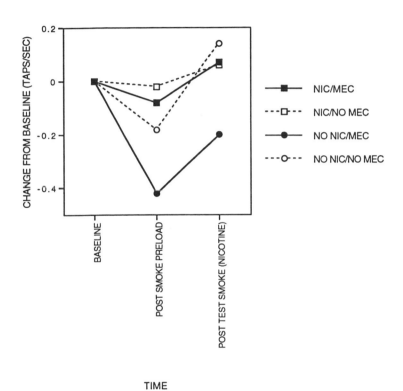

FIGURE 3. Effects of nicotine and mecamylamine on tapping speed.

Preliminary Two-Dimensional [1H]NMR Characterization of the Complex Formed between an 18-Amino Acid Peptide Fragment of the α-Subunit of the Nicotinic Acetylcholine Receptor and α-Bungarotoxin[a]

LISA N. GENTILE,[b] VLADIMIR J. BASUS,[c]
QING-LUO SHI,[b] AND EDWARD HAWROT[b]

[b]Section of Molecular and Biochemical Pharmacology
Brown University
Providence, Rhode Island 02912

[c]Department of Pharmaceutical Chemistry
University of California at San Francisco
San Francisco, California 94143

The nicotinic acetylcholine receptor (nAChR) is a ~250 kDa pentameric complex (subunit composition: $\alpha_2\beta\gamma\delta$) which is located at the neuromuscular junction and which is critically involved in eliciting skeletal muscle contraction. The nAChR is a transmembrane protein believed to have a topographical distribution as depicted in FIGURE 1.[1] The snake venom-derived, protein α-neurotoxins, such as α-bungarotoxin (BGTX), which behave as high-affinity, competitive antagonists at the nAChR are believed to bind to an extracellular site at or near the outermost edge or cusp of the receptor (FIG. 1). Although the four subunits are encoded by separate genes, they have a high degree of sequence homology and are presumed to share a common tertiary structure. Each of the subunits is predicted on the basis of primary sequence hydropathy analysis to contain four membrane-spanning regions (M1–M4), with M2 forming the lining of the nonselective cation channel. It is widely held that the first ~200 N-terminal residues from each subunit are extracellular. All four subunits contain a common disulfide loop involving Cys residues 128 and 142 (*Torpedo californica* α-subunit numbering scheme). The α-subunit alone contains an additional, extremely rare vicinal disulfide between residues 192 and 193. Many lines of investigation argue that the major determinants of the agonist and antagonist binding sites are located predominantly on the α-subunit although contributions from the γ- and δ-subunits also appear to be involved. At present, the exact role of the vicinal disulfide in the α-subunit is unclear, but it has been postulated that the conformational movements about the highly strained eight-membered disulfide ring

[a]This work was supported by National Science Foundation grants BNS 90-21227 (to E.H.) and DMB9104794 (to V.J.B.); National Institutes of Health grant GM32629 (to E.H.); and a NIH predoctoral training grant GM07601 (to L.N.G.). Some of the work presented here was done in partial fulfillment of the requirements for a Ph.D. degree (L.N.G.) from Brown University.

may be important in channel activation, subserving in some way the key signal transduction event at the neuromuscular junction.[2]

Small synthetic peptides corresponding to overlapping regions of the N-terminus (upstream of the M1 region) of the α-subunit have been examined in an attempt to localize the full extent of the α-BGTX binding site.[3-12] Several groups now agree that synthetic or recombinant peptide fragments of the α-subunit that contain residues 181–198 (*Torpedo californica*), and which notably include the 192–193 disulfide site within the native receptor, retain significant ability to bind BGTX as evidenced by a variety of binding assays. For example, the 18mer (α181-YRGWKHWVYYTCCP-DTPY-198) binds BGTX with an apparent K_d of 65 nM.[4,11]

Due to well-documented technical difficulties in preparing crystals of membrane proteins, an atomic resolution structure of the receptor or of its binding site is presently lacking. A 9 Å cryoelectron microscopy derived structure based on

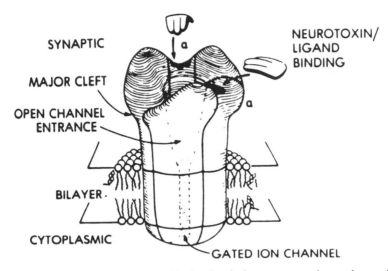

FIGURE 1. Schematic model of the nAChR showing the heteropentameric complex, as viewed in the membrane, as well as the α-bungarotoxin and ligand binding sites. (From Stroud.[1] Reproduced with permission.)

pseudocrystalline arrays of receptor in two-dimensional membranes is providing some moderate resolution insights into the three-dimensional configuration of the receptor in native membranes.[13] Our interests are in defining the nature of the contact zone between portions of the N-terminal domain of the α-subunit of the nAChR (specifically residues 181–198) and BGTX on an atomic level. We feel that the ability of the 18mer to bind so tightly to the toxin is a good indication that a significantly large fraction (at least 50%) of the actual binding site residues in the native receptor may be included in the 18mer peptide. Hence, we feel that a structural analysis of the receptor peptide fragments bound to BGTX will provide valuable information on the structure, at least that of the antagonist bound state, of the corresponding region in the intact receptor. Towards this end, we have been using two-dimensional [¹H]NMR techniques to characterize the solution structure of the receptor-antagonist complex.

Our current NMR investigations enlarge upon our earlier studies of the disulfide form of a 12-amino acid peptide fragment of the AChR (α185–196).[14] The view that is emerging from both of our NMR studies is one of a broad contact zone between the proteins. The 12mer has extensive contacts with the first two N-terminal loops of the toxin as evidenced by large chemical shift changes in bound versus free BGTX, as well as by intermolecular NOEs (nuclear Overhauser effects) between the peptide and toxin. The contact zone of the stoichiometric complex formed between BGTX and the 18mer (disulfide form of the peptide) also includes the first two loops of BGTX. The area of contact is extended further, however, on both the N- and C-terminus ends of the nested 12mer sequence (KHWVYYTCCPDT), and we now detect potential contacts with residues in the third loop and an additional residue within the C-terminal tail region.

METHODS

The 12-amino acid peptide corresponding to residues 185–196 (12mer), as well as the 18-amino acid peptide corresponding to residues 181–198 (18mer) of the α-subunit, were synthesized and purified by reverse-phase high-performance liquid chromatography (RP-HPLC) in the Protein and Nucleic Acid Facility, Yale University School of Medicine, New Haven, Connecticut. The primary amino acid sequences are those of the nAChR as isolated from *Torpedo californica* and are highly conserved among species in this region. Both amino and carboxyl termini were prepared unblocked, and the cysteines were prepared and maintained in the reduced state. Susceptibility to covalent alkylation by N-ethylmaleimide, as indicated by a marked increase in retention time on HPLC, confirmed that > 90% of the synthesized peptide contained cysteines in the free sulfhydryl form (see FIG. 2). The amino acid composition of the peptide was verified by ion exchange chromatographic analysis of performic acid oxidation products on a Beckman model 6300 amino acid analyzer. The mass of the 12mer (1,515) and the 18mer (2,340) were confirmed by fast-atom mass bombardment (FAB) techniques, at the Yale Cancer Center Mass Spectroscopy Facility.

Both the 12mer and the 18mer were oxidized with $K_3Fe(CN)_6$. In both cases, one equivalent of the reduced peptide in CH_3CN/H_2O (10% acetonitrile in the case of the 12mer and 50% acetonitrile for the less soluble 18mer) at 0.13 mM, pH 7.1 was added dropwise with a peristaltic pump to a solution containing 1.2 equivalents of potassium ferricyanide in 0.05 M ammonium bicarbonate at pH 7.1. These conditions, employing slow addition of peptide to oxidant and high dilution were required to prevent intermolecular disulfide formation. Completion of the reaction was monitored with a standard Ellman's reaction. Upon completion of oxidation of the 12mer, 5 mL of Biorad Affigel-501 in ammonium bicarbonate buffer was added to the reaction vessel, and the mixture was gently stirred overnight at 4 °C. After filtration and C18 Sep-Pak desalting, the peptide was purified by C18 RP-HPLC in 50% CH_3CN/H_2O (0.1% TFA). The preparation of the disulfide form of the 18mer did not require the organomercurial step for removal of the residual thiol-containing starting material, as the oxidized and reduced forms of the 18mer were resolvable upon RP-HPLC (see the elution profile in FIG. 3), unlike the oxidized and reduced 12mer, which co-eluted as one broad peak (note the observed retention times, t_R, of the oxidized and reduced 12mer in FIG. 2). The observed broadness of the 12mer-disulfide peak, relative to that of the starting material, suggests that the disulfide form of the peptide may exist as multiple conformers. Consistent with this interpretation is the fact that on the NMR time scale, we have observed, even at elevated temperatures, multiple low-energy conformers of the 12mer-disulfide peptide suggestive of conformational rearrangements involving the eight-membered disulfide ring (V. J. Basus, unpublished observations). The mass of both disulfide-containing peptides was confirmed by FAB mass spectroscopy, as above. Typical final yields for the 18mer oxidation were 70%, and those for the 12mer were about 50%.

All NMR spectra were acquired on a GE GN-500MHz spectrometer at the University of California at San Francisco. The 1:1 complex of 12mer:BGTX (5 mM each) in H_2O was prepared at pH 5.8, whereas the 1:1 complex of 18mer:BGTX (7.5 mM) was prepared at pH 4.0. In both cases, the peptide in solid form was added to a solution of BGTX.

The ROESY (rotating frame Overhauser spectroscopy) NMR experiment of a 2:1 mixture of BGTX:12mer ($K_d \approx$ 1.2–3 μM) was carried out at 45 °C, with a 100 ms

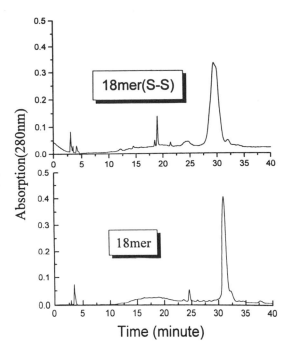

FIGURE 2. Reverse-phase (C-18) high-performance liquid chromatography profile of the reduced (*bottom*) and oxidized (*top*) 18mer (α181–198). The reduced 18mer has a longer retention time and is baseline separable from the disulfide form of the 18mer.

mixing time, at 7.5 mM BGTX, 3.75 mM peptide. Due to the tighter binding in the complex formed with the 18mer ($K_d \approx$ 65 nM) the experiment was performed at 65 °C, with a 50 ms mixing time. Usually, the NMR experiments involved an acquired data set of 512 × 2048 with 16 scans per t_1 increment, and the use of Doddrell's 1:3:3:1 water suppression scheme.

The NOESY (nuclear Overhauser spectroscopy) experiments of the 1:1 peptide: BGTX complex were carried out in temperature intervals of 10 ° beginning from the temperature used for the ROESY experiments of the complex (65 °C for the 18mer complex, 45 °C for the 12mer complex) down to 35 °C where the resonances of free BGTX were assigned. The usual NOESY experiment was run with a mixing time of 150 ms, the above-mentioned data set size, and a "jump and return" water suppression scheme. The HOHAHA (homonuclear Hartman Hahn) experiments were done with a 50 ms mixing time and a combination of SCUBA, DANTE, and 1:3:3:1 water suppression programs. The DQCOSY (double quantum filtered correlation spectroscopy) spectra were acquired with SCUBA water suppression. All experiments are run in H_2O and the reported chemical shifts are relative to sodium 3-(trimethylsilyl)propionate, or the water resonance.

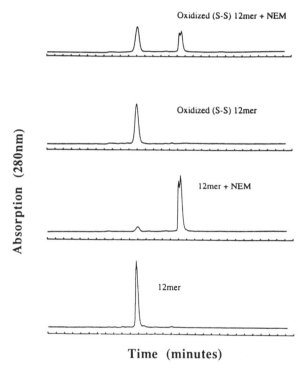

FIGURE 3. Reverse-phase high-performance liquid chromatography (HPLC) analysis of the 12mer (α185–196) before and after ferricyanide oxidation to the disulfide. As the reduced and oxidized form of the 12mer appear to co-migrate upon HPLC, the disulfide form was revealed only after N-ethylmaleimide (NEM) treatment. The NEM alkylates the available sulfhydryl only in the reduced form of the 12mer causing a shift in elution position as indicated. The elution position of the disulfide form of the 12mer is not affected by NEM.

RESULTS AND DISCUSSION

The first step in defining the contact zone between receptor and toxin is the complete ^{1}H assignment of the 1:1 complex (TABLE 1 for 12mer, TABLE 2 for 18mer). To accomplish this, we took advantage of two factors, the difference in sign of chemical exchange cross peaks versus NOE cross peaks in a ROESY experiment of a 2:1 BGTX:18mer complex, and the complete assignment of free BGTX. In the

TABLE 1. [^{1}H] Chemical Shifts[a] and Assignments for α-Bungarotoxin Complexed to the AChR Peptide Fragment α185–196, at 35 °C and pH 5.9

	NH	CαH	CβH	Others
BGTX Residue				
Ile-1	n.o.[b]	4.15	1.53	CγH$_3$ 0.91, CγH$_2$ 1.20, CδH$_3$ 0.74
Val-2	8.11	4.99	1.57	CγH$_3$ 0.57, 0.86
Cys-3	8.76	5.06	2.44, 3.01	
His-4	9.23	5.06	2.89, 2.54	CδH 6.34, CεH 7.71
Thr-5	9.06	5.29	4.03	CγH$_3$ 1.36
Thr-6	8.62	4.79	5.70	CγH$_3$ 1.27
Ala-7	9.23	4.53	1.42	
Thr-8	7.44	4.68	4.28	CγH$_3$ 1.06
Ser-9	8.28	n.a.[c]	3.79, 3.80	
Pro-10		4.80	2.05, 2.18	
Ile-11	8.85	3.95	1.50	CδH 0.25, CγH$_2$ 0.91, 1.37
Ser-12	7.66	4.93	3.73, 3.82	
Ala-13	8.19	5.09	0.88	
Val-14	8.90	4.66	2.09	CγH$_3$ 0.89, 0.86
Thr-15	8.47	4.45	4.01	CγH$_3$ 1.24
Cys-16	8.85	4.85	3.35, 3.00	
Pro-17		4.72	2.42, 1.78	
Pro-18		4.31	2.32, 1.88	CδH$_2$ 3.88, 3.60
Gly-19	8.74	4.28, 3.67		
Glu-20	7.91	4.28	2.14, 1.62	
Asn-21	7.82	4.98	3.00, 2.67	NδH$_2$ 6.98, 7.44
Leu-22	8.33	4.99	1.72, 1.54	CγH 1.58, CδH$_3$ 0.72, 0.79
Cys-23	8.73	5.96	3.27, 2.84	
Tyr-24	9.07	6.00	3.00, 2.66	CδH 6.65, CεH 6.75
Arg-25	9.00	5.25	n.a.	
Lys-26	9.95	5.81	2.11, 1.88	CγH 1.60, CεH 2.53
Met-27	9.21	6.24	2.00, 1.98	CγH$_2$ 2.43, 2.62, Cε 1.97
Trp-28	8.16	5.20	3.79, 3.45	CηH 7.35, Cξ^{3}H 6.80, Cξ^{2}H 7.58, CδH 7.07
Cys-29	9.55	5.40	3.48, 3.06	
Asp-30	8.10[d]	4.99	3.56, 2.75	
Ala-31	8.30	4.08	1.06	
Phe-32	8.82	4.83	3.10, 2.86	CδH 7.29, CξH 7.35
Cys-33	7.77	4.22	3.76, 3.12	
Ser-34	n.a.	n.a.	n.a.	
Ser-35	n.a.	4.29	n.a.	
Arg-36	8.33	4.57	n.a.	
Gly-37	7.25	4.58, 3.78		
Lys-38	7.81	n.a.	n.a.	
Val-39	8.71	3.79	0.15	CγH$_3$ 0.49, 0.33
Val-40	8.10	4.56	1.48	CγH$_3$ 0.53, 0.47
Glu-41	9.20	5.06	2.40, 2.08	
Leu-42	8.76	5.09	1.55	CδH 0.89, CγH 1.52
Gly-43	6.65	4.38, 4.09		
Cys-44	8.40	5.60	3.36, 3.18	
Ala-45	9.34	4.58	1.41	
Ala-46	8.75	4.85	1.54	
Thr-47	7.40	4.33	3.97	CγH$_3$ 1.13
Cys-48	9.00	4.55	n.a.	
Pro-49		4.11	1.52, 2.08	CδH$_2$ 3.57, 3.17, CγH$_2$ 1.0, 1.36
Ser-50	8.02	4.19	3.74, 3.79	
Lys-51	8.23	4.40	n.a.	

(*continued*)

TABLE 1. *(continued)*

	NH	CαH	CβH	Others
BGTX Residue (Cont'd)				
Lys-52	8.66	4.49	1.49, 0.79	CγH₂ 1.79, 1.32
Pro-53		4.21	1.82, 2.25	
Tyr-54	6.89	4.72	3.53, 3.13	CεH 6.76, CδH 7.12
Glu-55	7.66	5.11	2.13, 1.92	CγH 1.81
Glu-56	8.91	4.78	2.10, 2.00	CγH 2.29
Val-57	8.54	5.30	1.89	CγH₃ 0.95, 0.88
Thr-58	9.07	4.77	4.02	CγH₃ 1.23
Cys-59	9.13	5.65	3.73	
Cys-60	9.24	5.14	3.61, 3.39	
Ser-61	8.86	4.94	4.20, 3.82	
Thr-62	7.50	4.76	4.29	CγH₂ 1.21
Asp-63	8.29	4.82	2.51, 2.31	
Lys-64	9.99	3.17	0.29, 0.98	CδH 1.52, CγH₂ 1.73, 1.49
Cys-65	7.64	4.57	3.77, 3.56	
Asn-66	9.01	4.93	n.a.	NδH₂ 7.52, 7.96
Pro-67		3.58	n.a.	
His-68	8.34	3.90	2.82, 2.67	CδH 6.44, CεH 8.10
Pro-69		n.a.	n.a.	CδH₂ 3.13, 2.15, CγH 1.84
Lys-70	n.a.	4.24	n.a.	
Gln-71	8.10	4.29	2.04	
Arg-72	8.29	4.46	n.a.	
Pro-73		4.48	n.a.	
Gly-74	7.91	3.84		
12mer Residue				
His-186	n.a.	4.74	3.29, 3.09	CδH 7.78, CεH 8.11
Trp-187	8.79	4.47	3.30, 3.19	Cξ²H 7.28, CδH 7.01, CεH 7.65, 9.65
Val-188	8.22	4.08	1.96	CγH₃ 1.06, 0.68
Tyr-189	7.85	4.85	3.15, 2.79	CδH 7.31, CεH 6.91
Tyr-190	9.31	5.40	3.33, 2.87	CδH 7.01, CξH 6.95
Thr-191	n.a.	4.17	4.25	CγH₃ 1.19
Cys-192	n.a.	n.a.	n.a.	
Cys-193	7.42	5.34	3.62, 3.38	
Pro-194		n.a.	n.a.	CδH₂ 3.94, 3.31
Asp-195	8.20	4.73	2.84, 2.52	
Thr-196	7.73	4.08	4.19	CγH₃ 1.12

[a]Chemical shifts in ppm from internal TSP or water; accuracy +/− 0.01 ppm.
[b]n.o., not observed.
[c]n.a., not assigned.
[d]Note the correction in this shift from that reported erroneously in ref. 14.

ROESY experiment, there was an equimolar amount of "free" and 18mer "bound" BGTX in the sample, and the temperature of the sample (65 °C) was such that there was significant chemical exchange set up between free and bound BGTX. Knowing what the free shifts of BGTX were at the operating temperature (these had to be assigned anew, but were made significantly easier by the assignments that had already been carried out of the toxin at 35 °C and 45 °C) allowed the identification of significantly shifted peaks in the complex. This approach was only useful for resonances that underwent large chemical shift perturbations upon binding, because those nearly similar to those in free toxin were lost in the diagonal noise of the spectrum. Once these "perturbed" residues were assigned, the traditional assign-

TABLE 2. [^{1}H] Chemical Shifts^a and Assignments for α-Bungarotoxin Complexed to the AChR Peptide Fragment α181–198, at 35 °C and pH 4.0

	NH	CαH	CβH	Others
BGTX Residue				
Ile-1	n.o.^{b}	4.17	2.075	CγH$_3$ 0.76, CγH$_2$ 1.16, CδH$_3$ 0.71
Val-2	8.05	4.95	1.52	CγH$_3$ 0.57, 0.84
Cys-3	8.75	5.06	2.39, 2.94	
His-4	9.44	5.05	3.06, 2.95	CδH 6.42
Thr-5	8.87	5.18	3.99	CγH$_3$ 1.28
Thr-6	8.28	4.43	5.33	CγH$_3$ 1.40
Ala-7	9.19	4.54	1.45	
Thr-8	7.04	4.50	4.28	CγH$_3$ 1.04
Ser-9	8.28	4.54	3.85, 3.56	
Pro-10		n.a.^{c}	n.a.	
Ile-11	8.30	4.12	1.68	CγH$_3$ 1.00, CγH$_2$ 1.18, CδH$_3$ 0.88
Ser-12	7.61	4.98	3.95, 3.80	
Ala-13	8.28	5.04	0.83	
Val-14	8.83	4.65	2.05	CγH$_3$ 0.83, 0.86
Thr-15	8.50	4.41	3.98	CγH$_3$ 1.22
Cys-16	8.83	4.84	3.22, 3.00	
Pro-17		n.a.	n.a.	CδH$_2$ 3.49
Pro-18		4.40	2.31, 1.98	
Gly-19	8.77	4.27, 3.66		
Glu-20	7.85	4.28	2.31, 2.19	
Asn-21	8.04	4.94	3.00, 2.65	NδH$_2$ 6.96, 7.44
Leu-22	8.22	4.99	1.69, 1.45	CγH 1.57, CδH$_3$ 0.69, 0.77
Cys-23	8.75	5.93	3.17, 2.84	
Tyr-24	8.93	5.97	3.00, 2.65	CδH 6.66
Arg-25	9.03	5.24	n.a.	CδH$_2$ 3.01, NεH 7.16
Lys-26	9.79	5.78	2.15, 1.92	CγH$_2$ 1.60
Met-27	9.18	6.19	2.60, 2.01	CγH$_2$ 1.94, SCH$_3$ 1.97
Trp-28	8.19	5.20	3.71, 3.55	CηH 6.86, C^{ξ3}H 7.28, CεH 7.55, CδH 7.05, NεH 10.49
Cys-29	9.15	5.10	3.37, 3.03	
Asp-30	9.33	4.98	3.44, 2.73	
Ala-31	8.10	4.04	1.04	
Phe-32	8.80	4.68	3.39, 3.10	CδH 7.21, CξH 7.23
Cys-33	7.67	3.46	3.50, 3.09	
Ser-34	8.94	4.26	4.00, 3.94	
Ser-35	7.83	4.65	3.96	
Arg-36	8.23	4.54	1.87	CγH$_2$ 1.78, 1.65, NεH 7.19
Gly-37	7.19	4.70, 3.72		
Lys-38	8.08	4.20	1.61, 1.48	CγH 1.25, CεH$_2$ 2.97
Val-39	8.65	3.56	0.13	CγH$_3$ 0.38, 0.25
Val-40	8.09	4.48	n.a.	CγH$_3$ 0.46, 0.53
Glu-41	9.20	5.00	2.35	
Leu-42	8.75	5.07	n.a.	CγH$_2$ 1.52, CδH$_3$ 0.92, 0.84
Gly-43	6.62	4.28, 3.98		
Cys-44	8.44	5.53	3.21, 2.99	
Ala-45	9.36	4.58	1.39	
Ala-46	8.73	4.84	1.52	
Thr-47	7.40	4.33	3.95	CγH$_3$ 1.12
Cys-48	9.01	4.54	2.90	
Pro-49		4.15	2.10, 1.54	
Ser-50	8.01	4.13	3.79, 3.74	
Lys-51	8.48	4.24	n.a.	

(continued)

TABLE 2. (*continued*)

	NH	CαH	CβH	Others
BGTX Residue (Cont'd)				
Lys-52	8.10	4.56	0.96	CγH$_2$ 1.36, 1.31, CδH$_2$ 1.59, 1.47
Pro-53		4.18	2.15, 1.70	
Tyr-54	7.22	4.56	3.16	CεH 6.77
Glu-55	7.73	5.11	2.01	CγH$_2$ 1.90
Glu-56	8.92	4.78	2.37	CγH$_2$ 2.13
Val-57	8.52	5.25	0.87	CγH$_3$ 0.94, 0.85
Thr-58	9.08	4.75	3.99	CγH$_3$ 1.21
Cys-59	9.14	5.62	3.72, 3.02	
Cys-60	9.17	5.12	3.58, 3.41	
Ser-61	8.86	4.92	4.17, 3.79	
Thr-62	7.47	4.75	4.24	CγH$_3$ 1.20
Asp-63	8.37	4.73	2.31	
Lys-64	9.91	3.05	0.96, 0.25	CγH$_2$ 1.52, CδH$_2$ 1.52, 1.49
Cys-65	7.64	4.51	3.74, 3.50	
Asn-66	8.88	4.97	2.54	NδH$_2$ 7.62, 7.53
Pro-67		n.a.	n.a.	
His-68	n.a.	n.a.	n.a.	
Pro-69		n.a.	n.a.	
Lys-70	9.04	4.40	1.92	CεH$_2$ 3.00
Gln-71	8.29	4.32	1.97	
Arg-72	8.29	4.63	1.87, 1.75	CγH$_2$ 1.68, 1.46, CδH$_2$ 3.21, NεH 7.16
Pro-73		n.a.	n.a.	
Gly-74	7.96	3.91		
18mer Residue				
Lys-185	7.47	3.11		
Tyr-190	9.35	4.39	2.92	
Thr-191	8.05	4.48	n.a.	
Cys-192	7.69	4.81	n.a.	
Cys-193	7.29	5.42	n.a.	
Pro-194		n.a.	n.a.	CδH$_2$ 4.00, 3.30
Asp-195	8.20	4.90	2.89	
Thr-196	6.97	3.89	4.65	
Tyr-198	9.53	5.36	3.45, 3.06	

*a*Chemical shifts in ppm from internal TSP or water; accuracy +/− 0.01 ppm.
*b*n.o., not observed.
*c*n.a., not assigned.

ment protocol of NOESY, HOHAHA, and DQCOSY experiments were performed on the 1:1 complex. It is especially important to note here that we were able to form a true 1:1 complex between BGTX and the peptide fragments. In work recently described using an *Escherichia coli* trpE fusion protein containing an overlapping region of the receptor (α183–204), < 0.1% of the recombinant protein appeared to be in an active conformation capable of binding toxin.[15]

BGTX is a 74-amino acid long, α-neurotoxin from the venom of *Bungarus multicinctus*. BGTX has been extremely useful in pharmacological, electrophysiological, and biochemical studies of the nAChR because of its high selectivity for the receptor and its high binding affinity. Its tertiary structure is characterized by the presence of five disulfide bonds, found between cysteines 3–23, 16–44, 29–33, 48–59, and 60–65. Its high resolution structure has been solved both in solution by two-dimensional [¹H]NMR and in the crystalline state by X-ray crystallography.[16,17] FIGURE 4 shows that it is a relatively flat, hand-shaped structure composed of 3 loops: the N-terminal loop, the middle so-called toxic loop, and a third loop that leads into the C-terminal tail. The secondary and tertiary structure of BGTX is dominated by a triple-stranded, antiparallel β-sheet including residues Leu22 to Trp28, Val40 to Ala45, and Glu56 to Cys60. The boxed residues in FIGURE 4 are those that undergo large chemical shift perturbations upon binding the 12mer. The circled residues are those that undergo large chemical shift perturbations upon binding the 18mer. BGTX residues that were perturbed both in the binding to the 12mer and to the 18mer receptor fragments are shown as dark filled boxes. The BGTX residues that are most affected by either 12mer or 18mer binding appear to be concentrated in the region of the N-terminal and middle toxic loops.

Residues Asp30 and Arg36 in BGTX have been postulated to form an acetylcholine-mimetic site on the toxin, and might be expected to form the main contact zone with peptide fragments from the receptor.[18] TABLE 3 shows the residues in BGTX that are significantly perturbed ($\Delta\delta > 0.15$ ppm) upon complex formation, and indicates that the contact zone between receptor and toxin is quite extensive. In the solution structure of the 12mer/BGTX complex, as calculated from the interproton distance constraints, the receptor peptide appears to come very close to contributing an additional antiparallel β-strand to the central antiparallel β-sheet structure of the toxin.[14] It is interesting to note from TABLE 3 the number of evolutionarily conserved residues, especially Trp28, Asp30, and Phe32, that are in significantly different

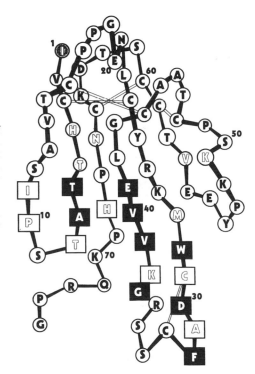

FIGURE 4. Schematic representation of α-bungarotoxin showing its three-looped antiparallel β-sheet structure. The light grey boxed residues are those perturbed significantly ($\Delta\delta > 0.15$ ppm) upon binding the 12mer (α185–196). The circled residues are those perturbed upon binding the 18mer (α181–198). Toxin residues perturbed both upon binding the 12mer and the 18mer are shown as dark filled boxes.

chemical environments after BGTX binding to the receptor fragment. The conservation of residues within the α-neurotoxin family of proteins suggests either an important structural or functional role for these conserved residues. The observation that several of the conserved residues are in regions significantly perturbed upon binding the receptor peptide fragments suggests that these residues may indeed play an important functional role in receptor recognition. Although the perturbations of the BGTX side chains may be more informative in terms of revealing direct interactions with the receptor peptide, in addition, a number of large main-chain

TABLE 3. α-Bungarotoxin Residues That Are Significantly Perturbed ($\Delta\delta > 0.15$) upon Binding of the AChR Peptide Fragment α181–198 at 35 °C and pH 4.0

Residue	Proton	Free α-BGTX (ppm)	Complexed α-BGTX (ppm)	Δ
1I	**βH**	1.87	2.07	0.20
4H	**NH**	9.18	9.44	0.26
	βH	2.67	3.05	0.38
5T	**NH**	9.03	8.87	0.16
6T	**αH**	4.73	4.43	0.30
	βH	5.02	5.32	0.30
7A	αH	4.32	4.54	0.22
27M	**αH**	6.00	6.19	0.19
28W	NH	8.63	8.19	0.44
30D	NH	8.35	9.33	0.98
	βH	3.17	3.44	0.27
32F	NH	8.27	8.81	0.54
37G	**NH**	7.71	7.17	0.54
	αH1	4.30	4.70	0.40
	αH2	3.92	3.72	0.20
39V	**βH**	0.52	0.14	0.38
	γH1	0.48	0.25	0.23
	γH2	0.58	0.39	0.19
40V	**NH**	8.09	7.70	0.39
	αH	4.78	4.48	0.30
41E	NH	9.38	9.19	0.19
51K	**NH**	8.30	8.48	0.18
57V	**βH**	1.87	0.87	1.00
66N	**δNH2**	7.82	7.62	0.20

NOTE: Residues shown in bold and underlined are those shifted upon binding the 18mer but not the 12mer.

perturbations exist, most especially those involving Asp30 NH. Histograms comparing the shifts of free versus bound BGTX in 12mer complex formation as well as 18mer complex formation of CαH residues are shown in FIGURES 5 and 6, respectively. The extension of the contact zone towards both the N- and C-termini of BGTX is quite easily seen in this analysis, the complexity and diversity of which would be difficult to explain solely on the basis of the 18mer being simply four residues longer on the N-terminus and two residues longer on the C-terminus than the 12mer. The interactions of the 18mer with the third loop are perhaps best

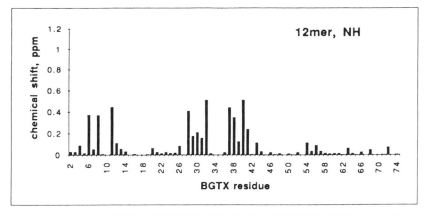

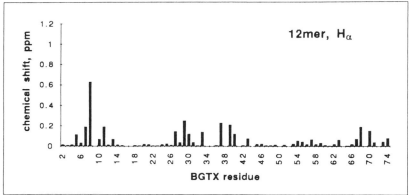

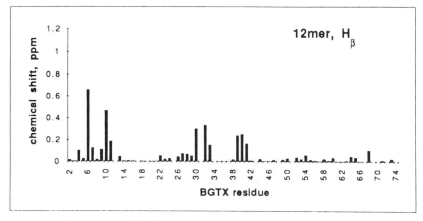

FIGURE 5. Histogram analysis of the chemical shift changes (as absolute values) between α-bungarotoxin "free" and when bound to the 12mer (α185–196). The *top* profile shows NH shifts, the *middle* one CαH shifts, and the *bottom* one CβH shifts.

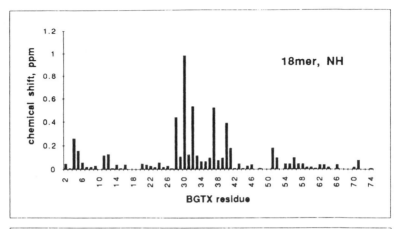

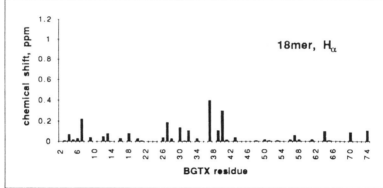

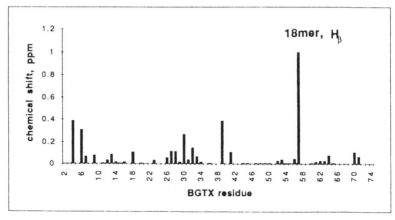

FIGURE 6. Histogram analysis of the chemical shift changes between α-bungarotoxin "free" and when bound to the 18mer (α181–198). The *top* profile shows NH shifts, the *middle* one CαH shifts, and the *bottom* one CβH shifts.

illustrated by the highly significant **1.00** ppm shift of the CβH proton of Val57. Intermolecular NOEs in this region, when available, will be very important in the positioning of the peptide residues near this region, especially the highly conserved Tyr198.

To refine the analysis of the 18mer:BGTX complex, more resonances of the peptide need to be unambiguously assigned. Although most of the main-chain BGTX resonances have been assigned, more than half of the peptide resonances, as seen in TABLE 2, remain obscured by overlap problems related to the large number of proton resonances in a complex of this size. A more complete assignment would allow for the identification of intermolecular NOEs between BGTX and 18mer, generating valuable distance constraints for molecular dynamics and molecular modeling studies based on the experimentally derived parameters. At present, however, the chemical shift overlap in the two dimensions available to us is prohibiting further rapid progress. It is also unfortunate that due to solubility reasons, we are not able to make a 2:1 peptide:toxin complex that might aid us in the assignment of the peptide residues through the use of the ROESY-based strategy that has been so useful in assigning the perturbed BGTX residues. Instead, to address the assignment problem, the chemical synthesis of a universally [15N]labeled synthetic 18mer is in progress. Such a labeled peptide would enable us to use additional powerful heteronuclear NMR experiments to complete the very difficult peptide assignments. Because of the limited supply of [15N]amino acids and the absence of "coupling ready" (fully protected) [15N]amino acids, this approach has been slow. First, complete synthetic protection schemes needed to be worked out for each individual amino acid in the peptide fragment, and then the details of bench-top peptide coupling needed to be explored at the scale determined in part by the cost of the labeled amino acids. The latter is important in this case, as the amount of amino acids needed in commercial peptide synthesizers are unreasonable in light of the cost of the [15N]amino acids. Nevertheless, a [15N]18mer would be extremely useful for [15N]filtered NOESY experiments that could be used to locate peptide residues, as well as to define the intermolecular NOEs between the toxin and the receptor fragment.

To circumvent the problems surrounding the preparation of [15N]labeled synthetic peptides and to provide further insights into the full extent of the contact zone between receptor and BGTX, we designed synthetic genes encoding several longer fragments (46 to 62 residues in length) from within the N-terminal portion of the α-subunit of the nAChR. The synthetic gene constructs are expressed in *E. coli*, and the conditions required for metabolic labeling of the expressed protein with relatively inexpensive [15N]NH₄Cl have also been established.[19] Upon final purification of the labeled recombinant protein fragments, we should be able to perform three-dimensional heteronuclear NMR experiments both in the presence and absence of BGTX. Such studies should greatly enhance our structural knowledge of the receptor in the vicinity of the ligand binding site. If the region to be studied is carefully chosen, a further advantage to working with longer receptor fragments is that longer sequences may encode enough protein folding information to begin to form autonomously folding structural elements or subdomains. It is widely recognized that small peptides (≤ about 25 residues) have little structure of their own, unless they are forming a distinct folding domain (i.e., a zinc finger) of a protein. The longer receptor fragments that we are now exploring should be more likely to fold into structures characteristic of those assumed by these regions in the native receptor. Assuming that the longer fragments remain soluble enough for the NMR investigations, we may therefore be in a good position to compare the structures of the receptor fragment both free in solution and when bound to BGTX or agonists.

We anticipate that the region we have chosen to concentrate our attentions on (α143–204) will likely form a distinct folding domain, based on the common exon organization in the native gene encoding the α-subunit in all species examined to date. One of the constructs that we are studying contains the 62-amino acid sequence, (α143–204), joined to the C-terminus region of a bacteriophage coat protein (T7 gene 9 protein). This fusion protein can be overexpressed in *E. coli* without the production of insoluble inclusion bodies. Preliminary BGTX binding studies of this fusion protein show that it binds BGTX with an affinity that is less than twofold reduced from that of intact membranes obtained from the electric organ of *Torpedo californica*. It appears, therefore, that nearly all of the BGTX recognition determinants of the native receptor can be found within this 62-residue region of the α-subunit.

Our NMR-based analysis of the BGTX/receptor peptide fragment contact zone has served to target residues on BGTX for site-directed mutagenesis in order to determine whether the affinity and possibly the specificity for the nAChR could be genetically manipulated. In a genetically engineered system similar to that being used for the overexpression of our receptor fragments, we also constructed a gene for BGTX which allows for the expression of wild-type and site-directed mutants of BGTX.[20] We began mutational analysis with an Ala scanning approach, and binding studies of the Asp30Ala mutant toxin have already been completed. Additional mutations involving Arg36Ala and Trp28Ala substitutions are currently being analyzed as well as the Asp30Ala-Arg36Ala double mutation. As the Asp30Ala mutant BGTX appears to bind native nAChR with normal affinity, it appears that the carboxylate side chain of Asp30 is not essential for receptor recognition.[20] It is unlikely therefore that the basis for receptor specificity resides in a role for Asp30 as a key component of an acetylcholine mimetic binding site on BGTX. We believe that the NMR structural analysis, together with the functional binding analysis of mutants constructed by site-directed mutagenesis, will continue to be a powerful combination of tools in elucidating the roles of individual residues in both structure and function of the nAChR and of the α-neurotoxin family of proteins.

REFERENCES

1. STROUD, R. M. 1981. Proceedings of the Second SUNYA Conversation in the Discipline of Biomolecular Stereodynamics R. H. Sarma, Ed. **2:** 55. Adenine Press. New York.
2. KAO, P. N. & A. KARLIN. 1986. J. Biol. Chem. **261:** 8085.
3. WILSON, P. T., T. L. LENTZ & E. HAWROT. 1985. Proc. Natl. Acad. Sci. USA **82:** 8790.
4. LENTZ, T. L., E. HAWROT & P. T. WILSON. 1988. Mol. Pharmacol. **34:** 643.
5. RADDING, W. P., W. R. CORFIELD, L. S. LEVINSON, G. A. HASSHIM & B. W. LOW. 1988. FEBS Lett. **231:** 212.
6. GOTTI, C., G. MAZZOLA, R. LONGHI, D. FORNASARI & F. CLEMENTI. 1987. Neurosci. Lett. **228:** 118.
7. RALSTON, S., V. SARIN, H. L. THANH, J. RIVIER, L. FOX & L. J. LINDSTROM. 1987. Biochemistry **26:** 3261.
8. NEUMANN, D., D. BARCHAN, M. FRIDKIN & S. FUCHS. 1986. Proc. Natl. Acad. Sci. USA **83:** 9250.
9. NEUMANN, D., D. BARCHAN, A. SAFRAN, J. M. GERSHONI & S. FUCHS. 1986. Proc. Natl. Acad. Sci. USA **83:** 3008.
10. CONTI-TRONCONI, B. M., F. TANG, B. M. DIETHELM, S. R. SPENCER, S. REINHARDT & A. MAELICKE. 1991. Biochemistry **30:** 2575.
11. PEARCE, S. F. A., P. PRESTON-HURLBURT & E. HAWROT. 1990. Proc. R. Soc. Lond. B **241:** 207.
12. TZARTOS, S. J. & M. S. REMOUNDOS. 1990. J. Biol. Chem. **265:** 21462.

13. UNWIN, N. 1993. Cell/Neuron **72/10:** 31.
14. BASUS, V. J., G. SONG & E. HAWROT. 1993. Biochemistry **32:** 12290.
15. FRAENKEL, Y., G. NAVON, A. ARONHEIM & J. M. GERSHONI. 1990. Biochemistry **29:** 2617.
16. BASUS, V. J., M. BILLETER, R. A. LOVE, R. M. STROUD & I. D. KUNTZ. 1988. Biochemistry **27:** 2763.
17. LOVE, R. A. & R. M. STROUD. 1986. Protein Eng. **1:** 37.
18. LOW, B. W., H. S. PRESTON, A. SATO, L. S. ROSEN, T. E. SEARLE, A. D. RUDKO & J. R. RICHARDSON. 1979. Proc. Natl. Acad. Sci. USA **73:** 2991.
19. GENTILE, L. N., C. A. VASLET, N. J. MESSIER & E. HAWROT. In preparation.
20. ROSENTHAL, J. A., S. H. HSU, N. J. MESSIER, D. SCHNEIDER, C. A. VASLET, L. N. GENTILE & E. HAWROT. 1994. J. Biol. Chem. **269:** 11178.

Nicotine Effects on Presynaptic Receptor Interactions[a]

H. SERSHEN,[b,c] E. TOTH,[b] A. LAJTHA,[b] AND E. S. VIZI[d]

[b]The N. S. Kline Institute for Psychiatric Research
Center for Neurochemistry
Orangeburg, New York 10962

[d]Institute of Experimental Medicine
Hungarian Academy of Sciences
Budapest, Hungary

Presynaptic regulation of neurotransmitter release in the brain influences the amount of neurotransmitter released. This may or may not depend on nerve activity, and can be modulated by other factors that influence the membrane potential of the nerve ending. The neurotransmitter itself can inhibit its own release by acting on presynaptic autoreceptors. Other transmitters can also act presynaptically to modulate transmitter outflow. The complexity and diversity of such modulatory mechanisms of release have received much attention. We were particularly interested in receptor–receptor interactions influenced by participation of nicotine in the central actions of nicotine involving presynaptic effects.

The mechanisms of these receptor–receptor interactions have been discussed recently. In receptor heteroregulation, neurotransmitters or neuromodulators, by binding to receptors on the neuronal membrane, may be able to regulate the characteristics and function of the recognition sites of another transmitter or modulator receptor. Whereas autoreceptor mechanisms regulate the sensitivity of a receptor and are modulated by the levels of its ligand through positive or negative feedback loops, heteroreceptor mechanisms include direct or indirect interactions between different receptors and transmitters, involving the plasma membrane, intracytoplasmatic loops, or interactions at the nuclear level.[1]

The mechanism of action of nicotine could be elucidated by the understanding of these different receptor–receptor and receptor–transmitter interactions.

NICOTINIC-CHOLINERGIC-DOPAMINERGIC INTERACTIONS

Our interest in studying the effect of nicotine on cerebral neurotransmitter systems was stimulated by a number of studies reporting that the incidence of smoking is lower in patients with Parkinson's disease, and that the lower rate of smoking may predate neurological symptoms.[2] A number of possible explanations exist including, for example, that nicotine affects dopaminergic cells. While examining nicotine-induced rotation after nigral lesions,[3] we found a dopamine-like action of nicotine, which indicated enhancement of activity in the intact nigrostriatal system, without any sign of tolerance after repeated nicotine administration. The nicotine antagonist mecamylamine blocked the induced rotation. There was no

[a]Supported in part by a grant from the Council for Tobacco Research-U.S.A., Inc.
[c]Corresponding author.

evidence of a change in dopamine metabolism after an acute challenge with nicotine or of a sustained change after repeated injection, although a modification of dopaminergic response to nicotine could not be excluded. Stimulation of nigrostriatal dopamine neurons by nicotine had been shown previously.[4] The nucleus accumbens seems to be a major site of nicotine action in the brain. After chronic nicotine treatment (6 weeks), the K_d and the B_{max} for domperidone in the accumbens increased severalfold with no changes observed in the caudate putamen or the frontal cortex. Such effects may influence preparkinsonian persons in regard to smoking.[5] The fairly specific effects in the accumbens suggest that the dopamine receptors in this structure differ from those in other areas. Pharmacological differences between the nigrostriatal and mesolimbic dopaminergic systems have been observed before.[6] After an acute dose of nicotine, we observed increased dopamine turnover in the accumbens;[7] changes in a number of other areas examined were minor. These results further indicated that the accumbens is an important site of nicotine action. Although nicotine receptors were shown to be located on norepinephrine and serotonin terminals as well,[8] little effect of nicotine on these neurotransmitters was observed in the accumbens;[7] however, changes in the cerebral levels of catecholamines and serotonin were observed after acute or chronic nicotine administration.[9] At high levels *in vitro* nicotine inhibited vesicular dopamine uptake,[10] an effect that may not occur under conditions of tobacco smoking. Nicotine effects on vesicular function may play a part in the effects on transmitter release.

Inasmuch as acetylcholine release in the striatum is inhibited by endogenous dopamine, nicotinic effects on the cholinergic system may be modulated by dopaminergic-nicotine effects. We found evidence for the presence of nicotinic receptors on striatal cholinergic interneurons. Nicotine increased acetylcholine release in striatal slices only after dopaminergic lesions;[11] in the control animals the nicotine-induced increased dopamine release probably inhibited the stimulation of acetylcholine release, because the release of acetylcholine is tonically inhibited by the endogenous dopamine. These findings indicate the existence of nicotinic receptors on striatal cholinergic interneurons, probably on the somatodendritic part. With electrically stimulated striatal slices, the effect of nicotine in releasing dopamine involved the participation of N-type calcium channels;[12] nicotine receptor-operated ion channels permeable to calcium and N-type voltage-sensitive Ca^{2+} channels may act additively to increase intracellular free Ca^{2+} levels. Thus, nicotine effects include those on the level of intracellular calcium. Increases in intracellular calcium by nicotine in cortical neuronal cultures have been reported.[13]

EFFECTS OF CHRONIC NICOTINE ADMINISTRATION

Because smoking exposes the nervous system to nicotine intermittently over a long period of time, chronic nicotine effects are of special interest. The increase in K_d and B_{max} of dopaminergic binding after nicotine has already been mentioned.[5] Chronic nicotine was reported to have a protective effect on dopamine cells in the substantia nigra after partial hemitransection.[14] We found that after an MPTP lesion nicotine had no effect on the lesion or on the rate of terminal recovery, as measured by the rate of recovery of striatal dopamine. This finding confirmed the report that exposure to cigarette smoke does not decrease the toxicity of MPTP.[15,16] In testing ligand binding at dopamine D_1 and D_2 receptors, MPTP was shown to increase binding at the D_2 site, and nicotine to increase binding at the D_1 site. The two effects were independent: nicotine did not influence MPTP effects, and MPTP did not alter nicotine effects. The fact that nicotine alters the ratio of D_1 to D_2 receptor binding activities is of interest because such ratio changes may play a role in the therapeutic effects of neuroleptics.[17]

Nicotine was shown to increase dopamine release in striatal slices of both untreated and chronic nicotine-treated mice. Dopamine metabolism was not affected. The higher rate of efflux of dopamine in response to electrical stimulation in the striatum of nicotine-pretreated animals suggests that chronic nicotine may affect the mechanism of transmitter release in dopaminergic nerve terminals. Dopamine agonists inhibited and antagonists enhanced the release—effects that were abolished in chronic nicotine-treated animals. The findings were interpreted to mean that increased release of dopamine by nicotine leads to dopamine D_2 autoreceptor subsensitivity, with nicotine possibly attenuating autoinhibition of dopaminergic neurotransmission in the striatum.[18] Although it is generally accepted that nicotine increases dopamine release by stimulating specific receptors on striatal dopaminergic nerve terminals, the identification of the subcellular dopamine pools involved in its action is controversial. Presynaptic autoreceptors respond to increased dopamine release with a reduced sensitivity. As a result of this impaired autoregulation of transmitter release, electrical stimulation induces higher dopamine efflux from striatum exposed to repeated nicotine. Such changes also result in altered behavioral responses, as evidenced by an attenuation of apomorphine-induced hypomotility in chronic nicotine-treated mice.[19] The resulting decrease in autoreceptor stimulation after chronic nicotine would reduce the response of D_2-autoreceptor agonists to decrease dopamine release, showing that both acute and chronic nicotine administration can induce changes in the dopaminergic system.

NICOTINE EFFECTS ON OTHER RECEPTORS

When nicotine was administered intracerebrally in specific regions via microdialysis, the release of other neurotransmitters in addition to that of dopamine could be observed. Extracellular levels of serotonin increased in the cingulate and frontal cortex, and norepinephrine increased in the substantia nigra, cingulate cortex, and pontine nucleus.[20] The effect of nicotine on dopamine release was inhibited by cholinergic, dopaminergic, and also glutamatergic antagonists, indicating that glutamic acid release participates in nicotine-induced dopamine release.[20] The extracellular levels of some other amino acids such as glycine and taurine were also increased. The levels of some nonneurotransmitter amino acids were increased as well, indicating that cell permeability or amino acid transport processes are also affected by high local nicotine concentrations. Thus, cholinergic, catecholaminergic, glutamatergic, and membrane effects of nicotine could be shown.[21] The effects of nicotine in the brain are mediated by several receptor systems, and nonreceptor effects are also involved; nicotine effects show significant regional heterogeneity.

Neuropeptides may have a role in the complex effects of nicotine. A significant reduction of substance P was observed after a single dose of nicotine, primarily in areas in which nicotine acts; this effect may be mediated by dopamine-induced release of substance P. Nicotinic antagonists abolish this effect.[22] The rapid and large changes in the level of substance P caused by nicotine indicate a significant role for this peptide in nicotine effects in the brain.

HETEROGENEITY OF NICOTINIC RECEPTORS

One aspect of the complexity of the effect of nicotine in the nervous system is that it affects several neurotransmitter systems—some directly, through nicotinic sites on different cells, such as cholinergic, glutamatergic, etc. The released neurotransmitters such as glutamate may in turn modulate other transmitter systems, representing the indirect effects of nicotine.

An additional complexity is represented by the heterogeneity of the nicotinic binding sites. The regional heterogeneity of nicotine effects already indicates a heterogeneous distribution of nicotinic binding sites. The pharmacological heterogeneity of nicotinic sites is shown, for example, by the differences between the postsynaptic somatodendritic and neuromuscular sites.[23] It is possible that presynaptic and postsynaptic nicotinic sites also show differences, including different pharmacological properties. In turn, the pools of acetylcholine released by nicotine may come from several distinct transmitter pools or compartments.[24]

NICOTINE-INDUCED DOPAMINE AND NOREPINEPHRINE RELEASE

The effect of nicotine on dopamine release is well established and has been shown in the brain both *in vivo*[3,7,18] and *in vitro;* for example, in synaptosomal[25] and in brain slice[26] preparations. The effect of nicotine on norepinephrine release is less clearly established. Early studies examining nicotine effects on sympathetic axons in brain vessels[27] and in rat heart[28] did not find any release of norepinephrine; such release could be detected in peripheral nerve fibers.[29] In guinea-pig vas deferens nicotine enhanced resting and stimulation-evoked release of norepinephrine in acute but not in chronic experiments, thus indicating the presence of inhibitory muscarinic and facilitatory nicotinic receptors on noradrenergic axon terminals.[30] The findings suggest that nicotinic receptors located on noradrenergic axon terminals are different from those located postsynaptically in striated muscle or ganglia, but are similar to those at cholinergic axon terminals at the neuromuscular junction.

EFFECT OF NICOTINE ON GLUTAMATE RELEASE

Many of the striatal neurons of cortical origin are glutamatergic. When *in vivo* nicotine was infused via microdialysis into the striatum of rats, the level of glutamate in the dialysate increased more than fourfold, indicating a release of glutamate into the extracellular fluid.[20] Kynurenic acid, a glutamic acid receptor antagonist, inhibited this release. Kynurenic acid also inhibited the nicotine-induced release of dopamine in the striatum. The fact that kynurenic acid completely blocked the nicotine-induced dopamine release indicates that nicotine does not act directly on dopamine release, but indirectly through glutamate release. Indeed, nicotine significantly increases the extracellular levels of glutamate in a calcium-dependent manner, as measured with microdialysis in the striatum;[21] muscarinic antagonists blocked the effect. We interpreted this finding to mean that at least in part the dopamine-releasing activity of nicotine is mediated through its effect on the glutamatergic system.[20] Nicotine administered intracerebrally via microdialysis increased the extracellular levels of aspartic acid as well. Dopaminergic or nicotinic antagonists did not affect the changes in amino acid levels, but muscarinic antagonists were inhibitory. This finding indicates that muscarinic cholinergic receptors also participate in the action of nicotine in the nervous system.

The interaction between glutamatergic and dopamine neurons is also reciprocal; in particular, dopamine receptors are located on striatal afferent fibers of cortical origin and dopamine agonists can inhibit the release of glutamate.[31] Such interactions would be regionally heterogeneous.

Adding to the complexity of the regulation of dopamine release, the presence of glutamate receptors on nondopamine neurons has been reported.[32] The nondopamine neurons mediating the effect of glutamate could be GABAergic interneurons. Glutamate appears to exert an inhibitory presynaptic control on dopamine release, indirectly mediated in part by GABAergic neurons, whereas glutamatergic neurons seem to have a facilitatory presynaptic influence on dopamine release. Thus, corticostriatal glutamatergic fibers have direct stimulatory and indirect inhibitory presynaptic effects on dopamine release.[32] A scheme of these receptor–receptor actions is given in FIGURE 1.

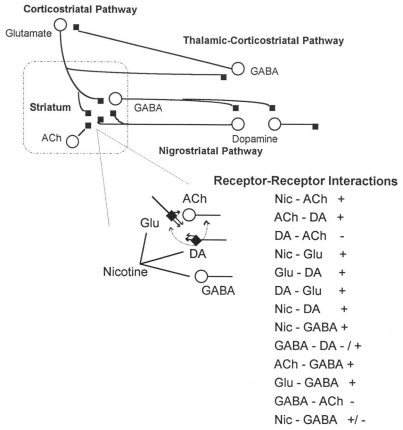

FIGURE 1. Schematic representation of nicotine-glutamatergic-dopaminergic presynaptic interactions.

INTERACTION OF RECEPTORS AT THE NUCLEAR LEVEL

Although not in the scope of this paper, interactions between receptors can also occur at the nuclear level because activation of one receptor coupled to adenylate cyclase can activate protein kinases, which in turn control the transcriptional activity of the gene for another receptor.[1] Nicotine may have direct effects on intracellular events involved in protein metabolism,[33] selectively affecting the metabolism of specific proteins only. The levels of some proteins were increased and others decreased. Thus, synthesis and breakdown are selectively affected. This reaction may involve effects at nuclear levels. Similar nuclear effects of nicotine are represented by the stimulation by nicotine of the expression of c-fos nuclear proteins, as early markers of neuronal activation.[34]

REFERENCES

1. AGNATI, L. F., K. FUXE, F. BENFENATI, G. VON EULER & B. FREDHOLM. 1993. Intramembrane receptor-receptor interactions: Integration of signal transduction pathways in the nervous system. Neurochem. Int. **22:** 213–222.
2. HAACK, D. G., R. J. BAUMANN, H. E. MCKEAN, H. D. JAMESON & J. A. TURBEK. 1981. Nicotine exposure and Parkinson's disease. Am. J. Epidemiol. **114:** 191–200.
3. LAPIN, E. P., H. S. MAKER, H. SERSHEN, Y. HURD & A. LAJTHA. 1987. Dopamine-like action of nicotine: Lack of tolerance and reverse tolerance. Brain Res. **407:** 351–363.
4. LICHTENSTEIGER, W., F. HEFTI, D. FELIX, T. HUWYLER, E. MELAMED & M. SCHLUMPF. 1982. Stimulation of nigrostriatal dopamine neurons by nicotine. Neuropharmacology **21:** 963.
5. REILLY, M. A., E. P. LAPIN, H. S. MAKER & A. LAJTHA. 1987. Chronic nicotine administration increases binding of [³H]domperidone in rat nucleus accumbens. J. Neurosci. Res. **18:** 621–625.
6. MISSALE, C., L. CASTELLETTI, S. GOVONI, P. F. SPANO, M. TRABUCCI & I. HANBAUER. 1985. Dopamine uptake is differentially regulated in rat striatum and nucleus accumbens. J. Neurochem. **45:** 51–56.
7. LAPIN, E. P., H. S. MAKER, H. SERSHEN & A. LAJTHA. 1989. Action of nicotine on accumbens dopamine and attenuation with repeated administration. Eur. J. Pharmacol. **160:** 53–59.
8. SCHWARTZ, R. D., J. LEHMANN & K. J. KELLAR. 1984. Presynaptic nicotinic cholinergic receptors labeled by [³H]acetylcholine on catecholamine and serotonin axons in brain. J. Neurochem. **42:** 1495.
9. WESTFALL, T. C., R. M. FLEMING, M. F. FUDGER & W. C. CLARK. 1967. Effect of nicotine and related substances upon amine levels in the brain. Ann. N.Y. Acad. Sci. **142:** 83–100.
10. KRAMER, H. K., H. SERSHEN, A. LAJTHA & M. E. A. REITH. 1989. The effect of nicotine on catecholaminergic storage vesicles. Brain Res. **503:** 296–298.
11. SANDOR, N. T., T. ZELLES, J. KISS, H. SERSHEN, A. TOROCSIK, A. LAJTHA & E. S. VIZI. 1991. Effect of nicotine on dopaminergic-cholinergic interaction in the striatum. Brain Res. **567:** 313–316.
12. HARSING, L. G., JR., H. SERSHEN, E. S. VIZI & A. LAJTHA. 1992. N-type calcium channels are involved in the dopamine releasing effect of nicotine. Neurochem. Res. **17:** 729–734.
13. FLUHLER, E. N., P. M. LIPIELLO, K. G. FERNANDES & R. J. REYNOLDS. 1989. Nicotine-evoked calcium changes in single fetal rat cortical neurons measured with the fluorescent probe. Soc. Neurosci. Abstr. **15:** 275.
14. JANSON, A. M., K. FUXE, I. KITAYAMA, A. HARFSTRAND & L. F. AGNATI. 1986. Morphometric studies on the protective action of nicotine on the substantia nigra dopamine nerve cells after partial hemitransection in the male rat. Trans. Eur. Winter Conf. Brain Res. Abstr.
15. SERSHEN, H., A. HASHIM, H. L. WIENER & A. LAJTHA. 1988. Effect of chronic oral nicotine on dopaminergic function in the MPTP-treated mouse. Neurosci. Lett. **93:** 270–274.
16. PERRY, T. L., S. HANSEN & K. JONES. 1987. Exposure to cigarette smoke does not decrease the neurotoxicity of N-methyl-4-phenyl-1,2,3,6-tetrahydropyridine in mice. Neurosci. Lett. **74:** 217–220.
17. WIENER, H. L., A. LAJTHA & H. SERSHEN. 1989. Dopamine D1 receptor and dopamine D2

receptor binding activity changes during chronic administration of nicotine in 1-methyl-4-phenyl-1,2,3,6-tetrahydropyridine-treated mice. Neuropharmacology 28: 535–537.

18. HARSING, L. G., JR., H. SERSHEN & A. LAJTHA. 1992. Dopamine efflux from striatum after chronic nicotine: Evidence for autoreceptor desensitization. J. Neurochem. 59: 48–54.

19. SERSHEN, H., A. HASHIM, L. HARSING & A. LAJTHA. 1991. Chronic nicotine-induced changes in dopaminergic system: Effect on behavioral response to dopamine agonist. Pharmacol. Biochem. Behav. 39: 545–547.

20. TOTH, E., H. SERSHEN, A. HASHIM, E. S. VIZI & A. LAJTHA. 1992. Effect of nicotine on extracellular levels of neurotransmitters assessed by microdialysis in various brain regions: Role of glutamic acid. Neurochem. Res. 17: 265–271.

21. TOTH, E., E. S. VIZI & A. LAJTHA. 1993. Effect of nicotine on levels of extracellular amino acids in regions of the rat brain in vivo. Neuropharmacology 32: 827–832.

22. NAFTCHI, N. E., H. MAKER, E. LAPIN, J. SLEIS, A. LAJTHA & S. LEEMAN. 1988. Acute reduction of brain substance P induced by nicotine. Neurochem. Res. 13: 305–309.

23. TOROCSIK, A., F. OBERFRANK, H. SERSHEN, A. LAJTHA, K. NEMESY & E. S. VIZI. 1991. Characterization of somatodendritic neuronal nicotinic receptors located on the myenteric plexus. Eur. J. Pharmacol. 202: 297–302.

24. ADAM-VIZI, V., Z. DERI, E. S. VIZI, H. SERSHEN & A. LAJTHA. 1991. Ca^{2+}-independent veratridine-evoked acetylcholine release from striatal slices is not inhibited by vesamicol (AH5183): Mobilization of distinct transmitter pools. J. Neurochem. 56: 52–58.

25. DE BELLEROCHE, J. S. & H. F. BRADFORD. 1978. Biochemical evidence for the presence of presynaptic receptors on dopaminergic nerve terminals. Brain Res. 142: 53–68.

26. GIORGUIEFF, M. F., M. L. LEFLOCH, T. C. WESTFALL, J. GLOWINSKI & M. J. BESSON. 1976. Nicotinic effect of acetylcholine on the release of newly synthesized ^{3}H-dopamine in rat striatal slices and cat caudate nucleus. Brain Res. 106: 117–131.

27. EDVINSSON, L., B. FALCK & C. H. OWMAN. 1977. Possibilities for cholinergic nerve action on smooth musculature and sympathetic axons in brain vessels mediated by muscarinic and nicotinic receptors. J. Pharmacol. Exp. Ther. 295: 225–230.

28. FUDER, H., R. SIEBENBORN & E. MUSCHOLL. 1982. Nicotine receptors do not modulate the ^{3}H-noradrenaline release from the isolated rat heart evoked by sympathetic nerve stimulation. Naunyn-Schmiedebergs Arch. Pharmacol. 318: 301–307.

29. LINDMAR, R., K. LOFFELHOLZ & E. MUSCHOLL. 1968. A muscarinic mechanism inhibiting the release of noradrenaline from peripheral adrenergic nerve fibers by nicotinic agents. Br. J. Pharmacol. Chemother. 32: 280–294.

30. TODOROV, L., K. WINDISCH, H. SERSHEN, A. LAJTHA, M. PAPASOVA & E. S. VIZI. 1990. Prejunctional nicotinic receptors involved in facilitation of stimulation-evoked noradrenaline release from the vas deferens of the guinea-pig. Br. J. Pharmacol. 102: 186–190.

31. GODUKHIN, O. V., A. D. ZHARIKOVA & A. YU. BUDANTSEV. 1984. Role of presynaptic dopamine receptors in regulation of the glutamatergic neurotransmission in rat neostriatum. Neuroscience 12: 377–383.

32. CHERAMY, A., R. ROMO, G. GODEHEU, P. BARUCH & J. GLOWINSKI. 1986. In vivo presynaptic control of dopamine release in the cat caudate nucleus II. Facilitatory or inhibitory influence of L-glutamate. Neuroscience 19: 1081–1090.

33. SERSHEN, H., M. BANAY-SCHWARTZ, D. S. DUNLOP, E. A. DEBLER & M. E. A. REITH. 1987. Nicotine-induced changes in the metabolism of specific brain proteins. Neurochem. Res. 12: 197–202.

34. MATTA, S. G., C. A. FOSTER & B. M. SHARP. 1993. Nicotine stimulates the expression of c-fos protein in the parvocellular paraventricular nucleus and brainstem catecholaminergic regions. Endocrinology 132: 2149–2156.

Acute and Chronic Nicotinic Interactions with Dopamine Systems and Working Memory Performance[a]

EDWARD D. LEVIN[b] AND JED E. ROSE

Department of Psychiatry
Duke University Medical Center Research Service
and
VA Medical Center
Durham, North Carolina 27710

Nicotine has been found to improve memory performance in a variety of studies (for review see ref. 1). Although this effect is not universally seen, studies have found nicotine to improve memory function in rats, monkeys, and humans. Improvements have been found in normal adults, but also in aged subjects and those with brain lesions. The neural bases for this effect is an active topic of study. Nicotine effects may critically involve interactions with other transmitter systems. We found important interactions of nicotinic systems with dopaminergic systems with regard to working memory performance in the radial-arm maze (RAM).

Nicotine has been found to promote the release of a wide variety of transmitters including dopamine (DA).[2] Nicotinic receptors are well represented in the midbrain dopamine nuclei, the substantia nigra (SN) and ventral tegmental area (VTA).[3–5] Acute nicotine administration increases the activity of DA cells in the substantia nigra and ventral tegmental area and promotes DA release in the striatum,[6–11] whereas nicotinic antagonist administration has been found to inhibit DA release from both striatal and mesolimbic structures.[12,13] However, with chronic administration nicotine-induced DA release becomes diminished.[14] There may even be some reversal of the effect, given that chronic nicotine administration has been found to increase DA receptor binding in the nucleus accumbens.[15] Nicotinic-DA interactions have been found in a variety of neurobehavioral studies.[16] The evidence from our laboratory concerning the importance of nicotine interactions with DA systems with regard to working memory function is the focus of this paper. Differences in nicotinic-DA relationships with acute and chronic nicotine treatment are discussed.

ACUTE MECAMYLAMINE STUDIES

The nicotinic antagonist mecamylamine has significant interactions with both dopaminergic agonists and antagonists with regard to working memory performance in the radial-arm maze. In six different studies we found that a high dose of 10 mg/kg of mecamylamine significantly impairs choice accuracy in the win-shift version of the radial-arm maze test.[17–22] This mecamylamine-induced deficit is reversed by the D_2/D_3 DA agonist quinpirole (FIG. 1), but not by a D_1 agonist SKF 38393.[18] A lower

[a] This research was supported by a grant from the Council for Tobacco Research-USA, Inc..
[b] Address correspondence to Dr. Edward D. Levin, Neurobehavioral Research Laboratory, Department of Psychiatry, Box 3412, Duke University Medical Center, Durham, NC 27710.

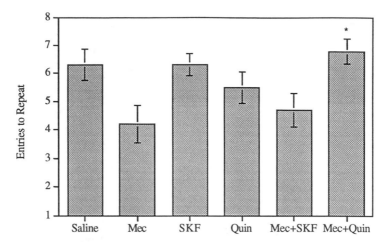

FIGURE 1. Reversal of the mecamylamine-induced radial-arm maze choice accuracy impairment with the D_2/D_3 agonist quinpirole (mean $\pm$ SEM). Mec = 10 mg/kg mecamylamine, SKF = 3 mg/kg SKF 38393, Quin = 0.05 mg/kg quinpirole (LY 171555). *$p < 0.05$ Mec versus Mec + Quin.[18]

dose of 2.5 mg/kg of mecamylamine does not by itself cause a deficit in radial-arm maze choice accuracy, but it does cause a significant choice accuracy deficit when given in conjunction with a low dose of the muscarinic antagonist scopolamine (0.05 mg/kg), which by itself does not cause a significant choice accuracy deficit.[19,20] This combined mecamylamine-scopolamine–induced deficit is also reversed by the D_2/D_3 agonist quinpirole.[20]

The opposite relationship is seen between mecamylamine and DA antagonists. A significant deficit can be elicited by a combination of the subthreshold dose of mecamylamine (2.5 mg/kg) and the mixed D_1/D_2 antagonist haloperidol.[22] The specific D_2 antagonist raclopride has the same effect of potentiating the mecamylamine-induced deficit, whereas addition of the D_1 antagonist SCH 23390 to this dose of mecamylamine has no significant effect.[23] Thus, D_2 receptors appear to have a consistent interaction with the nicotinic antagonist mecamylamine, with an agonist reversing the mecamylamine-induced memory deficit and an antagonist potentiating it. In contrast, D_1 receptors appear to have little interactive effect with mecamylamine.

ACUTE NICOTINE STUDIES

In several studies, we found that acute administration of 0.2 mg/kg of nicotine significantly improves working memory performance in the radial-arm maze.[24–26] This acute nicotine-induced improvement in choice accuracy is reversed by concurrent acute administration of the nicotinic antagonist mecamylamine or the muscarinic antagonist scopolamine.[25] The acute nicotine effect is not as robust as the chronic nicotine effect. In some experiments significant improvements were not seen with the 0.2 mg/kg dose when given alone. However, when the positive effects of acute nicotine are less obvious, they can be made more apparent by co-administration of dopaminergic agonists. The D_1/D_2 agonist pergolide[27] and the D_2/D_3 agonist quinpirole[26] (FIG. 2) both have mutually potentiating effects when given together with acute doses of 0.2 mg/kg of nicotine. Interestingly, we recently found that this dose of nicotine is effective in attenuating a choice accuracy deficit elicited by the D_1 agonist SKF 38393.[26] Thus, like the nicotinic antagonist mecamylamine, nicotine has significant interactions with dopaminergic drugs. The receptor subtype breakdown is not as clear with nicotine as it is with mecamylamine. As expected, nicotine had a mutually potentiating effect with quinpirole, which effectively reversed the mecamylamine-induced deficit. However, a significant interaction of nicotine with the D_1 agonist SKF 38393 was also found, which did not have a significant interaction with mecamylamine.

ACUTE LOCAL INFUSION STUDIES

Recently we began a series of studies of the effects of nicotinic drugs infused into the ventricles and local brain regions which are the sources or targets of DA systems. Intracerebroventricular (i.c.v.) infusion of nicotine significantly improved choice accuracy performance in rats with little training on the radial-arm maze. In rats trained to high levels of performance no such improvement was detected. However, in this set of rats nicotine was effective in reversing the impairment in performance caused by i.c.v. infusion of the nicotinic antagonist mecamylamine.[28]

Our initial study[29] examined the effects of local infusions of nicotinic agonist and antagonist drugs into the SN and VTA. Mecamylamine infused into either the SN or

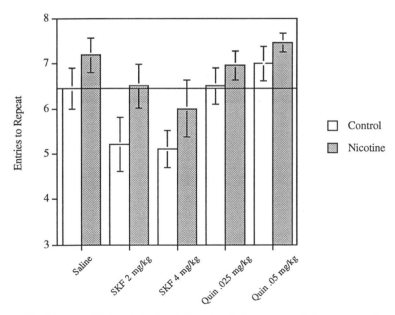

FIGURE 2. Additive effects of nicotine-induced radial-arm maze choice accuracy improvement with the effect of the D_2/D_3 agonist quinpirole and attenuation of the deficit caused by the D_1 agonist SKF 38393 (mean ± SEM).[26]

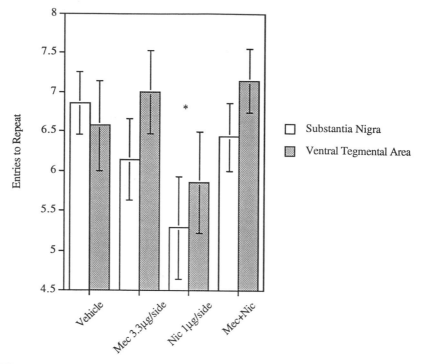

FIGURE 3. Local infusion of nicotine and mecamylamine into the substantia nigra and ventral tegmental area: Effects on choice accuracy in the radial-arm maze (mean ± SEM). *$p < 0.025$ versus vehicle and Nic + Mec.[29]

VTA significantly impaired choice accuracy in the radial-arm maze. Interestingly, both nicotine and the nicotinic agonist cytisine also showed signs of impairing choice accuracy. This may have been due to the local high concentrations causing desensitization, depolarization blockade or excessive stimulation of the nicotinic receptors in these areas. The results of the combination nicotine-mecamylamine experiment suggest that this may be true. Mecamylamine was effective in significantly reducing the nicotine-induced deficit (FIG. 3). In this same study we did not find any evidence of effects of local infusions of the muscarinic antagonist scopolamine or the muscarinic agonist pilocarpine.

CHRONIC SYSTEMIC STUDIES

We conducted seven different experiments examining the effect of chronic subcutaneous infusion of nicotine (approximately 12 mg/kg/day) via a glass and Silastic pellet or an osmotic minipump. We consistently found that chronic nicotine administration significantly improves memory performance in the radial-arm maze (FIG. 4).[21,30–33] This effect is reversed by concurrent administration of chronic mecamylamine.[32] Acute challenge with mecamylamine during the course of chronic nicotine causes an overall decline in choice accuracy in control and nicotine groups, but the enhanced performance of the nicotine group relative to controls is preserved.[21] In contrast, we found that acute administration with the muscarinic acetylcholine antagonist scopolamine eliminated the chronic nicotine-induced improvement in choice accuracy.

We have not found concurrent manipulations of D_2 DA receptors to affect significantly the chronic nicotine-induced improvement in choice accuracy. Chronic concurrent administration of either the D_2 agonist quinpirole or the D_2 antagonist raclopride had no discernible effect on the chronic nicotine-induced improvement of choice accuracy. Acute challenge with a range of doses of quinpirole likewise had no significant effect on the chronic nicotine-induced improvement in radial-arm maze choice accuracy. This stands in contrast to the significant interactions we saw between acute nicotine and quinpirole,[26] as well as the significant interactions between acute mecamylamine and both quinpirole and raclopride.[18,23]

LESION STUDIES

Chronic nicotine administration is effective in reversing the radial-arm maze working memory impairments caused by knife-cut lesions of the medial basalocortical projection or the fimbria-fornix, which connects the septum with the hippocampus. This points to the possible therapeutic use of chronic nicotine, but it also gives important information concerning the critical neural substrates for the chronic nicotine-induced improvement. Apparently, neither of these pathways is necessary for the chronic nicotine effects inasmuch as the effect was still seen in the lesioned rats. It may be the case that at least one of the lesioned pathways must be intact for the chronic nicotine-induced memory improvement or, alternatively, other pathways

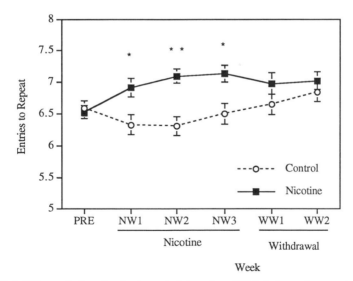

FIGURE 4. Effects of chronic nicotine administration (SC infusions of approximately 12 mg/kg/day) on choice accuracy in the radial-arm maze, averaged over seven experiments (mean ± SEM). During administration of nicotine: Control $n = 60$, nicotine $n = 63$; after withdrawal: Control $n = 51$, nicotine $n = 54$. *$p < 0.01$, **$p < 0.001$ control versus nicotine.

may subserve the nicotine effect. Interestingly, similar to the therapeutic effect of chronic nicotine, repeated doses of quinpirole also are effective in attenuating the radial-arm maze working memory deficit caused by knife-cut lesions of the medial basalocortical projection.[34] However, simple similarity of effect does not imply common mechanisms of action.

CONCLUSIONS

Nicotinic systems have direct interactions with dopaminergic systems which are important for the acute effects of nicotinic agonist and antagonist manipulations. D_2 receptors seem to be particularly important in this regard. In contrast, chronic nicotine-induced working memory improvement in the radial-arm maze seems to be relatively impervious to concurrent D_2 agonist or antagonist manipulations. The improvement is still seen after lesions to the basalocortical or septohippocampal pathways. Chronic mecamylamine co-administration does prevent the nicotine-induced improvement, but acute mecamylamine challenge does not reverse it. The only other manipulation we found to reverse the chronic nicotine-induced memory improvement is acute challenge with the muscarinic antagonist scopolamine. Nicotinic stimulation is necessary for the induction of the effect but not for the expression of it. In contrast, muscarinic stimulation may be necessary for the expression of the chronic nicotine-induced memory improvement.

Both acute and chronic nicotine administration improve memory performance in the win-shift working memory radial-arm maze task. The involvement of interactions with dopaminergic systems seems to be different for each.

SUMMARY

Nicotine has been found to improve memory performance in a variety of tests in rats, monkeys, and humans. Interactions of nicotinic systems with dopamine (DA) systems may be important for this effect. We conducted a series of studies of nicotinic agonist and antagonist interactions with DA systems using rats in a win-shift working memory task in the radial-arm maze. The working memory deficit caused by the nicotinic antagonist mecamylamine was potentiated by the D_1/D_2 DA antagonist haloperidol and the specific D_2 antagonist raclopride. In contrast, the mecamylamine-induced deficit was reversed by co-administration of the D_2/D_3 agonist quinpirole. Nicotine also has significant interactions with dopamine drugs with regard to working memory performance in the radial-arm maze. The DA agonist pergolide did not by itself improve radial-arm maze memory performance, but when given together with nicotine it produced an elevated dose-dependent increase in choice accuracy. The D_1 agonist SKF 38393 significantly impaired radial-arm maze choice accuracy. Nicotine was effective in reversing this deficit. When given together with nicotine, the D_2/D_3 agonist quinpirole improved RAM choice accuracy relative to either drug alone. Acute local infusion of mecamylamine to the midbrain DA nuclei effectively impairs working memory function in the radial-arm maze. In contrast to acute nicotinic manipulations, considerably less evidence exists that the effects of chronic nicotine administration are influenced by DA systems. This may be an example of the different neural substrates that underlie the memory improvement caused by acute and chronic nicotine.

ACKNOWLEDGMENTS

The authors would like to thank Sandra Briggs and Channelle Christopher for their help in performing these studies.

REFERENCES

1. LEVIN, E. D. 1992. Nicotinic systems and cognitive function. Psychopharmacology **108:** 417–431.
2. WONNACOTT, S., J. IRONS, C. RAPIER, B. THORNE & G. G. LUNT. 1989. Presynaptic modulation of transmitter release by nicotinic receptors. In Progress in Brain Research. A. Nordberg, K. Fuxe, B. Holmstedt & A. Sundwall, Eds.: 157–163. Elsevier Science Publishers, Amsterdam.
3. CLARKE, P. B. S. & A. PERT. 1985. Autoradiographic evidence for nicotine receptors on nigrostriatal and mesolimbic dopaminergic neurons. Brain Res. **348:** 355–358.
4. CLARKE, P. B. S., C. B. PERT & A. PERT. 1984. Autoradiographic distribution of nicotine receptors in rat brain. Brain Res. **323:** 390–395.
5. SCHWARTZ, R. D., J. LEHMANN & K. J. KELLAR. 1984. Presynaptic nicotinic cholinergic receptors labeled by (^{3}H)acetylcholine on catecholamine and serotonin axons in brain. J. Neurochem. **42:** 1495–1498.
6. CLARKE, P. B. S., D. W. HOMMER, A. PERT & L. R. SKIRBOLL. 1985. Electrophysiological actions of nicotine on substantia nigra single units. Br. J. Pharmacol. **85:** 827–835.
7. GRENHOFF, J., G. ASTON-JONES & T. H. SVENSSON. 1986. Nicotinic effects on the firing pattern of midbrain dopamine neurons. Acta Physiol. Scand. **128:** 151–158.
8. IMPERATO, A., A. MULAS & G. DI CHIARA. 1986. Nicotine preferentially stimulates dopamine release in the limbic system of freely moving rats. Eur. J. Pharmacol. **132:** 337–338.
9. LICHTENSTEIGER, W., F. HEFTI, D. FELIX, T. HUWYLER, E. MELAMED & M. SCHLUMPF. 1982. Stimulation of nigrostriatal dopamine neurons by nicotine. Neuropharmacology **21:** 963–968.
10. MERU, G., K. P. YOON, V. BOI, G. L. GESSA, L. NAES & T. C. WESTFALL. 1987. Preferential stimulation of ventral tegmental area dopaminergic neurons by nicotine. Eur. J. Pharmacol. **141:** 395–399.
11. RAPIER, C., G. G. LUNT & S. WONNACOTT. 1988. Stereoselective nicotine-induced release of dopamine from striatal synaptosomes: Concentration dependence and repetitive stimulation. J. Neurochem. **50:** 1123–1130.
12. AHTEE, L. & S. KAAKKOLA. 1978. Effect of mecamylamine on the fate of dopamine in striatal and mesolimbic areas of rat brain: Interaction with morphine and haloperidol. Br. J. Pharmacol. **62:** 213–218.
13. HAIKALA, H. & L. AHTEE. 1988. Antagonism of the nicotine-induced changes of the striatal dopamine metabolism in mice by mecamylamine and pempidine. Naunyn-Schmiedbergs Arch. Pharmakol. **338:** 169–173.
14. LAPIN, E. P., H. S. MAKER, H. SERSHEN & A. LAJTHA. 1989. Action of nicotine on accumbens dopamine and attenuation with repeated administration. Eur. J. Pharmacol. **160:** 53–59.
15. REILLY, M. A., E. P. LAPIN, H. S. MAKER & A. LAJTHA. 1987. Chronic nicotine administration increases binding of [^{3}H]domperidone in rat nucleus accumbens. J. Neurosci. Res. **18:** 621–625.
16. LEVIN, E. D. & J. E. ROSE. 1992. Cognitive effects of D_1 and D_2 interactions with nicotinic and muscarinic systems. In Neurotransmitter Interactions and Cognitive Function. E. D. Levin, M. W. Decker & L. L. Butcher, Eds.: 144–158. Berkhäuser. Boston, MA.
17. LEVIN, E. D., M. CASTONGUAY & G. D. ELLISON. 1987. Effects of the nicotinic receptor blocker, mecamylamine, on radial arm maze performance in rats. Behav. Neural Biol. **48:** 206–212.
18. LEVIN, E. D., S. R. McGURK, J. E. ROSE & L. L. BUTCHER. 1989. Reversal of a

mecamylamine-induced cognitive deficit with the D_2 agonist, LY 171555. Pharmacol. Biochem. Behav. **33:** 919–922.

19. LEVIN, E. D., S. R. McGURK, D. SOUTH & L. L. BUTCHER. 1989. Effects of combined muscarinic and nicotinic blockade on choice accuracy in the radial-arm maze. Behav. Neural Biol. **51:** 270–277.

20. LEVIN, E. D., J. E. ROSE, S. R. McGURK & L. L. BUTCHER. 1990. Characterization of the cognitive effects of combined muscarinic and nicotinic blockade. Behav. Neural Biol. **53:** 103–112.

21. LEVIN, E. D. & J. E. ROSE. 1990. Anticholinergic sensitivity following chronic nicotine administration as measured by radial-arm maze performance in rats. Behav. Pharmacol. **1:** 511–520.

22. McGURK, S. R., E. D. LEVIN & L. L. BUTCHER. 1989. Nicotinic-dopaminergic relationships and radial-arm maze performance in rats. Behav. Neural Biol. **52:** 78–86.

23. McGURK, S. R., E. D. LEVIN & L. L. BUTCHER. 1989. Radial-arm maze performance in rats is impaired by a combination of nicotinic-cholinergic and D_2 dopaminergic drugs. Psychopharmacology **99:** 371–373.

24. LEVIN, E. D., J. E. ROSE & L. ABOOD. 1995. Effects of nicotinic dimethylaminoethyl esters on working memory performance of rats in the radial-arm maze. Pharmacol. Biochem. Behav. In Press.

25. LEVIN, E. D. & J. E. ROSE. 1991. Nicotinic and muscarinic interactions and choice accuracy in the radial-arm maze. Brain Res. Bull. **27:** 125–128.

26. LEVIN, E. D. & B. EISNER. 1994. Nicotine interactions with D_1 and D_2 agonists: Effects on working memory function. Drug Dev. Res. **31:** 32–37.

27. LEVIN, E. D. 1995. Interactive effects of nicotinic and muscarinic agonists with the dopaminergic agonist pergolide on rats in the radial-arm maze. Pharmacol. Biochem. Behav. Submitted.

28. BRUCATO, F. H., E. D. LEVIN, J. E. ROSE & H. S. SWARTZWELDER. 1994. Intracerebroventricular nicotine and mecamylamine alter radial-arm maze performance in rats. Drug Dev. Res. **31:** 18–23.

29. LEVIN, E. D., S. J. BRIGGS, N. C. CHRISTOPHER & J. T. AUMAN. 1994. Working memory performance and cholinergic effects in the ventral tegmental area and substantia nigra. Brain Res. **657:** 165–170.

30. LEVIN, E. D., C. LEE, J. E. ROSE, A. REYES, G. ELLISON, M. JARVIK & E. GRITZ. 1990. Chronic nicotine and withdrawal effects on radial-arm maze performance in rats. Behav. Neural Biol. **53:** 269–276.

31. LEVIN, E. D., S. J. BRIGGS, N. C. CHRISTOPHER & J. E. ROSE. 1992. Persistence of chronic nicotine-induced cognitive facilitation. Behav. Neural Biol. **58:** 152–158.

32. LEVIN, E. D., S. J. BRIGGS, N. C. CHRISTOPHER & J. E. ROSE. 1993. Chronic nicotinic stimulation and blockade effects on working memory. Behav. Pharmacol. **4:** 179–182.

33. LEVIN, E. D., N. C. CHRISTOPHER, S. J. BRIGGS & J. E. ROSE. 1993. Chronic nicotine reverses working memory deficits caused by lesions of the fimbria or medial basalocortical projection. Cognit. Brain Res. **1:** 137–143.

34. McGURK, S. R., E. D. LEVIN & L. L. BUTCHER. 1992. Dopaminergic drugs reverse the impairment of radial-arm maze performance caused by lesions involving the cholinergic medial pathway. Neuroscience **50:** 129–135.

Structural and Functional Diversity of β-Adrenergic Receptors[a]

A. D. STROSBERG

Institut Cochin de Génétique Moléculaire
Laboratoire d'Immuno-Pharmacologie Moléculaire
CNRS UPR 0415
22, rue Méchain
75014 Paris, France

and

Université de Paris VII
Paris, France

Recent progress in molecular cloning techniques has allowed the identification of the genes or cDNAs corresponding to proteins homologous to but different from previously known receptors. These "new" sequences have since been shown to encode subtypes whose existence had been suspected on the basis of pharmacologic data. A good example of such a new subtype is the β_3-adrenergic receptor (β_3AR). Its gene was first cloned by homology using β_2- and β_1-specific DNA oligonucleotide probes.[1] Corresponding β_3 species homologues were later cloned from mouse[2] and rat.[3,4] After transfection of the β_3AR gene in Chinese hamster ovary (CHO) cells, which are devoid of such receptors, it was shown that the βAR subtype displays most of the ligand-binding and adenylyl cyclase-stimulating properties of previously described "atypical" receptors in various rodent species (reviewed in refs. 5 and 6).

The β_3AR is quite distinct from the β_1- and β_2AR and thus presents an interesting case of diversity in the small family of βARs.[7] In this paper we discuss the various features that distinguish the three βAR subtypes.

MOLECULAR DIFFERENCES

All three βARs belong to the R_7G superfamily of receptors coupled to GTP binding proteins.[8] They are thus composed of a single polypeptide chain with seven putative transmembrane domains, an extracellular glycosylated N-terminal and an intracellular C-terminal region (FIG. 1). As in all other R_7G proteins, the βAR subtypes vary extensively both in terms of length and sequences in the N- and C-terminal as well as in the third intracellular (i3) regions (FIG. 2). Although the seven hydrophobic regions are quite conserved, a number of differences may explain the striking pharmacologic and regulatory variations.

The degree of amino acid sequence identity between man, mouse, and rat β_3AR (FIG. 2) is much higher (80–90%) than that existing between different βAR subtypes

[a] This work was supported mainly by the Centre National de la Recherche Scientifique, the Institut National de la Santé et de la Recherche Médicale, the University of Paris VII, the Ministry for Research and Technology, and the Bristol-Myers-Squibb Company. We are also grateful for support from the Ligue Nationale contre le Cancer, the Fondation pour la Recherche Médicale Française, and, last but not least, the Association pour la Recherche contre le Cancer.

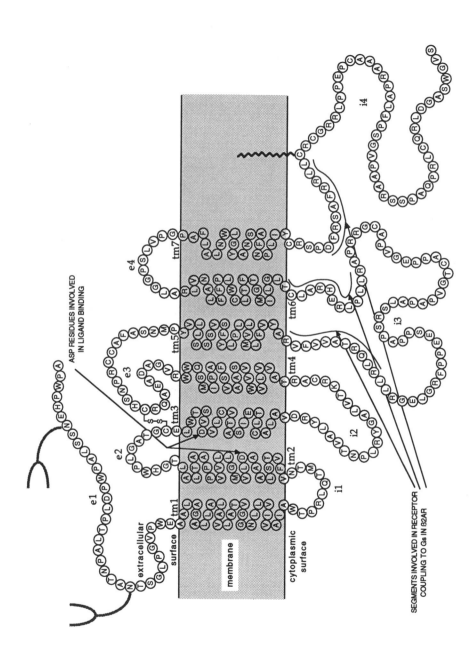

(40–50%) and is of the same order as that observed across species for a given receptor subtype. Several residues located in the functional domains of the receptor are specifically shared by the human, murine, and rat β₃AR, but are not found in the β₁- and β₂AR sequences. From a genetic point of view, the human, mouse, and rat β₃AR genes all display a similar genomic organization with an intron interrupting the 3' end of the coding block,[9–12] whereas the other βAR subtypes are intronless. Moreover, the human and mouse genes have been assigned to a chromosomal linkage group which is conserved between the two species.[2] Together, these structural data strongly suggest that the human, mouse, and rat genes do not encode distinct βAR subtypes,[4] but are species homologues of the β₃AR gene.

A number of residues found in human β₁- and β₂AR sequences are substituted by residues that appear to be specific for β₃. Most of these residues are found in all β₃-receptors sequenced so far. Some of the changes appear to be selective for the human β₃. Last, but not least, the Ser/Thr phosphorylation target sites documented for β₁- and β₂ARs are absent from all β₃. The rodent β₃ARs appear to share with the human β₃ quite a number of structural features that distinguish this subtype from β₁ and β₂. There are, in addition to these β₃-specific substitutions, a number of rodent specific changes, as is the case, for example, for the Val Ala Leu deletion in the tm1 domain.

PHARMACOLOGIC DIFFERENCES

Five major features distinguish the human β₃AR from those of β₁ and β₂: (1) atypically low affinity for conventional β-antagonists including reference radioligands; (2) atypically low stereoselectivity index for agonist and antagonist enantiomers; (3) atypically low potencies of reference agonists; (4) high potency of a novel class of compounds initially described as potent activators of lipolysis and thermogenesis in white and brown adipose tissues of rodents; and (5) partial agonistic activities of several β₁/β₂-antagonists, reflecting intrinsic sympathomimetic activities in tissues. Such general properties are reminiscent of those ascribed to the atypical βAR initially proposed to exist in adipose tissues and later in the digestive tract of rodents. These tissues are also the only ones where β₃ mRNA expression was unambiguously demonstrated both in human[13] and rodents.[2,3,12]

The rodent β₃ again appears to be somewhat different from the human subtype: in CMO β₃ cells, propranolol is an antagonist for mouse β₃, as it is for β₁ and β₂, in contrast to what is seen in the human β₃ where it behaves as a weak agonist. A number of other compounds display either stronger or weaker effects in the β₃ human or rodent without any obvious trend: BRL 37344, for example, approximately 30 times more potent in the rodent, whereas carazolol is about 8 times more potent in the human β₃.

FIGURE 1. Primary structure of the human β₃AR. The sequences are represented in the one-letter code for amino acids. The single polypeptide chain is arranged according to the model for rhodopsin. The disulfide bond essential for activity Cys[111] and Cys[109] is represented by -S-S-. The two N-glycosylation sites in the amino-terminal portion of the protein are indicated by ⤙. The palmitoylated Cys[360] residue in the N-terminus of the i4 loop is indicated by the symbol ⸯ .

```
                                              e1                      tm1              i1
MoB3    MAPWPHRNGSLALWSDAPTLDPSAANTSGLPGVPWAAALAGALLALA  TVGGNLLVIIAIARTPRLQTITNVFVTSL         76
RaB3    M-----K------F-------------------------------------------T--------------W-------M--------  76
HuB3    M-----E-S---P-P-L---A-NT------------------E--------VLA-----V-----W-----L--L-IM---          79
HuB1    MGAGVLVLGASEPGNLSSAAPLPDGAATAAR-LVPASP-ASLLPP-SESPE-LSQQWT-GM-L-M--IVLLI-A--V---V--K-----L-L-IM---  100
HuB2    MGQPGNGSAFLL--NGSHAPDHDVTQQRDEVWVV-M-IVMS-IVLAI-F--V---T---KFE---V--Y-I----                75

              tm2                       e2       tm3          ▼i2   tm4
MoB3    AAADLVVGLLVMPPGATLALTGHWPLGETGCELWTSVDVLCVTASIETLCALAVDRYLAVTNPLRYGTLVTKRRARAAVVLVWIVSAAVSFAPIMSQWWR  176
RaB3    -T-----------------------------------------------------------------------------------------------T-  176
HuB3    -------M----V--A----------A-------------------------------------A---C--T----V-------T--------------  179
HuB1    -S----M---V-F---IVVW-R-EY-SFF----------------------VI-L---I-S-F--QS-L-RA---GL-CT--AI--L---L--LMH---  200
HuB2    -C----M--A-V-F--AHI-MKM-TF-NFW--F---I-------------------VI--F-I-S-FK-QS-L--NK--VIILM----GLT--L--QMH-Y-  175

                  e3      ▼ tm5                                           i3
MoB3    VGADAEAQECHSNPRCCSFASNMPYALLSSVSFYLPLLVMLFVYARVFVWAKRQRHLLRRELGRF  SPEESPPSPSRSPSATGGTPAAPD      267
RaB3    ------------------------------------------------------------R------------V---T-S-                267
HuB3    ------R-------A-----V--------------------------------------P------------V--PV--C-P-E             270
HuB1    AES----RR-YND-K--D-VT-RA--IA--V---V--CI-A--L---RE-QK-VKKIDSCER--LGGPAR--S--P--V--PAPP-G--RPAAAAATA  299
HuB2    ATHQ---IN-YA-ET--DFFT-QA--IA--I----V--VI-V---S---QE---LQKIDKSE---HVQNL-QVEQ                      250

              i4                          tm6                    e4      tm7
MoB3    GVPPCGRRPARLLPLREHRALRTLGLIMGIFSLCWLPFFLANVLRALAGPSLVPSGVFIALNWLGYANSAFNPVIYCRSPDFRDAFRRLLCSYGGRG  364
RaB3    ---S-----G-----------------------------------------V-------------------L---------------S---RC-R-L  364
HuB3    ---A----------------C------------T-T-------------G----GPA-L------L----S---RC-R-L                  367
HuB1    PLANGRAGK---S--VA---QK--K---I--V-T-----------VK-FHRE---DRL-VFF---------I---K-QG---CARRAA          398
HuB2    DGRTGHGL--SSKF C-K--K---I----T-------------IV-IVHVIQDN-IRKE-Y-L---I--V-G---L----I--QE---LRRSSL     347

MoB3    PEEPRAVTFPASPVEARQSPPLNRFDGYEGARPFPT                                                            400
RaB3    ----V-------AS-NS-----E-----                                                                    400
HuB3    -P--C-AAR--LFPSGVPAARSSPAQPRLCQRLDGASWGVS                                                       408
HuB1    RRRHATHGDRPRASGCLARPGPP-SPGAASDDD-DDVGATPPARLLEPWAGCNGGAAADSDSSLDEPCRPGFASESKV                   477
HuB2    KAYGNGYSSNGNTGEQSGYHVEQEKENK-LCEDLPGTEDFVGHQGTVPSDNIDSQGRNCSTNDSLL                              413
```

DIFFERENT CONTROL MECHANISMS REGULATE β₁-, β₂- AND β₃AR EXPRESSION

A variety of control mechanisms regulate the function and expression of the βAR subtypes. We summarize in TABLE 1 a few of the factors involved in βAR regulation, and show the differences between β_1, β_2, and β_3.

Agonist-induced Desensitization

One of the most specific mechanisms is desensitization, a complex physiologic process, which prevents the hormonal overload of most G-protein–coupled receptors by impairing the signal-transmission pathway at receptor and/or postreceptor levels. An important clinical consequence of desensitization is the relative or complete resistance to pharmacologic agonists given for an extended period of time.

Several molecular mechanisms involved in receptor desensitization have been characterized for the β_2AR.[14] After a few minutes of receptor activation by an agonist, phosphorylation of β_2AR by protein kinase A (PKA) and βAR kinase (βARK) results in the rapid uncoupling of the receptor from the transducing pathway.[15,16] Phosphorylation by PKA is a negative-feedback of receptor activation, mediated by the raise of intracellular cAMP, which affects all β_2AR, whereas βARK phosphorylates only those receptors that are occupied by the agonist.[17] When receptor activation is sustained for longer periods of time (hours), protein degradation of preexisting receptors and destabilization of the receptor mRNA contribute to the reduction of the total number of β_2AR (i.e., receptor down-regulation).[18]

The three βAR subtypes are not equally sensitive to desensitization. The β_1AR, which has fewer potential phosphorylation sites and lacks the two tyrosine residues implicated in the down-regulation of the β_2AR,[19] is less prone than the β_2AR to both short-[20] and long-term[21] desensitization. The third βAR subtype is almost completely resistant to short-term desensitization,[12,22] primarily because this receptor does not undergo PKA- or βARK-induced phosphorylation—most likely because of the absence of the target sequences identified on the β_2AR.[8]

Other Mechanisms for βAR Modulation

A number of other compounds and factors have been shown to affect βAR expression and function (TABLE 1). These include glucocorticoids, butyrate, insulin adipocyte differentiation, and cold. The three βARs are diversely affected by these different conditions, again demonstrating that each subtype is regulated independently and may thus play a distinct physiologic role.

FIGURE 2. Amino acid sequence comparison of the mouse (Mo), rat (Ra), and human (Hu) β1-, β2-, and βARs. The seven transmembrane segments (tm1–tm7) are boxed and alternate with extracellular (e1–e4) and intracellular (i1–i4) domains. Gaps (*horizontal dashes*) have been introduced to maximize sequence alignment. Black triangles indicate residues conserved in all nine proteins, and dots represent classical substitution according to Daihoff PAM 250 matrix.

CONCLUSION

The comparison of the three βAR subtypes reveals both striking structural similarities, especially in the ligand binding and G-protein interacting sites, and important differences. These concern mainly regulatory control of receptor expression and function. It is likely that other families of receptor subtypes will display similar features by which binding of the same natural ligands may result in widely different functional effects, well controlled by developmentally or hormonally regulated mechanisms.

SUMMARY

The molecular and functional properties of the three βAR subtypes appear to be quite diversified even though all three bind the same natural catecholamines, adrenaline and noradrenaline, couple apparently to the same G_s transducer protein, and stimulate the same adenylyl cyclase effector. Binding characteristics for a variety of synthetic ligands thus encompass a wide range of K_d values, and a number of β_1/β_2 antagonists turn out to be potent agonists towards the β_3 subtype.

TABLE 1. Differential Up- and Down-Regulation of the Three β-Subtypes

Factors	β_1	β_2	β_3	References
Adipose differentiation	↓	↓	↑	Fève et al.[23,24]
cAMP	↓	↓	↑	Thomas et al.[25]
Dexamethasone	↓	↑	↓	Fève et al.[23,26]
Butyrate	↑	↑	↓	Krief et al.[27]
Insulin	→	→	↓	Fève et al.[28]

Regulatory mechanisms also appear to vary considerably from one β-receptor to another, with dexamethasone, for example, up-regulating the β_2 and down-regulating the β_1 and β_3 subtypes. Most striking is the fact that the β_3 subtype is resistant to the agonist-induced short-term desensitization initiated for the β_2 receptor by phosphorylation target sequences absent in β_3.

Diversity in the three βAR subtypes clearly suggests different functional roles in the various tissues in which these receptors are expressed together or alone.

REFERENCES

1. EMORINE, L. J., S. MARULLO, M. M. BRIEND-SUTREN, G. PATEY, K. TATE, C. DELAVIER-KLUTCHKO & A. D. STROSBERG. 1989. Molecular characterization of a new human β-adrenergic receptor involved in catecholamine control of metabolism. Science 245: 1118–1121.
2. NAHMIAS, C., N. BLIN, J-M. ELALOUF, M. G. MATTEI, A. D. STROSBERG & L. J. EMORINE. 1991. Molecular characterization of the mouse β_3-adrenergic receptor: Relationship with the atypical receptor of adipocytes. EMBO J. 10: 3721–3727.
3. GRANNEMAN, J. G., K. N. LAHNERS & A. CHAUDHRY. 1991. Molecular cloning and expression of the rat β_3-adrenergic receptor. Mol. Pharmacol. 40: 895–899.
4. MUZZIN, P., J. P. REVELLI, F. KUHNE, J. D. GOCAYNE, W. R. McCONBIE, J. G. VENTER,

J. P. GIACOBINO & M. C. FRASER. 1991. An adipose tissue-specific β-adrenergic receptor. Molecular cloning and down-regulation in obesity. J. Biol. Chem. **266:** 24053–24058.

5. BLIN, N., C. NAHMIAS, M-F. DRUMARE & A. D. STROSBERG. 1994. Mediation of most atypical effects by species homologues of the β3-adrenoceptor. Br. J. Pharmacol. **112:** 911–919.

6. EMORINE, L. J., N. BLIN & A. D. STROSBERG. 1994. The human β3-adrenoceptor: The search for a physiological function. Trends Pharmacol. Sci. **15:** 3–7.

7. BLIN, N., L. CAMOIN, B. MAIGRET & A. D. STROSBERG. 1993. Structural and conformational features determining selective signal transduction in the β3-adrenergic receptor. Mol. Pharmacol. **44:** 1094–1104.

8. STROSBERG, A. D. 1993. Structure, function and regulation of adrenergic receptors. Protein Sci. **12:** 1198–1209.

9. VAN SPRONSEN, A., C. NAHMIAS, S. KRIEF, M-M. BRIEND-SUTREN, A. D. STROSBERG & L. J. EMORINE. 1993. The human and mouse β3-adrenergic receptor genes: Promoter and intron/exon structure. Eur. J. Biochem. **213:** 1117–1124.

10. LELIAS, J. M., M. KAGHAD, M. RODRIGUEZ, P. CHALON, J. BONNIN, B. DUPRÉ, M. DELPECH, M. BENSAID, G. LEFUR, P. FERRARA & D. CAPUT. 1993. Molecular cloning of a human β3-adrenergic receptor cDNA. FEBS Lett. **324:** 127–130.

11. GRANNEMAN, J. G., K. N. LAHNERS & A. CHANDLEY. 1993. Characterization of the human β3-adrenergic receptor gene. Mol. Pharmacol. **44:** 264–270.

12. GRANNEMAN, J. G., LAHNERS & D. D. RAO. 1992. Rodent and human β3-adrenergic receptor genes contain an intron within the protein coding block. Mol. Pharmacol. **42:** 964–970.

13. KRIEF, S., F. LÖNNQVIST, J. RAIMBAULT, B. BAUDE, A. VAN SPRONSEN, P. ARNER, A. D. STROSBERG, D. RICQUIER & L. J. EMORINE. 1993. Tissue distribution of β3-adrenergic receptor mRNA in man. J. Clin. Invest. **91:** 344–349.

14. BENOVIC, J. L., M. BOUVIER, M. G. CARON & R. J. LEFKOWITZ. 1988. Regulation of the adenyl cyclase-coupled β-adrenergic receptors. Annu. Rev. Cell Biol. **4:** 405–408.

15. BOUVIER, M., P. HAUSSDORF, A. DEBLASI, B. F. O'DOWD, B. K. KOBILKA, M. G. CARON & R. J. LEFKOWITZ. 1988. Removal of phosphorylation sites from the β2-adrenergic receptor delays onset of agonist-promoted desensitization. Nature **333:** 370–373.

16. BENOVIC, J. L., A. DEBLASI, W. C. STONE, M. G. CARON & R. J. LEFKOWITZ. 1989. β-adrenergic receptor kinase: Primary structure delineates a multigene family. Science **246:** 235–240.

17. HAUSDORFF, W. P., M. HNATOWICH, B. F. O'DOWD, M. G. CARON & R. J. LEFKOWITZ. 1990. A mutation of the β2-adrenergic receptor impairs agonist activation of adenylate cyclase without affecting high affinity agonist-binding. J. Biol. Chem. **265:** 1388–1393.

18. HADCOCK, J. R. & C. C. MALBON. 1988. Down-regulation of β-adrenergic receptors: Agonist-induced reduction in receptor mRNA levels. Proc. Natl. Acad. Sci. USA **85:** 5021–5025.

19. VALIQUETTE, M., H. BONIN, M. HNATOWICH, M. G. CARON, R. J. LEFKOWITZ & M. BOUVIER. 1990. Involvement of tyrosine residues located in the carboxyl tail of the human β2-adrenergic receptor in its agonist-induced down-regulation. Proc. Natl. Acad. Sci. USA **87:** 5089–5093.

20. ZHOU, X. M. & P. H. FISHMAN. 1991. Desensitization of the human β1-adrenergic receptor: Involvement of the cyclic cAMP-dependent but not receptor-specific protein kinase. J. Biol. Chem. **266:** 7462–7468.

21. SUZUKI, T., C. T. N'GUYEN, F. NANTEL, H. BONIN, M. VALIQUETTE, T. FRIELLE & M. BOUVIER. 1992. Distinct regulation of β1- and β2-adrenergic receptors in Chinese hamster fibroblasts. Mol. Pharmacol. **41:** 542–548.

22. NANTEL, F., H. BONIN, L. J. EMORINE, V. ZILBERFARB, A. D. STROSBERG, M. BOUVIER & S. MARULLO. 1993. The human β3-adrenergic receptor is resistant to short-term agonist-promoted desensitization. Mol. Pharmacol. **43:** 548–555.

23. FÈVE, B., L. J. EMORINE, M.-M. BRIEND-SUTREN, F. LASNIER, A. D. STROSBERG & J. PAIRAULT. 1990. Differential regulation of β1- and β2-adrenergic receptor protein and

mRNA levels by glucocorticoids during 3T3-F442A adipose differentiation. J. Biol. Chem. **265:** 16343–16349.

24. FÈVE, B., L. J. EMORINE, F. LASNIER, N. BLIN, B. BAUDE, C. NAHMIAS, A. D. STROSBERG & J. PAIRAULT. 1991. Atypical β-adrenergic receptor in 3T3-F442A adipocytes: Pharmacological and molecular relationship with the human β3-adrenergic receptor. J. Biol. Chem. **266:** 20329–20336.

25. THOMAS, R. F., B. D. HOLT, D. A. SCHWINN & S. B. LIGGETT. 1992. Long-term agonist exposure induces up-regulation of β3AR exposure via multiple cAMP response elements. Proc. Natl. Acad. Sci. USA **89:** 4490–4494.

26. FÈVE, B., B. BAUDE, S. KRIEF, A. D. STROSBERG, J. PAIRAULT & L. J. EMORINE. 1992. Dexamethasone down-regulates β3-adrenergic receptors in 3T3-F442A adipocytes. J. Biol. Chem. **267:** 15909–15915.

27. KRIEF, S., B. FÈVE, B. BAUDE, V. ZILBERFARB, A. D. STROSBERG, J. PAIRAULT & L. J. EMORINE. 1994. Transcriptional modulation by n-butyric acid of β1-, β2, and β3-adrenergic receptor balance in 3T3-F442A adipocytes. J. Biol. Chem. **269:** 6664–6670.

28. FÈVE, B., K. ELHADRI, A. QUIGNARD-BOULANGÉ & J. PAIRAULT. 1994. Transcriptional down-regulation by insulin of the β3-adrenergic receptor expression in 3T3-F442A adipocytes: A novel mechanism for repressing the cAMP signalling pathway. Proc. Natl. Acad. Sci. USA **91:** 5677–5681.

Inverse Regulation of Hepatic α_{1B}- and β_2-Adrenergic Receptors

Cellular Mechanisms and Physiological Implications

GEORGE KUNOS, EDWARD J. N. ISHAC, BIN GAO,
AND LIU JIANG

Departments of Pharmacology & Toxicology and Medicine
Medical College of Virginia
Virginia Commonwealth University
Richmond, Virginia 23298

Catecholamines control a wide variety of metabolic processes in the liver, including key steps in carbohydrate, lipid, and amino acid metabolism. Of the multiple subtypes of adrenergic receptors (AR) that have been identified by pharmacological means as well as by molecular cloning, two subtypes have major roles in the control of hepatic functions in the rat: calcium-linked α_{1B}AR and cAMP-linked β_2AR. The most extensively studied metabolic effect of catecholamines in the liver is glycogenolysis and the subsequent release of glucose, brought about by activation of the rate-limiting enzyme, glycogen phosphorylase. Phosphorylase activation by catecholamines can occur both through α_{1B}AR and β_2AR.[1]

In addition to their short-term metabolic effects, catecholamines also influence hepatocyte growth and differentiation: they increase hepatic DNA synthesis and are involved in the early phases of the regenerative response after hepatic injury or partial hepatectomy.[2] The mammalian liver displays unusual plasticity in that it can fully regenerate after extensive tissue loss or injury. This regenerative response is preceded by a temporary dedifferentiation of liver cells, characterized by the loss of liver-specific functions and gene products, such as the synthesis of albumin and transferrin, and the parallel rapid emergence of growth-related gene products, such as the protooncogenes c-*myc*, c-*jun*, c-*fos*, and h-*ras*.[3] Adrenergic receptors also display a unique form of plasticity, best exemplified by studies of hepatic glycogenolysis in the rat. Although in the normal, adult, male rat this response is mediated exclusively by α_{1B}AR, after partial hepatectomy the same response is rapidly converted to a predominantly β_2AR-mediated event.[4] Interestingly, a similar conversion from α_1- to β_2-type response occurs in a number of other conditions, including glucocorticoid deficiency,[5] hypothyroidism,[6] toxin-induced liver regeneration,[7] malignant transformation,[8] cholestasis,[4] fetal versus adult state,[9] and dissociation of hepatocytes by enzymatic digestion.[10–14] In many of these conditions, a corresponding decrease in the expression of α_{1B}AR and increase in the expression of β_2AR have also been noted.[4,6,12–17] Furthermore, the effects of other glycogenolytic hormones acting through calcium, such as vasopressin, or acting though cAMP, such as glucagon, were either unaffected or changed in a different direction than the corresponding AR response.[11] This strongly suggests that the conversion from α_{1B}- to β_2-adrenergic control of glycogenolysis is related to corresponding inverse changes in the expression of the α_{1B}AR and β_2AR genes,[16,17] although it is clear that additional mechanisms, such as selective changes in the coupling of α_{1B}AR and β_2AR to their respective G-proteins,[1,10] or changes in G-protein expression[18] are also involved. Because cellular dedifferentiation appears to be a common denominator among the

261

conditions associated with a switch from α_1- to β_2-adrenergic glycogenolysis, there may be common underlying mechanisms. Acute dissociation of liver cells by enzymatic digestion and partial hepatectomy both result in very rapid changes in the AR response[4,10,11] and receptor gene expression,[12–14,17] and they also share the pattern of altered expression of other affected genes, such as various protooncogenes.[3,19] We used these two models in further studies to explore the underlying mechanisms.

MECHANISMS INVOLVED IN THE RAPID CONVERSION FROM α_1- TO β_2-ADRENERGIC GLYCOGENOLYSIS IN ACUTELY DISSOCIATED HEPATOCYTES

Role of Cyclooxygenase Products

Hepatocytes were isolated from adult, male rats by a collagenase perfusion protocol and were maintained in serum-free Krebs buffer containing 1.5% gelatin for improved viability. FIGURE 1 illustrates the typical α_1-adrenergic response pattern of freshly isolated cells and β_2-adrenergic response pattern of cells preincubated for 4 h, by the time-dependent decrease in the effectiveness of the α_1-agonist, phenylephrine, and the parallel emergence of an effect of the β-agonist, isoproterenol, on phosphorylase a activity. The effects of dibutyryl cAMP and of the calcium ionophore, A23187, were unaffected, indicating that the inverse changes in the α_1- and β-adrenergic response must have occurred before the generation of the second messengers cAMP and calcium, respectively. Whereas prolonged *in vitro* incubation of hepatocytes was shown to be associated with up-regulation of β_2AR and down-regulation of α_1AR,[12–14] the altered adrenergic activation of phosphorylase develops faster than the corresponding changes in receptor density. This suggests that, at least initially, the altered response must be due to inverse changes in the coupling of receptors to their respective post-receptor pathways.

Glucocorticoids are thought to produce most of their biological effects by inhibition of the breakdown of membrane phospholipids via phospholipase A_2 (PLA$_2$) and the subsequent generation of arachidonic acid (AA) metabolites, predominantly through the cyclooxygenase pathway. Inasmuch as glucocorticoid deficiency is one of the conditions associated with a conversion from α_1- to β_2-adrenergic glycogenolysis in rat liver,[5] we hypothesized that increased activity of the PLA$_2$/cyclooxygenase system is a common pathway involved in the α/β change induced by various stimuli, including the acute dissociation of hepatocytes. Evidence in support of this hypothesis was provided by experiments using a "lipid trap" paradigm.[20] Defatted bovine serum albumin (BSA), which avidly binds fatty acids, can be used to trap fatty acids released by cells, whereas regular BSA, which is saturated with fatty acids, has no such effect. Hepatocytes incubated in their regular Krebs medium, or in medium in which 0.5% gelatin was replaced with 0.5% regular BSA, displayed a similar switch from α_1- to β-adrenergic glycogenolysis after *in vitro* incubation for 4 h.[20] However, when the medium contained 0.5% defatted BSA, the cells that had been incubated for 4 h retained the α_1-adrenergic response pattern observed in the freshly isolated cells.[20] Because the predominant fatty acid in the sn-2 position of membrane phospholipids is AA, we tested the effect of exogenous AA on the adrenergic activation of liver glycogen phosphorylase. A 20-min exposure of freshly isolated hepatocytes to 10 μM AA, but not to stearic or palmitic acids, caused an acute shift in the receptor response from α_1- to mixed α_1/β-type, and this change could be prevented by simultaneous exposure of the cells to the cyclooxygenase inhibitor, ibuprofen, but not the lipoxygenase inhibitor, nordihydroguaiaretic acid.[20]

Ibuprofen also prevented the time-dependent shift of the receptor response from α_1- to β-type.[20] Incubation of hepatocytes for 4 h with actinomycin D, which blocks the time-dependent conversion of the AR response, does not prevent the similar but more acute change caused by exogenous AA.[20] The rapid onset of the effects of exogenous AA and its independence from mRNA synthesis suggest that it is the coupling of AR rather than their expression that may be regulated by an AA metabolite. The more pronounced effects of AA and ibuprofen on the β- than on the α_1-adrenergic response[1,20] suggest that the primary target is the β_2AR system. In agreement with this possibility, we found that hepatocytes from rats raised on an essential fatty acid (EFA)-free diet, which have markedly reduced AA and linoleic acid contents, fail to develop a β-adrenergic response on prolonged *in vitro* incubation.[21]

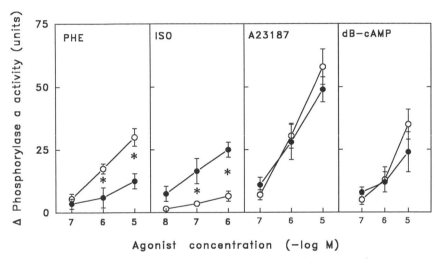

FIGURE 1. The effect of prolonged (4 h) incubation of isolated rat hepatocytes on the glycogenolytic response to various agonists. Phosphorylase *a* activity was determined[6] in aliquots of freshly isolated cells (*open circles*) or cells preincubated for 4 h (*filled circles*) in the absence of drugs or after a 3-min exposure to the indicated concentrations of phenylephrine (PHE), isoproterenol (ISO), the calcium ionophore A23187, or dibutyryl cAMP (dB-cAMP). Means ± SE from five experiments are shown. Asterisk indicates significant difference between corresponding 0-h and 4-h values ($p < 0.05$). Baseline phosphorylase *a* activity was 20–25 units.[6]

Role of Protein Kinase C

Protein kinase C (PKC) plays a key role in signal transduction as well as in cell proliferation.[22] Activation of PKC by phorbol esters inhibits differentiation and promotes growth in various tissues, including the liver where regeneration after partial hepatectomy is associated with activation of PKC.[23] Acute exposure of rat hepatocytes was also shown to selectively inhibit α_1-receptor-mediated glycogenolysis,[24,25] probably due to phosphorylation of the α_{1B}AR,[25] and in certain cell types phorbol esters potentiate βAR-mediated cAMP accumulation.[26,27] These findings

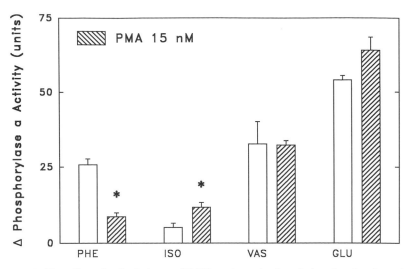

FIGURE 2. The effect of a phorbol ester (PMA) on the activation of phosphorylase by various glycogenolytic agents in freshly isolated hepatocytes. Cells were incubated for 10 min with vehicle (*open columns*) or 15 nM PMA (*cross-hatched columns*), and then exposed to vehicle (baseline) or to maximally effective concentrations of phenylephrine (10 μM), isoproterenol (1 μM), vasopressin (10 nM) or glucagon (10 nM). Means ± SE from six experiments are shown.

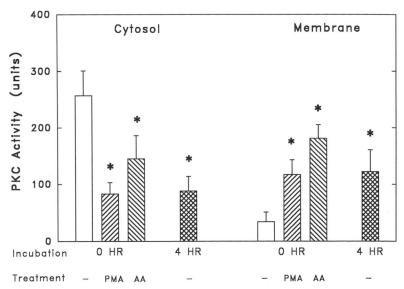

FIGURE 3. Drug-induced or time-dependent translocation of protein kinase C (PKC) activity in isolated hepatocytes. PKC activity was determined[29] in cytosol and plasma membrane fractions prepared from freshly isolated cells (0 h) or cells preincubated for 4 h. Zero-hour cells were exposed for 10 min to vehicle only, 15 nM PMA or 10 μM AA before preparation of the subcellular fractions. Four-hour cells were exposed to vehicle only. Asterisk indicates significant difference from corresponding value in untreated 0-h cells ($p < 0.05$). Columns and bars represent means + SE ($n = 4$).

could suggest that activation of PKC is also involved in the conversion of the adrenergic activation of phosphorylase from an $\alpha_1 AR$- to a $\beta_2 AR$-mediated event. The following observations support this possibility.

Effects of phorbol esters. Exposure of freshly isolated hepatocytes to 5 nM of phorbol 12-myristate, 13-acetate (PMA) for 4 min resulted in a marked reduction in the effect of phenylephrine, and a small but significant increase in the effect of isoproterenol on phosphorylase *a* activity, with no change in the effects of vasopressin or glucagon (FIG. 2). Exposure to phorbol 12-monoacetate, which does not activate PKC, had no effect on any of these drug responses. Experiments with different concentrations of PMA indicated that the EC_{50} of PMA for both the increase in the isoproterenol response and the decrease in the phenylephrine response was 3 nM, which is close to the K_d of PMA for PKC.[28]

Time-dependent activation of PKC in isolated hepatocytes. It is well established that activation of PKC in various cell types is associated with its translocation from the cytosol to the membrane. We quantified PKC activity in cytosol and membrane

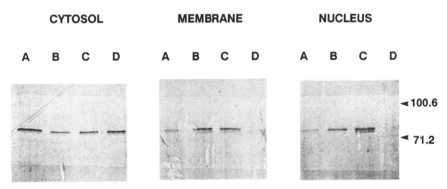

FIGURE 4. Translocation of PKC from cytosol to membrane and nucleus induced by PMA or by 4-h incubation. Detergent-solubilized subcellular fractions were prepared from cells pretreated as indicated, and were size-fractionated by electrophoresis on a 8% polyacrylamide gel, transferred to nitrocellulose, and blotted with a nonsubtype selective polyclonal antibody against PKC (UBI). Bands were visualized by biotinylated streptavidin. Band **A**, 0-h control; **B**, 0 h + PMA; **C**, 4-h control; **D**, 4 h + calphostin. For further explanation, see text.

fractions prepared from isolated hepatocytes, by measuring histone phosphorylation in the presence and absence of added phospholipids.[29] As illustrated in FIG. 3, in freshly isolated cells most of the activity was in the cytosol and very little in the membrane fraction. A 20-min exposure of these cells to 15 nM PMA resulted in a significant decrease in PKC activity in the cytosolic fraction and an increase in the membrane fraction, and a similar translocation occurred in cells incubated for 4 h in the absence of PMA. Interestingly, a similar cytosol-to-membrane translocation of PKC was induced by exposure of freshly isolated hepatocytes for 20 min to 10 μM AA (FIG. 3). In these experiments, the increase in PKC activity in the membrane fraction was less than the decrease in the cytosolic fraction, suggesting that translocation may occur to additional sites, such as the nucleus. We examined the presence of immunoreactive PKC in cytosolic, membrane, and nuclear fractions by Western blotting, using a nonisoform selective polyclonal antibody against PKC. FIGURE 4 illustrates the results of an experiment with subcellular fractions prepared from freshly isolated hepatocytes exposed for 20 min to vehicle (A lanes) or to 15 nM

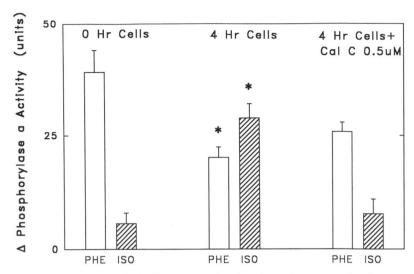

FIGURE 5. Calphostin C (Cal C) prevents the time-dependent conversion from α_1- to β_2-adrenergic glycogenolysis in isolated hepatocytes. Columns and bars (means + SE, $n = 5$) represent the increase in phosphorylase a activity caused by 10 μM phenylephrine or 1 μM isoproterenol under the indicated conditions.

PMA (B lanes), and from cells preincubated for 4 h with no drug added (C lanes) or incubated in the presence of 0.5 μM calphostin C, a potent and selective inhibitor of PKC (D lanes). As seen in FIGURE 4, exposure of freshly isolated cells to PMA, or their *in vitro* incubation for 4 h without PMA, caused similar decreases in PKC in the cytosolic fraction and increases in both the membrane and nuclear fractions. The presence of calphostin throughout the 4-h incubation prevented the translocation of PKC to either membrane or nucleus.

Effects of calphostin C. Calphostin C not only prevented the time-dependent translocation of PKC, as evidenced by Western blots (see FIG. 4), but also prevented the parallel decrease in the α-adrenergic and increase in the β-adrenergic activation of phosphorylase (FIG. 5).

Together these observations strongly suggest that loss of cell-to-cell contact in the liver leads to activation of PKC, resulting in its translocation to both the plasma membrane and the nucleus, which is involved in the conversion from α_1- to β_2-adrenergic glycogenolysis.

TRANSCRIPTIONAL REGULATION OF HEPATIC α_{1B}-ADRENERGIC RECEPTORS

The rapidity with which changes in the adrenergic receptor response develop upon activation of PKC or of the AA pathway suggests that these changes occur at the level of the coupling of the receptors to their respective signal transduction pathways. However, primary culturing of hepatocytes also leads to corresponding, although more slowly developing, inverse changes in the expression of α_1AR and β_2AR.[12-14] In order to obtain a more direct indicator of receptor synthesis, we quantified the steady-state levels of $\alpha_{1B}AR$ and β_2AR mRNAs in total RNA extracted from 0-h and 4-h cells. Of the three α_1AR subtypes cloned to date, only the $\alpha_{1B}AR$ has been found to be expressed in the rat liver.[30] mRNA levels were measured by highly sensitive DNA excess solution hybridization assays with sensitivity limits in the range of 0.1 amol mRNA/μg total RNA.[16,17] A progressive decrease in $\alpha_{1B}AR$ mRNA and a parallel increase in β_2AR mRNA were detected which reached statistical significance by the second to third hour of incubation.[17] FIGURE 6 illustrates a similar inverse change, that is, a decrease in $\alpha_{1B}AR$ mRNA and an increase in β_2AR mRNA induced by partial hepatectomy, as detected in a Northern blot of poly A^+ RNA prepared from the livers of sham-operated animals, or from the residual liver tissue within 2 or 6 h of 2/3 partial hepatectomy. The $\alpha_{1B}AR$ and β_2AR mRNAs were detected in mRNA preparations obtained from the same livers. FIGURE 6 also illustrates the well-documented mRNA heterogeneity for both receptors. The $\alpha_{1B}AR$ has three mRNAs: a major 2.7 kb and two minor bands at 3.3

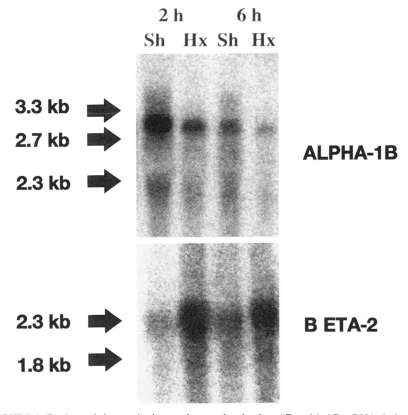

FIGURE 6. Reciprocal changes in the steady-state levels of $\alpha_{1B}AR$ and β_2AR mRNAs induced by partial hepatectomy. Northern blots of poly A^+ RNA prepared from the liver of sham-operated (Sh) or 2/3 hepatectomized (Hx) male Sprague-Dawley rats (200 g) 2 or 6 h following surgery were hybridized with [^{35}S]labeled cDNA probes for $\alpha_{1B}AR$ and β_2AR. Bands were visualized by a phosphorimager. Note the Hx-induced opposite changes in the $\alpha_{1B}AR$ and β_2AR message in the same RNA preparations.

and 2.3 kb,[31] of which the 3.3 kb species is only expressed in liver.[32] Nuclear run-on assays indicate that these inverse changes in steady-state mRNA levels can be accounted for by a corresponding increase or decrease in the rate of transcription of the β_2AR or α_{1B}AR genes, respectively (not shown).

Once it is clear that a major mechanism by which receptor expression is regulated is at the level of the transcription of the receptor gene, it is apparent that further analysis requires a detailed characterization of the regulatory domains of the gene and of the *trans*-acting factors that interact with these domains to direct or to modulate the rate of transcription. As a first step towards this goal, we isolated and sequenced the gene encoding the rat α_{1B}AR and characterized its 5'-flanking region.[33] Unlike the intronless genes encoding other adrenergic receptors cloned to date, the rat α_{1B}AR has a large intron (> 16 kb) interrupting its coding region. A similar feature is evident in the human α_{1B}AR gene, published at the time this work was completed.[34] Analysis of the sequence of the first 1000 bp immediately upstream from the coding region indicates the absence of TATA and CCAAT boxes and G + C rich regions, features characteristic of housekeeping genes.[33] The 5'-flanking region also contains consensus sequences that are recognized response elements for various *trans*-acting factors, such as AP1, cAMP (CRE), glucocorticoid receptor (GRE), and thyroid hormone receptor (TRE),[33] which are probably involved in the well-documented regulation of α_{1B}AR by PKC,[35] cAMP,[36] corticosteroids,[37] and thyroid hormones,[6,15,16] respectively. Primer extension analyses using 5' upstream primers identified transcription start points at -54 and -57 bp (tsp1), -443 bp (tsp2), and a cluster between -1035 and -1340 bp (tsp3).[31] Further analysis by transient transfections of putative promoter/CAT constructs revealed that the α_{1B}AR gene has three independent promoters, as illustrated in FIGURE 7.[31] The promoters are located at -49 to -127 bp (P1), -432 to -813 bp (P2), and -1107 to -1363 bp (P3) upstream from the start codon, and they direct transcription from tsp1, tsp2, and tsp3, to generate mRNA species of 2.3, 2.7, and 3.3 kb in length, respectively.[31] Analysis of the structure of these promoters indicate that P1 and P2 have features of housekeeping-like promoters, whereas P3, which is responsible for the liver-specific expression of the 3.3 kb mRNA species,[32] has both a TATA and a CCAAT box and is flanked by recognition sites for liver specific transcription factors, such as the CCAAT/enhancer binding protein and hepatocyte nuclear factor-5.[31] These findings suggest that differential control of these promoters may underlie the well-documented developmental and tissue-specific regulation of the α_{1B}AR. Studies are in progress to identify the regulatory domains of the β_2AR and the *trans*-acting factors that bind to these domains to control transcription of the β_2AR gene. If liver-specific factors are found that are involved in the transcriptional regulation of both the α_{1B}AR and the β_2AR, the possible role of such factors in the inverse regulation of these two receptors will be further explored.

POSSIBLE BIOLOGICAL SIGNIFICANCE

Inverse regulation of α_{1B}AR and β_2AR has been observed in a number of different physiological and pathological conditions (see introduction). The remarkable similarity of the altered receptor response pattern under these various conditions suggests that it represents a unique form of receptor regulation. Furthermore, the close parallel between conversion of the AR response and a shift from liver-specific to growth-related functions suggests a role in the process of liver cell growth and differentiation. Whereas the activation of glycogenolysis by $\alpha_{1B}AR$ or β_2AR is a unidirectional response, this is not the case for effects on liver cell proliferation. $\alpha_{1B}AR$ are mitogenic, increase DNA synthesis, and their inhibition can prevent liver

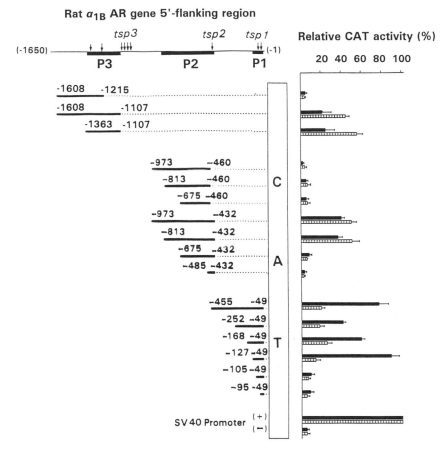

FIGURE 7. Characterization of three promoters of the $\alpha_{1B}AR$ receptor gene. The left side is a schematic representation of the pCAT constructs used in cell transfection experiments; the right side shows CAT activity in Hep3B (*solid bars*) and in DDT₁ MF-2 cells (*cross-hatched bars*), expressed as % of positive control. CAT activities (means ± SE, $n = 3$) are corrected for transfection efficiency, as described.[31] The positions of the three promoters (*horizontal bars*) and 3 tsp (*arrows*) are indicated on the line representing the 5′-flanking region. (From Gao and Kunos.[31] Reproduced, with permission, from the *Journal of Biological Chemistry*.)

regeneration after partial hepatectomy.[2] In contrast, in proliferating liver cells, stimulation of β_2AR strongly inhibits the G1-S transition,[38] and thus may be involved in the termination of the proliferative process. This, and the reported shift from α_1AR to β_2AR in human hepatocellular carcinoma,[39] could suggest that this form of regulation is important in terminating hepatocyte proliferation in order to prevent its potential progression into malignant transformation.

REFERENCES

1. KUNOS, G. & E. J. N. ISHAC. 1987. Biochem. Pharmacol. **36:** 1185–1191.
2. MICHALOPOULOS, G. K. 1990. FASEB J. **4:** 176–181.
3. FAUSTO, N. & J. E. MEAD. 1989. Lab. Invest. **60:** 4–13.
4. AGGERBECK, M., N. FERRY, E. S. ZAFRANI, M. C. BILLON, R. BAROUKI & J. HANOUNE. 1983. J. Clin. Invest. **71:** 476–486.
5. CHAN, T. M., P. F. BLACKMORE, K. E. STEINER & J. H. EXTON. 1979. J. Biol. Chem. **254:** 2428–2433.
6. PREIKSAITIS, H. G., W. H. KAN & G. KUNOS. 1982. J. Biol. Chem. **257:** 4321–4327.
7. GARC[DI]A-SÁINZ, J. A. & A. NÁJERA-ALVARADO. 1986. Biochem. Biophys. Acta **885:** 102–109.
8. CHRISTOFFERSEN, T. & T. BERG. 1975. Biochem. Biophys. Acta **381:** 72–77.
9. BLAIR, J. B., M. E. JAMES & J. L. FOSTER. 1979. J. Biol. Chem. **254:** 7579–7584.
10. ITOH, H., F. OKAJIMA & M. UI. 1984. J. Biol. Chem. **259:** 15464–15473.
11. KUNOS, G., F. HIRATA, E. J. N. ISHAC & L. TCHAKAROV. 1984. Proc. Natl. Acad. Sci. USA **81:** 6178–6182.
12. NAKAMURA, T., A. TOMOMURA, S. KATO, C. NODA & A. ICHIHARA. 1984. J. Biochem. **96:** 127–136.
13. SCHWARTZ, K. R., S. M. LANIER, E. A. CARTER, C. J. HOMCY & R. M. GRAHAM. 1985. Mol. Pharmacol. **27:** 200–209.
14. SANDNES, D., T. E. SAND, G. SAGER, G. O. BRNSTAD, M. R. REFSNES, I. P. GLADHAUG, S. JACOBSEN & T. CHROSTOFFERSEN. 1986. Exp. Cell Res. **165:** 117–126.
15. MALBON, C. C. 1980. J. Biol. Chem. **255:** 8692–8699.
16. LAZAR-WESLEY, E., J. R. HADCOCK, C. C. MALBON, G. KUNOS & E. J. N. ISHAC. 1991. Endocrinology **129:** 1116–1119.
17. ISHAC, E. J. N., E. LAZAR-WESLEY & G. KUNOS. 1992. J. Cell. Physiol. **152:** 79–86.
18. RAPIEJKO, P. J., D. C. WATKINS, M. ROS & C. C. MALBON. 1989. J. Biol. Chem. **264:** 16183–16189.
19. KRUIJER, W., H. SKELLY, F. BOTTERI, H. VAN DEN PUTTEN, J. R. BARBER, I. M. VERMA & H. L. LEFFERT. 1986. J. Biol. Chem. **261:** 7929–7936.
20. ISHAC, E. J. M. & G. KUNOS. 1986. Proc. Natl. Acad. Sci. USA **83:** 53–57.
21. GROJEC, M. S., E. J. N. ISHAC, J. KAPOCSI & G. KUNOS. 1990. Arch. Biochem. Biophys. **283:** 34–39.
22. NISHIZUKA, Y. 1988. Nature **334:** 661–665.
23. BUCKLEY, A. R., C. W. PUTNAM, R. EVANS, H. E. LAIRD, G. N. SHAH, D. W. MONTGOMERY & D. H. RUSSELL. 1987. Life Sci. **41:** 2827–2834.
24. LYNCH, C. J., R. CHAREST, S. B. BOCCKINO, J. H. EXTON & P. F. BLACKMORE. 1985. J. Biol. Chem. **260:** 2844–2851.
25. LEEB-LUNDBERG, L. M. F., S. COTECCHIA, J. W. LOMASNEY, J. F. DEBERNARDIS, R. J. LEFKOWITZ & M. G. CARON. 1985. Proc. Natl. Acad. Sci. USA **82:** 5651–5655.
26. SUGDEN, D., J. VANECEK, D. C. KLEIN, T. P. THOMAS & W. B. ANDERSON. 1985. Nature **314:** 359–361.
27. CHAMBAUT-GUERIN, A. M. & P. THOMOPOULOS. 1987. Eur. J. Biochem. **170:** 381–387.
28. NISHIZUKA, Y. 1984. Nature **308:** 693–697.
29. KRAFT, A. S. & W. B. ANDERSON. 1983. J. Biol. Chem. **258:** 9178–9183.
30. LOMASNEY, J. W., S. COTECCHIA, R. J. LEFKOWITZ & M. G. CARON. 1991. Biochim. Biophys. Acta **1095:** 127–139.
31. GAO, B. & G. KUNOS. 1994. J. Biol. Chem. **269:** 15762–15767.
32. MCGEHEE, R. E., JR., S. P. ROSSBY & L. E. CORNETT. 1990. Mol. Cell. Endocrinol. **74:** 1–9.
33. GAO, B. & G. KUNOS. 1993. Gene **131:** 243–247.

34. RAMARAO, C. S., J. M. KINCAID-DENKER, D. M. PEREZ, R. J. GAIVIN, R. P. RIEK & R. M. GRAHAM. 1992. J. Biol. Chem. **267:** 21936–21945.
35. HU, Z. W., X. Y. SHI, M. SAKAUE & B. B. HOFFMAN. 1993. J. Biol. Chem. **268:** 3610–3615.
36. MORRIS, G. M., J. R. HADCOCK & C. C. MALBON. 1991. J. Biol. Chem. **266:** 2233–2238.
37. SAKAUE, M. & B. B. HOFFMAN. 1991. J. Clin. Invest. **88:** 385–389.
38. REFSNES, M. R., G. H. THORESEN, G. SANDNES, O. F. DAJANI, L. DAJANI & T. CHRISTOFFERSEN. 1992. J. Cell. Physiol. **125:** 164–171.
39. BEVILACQUA, M., G. NORBIATO, E. CHEBAT, G. BALDI, P. BERTORA, E. REGALIA, G. COLELLA, L. GENNARI & T. VAGO. 1991. Cancer **67:** 2543–2551.

Induced Hibernation
by α2-Adrenoceptor Agonists[a]

N. ERIC NAFTCHI

Laboratory of Biochemical Pharmacology
New York University Medical Center
New York, New York 10016

Guanabenz, 2,6-dichlorobenzylidene aminoguanidine, is an α2-adrenoceptor agonist and is considered to act similarly to clonidine with respect to side effects and efficacy.[1,2] Three adrenoceptor agonists—two commonly employed anesthetic agents and a minor tranquilizer—were compared in the rat for their effect on core temperature and behavior. At an ambient temperature of 26.7 °C, 24 Sprague-Dawley rats were injected intraperitoneally (i.p.) with guanabenz acetate (12 mg/kg), clonidine (0.3 mg/kg), pentobarbital (35 mg/kg), ketamine HCl (35 mg/kg), diazepam (6 mg/kg), guanabenz + diazepam, and guanabenz + pentobarbital at the same stated doses. Within 45 min, the mean core temperature of the animals dropped to 27.5, 34.5, 33.6, 35.1, 37.6, 30.7, and 29.3 °C, respectively. The patellar, urethroanal (bulbocavernosus), startle, and corneal reflexes were absent after 30 min in guanabenz, pentobarbital, ketamine, guanabenz plus pentobarbital treated animals, but they were present in clonidine and diazepam treated animals. In the rats treated with a combination of the drugs guanabenz and diazepam, the urethroanal and vestibular reflexes were absent. Other reflexes, that is, withdrawal, tail compression response, corneal, and righting were present in these animals. Those rats receiving clonidine showed signs of sedation but not anesthesia. Guanabenz (12 mg/kg) was administered i.p. to eight Sprague-Dawley rats. Immediately after injection of guanabenz or the vehicle (5% aqueous dextrose) into another group of eight animals, the rats were placed in a temperature-controlled room (4–5 °C). The brain temperature of the control rats did not drop appreciably. By comparison, the brain temperature of the guanabenz-treated rats dropped by 10 °C (from 37 °C to 27 °C) whereas colonic temperature dropped from 37 °C to 17 °C, representing a fall of 20 °C in core temperature (FIG. 1). Three sets of two rats each were kept in the state of hibernation for 3, 7, and 12 days, with repeated injections, without any untoward effect. One of the rats carried for 12 days, however, did not recover from anesthesia because of an overdose, extensive weight loss, or most probably because of both. The state of anesthesia/hypothermia induced by guanabenz was not reversed by naloxone, but was selectively reversed by yohimbine, tolazoline, and xanthines.

A group of four rats were administered i.p. guanabenz (12 mg/kg) at an ambient temperature of 21.3 °C. Their colonic, intraperitoneal, hepatic, and renal temperatures were compared with the vehicle-treated rats housed at a room temperature of 21.5 °C. The core temperature was measured at 5-min intervals, and that of the vital organs at 15 and 25 min post injection.

Stabilized at 25 min after guanabenz injection, the mean intraperitoneal temperature of 31.5 °C and that of the liver and kidneys (33.5 °C), respectively, were 9.5 °C

[a] This work was supported by Edmund A. Guggenheim, and Murray and Leonie Guggenheim Clinical Research Endowment Funds.

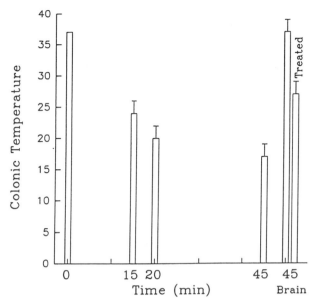

FIGURE 1. Guanabenz-induced hypothermia. Guanabenz (12 mg/kg) was administered i.p. to eight Sprague-Dawley rats. Immediately after injection of guanabenz or the vehicle (5% aqueous dextrose), the rats were placed in a temperature controlled room (4–5 °C). No change occurred in colonic and brain temperatures of the control rats. The brain temperature of guanabenz-treated rats, however, dropped by 10 °C (from 37 to 27 °C), and the colonic temperature dropped by 20 °C from 37 to 17 °C. Note that the onset of hypothermia is very rapid; the major drop in core temperature occurs within 15–20 min.

and 11.5 °C higher than the comparable mean colonic temperature of 22.5 °C for vehicle controls (TABLE 1).

Previous studies by Franz and co-workers[3,4] have shown that clonidine, an α2-adrenoceptor agonist, depresses transmission through sympathetic preganglionic neurons in the spinal cord, at least in part by reducing neuronal levels of cyclic AMP.

The state of anesthesia can be reversed by aminophyline, a cyclic-AMP phosphodiesterase inhibitor. Any agent that crosses the blood-brain barrier and increases cyclic-AMP levels, such as xanthines and β-adrenoceptor agonists, can reverse the state of anesthesia, but not that of hypothermia.

TABLE 1. Temperature Alterations after Treatment with Guanabenz[a]

Drug	Time (min)	Core	Intraperitoneal	Abdominal Cavity	Liver	Kidney
Guanabenz	15	22	31.5	31.5	33.5	33.5
	25	22.5	31.5	31.5	34.5	33.5
	40	22.6				
Vehicle (control)	15	35.5	37.0	37.8	38.2	38.2
	25	35.8	36.4	37.3	37.8	38.2
	40	35.6				

[a]Mean core temperature before injection was 35.5 ± 0.1 °C; ambient temperature, 21 °C.

The mechanism of action of α2-adrenoceptor agonists is believed to be stimulation of α2-adrenoceptors in the central nervous system. Stimulation of the receptors in the vasomotor center and the nucleus tractus solitarii decreases the sympathetic nerve impulses to the peripheral organs, such as the heart and blood vessels, with consequent reduction of peripheral vascular resistance and hypotension. Thus, the induced state of anesthesia and deep hypothermia mimics hibernation and is compatible with life; if an adequate nutritional balance is maintained, a mammal can be kept in a hibernation-like state for a long period of time without any untoward effect.

ACKNOWLEDGMENTS

This paper is dedicated to Jude Sleis for her years of impeccable service. Her untimely departure leaves a great void. The support and help of my colleagues, Drs. Abel Lajtha, Henry Sershen, and Audrey Hashim, are gratefully acknowledged.

REFERENCES

1. NAFTCHI, N. E. 1991. Int. J. Dev. Neurosci. **9:** 113–126.
2. NAFTCHI, N. E. 1982. Science **217:** 1042–1044.
3. FRANZ, D. A. & P. W. MADSEN. 1982. Eur. J. Pharmacol. **78:** 53–59.
4. FRANZ, D. A. & P. W. MADSEN. 1982. Neurosci. Lett. **28:** 211–216.

Effect of Hypothermia/Anesthesia Induced by α2-Adrenoceptor Agonist on Monoamine Turnover and Neurotensin Concentrations in the Rat Brain[a]

N. ERIC NAFTCHI,[b] HENRY SERSHEN,[c]
AUDREY HASHIM,[c] AND GARTH BISSETT[d]

[b]Laboratory of Biochemical Pharmacology
New York University Medical Center
New York, New York 10016

[c]The N.S. Kline Institute for Psychiatric Research
Center for Neurochemistry
Orangeburg, New York 10962

[d]Duke University Medical Center
Durham, North Carolina

Treatment of 13 Sprague-Dawley rats with guanabenz (4 mg/kg), an α-adrenoceptor agonist, significantly reduced neurotensin (NT) concentrations only in the caudate nucleus. The NT content of the hypothalamus, olfactory tubercles, and nucleus accumbens was also reduced, but not at the 95% significant level; that of the frontal cortex, preoptic nucleus, septum or ventral tegmental area/substantia nigra was unchanged. Twenty minutes after treatment with guanabenz (4 mg/kg) at a room temperature of 22.5–23.0 °C, the core temperature of the treated rats (warm treated, WT) was 34.7 ± 0.4 °C and that of the control rats (warm control, WC) was 36.7 ± 0.75 °C. The core temperature of the control animals (cold control, CC) in a temperature-controlled room (3.7 °C) was 35.4 ± 0.12 °C compared with 30.8 ± 0.3 °C for guanabenz-treated animals (cold treated, CT). In the cold-control animals, 5-hydroxyindoleacetic acid (5-HIAA) increased significantly in the frontal cortex, hippocampus, and corpus striatum (TABLE 1). The concentration of 5-HT was also significantly reduced in the striatum. These results indicate that, with the exception of the hippocampus, an increase occurs in the turnover of 5-HT in the frontal cortex and the corpus striatum due to cold stress. Dopamine turnover also increased significantly in the frontal cortex and the corpus striatum. The induced changes in 5-HT and dopamine turnover by cold stress were abolished by guanabenz treatment; no significant difference was found in the striatal dopamine and 5-HT among the vehicle-treated control rats in the room ambiance (WC) and rats treated in the cold (CT) temperatures (TABLE 1). Anterior hypothalamus showed no change either due to cold or to treatment with guanabenz in any of the neurotransmitters and their metabolites (TABLE 2).

The regions of caudate nucleus and preoptic hypothalamus where NT concentra-

[a]This work was supported by Edmund A. Guggenheim, and Murray and Leonie Guggenheim Clinical Research Endowment Funds.

TABLE 1. Effect of Guanabenz on Catecholamine Levels in the Brain

	5HIAA	5HT	DOPAC + HVA/DA	5HIAA/ 5HT
Frontal Cortex				
WC	1.95 ± 0.11	4.71 ± 0.20	0.69 ± 0.11	0.38
WT	1.58 ± 0.24	3.96 ± 0.50	1.02 ± 0.08	0.37
CC	3.13 ± 0.36^a	4.65 ± 0.53	1.09 ± 0.11^b	0.62^a
CT	1.93 ± 0.13^a	5.12 ± 0.53	0.71 ± 0.14^b	0.35
Hippocampus				
WC	2.42 ± 0.10	2.10 ± 0.11	0.54 ± 0.12	1.06
WT	2.80 ± 0.10	2.75 ± 0.12	0.55 ± 0.13	0.94
CC	4.43 ± 0.20^a	3.77 ± 0.99	0.59 ± 0.07	1.08
CT	3.76 ± 0.06^a	3.88 ± 0.21	0.72 ± 0.28	0.89
Striatum				
WC	3.04 ± 0.17	3.27 ± 0.29	0.19 ± 0.01	0.86
WT	3.54 ± 0.19	4.18 ± 0.16	0.17 ± 0.01	0.78
CC	4.00 ± 0.33^b	2.85 ± 0.27^b	0.25 ± 0.02^a	1.29^a
CT	2.70 ± 0.25^a	3.36 ± 0.53	0.17 ± 0.01^a	0.74

NOTE: Animals killed 20 min after injection. Values are expressed as ng/mg protein, mean ± SEM, n = 4–5; significant comparisons are between WC versus CC and CC versus CT.
$^a p < 0.01$ and $^b p < 0.05$.
Legend: Warm control (WC): Saline; room temperature, 22.5–23 °C; body temperature, 36.7 ± 0.75 °C. Warm treated (WT): Guanabenz, body temperature, 34.7 ± 0.04 °C. Cold control (CC): Saline, room temperature, 3.7 °C; body temperature 35.4 ± 0.12 °C. Cold treated (CT): Guanabenz, body temperature 30.8 ± 0.3 °C.

tions were significantly altered contain intrinsic NT neurons, as well as some terminals that project from other regions.[1,2] Neurotensin is known to be colocalized with dopamine in neurons of the ventral tegmental area that project to the nucleus accumbens and frontal cortex. The lack of significant NT changes in any of these regions, however, would argue against this source of NT being affected by the treatment with guanabenz. Administration of neuroleptic drugs is known to increase NT levels[3] and NT messenger RNA concentrations[4] in the caudate nucleus and nucleus accumbens. This effect of neuroleptics is still present after destruction of dopamine neurons innervating these regions.[5] Thus, guanabenz may be decreasing NT concentrations by affecting local circuit neurons containing NT in the caudate nucleus and preoptic hypothalamus.

The results demonstrate that cold stress induces a marked increase in the turnover of 5-HT in the frontal cortex, hippocampus, and corpus striatum; there is a decrease in the [5-HT/5-HIAA] ratio. With the exception of the hippocampus, the same results also apply to dopamine turnover; the ratio of [DOPAC + HVA)/DA]

TABLE 2. Effect of Anesthesia/Hypothermia Induced by Guanabenz-Acetate on Biogenic Amine Levels in Anterior Hypothalamus[a]

	DOPAC	DA	HVA	5HT	5HIAA	NE
Control	0.65 ± 0.2	4.5 ± 1.0	0.42 ± 0.07	9.8 ± 0.9	3.9 ± 0.4	13.5 ± 2.7
Cold Control	0.65 ± 0.2	4.8 ± 1.6	0.43 ± 0.08	8.2 ± 2.0	4.1 ± 1.0	14.4 ± 3.0
Treated	0.79 ± 0.3	6.3 ± 2.0	0.55 ± 0.16	10.2 ± 1.6	2.8 ± 0.5^b	13.8 ± 1.7

[a]Results expressed as ng/mg protein, mean ± SD.
$^b p < 0.05$.

increased significantly in the frontal cortex and the corpus striatum (TABLE 1). Alterations in 5-HT and dopamine turnover induced by cold stress are abolished by guanabenz treatment. Guanabenz, therefore, attenuates the increase in dopamine and 5-HT turnover induced by cold stress in the brain (TABLE 1 compares results in ambient temperature, WC versus CT).

Anterior hypothalamus showed no change in any of the neurotransmitters and their metabolites either due to cold or to treatment with guanabenz (4 mg/kg). In this discrete area of the brain, including the posterior hypothalamus, a significant reduction in 5-HIAA occurred, which may be a sign of decreased activity of membrane-bound intraneuronal monoamine oxidase, but not that of extraneuronal catechol-*O*-methyl transferase in this tissue (TABLE 2).

ACKNOWLEDGMENTS

This paper is dedicated to Jude Sleis for her years of impeccable service. Her untimely departure leaves a great void.

REFERENCES

1. EMSON, P. C., M. GOEDERT & P. W. MANTYH. 1985. *In* Handbook of Chemical Neuroanatomy, Vol. 4. A. Bjorkland & T. Hokfelt, Eds.: 355–405. Elsevier. Amsterdam.
2. ZAHMS, D. S. 1987. Neurosci. Lett. **81:** 41–47.
3. KILTS, C. D., C. M. ANDERSON, G. BISSETTE, T. D. ELY & C. B. NEMEROFF. 1988. Biochem. Pharmacol. **37:** 1547–1554.
4. MERCHANT, K. M., P. R. DOBNER & D. M. DORSA. 1992. J. Neurosci. **12:** 652–663.
5. BISSETTE, G., K. DOLE, M. JOHNSON, *et al.* 1988. Soc. Neurosci. Abstr. **14:** 1211.

Novel Dopamine Receptor Subtypes as Targets for Antipsychotic Drugs

PIERRE SOKOLOFF,[a] JORGE DIAZ,[b]
DANIEL LEVESQUE,[a] CATHERINE PILON,[a]
VIOLETTA DIMITRIADOU,[b] NATHALIE GRIFFON,[a]
CLAAS H. LAMMERS,[a] MARIE-PASCALE MARTRES,[a]
AND JEAN-CHARLES SCHWARTZ[a]

[a] Unité de Neurobiologie et de Pharmacologie de l'INSERM
Centre Paul Broca
2ter rue d'Alésia
75014 Paris, France

[b] Laboratoire de Physiologie
Université René Descartes
4 avenue de l'Observatoire
75006 Paris, France

Dopamine (DA) is an important neurotransmitter involved in diverse cerebral functions, among which those that control hormone secretion, emotions, and motor and motivated behaviors. The molecular diversity of dopamine receptors, recently revealed by the approaches of molecular biology, indicates that DA may mediate its various functions by interacting with at least five different DA receptors encoded by different genes. These receptors can be classified in D_1-like and D_2-like subfamilies according to primary sequence homology, gene organization, pharmacology, and, to some extent, intracellular signaling.[1-4] Thus, the D_1-like receptors—D_1 and D_5 receptors—are encoded by intronless genes and coupled to G_s-mediated activation of adenylyl cyclase. D_2-like receptors—D_2, D_3, and D_4 receptors—are encoded by genes with their coding sequence interrupted by introns and mediate G_i and G_o-mediated inhibitory responses, namely, inhibition of cAMP formation.[5-8]

Disturbances in DA neurotransmission have been implicated in several neuropsychiatric disorders, such as schizophrenia. Although it has been recognized that antipsychotic drugs primarily block DA receptors, it is unclear which precise target is involved. Moreover, the favorable therapeutic effect of currently used antipsychotic drugs is often impaired by severe motor and/or endocrine side effects, which limit their use. The possibility that blockade of the various DA receptors results in different clinical effects, for instance, favorable versus adverse effects, affords opportunities to select more efficient and safer drugs.

The DA receptor subtypes cannot be simply classified into two independent and functionally opposing receptor families, because they have been shown to participate in various reciprocal interactions. In spite of their opposite effects on cyclic AMP formation, D_1-like and D_2-like receptors display cooperative interactions that have been extensively illustrated in animal behavioral models.[9] More recently, functional communication at the molecular level between D_1 and D_2 receptors has been suggested,[10] possibly involving G-protein interactions.[11] Likewise, the D_1 receptor-induced facilitation of D_2 receptor-mediated arachidonate release[12] indicates that interacting intracellular signaling pathways may also account for synergism between

278

the receptors of the two subfamilies. On the other hand, D_2-like receptors cannot be regarded just as homologous and functionally related isoreceptors. We show here that D_2 and D_3 receptors, two likely targets for antipsychotic drugs, display distinct intracellular signaling pathways in a heterologous expression system and distinct modes of regulation in brain after interruptions of DA neurotransmission. Furthermore, D_2 and D_3 receptors blockaded by antipsychotic drugs have the opposite effects on neurotensin/neuromedin N expression in nucleus accumbens, a putative biochemical index of antipsychotic drug effects.[13-15] The therapeutical consequences of this dual interaction and differential regulation will be examined in the treatment of schizophrenia, a disorder characterized by the occurrence of both positive symptoms (hallucinations, delusions) and negative symptoms (impoverished thought and affect).

TABLE 1. Comparison of Dissociation Constants of Antipsychotic Drugs at Cloned Dopamine Receptor Subtypes[a]

Antipsychotic Agent	D_1-like Receptors		D_2-like Receptors		
	D_1 Receptor	D_5 Receptor	D_2 Receptor	D_3 Receptor	D_4 Receptor
Haloperidol	30	40	0.6	3	5
Chlorpromazine	16	33	2	6	37
Thioproperazine	–	–	0.5	1	50
Thioridazine	–	–	5	8	12
Pimozide	–	–	10	11	43
Sulpiride	40,000	80,000	10	20	1,000
Raclopride	10,000	–	2	4	1,500
Clozapine	140	250	70	300	9
(+) UH232	–	–	40	10	–

[a]Dissociation constants expressed in nM. Values taken from references 26–28, 30, 34, 36, and 76.

D_2 AND D_3 RECEPTORS AS COMMON TARGETS FOR ANTIPSYCHOTIC DRUGS

Before the advent of molecular biology, it had been generally assumed that neuroleptics derived their antipsychotic activity from the blockade of a single D_2 receptor.[16] Nevertheless, the recent discovery of several D_2-like receptor subtypes raises the possibility that the antipsychotic effects result in more discrete blockade of a particular subtype. We compared the binding data in the literature for several antipsychotic drugs used in clinical practice, at cloned DA receptors subtypes (TABLE 1). The low affinities at D_1-like receptors of several antipsychotic drugs, including the substituted benzamides sulpiride and raclopride and to a lesser extent haloperidol, indicate that blocking of these receptors is probably not achieved during treatment. In contrast, antipsychotic drugs have higher affinities at receptors of the D_2-like subfamily: D_2 and D_3 receptors appear to represent common targets for these drugs, with affinities in the nanomolar range for all compounds listed. This suggests that antipsychotic drugs produce their clinical effects primarily by blocking D_2 and D_3 receptors. In agreement, the study of brain of schizophrenic patients by positron emission tomography[17] indicates that antipsychotic drugs at clinically active dosages occupy the striatal D_2 receptor by 65–85%, a figure which is probably not very different for the D_3 receptor, given its similar pharmacological properties. In addition, both D_2 and D_3 receptors are highly expressed in brain limbic structures,[18] where DA is involved in various aspects of behavior, mood, and cognition through a feedback with cortical activities. Disturbances at this level may participate in the etiology of schizophrenia. However, the D_2 receptor, unlike the D_3 receptor, is also highly expressed in dorsal striatum, a region implicated in the control of motor activity and in the pituitary, where DA controls prolactin release. This suggests that blockade of the D_2 receptor also produces the motor and endocrine adverse effects of antipsychotic drugs, a drawback of present antipsychotic medication that would not meet D_3 receptor blockers. Clinical assessment of putative antipsychotic compounds with D_3-preferring affinity, such as (+) UH232 (TABLE 1), will allow us to evaluate this hypothesis.

The D_4 receptor also recognizes antipsychotic drugs with high affinity, but with a much higher variability. Particularly, it seems unlikely that the antipsychotic properties of raclopride[19–21] are due to the blockade of the D_4 receptor, for which this compound has a very low affinity.[22] Nonetheless, clozapine, an atypical antipsychotic drug that is relatively free of the adverse effects of drug-induced parkinsonism and tardive dyskinesia, binds to the D_4 receptor with an affinity at least 10 times higher than to other DA receptor subtypes. However, it should be noted that antipsychotic drugs, such as clozapine, have additional serotonergic,[23] muscarinic,[24] and α-1 adrenergic[25] properties, which may be responsible for peculiar pharmacological profiles and/or atypical properties. Hence, it has been suggested that atypical properties correlate with dopamine D_2/serotonin 5HT$_2$ receptor pKi ratios.[23] This latter hypothesis is not inconsistent with the concept of a specific involvement of D_2 or D_3 receptors in antipsychotic drug effects, if serotonin or another transmitter interacting with extrapyramidal systems may prevent the negative consequences of dopamine receptor blockade on motor control. It seems likely, indeed, that an atypical antipsychotic profile may be achieved in more than one way.

A FUNCTIONAL *IN VITRO* MODEL FOR D$_3$ RECEPTOR ACTIVATION

Heterologous cell expression systems allowed us to identify the second messenger pathways of DA receptor subtypes. The D_1[26–29] and D_5[30,31] receptors stimulate adenylyl cyclase, whereas D_2[5,6] and D_4[7] inhibit this enzyme activity. The D_2 receptor is functionally coupled to additional effector systems: it decreases Ca^{2+} influx by activating K^+ channels and activates phospholipase C in some cells,[6] but it inhibits this enzyme in other systems.[32] D_2 receptors also activate arachidonic acid release, provided that phospholipase A_2 is stimulated by raised intracellular Ca^{2+}.[12,33] All these effects are mediated via GTP-binding proteins (G-proteins) of the G_i/G_o group, which also regulate agonist binding. Binding at D_2 and D_4 receptors in membranes is generally described as occurring in two affinity states, the high-affinity state being converted into the low-affinity state by GTP.[5,34]

The evidence for such coupling of the D_3 receptor to G-proteins has long been lacking. In various transfected cells, including transfected fibroblasts such as Chinese hamster ovary cells (CHO), no[35] or little[8,36–38] GTP-induced shift in agonist affinity could be observed, as well as inconsistent inhibition of adenylyl cyclase[36,37] and weak activation of phospholipases.[12,38] The lack of indication of D_3 receptor coupling to conventional effector systems may seem paradoxical, given the sequence homology this receptor displays with the D_2 receptor in the third intracytoplasmic loop, a part of the receptor presumably implicated in coupling to G-proteins. One possible explanation is that the recipient cells used in previous studies may not be appropriate, either because the D_3 receptor is incorrectly processed or integrated in the membrane or because the cells do not express the adequate G-protein or effector system.

In view of the above considerations, we sought an appropriate recipient cell by transfecting a neuroblastoma-glioma NG 108-15 hybrid cell line,[39] the neuronal origin of which may afford a repertoire of effectors and G-proteins more extended than in the CHO cell line. Accordingly, in contrast with this latter cell line, the transfected NG108-15 cell line expresses a D_3 receptor that exists in two affinity states, interconverting by GTP analogs. The relevance of this observation to the physiological function of the D_3 receptor is indicated by the similar effects found under identical experimental conditions in membranes of lobules 9 and 10 of rat cerebellum, where a pure population of constitutive D_3 receptors can be studied.[40] Previous observations (P. Sokoloff and C. Pilon, unpublished results) indicated that

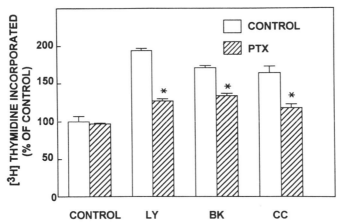

FIGURE 1. Effect of pertussis toxin (PTX) on stimulation of mitogenesis induced by quinpirole, bradykinin or carbamylcholine. Cells were pretreated for 24 h with pertussis toxin (200 ng/mL) stimulated by 0.1 μM quinpirole (LY), 1 μM bradykinin (BK) or 100 μM carbamylcholine (CC). Mitogenesis was evaluated by measuring [³H]thymidine incorporation. Results are expressed as percent of radioactivity incorporated in unstimulated cells. Asterisk denotes a significant difference ($p < 0.002$) in treated cells (*hatched columns*) versus untreated cells (*open columns*) by the Student's *t* test.

the effects of guanylnucleotides on agonist binding at D_3 receptors expressed by transfected CHO cells were enhanced by co-transfecting an α-subunit cDNA of G_o. This suggested that a G-protein of this type, constitutively expressed in the NG 108-15,[39] but not in the CHO cell line,[41] naturally couples to the D_3 receptor. Accordingly, D_3 receptor stimulation in transfected NG 108-15 increases mitogenesis through a pertussis toxin-sensitive mechanism (FIG. 1). Previous studies have indicated that a wide variety of G-protein-coupled receptors are able to induce mitogenesis.[42] Several concomitant and interacting mechanisms probably contribute to mitogenesis, which complicates the identification of the initial second messenger signaling pathways involved. Potent mitogenic factors either increase phosphatidylinositol turnover (see FIG. 2 for effects of bradykinin) or inhibit forskolin-stimulated cyclic AMP formation (see FIG. 2 for effects of carbamylcholine). The D_3 receptor-

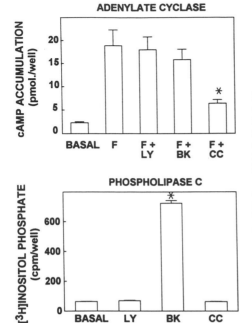

FIGURE 2. Effects of quinpirole, bradykinin or carbamylcholine on cyclic AMP (*top*) and inositol phosphate (*bottom*) accumulations. Cyclic AMP accumulation was measured in cells stimulated by 10 μM forskolin (F). LY, quinpirole, 0.1 μM; BK, bradykinin, 1 μM; CC, carbamylcholine, 100 μM. Asterisk denotes a significant difference ($p < 0.01$) versus forskolin alone (*top panel*) and ($p < 0.001$) versus basal level (*bottom panel*), by the Student's t test.

induced mitogenesis does not appear to result from increased production of phospholipase C-associated second messengers or inhibition of adenylyl cyclase (FIG. 2). Nevertheless, the increase of diacylglycerol, one of the phospholipase C products that activates protein kinases, may play a significant role in D_3 receptor-induced mitogenesis because a phorbol ester, which also activates protein kinases, potentiated the response.[39] Thus, the D_3 receptor produces mitogenesis by affecting a still unidentified pathway, through a pertussis toxin-sensitive mechanism.

This functional model now allows us to pharmacologically characterize a D_3 receptor-mediated response (TABLE 2). DA receptor agonists, some of which were previously classified as autoreceptor-selective agonists, appear as full D_3 receptor agonists with subnanomolar potencies; (+)UH 232, a partially selective D_3 receptor compound, which displays in animal models a peculiar behavioral pattern ascribed to selective autoreceptor blockade,[43] appears now as a reversible D_3 receptor antagonist.[39] Taken together, these pharmacological data support the previously suggested hypothesis[35,36] that some behavioral and biochemical actions of DA agonists in low dosages, which had been attributed to their selective interaction with DA autoreceptors, may actually involve the D_3 receptor. In addition, not only the D_3 receptor but also the D_2 receptor is able to enhance mitogenesis, which allowed us to compare the efficacies of various agonists at these two receptors in similar experimental conditions (TABLE 2). It is clear from this comparison that the D_3 receptor selectivity is much lower when biological activity is considered than in binding studies. For example, quinpirole and 7-hydroxy dipropyl aminotetralin (7OH-DPAT), which are 50 times more potent in binding studies, have only a 2–7 times higher potency in the functional models. This suggests that attempts to identify an *in vivo* D_3 receptor-mediated response by using only agonists that displayed D_3 receptor binding selectivity[44–46] should be considered with caution. It remains that the mitogenic response in transfected NG108-15 is a suitable model for identifying selective D_3 receptor agonists, which may prove useful in the characterization of the D_3 receptor function in the brain.

D_3 RECEPTOR-MEDIATED ACTIVATION OF THE c-*FOS* GENE IN TRANSFECTED CELLS

In transfected NG 108-15, the D_3 receptor stimulates the transcription of the proto-oncogene c-*fos,* as measured by the appearance of Fos immunoreactivity (FIG. 3). The c-*fos* gene, an immediate early gene rapidly and transiently expressed in a wide variety of cell types including neurons,[47,48] constitutes a marker for cell activity and is involved in cell differentiation-proliferation balance.[49-51] Activation of c-*fos* by a DA receptor has never been reported in transfected cells, even though c-*fos* is activated in brain upon administration of indirect DA agonists in normal animals[52,53] and of D_1 receptor agonists following 6-hydroxydopamine-induced denervation of the striatum.[54] Antagonists of D_2-like receptors, that is, D_2, D_3 or D_4 receptors, also activate c-*fos* transcription *in vivo,*[55-58] suggesting that a receptor of this type is tonically and negatively coupled to c-*fos* expression. D_2 receptor-mediated inhibition of both cAMP formation and of calcium channel activity presumably contributes to this negative coupling by counteracting the action of a calcium/cAMP-dependent responsive enhancer in the c-*fos* gene.[48] Such a process may be relevant for the therapeutic action of neuroleptics, because, although "typical" antipsychotic drugs induce c-*fos* in various parts of the striatal complex, including those involved in motor functions, the action of the "atypical" ones, such as clozapine and remoxipride, is

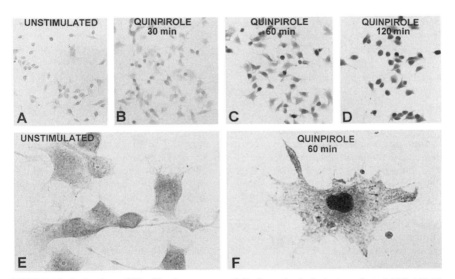

FIGURE 3. Induction of Fos-like immunoreactivity by quinpirole in transfected NG 108-15 cells. Monolayer cell preparations were incubated without (**A**) or with 0.1 μM quinpirole for 30 (**B**), 60 (**C**), or 120 min (**D**). The cells were stained with an anti-Fos polyclonal antibody. The intense coloration appearing after a 30-min stimulation by quinpirole is restricted to nucleus, which is shown in (**F**) at a higher magnification, whereas unstimulated cells display only a diffuse and weak coloration (**E**).

restricted to the limbic parts of the nucleus accumbens,[56-58] a region in which disturbances in DA neurotransmission have been implicated in schizophrenia.

OPPOSING ROLES FOR D_2 AND D_3 RECEPTORS ON NEUROTENSIN EXPRESSION IN NUCLEUS ACCUMBENS

In contrast with the D_2 receptor, the D_3 receptor is almost absent from the dorsal striatum; like the D_2 receptor, however, it is well expressed in the ventral striatum, particularly the nucleus accumbens.[18,40,59] We have compared the distributions of D_2 and D_3 receptor mRNAs in accumbal subterritories, namely, shell and core, known to display distinct cytochemical features, connections, and therefore functions.[60-62] At the level of the island of Calleja major, where shell and core subterritories can be easily distinguished, D_3 receptor mRNA was almost exclusively expressed in the shell, whereas the D_2 receptor mRNA was expressed in the core as well as in restricted parts of the shell (FIG. 4). As in other brain areas,[18] however, no overlap is found between D_2 and D_3 receptor mRNA distributions in the shell. The D_3 receptor mRNA is expressed in the ventromedial area of the shell (ShV), which also expresses neurotensin (NT) but not D_2 receptor mRNA, whereas the reverse situation exists in a more dorsal shell area called the "cone" (ShC in FIG. 4).

Since the distribution of D_3 receptor and NT/neuromedin N mRNAs within the ShV seemed to overlap, a possible colocalization of these markers was assessed in thin successive sections (FIG. 4D and E). Both mRNAs are coexpressed in a subpopulation of medium-sized neurons representing 42% of the NT neurons in the analyzed area. In comparison, D_3 receptor mRNA is expressed by 43% of the total cellular population of the same area. Because these values are inherently minimized, it can be safely concluded that a major proportion of NT neurons of this area of the shell, known to heavily project to the ventromedial pallidum, express the D_3 rather than the D_2 receptor.

In the absence of highly selective D_3 receptor ligands, the role of dopamine innervation on these neurons was assessed by simultaneous blockade of D_2 and D_3 receptors using haloperidol or sulpiride, two antipsychotic drugs with a D_2-like selectivity (FIG. 5), in rather high dosage. In agreement with previous studies[13-16,63] these agents induced a marked activation of NT gene expression in D_2-receptor rich areas, for example, the dorsal striatum (not shown) or shell cone of accumbens (FIG. 5). In contrast, both agents induced opposite changes in the ShV, an area selectively expressing D_3 receptors; therein the preexisting NT mRNA signal is markedly decreased. Thus, in subterritories of the shell, the D_3 receptor may exert a tonic stimulation of NT expression, an effect opposite to that exerted in other striatal divisions by the D_2 receptor. Nevertheless, the increase in striatal NT expression triggered by antipsychotic drugs may be the indirect result of an increase in DA neuron firing, mediated by D_2 autoreceptor blockade, which results in a higher availability of DA at D_1 receptors.[64] However, SCH 23390, a selective dopamine D_1-like receptor antagonist, do not induce any change in NT mRNA in any region of the nucleus accumbens (FIG. 4C), supporting the involvement of a D_2-like receptor in the regulation of NT expression in this region.

A D_3 receptor-mediated activation of NT gene transcription is consistent with the detection of spontaneously expressed transcripts in D_3 receptor-rich accumbal areas. In addition, it is also consistent with the observations that, in transfected NG-108-15 cells, the D_3 receptor promotes activation of the c-*fos* gene, whose product is known to activate transcription of the NT gene via binding to its AP1 site.[65]

REGULATION OF THE D₃ RECEPTOR AFTER INTERRUPTION OF DOPAMINERGIC TRANSMISSION

Prolonged interruption of DA neurotransmission in animals by chronic treatment with neuroleptics or by lesions of dopaminergic neurons results in behavioral

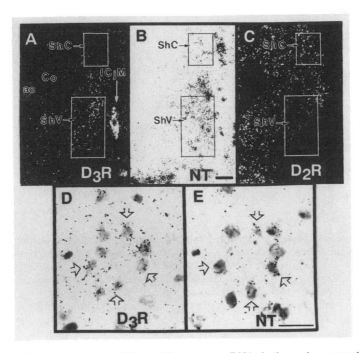

FIGURE 4. Expression patterns of D_2 and D_3 receptor mRNAs in the nucleus accumbens and comparison with neurotensin (NT) mRNA in control animals. Frontal sections performed at 1.2 mm to bregma (Paxinos and Watson, 1982) were hybridized with [^{35}S]labeled cRNA probes for D_3 receptor (**A**), NT (**B**) or D_2 receptor (**C**). Note that in the ventromedial part of the shell subdivision of the nucleus accumbens (area surrounded by a rectangle and subsequently referred to as ShV) the distribution of D_3 receptor mRNA matched the expression pattern of NT mRNA. In **A** and **C,** the microphotographs were obtained under darkfield illumination. On brightfield microphotographs of 3-μm sections hybridized with the D_3 receptor cRNA probe (**D**) and corresponding adjacent section hybridized with the NT cRNA probe (**E**) at the level of the ventromedial part of the shell, four of nine neurons present on both sections showed hybridization signals with both probes and are indicated by arrows. Ac, anterior commissura; Co, core subdivision of the nucleus accumbens; ICj M, island of Calleja Major; ShC, cone part of the shell subdivision; ShV, ventromedial part of the shell. Bars = 250 μm (**A–C**), 20 μm (**D, E**).

supersensitivity to DA agonists and increased number of receptors.[66] Elevated D_2 receptor mRNA[67–70] after chronic neuroleptic treatment is direct evidence that enhanced responsiveness via increase in DA receptor number results from increased rate of synthesis through activation of gene transcription. It is generally recognized, however, that no tolerance to the antipsychotic activity of neuroleptics develops in

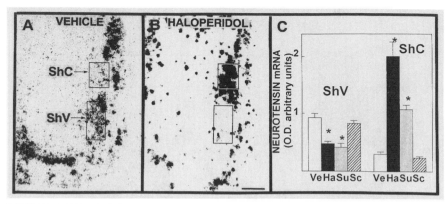

FIGURE 5. Effects of dopamine receptor antagonists on expression of NT mRNA. Frontal sections taken from a vehicle (**A**) or a haloperidol-treated animal (**B**) were hybridized with a neurotensin (NT) cRNA probe. Bar = 500 μm. In **C**, autoradiograms obtained as in **A** and **B** from 6–8 sections from groups of 3–5 rats receiving vehicle (Ve), haloperidol (Ha), sulpiride (Su) or SCH 23390 (Sc) were analyzed with an image analyzer (Biocom 2000, Les Ulis, France) in the ShV and the ShC subdivisions as delineated in **A**. *$p < 0.005$ as compared to vehicle-treated rats by the Mann-Whitney U test.

schizophrenic subjects,[71] whereas tolerance to the motor side effects is progressively setting in the course of the treatment. We compared the effects of dopaminergic neuron ablation and of a chronic haloperidol treatment on levels of D_2 and D_3 receptor binding and mRNA in the nucleus accumbens. After a two-week treatment, haloperidol enhanced both D_2 receptor mRNA and D_2 receptor binding in the nucleus accumbens. In contrast, neither D_3 receptor binding nor mRNA changed, suggesting that no up-regulation develops at this receptor. In addition, no tolerance to the effect of haloperidol on the activation of NT expression was seen (TABLE 3). This observation might be regarded as further support to the idea that the D_3 receptor is an important target for antipsychotic drugs. In contrast with their motor

TABLE 2. Comparison of Potencies of Dopamine and Agonists at D_2 and D_3 Receptors in Binding and Functional Studies

	Receptor Binding (K_i values, nM)			Stimulation of [³H]Thymidine Incorporation (EC_{50} values, nM)		
Agonist	D_2 Receptor	D_3 Receptor	$K_i(D_2)/$ $K_i(D_3)$	D_2 Receptor	D_3 Receptor	$EC_{50}(D_2)/$ $EC_{50}(D_3)$
Dopamine	544	23	24	20	1.4	15
Apomorphine	63	73	0.87	2.3	2.2	1.1
Quinpirole	1,400	39	36	2.8	0.86	3.3
(+)7OH-DPAT	103	2.1	49	2.7	0.39	7.0

NOTE: Data for binding at D_2 and D_3 receptors were taken from Sokoloff *et al.*[36] EC_{50} values for D_2 and D_3 receptor-mediated stimulation of mitogenesis, evaluated by measuring the incorporation of [³H]thymidine, were calculated from concentration-response curves obtained using Chinese hamster ovary cells transfected with the rat dopamine D_2 receptor cDNA and with NG 108-15 cells transfected with the human D_3 receptor cDNA, respectively.

side effects which progressively diminish during long-term treatments, their therapeutic effects do not show any impairment.

Interestingly, ablation of dopamine neurons by 6-hydroxydopamine injection (TABLE 3) or medial forebrain lesions (not shown) results in a dramatic but paradoxical decrease in D_3 receptor expression in the nucleus accumbens. This effect is not reproduced by blockade of D_1-like, D_2-like, and cholecystokinin receptors, suggesting that a messenger molecule released from catecholamine neurons—but distinct from dopamine or its co-transmitter[72] cholecystokinin—is necessary to maintain the expression of the D_3 receptor in the accumbens. This not only confirms that D_2 and D_3 receptors are opposites in terms of their roles and regulation, but also suggests that identification of this putative messenger molecule will constitute a heuristic research area in the pathophysiology of schizophrenia.

TABLE 3. Effects of Interruptions of Dopamine Neurotransmission on the Expression of D_2 and D_3 Receptors and Neurotensin in the Nucleus Accumbens

Treatment	D_2 Receptor		D_3 Receptor	
	mRNA	Binding	mRNA	Binding
6-OHDA	$+59 \pm 25^a$	$+42 \pm 6^a$	-52 ± 10^a	-54 ± 12^b
Haloperidol	$+61 \pm 12^b$	$+66 \pm 6^b$	-3 ± 20 ns	$+7 \pm 4$ ns

NOTE: In the lesion study, animals received a unilateral injection of 6-hydroxydopamine (6-OHDA) and were killed 3 weeks later. Treatment with haloperidol consisted of either twice daily injections of haloperidol (20 mg/kg) for 2 weeks when D_2 and D_3 receptor mRNA and binding were measured or a single injection when neurotensin mRNA was measured. Binding at D_2 and D_3 receptors was measured using [^{125}I]iodosulpride[74] and [^{3}H]7OH-DPAT,[42] respectively. mRNA levels were measured by quantitative PCR with internal standards.[75] Neurotensin mRNA was measured by analyzing with an image analyzer the autoradiograms obtained in *in situ* hybridization experiments. Data are percent changes over mean values obtained in contralateral side or in vehicle-treated animals.
$^a p < 0.05$ and $^b p < 0.01$ by the Mann-Whitney test; ns, not significant.

CONCLUSIONS

Neuroleptics share the common characteristic of being recognized by the DA D_2 and D_3 receptors, the blockade of which may, therefore, represent the primary mechanism for antipsychotic drug action. In addition, both D_2 and D_3 receptors are well expressed in brain limbic structures, such as the shell part of the nucleus accumbens, where DA neurotransmission is involved in various aspects of behavior, mood, and cognition through a feedback loop controlling cortical activities. Disturbances at this level may participate in the etiology of schizophrenia.

From functional studies, it appears, however, that D_2 and D_3 receptors act in opposite directions in the nucleus accumbens. The D_2 receptor exerts a tonic inhibition on c-*fos* and neurotensin gene transcriptions, whereas the D_3 receptor tonically stimulates NT expression and promotes c-*fos* expression, at least in transfected cells. That DA exerts opposite effects via D_2 and D_3 receptors is consistent with data of behavioral studies in rodents: agonists of D_2-like receptors (i.e., D_2, D_3 or D_4) induce contrasting locomotor manifestations when applied in different accumbal subterritories.[73] In addition, the partially selective D_3-antagonists AJ 76 and UH 232 trigger, in low dosage, paradoxical behavioral activations, originally attributed to autoreceptor blockade because they resemble those of D_2 agonists; at somewhat higher dosage, however, these experimental drugs induce the behavioral disruptions characteristic of neuroleptics, that is, preferential D_2 antagonists.[43]

The discovery of the dual effects of D_2 and D_3 receptors on accumbal neurons may be relevant to the treatment of schizophrenia. Neuroleptics are more effective in alleviating positive symptoms such as hallucinations and delusions, than negative symptoms such as impoverished thought and affect. The occurrence of these distinct, even opposite, manifestations, sometimes in the same patient, suggests that the antipsychotic drugs do not normalize a single overactive dopaminergic pathway. The D_2 preference of the drugs presently available might be responsible for the greater efficiency of these drugs against positive rather than negative symptoms, and, in addition, with the drug-induced deficient symptoms in patients with schizophrenia. The forthcoming introduction in clinics of first D_3 receptor selective antagonists may contribute to the improvement in the treatment of schizophrenia.

REFERENCES

1. SCHWARTZ, J.-C., B. GIROS, M.-P. MARTRES & P. SOKOLOFF. 1992. The dopamine receptor family: Molecular biology and pharmacology. Sem. Neurosci. **4:** 99–108.
2. CIVELLI, O., J. R. BUNZOW & D. K. GRANDY. 1993. Molecular diversity of dopamine receptors. Annu. Rev. Pharmacol. Toxicol. **33:** 281–307.
3. GINGRICH, J. A. & M. G. CARON. 1993. Recent advances in the molecular biology of dopamine receptors. Annu. Rev. Neurosci. **16:** 299–321.
4. SIBLEY, D. R. & F. J. MONSMA, JR. 1992. Molecular biology of dopamine receptors. Trends Pharmacol. Sci. **13:** 61–65.
5. NEVE, K. A., R. A. HENNINGEN, J. R. BUNZOW & O. CIVELLI. 1985. Functional characterization of a rat dopamine D_2 cDNA expressed in a mammalian cell line. Mol. Pharmacol. **36:** 446–451.
6. VALLAR, L., C. MUCA, M. MAGNI, P. ALBERT, J. BUNZOW, J. MELDOLESI & O. CIVELLI. 1990. Differential coupling of dopaminergic D_2 receptors expressed in different cell types. J. Biol. Chem. **265:** 10320–10326.
7. COHEN, A. I., R. D. TODD, S. HARMON & K. L. O'MALLEY. 1992. Photoreceptors of mouse retinas possess D_4 receptors coupled to adenylate cyclase. Proc. Natl. Acad. Sci. USA **89:** 12093–12097.
8. CHIO, C. L., M. E. LAJINESS & R. M. HUFF. 1984. Activation of heterologously expressed D_3 dopamine receptors: Comparison with D_2 dopamine receptors. Mol. Pharmacol. **45:** 51–60.
9. CLARK, D. & F. J. WHITE. 1987. D_1 dopamine receptor; the search for a function: A critical evaluation of the D_1/D_2 dopamine receptor classification and its functional implications. Synapse **1:** 347–388.
10. SEEMAN, P., H. B. NIZNIK, H.-C. GUAN, G. BOOTH & C. ULPIAN. 1989. Link between D_1 and D_2 dopamine receptors is reduced in schizophrenia and Huntington diseased brain. Proc. Natl. Acad. Sci. USA **86:** 10156–10160.
11. NIZNIK, H. B. 1994. Molecular mechanisms of dopamine D_1 and D_2 receptor interactions. In Schizophrenia, Dopamine Receptor Subtypes and Antipsychotics. J. Gerlach, Ed. Springer. Berlin. In press.
12. PIOMELLI, D., C. PILON, B. GIROS, P. SOKOLOFF, M.-P. MARTRES & J.-C. SCHWARTZ. 1991. Dopamine activation of the arachidonic acid cascade *via* a modulatory mechanism as a basis for D_1/D_2 receptor synergism. Nature **353:** 164–167.
13. KILTS, C. D., C. M. ANDERSON, G. BISSETTE, T. D. ELY & C. B. NEMEROFF. 1988. Differential effects of antipsychotic drugs on the neurotensin concentration of discrete rat brain nuclei. Biochem. Pharmacol. **37:** 1547–1554.
14. LEVANT, B., G. BISSETTE, E. WIDERLÖV & C. B. NEMEROFF. 1991. Alterations in regional brain neurotensin concentrations produced by atypical antipsychotic drugs. Regul. Pept. **32:** 193–201.

15. MERCHANT, K. M. & D. M. DORSA. 1993. Differential induction of neurotensin and c-fos gene expression by typical versus atypical antipsychotics. Proc. Natl. Acad. Sci. USA **90:** 3447–3451.
16. SEEMAN, P. 1980. Brain dopamine receptors. Pharmacol. Rev. **32:** 229–313.
17. FARDE, L., F. A. WIESEL, H. HALL, C. HALLDIN, S. STONE-ELANDER & G. SEDVALL. 1987. No D_2 receptor increase in PET study of schizophrenia. Arch. Gen. Psychiatry **44:** 671–672.
18. BOUTHENET, M.-L., E. SOUIL, M.-P. MARTRES, P. SOKOLOFF, B. GIROS & J.-C. SCHWARTZ. 1991. Localization of dopamine D_3 receptor mRNA in the rat brain using in situ hybridization histochemistry: Comparison with dopamine D_2 receptor mRNA. Brain Res. **564:** 203–219.
19. COOKSON, J. C., B. NATORF, N. HUNT, T. SILVESTONE & G. UPPFELDT. 1989. Efficacy, safety and tolerability of raclopride, a specific D_2 receptor blocker, in acute schizophrenia. Int. Clin. Psychopharmacol. **4:** 61–70.
20. HIRSCH, S. R., T. R. E. BARNES, M. DICKINSON, M. FERNANDEZ, A. JOLLEY, L. PUSAVAT, M. RICCIO, J. SPELLER, R. G. MCREADIE & M. STEWART. 1992. A double-blind comparison of raclopride and haloperidol in the acute phase of schizophrenia. Acta Psychiatr. Scand. **86:** 391–398.
21. FARDE, L., F. A. WIESEL, G. UPPFELDT, A. WAHLEN & G. SEDVALL. 1988. An open trial of raclopride in acute schizophrenia. Confirmation of D_2-dopamine occupancy by PET. Psychopharmacology **94:** 1–7.
22. SEEMAN, P., H. C. GUAN & H. H. M. VAN TOL. 1993. Dopamine D_4 receptors elevated in schizophrenia. Nature **365:** 441–445.
23. MELTZER, H. Y., S. MATSUBARA & J.-C. LEE. 1989. Classification of typical and atypical antipsychotic drugs on the basis of dopamine D_1, D_2 and serotonin$_2$ pK_i values. J. Pharmacol. Exp. Ther. **251:** 238–246.
24. MILLER, R. J. & C. R. HILEY. 1974. Anti-muscarinic properties of neuroleptics and drug-induced parkinsonism. Nature **248:** 596–597.
25. COHEN, B. M. & J. F. LIPINSKI. 1986. *In vivo* potencies of antipsychotic drugs in blocking alpha$_1$ noradrenergic and dopamine D_2 receptors: Implications for drug mechanisms of action. Life Sci. **39:** 2571–2580.
26. DEARRY, A., J. A. GINGRICH, P. FALARDEAU, R. T. FREMEAU, M. D. BATES & M. G. CARON. 1990. Molecular cloning and expression of the gene for a human D_1 dopamine receptor. Nature **347:** 72–76.
27. ZHOU, Q. Z., D. K. GRANDY, L. THAMBI, J. A. KUSHNER, H. H. M. VAN TOL, R. CONE, D. PRIBNOW, J. SALON, J. R. BUNZOW & O. CIVELLI. 1990. Cloning and expression of human and rat D_1 dopamine receptors. Nature **347:** 76–80.
28. SUNAHARA, R. K., H. B. NIZNIK, D. M. WEINER, T. M. STORMANN, M. R. BRANN, J. L. KENNEDY, J. E. GELERNTER, R. ROZMAHEL, Y. YANG, Y. ISRAEL, P. SEEMAN & B. F. O'DOWD. 1990. Human dopamine D_1 receptor encoded by an intronless gene on chromosome 5. Nature **347:** 80–83.
29. MONSMA, F. J., L. C. MAHAN, L. D. MCVITTIE, C. R. GERFEN & D. R. SIBLEY. 1990. Molecular cloning and expression of a D_1 dopamine receptor linked to adenylyl cyclase activation. Proc. Natl. Acad. Sci. USA **87:** 6723–6727.
30. SUNAHARA, R. K., H. C. GUAN, B. F. O'DOWD, P. SEEMAN, L. G. LAURIER, G. NG, S. R. GEORGE, J. TORCHIA, H. H. M. VAN TOL & H. B. NIZNIK. 1991. Cloning of the gene for a human dopamine D_5 receptor with higher affinity for dopamine than D_1. Nature **350:** 614–619.
31. TIBERI, M., K. R. JARVIE, C. SILVIA, P. FALARDEAU, J. A. GINGRICH, N. GODINOT, L. BERTRAND, T. L. YANG-FENG, R. T. FREMEAU JR. & M. G. CARON. 1991. Cloning, molecular characterization, and chromosomal assignment of a gene encoding a second D_1 dopamine receptor subtype: Differential expression pattern in rat brain compared with the D_{1A} receptor. Proc. Natl. Acad. Sci. USA **88:** 7491–7495.
32. ENJALBERT, A., F. SLADECZEK, G. GUILLON, P. BERTRAND, C. SHU, J. EPELBAUM, A. GARCIA-SAINZ, C. LOMBARD, C. KORDON & J. BOCKAERT. 1986. Angiotensin II and dopamine modulate both cAMP and inositol phosphate productions in anterior pituitary cells. J. Biol. Chem. **261:** 4071–4075.

33. KANTERMAN, R. Y., L. C. MAHAN, E. M. BRILEY, F. J. MONSMA, D. R. SIBLEY, J. AXELROD
    & C. C. FELDEK. 1991. Transfected D_2 dopamine receptors mediate the potentiation of
    arachidonic release in Chinese hamster ovary cells. Mol. Pharmacol. **39:** 364–369.
34. VAN TOL, H. H. M., J. R. BUNZOW, H. C. GUAN, R. K. SUNAHARA, P. SEEMAN, H. B.
    NIZNIK & O. CIVELLI. 1991. Cloning of the gene for a human dopamine D_4 receptor
    with high affinity for the antipsychotic clozapine. Nature **350:** 610–614.
35. SOKOLOFF, P., B. GIROS, M.-P. MARTRES, M.-L. BOUTHENET & J.-C. SCHWARTZ. 1990.
    Molecular cloning and characterization of a novel dopamine receptor (D_3) as a target
    for neuroleptics. Nature **347:** 146–151.
36. SOKOLOFF, P., M. ANDRIEUX, R. BESANÇON, C. PILON, M.-P. MARTRES, B. GIROS & J.-C.
    SCHWARTZ. 1992. Pharmacology of human D_3 dopamine receptor expressed in a
    mammalian cell line: Comparison with D_2 receptor. Eur. J. Pharmacol. Mol. Pharma-
    col. Sect. **225:** 331–337.
37. CASTRO, S. W. & P. G. STRANGE. 1993. Differences in the ligand binding properties of the
    short and long versions of the D_2 dopamine receptor. J. Neurochem. **60:** 372–375.
38. SEABROOK, G. R., S. PATEL, R. MARWOOD, F. EMMS, M. R. KNOWLES, S. B. FREEDMAN &
    G. MCALLISTER. 1992. Stable expression of human D_3 dopamine receptors in GH_4 Ci
    pituitary cells. FEBS Lett. **312:** 123–126.
39. PILON, C., D. LEVESQUE, V. DIMITRIADOU, N. GRIFFON, M.-P. MARTRES, J.-C. SCHWARTZ
    & P. SOKOLOFF. 1994. Functional coupling of the human dopamine D_3 receptor in a
    transfected NG 108-15 neuroblastoma-glioma hybrid cell line. Eur. J. Pharmacol. Mol.
    Pharmacol. Sect. **268:** 129–139.
40. LEVESQUE, D., J. DIAZ, C. PILON, M.-P. MARTRES, B. GIROS, E. SOUIL, D. SCHOTT, J.-L.
    MORGAT, J.-C. SCHWARTZ & P. SOKOLOFF. 1992. Identification, characterization and
    localization of the dopamine D_3 receptor in rat brain using 7-[^{3}H]hydroxy-N,N-di-n-
    propyl-2-aminotetralin. Proc. Natl. Acad. Sci. USA **89:** 8155–8159.
41. LANG, J. 1983. Purification and characterization of subforms of the guanine-nucleotide-
    binding proteins $G\alpha_i$ and $G\alpha_o$. Eur. J. Biochem. **183:** 687–692.
42. JULIUS, D. 1990. Molecular biology of serotonin receptors. Annu. Rev. Neurosci. **14:** 335–
    360.
43. SVENSSON, K., A. M. JOHANSSON, T. MAGNUSSON & A. CARLSSON. 1986. (+)-AJ 76 and
    (+)-UH 232: Central stimulants acting as preferential dopamine autoreceptor antago-
    nists. Naunyn-Schmiedebergs Arch. Pharmakol. **334:** 234–245.
44. DALY, S. A. & J. L. WADDINGTON. 1993. Behavioural effects of the putative D_3 dopamine
    receptor agonist 7-OH-DPAT in relation to other "D_2-like" agonists. Neuropharmacol-
    ogy **32:** 509–510.
45. CAINE, S. B. & G. F. KOOB. 1993. Modulation of cocaine self-administration in the rat
    through D_3 dopamine receptors. Science **260:** 1815–1816.
46. MELLER, E., K. BOHMAKER, M. GOLDSTEIN & B. A. BASHAM. 1993. Evidence that striatal
    synthesis-inhibiting autoreceptors are dopamine D_3 receptors. Eur. J. Pharmacol.
    249: R5–R6.
47. MORGAN, J. I. & T. CURRAN. 1989. Stimulus-transcription coupling in neurons: Role of
    cellular immediate-early genes. Trends Neurosci. **12:** 459–462.
48. SHENG, M. & M. E. GREENBERG. 1990. The regulation and function of c-fos and other
    immediate early genes in the central nervous system. Neuron **4:** 477–485.
49. MÜLLER, R. 1986. Cellular and viral *fos* genes: Structure, regulation of expression and
    biological properties of their encoded products. Biochim. Biophys. Acta **823:** 207–225.
50. MARX, J. L. 1987. The fos gene as "master switch." Science **237:** 854–856.
51. CURRAN, T. 1988. The fos oncogene. *In* The Oncogene Handbook. E. P. Reddy, A. M.
    Shalka & T. Curran, Eds.: 307–554. Elsevier Science Publishers. Amsterdam.
52. GRAYBIEL, A. M., R. MORATALLA & H. A. ROBERTSON. 1990. Amphetamine and cocaine
    induce drug-specific activation of the c-fos gene in striosome-matrix and limbic subdivi-
    sions of the striatum. Proc. Natl. Acad. Sci. USA **87:** 6912–6916.
53. YOUNG, S. T., L. J. PORRINO & M. J. IADOROLA. 1991. Cocaine induces striatal c-fos
    immunoreactivity proteins via dopaminergic D_1 receptors. Proc. Natl. Acad. Sci. USA
    88: 1291–1295.
54. ROBERTSON, H. A., M. R. PETERSON, K. MURPHY & G. S. ROBERTSON. 1989. D_1-dopamine

receptor agonists selectively activate c-fos independent of rotational behavior. Brain Res. **503:** 346–349.

55. MILLER, J. C. 1990. Induction of c-fos mRNA expression in rat striatum by neuroleptic drugs. J. Neurochem. **54:** 1453–1455.
56. NGUYEN, T. V., B. KASOFSKY, B. BIRNBAUM, B. COHEN & S. E. HYMAN. 1992. Differential expression of c-fos and Zif 268 in rat striatum following haloperidol, clozapine and amphetamine. Proc. Natl. Acad. Sci. USA **89:** 4270–4274.
57. ROBERTSON, G. S. & H. C. FIBIGER. 1992. Neuroleptics increase c-*fos* expression in the forebrain: Contrasting effects of haloperidol and clozapine. Neuroscience **46:** 315–328.
58. DEUTCH, A. Y., M. C. LEE & M. J. IADOROLA. 1992. Regionally specific effects of atypical antipsychotic drugs on striatal Fos expression: The nucleus accumbens shell as a locus of antipsychotic action. Mol Cell. Neurosci. **3:** 332–341.
59. MANSOUR, A., J. H. MEADOR-WOODRUFF, J. R. BUNZOW, O. CIVELLI, H. AKIL & S. J. WATSON. 1990. Localization of dopamine D_2 receptor mRNA and D_1 and D_2 receptor binding in the rat brain and pituitary. An in situ hybridization autoradiographic analysis. J. Neurosci. **10:** 2587–2601.
60. GROENEWEGEN, H. J., H. W. BERENDSE, G. E. MEREDITH, S. N. HABER, P. VOORN, J. G. WOLTERS & A. H. M. LOHMAN. 1991. Functional anatomy of the ventral limbic system-innervated striatum. *In* The mesolimbic dopamine system: From motivation to action. P. Willner & J. Scheel-Kruger, Eds.: 19–59. John Wiley. Chichester, UK.
61. ZAHM, D. S. & J. S. BROG. 1992. On the significance of subterritories in the "accumbens" part of the rat ventral striatum. Neuroscience **50:** 751–767.
62. DEUTCH, A. Y. & D. S. CAMERON. 1992. Pharmacological characterization of dopamine systems in the nucleus accumbens core and shell. Neuroscience **46:** 49–56.
63. GOVONI, S., J. S. HONG, H.-Y. T. YANG & E. COSTA. 1980. Increase of neurotensin content elicited by neuroleptics in nucleus accumbens. J. Pharmacol. Exp. Ther. **215:** 413–417.
64. CASTEL, M. N., P. MORINO & T. HÖKFELT. 1993. Modulation of the neurotensin striato-nigral pathway by D_1 receptors. Mol. Neurosci. **5:** 281–284.
65. KISLAUKIS, E. & P. R. DOBNER. 1990. Mutually dependent response elements in the *cis*-regulated region of the neurotensin/neuromedin N gene integrate environmental stimuli in PC 12 cells. Neuron **4:** 783–795.
66. CREESE, I. & D. R. SIBLEY. 1980. Receptor adaptations to centrally acting drugs. Annu. Rev. Pharmacol. Toxicol. **21:** 357–391.
67. AUTELITANO, D. J., L. SNYDER, S. C. SEALFON & J. L. ROBERTS. 1989. Dopamine D_2-receptor messenger RNA is differentially regulated by dopaminergic agents in rat anterior and neurointermediate pituitary. Mol. Cell. Endocrinol. **67:** 101–105.
68. MARTRES, M.-P., P. SOKOLOFF, B. GIROS & J.-C. SCHWARTZ. 1992. Effects of dopaminergic transmission interruption on the D_2 receptor isoforms in various cerebral tissues. J. Neurochem. **58:** 673–679.
69. BERNARD, V., C. LE MOINE & B. BLOCH. 1991. Striatal neurons express increased level of dopamine D_2 receptor mRNA in response to haloperidol treatment: A quantitative *in situ* hybridization study. Neuroscience **45:** 117–126.
70. KOPP, J., N. LINDEFORS, S. BRENE, H. HALL, H. PERSSON & G. SEDVALL. 1992. Effect of raclopride on dopamine D_2 receptor mRNA expression in rat brain. Neuroscience **47:** 771–779.
71. CROW, T. J., A. J. CROSS, E. C. JOHNSTONE, A. LONGDEN, F. OWEN & R. M. RIDLEY. 1980. Time course of the antipsychotic effect in schizophrenia and some changes in postmortem brain and their relation to neuroleptic medication. Adv. Biochem. Psychopharmacol. **24:** 495–503.
72. HÖKFELT, T., J. F. REHFELD, L. SKIRBOLL, B. IVEMARK, M. GOLSTEIN & K. MARKEY. 1980. Evidence for coexistence of dopamine and CCK in meso-limbic neurons. Nature **225:** 476–478.
73. ESSMAN, W. D., P. MCGONIGLE & I. LUCKI. 1993. Anatomical differentiation within the nucleus accumbens of the locomotor stimulatory actions of selective dopamine agonists and d-amphetamine. Psychopharmacology **112:** 233–241.
74. MARTRES, M.-P., M.-L. BOUTHENET, N. SALES, P. SOKOLOFF & J-C. SCHWARTZ. 1985.

Widespread distribution of brain dopamine receptors evidenced with [^{125}I]iodosulpride, a highly selective ligand. Science **228:** 752–755.
75. SIEBERT, P. D. & J. W. LARRICK. 1992. Competitive PCR. Nature **359:** 557–558.
76. PEDERSEN, U. B., B. NORBY, A. A. JENSEN, M. SCHIODT, A. HANSEN, P. SUHR-JENSSEN, M. SCHEIDELER, O. THASTRUP & P. H. ANDERSEN. 1994. Characteristics of stably expressed human dopamine D_{1a} and D_{1b} receptors: Atypical behavior of the dopamine D_{1b} receptor. Eur. J. Pharmacol. Mol. Pharmacol. Sect. **267:** 85–93.

Denervation, Hyperinnervation, and Interactive Regulation of Dopamine and Serotonin Receptors[a]

TOMÁS A. READER,[b,c] FATIHA RADJA,[b]
KAREN M. DEWAR,[d] AND LAURENT DESCARRIES[e]

Departments of Physiology,[b] Psychiatry,[d] and Pathology[e]
Faculty of Medicine
University of Montreal
Montreal, Quebec, Canada H3C 3J7

In adult experimental animals, 6-hydroxydopamine (6-OHDA) lesions that destroy more than 90% of the dopamine (DA) afferents to the neostriatum entail a dysfunction that resembles Parkinson's disease in the human.[1-3] Sensorimotor neglect, aphagia, adipsia, cognitive alterations, and postural abnormalities characterize this behavioral syndrome in the rat.[3-6] Even after severe nigrostriatal 6-OHDA lesion, however, considerable recovery of function may occur, associated with modified properties of the spared DA terminals, including increases in DA synthesis,[7,8] metabolism,[9] and fractional efflux.[10] Supersensitivity of neostriatal DA receptors has also been reported as another factor accounting for some functional recovery in these animals.[11,12] On the other hand, following an extensive nigrostriatal DA denervation but carried out in neonates, the rats reach adulthood with only minor sensorimotor and ingestive deficits.[13,14] Among mechanisms underlying this functional sparing, a major reorganization of the neostriatal circuitry has been proposed, as evidenced by the striking serotonin (5-HT) hyperinnervation that then takes place in the rostral half of neostriatum.[15-18] Moreover, autoradiographic evidence has been obtained suggesting that up-regulation of certain 5-HT receptors might be associated with 5-HT hyperinnervation, at least in conditions of homotypic sprouting after 5,7-dihydroxytryptamine lesion.[19] The neonatally 6-OHDA-lesioned rat provided an interesting opportunity to investigate eventual regulations of DA and 5-HT receptors in a model of 5-HT hyperinnervation unassociated with prior lesion of the 5-HT system. We therefore used quantitative autoradiography after radioligand binding to examine the amount and distribution of dopamine D_1 and D_2 receptors and of some of the known 5-HT receptor subtypes, in the brain of adult rats subjected to nigrostriatal 6-OHDA lesion soon after birth. It was expected that the localization of eventual changes within the neostriatum and related brain regions

[a]This research was supported by grants from the Medical Research Council of Canada to Drs. Tomás A. Reader and Laurent Descarries, and from the Banting Research Foundation and the Natural Sciences and Engineering Research Council of Canada to Dr. Karen M. Dewar. K.M.D. was supported by a Chercheur-boursier Junior 1 scholarship from the FRSQ, and F.R. by postdoctoral fellowships from the FRSQ and the Groupe FCAR de recherche sur le système nerveux central.

[c]Address correspondence to Dr. Tomás A. Reader, Département de physiologie, Faculté de Médecine, Université de Montréal, C.P. 6128, Succ. Centre-ville, Montréal, Québec H3C 3J7, Canada.

TABLE 1. Concentration of Aromatic Monoamines in the Neostriatum of Control and Neonatally 6-OHDA-Lesioned Rats[a]

Region	DA	DOPAC	HVA	3-MT	5-HT	5-HIAA
Rostral neostriatum						
Control	99.7 ± 8.8	65.6 ± 9.7	14.3 ± 1.0	6.6 ± 1.0	3.3 ± 0.40	7.4 ± 1.0
Lesioned	0.04 ± 0.02	0.50 ± 0.20	0.10 ± 0.02	0.10 ± 0.01	5.2 ± 0.80	24.1 ± 3.0
% change	↓ 99.9	↓ 99.3	↓ 99.2	↓ 98.6	↑ 56.8	↑ 225.7
	***	***	***	***	***	**
Caudal neotriatum						
Control	46.3 ± 9.3	34.7 ± 7.2	7.4 ± 1.0	4.2 ± 0.70	9.5 ± 1.2	24.4 ± 5.0
Lesioned	0.30 ± 0.09	0.90 ± 0.20	0.10 ± 0.05	0.10 ± 0.02	9.8 ± 1.3	30.7 ± 3.5
% change	↓ 99.3	↓ 97.4	↓ 98.1	↓ 98.6	↑ 3.2	↑ 25.8
	***	***	***	***	n.s.	n.s.

Abbreviations: DA, dopamine; DOPAC, 3,4-dihydroxyphenylacetic acid; HVA, homovanillic acid; 3-MT, 3-methoxytyramine; 5-HT, serotonin; 5-HIAA, 5-hydroxyindole-3-acetic acid.

[a]Values are mean ± SEM ($n = 12$) in ng/mg protein. $*p < 0.05$, $**p < 0.01$ and $***p < 0.001$, by Student's t test. (From Dewar *et al.*[22] Reproduced, with permission, from *Brain Research*.)

might shed some light on possible regulatory interactions between these two monoamine systems.

MONOAMINE METABOLISM IN ADULT NEOSTRIATUM AFTER NEONATAL 6-OHDA LESION

The experiments were carried out in normal control and adult Sprague-Dawley rats (Charles River, Montreal) subjected as 3-day-old pups to bilateral intraventricular administration of 6-OHDA (Sigma, St. Louis, MO). Fifty micrograms of 6-OHDA (free base) were delivered in each lateral ventricle after systemic pretreatment with desipramine (Sigma), to protect noradrenaline neurons.[20–22]

As repeatedly demonstrated after such lesions,[3,15,17,22–26] the neonatal administration of 6-OHDA led to profound depletions of DA and its metabolites, 3,4-dihydroxyphenylacetic acid (DOPAC), homovanillic acid (HVA), and 3-methoxytyramine (3-MT), in the neostriatum of 3-month-old animals (TABLE 1). Similar depletions have been documented as early as 15 and 30 days after the 6-OHDA lesion[22] and up to six months.[23] It has also been shown recently that, 1 and 3 months after the lesion, specific [³H]BTCP binding to neostriatal DA uptake sites was reduced to about 2% of control values.[26] In this latter study, the synthesis rate of DA in the spared DA terminals was confirmed to be greatly enhanced, and neostriatal DA catabolism was found to be switched from DOPAC to HVA production.[26] Such a change presumably reflected a higher fractional release of DA together with a decreased capacity for DA uptake, leading to preferential extracellular DA degradation by catechol-O-methyltransferase as opposed to intraneuronal degradation by MAO.

In the case of 5-HT, considerable increases in 5-HT and 5-hydroxyindole-3-acetic acid (5-HIAA) levels were measured in the rostral but not the caudal half of neostriatum (TABLE 1). In two separate studies, such increases were already apparent after 1 month of survival.[22,26] The most recent investigation also indicated that neither 5-HT synthesis nor catabolism were concomitantly affected. However, inasmuch as the density of specific [³H]citalopram binding to neostriatal 5-HT uptake sites was not significantly increased at 1 month, and less increased than the corresponding 5-HT concentration at 3 months, it was concluded that the amount of 5-HT per 5-HT terminal had risen prior to the 5-HT hyperinnervation and remained elevated thereafter.[26] Such an elevation in 5-HT steady-state level might be the result of an inhibition of 5-HT release mediated by 5-HT$_{1B}$ autoreceptors (see below).

QUANTITATIVE DISTRIBUTION OF MONOAMINE RECEPTORS AFTER NEONATAL 6-OHDA LESION

Dopamine D_1 and D_2 as well as serotonin 5-HT$_{1A}$, 5-HT$_{1B}$, 5-HT$_{1nonAB}$, and 5-HT$_{2A}$ receptors were measured by quantitative autoradiography according to well-established procedures.[23,24,27–29] Details of the respective incubation conditions are given in legends of TABLES 2–7. After incubation, film autoradiographs were prepared and exposed for 3 days ([¹²⁵I] ligands) or several weeks ([³H] ligands). Densitometric readings were made with an image analysis system (MCID™, Imaging Research, Ontario, Canada), in the rostral and caudal halves of neostriatum and other anatomical regions of interest, selected according to the receptor subtype investigated. Standard curves from [³H]- or [¹²⁵I]-Microscales™ were used to convert density levels into femtomoles per milligram of protein (fmol/mg protein). Multiple readings (6–25) were made in each region and the mean density measured from at least six sections per region per rat. Nonspecific binding was determined in adjacent sections and specific binding obtained by subtracting nonspecific from total binding. Each receptor subtype was studied in at least four lesioned and four control rats.

Decreased D_1 Receptors in Rostral Neostriatum

As shown in FIGURE 1, the density of dopamine D_1 receptors labeled with [³H]SCH 23390 was uniformly high in the rostral neostriatum of controls, and less dense in the caudal half of neostriatum as well as in the substantia nigra. In neonatally lesioned rats, there was a slight decrease by comparison to control in the rostral but not the caudal half of neostriatum (TABLE 2). A slight apparent increase in the substantia nigra of lesioned rats was not statistically significant.

TABLE 2. D_1 Receptors in Control and Neonatally 6-OHDA-Lesioned Rats

Region	Control	Lesioned	% Change
Rostral neostriatum	1912 ± 37[a]	1500 ± 213	↓ 22[b]
Caudal neostriatum	1299 ± 135	1300 ± 141	0
Substantia nigra	1304 ± 229	1484 ± 175	↑ 13

NOTE: Dopamine D_1 receptors were labeled with the antagonist [³H]SCH 23390 (DuPont, 73.2 Ci/mmol). Sections were first preincubated at 25 °C for 30 min in 50 mM Tris-HCl buffer (pH 7.4) containing 120 mM NaCl, 5 mM KCl, 2 mM CaCl$_2$, and 1 mM MgCl$_2$. They were then incubated for 60 min in the same buffer with 1 nM [³H]SCH 23390 in the presence of 100 nM of ketanserin to prevent binding of the ligand to 5-HT receptors. Nonspecific binding was determined in adjacent sections incubated with the radioligand in the presence of 30 μM (±)SKF 38393 hydrochloride.[22,24] Values are mean ± SD in fmol/mg protein. (From Radja *et al.*[24] Reproduced, with permission, from *Neuroscience.*)

[a]$p < 0.001$ for caudal versus rostral neostriatum, by Student's *t* test.
[b]$p < 0.01$ for lesioned versus control, by Student's *t* test.

TABLE 3. D_2 Receptors in Control and Neonatally 6-OHDA-Lesioned Rats

Region	Control	Lesioned	% Change
Rostral neostriatum			
Dorsolateral	175 ± 13	223 ± 20	↑ 27[a]
Dorsomedian	126 ± 10	162 ± 25	↑ 28[b]
Ventrolateral	161 ± 8	198 ± 8	↑ 23[a]
Ventromedian	116 ± 9	128 ± 26	↑ 10
Caudal neostriatum			
Dorsal	84 ± 9	102 ± 10	↑ 21[b]
Medial	120 ± 6	153 ± 22	↑ 27[b]
Ventral	137 ± 25	193 ± 29	↑ 41[b]
Substantia nigra	66 ± 4	13 ± 7	↓ 80[a]

NOTE: Dopamine D_2 receptors were labeled with the antagonist [³H]raclopride (DuPont, 63 Ci/mmol). Sections were first preincubated at 25 °C for 30 min in 50 mM Tris-HCl buffer (pH 7.4) containing 120 mM NaCl and 5 mM KCl, and then incubated for 60 min in the same buffer with 2 nM [³H]raclopride. Nonspecific binding was determined in adjacent sections incubated with the radioligand in the presence of 300 μM (±)sulpiride.[22,24] Values are mean ± SD in fmol/mg protein. (From Radja et al.[24] Reproduced, with permission, from *Neuroscience*.)
[a]$p < 0.01$ and [b]$p < 0.05$, by Student's t test.

Earlier measurements of D_1 receptor binding after neonatal 6-OHDA lesion had shown slight increases,[30] dramatic losses[31] or no changes[32,33] in the number of these sites, perhaps due to different degrees of DA denervation in different conditions of 6-OHDA administration. In those and the present studies, the DA lesions were carried out at a postnatal age when the rat neostriatum is only partly DA inner-vated,[34] but a significant proportion (60–70%) of D_1 receptors is already measur-able.[30,35] Since major developmental changes occur during the first postnatal week, not only in the patch and matrix distribution of the neostriatal DA innervation, but also in that of its D_1 receptors,[35] the exact time when the lesion is made may be a critical factor. After 6-OHDA lesions in the adult, significant losses of D_1 receptors were observed in the neostriatum, as if independent from the presence of DA.[36] These losses are not reversed by transplantation of fetal mesencephalic neurons. On the other hand, disruption[37] or blockade[38] of DA neurotransmission has been shown to increase D_1 receptor number. The D_1 receptor decrease after neonatal DA denervation might therefore be imputable to removal of the DA innervation itself rather than to the decreased availability of DA. Indeed, following neonatal DA

TABLE 4. 5-HT_{1A} Receptors in Control and Neonatally 6-OHDA-Lesioned Rats

Region	Control	Lesioned	% Change
Lateral septum	672 ± 39	705 ± 44	↑ 5
CA1 region	530 ± 83	564 ± 67	↑ 6
Dentate gyrus	716 ± 45	666 ± 79	↓ 7
Dorsal raphe	390 ± 36	411 ± 14	↑ 6

NOTE: Serotonin 5-HT_{1A} receptors were labeled with the agonist [³H]8-hydroxy-2-(di-n-propylamino)tetralin or [³H]8-OH-DPAT (DuPont, 63 Ci/mmol). Sections were first preincu-bated at 25 °C for 30 min in 170 mM Tris-HCl buffer (pH 7.6) and then incubated for 60 min in the same buffer with 2 nM [³H]8-OH-DPAT. Nonspecific binding was determined in adjacent sections incubated with the radioligand in the presence of 10 μM unlabeled 5-HT creatinine sulfate.[23] Values are mean ± SD in fmol/mg protein. (From Radja et al.[23] Reproduced, with permission, from *Brain Research*.)

TABLE 5. 5-HT$_{1B}$ Receptors in Control and Neonatally 6-OHDA-Lesioned Rats

Region	Control	Lesioned	% Change
Rostral neostriatum	8.2 ± 1.8	10.8 ± 1.0	↑ 32[a]
Caudal neostriatum	7.2 ± 1.9	9.6 ± 0.9	↑ 33[a]
Substantia nigra	24.7 ± 3.5	38.0 ± 7.4	↑ 54[b]
Globus pallidus	22.7 ± 4.2	30.3 ± 4.3	↑ 33[a]
Dorsal subiculum	28.1 ± 6.5	28.2 ± 5.1	0

NOTE: Serotonin 5-HT$_{1B}$ receptors were labeled with the antagonist [^{125}I]cyanopindolol (DuPont, 2200 Ci/mmol). Sections were first preincubated at 25 °C for 30 min in 170 mM Tris-HCl buffer (pH 7.6) containing 150 mM NaCl, and then incubated for 120 min in the same buffer with 12 pM [^{125}I]cyanopindolol. Nonspecific binding was determined in adjacent sections incubated with the radioligand in the presence of 10 μM unlabeled 5-HT creatinine sulfate.[23] Values are mean ± SD in fmol/mg protein (From Radja *et al.*[23] Reproduced, with permission, from *Brain Research.*)
[a]$p < 0.05$ and [b]$p < 0.01$, by one-way analysis of variance.

denervation, extracellular levels of neostriatal DA have been shown to exceed the values expected from the extent of DA denervation,[39] in keeping with the increased turnover within the residual or spared DA terminals.[26]

The localization of the D$_1$ receptor decrease to the rostral half of neostriatum suggested a relationship with the 5-HT hyperinnervation that also predominates in this part of neostriatum. 5-HT has already been shown to inhibit DA release from neostriatal slices acting through 5-HT2A receptors,[40] which are known to be subject to up-regulation in the present model (see below). However, because the present D$_1$ receptor decrease takes place in a DA-denervated neostriatum, a more direct interaction at the level of the expression of the two receptors is likely to be the cause.

Although the changes in D$_1$ receptors were restricted to the rostral neostriatum and consisted of a down-regulation,[22,24] they were accompanied by an *increase* in the responsiveness of neostriatal neurons to the iontophoretic application of DA or the D$_1$ receptor agonist, SKF 38393.[24] Such an observation was in agreement with previous demonstrations of behavioral hypersensitivity to DA[41,42] and to a D$_1$ receptor agonist[43] after adult lesions, presumably unassociated with D$_1$ receptor increases.[22,32,33] This lack of correlation between sensitivity to receptor agonists and

TABLE 6. 5-HT$_{1nonAB}$ Receptors in Control and Neonatally 6-OHDA-Lesioned Rats

Region	Control	Lesioned	% Change
Rostral neostriatum	163 ± 19	225 ± 37	↑ 38[a]
Caudal neostriatum	160 ± 33	216 ± 26	↑ 35[a]
Substantia nigra	262 ± 17	394 ± 48	↑ 50[b]
Globus pallidus	197 ± 45	212 ± 35	↑ 8
Choroid plexus	699 ± 35	727 ± 14	↑ 4

NOTE: Other serotonin 5-HT$_1$ receptors (nonAB) were labeled with [^{3}H]serotonin creatinine sulfate or [^{3}H]5-HT (Amersham, 25 Ci/mmol). Sections were first preincubated at 25 °C for 30 min in 170 mM Tris-HCl buffer (pH 7.6), and then incubated for 60 min in the same buffer with 2 nM [^{3}H]5-HT, 0.01% ascorbic acid, 10 μM fluoxetine and 10 μM pargyline. Pindolol (1 μM) was added to occlude 5-HT$_{1A}$ and 5-HT$_{1B}$ sites. Nonspecific binding was determined in adjacent sections incubated with the radioligand in the presence of 10 μM unlabeled 5-HT creatinine sulfate.[23] Values are mean ± SD in fmol/mg protein. (From Radja *et al.*[23] Reproduced, with permission, from *Brain Research.*)
[a]$p < 0.05$ and [b]$p < 0.01$, by one-way analysis of variance.

TABLE 7. 5-HT$_{2A}$ Receptors in Control and Neonatally 6-OHDA-Lesioned Rats

Region	Control	Lesioned	% Change
Claustrum	3.3 ± 0.4	3.5 ± 0.7	↑ 6
Nucleus accumbens	2.3 ± 0.7	2.4 ± 0.2	↑ 4
Rostral neostriatum	1.5 ± 0.3	2.4 ± 0.2	↑ 60[a]
Caudal neostriatum	2.3 ± 0.3[b]	2.4 ± 0.8	↑ 4
Choroid plexus	1.2 ± 0.1	1.3 ± 0.2	↑ 8

NOTE: Serotonin 5-HT$_{2A}$ receptors were labeled with the agonist [^{125}I](2,5-dimethoxy-4-iodophenyl)-2-aminopropane or [^{125}I]DOI (DuPont, 2200 Ci/mmol). Sections were first preincubated at 25 °C for 30 min in 50 mM Tris-HCl buffer (pH 7.4) containing 4 mM CaCl$_2$, 0.1% ascorbic acid and 0.1% bovine serum, and then incubated for 90 min in the same buffer with 200 pM [^{125}I]DOI in the presence of 30 nM unlabeled 5-HT to occlude 5-HT$_{2C}$ sites. Nonspecific binding was determined in adjacent sections incubated with the radioligand in the presence of 4 mM unlabeled 5-HT creatinine sulfate.[23] Values are mean ± SD in fmol/mg protein. (From Radja *et al.*[23] Reproduced, with permission, from *Brain Research.*)
[a] $p < 0.01$, by one-way analysis of variance.
[b] $p < 0.05$, for caudal versus rostral neostriatum.

receptor measurements by radioligand binding might be attributed to the presence of spare receptors in neostriatum.[44] On the other hand, the effect of DA receptor agonists might be more dependent on the efficiency of the transducing mechanisms in target neurons than on binding parameters, that is, number and/or affinity of the receptors.[45] It has also been shown that after neonatal DA denervation, D$_1$ receptor-coupled adenylyl cyclase activity may be enhanced without changes in the number or affinity of D$_1$ receptors.[33] Earlier studies in adult rats chronically treated with the D$_1$ receptor antagonist SCH 23390 have also shown considerable increases in adenylyl cyclase activity dissociated from a slight increase in D$_1$ receptors.[46] Conversely, and within a broader perspective, chronic lithium in adult rats causes important decreases in DA-activated adenylyl cyclase activity unassociated with changes in receptor antagonist binding parameters.[47] Altogether, these and other studies indicate a clear-cut dissociation between functional properties and receptor binding parameters. In the case of the neonatally 6-OHDA-lesioned rat, the behavioral hypersensitivity to DA receptor agonists,[32,48,49] as well as the increased responsiveness of neostriatal neurons to iontophoresed DA and SKF 38393,[24] may therefore depend on the activation of transducing mechanisms rather than on the properties of the receptor-ligand recognition site.

Increased D$_2$ Receptors

Considerable increases were found in specific [^{3}H]raclopride binding throughout the neostriatum of neonatally 6-OHDA-lesioned versus control rats (FIG. 2). These increases were significant in three quadrants of the rostral neostriatum (dorsolateral, dorsomedial, and ventrolateral) as well as in the dorsal, medial, and ventral thirds of the caudal neostriatum (TABLE 3). As previously shown,[50] medial-to-lateral and dorsal-to-ventral increasing gradients in [^{3}H]raclopride binding are respectively observed in the rostral and caudal neostriatum of controls. These gradients were still present in the increasingly labeled neostriatum after neonatal lesion. The density of D$_2$ receptors in the substantia nigra of the controls was about half that of the neostriatum as a whole. As already described, after adult lesions[51] [^{3}H]raclopride binding was markedly decreased in the substantia nigra of the neonatally 6-OHDA-lesioned rats.

The increase in neostriatal [³H]raclopride binding, presumably representing an up-regulation of D_2 receptors,[22] was approximately the same in all portions of the rostral and caudal neostriatum (TABLE 3), suggesting a common regulatory mechanism operating at the same rate and/or efficiency throughout this part of brain. The widespread neostriatal increase in D_2 receptor density was not accompanied by a parallel elevation of D_2 receptor mRNA levels.[24] At the time of the lesion, D_2 receptor mRNA levels had presumably reached only 60–75% of control adult values.[52] The development of a normal expression level of D_2 receptor mRNA in the absence of DA innervation has already been reported.[53,54]

In adult rat, increases in neostriatal D_2 receptors with no changes in mRNA levels have been measured after chronic blockade with haloperidol.[52,55,56] Decreases in D_2 receptors in the neostriatum of aged rats are less severe than decreases in mRNA levels,[57] and could reflect posttranscriptional changes of this receptor.[58] Although the mechanisms implicated in the regulation of mRNA and protein levels for the D_2 receptor remain to be clarified, one can propose that the turnover of the D_2 receptor protein in the neonatal DA-denervated neostriatum is reduced, perhaps

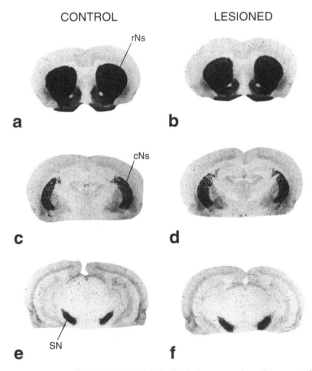

FIGURE 1. D_1 receptors ([³H]SCH 23390 binding) in control and neonatally 6-OHDA-lesioned rats. In the control (**a, c, e**), D_1 receptors are abundant in both the rostral (rNS) and caudal (cNS) halves of neostriatum as well as in the substantia nigra (SN). After neonatal 6-OHDA lesion (**b, d, f**), slight but significant decreases are observed in the rNS. See TABLE 2 for quantitative data.

because of a lack of internalization and/or removal from the neuronal membrane surface. If D_2 receptor protein synthesis is maintained at a normal rate but removal or inactivation is diminished, this will lead to an increased density of surface D_2 receptors, although not necessarily to an increased number of functional receptors, because these have to be adequately coupled to their transducing mechanisms.

The D_2 receptor increase in the neonatally DA-denervated neostriatum did not modify the sensitivity of neostriatal neurons to iontophoresed PPHT, a potent D_2 receptor agonist.[24] This observation was consistent with earlier studies showing only

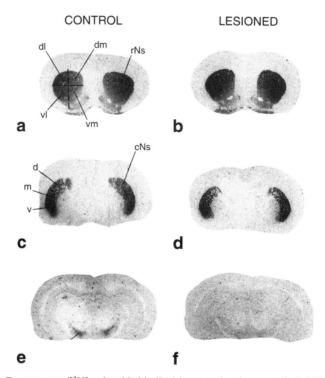

FIGURE 2. D_2 receptors ([^{3}H]raclopride binding) in control and neonatally 6-OHDA-lesioned rats. In the control (**a, c, e**), D_2 receptors show a medial-to-lateral increasing gradient in the rostral half of neostriatum (rNS), and a dorsal-to-ventral increasing gradient in its caudal half (cNS). A lower binding density is observed in the substantia nigra (SN). After neonatal 6-OHDA lesion (**b, d, f**), moderate increases are observed in both the rNS and cNS, and a marked decrease in the SN. See TABLE 3 for quantitative data.

slight increases in the behavioral sensitivity of these rats to the administration of the D_2 receptor agonist, quinpirole.[48,59] In contrast, in rats DA-denervated with 6-OHDA as adults, this same treatment was accompanied by a marked behavioral supersensitivity witnessed by locomotor changes.[32,48,49] The lesser electrophysiological and behavioral responsiveness to D_2 receptor agonists observed in the neonatally versus adult DA-denervated neostriatum raises the possibility that at least some of the increased D_2 binding sites measured after neonatal lesion are not primarily devoted to DA-mediated function. This suggestion would be in line with the former hypothesis that some of the increased D_2 receptors available for binding might not be efficiently coupled to their appropriate transducing mechanisms.

Unchanged 5-HT$_{1A}$ Receptors

Specific [^{3}H]8-OH-DPAT binding was relatively high in brain regions already described as rich in 5-HT$_{1A}$ receptors.[27,60,61] In none of these regions was this binding significantly altered 6 months after neonatal 6-OHDA lesion (TABLE 4). In the neostriatum of lesioned as well as control rats, no detectable [^{3}H]8-OH-DPAT binding was found. At the concentration of radioligand used (2 nM), only high-affinity 5-HT$_{1A}$ receptors were labeled;[62] therefore, the presence and/or changes in the number of low-affinity [^{3}H]8-OH-DPAT binding sites could not be ruled out.

The absence of 5-HT$_{1A}$ receptors in rat neostriatum[60,62] has been confirmed by immunocytochemistry[63,64] and *in situ* hybridization of the mRNA.[63,65] Homogenate binding studies have revealed the presence in neostriatum of a [^{3}H]8-OH-DPAT binding site pharmacologically distinct from the 5-HT$_{1A}$ receptor and which was decreased after 5,7-dihydroxytryptamine lesion.[66,67] In spite of the 5-HT hyperinnervation, this [^{3}H]8-OH-DPAT binding site was undetectable under the present autoradiographic conditions.

Increased 5-HT$_{1B}$ Receptors

Considerable increases were found in [^{125}I]cyanopindolol binding to 5-HT$_{1B}$ receptors in the neonatally 6-OHDA-lesioned versus control rats (FIG. 3). In the controls, as previously reported,[68,69] relatively high densities of 5-HT$_{1B}$ sites were measured in the dorsal subiculum, followed by those in the substantia nigra and globus pallidus (TABLE 5). Lower densities were detected in both the rostral and caudal neostriatum. All these regions, except for the dorsal subiculum, showed significant increases in [^{125}I]cyanopindolol binding, three months after the neonatal lesion. The highest increase was in the substantia nigra, followed by those in the globus pallidus and the two portions of neostriatum. No significant differences were present in either lesioned or control rats between rostral versus caudal halves of neostriatum. Apparent increases in the hippocampus, periaqueductal gray, and superior colliculus were not quantified, because the latter regions contain 5-HT$_{1A}$ as well as 5-HT$_{1B}$ receptors, and cyanopindolol does not distinguish between these two subtypes.[70]

The increased number of 5-HT$_{1B}$ receptors in the neonatally DA-denervated neostriatum was documented throughout this anatomical region, and similar increases were also measured in the substantia nigra and the globus pallidus, that is, the two major territories of projection of neostriatum. In rat brain, the 5-HT$_{1B}$ receptors act as terminal autoreceptors as well as receptors postsynaptic to 5-HT neurons. The neostriatal increase in 5-HT$_{1B}$ binding could not be attributed solely to an augmented number of 5-HT terminals, because it extended to both caudal and rostral halves of the neostriatum, and neither the substantia nigra nor the globus pallidus could be assumed to be 5-HT hyperinnervated.[71] The anatomical distribution of these increases suggested an up-regulation of the 5-HT$_{1B}$ receptors in the somata/dendrites of neostriatal projection neurons, accompanied by an increase of their axonal transport to both territories of projection. This interpretation was consistent with *in situ* hybridization data demonstrating the presence of mRNA for this receptor in the rat neostriatum, but not in substantia nigra nor globus pallidus.[72]

CONTROL   LESIONED

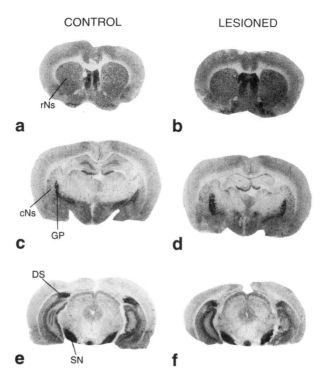

FIGURE 3. 5-HT$_{1B}$ receptors ([^{125}I]cyanopindolol binding) in control and neonatally 6-OHDA-lesioned rats. In the control (**a, c, e**), these receptors are relatively abundant in the rostral (rNS) and caudal (cNS) neostriatum, globus pallidus (GP), dorsal subiculum (DS), and substantia nigra (SN). After neonatal 6-OHDA lesion (**b, d, f**), all these regions except the dorsal subiculum show significant increases. See TABLE 5 for quantitative data.

Inasmuch as the 5-HT$_{1B}$ increase involved the entire neostriatum, it appeared more closely related to the overall DA denervation than the rostrally predominant 5-HT hyperinnervation, suggesting that DA afferents might normally regulate 5-HT$_{1B}$ receptor expression, at least during early postnatal ontogenesis. However, similar DA denervations produced by 6-OHDA in adult rat do not lead to appreciable changes in 5-HT$_{1B}$ receptors, as assessed either by quantitative autoradiography or by the electrophysiological responsiveness to the iontophoretic application of 5-HT or the 5-HT$_{1B/2C}$ receptor agonist, m-CPP.[73]

Increased 5-HT$_{1nonAB}$ Receptors

In the present autoradiographic studies, 5-HT$_1$ sites labeled with [^{3}H]5-HT—but distinct from the 5-HT$_{1A}$ and 5-HT$_{1B}$ subtypes—were designated as nonAB, because both 5-HT$_{1C}$ (now called 5-HT$_{2C}$) and 5-HT$_{1D}$, as well as other 5-HT$_1$ subtypes, could be labeled altogether.[74-77] No attempts were made to label 5-HT$_{2C}$ or 5-HT$_{1D}$ sites with [^{3}H]mesulergine or [^{3}H]5-HT in the presence of appropriate blockers, respectively, because the former compound also binds to 5-HT$_{2A}$ receptors,[61] and the density of 5-HT$_{1D}$ sites in rat brain is too low for autoradiographic detection.[77] Strong labeling of the choroid plexus did indicate binding of [³H]5-HT to 5-HT$_{2C}$ receptors,[74,75] which have been reported to be present in the neostriatum, substantia nigra, and globus pallidus of adult rat.[78] 5-HT$_{1D}$ receptors have also been demonstrated in rat brain,[77] and appear to be postsynaptic to 5-HT fibers in the neostriatum as well as in the cerebral cortex.[77]

In the controls, 5-HT$_{1nonAB}$ binding was the highest in the choroid plexus, followed by that in the substantia nigra, globus pallidus, and neostriatum (FIG. 4). No binding was detected in the dorsal subiculum. In the neonatally lesioned rats, both the neostriatum and the substantia nigra, but not the globus pallidus nor the choroid plexus, showed post lesion increases (TABLE 6). As was the case for 5-HT$_{1B}$ sites, the highest increase was that in the substantia nigra, followed by equal increases in the two halves of neostriatum. However, in contrast to the 5-HT$_{1B}$ binding, the 5-HT$_{1nonAB}$ binding was not increased in the globus pallidus, suggesting a preferential localization of the corresponding receptors on striatonigral, as opposed to striatopallidal, projection neurons.[79] Interestingly, a similar differential localization of DA receptor subtypes on striatonigral as opposed to striatopallidal neurons has been documented

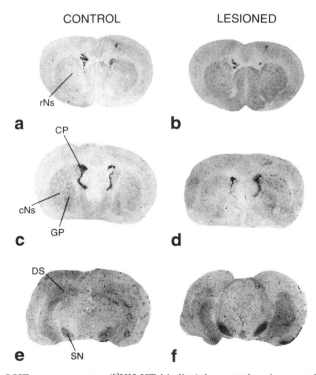

CONTROL　　　　LESIONED

FIGURE 4. 5-HT$_{1nonAB}$ receptors ([³H]5-HT binding) in control and neonatally 6-OHDA-lesioned rats. In the control (**a, c, e**), this binding is moderate in the rostral (rNS) and caudal (cNS) neostriatum and also detectable in the globus pallidus (GP) and substantia nigra (SN). It is also prominent in the choroid plexus (CP), indicating the presence of 5-HT$_{2C}$ receptors. After neonatal 6-OHDA lesion (**b, d, f**), increases are noticeable in the two halves of NS and in the SN. See TABLE 6 for quantitative data.

for D_1 versus D_2 receptors, respectively.[80] It is noteworthy that, in the neonatally lesioned rats, an increase in 5-HT_{2C} receptors, together with the 5-HT_{1B} receptor increase, could account for enhanced electrophysiological responses of neostriatal neurons to the iontophoretic application of 5-HT or its receptor agonist, m-CPP.[73]

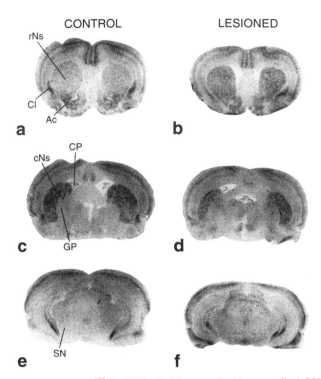

FIGURE 5. 5-HT_{2A} receptors ($[^{125}\text{I}]\text{DOI}$ binding) in control and neonatally 6-OHDA-lesioned rats. In the control (**a, c, e**), this binding is the strongest in the claustrum (Cl) and nucleus accumbens (Ac), but also high in the caudal neostriatum (cNS); it is weaker in the rostral neostriatum (rNS) and undetectable in the globus pallidus (GP) and the substantia nigra (SN). After neonatal 6-OHDA lesion (**b, d, f**), a considerable increase is seen in the rNS. Note the moderate binding of the CP in both control and lesioned rats. See TABLE 7 for quantitative data.

Increased 5-HT_2A Receptors

Specific $[^{125}\text{I}]\text{DOI}$ binding to 5-HT_{2A} receptors was markedly increased by comparison to control in the rostral neostriatum of the neonatally 6-OHDA-lesioned rats (FIG. 5). In all regions examined, this binding was relatively weak. As previously reported,[81] a laminar distribution could be observed in the neocortex, with a conspicuous band of relatively high density at the level of layer Va (FIG. 5). Overall, $[^{125}\text{I}]\text{DOI}$ binding was the highest in the claustrum, moderate in the caudal neostriatum and nucleus accumbens, and low in the rostral neostriatum and choroid plexus (TABLE 7). Three months after the neonatal 6-OHDA lesion, a similar distribution was found in all regions except the rostral neostriatum. Both the lesioned and control rats showed some preferential labeling of the choroid plexus, in spite of incubation in the presence of 30 nM cold serotonin. However, there was no significant labeling in the substantia nigra.

The increased [^{125}I]DOI binding associated with 5-HT hyperinnervation in the rostral neostriatum eliminated the normally observed caudorostral decreasing gradient in the density of these receptors.[82] This increase of 5-HT$_{2A}$ receptors was reminiscent of that reported in the 5-HT-hyperinnervated inferior olivary complex of adult rat following cytotoxic lesioning of its 5-HT innervation with 5,6-dihydroxytryptamine.[19] In this latter study, the 5-HT$_{2A}$ receptor increase had been interpreted as the result of an up-regulation without ruling out the possibility that it be due to the initial 5-HT denervation rather than to the ensuing 5-HT hyperinnervation. In the present study, 5-HT denervation was no longer in question as the cause of the 5-HT$_{2A}$ up-regulation. Moreover, the 5-HT$_{2A}$ receptor increase seemed tightly related to the 5-HT hyperinnervation in view of the coinciding anatomical distribution of the two phenomena. Yet, it is well established that this 5-HT receptor subtype is essentially postsynaptic to 5-HT neurons.[83] Interestingly, it was also demonstrated recently that in the neonatally DA-denervated and 5-HT–hyperinnervated neostriatum—but not in its adult counterpart without 5-HT hyperinnervation nor changes in 5-HT$_{2A}$ receptor binding—neuronal responsiveness to the iontophoretic application of 5-HT and to the 5-HT$_{2A/2C}$ receptor agonist DOI is also considerably increased.[71]

SUMMARY AND CONCLUSIONS

The destruction of nigrostriatal DA neurons by cerebroventricular injection of 6-OHDA in newborn rat results in a nearly complete DA denervation of the neostriatum and its subsequent 5-HT hyperinnervation. Three to six months after the neonatal 6-OHDA lesion, D$_1$ receptors are slightly decreased in the rostral neostriatum, but unchanged in its caudal half and in the substantia nigra. In contrast, D$_2$ receptors are increased throughout the neostriatum and decreased in the substantia nigra. Interestingly, no parallel changes occur in D$_2$ receptor mRNA, as measured by *in situ* hybridization on adjacent sections. The responsiveness to DA and to the D$_1$ receptor agonist SKF 38393 is markedly enhanced,[24] suggesting that this hypersensitivity is independent of the ligand recognition sites but rather mediated by postreceptor transducing mechanisms. On the other hand, the responsiveness to the D$_2$ receptor agonist PPHT remains unchanged in spite of the up-regulation of neostriatal D$_2$ receptors. The opposite changes in the number of D$_1$ and D$_2$ binding sites, dissociated from the expression of D$_2$ receptor mMRNA and from the sensitivity to DA receptor agonists, suggest independent adaptations triggered by the neonatal DA denervation and/or ensuing 5-HT hyperinnervation.

With regard to 5-HT receptor subtypes, a considerable increase in 5-HT$_{1B}$ binding sites in both the rostral and caudal neostriatum, as well as in the substantia nigra and globus pallidus, suggests an up-regulation and increased axonal transport of these receptors in neostriatal projection neurons. A similar increase in the density of 5-HT$_{1nonAB}$ receptors in the two halves of neostriatum and in the substantia nigra, but not the globus pallidus, presumably reflects an up-regulation and transport of the corresponding receptors (5-HT$_{1D}$ and 5-HT$_{2C}$) in striatonigral projection neurons only. The distribution of both increases throughout the neostriatum suggest a possible role of DA in the regulation of these receptors during ontogenesis. The even greater increase in 5-HT$_{2A}$ receptor density predominating in the rostral, 5-HT-hyperinnervated part of the neostriatum suggests a tight relationship with the 5-HT innervation. These increases in 5-HT receptor density can be correlated with an enhanced electrophysiological responsiveness[72] of neostriatal neurons to the iontophoretic application of 5-HT and its agonists, mCPP and DOI. They could therefore account for an enhancement of 5-HT neurotransmission in the neonatally 6-OHDA-denervated and 5-HT-hyperinnervated neostriatum, even if basal extracellular 5-HT levels remain normal under these conditions due to an increased number of 5-HT uptake sites.[84]

ACKNOWLEDGMENTS

We thank Daniel Cyr, Giovanni Battista Filosi, and Claude Gauthier for preparing the illustrations.

REFERENCES

1. UNGERSTEDT, U. 1968. 6-Hydroxy-dopamine induced degeneration of central monoamine neurons. Eur. J. Pharmacol. **5:** 107–110.
2. URESTKY, N. J. & L. L. IVERSEN. 1970. Effects of 6-hydroxydopamine on catecholamine containing neurons in the rat brain. J. Neurochem. **17:** 269–278.
3. ZIGMOND, M. J. & E. M. STRICKER. 1989. Animal models of parkinsonism using selective neurotoxins: Clinical and basic implications. Int. Rev. Neurobiol. **31:** 1–79.
4. UNGERSTEDT, U. 1971. Adipsia and aphagia after 6-hydroxydopamine induced degeneration of the nigro-striatal dopamine system. Acta Physiol. Scand. Suppl. **367:** 95–122.
5. ZIGMOND, M. J. & E. M. STRICKER. 1972. Deficits in feeding behaviour after intraventricular injection of 6-hydroxydopamine in rats. Science **177:** 1211–1214.
6. ZIGMOND, M. J., T. G. HASTINGS & E. D. ABERCROMBIE. 1992. Neurochemical responses to 6-hydroxydopamine and L-dopa therapy: Implications for Parkinson's disease. Ann. N. Y. Acad. Sci. **648:** 71–86.
7. AGID, Y., F. JAVOY & J. GLOWINSKI. 1973. Hyperactivity of remaining dopaminergic neurons after partial destruction of the nigro-striatal dopaminergic system in the rat. Nature **245:** 150–151.
8. ZIGMOND, M. J., A. L. ACHESON, M. K. STACHOWIAK & E. M. STRICKER. 1984. Neurochemical compensation after nigrostriatal bundle injury in an animal model of preclinical parkinsonism. Arch. Neurol. **41:** 856–861.
9. ALTAR, C. A., M. R. MARIEN & J. F. MARSHALL. 1987. Time course of adaptations in dopamine biosynthesis, metabolism, and release following nigrostriatal lesions: Implications for behavioural recovery from brain injury. J. Neurochem. **48:** 390–399.
10. STACHOWIAK, M. K., R. W. KELLER, JR., E. M. STRICKER & M. J. ZIGMOND. 1987. Increased dopamine efflux from striatal slices during development and after nigrostriatal bundle damage. J. Neurosci. **7:** 1648–1654.
11. CREESE, I., D. R. BURT & S. H. SNYDER. 1977. Dopamine receptor binding enhancement accompanies lesion-induced behavioural supersensitivity. Science **197:** 596–598.
12. NEVE, K. A., M. R. KOZLOWSKI & J. F. MARSHAL. 1982. Plasticity of neostriatal dopamine receptors after nigrostriatal injury: Relationship to recovery of sensorimotor functions and behavioural supersensitivity. Brain Res. **244:** 33–44.
13. BREESE, G. R., A. A. BAUMEISTER, T. J. McCOWN, S. G. EMERICK, G. D. FRYE, K. CROTTY & R. A. MUELLER. 1984. Behavioural differences between neonatal and adult 6-hydroxy-dopamine-treated rats to dopamine agonist: Relevance to neurological symptoms in clinical syndromes with reduced brain dopamine. J. Pharmacol. Exp. Ther. **231:** 343–354.
14. JOHNSON, B. J. & J. P. BRUNO. 1990. D_1 and D_2 receptor contributions to ingestive and locomotor behavior are altered after dopamine depletions in neonatal rats. Neurosci. Lett. **118:** 120–123.

15. STACHOWIAK, M. K., J. P. BRUNO, A. M. SNYDER, E. M. STRICKER & M. J. ZIGMOND. 1984. Apparent sprouting of striatal serotonergic terminals after dopamine-depleting brain lesions in neonatal rats. Brain Res. **291:** 164–167.

16. BERGER, T. W., S. KAUL, E. M. STRICKER & M. J. ZIGMOND. 1985. Hyperinnervation of the striatum by dorsal raphe afferents after dopamine-depleting brain lesions in neonatal rats. Brain Res. **336:** 354–358.

17. SNYDER, A. M., M. J. ZIGMOND & R. D. LUND. 1986. Sprouting of serotoninergic afferents into striatum after dopamine-depleting lesions in infant rats: A retrograde transport and immunocytochemical study. J. Comp. Neurol. **245:** 274–281.

18. DESCARRIES, L., J.-J. SOGHOMONIAN, S. GARCIA, G. DOUCET & J. P. BRUNO. 1992. Ultrastructural analysis of the serotonin hyperinnervation in adult rat neostriatum following neonatal dopamine denervation with 6-hydroxydopamine. Brain Res. **569:** 1–13.

19. PARÉ, M., L. DESCARRIES & R. QUIRION. 1992. Up-regulation of 5-hydroxytryptamine-2 and NK-1 receptors associated with serotonin/substance P hyperinnervation in the rat inferior olive. Neuroscience **51:** 97–106.

20. BREESE, G. R. & T. D. TRAYLOR. 1971. Depletion of brain noradrenaline and dopamine by 6-hydroxydopamine. Br. J. Pharmacol. **42:** 88–99.

21. BRUNO, J. P., E. M. STRICKER & M. J. ZIGMOND. 1985. Rats given dopamine-depleting brain lesions as neonates are subsensitive to dopaminergic antagonists as adults. Behav. Neurosci. **99:** 771–775.

22. DEWAR, K. M., J.-J. SOGHOMONIAN, J. P. BRUNO, L. DESCARRIES & T. A. READER. 1990. Elevation of dopamine D_2 but not D_1 receptors in adult rat neostriatum after neonatal 6-hydroxydopamine denervation. Brain Res. **536:** 287–296.

23. RADJA, F., L. DESCARRIES, K. M. DEWAR & T. A. READER. 1993. Serotonin 5-HT$_1$ and 5-HT$_{2A}$ receptors in adult rat brain after neonatal destruction of nigrostriatal dopamine neurons: A quantitative autoradiographic study. Brain Res. **606:** 271–285.

24. RADJA, F., M. EL MANSARI, J.-J. SOGHOMONIAN, K. M. DEWAR, A. FERRON, T. A. READER & L. DESCARRIES. 1993. Changes of D_1 and D_2 receptors in adult rat neostriatum after neonatal dopamine denervation: Quantitative data from ligand binding, in situ hybridization and iontophoresis. Neuroscience **57:** 635–648.

25. MOLINA-HOLGADO, E., K. M. DEWAR, L. GRONDIN, N. M. VAN GELDER & T. A. READER. 1993. Amino acid levels and GABA$_A$ receptors in rat neostriatum, cortex and thalamus after neonatal 6-hydroxydopamine lesion. J. Neurochem. **60:** 23–32.

26. MOLINA-HOLGADO, E., K. M. DEWAR, L. DESCARRIES & T. A. READER. 1994. Altered dopamine and serotonin metabolism in the dopamine-denervated and serotonin-hyperinnervated neostriatum of adult rat after neonatal 6-hydroxydopamine. J. Pharmacol. Exp. Ther. **270:** 713–721.

27. MARCINKIEWICZ, M., D. VERGÉ, H. GOZLAN & M. HAMON. 1984. Autoradiographic evidence for the heterogeneity of 5-HT$_1$ sites in rat brain. Brain Res. **291:** 159–163.

28. KÖHLER, C. & A. C. RADESTÄTER. 1986. Autoradiographic visualization of D_2 receptors in monkey brain using the selective benzamide drug [^{3}H]raclopride. Neurosci. Lett. **66:** 85–90.

29. DAWSON, T. M., D. R. GEHLERT, H. I. YAMAMURA, A. BARNETT & J. K. WAMSLEY. 1985. D_1 dopamine receptors in the rat brain: Autoradiographic localization using [^{3}H]SCH 23390. Eur. J. Pharmacol. **108:** 323–325.

30. BROADDUS, W. C. & J. P. BENNETT. 1990. Postnatal development of striatal dopamine function. II. Effects of postnatal 6-hydroxydopamine treatment on D_1 and D_2 receptors, adenylate cyclase activity and postsynaptic dopamine function. Dev. Brain Res. **52:** 273–277.

31. GELBARD, H. A., M. H. TEICHER, R. J. BALDESSARINI, A. GALITANO, E. R. MARSH, J. ZORC & G. FAEDDA. 1990. Dopamine D_1 receptor development depends on endogenous dopamine. Dev. Brain Res. **56:** 137–140.

32. BREESE, G. R., G. R. DUNCAN, T. C. NAPIER, S. C. BONDY, L. C. IORIO & R. A. MUELLER. 1987. 6-Hydroxydopamine treatment enhances behavioral responses to intracerebral microinjection of D_1 and D_2-dopamine agonists into nucleus accumbens and striatum without changing dopamine antagonist binding. J. Pharmacol. Exp. Ther. **240:** 167–176.

33. LUTHMAN, J., E. LINDQVIST, D. YOUNG & R. COWBURN. 1990. Neonatal dopamine lesion

in the rat results in enhanced adenylate cyclase activity without altering dopamine receptor binding or dopamine- and adenosine-3':5'-monophosphate-regulated phosphoprotein (DARP-32) immunoreactivity. Exp. Brain Res. **83:** 85–95.

34. VOORN, P., A. KALSBECK, B. JORRITSMA-BYHAM & H. J. GROENEWEGEN. 1988. The pre- and postnatal development of the dopaminergic cell groups in the ventral mesencephalon and the dopaminergic innervation of the striatum of the rat. Neuroscience **25:** 857–887.

35. MURRIN, L. C. & W. ZING. 1989. Dopamine D$_1$ receptor development in the rat striatum: Early localization in striosomes. Brain Res. **480:** 170–177.

36. BLUNT, S. B., P. JENNER & C. D. MARSDEN. 1992. Autoradiographic study of striatal D$_1$ and D$_2$ dopamine receptors in 6-OHDA-lesioned rats receiving foetal ventral mesencephalic grafts and chronic treatments with L-DOPA and carbidopa. Brain Res. **582:** 299–311.

37. JOYCE, J. N. 1991. Differential response of striatal dopamine and muscarinic cholinergic receptor subtypes to the loss of dopamine. II. Effects of 6-hydroxydopamine or colchicine microinjections into TVA or reserpine treatment. Exp. Neurol. **113:** 277–290.

38. SAVASTA, M., A. DUBOIS, J. BENAVIDEZ & B. SCATTON. 1988. Different plasticity changes in D$_1$ and D$_2$ receptors in rat striatal subregions following impairment of dopaminergic neurotransmission. Neurosci. Lett. **85:** 119–124.

39. CASTAÑEDA, E., I. Q. WHISHAW, L. LERMER & T. E. ROBINSON. 1990. Dopamine depletion in neonatal rats: Effects on behavior and striatal dopamine release assessed by intracerebral microdialysis during adulthood. Brain Res. **508:** 30–39.

40. ENNIS, C., J. D. KEMP & B. COX. 1981. Characterization of inhibitory 5-hydroxytryptamine receptors that modulate dopamine release in the striatum. J. Neurochem. **36:** 1515–1520.

41. FELTZ, P. & J. DE CHAMPLAIN. 1972. Enhanced sensitivity of caudate neurones to microiontophoretic injections of dopamine in 6-hydroxydopamine treated cats. Brain Res. **43:** 601–605.

42. SCHULTZ, W. & U. UNGERSTEDT. 1978. Striatal cell supersensitivity to apomorphine in dopamine-lesioned rats correlated to behavior. Neuropharmacology **17:** 180–186.

43. HU, X. T. & R. W. WANG. 1988. Comparison of effects of D$_1$ and D$_2$ dopamine receptor agonists on neurons in the rat caudate putamen: An electrophysiological study. J. Neurosci. **8:** 4340–4348.

44. BATTAGLIA, G., A. B. NORMAN, E. J. HESS & I. CREESE. 1986. Functional recovery of D$_1$ dopamine receptor-mediated stimulation of rat striatal adenylate cyclase activity following irreversible receptor modification by N-ethoxycarbonyl-2-ethoxy-1,2-dihydroquinoline (EEDQ): Evidence for spare receptors. Neurosci. Lett. **69:** 290–295.

45. PARENTI, M., S. GENTLEMAN, C. OLIANAS & N. H. NEFF. 1982. The dopamine receptor adenylate cyclase complex: Evidence for post recognition site involvement for development of supersensitivity. Neurochem. Res. **7:** 115–124.

46. HESS, E. J., L. J. ALBERS, L. HOANG & I. CREESE. 1986. Effects of chronic SCH 23390 treatment on the biochemical and behavioral properties of D$_1$ and D$_2$ dopamine receptors: Potentiated behavioral responses to a D$_2$ dopamine agonist after selective D$_1$ dopamine receptor upregulation. J. Pharmacol. Exp. Ther. **238:** 846–851.

47. CARLI, M., M. B. ANAND-SRIVASTAVA, E. MOLINA-HOLGADO, K. M. DEWAR & T. A. READER. 1994. Effects of chronic lithium treatments on central dopaminergic receptor systems: G proteins as possible targets. Neurochem. Int. **24:** 13–22.

48. BREESE, G. R., A. BAUMEISTER, T. C. NAPIER, G. D. FRYE & R. A. MUELLER. 1985. Evidence that D$_1$-dopamine receptors contribute to the supersensitive behavioral responses induced by L-dihydroxyphenylalanine in rats treated neonatally with 6-hydroxydopamine. J. Pharmacol. Exp. Ther. **253:** 287–288.

49. CRISWELL, H., R. A. MUELLER & G. R. BREESE. 1989. Priming of D$_1$-dopamine receptor responses: Long-lasting behavioral supersensitivity to a D$_1$-dopamine agonist following repeated administration to neonatal 6-OHDA-lesioned rats. J. Neurosci. **9:** 125–133.

50. SAVASTA, M., A. DUBOIS, C. FEUERSTEIN, M. MANIER & B. SCATTON. 1987. Denervation

super-sensitivity of striatal D2 dopamine receptors is restricted to the ventro- and dorsolateral regions of the striatum. Neurosci. Lett. **74:** 180–186.

51. MORELLI, M., T. MENNINI & G. DI CHIARA. 1988. Nigral dopamine autoreceptors are exclusively of the D_2 type: Quantitative autoradiography of [^{125}I]iodosulpiride and [^{125}I]SCH 23982 in adjacent brain sections. Neuroscience **27:** 865–870.

52. CREESE, I., D. R. SIBLEY & S. X. XU. 1992. Expression of rat striatal D_1 and D_2 dopamine receptor mRNAs: Ontogenic and pharmacological studies. Neurochem. Int. **20:** 45S–48S.

53. CHEN, J. F. & B. WEISS. 1986. Ontogenetic expression of D_2 dopamine receptor mRNA in rat corpus striatum. Dev. Brain. Res. **63:** 95–104.

54. SOGHOMONIAN, J.-J. 1993. Effects of 6-hydroxydopamine injections on glutamate-decarboxylase, prepro-enkephalin and dopamine D_2 receptor mRNAs in the adult rat striatum. Brain Res. **621:** 249–259.

55. GOSS, J. R., A. B. KELLY, S. A. JOHNSON & D. G. MORGAN. 1991. Haloperidol treatment increases D_2 dopamine receptor protein independently of RNA levels in mice. Life Sci. **48:** 1015–1022.

56. VAN TOL, H. H. M., M. RIVA, O. CIVELLI & I. CREESE. 1990. Lack of effect of chronic dopamine receptor blockade on D_2 dopamine receptor mRNA level. Neurosci. Lett. **111:** 303–308.

57. MESCO, E. R., J. A. JOSEPH, M. J. BLAKE & G. S. ROTH. 1991. Loss of D_2 receptors during aging is partially due to decreased levels of mRNA. Brain Res. **545:** 355–358.

58. SAKATA, M., S. M. FAROOQUI & C. PRASAD. 1992. Post-transcriptional regulation of loss of rat striatal D_2 dopamine receptors during aging. Brain Res. **575:** 309–314.

59. KOSTRZEWA, R. M. & L. GONG. 1991. Supersensitized D_1 receptors mediated enhanced oral activity after neonatal 6-OHDA. Pharmacol. Biochem. Behav. **39:** 677–682.

60. VERGÉ, D., G. DAVAL, M. MARCINKIEWICZ, A. PATEY, S. EL MESTIKAWY, H. GOZLAN & M. HAMON. 1986. Quantitative autoradiography of multiple 5-HT$_1$ receptor subtypes in the brain of control or 5,7-dihydroxytryptamine-treated rats. J. Neurosci. **12:** 3474–3482.

61. PAZOS, A., D. HOYER, M. M. DIETL & J. M. PALACIOS. 1988. Autoradiography of serotonin receptors. In Neuronal Serotonin. N. N. Osborne & M. Hamon, Eds.: 507–543. Wiley. Chichester, UK.

62. NÉNONÉNÉ, E. K., F. RADJA, M. CARLI, L. GRONDIN & T. A. READER. 1994. Heterogeneity of cortical and hippocampal 5-HT$_{1A}$ receptors: A reappraisal of homogenate binding with 8-[^{3}H]hydroxypropyl-aminotetralin. J. Neurochem. **62:** 1822–1834.

63. MIQUEL, M. C., E. DOUCET, C. BONI, S. EL MESTIKAWY, L. MATTHIESSEN, G. DAVAL, D. VERGÉ & M. HAMON. 1991. Central serotonin$_{1A}$ receptors: Respective distribution of encoding mRNA, receptor protein and binding sites by in situ hybridization histochemistry, radioimmunohistochemistry and autoradiographic mapping in the rat brain. Neurochem. Int. **19:** 453–465.

64. RIAD, M., S. EL MESTIKAWI, D. VERGÉ, H. GOZLAN & M. HAMON. 1991. Visualization and quantification of central 5-HT$_{1A}$ receptors with specific antibodies. Neurochem. Int. **19:** 413–423.

65. CHALMERS, D. T. & S. J. WATSON. 1991. Comparative anatomical distribution of 5-HT$_{1A}$ receptor mRNA and 5-HT$_{1A}$ binding in rat brain—A comparative in situ hybridization/in vitro receptor autoradiographic study. Brain. Res. **561:** 51–60.

66. GOZLAN, H., S. EL MESTIKAWI, M. B. EMERIT, L. PICHAT, J. GLOWINSKI & M. HAMON. 1983. Identification of presynaptic 5-HT autoreceptors using a new ligand: [^{3}H]PAT. Nature **305:** 140–142.

67. HALL, M. D., S. EL MESTIKAWI, M. B. EMERIT, L. PICHAT, M. HAMON & H. GOZLAN. 1985. [^{3}H]-8-hydroxy-2-(di-n-propylamino)tetralin binding to pre- and postsynaptic 5-hydroxytryptamine sites in various regions of rat brain. J. Neurochem. **44:** 1685–1696.

68. PAZOS, A., G. ENGEL & J. M. PALACIOS. 1985. α-Adrenoceptor blocking agents recognize a subpopulation of serotonin receptors in brain. Brain Res. **343:** 205–230.

69. RADJA, F., G. DAVAL, M. B. EMERIT, M. C. GALLISSOT, M. HAMON & D. VERGÉ. 1989. Selective irreversible blockade of 5-hydroxytryptamine$_{1A}$ and 5-hydroxytryptamine$_{1A}$

receptor binding sites in the rat brain by 8-MeO-2'-chloro-PAT: A quantitative autoradiographic study. Neuroscience **31:** 723–733.

70. PEROUTKA, S. J., A. W. SCHMIDT, A. J. SLEIGHT & M. A. HARRINGTON. 1990. Serotonin receptor "families" in the central nervous system: An overview. *In* The Neuropharmacology of Serotonin. P. M. Whittaker-Azmitia & S. J. Peroutka, Eds. Ann. N.Y. Acad. Sci. **600:** 104–112.

71. LUTHMAN, J., B. BOLIOLI, T. TSUTSUMI, A. VERHOFSTAD & G. JONSSON. 1987. Sprouting of striatal serotonin nerve terminals following selective lesions of nigro-striatal dopamine neurons in neonatal rat. Brain Res. Bull. **19:** 269–274.

72. VOIGT, M. M., D. J. LAURIE, P. H. SEEBURG & A. BACH. 1991. Molecular cloning and characterization of a rat brain cDNA encoding a 5-hydroxytryptamine$_{1B}$ receptor. EMBO J. **10:** 4017–4023.

73. EL MANSARI, M., F. RADJA, A. FERRON, T. A. READER, E. MOLINA-HOLGADO & L. DESCARRIES. 1994. Hypersensitivity to serotonin and its agonists in serotonin-hyperinnervated neostriatum after neonatal dopamine denervation. Eur. J. Pharmacol. **261:** 171–178.

74. HOYER, D., S. SRIVASTA, A. PAZOS, G. ENGEL & J. M. PALACIOS. 1986. [^{125}I]LSD labels 5-HT$_{1C}$ recognition sites in pig choroid plexus membranes. Comparison with [^{3}H]mesulergine and [^{3}H]5-HT binding. Neurosci. Lett. **69:** 269–274.

75. PAZOS, A. & J. M. PALACIOS. 1985. Quantitative autoradiographic mapping of serotonin receptors in rat brain. I. Serotonin-1 receptors. Brain Res. **346:** 205–230.

76. FRAZER, A., S. MAAYANI & B. B. WOLFE. 1988. Subtypes of receptors for serotonin. Annu. Rev. Pharmacol. Toxicol. **30:** 307–348.

77. HERRICK-DAVIS, K., I. M. MAISONNEUVE & M. TITILER. 1989. Postsynaptic localization and up-regulation of serotonin 5-HT$_{1D}$ receptors in rat brain. Brain Res. **483:** 155–157.

78. MENGOD, G., H. NGUYEN, C. WAEBER, H. LÜBBERT & J. M. PALACIOS. 1990. The distribution and cellular localization of the serotonin$_{1C}$ receptor mRNA in the rodent brain examined by in situ hybridization histochemistry. Comparison with receptor binding distribution. Neuroscience **35:** 577–591.

79. LOOPUIJT, L. D. & D. VAN DER KOOY. 1985. Organization of the striatum: Collateralization of its efferent axons. Brain Res. **348:** 86–99.

80. GERFEN, C. R., T. M. ENGBER, L. C. MAHAN, Z. SUSEL, T. N. CHASE, F. J. MOMSMA & D. R. SIBLEY. D$_1$ and D$_2$ dopamine receptor-regulated gene expression of striatonigral and striatopallidal neurons. Neuroscience **250:** 1429–1432.

81. MCKENNA, D. J., A. J. NAZARALI, A. J. HOFFMSN, D. E. NICHOLS, C. A. MATHIS & J. M. SAAVEDRA. 1989. Common receptors for hallucinogens in rat brain: A comparative autoradiographic study using [^{125}I]LSD and [^{125}I]DOI, a new psychotomimetic radioligand. Brain Res. **476:** 45–56.

82. BLUE, M. E., K. A. YAGALOFF, L. A. MAMOUNAS, P. R. HARTIG & M. E. MOLLIVER. 1988. Correspondence between 5-HT$_2$ receptors and serotonergic axons in rat neocortex. Brain Res. **453:** 315–328.

83. MENGOD, G., M. POMPEINANO, I. MARTINEZ-MIR & J. M. PALACIOS. 1990. Localization of the mRNA for 5-HT$_2$ receptor by in situ hybridization histochemistry. Correlation with the distribution of receptor sites. Brain Res. **524:** 139–143.

84. JACKSON, D. & E. D. ABERCROMBIE. 1992. In vivo neurochemical evaluation of striatal serotonergic hyperinnervation in rats depleted of dopamine at infancy. J. Neurochem. **58:** 890–897.

Aspartame Does Not Affect Aminergic and Glutamatergic Receptor Kinetics in Rat Brain

M. A. REILLY AND A. LAJTHA

The Nathan S. Kline Institute for Psychiatric Research
Center for Neurochemistry
Orangeburg, New York 10962

Chronic exposure to antagonists or agonists can up- or down-regulate neurotransmitter receptors. Changes in central nervous system (CNS) receptors might underlie adverse reactions anecdotally attributed to ingestion of aspartame (L-aspartyl-L-phenylalanine methyl ester, APM). This high-intensity sweetener is metabolized in the gastrointestinal tract to phenylalanine, which is absorbed into the circulatory system. Phenylalanine and its metabolite tyrosine compete with tryptophan for transport into brain, where levels of these amino acids may influence the synthesis of the aminergic neurotransmitters norepinephrine, dopamine, and serotonin. The influence of postprandial variation in plasma amino acid levels and of food additives that may affect plasma amino acids deserves careful attention. Phenylalanine with its known neurotoxic effects in phenylketonuria is of special interest. This paper reports the lack of effect of prolonged APM administration on several neurotransmitter receptor systems in brain of adult and weanling rats.

Adult male Sprague-Dawley rats were given APM 500 mg/kg daily for 30 days in their drinking water. By use of standard receptor assay procedures,[1,2] the binding kinetics for adrenergic, dopaminergic, and serotonergic receptors were determined in appropriate brain regions (TABLE 1). Brain content of the aminergic neurotransmitters and their major metabolites was measured, as were plasma and brain levels of phenylalanine, tyrosine, and other amino acids. Neither receptor kinetics (K_d and B_{max}), nor neurotransmitter content, nor levels of phenylalanine and tyrosine in plasma and brain were altered.

Because of the greater vulnerability of the immature mammalian brain to neurotoxic action, studies for possible effects of APM on CNS aminergic function were done in weanlings (age 20 to 22 days) born to rats given APM 500 mg/kg daily in drinking water throughout gestation and lactation. No CNS effect of APM exposure was found in aminergic receptor kinetics (TABLE 1), levels of neurotransmitters and metabolites, or in plasma levels of phenylalanine and tyrosine. However, small but significant decreases occurred in brain levels of phenylalanine, glutamate, and aspartate. Subsequent studies found no effect of perinatal APM exposure on glutamatergic binding (TABLE 1), but decreases in glutamate and aspartate were again observed in cerebral cortex and hippocampus (TABLE 2). After termination of exposure to APM, levels of these excitatory amino acids returned to normal within three weeks (TABLE 2).

We conclude that prolonged high-dose APM ingestion does not affect aminergic or glutamatergic receptor kinetics in brain of adult or weanling rats.[1-3] Because APM administration in drinking water did not significantly affect plasma levels of amino acids, these findings are not unexpected. However, the possible effects of significant changes in plasma amino acids levels require close attention.

TABLE 1. Receptor Kinetics in Rat Brain

Receptor	Control K_d	Control B_{max}	APM (500 mg/kg daily) K_d	APM (500 mg/kg daily) B_{max}
Adrenergic α_1				
Adult	0.05 ± 0.01	203 ± 11	0.04 ± 0.01	215 ± 25
Adrenergic α_2				
Adult	2.5 ± 0.02	115 ± 8	2.1 ± 0.3	100 ± 11
Weanling	1.5 ± 0.04	143 ± 3	1.4 ± 0.06	29 ± 5
Serotonergic 5-HT$_{1A}$				
Adult	2.2 ± 0.3	167 ± 14	2.6 ± 0.6	163 ± 20
Serotonergic 5-HT$_2$				
Adult	0.54 ± 0.04	269 ± 34	0.60 ± 0.10	284 ± 54
Weanling	0.33 ± 0.02	281 ± 15	0.43 ± 0.04	251 ± 17
Dopaminergic D$_1$				
Adult	0.55 ± 0.1	780 ± 54	0.60 ± 0.10	806 ± 62
Weanling	0.35 ± 0.03	852 ± 54	0.39 ± 0.01	743 ± 55
Dopaminergic D$_2$				
Adult	0.06 ± 0.01	279 ± 15	0.05 ± 0.01	250 ± 17
Weanling	0.056 ± 0.005	307 ± 15	0.054 ± 0.004	256 ± 22
Glutamatergic NMDA				
Weanling				
Cortex	2.64 ± 0.17	1.82 ± 0.12	3.11 ± 0.48	1.58 ± 0.17
Hippocampus	4.49 ± 0.88	1.99 ± 0.20	5.41 ± 0.57	1.80 ± 0.20
Total glutamatergic				
Weanling				
Cortex	843 ± 119	35.3 ± 6.7	941 ± 60	37.3 ± 3.2
Hippocampus	537 ± 54	18.9 ± 3.1	534 ± 42	26.9 ± 3.5

NOTE: Dopaminergic receptor kinetics were determined in striatum, serotonergic 5-HT$_{1A}$ kinetics in hippocampus, and all others in cerebral cortex except where otherwise noted.

K_d values represent nM concentrations; B_{max} values are fmol/mg protein, except for glutamatergic receptors, where B_{max} values are pmol/mg protein. Values are means ± SEM for 6–12 individual Scatchard determinations. No K_d or B_{max} comparison (APM treatment vs. control) yielded a statistically significant difference (Student's t test).

Abbreviations: APM, L-aspartyl-L-phenylalanine methyl ester; NMDA, N-methyl-D-aspartate.

TABLE 2. Glutamate and Aspartate Levels in Cerebral Cortex and Hippocampus of Rats after Maternal Ingestion of Aspartame[a]

Region	Amino Acid	Control	APM Exposure
At time of weaning			
Cerebral cortex	Glutamate	9050 ± 260	8230 ± 210[b]
	Aspartate	3720 ± 100	3240 ± 130[b]
Hippocampus	Glutamate	9250 ± 230	8370 ± 290[b]
	Aspartate	2870 ± 140	2790 ± 110
Three weeks after termination of aspartame			
Cerebral cortex	Glutamate	8820 ± 260	10400 ± 1080
	Aspartate	2950 ± 130	3150 ± 83
Hippocampus	Glutamate	8470 ± 540	9640 ± 870
	Aspartate	2010 ± 170	2010 ± 53

[a]Values represent nmol/g tissue and are means ± SEM of five rats per group.
[b]Differences in control vs. APM, $p < 0.05$; all others, not significant (Student's t test). APM, L-aspartyl-L-phenylalanine methyl ester.

REFERENCES

1. REILLY, M. A., E. A. DEBLER, A. FLEISCHER & A. LAJTHA. 1989. Biochem. Pharmacol. **38:** 4339–4341.
2. REILLY, M. A., E. A. DEBLER & A. LAJTHA. 1990. Res. Commun. Psychol. Psychiatry Behav. **15:** 141–159.
3. REILLY, M. A., E. A. DEBLER, A. FLEISCHER & A. LAJTHA. 1989. Res. Commun. Psychol. Psychiatry Behav. **14:** 287–303.

Molecular and Functional Diversity of Histamine Receptor Subtypes

JEAN-MICHEL ARRANG, GUILLAUME DRUTEL,
MONIQUE GARBARG, MARTIAL RUAT,
ELISABETH TRAIFFORT,
AND JEAN-CHARLES SCHWARTZ

Unité de Neurobiologie et Pharmacologie (U. 109) de l'INSERM
Centre Paul Broca
2ter rue d'Alésia
75014 Paris, France

The early history of histamine is largely associated with allergy. The major actions of histamine were described at the beginning of this century by Sir Henry Dale and his colleagues after its isolation from ergot extracts. Namely, its potent contractile effects on smooth muscles and the capillary dilation it induces, which mimic some initial manifestations of the anaphylactic shock, were identified by these scientists. They also detected the presence of the amine in many tissues, but it was another British scientist, Feldberg, who clearly demonstrated that histamine was released from the lung during the anaphylactic response and that it induced a marked bronchoconstriction.

The idea that histamine exerts its various biological effects via interaction with several distinct receptor subtypes progressively arose mainly with the design of subtype-selective antagonists. It was first realized that the "antihistamines" (now termed H_1-receptor antagonists), the first of which were designed by Bovet and Staub,[1] did not block uniformly all actions of histamine, leaving, for instance, gastric secretion unaffected. On this basis as well as on the differential action of agonists, Ash and Schild[2] clearly postulated the existence of a second receptor subtype. The existence of the H_2 receptor was definitively established with the design of burimamide, a selective (non-H_1) antagonist, as well as of several relatively selective agonists.[3]

Arrang *et al.*[4] proposed the existence of the third receptor subtype, an autoreceptor controlling the synthesis and release of histamine in cerebral neurons. Four years later, the existence of the H_3 receptor was definitively established with the design of highly potent and selective agonists and antagonists.[5]

The fields of histamine receptor pharmacology and biochemistry were recently reviewed in an extensive manner.[6-8] However, the very recent cloning of the genes encoding the histamine H_1- and H_2-receptor subtypes has notably enlarged our knowledge of these receptors. Although the histamine H_3 receptor has not yet been cloned, all three seem to belong to the superfamily of receptors with seven transmembrane domains and coupled to guanylnucleotide-sensitive G-proteins (TABLE 1).

THE HISTAMINE H_1 RECEPTOR

The H_1-receptor pharmacology was initially defined in functional assays such as smooth muscle contraction, with the design of potent antagonists, the so-called antihistamines, most of which are known to interfere with central histaminergic transmissions and display prominent sedative properties. Biochemical and localization studies of the H_1 receptor were made feasible with the design of reversible and irreversible radiolabeled probes such as [³H]mepyramine, [¹²⁵I]iodobolpyramine, and [¹²⁵I]iodoazidophenpyramine (reviewed in refs. 9–11).

Initial biochemical studies indicated that the cerebral guinea pig H_1 receptor was a glycoprotein of apparent molecular mass of 56 kDa with critical disulfide bonds and that agonist binding was regulated by guanyl nucleotides, implying that the receptor

TABLE 1. Properties of Three Histamine Receptor Subtypes

Property	H_1	H_2	H_3
Coding sequence	491 a.a. (bovine) 488 a.a. (guinea pig) 486 a.a. (rat) 487 a.a. (human)	358 a.a. (rat) 359 a.a (dog, human)	?
Chromosome localization	Chromosome 3	?	?
Highest brain densities	Thalamus Cerebellum Hippocampus	Striatum Cerebral cortex Amydgala	Striatum Frontal cortex Substantia nigra
Autoreceptor	No	No	Yes
Affinity for histamine	Micromolar	Micromolar	Nanomolar
Characteristic agonists	2(m. chlorophenyl) histamine	Impromidine Sopromidine	(R)α-methylhistamine Imetit
Characteristic antagonists	Mepyramine (pyrilamine)	Cimetidine	Thioperamide
Radioligands	[³H]Mepyramine [¹²⁵I]Iodobolpyramine	[³H]Tiotidine [¹²⁵I]Iodoamino- potentidine	[³H](R)α-methylhistamine [¹²⁵I]Iodophenpropit [¹²⁵I]Iodoproxyfan
Second messengers	Inositol phosphates (+) Arachidonic acid (+) cAMP (potentiation)	cAMP (+) Arachidonic acid (−) Ca^{2+} (+)	Inositol phosphates (−)

a.a.: Amino acid residue.

belongs to the superfamily of receptors coupled to G-proteins. In addition, various intracellular responses were found to be associated with H_1-receptor stimulation, for example, inositol phosphate release, increase in Ca^{2+} fluxes, cyclic AMP and cyclic GMP accumulation in whole cells, arachidonic acid release.[6,12] It was not known, however, whether such a variety of responses corresponds to a single receptor or to distinct isoreceptors. Indeed several photoaffinity-labeled proteins of slightly different masses, but similar H_1 pharmacology, were detected in some tissues.[10]

In spite of preliminary attempts using affinity columns with a mepyramine derivative, the H_1 receptor was never purified to homogeneity. Nevertheless, the deduced amino acid sequence of a bovine H_1 receptor was recently disclosed after expression cloning of a corresponding cDNA. The latter was based upon the detection of a Ca^{2+}-dependent Cl^- influx into microinjected *Xenopus* oocytes. Following the transient expression of the cloned cDNA into COS-7 cells, the identity of the protein as an H_1 receptor was confirmed by binding studies.[13] More recently, by using the cloned bovine cDNA as a probe, the gene encoding the H_1 receptor was isolated in rats,[14] guinea pigs,[15,16] and humans.[17]

We recently cloned a guinea-pig cDNA encoding an H_1 receptor in order to identify the signaling systems of the H_1 receptor in a well-studied animal species, as well as to assess the possible existence of isoreceptors.[16] It encodes a glycoprotein of 488 amino acids (FIG. 1) with a calculated M_r of 56 kDa, in good agreement with the apparent size of the photoaffinity-labeled receptor from guinea-pig brain or heart, as determined by SDS/PAGE analysis.[18,19] Northern blot analysis of a variety of guinea-pig peripheral or cerebral tissues identified, in most cases, a single transcript of 3.3 kb. However, in some tissues—for example, ileum or lung—a second transcript of 3.7 kb was generated, possibly by the use of distinct promoter or polyadenylation sites or corresponding to a transcript from a distinct gene.[15,16]

In situ hybridization studies showed a highly contrasted expression of the H_1 receptor gene transcript in guinea-pig brain.[16] When compared with the autoradiographic localization of the corresponding receptor protein,[20] consistent as well as complementary information was provided. For instance, the mRNAs were found in high levels in cerebellar Purkinje cells and hippocampal pyramidal cells, whereas dense [^{125}I]iodobolpyramine binding sites are found in the molecular layers of both areas. This presumably reflects the synthesis of the receptor in perikarya and its final insertion in membranes of the abundant dendritic trees of both cell types.

Transfection with the guinea-pig gene followed by stable expression of H_1 receptors by a CHO cell line allowed the characterization of multiple signaling pathways.[21] In each case the involvement of a G_i/G_o protein with pertussis toxin (PTX), the influence of extracellular Ca^{2+} and of protein kinase C (PKC) activation by phorbol 12-myristate 13-acetate (PMA) were assessed.

Histamine induced in a PTX- and PMA-insensitive manner a biphasic increase in intracellular Ca^{2+} level of which only the second, sustained phase was dependent on the extracellular Ca^{2+} level. In addition, histamine also caused a threefold elevation of inositol phosphate production, which was PTX-insensitive but slightly inhibited by PMA and reduced by 75% in the absence of extracellular Ca^{2+}.

Histamine also caused a massive release of arachidonic acid (AA), occurring in a Ca^{2+}- and PMA-sensitive manner, probably through the activation of a cytosolic phospholipase A_2, which partly involves coupling to a PTX-sensitive G-protein. In comparison, in HeLa cells endowed with a native H_1 receptor, the histamine-induced arachidonic acid release was also Ca^{2+}- and PMA-sensitive, but totally PTX-insensitive.

Finally, in the same CHO(H_1) cell line, histamine in very low concentrations potentiated the cyclic AMP accumulation induced by forskolin. This response appeared to be insensitive to PTX, extracellular calcium, and PMA.

These various observations show that stimulation of a single receptor subtype, the guinea-pig H_1 receptor, can trigger four major intracellular signals, presumably through coupling to several G-proteins, which are variously modulated by extracellular Ca^{2+} and PKC activation.

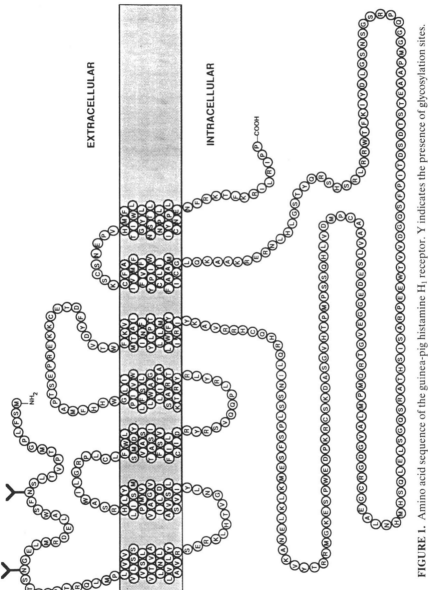

FIGURE 1. Amino acid sequence of the guinea-pig histamine H_1 receptor. Y indicates the presence of glycosylation sites.

THE HISTAMINE H₂ RECEPTOR

Until recently, the information on H_2 receptors was mainly derived from the physiological and biochemical responses they mediate, and molecular properties of the H_2 receptor have remained largely unknown for a long time. For instance, only recently the reversible labeling of the H_2 receptor was achieved using [³H]tiotidine[22] or, more reliably, [¹²⁵I]iodoaminopotentidine.[23] Irreversible labeling, achieved with a photoaffinity probe, followed by SDS-PAGE, led to the identification of H_2 receptor peptides from the guinea-pig brain.[23]

By screening cDNA or genomic libraries with homologous probes, the gene encoding the H_2 receptor was first identified in dogs[24] and, subsequently, in rats[25] and humans.[26] Comparison of these proteins shows that they display a high degree of homology, that is, 82% between the rat and dog receptor (FIG. 2), whereas the degree of homology between the H_1 and H_2 receptors is limited. The H_2 receptor is organized like other receptors positively coupled to adenylyl cyclase; that is, it displays a short (30 amino acids) third intracellular loop and a long (71 amino acids) C-terminal cytoplasmic tail (FIG. 2).

Consistent with their histamine binding function, the H_2 receptors display in the third transmembrane helix (TM3) an aspartate residue (Asp^{98}) likely to bind the ammonium group of the endogenous ligand, because it is found in all other aminergic receptors. In the TM5, an aspartate and a threonine residue (Asp^{185} and Thr^{189} in the rat and Asp^{186} and Thr^{190} in the dog) seemed responsible for hydrogen bonding with the nitrogen atoms of the imidazole ring of histamine. This was partially confirmed by site-directed mutagenesis.[27]

A potential regulation of the rat H_2 receptor by phosphorylation is suggested by the presence of three consensus sites for protein kinase C.

Northern blot analysis of various tissues using a probe derived from the rat cDNA sequence revealed the presence of a single major transcript of 6.0 and 4.5 kb in rat and guinea pig, respectively.[25,28] The distribution of the mRNAs in these two species was consistent with the known distribution of the receptor as mainly established using the sensitive probe [¹²⁵I]iodoaminopotentidine.[23]

Transfected CHO cells were found to express a high level of rat H_2 receptors.[28] In these cells, histamine, in low concentration, induced an accumulation of cAMP, confirming the association of the H_2 receptor with adenylyl cyclase. In addition, in the same cells, histamine potently inhibited the release of arachidonic acid induced by stimulation of constitutive purinergic receptors or by application of a Ca^{2+} ionophore. This inhibition was independent of either cAMP or Ca^{2+} levels in the cells. The results indicate that a single H_2 receptor may be linked not only to adenylyl cyclase activation but also to reduction of phospholipase A_2 activity. Because H_1 receptors have been reported to stimulate arachidonic acid release, inhibition of this release, an unexpected signaling pathway for H_2 receptors, may account for the opposite physiological responses elicited in many tissues by activation of these two receptor subtypes.

In rat hepatoma-derived cells transfected with the canine H_2-receptor gene, histamine induced an increase in intracellular cAMP and Ca^{2+} concentrations, revealing in this system a positive coupling of the H_2 receptor to two signaling mechanisms.[29]

THE HISTAMINE H₃ RECEPTOR

The H_3 receptor was initially detected as an autoreceptor controlling histamine synthesis and release in brain.[4] It was thereafter shown to inhibit presynaptically the

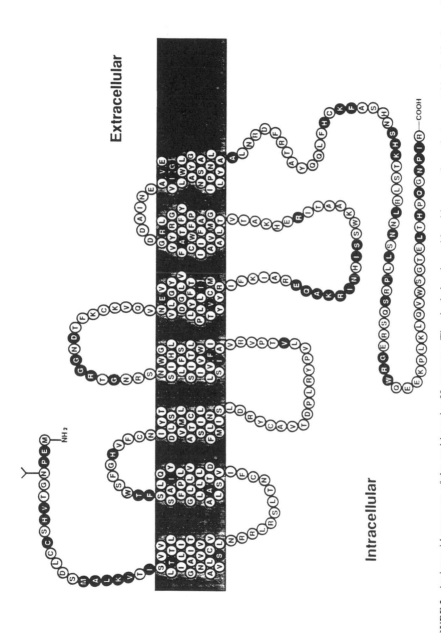

FIGURE 2. Amino acid sequence of the rat histamine H_2 receptor. The shaded amino acid residues are those that are not identical in dog and rat.

release of other monoamines in brain and peripheral tissues as well as of neuropeptides from unmyelinated C-fibers (reviewed in refs. 7 and 30).

The molecular structure of the H_3 receptor remains to be established. Reversible labeling of this receptor was first achieved using a highly selective agonist [³H](R)α-methylhistamine.[5] [³H]Nα-methylhistamine, a less selective agonist,[31] and, more recently, [¹²⁵I]iodoproxyfan, a selective antagonist,[32] were also proposed. It appears that the binding of [³H](R)α-methylhistamine is regulated by guanyl nucleotides, strongly suggesting that the H_3 receptor, like the other histamine receptors, belongs to the superfamily of receptors coupled to G-proteins.[33] Constitutive H_3 receptors in a gastric cell line appear to be negatively coupled to phospholipase C.[34] In the vascular smooth muscle, H_3 receptors mediate voltage-dependent Ca^{2+}-channel stimulation via a pertussis-insensitive G-protein.[35]

During the last few years several potent and highly selective H_3 receptor agonists were designed.[36] Among them, (R)α-methylhistamine[5] and (R)α,(S)β-dimethylhistamine[36] display a high degree of stereoselectivity, imetit[37] being a nonchiral and very potent H_3-receptor agonist (TABLE 2).

TABLE 2. Potent and Selective H_3-Receptor Agonists

Compound	Relative Potency at Receptors		
	H_1	H_2	H_3
Histamine: Im-CH$_2$-CH$_2$-NH$_2$	100	100	100
(R)α-methylhistamine:lm-CH$_2$-CH(CH$_3$)-NH$_2$	0.5	1	1,500
(R)α,(S)β-dimethylhistamine: lm-CH(CH$_3$)-CH(CH$_3$)-NH$_2$	0.03	0.2	1,800
Imetit: Im-CH$_2$-CH$_2$-SC(NH$_2$)=NH	<0.1	0.6	6,200

By the use of these compounds as well as the prototypic H_3-receptor antagonist thioperamide,[5] several effects and physiological roles of histamine could be unraveled or confirmed.

In the brain, H_3-receptor ligands have largely confirmed the role played by histaminergic neurons in cortical activation and arousal mechanisms.[7,38] In the respiratory tract, H_3 receptors inhibit both acetylcholine release from the vagus nerve and the release of neuropeptides from sensory nerves.[39] In the digestive system, similar prejunctional H_3 receptors are involved in the regulation of gastrointestinal functions.[40] However, a direct stimulation of H_3 receptors on enterochromaffin-like cells in the effector organs has also been reported.[41,42] Both populations of H_3 receptors are likely to be involved in the regulation of gastric acid secretion.[40,43]

CONCLUSIONS

All three histamine receptor subtypes presently known were identified through the classical strategy based upon the design of a suitable bioassay and the synthesis of new chemical drugs. In the case of H_1 and H_2 receptors, this strategy has led not only to fundamental discoveries in the field of receptors, but, at the same time, to very useful drugs for treating life-threatening allergic and gastrointestinal disorders. A similar classical process involving a collaboration between pharmacologists and chemists led us to define the H_3 receptor. Although this remains to be firmly established, it can be anticipated that some of the H_3-receptor ligands will constitute novel drugs for the treatment of central or peripheral disorders in humans. The molecular biology approach has already allowed to complement, in greater detail, information about H_1 and H_2 receptors. It would be surprising if this cloning strategy, which has been so fruitful for the discovery of new isoforms or receptor subtypes, does not lead, as in other areas, to an expansion of the histamine receptor family during the coming years.

REFERENCES

1. BOVET, D. & A. M. STAUB. 1937. Action protectrice des éthers phénoliques au cours de l'intoxication histaminique. C. R. Soc. Biol. Paris **124:** 547–549.
2. ASH, A. S. F. & H. O. SCHILD. 1966. Receptors mediating some actions of histamine. Br. J. Pharmacol. Chemother. **27:** 427–439.
3. BLACK, J. W., W. A. M. DUNCAN, C. J. DURANT, C. R. GANELLIN & M. E. PARSONS. 1972. Definition and antagonism of histamine H_2-receptors. Nature **236:** 385–390.
4. ARRANG, J. M., M. GARBARG & J-C. SCHWARTZ. 1983. Autoinhibition of histamine release mediated by a novel class (H_3) of histamine receptor. Nature **302:** 832–837.
5. ARRANG, J. M., M. GARBARG, J-C. LANCELOT, J. M. LECOMTE, H. POLLARD, M. ROBBA, W. SCHUNACK & J-C. SCHWARTZ. 1987. Highly potent and selective ligands for histamine H_3-receptors. Nature **327:** 117–123.
6. HILL, S. J. 1990. Distribution, properties and functional characteristics of three classes of histamine receptors. Pharmacol. Rev. **42:** 46–83.
7. SCHWARTZ, J-C., J. M. ARRANG, M. GARBARG, H. POLLARD & M. RUAT. 1991. Histaminergic transmission in the mammalian brain. Physiol. Rev. **71:** 1–51.
8. SCHWARTZ, J. C. & H. L. HAAS, EDS. 1992. The Histamine Receptor. Vol. 16. Receptor Biochemistry and Methodology. Wiley Liss Inc. New York.
9. GARBARG, M., E. TRAIFFORT, M. RUAT, J. M. ARRANG & J-C. SCHWARTZ. 1992. Reversible labelling of H_1, H_2 and H_3-receptors. *In* The Histamine Receptor. J-C. Schwartz & H. L. Haas, Eds.: 73–95. Wiley Liss Inc. New York.
10. RUAT, M., E. TRAIFFORT & J-C. SCHWARTZ. 1992. Biochemical properties of histamine receptors. *In* The Histamine Receptor. J-C. Schwartz & H. L. Haas, Eds.: 97–107. Wiley Liss Inc. New York.
11. POLLARD, H. & M. L. BOUTHENET. 1992. Autoradiographic visualization of the three histamine receptor subtypes. *In* The Histamine Receptor. J-C. Schwartz & H. L. Haas, Eds.: 179–192. Wiley Liss Inc. New York.
12. HILL, S. J. & J. DONALDSON. 1992. The H_1 receptor and inositol phospholipid hydrolysis. *In* The Histamine Receptor. J-C. Schwartz & H. L. Haas, Eds.: 109–128. Wiley Liss Inc. New York.
13. YAMASHITA, M., H. FUKUI, K. SUGAWA, Y. HORIO, S. ITO, H. MIZUGUCHI & H. WADA. 1991. Expression cloning of a cDNA encoding the bovine histamine H_1 receptor. Proc. Natl. Acad. Sci. USA **88:** 11515–11519.
14. FUJIMOTO, K., Y. HORIO, K. SUGAMA, S. ITO, Y. Q. LIU & H. FUKUI. 1993. Genomic cloning of the rat histamine H_1 receptor. Biochem. Biophys. Res. Commun. **190:** 294–301.
15. HORIO, Y., Y. MORI, I. IGUCHI, K. FUJIMOTO, S. ITO & H. FUKUI. 1993. Molecular cloning of the guinea pig histamine H_1 receptor gene. J. Biochem. **114:** 408–414.
16. TRAIFFORT, E., R. LEURS, J. M. ARRANG, J. TARDIVEL-LACOMBE, J. DIAZ, J-C. SCHWARTZ & M. RUAT. 1994. Guinea pig histamine H_1 receptor. I. Gene cloning, characterization and tissue expression revealed by *in situ* hybridization. J. Neurochem. **62:** 507–518.

17. DE BACKER, M. D., W. GOMMEREN, H. MOEREELS, G. NOBELS, P. VAN GOMPEL, J. E. LEYSEN & W. H. M. L. LUYTEN. 1993. Genomic cloning, heterologous expression and pharmacological characterization of a human histamine H_1 receptor. Biochem. Biophys. Res. Commun. **197:** 1601–1608.

18. RUAT, M., M. KÖRNER, M. GARBARG, C. GROS, J-C. SCHWARTZ, W. TERTIUK & C. R. GANELLIN. 1988. Characterization of histamine H_1-receptor binding peptides in guinea pig brain using [^{125}I]iodoazidophenpyramine, an irreversible specific photoaffinity probe. Proc. Natl. Acad. Sci. USA **85:** 2743–2747.

19. RUAT, M., M. L. BOUTHENET, J-C. SCHWARTZ & C. R. GANELLIN. 1990. Histamine H_1 receptor in heart: Unique electrophoretic mobility and autoradiographic localization. J. Neurochem. **55:** 379–385.

20. BOUTHENET, M. L., M. RUAT, N. SALES, M. GARBARG & J-C. SCHWARTZ. 1988. A detailed mapping of histamine H_1-receptors in guinea pig central nervous system established by autoradiography with [^{125}I]iodobolpyramine. Neuroscience **26:** 553–600.

21. LEURS, R., E. TRAIFFORT, J. M. ARRANG, J. TARDIVEL-LACOMBE, M. RUAT & J-C. SCHWARTZ. 1994. Guinea pig histamine H_1 receptor. II. Stable expression in Chinese hamster ovary cells reveals the interaction with three major signal transduction pathways. J. Neurochem. **62:** 519–527.

22. GAJTKOWSKI, G. A., D. B. NORRIS, T. J. RISING & T. P. WOOD. 1983. Specific binding of ^{3}H-tiotidine to histamine H_2-receptors in guinea pig cerebral cortex. Nature **304:** 65–67.

23. RUAT, M., E. TRAIFFORT, M. L. BOUTHENET, J-C. SCHWARTZ, J. HIRSCHFELD, A. BUSCHAUER & W. SCHUNACK. 1990. Reversible and irreversible labeling and autoradiographic localization of the cerebral histamine H_2 receptor and [^{125}I]iodinated probes. Proc. Natl. Acad. Sci. USA **87:** 1658–1662.

24. GANTZ, I., M. SCHAFFER, J. DELVALLE, C. LOGSDON, V. CAMPBELL, M. UHLER & T. YAMADA. 1991. Molecular cloning of a gene encoding the histamine H_2 receptor. Proc. Natl. Acad. Sci. USA **88:** 429–433.

25. RUAT, M., E. TRAIFFORT, J. M. ARRANG, R. LEURS & J-C. SCHWARTZ. 1991. Cloning and tissue expression of a rat histamine H_2-receptor gene. Biochem. Biophys. Res. Commun. **179:** 1470–1478.

26. GANTZ, I., G. MUNZERT, T. TASHIRO, M. SCHAFFER, L. WANG, J. DELVALLE & T. YAMADA. 1991. Molecular cloning of the human histamine H_2 receptor. Biochem. Biophys. Res. Commun. **178:** 1386–1392.

27. GANTZ, I., J. DELVALLE, L. D. WANG, T. TASHIRO, G. MUNZERT, Y. J. GUO, Y. KONDA & T. YAMADA. 1992. Molecular basis for the interaction of histamine with the histamine H_2 receptor. J. Biol. Chem. **267:** 20840–20843.

28. TRAIFFORT, E., M. RUAT, J. M. ARRANG, R. LEURS, D. PIOMELLI & J-C. SCHWARTZ. 1992. Expression of a cloned rat histamine H_2 receptor mediating inhibition of arachidonate release and activation of cAMP accumulation. Proc. Natl. Acad. Sci. USA **89:** 2649–2653.

29. DELVALLE, J., L. WANG, I. GANTZ & T. YAMADA. 1992. Characterization of H_2-histamine receptor: Linkage to both adenylate cyclase and [Ca^{2+}]i signaling systems. Am. J. Physiol. **263:** G967–G972.

30. ARRANG, J. M., M. GARBARG & J-C. SCHWARTZ. 1992. H_3-receptor and control of histamine release. *In* The Histamine Receptor. J-C. Schwartz & H. L. Haas, Eds.: 145–159. Wiley Liss Inc. New York.

31. KORTE, A., J. MYERS, N. Y. SHIH, R. W. EGAN & M. A. CLARK. 1990. Characterization and tissue distribution of H_3-histamine receptors in guinea pigs by N^{α}-methylhistamine. Biochem. Biophys. Res. Commun. **168:** 979–986.

32. LIGNEAU, X., M. GARBARG, M. L. VIZUETE, J. DIAZ, K. PURAND, H. STARK, W. SCHUNACK & J. C. SCHWARTZ. 1994. [^{125}I]Iodoproxyfan, a new antagonist to label and visualize cerebral histamine H_3 receptors. J. Pharmacol. Exp. Ther. **271:** 452–459.

33. ARRANG, J. M., J. ROY, J. L. MORGAT, W. SCHUNACK & J-C. SCHWARTZ. 1990. Histamine H_3-receptor binding sites in rat brain membranes: Modulation by guanine nucleotides and divalent cations. Eur. J. Pharmacol. **188:** 219–227.

34. CHERIFI, Y., C. PIGEON, M. LE ROMANCER, A. BADO, F. REYL-DESMARS & M. J. M. LEWIN. 1992. Purification of a histamine H_3 receptor negatively coupled to phosphoinositide turnover in the human gastric cell line HGT1. J. Biol. Chem. **267:** 25315–25320.
35. OIKE, M., K. KITAMURA & H. KURIYAMA. 1992. Histamine H_3-receptor activation augments voltage-dependent Ca^{2+} current via GTP hydrolysis in rabbit saphenous artery. J. Physiol. **448:** 133–152.
36. LIPP, R., H. STARK & W. SCHUNACK. 1992. Pharmacochemistry of H_3 receptors. *In* The Histamine Receptor. J-C. Schwartz & H. L. Haas, Eds.: 57–72. Wiley Liss Inc. New York.
37. GARBARG, M., J. M. ARRANG, A. ROULEAU, X. LIGNEAU, M. DAM TRUNG TUONG, J-C. SCHWARTZ & C. R. GANELLIN. 1992. S-[2-(4-Imidazolyl)ethyl]isothiourea, a highly specific and potent histamine H_3-receptor agonist. J. Pharmacol. Exp. Ther. **263:** 304–310.
38. LIN, J. S., K. SAKAI, G. VANNI-MERCIER, J. M. ARRANG, M. GARBARG, J-C. SCHWARTZ & M. JOUVET. 1990. Involvement of histaminergic neurons in arousal mechanisms demonstrated with H_3-receptor ligands in the cat. Brain Res. **523:** 325–330.
39. BARNES, P. J. 1992. Histamine receptors in the respiratory tract. *In* The Histamine Receptor. J-C. Schwartz & H. L. Haas, Eds.: 253–270. Wiley Liss Inc. New York.
40. BERTACCINI, G. & G. CORUZZI. 1992. Histamine receptors in the digestive system. *In* The Histamine Receptor. J-C. Schwartz & H. L. Haas, Eds.: 193–230. Wiley Liss Inc. New York.
41. SCHWÖRER, H., S. KATSOULIS & K. RACKE. 1992. Histamine inhibits 5-hydroxytryptamine release from the porcine small intestine: Involvement of H_3 receptors. Gastroenterology **102:** 1906–1912.
42. PRINZ, C., M. KAJIMURA, D. R. SCOTT, F. MERCIER, H. F. HELANDER & G. SACHS. 1993. Histamine secretion from rat enterochromaffinlike cells. Gastroenterology **105:** 449–461.
43. BADO, A., L. MOIZO, J. P. LAIGNEAU & M. J. LEWIN. 1992. Pharmacological characterization of histamine H_3 receptors in isolated rabbit gastric glands. Am. J. Physiol. **262:** G56–G61.

Recent Studies on a μ-Opioid Receptor Purified from Bovine Striatum[a]

ERIC J. SIMON

Departments of Psychiatry and Pharmacology
New York University Medical Center
550 First Avenue
New York, New York 10016

The existence of opioid receptors in the central nervous system of animals and humans was demonstrated in 1973 simultaneously by our laboratory[1,2] and by others.[3-5] At the time only a single type of receptor was postulated, a concept soon challenged and disproved. Martin and co-workers[6,7] postulated three classes of receptors, μ, κ, and σ. The latter is now not considered to be an opioid receptor and has been replaced by δ, an enkephalin-preferring receptor first demonstrated by Kosterlitz and co-workers[8] in the mouse vas deferens. The enkephalins are therefore thought to be the endogenous ligands of the δ-receptor, dynorphin that of the κ-receptor, and β-endorphin a putative endogenous ligand of the μ-receptor. Within the past year the δ-opioid receptor was cloned independently by two laboratories.[9,10] This was rapidly followed by the cloning of a μ- and a κ-receptor,[11,12] by low stringency screening of cDNA libraries with oligonucleotide probes from the structure of the δ-receptor cDNA. The protein sequences derived from the cDNA's coding for the three types of opioid receptors are shown in FIGURE 1. All of them have structures with seven putative transmembrane domains, typical of G-protein coupled receptors. The figure shows a number of receptors of this class that exhibit high levels of homology with the opioid receptors.

There is considerable pharmacological evidence for the existence of multiple subtypes of each of the three major types of opioid receptors. However, no evidence for subtypes has yet been provided by cloning, that is, all clones of a given type, obtained in a number of laboratories,[13-19] including those cited earlier, have proved to be identical to the cDNA's coding for the μ-, δ-, and κ-opioid receptors originally published. The present paper discusses recent work in our laboratory on a μ-opioid receptor purified to homogeneity from bovine striatal membranes, which appears to be a subtype, different from the μ-receptor which has been cloned in several laboratories. Final proof of this hypothesis awaits the cloning of this protein.

RESULTS AND DISCUSSION

Purification and Microsequencing of a μ-Opioid Binding Protein

The purification to homogeneity of an opioid binding protein (OBP) from bovine striatal membrane was reported by us some years ago.[20] Briefly, the membranes were solubilized with digitonin and the extract was purified in two steps, ligand-affinity chromatography, which provided the major purification (4000- to 5000-fold) fol-

[a]This work is supported by National Institute on Drug Abuse grant DA-00017. Gifts from Hoffmann-LaRoche, Inc. and Pfizer Central Research are gratefully acknowledged.

lowed by lectin-affinity chromatography on wheat-germ agglutinin agarose, for a total purification of 60,000- to 70,000-fold. The purified protein gives a single band on SDS-polyacrylamide gel electrophoresis corresponding to a molecular mass of 65 kDa. It is highly glycosylated and enzymatic deglycosylation yields a protein of 40–43 kDa (Gioannini and Simon, unpublished results). Preliminary evidence suggested that it has the properties of a μ-opioid receptor, though this evidence was not conclusive because the purified protein bound only antagonists with high affinity. High-affinity agonist binding was lost, presumably due to uncoupling of the receptor protein from G-protein (see below for experiments on reconstitution).

The purified protein was found to be blocked at the N-terminus. Fragmentation with cyanogen bromide and sequencing, kindly carried out for us by Dr. C. Strader at Merck & Co., provided two peptides, one 20 amino acids long (peptide I) and the other 13 amino acids in length (peptide II). More recently, Dr. W. Burgess at the American Red Cross obtained four more short peptides for us by fragmentation with lysine-protease and microsequencing. It is of interest that none of these six peptides are found in the published amino acid sequence of the cloned μ-opioid receptor. Attempts to clone OBP, using PCR and direct screening of cDNA and genomic libraries, with oligonucleotides made from peptides I and II, have not yet succeeded. This is presumably because of the relatively short peptides and their high level of degeneracy. The following sections deal with studies using antibodies (Abs) made against portions of peptide I and II and with the reconstitution of purified OBP with G-proteins in liposomes.

Antisera against Peptides Derived from Opioid Binding Protein

Our laboratory, in collaboration with Dr. Lawrence Taylor in Dr. Huda Akil's laboratory, University of Michigan (Ann Arbor, MI), produced six antisera against three sequences in peptides I and II.[21] TABLE 1 lists the sequences and the Abs made against them. All of these Abs recognized the purified OBP, from which the peptides are derived. They immunoprecipitate the bulk of radioiodine-labeled OBP and give an immunoblot at the correct molecular mass of 65 kDa. Most of our subsequent studies were done with antiserum Ab165, made against the N-terminal 12 amino acids of peptide I, the most immunoreactive serum against the OBP protein. In studies of a number of bovine brain regions, positive immunoblot signals corresponding to a molecular mass of 65 kDa were obtained with regions known to be rich in μ-receptors, whereas no signal was obtained with regions devoid of opioid receptors, such as white matter. Similarly, the cell line, SK-N-SH, high in μ-opioid receptors, gave a band in immunoblots, whereas HeLa and C-6 cells, devoid of opioid receptors, do not. Interestingly, NG-108-15 cells that express only δ-receptors give a band at a slightly lower molecular mass of approximately 60 kDa with Ab165. This may suggest cross-reaction with δ-receptors or the presence of silent and slightly different μ-receptors. Experiments are in progress to determine which of these alternatives is correct.

More recently, Dr. Hiller in our laboratory has carried out immunocytochemical studies of the distribution of Ab165-reactive material in the rat central nervous system.[22] It parallels very closely the distribution of μ-opioid receptor binding, determined by autoradiography by ourselves[23] and by others.[24,25] FIGURE 2 shows an example of these studies, namely, the spinal cord distribution of immunoreactivity to Ab165.

```
            1              #                            #        #   50
MUOR1    .......MDS STGPGNTSDC SDPLAQASCS PAPGSWLNLS HVDGNQSDPC
DOR1     .......... .......... MELVPSARAE LQSSPLVNLS ....DAFPS
KOR1     .......... ......MESP IQIFRGDPGP TCSPSACLLP NSSSWFPNWA
SOMR1    .......... ..MFPNAPPP LPHSSPSSSP GGCGEGVCSR
FPEP     .......... .......... .......... .......... ..........
OBPR     .......... ......MASP AGNLS.AWPG WGWPP..PAA LRNLTSSPAP
NEUROKR  MASVPRGENW TDGTVEVGTH TGNLSSALGV TEWLALQAGN FSSALGLPAT
B2AR     .......... .......... .......... .......... .......MEP

            51#          #                                    100
MUOR1    GLNRTGLGGN DSLCPQTG.. ...SPSMVTA ITIMALYSIV CVVGLFGNFL
DOR1     AFPSAGANAS GSPGARSA.. ...S.SLALA IAITALYSAV CAVGLLGNVL
KOR1     ESDSNGSVGS EDQQLESA.. ...HISPAIP VIITAVYSVV FVVGLVGNSL
SOMR1    GPGSGAADGM EEPGRNSSQN GTLSEGQGSA ILISFIYSVV CLVGLCGNSM
FPEP     ....METNSS LPTNISGGTP AVSAGYLFLD IITYLVFAVT FVLGVLGNGL
OBPR     TASPSPAPSW TPSPRPGPAH PFLQPPWAVA LWSLA.YGAV VAVAVLGNLV
NEUROKR  TQAPSQV... ....RANLTN QFVQPSWRIA LWSLA.YGLV VAVAVFGNLI
B2AR     HGNDSDFLLA PNGSRAPGHD ITQERDEAWV VGMAILMSVI VLAIVFGNVL

            101                                              150
MUOR1    VMYVIVRYTK MKTATNIYIF NLALADALAT STLP.FQSVN YLMGTWPFGT
DOR1     VMFGIVRYTK LKTATNIYIF NLALADALAT STLP.FQSAK YLMETWPFGE
KOR1     VMFVIIRYTK MKTATNIYIF NLALADALVT TTMP.FQSAV YLMNSWPFGD
SOMR1    VIYVILRYAK MKTATNIYIL NLAIADELLM LSVP.FLVTS TLLRHWPFGA
FPEP     VIWVAG.FRM THTVTTISYL NLAVADFCFT STLPFFMVRK AMGGHWPFGW
OBPR     VIWIVLAHKR MRTVTNSFLV NLAFADAAMA ALNALVNFIY ALHGEWYFGA
NEUROKR  VIWIILAHKR MRTVTNYFLV NLAFSDASVA AFNTLINFIY GLHSEWYFGA
B2AR     VITAIAKFER LQTVTNYFIT SLACADLVMG LAVVPFGASH ILMKMWNFGN

            151                                              200
MUOR1    ILCKIVISID YYNMFTSIFT LCTMSVDRYI AVCHPVKALD FRTPRNAKIV
DOR1     LLCKAVLSID YYNMFTSIFT LTMMSVDRYI AVCHPVKALD FRTPAKAKLI
KOR1     VLCKIVISID YYNMFTSIFT LTMMSVDRYI AVCHPVKALD FRTPLKAKII
SOMR1    LLCRLVLSVD AVNMFTSIYC LTVLSVDRYV AVEHPIKAAR YRRPTVAKVV
FPEP     FLCKFVFTIV DINLFGSVFL IALIALDRCV CVLHPVWTQN HRTVSLAKKV
OBPR     NYCRFQNFFP ITAVFASIYS MTAIAVDRYM AIIDPLKPR. .LSATATRIV
NEUROKR  NYCRFQNFFP ITAVFASIYS MTAIAVDRYM AIIDPLKPR. .LSATATKIV
B2AR     FWCEFWTSID VLCVTASIET LCVIAVDRYV AITSPFKYQS LLTKNKARVV

            201                                              250
MUOR1    NVCNWILSSA IGLPVMFMAT TKYRQ..GSI DCTL.TFS.. ..HPTW.YWE
DOR1     NICIWVLASG VGVPIMVMAV TQPRD..GAV VCML.QFP.. ..SPSW.YWD
KOR1     NICIWLLASS VGISAIVLGG TKVREDVDVI ECSL.QFP.. ..DDEYSWWD
SOMR1    NLGVWVLSLL VILPIVVFSR TAANSD.GTV ACNM.LMP.. ..EPAQRWLV
FPEP     IIGPWVMALL LTLPVIIRVT T.VPGKTGTV ACTF.NFSPW TNDPKERIKV
OBPR     IGSIWILAFL LAFPQCLYSK IKVMP..... GRTL.CYVQW PEGSRQHFTY
NEUROKR  IGSIWILAFL LAFPQCLYSK IKVMP..... GRTL.CYVQW PEGPKQHFTY
B2AR     ILMVWIVSGL TSFLPIQMHW YRATHKQAID CYAKETCCDF FTNQAYAIAS
```

FIGURE 1. Amino acid sequences of the cloned µ-, δ-, and κ-opioid receptors and homologous G-protein coupled receptors. Boldface type and shading: putative transmembrane domains. *, putative sites for phosphorylation by protein kinase A. #, putative sites for N-linked glycosylation. (Adapted from Wang et al.,[13] with permission.)

Reconstitution of High-Affinity Binding and GTPase Stimulation

Our preliminary pharmacological evidence as well as the results of the studies with the antibodies directed against peptides from purified OBP suggested that OBP was a µ-binding protein. However, these results were not conclusive, and real proof had to come from pharmacological studies of the reconstituted purified OBP, described in this section.

```
         251                                                 300
MUOR1    NLLKICV.......FIFAFIMP VLIITVCYGL MILRLKSVRM LSGS....KE
DOR1     TVTKICV... ..FLFAFVVP ILIITVCYGL MLLRLRSVRL LSGS....KE
KOR1     LFMKICV... ..FVPAFVIP VLIIIVCYTL MILRLKSVRL LSGS....RE
SOMR1    GFV.LYT... ..FLMGFLLP VGAICLCYVL IIAKMRMVPS RPAG.....S
FPEP     AVAMLTVRGI IRFIIGFSAP MSIVAVSYGL IATKIHKQGL IKSS......
OBPR     HMIVI..... ...VLVYCFP LLIMGITYTI VGITLWGGEI PGDTCDKYQE
NEUROKR  HIIVI..... ...ILVYCFP LLIMGVTYTI VGITLWGGEI PGDTCDKYHE
B2AR     SIVSFYVP.. .LVVMVFVYS RVFQVAKRQL QKIDKSEGRF HAQNLSQVEQ

         301               *                               350
MUOR1    KDRN...... ......LRR ITRMVLVVVA VFIYCWTPIH IYVIIKALIT
DOR1     KDRS...... .......LRR ITRMVLVVVG AFVVCWAPIH IFVIVWTLVD
KOR1     KDRN...... .......LRR ITKLVLVVVA VFIICWTPIH IFILVEALGS
SOMR1    TQRS...... .......ERK ITLMVMMVVM VFVICWMPFY VVQLVNVFAE
FPEP     .......... ........R PLRVLSFVAA AFFLCWSPYQ VVALIATVRI
OBPR     QLKA...... .......KRK VVKMMIIVVV TFAICWLPYH IYFILTAIYQ
NEUROKR  QLKA...... .......KRK VVKMMIIVVV TFAICWLPYH VYFILTAIYQ
B2AR     DGRSGHGLRS SSKFCLKEHK ALKTLGIIMG TFTLCWLPFF IVNIVHVIRA

         351                                                 400
MUOR1    I.PETTF... QTVSWHFCIA LGYTNSCLNP VLYAFLDENF KRCF.REF..
DOR1     INRRDPL... VVAALHLCIA LGYANSSLNP VLYAFLDENF KRCF.RQL..
KOR1     TSHSTA.... ALSSYYFCIA LGYTNSSLNP VLYAFLDENF KRCF.RDF..
SOMR1    QDDATVS... Q......LSVI LGYANSCANP ILYGFLSDNF KRSFQRIL..
FPEP     RELLQGMYKE IGIAVDVTSA LAFFNSCLNP MLYVFMGQDF RERLIHAL..
OBPR     QLNRWKYIQQ VYLASFW... LAMSSTMYNP IIYCCLNKRF RAGFKRAFRW
NEUROKR  QLNRWKYIQQ VYLASFW... LAMSSTMYNP IIYCCLNKRF RAGFKRAFRW
B2AR     NLIPKEV... ....YILLNW LGYVNSAFNP LIYC.RSPDF RIAFQELLCL

         401                                                 450
MUOR1    .CIPTSSTIE QQNSTRVRQN TREHPSTANT VDRTNHQLEN LEAETAPLP.
DOR1     .CRTPCGRQE PGSLRRPRQA TTRERVTACT PS.....DG PGGGAAA...
KOR1     .CFPIKMRME RQSTNRVR.N TVQDPASMRD VGGMNKPV.. ..........
SOMR1    .CL...SWMD NAAEEPVDYY ATALKSRAYS VEDFQPENLE SGGVFRNGTC
FPEP     .PASLERALT EDSTQTSDTA TNSTLPSAEV ALQAK..... ..........
OBPR     CPFIHVSSYD ELELKATRLH PMRQ.SSLYT VTRMESMSVV FDSNDGDSAR
NEUROKR  CPFIQVSSYD ELELKTTRFH PTRQ.SSLYT VSRMESVTVL FDPNDGDPTK
B2AR     RRSSSKTYGN GYSSNSNGRT DYTGEQSAYQ LGQEKENELL CEEAPGMEGF
```

FIGURE 1. (*continued*)

Our initial studies of OBP reconstitution[26] were done with a CHAPS extract of bovine striatal membranes as the source of phospholipids and G-proteins. This extract was prepared in the absence of sodium, which we have shown to solubilize very little opioid binding activity. Heating the extract at 37 °C for 30 min removed any residual traces of opioid binding activity. Lipid vesicles were formed by adding purified OBP to the CHAPS extract, precipitating the mixture with polyethyleneglycol-6000 (PEG), and resuspending the pellet, resulting from centrifugation at 12,000 × g,

TABLE 1. Antipeptide Antibodies Generated against Amino Acid Sequences Derived from Opioid Binding Protein

	Amino Acid Sequence	Antibody
Peptide I-1-12	IRNLRQDRSKYY[X]	165, 166, 6639
Peptide I-14-20	NFFYKRL	163
Peptide II-1-9	Y*SNNVLFVSH[XFND]	161, 162

NOTE: *, Y-tyrosine added to permit iodination of peptide. X indicates unknown amino acids. Bracketed residues were not used for antibody production. (Gioannini et al.[21] Reprinted with permission from *Molecular Pharmacology*.)

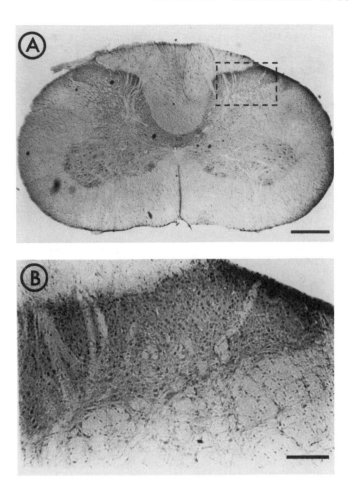

FIGURE 2. Distribution of immunoreactivity against Ab165 in the rat lumbar spinal cord. (**A**) High levels of immunoreactivity are seen in lamina I and lamina II (substantia gelatinosa) of the dorsal horn. Higher magnification of the (boxed area of panel **A**) dorsal horn, shown in panel **B**, reveals that laminae III and IV also contain immunopositive cell bodies. An edge artifact is apparent in the upper right corner of this panel. It is also notable that the median aspect of lamina V (see panel **A**) contains a substantial level of immunoreactivity. In the ventral horns, large individual immunopositive cell bodies can be seen in laminae VII, VIII, and IX. Bars, **A** = 100 μm, **B** = 2 μm. (Hiller *et al.*[22] Reproduced, with permission, from *Neuroscience.*)

in Tris buffer containing magnesium. FIGURE 3 shows that the selective μ-agonist DAGO binds saturably and with high affinity to the reconstituted OBP. The figure also shows that the binding is very selective for μ-ligands. The binding was found to be completely dependent on and proportional to the amount of OBP added to the liposomes, stereospecific and sensitive to inhibition by GTPγS. The affinity of DAGO for reconstituted OBP was 1.5 nM, identical to that seen in the membrane-bound μ-opioid receptor in bovine striatal membranes. Functional coupling was also achieved, as evidenced by the stimulation of low K_m GTPase by μ-agonists, such as DAGO.

More recently we repeated the reconstitution studies with purified G-proteins in collaboration with the laboratory of John Hildebrandt at South Carolina Medical University (Charleston, SC) (manuscript in preparation). Most of the studies were carried out with a highly purified mixture of brain G-proteins, composed of 80% G_{oA}, 7–8% G_{oC}, 7–8% G_{il} and small amounts of G_{oB} and G_{i2}. Some experiments with individual purified G-proteins were also done. The liposomes were formed by mixing purified OBP with phosphatidylcholine (Sigma) and the purified G-proteins. The mixture was precipitated with PEG, centrifuged at $100,000 \times g$, and resuspended in Tris buffer containing magnesium. Functional reconstitution was assessed by measurement of opioid stimulation of low K_m GTPase. Opioid ligands, such as DAGO, morphine and levorphanol, produced up to 100% stimulation of GTPase. The stimulation was reversed by ($-$)naloxone but not by ($+$)naloxone. Dextrorphan, the inactive enantiomer of levorphanol, was inactive and so were δ- and κ-agonists.

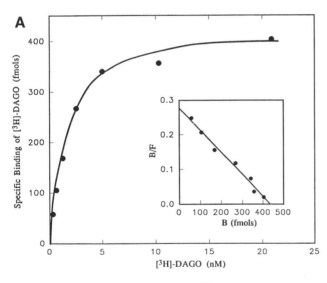

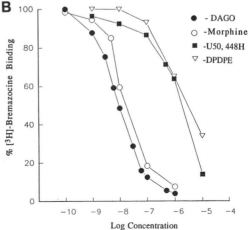

FIGURE 3. Studies of opioid binding to purified opioid binding protein (OBP) reconstituted in liposomes with CHAPS extract of bovine striatal membranes. (**A**) Saturation curve and Scatchard plot of [³H]DAGO binding to reconstituted purified OBP. (**B**) Competition of μ (DAGO), δ (DP-DPE), and κ (U50, 488H) agonist ligands for [³H]bremazocine (1.5 nM) binding to reconstituted OBP. (Gioannini *et al.*[26] Reproduced, with permission, from *Biochemical and Biophysical Research Communications.*)

Studies of μ-agonist binding showed that high-affinity binding to reconstituted OBP was obtained both with the G-protein mixture and with individual G-proteins. Differences in efficacy between different G-proteins were not significant. Furthermore, differences observed could represent differences in incorporation of various G-proteins into liposomes, in the level of coupling or in the nature of the coupling. We cannot at present distinguish among these possibilities. It should be noted that the level of binding and the affinity of DAGO binding were lower than those seen when CHAPS extract was the source of G-proteins. This could be due to the absence of factors that favor high levels of incorporation into liposomes and coupling or to some slight denaturation of G-proteins during purification. The affinity of DAGO binding, K_d = 7 nM, was high but somewhat lower than the K_d = 1.5 nM we observed in the experiments with CHAPS extract.

In summary, our studies with antisera directed against peptides derived from purified OBP, and especially the pharmacological profile obtained with reconstituted OBP, demonstrate clearly that we have purified a μ-receptor. The fact that none of the peptides obtained from the purified protein are identical to sequences in the cloned μ-receptor suggests that it may be a different subtype, the predominant subtype in bovine brain. Our results suggest that this subtype of μ-opioid receptor is also present in rat brain or, alternatively, that sequence homology and/or tertiary structural similarities give rise to cross-reactivity of the rat μ-receptor with our antisera.

REFERENCES

1. SIMON, E. J., J. M. HILLER & I. EDELMAN. 1973. Stereospecific binding of the potent narcotic analgesic ^{3}H-etorphine to rat brain homogenate. Proc. Natl. Acad. Sci. USA **70:** 1947–1949.
2. HILLER, J. M., J. PEARSON & E. J. SIMON. 1973. Distribution of stereospecific binding of the potent narcotic analgesic etorphine in the human brain: Predominance in the limbic system. Res. Commun. Chem. Pathol. Pharmacol. **6:** 1052–1062.
3. PERT, C. B. & S. H. SNYDER. 1973. Opiate receptor: Demonstration in nervous tissue. Science **179:** 1011–1014.
4. KUHAR, M. J., C. B. PERT & S. H. SNYDER. 1973. Regional distribution of opiate receptor binding in monkey and human brain. Nature **245:** 447–451.
5. TERENIUS, L. 1973. Stereospecific interaction between narcotic analgesics and a synaptic plasma membrane fraction of rat cerebral cortex. Acta Pharmacol. Toxicol. **32:** 317–320.
6. MARTIN, W. R., C. G. EADES, J. A. THOMPSON, R. E. HUPPLER & P. E. GILBERT. 1976. The effects of morphine- and nalorphine-like drugs in the nondependent and morphine-dependent chronic spinal dog. J. Pharmacol. Exp. Ther. **197:** 517–532.
7. GILBERT, P. E. & W. R. MARTIN. 1976. The effects of morphine- and nalorphine-like drugs in the nondependent morphine-dependent and cyclazocine-dependent chronic spinal dog. J. Pharmacol. Exp. Ther. **198:** 66–82.
8. LORD, J. A. H., A. A. WATERFIELD, J. HUGHES & H. W. KOSTERLITZ. 1977. Endogenous opioid peptides: Multiple agonists and receptors. Nature **267:** 495–499.
9. EVANS, C. J., D. E. KEITH, JR., H. MORRISON, K. MAGENDZO & R. H. EDWARDS. 1992. Cloning of a delta opioid receptor by functional expression. Science **258:** 1952–1955.
10. KIEFFER, B. L., K. BEFORT, C. GAVERIAUX-RUFF & C. G. HIRTH. 1992. The delta-opioid receptor: Isolation of a cDNA by expression cloning and pharmacological characterization. Proc. Natl. Acad. Sci. USA **89:** 12048–12052.
11. CHEN, Y., A. MESTEK, J. LIU, J. A. HURLEY & L. YU. 1993. Molecular cloning and functional expression of a mu-opioid receptor from rat brain. Mol. Pharmacol. **44:** 8–12.
12. YASUDA, K., K. RAYNOR, H. KONG, C. D. BREDER, J. TAKEDA, T. REISINE & G. I. BELL.

1993. Cloning and functional comparison of kappa and delta opioid receptors from mouse brain. Proc. Natl. Acad. Sci. USA **90:** 6736–6740.

13. WANG, J. B., Y. IMAI, C. M. EPPLER, P. GREGOR, C. E. SPIVAK & G. R. UHL. 1993. Mu opiate receptor: cDNA cloning and expression. Proc. Natl. Acad. Sci. USA **90:** 10230–10234.

14. MINAMI, M., T. TOYA, Y. KATAO, K. MAEKAWA, S. NAKAMURA, T. ONOGI, S. KANEKO & M. SATOH. 1993. Cloning and expression of a cDNA for the rat kappa-opioid receptor. FEBS Lett. **329:** 291–295.

15. NISHI, M., H. TAKESHIMA, K. FUKUDA, S. KATO & K. MORI. 1993. cDNA cloning and pharmacological characterization of an opioid receptor with high affinities for kappa-subtype-selective ligands. FEBS Lett. **330:** 77–80.

16. FUKUDA, K., S. KATO, K. MORI, M. NISHI & H. TAKESHIMA. 1993. Primary structures and expression from cDNAs of rat opioid receptor delta and mu-subtypes. FEBS Lett. **327:** 311–314.

17. MENG, F., G. X. XIE, R. C. THOMPSON, A. MANSOUR, A. GOLDSTEIN, S. J. WATSON & H. AKIL. 1993. Cloning and pharmacological characterization of a rat kappa opioid receptor. Proc. Natl. Acad. Sci. USA **90:** 9954–9958.

18. THOMPSON, R. C., A. MANSOUR, H. AKIL & S. J. WATSON. 1993. Cloning and pharmacological characterization of a rat mu opioid receptor. Neuron **11:** 903–913.

19. BZDEGA, T., H. CHIN, H. KIM, H. H. JUNG, C. A. KOZAK & W. A. KLEE. 1993. Regional expression and chromosomal localization of the delta opiate receptor gene. Proc. Natl. Acad. Sci. USA **90:** 9305–9309.

20. GIOANNINI, T. L., A. D. HOWARD, J. M. HILLER & E. J. SIMON. 1985. Purification of an active opioid binding protein from bovine striatum. J. Biol. Chem. **260:** 15117–15121.

21. GIOANNINI, T. L., Y-H. YAO, J. M. HILLER, L. P. TAYLOR & E. J. SIMON. 1993. Antisera against peptides derived from a purified mu-opioid binding protein recognize the protein as well as mu-opioid receptors in brain regions and a cell line. Mol. Pharmacol. **44:** 796–801.

22. HILLER, J. M., Y. ZHANG, G. BING, T. L. GIOANNINI, E. A. STONE & E. J. SIMON. 1994. Immunohistochemical localization of mu opioid receptors in rat brain using antibodies generated against a peptide sequence present in a purified mu opioid binding protein. Neuroscience **62:** 829–841.

23. PEARSON, J., L. BRANDEIS, E. J. SIMON & J. M. HILLER. 1980. Radioautography of binding of tritiated diprenorphine to opiate receptors in the rat. Life Sci. **26:** 1047–1052.

24. WAMSLEY, J. K. 1983. Opioid receptors: Autoradiography. Pharmacol. Rev. **35:** 69–83.

25. MANSOUR, A. & S. J. WATSON. 1993. Anatomical distribution of opioid receptors in mammalians: An overview. *In* Opioids I, Handbook of Experimental Pharmacology. Vol. 104/I. A. Herz, H. Akil & E. J. Simon, Eds.: 79–105. Springer-Verlag. Heidelberg.

26. GIOANNINI, T. L., L. Q. FAN, L. HYDE, D. OFRI, Y-H. YAO, J. M. HILLER & E. J. SIMON. 1993. Reconstitution of a purified mu-opioid binding protein in liposomes: Selective, high affinity, GTP-γS-sensitive mu-opioid agonist binding is restored. Biochem. Biophys. Res. Commun. **194:** 901–908.

Correlating the Pharmacology and Molecular Biology of Opioid Receptors

Cloning and Antisense Mapping a Kappa₃-related Opiate Receptor[a]

G. W. PASTERNAK,[b–d] Y-X. PAN,[c] AND J. CHENG[c]

[c]The Cotzias Laboratory of Neuro-Oncology
Memorial Sloan-Kettering Cancer Center
and
[d]Departments of Neurology & Neuroscience and Pharmacology
Cornell University Medical College
New York, New York 10021

The pharmacology of the various opioid receptors has been well studied. The availability of large numbers of selective agonists and antagonists has permitted the correlation of specific receptors defined in binding assays with selected pharmacological actions (TABLE 1).[1,2] Virtually all the established opioid receptor subtypes elicit analgesia, although the localization of their actions varies. In addition to regional differences, highly selective agonists for the different subtypes do not demonstrate cross-tolerance, implying that they are activating distinct systems leading to a common response. A full understanding of these receptor mechanisms requires the elucidation of the molecular mechanisms responsible for these *in vivo* actions. We now review recent work from our laboratory correlating the molecular biology and pharmacology of opioid receptors.

DELTA RECEPTORS

Recently, two groups reported the cloning of the delta opioid receptor (DOR-1).[3,4] This seminal discovery was soon followed by the identification of clones encoding mu and kappa receptors.[5–14] All three families show high degrees of homology, but are quite distinct in both the N- and C-termini, as well as the second extracellular loop and the two adjacent transmembrane regions. When expressed, all three families show the anticipated binding selectivities towards large numbers of opioids. Functionally they are active, inhibiting stimulated adenylyl cyclase.

Although these studies all inferred that the cloned receptors corresponded to those identified pharmcologically, a formal connection had not been demonstrated.

[a]This work was supported by grants from the National Institute on Drug Abuse to G.W.P. (DA02615 and DA07242). G.W.P. is a recipient of a Research Scientist Award from the National Institute on Drug Abuse (DA00220), and Y-X.P. is supported by a fellowship from the Aaron Diamond Foundation.

[b]Address correspondence to Dr. Gavril W. Pasternak, Department of Neurology, Memorial Sloan-Kettering Cancer Center, 1275 York Avenue, New York, NY 10021.

We addressed this question using an antisense approach to selectively down-regulate naturally occurring mRNA by using short oligodeoxynucleotides (ODN) with complementary sequences. We first examined delta receptors in a tissue culture model (FIG. 1).[15] We tested a series of antisense ODN directed at different regions of cDNA encoding the receptor, including both coding and untranslated regions. All the antisense ODN successfully down-regulated delta binding in the NG108-15 cells.[15] This indicated that the location of the antisense ODN on the mRNA was not critical. As controls, we also examined additional ODN. The sense ODN, which cannot anneal to the mRNA, was inactive. However, mixing the sense ODN with its corresponding antisense ODN prior to their addition to the cultured cells prevented the actions of the antisense ODN. A mismatch ODN, in which we scrambled the sequence of four bases without changing the overall base composition, also was without effect, demonstrating the stringent specificity of the response.

TABLE 1. Tentative Classification of Opioid Receptor Subtypes and Their Actions[a]

Receptor	Analgesia	Other
Mu		
Mu_1	Supraspinal	Prolactin release
		Feeding
		Acetylcholine release in the brain
Mu_2	Spinal	Respiratory depression
		Gastrointestinal transit
		Dopamine turnover in the brain
		Feeding
		Guinea-pig ileum bioassay
		Most cardiovascular effects
Kappa		
$Kappa_1$	Spinal	Diuresis
		Feeding
$Kappa_2$	Unknown	
$Kappa_3$	Supraspinal	
Delta		Mouse vas deferens bioassay
$Delta_1$	Supraspinal	
$Delta_2$	Spinal and supraspinal	

[a]Modified from Pasternak.[1,2]

We then examined antisense ODN *in vivo.* The paradigm employed injections of antisense on days 1, 3, and 5, followed by analgesic testing on day 6. We used this approach to obtain a prolonged down-regulation of receptor synthesis. The turnover of opioid receptors is approximately 3–4 days.[16] By maintaining the antisense treatments over 5 days, we permitted the loss of preexisting receptors. Administered intrathecally, the antisense selectively blocked the analgesic actions of two delta agonists, DPDPE and deltorphin II, without affecting either mu or delta actions (FIG. 2).[15] Again, the mismatch controls were without effect.

Additional controls explored the stability of the ODN under these conditions. In tissue culture studies, radiolabeled ODN is rapidly taken up by cells and as much as 5% remains associated with the cells as intact ODN, as indicated by its size on gels. Similar results were seen *in vivo.* Measurements of mRNA levels of DOR-1 following intrathecal administration reveal a 30% reduction, similar to the loss of binding.

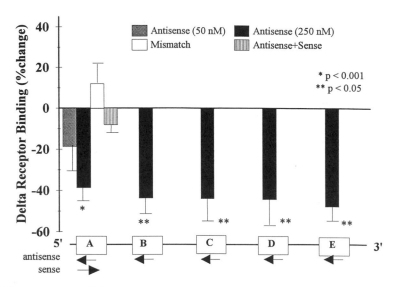

FIGURE 1. Down-regulation of delta receptor binding by antisense to DOR-1. NG108-15 cells were incubated for 5 days with the various antisense oligodeoxynucleotides at 250 nM, unless stated otherwise. The schematic representation provides an indication of the location of the antisense. Antisense A is directed at the N-terminus of the receptor, whereas the others are directed at regions downstream, including the 3'-untranslated region. (From Standifer *et al.*[15] Reproduced, with permission, from *Neuron.*)

Although the mechanism through which these ODN act remains unclear, they do down-regulate both receptors and mRNA.

MU AND KAPPA$_1$ RECEPTORS

Using similar approaches, we also demonstrated that antisense ODN directed against kappa$_1$[17] and mu receptors[18] also down-regulated the receptors against which they were designed. In all cases, our studies demonstrated remarkable selectivity among the various receptor classes, suggesting that this antisense approach might be a reasonable approach for selectively screening partial sequences of novel proteins for pharmacological activity without having to clone, sequence, and express complete gene products.

CLONING A KAPPA$_3$-RELATED OPIOID RECEPTOR

Several years ago, we identified a novel opioid binding site, termed kappa$_3$[19–23] (for review, see ref. 1). Although it fit the traditional criteria for inclusion in the opioid receptor family, it is readily distinguished from the mu, delta or kappa$_1$ receptors. Kappa$_3$ receptors display a binding profile that differs from the others and, in tissue culture studies, inhibits adenylyl cyclase through mechanisms independent of mu, delta or kappa$_1$ receptors.[24] *In vivo,* kappa$_3$ receptors elicit analgesia supraspinally. Additionally, it demonstrates no cross-tolerance with the other subtypes, particularly mu receptors.[20,22]

The cloning of delta receptors quickly led to the identification of clones corresponding to the mu and kappa$_1$ receptors using reverse transcriptase-polymerase

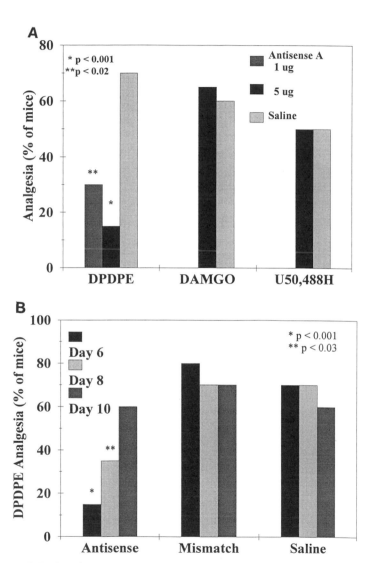

FIGURE 2. Selective blockade of delta analgesia by a DOR-1 antisense. (**A**) Groups of mice received Antisense A on days 1, 3, and 5 intrathecally, and analgesia was assessed on day 6 in the tailflick assay 15 min after receiving DPDPE (0.5 µg), DAMGO (8 ng) or U50,488H (25 µg). (**B**) Groups of mice received Antisense A, Mismatch A or saline on days 1, 3, and 5 and were tested for analgesia with DPDPE (0.5 µg) on days 6, 8, and 10. (From Standifer *et al.*[15] Reproduced, with permission, from *Neuron.*)

chain reaction. Anticipating strong homology among the opioid receptor classes, we employed a similar approach based upon the sequence of the delta receptors in an effort to look for additional opioid receptor subtypes. In addition to the sequences corresponding to delta, mu, and kappa$_1$ clones, we isolated a unique sequence of approximately 500 bases which had high homology to the others. Before committing our resources to cloning the full-length cDNA, we explored the pharmacological relevance of the PCR employing the antisense approach described above. We designed a 20 mer complimentary to the PCR product and administered it intracerebroventricularly (i.c.v.) in mice using the same paradigm we had previously developed for the other opioid receptors.[15,17,18] For a control, we designed a mismatch ODN in which the sequence of four bases of the 20 were scrambled, preventing effective annealing with the mRNA of interest. When we examined the animals, antisense treatment had little effect on the analgesic actions of morphine, DPDPE or U50,488, implying that the sequence was not encoding a mu, delta or kappa$_1$ receptor, respectively. However, the antisense effectively blocked the analgesic actions of the kappa$_3$ analgesic naloxone benzoylhydrazone (NalBzoH).[14,25] The mismatch control was inactive. This selective down-regulation of kappa$_3$ analgesia indicated that this clone was closely related to the kappa$_3$ receptor.

With this information, we proceeded to clone the full-length cDNA, which we termed KOR-3 (GenBank accession number U09421). The sequence is very similar to a recently reported putative opioid receptor, ORL1,[26] and has been observed by other groups.[14] Once we had obtained the full sequence, we expressed the clone in COS-7 cells. Using a monoclonal antibody (mAb8D8) directed against native kappa$_3$ receptors in BE(2)-C neuroblastoma cells, we found that on Western analysis the monoclonal antibody recognizes our clone expressed in COS-7 cells (Brooks, A., et al., in preparation). Control COS-7 cells transfected with the vector lacking our clone did not display any immunoreactivity. The in vitro translation product of the clone also was recognized by Western using mAb8D8. Thus, both the antibody and antisense approaches closely associate the cloned receptor with the kappa$_3$ site. However, the expressed receptor was not functionally active. Attempts to demonstrate binding were ambiguous. Although we occasionally could see cyclase effects, they were not very robust.

In view of the difficulty in expressing a functional receptor, we returned to the antisense approach to determine whether the clone truly corresponds to the kappa$_3$ receptor. We chose five regions in the open reading frame as well as two regions in the 3'-untranslated region, designed antisense ODN, and performed antisense studies on analgesia. In all cases, the antisense blocked the kappa$_3$ analgesic response in vivo. Southern analysis indicates a single gene encoding the sequence, further strengthening the implication that the KOR-3 clone derives from the gene encoding the kappa$_3$ receptor. However, it does not explain the difficulty obtaining a functionally active receptor. A number of possibilities exist. Although the receptor is expressed in the COS-7 cells, needed posttranslational changes may not be taking place or necessary transduction systems or G-proteins may not be available. Alternatively, there may be splice variants of the receptor. Indeed, we have evidence for a splice site between the first and second transmembrane domains, a location similar to that seen with the other opioid receptor classes. Although we have not yet identified alternative sequences in the coding region, alternative splicing remains a strong possibility. All the antisense ODN directed downstream from the splice site down-regulated NalBzoH analgesia as noted above. However, when we examined an additional series of ODN directed at sequences upstream from the splice, only one of the five blocked NalBzoH analgesia. This response is quite distinct from that seen with the ODN designed against the regions downstream, all of which block kappa$_3$

analgesia. Although indirect, this may be an indication that the kappa₃ receptor is a splice variant of our clone, both being derived from a single gene.

SUMMARY

We cloned a kappa₃-related opioid receptor, and although it is still not clear whether this clone corresponds to the kappa₃ receptor itself or is a related gene product, the extensive antisense mapping and the antibody immunoreactivity strongly associate this clone with the kappa₃ receptor. Our approach also indicates the usefulness of antisense approaches in mapping and identifying orphan receptors. Perhaps it is most effective in identifying partial sequences prior to cloning them in their entirety. It also provides a mechanism of identifying proteins that are not expressed functionally.

REFERENCES

1. PASTERNAK, G. W. 1993. Clin. Neuropharmacol. **16:** 1–18.
2. PASTERNAK, G. W. 1988. J. Am. Med. Assoc. **259:** 1362–1367.
3. EVANS, C., D. KEITH, H. MORRISON, K. MAGENDZO & R. EDWARDS. 1992. Science **258:** 1952–1955.
4. KIEFFER, B., K. BEFORT, C. GARERIAUX-RUFF & C. HIRTH. 1992. Proc. Natl. Acad. Sci. USA **89:** 12048–12052.
5. CHEN, Y., A. MESTEK, J. LIU, A. HURLEY & L. YU. 1993. Mol. Pharmacol. **44:** 8–12.
6. FUKUDA, K. S. KATTO, K. MORI, M. NISHI & H. TAKESHIMA. 1993. FEBS Lett. **327:** 311–314.
7. MENG, F., G. XIE, R. THOMPSON, A. MANSOUR, A. GOLDSTEIN, S. WATSON, & H. AKIL. 1993. Proc. Natl. Acad. Sci. USA **90:** 9954–9958.
8. MINAMI, M., T. TOYA, Y. KATAO, K. MAEKAWA, S. NAKAMURA, T. ONOGI, S. KANEKO & M. SATOH. 1993. FEBS Lett. **329:** 291–295.
9. TAKESHIMA, H., K. FUKUDA, S. KATO & K. MORI. 1993. FEBS Lett. **330:** 77–80.
10. THOMPSON, R. C., A. MANSOUR, H. AKIL & S. J. WATSON. 1993. Neuron **11:** 1–20.
11. WANG, J.-B., Y. MEI, C. M. EPPLER, P. GREGOR, C. E. SPIVAK & G. R. UHL. 1993. Proc. Natl. Acad. Sci. USA **90:** 10230–10234.
12. YASUDA, K., K. RAYNOR, H. KONG, C. BREDER, J. TAKEDA, T. REISINE & G. BELL. 1993. Proc. Natl. Acad. Sci. USA **90:** 6736–6740.
13. REISINE, T. & G. BELL. 1993. Molecular biology of opioid receptors. Trends Neurosci. **16:** 506–510.
14. UHL, G. R., S. R. CHILDERS & G. W. PASTERNAK. 1994. Trends Neurosci. **17:** 89–93.
15. STANDIFER, K. M., C.-C. CHIEN, C. WAHLESTEDT, G. P. BROWN & G. W. PASTERNAK. 1994. Neuron **12:** 805–810.
16. PASTERNAK, G. W., S. R. CHILDERS & S. H. SNYDER. 1980. J. Pharmacol. Exp. Ther. **214:** 455–462.
17. CHIEN, C. C., G. P. BROWN, Y. X. PAN & G. W. PASTERNAK. 1994. Eur. J. Pharmacol. **253:** R7–8.
18. ROSSI, G., Y. X. PAN, J. CHENG & G. W. PASTERNAK. 1994. Life Sci. **54:** PL375–379.
19. PRICE, M., M. A. GISTRAK, Y. ITZHAK, E. F. HAHN & G. W. PASTERNAK. 1989. Mol. Pharmacol. **35:** 67–74.
20. GISTRAK, M. A., D. PAUL, E. F. HAHN & G. W. PASTERNAK. 1989. J. Pharmacol. Exp. Ther. **251:** 469–476.
21. CLARK, J. A., L. LIU, M. PRICE, B. HERSH, M. EDELSON & G. W. PASTERNAK. 1989. J. Pharmacol. Exp. Ther. **251:** 461–468.
22. PAUL, D., J. A. LEVISON, D. H. HOWARD, C. G. PICK, E. F. HAHN & G. W. PASTERNAK. 1990. J. Pharmacol. Exp. Ther. **255:** 769–774.

23. PAUL, D., C. G. PICK, L. A. TIVE & G. W. PASTERNAK. 1991. J. Pharmacol. Exp. Ther. **257:** 1–7.
24. STANDIFER, K. M., J. CHENG, A. BROOKS, W. SU, C. HONRADO & G. W. PASTERNAK. 1994. J. Pharmacol. Exp. Ther. **270:** 1246–1255.
25. PAN, Y. X., J. CHENG & G. W. PASTERNAK. 1994. Regul. Pept. **54:** 217–218.
26. MOLLEREAU, C., M. PARMENTIER, P. MAILLEUX, J. L. BUTOUR, C. MOISAND, P. CHALON, D. CAPUT, G. VASSART & J. C. MEUNIER. 1994. FEBS Lett. **341:** 33–38.

Characterization of Opioid Receptor Types and Subtypes with New Ligands

ANNA BORSODI[a] AND GÉZA TÓTH[b]

[a]Institute of Biochemistry and [b]Isotope Laboratory
Biological Research Center of the Hungarian Academy of Sciences
P.O. Box 521
H-6701 Szeged, Hungary

Opioid drugs and opioid peptides produce their behavioral effects, including antinociception, by interactions with opioid receptors in the central nervous system. The existence of specific opioid receptors was originally suggested by behavioral and clinical studies, and was confirmed by biochemical identification (*in vitro* binding experiments) in 1973.[1-4] Since then, extensive studies have been undertaken on their localization, biochemical, and pharmacological characterization.

The classical ligand for opioid receptors is morphine (FIG. 1.), originating from the alkaloids of the poppy plant. As far as the structural requirements are concerned for opiate action, the presence of a phenolic OH group in position 3 is important. The substitutions on the nitrogen determine the agonistic or antagonistic character of the ligand: the methyl group results in agonistic properties, whereas the allyl group (as in the case of naloxone) leads to an antagonistic character. The allyl group can be replaced by a cyclopropylmethyl or propyl group (e.g., naltrexone, or N-propyl-noroxymorphone).

The first endogenous ligands for opioid receptors were identified in 1975 by Hughes *et al.*[5] The N-terminal Tyr of the two pentapeptides (methionine- and leucine-enkephalins) correspond to the A ring of morphine (FIG. 2). A number of other endogenous opioid peptides have been described in the meantime. A representative list of them is outlined in TABLE 1. Most of these compounds show a homology—the first four residues are identical. This structural arrangement can easily be explained by the "message-address" concept, originally described by Schwyzer.[6] The N-terminal region of the molecule is constant and carries the message, whereas the other part is fairly variable, resulting in functional heterogeneity. Accordingly, opioid receptors are heterogeneous, consisting of at least three major types, mu, delta, and kappa. The major opioid receptor types, their representative ligands, and effects are shown in TABLE 2. The three opioid receptors exhibit different ligand selectivity profiles. Most endogenous opioids and synthetic ligands do not possess absolute specificity for a given receptor type, but can interact with more than one opioid receptor type. The situation is further complicated by the fact that multiple receptor types may coexist within a single tissue, or even in a cell. Although the multiplicity of opioid receptors is generally accepted, the molecular basis of the heterogeneity is not completely understood. The primary structure of the delta receptor cDNA was reported simultaneously by Kieffer *et al.*[7] and Evans *et al.*[8] at the end of 1992, followed by the cloning of the mu[9] and kappa receptor.[10] Further multiplicity has not been proved yet by the cloning experiments.[11,12]

A better understanding of multiple opioid receptor structures and functions is of great importance for both the theoretical and practical points of view. Over the last few years increasing attention has been focused on studies of the heterogeneity of the opioid receptor types and especially on opioids acting at the delta as well as the

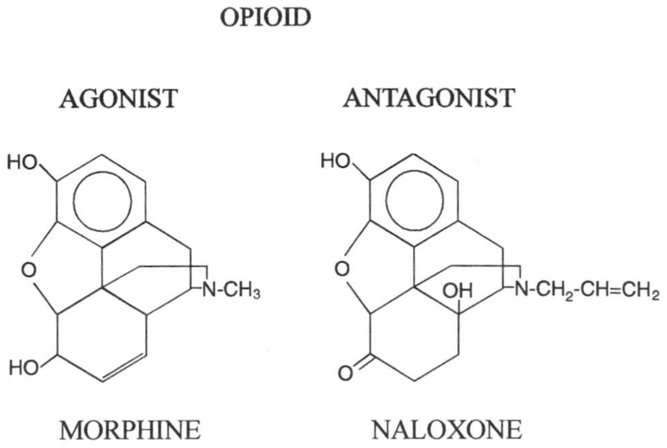

OPIOID

AGONIST   ANTAGONIST

MORPHINE   NALOXONE

FIGURE 1. Chemical structure of classical opioid ligands.

kappa receptors, because these compounds may cause fewer side effects than mu opioid receptor agonists do, thereby possibly providing an attractive alternative to the currently used opioid analgesics. Several attempts are presently under way in various laboratories to develop highly specific compounds, the use of which is crucial for understanding the mechanism of opioid action at the level of the endogenous system, in neurochemical processes in various mental diseases and pain states, and will be of direct benefit in improved therapy.

Opioid drugs are, and will continue to be, essential therapeutic agents. They provide the ultimate treatment for pain, but their use is complicated by many other effects. The most notable ones include respiratory depression, sedation, and gastro-intestinal dysfunction. Chronic use of opioids can also result in addiction and physical dependence. The production of compounds selective for the opioid receptor types/subtypes may provide the means to safe analgesics. The observation that distinct receptor types may mediate different nonanalgesic effects opened the possibility that some opioid side effects might be avoided by more selective drugs acting on different opioid receptor populations. The important role of mu opioid receptors in the development of opioid tolerance and physical dependence is well

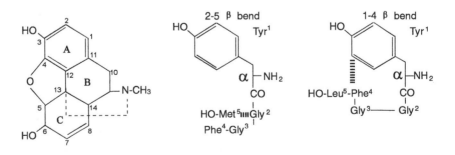

MORPHINE   MET-ENKEPHALIN   LEU-ENKEPHALIN

FIGURE 2. Comparison of structure of ± morphine and enkephalins.

TABLE 1. Endogenous Opioid Peptides

Precursor	Opioid-Peptide	Structure	Selectivity
Pro-opio-melanocortin (POMC)	β-Endorphin	Tyr-Gly-Gly-Phe-Met-Thr-Ser-Gln-Thr-Pro-Leu-Val-Thr-Leu-Phe-Lys-Asn-Ala-Ile-Ile-Lys-Asn-Ala-Tyr-Lys-Lys-Gly-Glu	$\mu, \epsilon > \delta \gg \kappa$
Pro-enkephalin A	[Leu^5]Enkephalin	Tyr-Gly-Gly-Phe-Leu	$\delta > \mu \gg \kappa$
	[Met^5]Enkephalin	Tyr-Gly-Gly-Phe-Met	$\mu \sim \delta \gg \kappa$
	[Met^5]Enkephalin-Arg^6-Phe^7	Tyr-Gly-Gly-Phe-Met-Arg-Phe	κ_2
Pro-dynorphin (Pro-enkephalin B)	Dynorphin A (1–17)	Tyr-Gly-Gly-Phe-Leu-Arg-Arg-Ile-Arg-Pro-Lys-Leu-Lys-Trp-Asp-Asn-Gly	$\kappa \gg \mu > \delta$
	Dynorphin A (1–13)	Tyr-Gly-Gly-Phe-Leu-Arg-Arg-Ile-Arg-Pro-Lys-Leu-Lys	$\kappa > \delta \sim \mu$
	Dynorphin A (1–8)	Tyr-Gly-Gly-Phe-Leu-Arg-Arg-Ile	$\kappa > \delta \sim \mu$
Others			
β-Casein derivatives	Morphiceptin	Tyr-Pro-Phe-Pro-NH$_2$	μ
	β-Casomorphin	Tyr-Pro-Phe-Pro-Gly-Pro-Ile	
α-Gliadin derivatives	Gliadorphin	Tyr-Pro-Gln-Pro-Gln-Pro-Phe	
Frog brain peptides			
Phyllomedusa sauvagei	Dermorphin	Tyr-D-Ala-Phe-Gly-Tyr-Pro-Ser-NH$_2$	μ
	Deltorphin	Tyr-D-Met-Phe-His-Leu-Met-Asp-NH$_2$	δ
Phyllomedusa bicolor	[D-Ala^2]Deltorphin I	Tyr-D-Ala-Phe-Asp-Val-Val-Gly-NH$_2$	δ
	[D-Ala^2]Deltorphin II	Tyr-D-Ala-Phe-Glu-Val-Val-Gly-NH$_2$	δ

documented. However, the involvement of delta opioid receptors in the development of these adaptive phenomena is less known.

Opioid antagonists have been indispensable pharmacological tools for identifying receptor types involved in the interaction with endogenous and synthetic opioid agonists. The antagonists are especially useful in the case, when the pharmacological endpoints are identical (e.g., antinociception or inhibition of a smooth muscle preparation by agonists), and it is not easy to distinguish among mu, delta, and kappa opioid receptor mediated agonist effects.

In this review we describe the characteristics of a number of new opioid ligands prepared in normal and in tritiated forms.

TABLE 2. Heterogeneity of Opioid Receptors

	μ	δ	κ
	β-end > dynA > met > leu[a]	met = leu > β-end > dynA[a]	dynA ≫ β-end > leu = met[a]
Selective agonists	DAMGO Sufentanyl PLO17	DPDPE DSBULET [D-Ala^2]Deltorphins	U69593 CI977 ICI197067
Selective antagonists	CTAP Cyprodime	ICI174864 Naltrindole TIPP	Nor-binaltorphimine
Radioligands	[^{3}H]DAMGO [^{3}H]PLO17	[^{3}H]DPDPE [^{3}H]TIPP [^{3}H]Naltrindole	[^{3}H]U69596 [^{3}H]CI977
Predominant effectors	cAMP ↓ K$^+$ channel ↑ Ca^{2+} channel ↓	cAMP ↓ K$^+$ channel ↑ Ca^{2+} channel ↓	cAMP ↓ K$^+$ channel ↑ Ca^{2+} channel ↓
Structural information	398 aa rat, mouse 7TM	372 aa rat, mouse 7TM	380 aa rat, mouse 7TM

[a]Potency order.

MU RECEPTOR-SPECIFIC LIGANDS

Agonists

The clinically employed opioid alkaloids (morphine, methadone, fentanyl, etc.) preferentially bind to the mu receptors and have a high potential for abuse. Their use is limited because of the development of tolerance and dependence and other side effects (respiratory depression, etc.). Recently a new group of compounds (14-alkoxymetopon derivatives) was described[13] with reduced dependence liability. It was shown earlier that the introduction of a 14-metoxy group to N-methylmorphinane-6-ones leads to a dramatic increase in antinociceptive potency.[14] A number of 14-alkoxymetopons (FIG. 3) have been tested in biochemical and pharmacological assays.[13] It was shown that the new ligands exhibited high affinity towards the mu sites. The sodium indexes were found to be extremely high (between 41–133), reflecting the agonist property of the compounds. This was further proved on isolated guinea-pig longitudinal muscle preparation, where the relative potency was 48–72 times higher than that of normorphine. The naloxone reversible antinocicep-

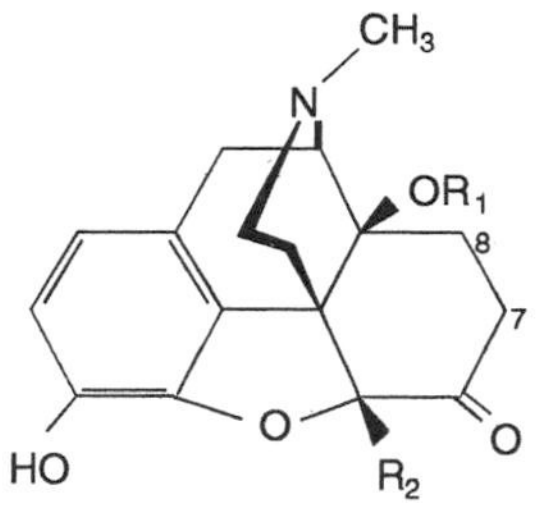

FIGURE 3. Chemical structure of alkoxymetopons.

(1) $R_1 = R_2 = H$ (OXYMORPHONE);
(2) $R_1 = R_2 = CH_3$ (14-METHOXYMETOPON);
(3) $R_1 = C_2H_5, R_2 = CH_3$ (14-METHOXYMETOPON);
(4) $R_1 = R_2 = CH_3$, 7.8 (14-METHOXY-5-METHYLMORPHINON).

tive effects in rats and mice were 130–300 times higher than in the case of morphine. Moreover, the dependence liability of the 14-alkoxymetopon derivatives in the withdrawal jumping tests was less pronounced than that of morphine (38–78% of control) in both species. It was also found by our laboratory that a major side effect—respiratory depression—of opioid alkaloids is diminished using certain codeinone analogues, which still hold analgesic potency[15] (see more detailed description under the section AFFINITY LABELING).

Antagonists

Cyprodime is known to be the only pure mu antagonist ligand among the heterocyclic compounds.[16] It has recently been radiolabeled[17] according to the schematic representation shown in FIGURE 4. This enzymatic procedure was chosen to have a brominated precursor that was followed by catalytic dehalogenation methods to obtain theoretical specific radioactivity. The complete biochemical characterization of this radioligand is currently in progress.

DELTA RECEPTOR-SPECIFIC LIGANDS

Pharmacological evidence has suggested the existence of two delta receptor subtypes in the brain in 1991.[18,19] From the known delta-specific ligands, [D-Pen²,

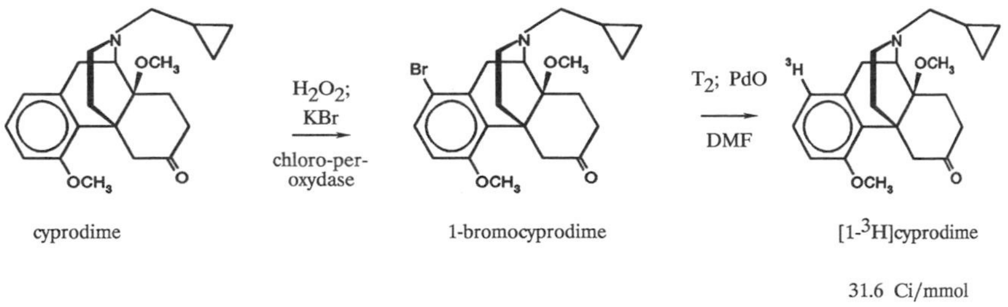

cyprodime   1-bromocyprodime   [1-³H]cyprodime

31.6 Ci/mmol

FIGURE 4. Tritiation of the mu-specific opioid antagonist, cyprodime.

TABLE 3. Selected Endogenous Opioid Peptides from Frog Skin

Dermorphin
TYR-D-ALA-PHE-GLY-TYR-PRO-SER-NH$_2$
Deltorphin
TYR-D-MET-PHE-HIS-LEU-MET-ASP-NH$_2$
Deltorphin I
TYR-D-ALA-PHE-ASP-VAL-VAL-GLY-NH$_2$
Deltorphin II
TYR-D-ALA-PHE-GLU-VAL-VAL-GLY-NH$_2$

D-Pen^5]enkephalin (DPDPE)[20] is thought to be primarily an agonist at the opioid delta$_1$ subtype, whereas [D-Ala^2, Glu^4]deltorphin is a selective agonist at the delta$_2$ subtype.[21]

Agonists

Linear hexapeptides, DSLET (Tyr-D-Ser-Gly-Phe-Leu-Thr) and DTLET (Tyr-D-Thr-Gly-Phe-Leu-Thr), have been developed in the laboratory of Roques. Recently, researchers there introduced the lipophylic and bulky tert-butyl group on the Ser^2 and Thr^6 amino acids of DSLET. The resulting new compounds (DSBULET, BUBU, and BUBUC) are highly potent and selective full agonist of delta receptors.[22] BUBU and BUBUC are protected from peptidases and were recently shown to be able to enter the brain,[23] allowing for the first time the effects resulting from delta receptor stimulation to be investigated after systematic administration, that is, in clinically relevant conditions. BUBU and BUBUC display interesting antidepressant-like properties and spinal analgesic activity without cross tolerance to morphine in chronic pain.

Following the isolation of a heptapeptide (dermorphin) from frog (*Phyllomedusa sauvagei*) skin,[24] another peptide (deltorphin) was found by recombinant DNA technology.[25] Both peptides derive from a common, larger precursor and contain amino acids with a D configuration in position 2. Later two more peptides were identified in another frog species[26] and were found to be excellent ligands for the delta receptor type (TABLE 3).

Buzas *et al.* prepared and characterized the delta$_2$-specific peptide deltorphin II in tritiated form.[27] Lately, new analogues of this compound have been synthesized, with the aim of better specificity and affinity towards the delta$_2$ sites. Among them, Ile residues were incorporated into the fifth and sixth position of deltorphin II[28] and radiolabeled using a diiodo-Tyr containing precursor peptide[29] (TABLE 4). The presence of the more hydrophobic residues resulted in an increased affinity (K$_d$: 0.4

TABLE 4. Tritiation of Deltorphin Analogues

Peptide	Catalyst	Specific Activity	
		GBq/mmol	(Ci/mmol)
[p-^{3}H-Phe^3]Deltorphin II	PdO	726	(20.6)
[p-^{3}H-Phe^3]Deltorphin II	PdO/BaSO$_4$	908	(24.5)
[3',5'-^{3}H-Tyr^1,Ile^{5,6}]Deltorphin II	PdO/BaSO$_4$	2364	(63.9)

nM) and selectivity (selectivity ratios: mu/delta 2400; kappa/delta 18000). Further advantages of this ligand are low, nonspecific binding ($<25\%$) and high, specific radioactivity (TABLE 4).

Antagonists

A potent and moderately selective delta antagonist was developed by Portoghese. The structure of naltrindole (NTI)[30] is based on the morphinane skeleton to which an indole nucleus is fused (FIG. 5). It was shown that the development of acute tolerance and dependence in mice pretreated with NTI before induction of tolerance and dependence with morphine sulfate was markedly suppressed. The use of this delta-opioid receptor antagonist allows a way to prevent opioid tolerance and physical dependence without compromising the antinociceptive activity of mu-opioid receptor agonists such as morphine. Recently, NTI was radiolabeled in one[31] and subsequently in two positions.[32,33] This ligand binds to delta opioid receptors with high affinity, but a significant proportion of the binding becomes wash-resistant. This pseudoirreversible nature of the binding might be due to the hydrophobic property of the ligand.

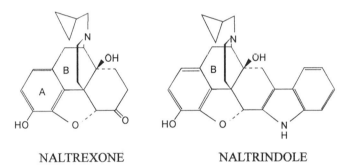

NALTREXONE   NALTRINDOLE

FIGURE 5. Structure of selected opioid antagonists.

Relatively selective antagonists have recently been obtained: [D-Ala^2, Leu^5, Cys^6]enkephalin (DALCE)[34] for delta$_1$ and naltrindole-5'-isothiocyanate (5'-NTII)[18] for delta$_2$ receptor subtypes. Several studies suggest that both delta$_1$ and delta$_2$ opioid receptors mediate antinociception in mice.[18,35] Both receptor subtypes appear to mediate antinociception at the supraspinal level, whereas the delta$_2$ receptor is involved in antinociception at the spinal level. Finally, cold-water swim stress produces an opioid mediated antinociceptive response, which appears to be antagonized by 5'-NTII, but not by DALCE. Thus, on the basis of the recent data the existence of different subtypes of delta receptors in the central nervous system of rodents is suggested. This heterogeneity of delta receptors was further supported by ligand binding assays,[19] although not yet supported by the cloning of a single opioid receptor gene coding a protein which binds a delta subtype selective ligand.

Schiller recently designed a peptide antagonist for the delta receptor, based on conformational restriction.[36] The tetrapeptide H-Tyr-Tic-Phe-Phe-OH (TIPP) (FIG. 6) and its tritiated form displays high delta-receptor affinity, unprecedented delta selectivity, high potency as a delta antagonist, and, unlike other delta antagonists, shows no mu-antagonist properties and is 80 times more selective than NTI.[37] It was shown earlier that antagonists may be obtained by the reduction of the peptide bond (CH$_2$NH) in the case of mu ligands. This approach has been applied also to obtain delta antagonists. A chemically and enzymatically stable, more potent and more selective analogue, Tyr-TicΨ(CH$_2$NH)Phe-OH (TIPP[Ψ]) was described very recently[38] and has already been radiolabeled (manuscript submitted).

KAPPA RECEPTOR-SPECIFIC LIGANDS

Kappa opioid receptors play a role in various pharmacological and physiological functions, such as analgesia, behavioral and autonomic effects, regulation of neurotransmitter and hormone release and synthesis, modulation of membrane ion-channels and calcium uptake.[39] The kappa agonists are considered to be advantageous drugs in producing spinal analgesia, for treating rheumatic fever disease, strokes, in reducing chemical, visceral, and thermal stimuli. Moreover, they do not

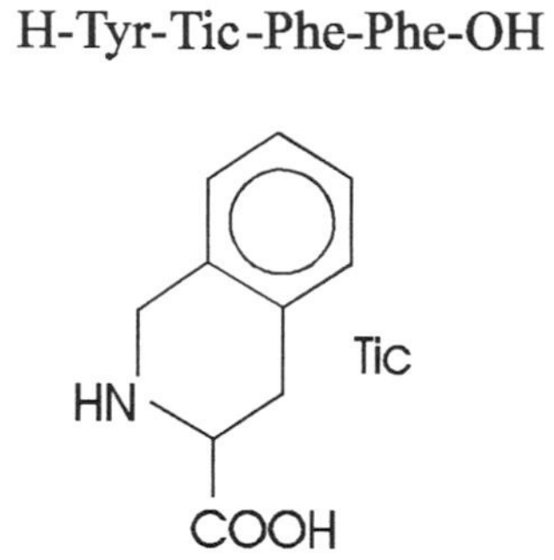

H-Tyr-Tic-Phe-Phe-OH

FIGURE 6. Structure of the delta opioid receptor specific peptide, H-Tyr-Tic-Phe-Phe-OH (TIPP). Tic: Tetrahydro isoquinoline-3-carboxylic acid.

induce dependence as mu-specific ligands do, but they may produce dysphoria. In the last few years the heterogeneity among kappa receptors became evident, although their exact role still has to be elucidated. Presently, U-69,593 is considered to be one of the best kappa$_1$ selective ligands. It is important to design, synthesize, and label new ligands for other subtypes as well.

Agonists

Met-enkephalin-Arg^6-Phe^7 was earlier considered as a kappa$_2$ selective ligand.[40] Therefore, [^{3}H]Met-enkephalin-Arg^6-Phe^7 has been synthesized and labeled from a diiodo-Tyr containing precursor peptide with dehalotritiation.[41] The results obtained with this radioligand raised the possibility of further heterogeneity. On the other hand its application is presently limited because of its sensitivity toward various peptidases. For this reason the synthesis of more stable analogues is feasible.

Previous pharmacological experiments showed the analgesic efficacy of the 2,4-dipyridine substituted dimethyl-3,7-diazabicyclo[3,3,1]nonan-9-on-1,5-dicarboxylate[42] (FIG. 7.). It was thought that the effect was the result of an interaction of this

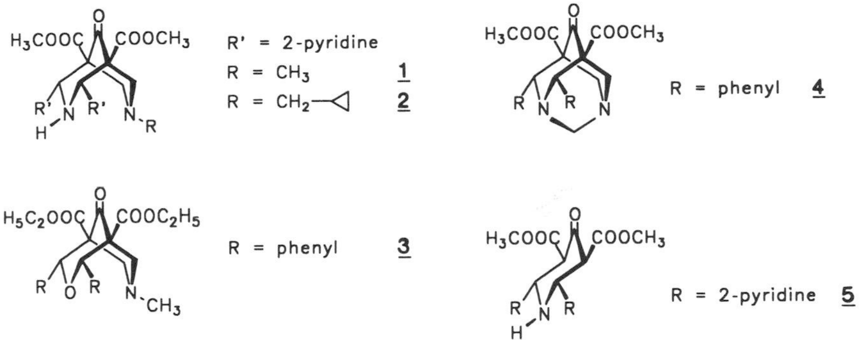

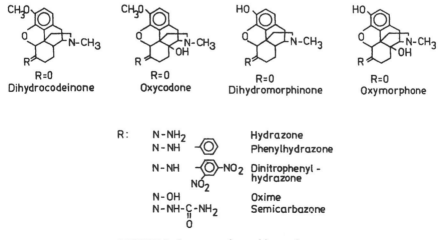

FIGURE 7. Structure of heterocyclic bicyclo[3,3,1]nonan-9-ones.

unusual compound with the kappa receptor type. Very recently it was found that this new class of ligands (heterocyclic bicyclo[3,3,1]nonan-9-ones) exhibit kappa_1 agonist selectivity.[43] The further modification of these ligands will allow the study of the structure of opioid kappa receptor subtypes. Besides the theoretical importance, this new group of ligands may lead to the development of potent analgesic drugs with reduced side effects.

AFFINITY LABELING OF OPIOID RECEPTORS

Structural studies of many receptor binding sites were greatly facilitated by the use of affinity ligands, which are capable of covalent interactions with the receptors. A number of these compounds have been developed in normal and tritiated form for the identification of opioid receptors. A large percentage of their binding became

FIGURE 8. Structure of morphinane-6-ones.

irreversible under the proper conditions in rat and frog brain. The new ligands were either modifications of morphine or derivatives of enkephalins.

Morphine Derivatives

Hydrazone, phenylhydrazone, and dinitrophenylhydrazone, oxime, and semicarbazone derivatives of dihydromorphone and oxymorphone (FIG. 8) were prepared.[15] The N-substituted hydrazones and the oxime and semicarbazone derivatives were all capable of irreversible inhibition of [^{3}H]naloxone binding. The blockade concerned mainly the high-affinity binding site. We concluded that the C = N double bond was responsible for the irreversible binding. A tritiated form of oxymorphazone has also been prepared and characterized in binding assays.[44] Among the corresponding codeinone derivatives, selective blockers were found for the low-affinity site. This observation has a pharmaceutical significance, because this site is thought to be responsible for the respiratory depression, which is the major side effect of the

TABLE 5. Effect of C-6 Substituted Oxycodone Derivatives on Arterial Blood Gases in Anesthetized Rats

Treatment	Arterial Blood Gas Measurements	
	pO$_2$	pCO$_2$
Fentanyl (10 μg/kg)		
Baseline values	111.1	28.9
Change (%)	−35.5	+69.5
Oxycodone oxime (10 μg/kg)		
Baseline values	107.5	27.1
Change (%)	+42.0	−44.3
Oxycodone oxime + fentanyl		
Baseline values	108.5	31.3
Change (%)	+5.2	−15.5

morphine derivatives.[45] It was suspected that different mu receptor subtypes may mediate antinociceptive, gastrointestinal, and respiratory actions induced by C-6 substituted oxicodone derivatives. It was found that low doses of the oxycodone derivatives failed to inhibit gastrointestinal transit and resulted in a slight increase of respiratory function. Moreover, the respiratory depression induced by mu agonists (e.g. morphine, fentanyl) was prevented in conscious rabbit and narcotized rat (TABLE 5).

Enkephalin Derivatives

A number of enkephalin analogues were elongated with a chloromethyl ketone group at the C-terminus, which led to a shift in their specificity from the delta towards the mu site. They were all able to irreversibly block the high-affinity naloxone binding site. A tritiated form of D-Ala^2-Leu^5 enkephalin chloromethyl ketone (DALECK) was used first for the identification of opioid binding sites in rat and frog brain.[46,47] Later, a highly mu-selective compound was prepared from Tyr-D-Ala^2-Gly-(Me)Phe-Gly-ol (DAMGO). The new chloromethyl-ketone derivative and its radiolabeled form were synthesized by a fragment condensation method (spec. radioactivity 56 Ci/mmol).[48] More recently, hydrodynamic parameters of mu opioid receptors were measured on [³H]DAMCK-prelabeled preparations of rat brain under nondenaturing conditions. The apparent M_r on SDS-PAGE followed by fluorography was found to be 58 kDa,[49] confirming previous data reported in the literature.[50] This size of the mu receptor was shown in various species including rat, guinea pig, rabbit, and chicken (unpublished results). The molecular mass of the cloned receptors is much less because of the lack of the polysaccharide chains. (All of the major opioid receptor types are glycoproteins.)

FUTURE DIRECTIONS

The recently described cloning of the major opioid receptor types will facilitate the development of new compounds (agonists and antagonists) for use in further detailed structural analysis of the receptors and for new clinically useful opioids. The use of antagonists will be inevitable in the transfected cells (containing different second messenger systems); besides, it is likely that there are distinct binding sites for agonists and antagonists.

The novel ligands will be tested by chemical, biochemical, and pharmacological assays for a better understanding of the interactions between the ligands and receptor types/subtypes. The determination of the subtype selectivity of the new compounds is expected, for example, in the case of delta receptors. As stated earlier in this review, there is evidence for the existence of two receptor subtypes, delta₁ and delta₂. Neither NTI nor TIPP seems to discriminate between these two receptor subtypes. Although structural modification of naltrindole has resulted in antagonists with some preference for both delta₁ and delta₂ receptors, pure antagonist compounds with further improved selectivity for either delta₁ and delta₂ receptors have to be developed. Such compounds are absolutely necessary for the definition of the distinct functional roles of these two receptor subtypes. In particular, the availability of such receptor subtype-specific antagonists would also allow us to examine the important question of whether the effect on morphine tolerance and dependence observed with naltrindole is mediated by the delta₁ or the delta₂ receptor or by both. On the basis of results obtained from *in vitro* binding assays, isolated tissue preparations and pharmacological tests, more candidates will be selected for radiolabeling.

Different chemical approaches can be applied to avoid enzymatic degradation of the peptides and to achieve improved bioavailability. Some of the peptides are modified for obtaining pseudopeptides with restricted conformations and enhanced resistance to peptidases. Based on NMR and theoretical conformational studies, we expected to obtain new information for defining the pharmacophore, corresponding to the receptor types (subtypes). Some of the modifications will lead to radiolabeled compounds, allowing the use of autoradiographic and electronmicroscopic techniques also for studying the receptors. For obtaining ligands with better selectivity and antagonist property, a further reduction in conformational freedom is required. It is expected that conformational restriction of peptides in some cases might also reduce or even totally abolish their intrinsic activity ("efficacy") and, thus, may produce partial agonists or antagonists.

Besides the functional assays, the application of molecular biological techniques (including site-directed mutagenesis and hybridization experiments) will be required for the complete understanding of the effects of the novel ligands in the opioid system. Studies of the regulation of opioid gene expression by the specially synthesized analogues will certainly be performed. It is also expected that regulation of opioid receptor mRNAs will be estimated by *in situ* hybridization histochemistry. All these approaches are necessary in order to fulfill the ultimate goal of developing new therapeutic agents with fewer undesired side effects.

REFERENCES

1. PERT, C. B. & S. H. SNYDER. 1973. Opiate receptor: Demonstration in nervous tissue. Science **179**: 1011–1014.
2. SIMON, E. J., J. M. HILLER & I. EDELMAN. 1973. Stereospecific binding of the potent narcotic analgesic [³H]etorphine to rat brain homogenates. Proc. Natl. Acad. Sci. USA **70**: 1947–1949.
3. TERENIUS, L. 1973. Stereospecific interaction between narcotic analgesics and a synaptic plasma membrane fraction of the guinea-pig ileum. Acta Pharmacol. Toxicol. **32**: 317–320.
4. GOLDSTEIN, A., L. I. LOWNEY & B. K. PAL. 1971. Stereospecific and nonspecific interactions of the morphine congener levorphanol in subcellular fractions of mouse brain. Proc. Natl. Acad. Sci. USA **68**: 1742–1747.
5. HUGHES, J., T. W. SMITH, H. W. KOSTERLITZ, L. A. FOTHERGILL, B. A. MORGAN & H. R. MORRIS. 1975. Identification of two related pentapeptides from the brain with potent opiate agonist activity. Nature **285**: 577–579.
6. SCHWYZER, R. 1977. ACTH: A short introductory review. Ann. N.Y. Acad. Sci. **247**: 3–26.
7. KIEFFER, B. L., K. BEFORT, C. GAVERIAUX-RUFF & C. G. HIRTH. 1992. The delta-opioid receptor: Isolation of a cDNA by expression cloning and pharmacological characterization. Proc. Natl. Acad. Sci. USA **89**: 12048–12052.
8. EVANS, C. J., D. E. KEITH, JR., H. MORRISON, K. MAGENDZO & R. H. EDWARDS. 1993. Cloning of a delta opioid receptor by functional expression. Science **258**: 1952–1955.
9. FUKUDA, K., S. KATO, K. MORI, M. NISHI & H. TAKESHIMA. 1993. Primary structures and expression from cDNAs of rat opioid receptor delta- and mu-subtypes. FEBS Lett. **327**: 311–314.
10. REISINE, T. & G. I. BELL. 1993. Molecular biology of opioid receptors. TINS **16**: 506.
11. RAYNOR, K., H. KONG, Y. CHEN, K. YASUDA, L. YU, G. I. BELL & T. REISINE. 1994. Pharmacological characterization of the cloned kappa, delta and mu opioid receptors. Mol. Pharmacol. **45**: 330–334.
12. UHL, G. R., S. CHILDERS & G. PASTERNAK. 1994. An opiate-receptor gene family reunion. TINS **17(3)**: 89–93.
13. FURST, ZS., B. BUZAS, T. FRIEDMANN, H. SCHMIDHAMMER & A. BORSODI. 1993. Highly potent novel opioid receptor agonist in the 14-alkoxymetopon series. Eur. J. Pharmacol. **236**: 209–215.
14. SCMIDHAMMER, H., L. AEPPLI, L. ATWELL, F. FRITSCH, A. E. JACOBSON, M. NEBUCHLA & G. SPERK. 1984. Synthesis and biological evaluation of 14-alkoxymorphinans. 1. Highly potent opioid agonist in the series of (-)-14-methoxy-*N*-methyl-morphinan-6-ones. J. Med. Chem. **27**: 1575–1579.
15. KRIZSÁN, D., E. VARGA, S. HOSZTAFI, S. BENYHE, M. SZUCS & A. BORSODI. 1991. Irreversible blockade of the high and low affinity [³H]naloxone binding sites by C-6 derivatives of morphinane-6-ones. Life Sci. **48**: 439–451.
16. SCHMIDHAMMER, H., W. P. BURKARD, L. EGGSTEIN-AEPPLI & C. F. C. SMITH. 1989. Synthesis and biological evaluation of 14-alkoxymorphinans. 2. (-)-N-cyclorpopylmethl-4,14-dimethoxymorphinan-6-one, a selective μ-opioid receptor antagonist. J. Med. Chem. **32**: 418–421.
17. ÖTVÖS, F., G. TOTH & H. SCHMIDHAMMER. 1992. Tritium labeling of cyprodime (= (-)-17-(cyclopropylmethyl)-4,14-dimethoxymorpinane-6-one), a mu receptor-selective opioid antagonist. Helv. Chim. Acta **75**: 1718–1720.
18. JIANG, Q., A. E. TAKEMORI, M. SULTANA, P. S. PORTOGHESE, W. D. BOWEN, H. I. MOSBERG & F. PORRECA. 1991. Differential antagonism of opioid antinociception by [D-Ala², Leu⁵, Cys⁶]enkephalin and naltrindole 5'-isothiocyanate: Evidence for delta subtypes. J. Pharmacol. Exp. Ther. **257**: 1069–1075.

19. NEGRI, L., R. L. POTENZA, R. CORSI & P. MELCHIORRI. 1991. Evidence for two subtypes of delta opioid receptor in rat brain. Eur. J. Pharmacol. **196:** 335–336.
20. MOSBERG, H. I., R. HURST, V. J. HRUBY, K. GEE, H. I. YAMAMURA, J. J. GALLIGAN & T. F. BURKS. 1983. Bis-penicillamine enkephalins possess highly improved specificity towards delta opioid receptors. Proc. Natl. Acad. Sci. USA **80:** 5871.
21. KREIL, G. D. BARRA, M. SIMMACO, V. ERSPAMER, G. FALCONERI-ERSPAMER, L. NEGRI, C. SEVERINI, R. CORSI & P. MELCHIORRI. 1989. Deltophin, a novel amphibian skin peptide with high selectivity and affinity for delta opioid receptors. Eur. J. Pharmacol. **162:** 123–128.
22. GACEL, G. A., E. FELLION, A. BAAMONDE, V. DAUGE & P. B. ROQUES. 1990. Synthesis, biochemical and pharmacological properties of BUBUC, a highly selective and systematically active agonist for in vivo studies of delta opioid receptors. Peptides **11:** 983.
23. DELAY-GOYET, P., M. RUIZGAYO, A. BAAMONDE, G. A. GACEL, J. L. MORGAT & P. B. ROQUES. 1991. Brain passage of BUBU, a highly selective and potent agonist for delta opioid receptors. In vivo binding and mu receptors versus delta receptors occupancy. Pharmacol Biochem. Behav. **38:** 155.
24. MONTECUCCHI, P. C., R. CASTIGLIONE, S. PIANI, L. GOZZINI & V. ERSPAMER. 1981. Aminoacid composition and sequence of dermorphin, a novel opiate-like peptide from the skin of *Phylomedusa sauvagei.* Int. J. Pept. Protein Res. **17:** 275–283.
25. MELCHIORI, P., L. NEGRI, G. FALCONERI-ERSPAMER, C. SEVERINI, R. CORSI, M. SOAJE, V. ERSPAMER & D. BARRA. 1991. Structure-activity relationships of the delta-opioid-selective agonists, deltorphins. Eur. J. Pharmacol. **195:** 201–207.
26. SALVADORI, S., M. MARASTONI, G. BALBONI, P. A. BOREA, M. MORARI & R. TOMATIS. 1991. Synthesis and structure-activity relationships of deltorphin analogues. J. Med. Chem. **34:** 1656–1661.
27. BUZÁS, B., G. TOTH, S. CAVAGNERO, V. J. HRUBY & A. BORSODI. 1992. Synthesis and binding characteristics of the highly delta-specific new tritiated opioid peptide, [³H]deltorphin II. Life Sci. **50:** PL75–PL78.
28. SASAKI, Y., A. AMBO & K. SUZUKI. 1991. [D-Ala²]Deltorphin analogues with high affinity and selectivity for delta-opioid receptors. Biochem. Biophys. Res. Commun. **180:** 822–827.
29. NEVIN, S. T., L. KABASAKAL, F. OTVOS, G. TOTH & A. BORSODI. 1993. Characterisation of the novel delta opioid agonist Ile⁵,⁶deltorphin II. Neuropeptides **26:** 261–265.
30. PORTOGHESE, P. S., M. SULTANA, H. NAGASE & A. E. TAKEMORI. 1988. Application of the message-address concept in the design of highly potent and selective non-peptide delta opioid receptor antagonists. J. Med. Chem. **31**(2)**:** 281–282.
31. YAMAMURA, M. S., R. HORVATH, G. TOTH, F. OTVOS, E. MALATYNSKA, R. J. KNAPP, F. PORRECA, V. J. HRUBY & H. I. YAMAMURA. 1992. Characterization of [³H]naltrindole binding to delta opioid receptors in rat brain. Life Sci. **50:** PL119–PL124.
32. BORSODI, A., G. OZDEMIRLER, S. T. NEVIN, L. KABASAKAL, F. OTVOS & G. TOTH. 1993. Binding characteristics of the delta antagonist [1,5'-³H]naltrindole in rat brain membranes. Br. J. Pharmacol. **109:** 17.
33. DORN, C. R., C. S. MARKOS, M. S. DAPPEN & B. S. PITZELE. 1992. Synthesis of [³H]naltrindole. J. Label. Comp. Radiopharm. **31**(5)**:** 375–380.
34. BOWEN, W. H., S. B. HELLWELL, M. KELEMEN, R. HUEY & D. STEWART. 1987. Affinity labeling of delta opiate receptors using [D-Ala²,Leu⁵,Cys⁶]enkephalin. J. Biol Chem. **262:** 13434.
35. MATTIA, A., S. C. FARMER, A. E. TAKEMORI, M. SULTANA, P. S. PORTOGHESE, H. I. MOSBERG, W. D. BOWEN & F. PORRECA. 1992. Spinal opioid delta antinociception in the mouse. Mediation by 5' NTII-sensitive delta receptor subtype. J. Pharmacol. Exp. Ther. **260:** 518–525.
36. SCHILLER, P. W., T. M.-D. NGUYEN, G. WELTROWSKA, B. C. WILKES, B. J. MARSDEN, C. LEMIEUX & N. N. CHUNG. 1992. Differential stereochemical requirements of mu versus delta receptors for ligand binding and signal transduction: Development of a new class of potent and highly selective delta-selective peptide antagonists. Proc. Natl. Acad. Sci. USA **89:** 11871–11875.
37. NEVIN, S. T., G. TOTH, T. M.-D. NGUYEN, P. W. SCHILLER & A. BORSODI. 1993. Synthesis

and binding characteristics of the highly specific, tritiated delta opioid antagonist [³H]TIPP. Life Sci. **53:** PL57–62.

38. SCHILLER, P. W., G. WELTROWSKA, T. M.-D. NGUYEN, B. C. WILKES, N. N. CHENG & C. LEMIEUX. 1993. TIPP[Y]: A highly potent and stable pseudopeptide delta opioid receptor antagonist with extraordinary delta selectivity. J. Med. Chem. **36:** 3182–3187.

39. WOLLEMANN, M., S. BENYHE & J. SIMON. 1993. The kappa-opioid receptor: Evidence for the different subtypes. Life Sci. **52:** 599–611.

40. BENYHE, S., E. VARGA, J. HEPP, A. MAGYAR, A. BORSODI & M. WOLLEMANN. 1990. Characterization of kappa$_1$ and kappa$_2$ opioid binding sites in frog (*Rana esculenta*) brain membrane preparation. Neurochem. Res. **15:** 899–904.

41. WOLLEMANN, M., J. FARKAS, G. TOTH & S. BENYHE. 1994. Characterization of [³H]Met-enkephalin-Arg[6]-Phe[7] binding to opioid receptors in frog brain membrane preparations. J. Neurochem. **63:** 1460–1465.

42. ERCIYAS, E. & U. HOLZGRABE. 1992. Synthese und stereochemie potentiell stark analgetischer 2,4-m-diarylsubstituierter 3,7-diazabicyclo[3,3,1]nonan-9-on-1,5-diester. Arch. Pharm. (Weinheim) **325:** 657–663.

43. BENYHE, S., A. BORSODI, U. HOLZGRABE & C. NACHTSHEIM. 1994. A novel class of kappa opioid receptor agonists with a heterocyclic bicyclo [3,3,1]nonan-9-one skeleton. Regul. Pept. **54(1):** 27–28.

44. VARGA, E., G. TOTH, S. BENYHE, S. HOSZTAFI & A. BORSODI. 1987. Synthesis and binding of [³H]oxymorphazone to rat brain membrane. Life Sci. **40:** 1579–1588.

45. FURST, ZS., T. FRIEDMANN, S. HOSZTAFI & A. BORSODI. 1994. Different mu opioid receptor subtypes may mediate C-6 substituted oxycodone derivatives induced antinociceptive, gastrointestinal and respiratory actions. Regul. Pept. Suppl. **1:** S105–S106.

46. SZUCS, M., M. BELCHEVA, J. SIMON, S. BENYHE, G. TOTH, J. HEPP, M. WOLLEMANN & K. MEDZIHRADSZKY. 1987. Covalent labeling of opioid receptors with [³H]-D-Ala2-Leu5-enkephalin chlorometyl ketone: I. Binding characteristics in rat brain membranes. Life Sci. **41:** 177–184.

47. SIMON, J., M. SZUCS, S. BENYHE, G. TOTH, J. HEPP, A. BORSODI, M. WOLLEMANN & K. MEDZIHRADSZKY. 1987. Covalent labelling of opioid receptors with 3H-D-Ala2-Leu5-enkephalin chloromethyl ketone. II. Binding characteristics in frog brain membranes. Life Sci. **41:** 185–192.

48. VARGA, E., G. TOTH, J. HEPP, S. BENYHE, J. SIMON, K. MEDZIHRADSZKY & A. BORSODI. 1988. Synthesis of [³H]-Tyr-D-Ala-Gly-N(Me)Phe-chloromethyl ketone. Neuropeptides **12:** 135–139.

49. OKTEM, H. A., J. MOITRA, S. BENYHE, G. TOTH, A. LAJTHA & A. BORSODI. 1991. Opioid receptor labeling with the chloromethyl ketone derivative of [³H]-Tyr-D-Ala-Gly-(Me)Phe-Gly-ol (DAGO) II. Covalent labeling of mu opioid binding site by [³H]-Tyr-D-Ala-Gly-(Me)Phe chloromethyl ketone. Life Sci. **48:** 1763–1768.

50. NEWMAN, E. L., A. BORSODI, G. TOTH, J. HEPP & E. A. BARNARD. 1986. Mu-receptor specificity of the opioid peptide irreversible reagent [³H] DALECK. Neuropeptides **8:** 305–315.

Opiate Receptor Changes after Chronic Exposure to Agonists and Antagonists

JAMES E. ZADINA, ABBA J. KASTIN,
LAURA M. HARRISON, LIN-JUN GE,
AND SULIE L. CHANG[a]

*VA Medical Center and
Department of Medicine and Neuroscience Training Program
Tulane University School of Medicine
New Orleans, Louisiana 70146*

*[a]Department of Biology
Seton Hall University
South Orange, New Jersey 07079*

The responsiveness of neurotransmitter receptor systems is altered by exposure of the receptors to their agonist and antagonist ligands. For agonists, the physiological response declines over time despite the presence of a constant stimulus. Chronic exposure to antagonists, by contrast, can lead to an increase in the number of receptors, often accompanied by increased responsiveness to agonists.

Three major processes, characterized in greatest detail in the β-adrenergic system[1] but typical of many transmitter systems, have been postulated to contribute to the loss of responsiveness to agonists. These include (1) loss of coupling to G-proteins, (2) sequestration of receptors, and (3) down-regulation. Each of these mechanisms may contribute to different aspects of the loss of responsiveness, particularly with regard to the time course of the change. Thus, uncoupling occurs in seconds to minutes, sequestration in minutes, and down-regulation in minutes to hours. Phosphorylation events appear critical for some and may be involved in all of these processes.

For opiates and their receptors, the role of these processes in the changing responsiveness observed after exposure to opiates has remained elusive. As discussed in the concluding section, *in vivo* studies on changes in receptors in adult animals have yielded numerous, often conflicting results, including down-regulation, up-regulation, and no change. Studies in developing animals indicate that the perinatal period is a time when the animal is particularly sensitive to regulation of receptors by agonists. *In vitro* cell lines, however, have been particularly useful for studying changes in receptors and responsiveness after exposure to agonists. Homogeneity of cell type and receptor content as well as control of the kinetics of drug exposure are among the advantages of these preparations.

This report focuses primarily on recent findings in opiate-receptor–containing neuronal cell lines as models of cellular responses to chronic exposure to opiates and their antagonists. The conclusion will discuss implications of these cellular models for studies *in vivo*.

REGULATION OF MU RECEPTORS *IN VITRO* BY MORPHINE

Many aspects of opiate receptor regulation were characterized with cell lines containing delta receptors such as N4TG1[2] and NG108-15[3] cells. However, morphine and other compounds with clinical usefulness, as well as abuse potential, act primarily at the mu receptor, which is not present on these cells. Characterization of the mu receptor and its response to prolonged exposure to agonists *in vitro* is therefore of considerable importance. One of the first cell lines used in studying mu receptors was derived from pituitary cells. In these 7315c cells, Puttfarcken *et al.*[4,5] demonstrated that, as with delta sites in the NG108-15 cells,[3] two components of the response to morphine could be characterized: an early desensitization, or decline in the inhibitory effect of an acute exposure to morphine on cyclic AMP accumulation, followed at later times and higher doses by down-regulation of the mu receptors.

The first neuronal cell line in which mu receptor down-regulation by morphine was demonstrated[6,7] was the SH-SY5Y cell line, a subclone of SK-N-SH cells. When SH-SY5Y cells were exposed to morphine and the mu receptors were measured with the mu-selective ligand [^{3}H]-DAMGO (Tyr-D-Ala-Gly-N-Me-Phe-Gly-ol), the maximum number of receptors (B_{max}) decreased over time. At about 3 h, the decrease was half-maximal. Maximal down-regulation occurred at about 24 h with little further change up to 72 h. The effect was reversible by naloxone and was dose-dependent, with half-maximal decreases occurring at a concentration of about 0.5 μM morphine. This dose corresponds well to the reported apparent affinity of morphine for the mu receptor in these culture conditions,[8] which is expected to be lower than that observed with isolated membranes. This difference is due to several factors including the reduction of agonist affinity by salts in the media and loss of agonist due to internalization.

The morphine-induced decrease in mu binding was also shown to be temperature-dependent, consistent with down-regulation processes in other systems. Muscarinic receptors present on SH-SY5Y cells were not affected by morphine, indicating that the down-regulation was homologous for opiate receptors.

Earlier studies with NG108-15 cells[3] had shown that efficacy of agonists correlated with the capacity to induce down-regulation of delta receptors and that partial agonists were ineffective at producing down-regulation. This idea was supported for mu receptors in SH-SY5Y cells because pentazocine, a partial agonist in SH-SY5Y cells,[9] failed to down-regulate the mu receptor.[7] Recent studies involving site-directed mutagenesis of adrenergic receptors[1] indicate that physical and functional receptor/G-protein coupling may not be identical, and that physical coupling may be more important for down-regulation. In general, however, efficacy should correlate with coupling, and thus with the capacity for down-regulation.

REGULATION BY MORPHINE OF MU AND DELTA RECEPTORS IN THE SAME CELLS

Although the SH-SY5Y cell line was initially used as a neuronal model for studying mu receptor responses to morphine, the cells also express delta receptors. The ratio of mu to delta receptors has been estimated to be from 2:1[10] to 5:1.[11,12] Under our culture conditions, the ratio of mu to delta sites is about 1.4 to 1.[13] Morphine binds preferentially to mu receptors, but it can also bind with lower affinity to delta receptors. We therefore tested whether morphine affected the delta sites that are coexpressed with mu sites in SH-SY5Y cells. The delta sites were also down-regulated by morphine. This would not be expected from studies in NG108-15 cells, because morphine did not affect receptors in those cells, which contain only delta sites.[3] Furthermore, morphine down-regulated delta receptors to a greater extent than mu receptors in the SH-SY5Y cells. Thus, the regulation of delta sites was qualitatively different in NG108-15 and SH-SY5Y cells. One possibility is that morphine could be down-regulating delta sites by acting at the mu sites present in SH-SY5Y cells but not NG108-15 cells. Baumhaker *et al.*[14] suggested such a mechanism for the parent cell line, SK-N-SH, based on studies with the mu receptor alkylating agent β-funaltrexamine.

To test whether mu and delta sites could be separately regulated by morphine, we combined treatment with morphine and the mu antagonist CTAP (D-Phe-Cys-Tyr-D-Trp-Arg-Thr-Pen-Thr-NH$_2$) or the delta antagonist ICI 174864 (N,N-diallyl-Tyr-Aib-Aib-Phe-Leu-OH). CTAP reversed the effects of morphine on the mu (but not the delta) receptor and ICI 174864 reversed the down-regulation of the delta (but not the mu) receptor. The pharmacology of both antagonists, however, proved complex. CTAP alone at high doses down-regulated the delta receptor. This action was found to result from delta agonist activity because ICI 174864 could reverse the CTAP-induced down-regulation. CTAP could, therefore, have blocked morphine-induced mu receptor–mediated down-regulation of delta sites, and simultaneously caused down-regulation by its activity as a delta agonist. A dose (300 nM) was found, however, that completely antagonized morphine-induced down-regulation of mu receptors without down-regulating delta receptors by itself or blocking the morphine-induced down-regulation of delta receptors. Taken together with the reversal of morphine-induced down-regulation by the delta antagonist ICI 174864, these results indicate that activation of the mu receptor is not required, but activation of delta sites is required, for morphine to down-regulate the delta receptor in SH-SY5Y cells. These qualitative differences between NG108-15 and SH-SY5Y cells in the regulation of delta receptors may help elucidate critical mechanisms in their regulatory processes.

REGULATION OF MU AND DELTA SITES *IN VITRO* BY ANTAGONISTS

Shortly after establishment of opiate binding assays, it was shown that chronic administration of opiate antagonists up-regulated opiate receptors,[15] and the phenomenon has been well-characterized since then.[16–19] *In vitro* models of this process have not been as consistent: Barg *et al.*[20] found increased [³H]-DADL binding in NG108-15 cells but Law *et al.*[3] found no effect on [³H]-diprenorphine binding.

We found[7,13] that 24-h exposure to naloxone increased both mu and delta receptors by about 40% in SH-SY5Y cells, indicating that these cells provide a good model for antagonist effects on neuronal opiate receptors. Antagonists selective for receptor subtypes were also able to up-regulate their respective receptors, but the effects were not as robust as with naloxone. CTAP increased mu receptors by 20–30%. As described above, however, it also down-regulated delta sites at high doses as a result of its delta agonist activity.

The up-regulation by CTAP was observed even in the presence of high concentrations of morphine. This is consistent with studies of Yoburn *et al.*[18] showing that up-regulation is resistant to blockade by agonists and that less than full occupancy of the receptors is required for up-regulation.

Indirect mechanisms are generally thought to be involved in antagonist-induced up-regulation. Thus, a basal "tone" of agonist activity is thought to induce partial down-regulation that is "unmasked" in the presence of antagonist. In culture, the source of this agonist activity could be the cells themselves or the media. SH-SY5Y cells are known to express both proenkephalin[21] and POMC.[22] In addition, we found that fetal calf serum contains material that can displace [^{3}H]-diprenorphine binding.[13] Several potential sources for "basal agonist activity" are therefore present in the cell culture and could be part of the mechanism of up-regulation by antagonists. Studies with the delta antagonist ICI 174,864, however, raised the additional possibility of a direct mechanism of antagonist-induced up-regulation.

ICI 174,864 up-regulated delta receptors in the SH-SY5Y cells, but unexpectedly it also up-regulated mu receptors.[13] It is not known to have mu antagonist properties,[23] but has been shown to exhibit "negative intrinsic activity" at the high-affinity GTPase associated with delta receptors in NG108-15 cells.[24] One of the earliest events in the signaling cascade for agonists acting at G-protein-coupled receptors is stimulation of GTPase. In NG108-15 cells, there is a low basal activation of GTPase even in the absence of agonist ligands, and ICI 174,864 can inhibit this activity. This inhibition, or "negative intrinsic activity," could be involved in a direct mechanism by which the antagonist up-regulates receptors. If this pathway is shared by mu and delta receptors, it could contribute to the combined up-regulation by ICI 174,864 of mu and delta sites in SH-SY5Y cells.

REGULATION OF RECEPTORS BY SELECTIVE AGONISTS

The use of agonists selective for mu or delta receptors permitted selective down-regulation of each of the sites.[13] PL017, an analogue of both β-casomorphin and morphiceptin that is one of the most selective mu agonists available, down-regulated mu sites in SH-SY5Y cells with an IC_{50} of 180 nM, but did not alter delta sites at concentrations up to 10,000 nM. The mu selective enkephalin analogue DAMGO also selectively down-regulated the receptors. Conversely, the highly selective delta agonist DPDPE down-regulated delta, but not mu receptors. DPDPE was effective at very low (sub-nanomolar) doses, indicating either high efficacy for the compound or relatively high sensitivity of the SH-SY5Y cells to delta receptor down-regulation.

REGULATION OF RECEPTORS BY DIFFERENTIATING AGENTS

One advantage of cell line models is that in the undifferentiated state the cells proliferate rapidly to permit generation of a sufficient population for experimentation. The cells may then be induced to differentiate into a number of morphologically and biochemically distinct phenotypes, some of which may include high concentrations of the receptor of interest. In SH-SY5Y cells, Yu et al.[11] showed that differentiation with retinoic acid (RA) increased mu receptors by 60%. We further characterized the effects of differentiating agents on opiate receptors[13] and found that delta receptors are also increased to about the same extent as mu receptors by RA, but differentiation with the phorbol agent TPA increased mu receptors without changing delta sites. Thus, the phenotypic ratio of mu to delta sites can be manipulated by the choice of differentiating agents.

IMPLICATIONS OF *IN VITRO* MODELS OF OPIATE RECEPTOR REGULATION FOR *IN VIVO* STUDIES AND FOR MODELS OF OPIATE TOLERANCE

Inasmuch as activation of a receptor is the first step in the signal transduction cascade, changes in receptor number or affinity provide an attractive mechanism to explain the loss of responsiveness that is characteristic of tolerance. It is most likely, however, that tolerance involves several molecular sites of action, including receptor, G-protein, and effector proteins as well as phosphorylation and transcriptional events. Measurable changes in the receptor number or affinity cannot fully account for the profile of changes that characterize tolerance. Indeed, tolerance can develop independently of down-regulation. Changes in any other single step in the signal cascade, however, are also unlikely by themselves to account for the full profile of tolerance. Further understanding of changes at each step is needed to develop a comprehensive understanding of the cellular mechanisms of tolerance.

Regulation by agonists of opiate receptors in brain has been highly variable. After *in vivo* chronic administration of agonists, opiate receptors in brain have been reported to decrease,[25,26] increase,[27–31] not change,[32,33] change in some, but not other brain areas,[34] and to change in amount and direction depending on the dose, efficacy, and selectivity of the agonist.[35,36]

The suggestion that efficacy is a critical factor in whether an agonist will induce down-regulation was first proposed on the basis of *in vitro* studies with the delta receptor[3] and was confirmed *in vitro* for the mu receptor.[7] Studies *in vivo* are consistent with this idea because the highly efficacious agonist etorphine decreased mu and delta receptors in rat brain[26] and mu receptors in mouse brain.[35] The high efficacy mu agonist DAMGO down-regulated mu receptors in SH-SY5Y cells[13] and in three of eight areas examined in rat brain.[34]

The dose of agonist required for down-regulation may also offer clues to differences in receptor regulation *in vivo* and *in vitro,* and perhaps between developing and adult animals. It has been suggested[35,37] that one reason that down-regulation is readily observed in culture but not *in vivo* is that the doses required are toxic *in vivo,* whereas cell cultures can handle higher doses. In the mu receptor-containing 7315c pituitary cells,[4,5] down-regulation required considerably higher doses and longer exposure times than did desensitization, which is the relatively rapid loss of the cyclicAMP response to the acute exposure to agonist. Desensitization and down-regulation were also found to be separable phenomena in earlier studies on the delta sites in NG108-15 cells.[3]

In the NG108-15 cells, however, low doses of highly efficacious compounds were capable of down-regulating receptors. In SH-SY5Y cells it has been reported that desensitization and down-regulation were not clearly distinguishable.[38] In addition, doses of morphine (IC_{50}) required for down-regulation of receptors in SH-SY5Y cells[7,13] were comparable to both the K_i for binding in culture and the doses required to induce the major functional response to morphine, which is inhibition of adenylate cyclase. Furthermore, the delta agonist DPDPE down-regulated the delta receptor at very low concentrations.[13] Thus, down-regulation does not necessarily require unusually high doses of agonists. Again, the efficacy of the agonist is an important consideration, and, for the *in vivo* situation, its pharmocokinetics and the relative sensitivity of systems mediating toxic effects of the drug may all contribute to differences observed *in vivo* and *in vitro.*

In the developing animal, related issues may contribute to the phenomenon observed by us and others[39–42] that agonist-induced down-regulation is more readily apparent during the perinatal period than in later life. Less activity of metabolic enzymes[43] or a less developed blood-brain barrier could contribute to greater delivery of opiates to the brain in young animals. Other factors must also be considered, however, including differential ontogenic expression of opiate receptor types or components of their signaling processes, and homeostatic mechanisms (described below) that may counteract the effects of opiates.

Like all models of tolerance, those involving changes in receptors have certain strengths and weaknesses.[44] Changes in receptors would account for certain aspects of cross-tolerance, in which agonists selective for one receptor type do not induce tolerance at other receptors. Theoretical models must accommodate the observation that opiates can induce changes in responsiveness of great magnitude (e.g., 1000 fold).[45] Considerable loss of response can be explained by receptor models. However, changes resulting from loss of receptor number alone would tend to be abrupt rather than continuous as is typical of tolerance. The concept of "receptor reserve," in which only a small proportion of receptors need to be occupied for a maximal response, in general does not specify the locus of the reserve. Excess of receptors or components of their signaling systems could mediate the reserve. Because chronic exposure to agonist diminishes the reserve, the dose-response curve of the agonist is first shifted to the right (greater doses are required for a given effect), and then the maximal response declines when the reserve is lost and the receptor number declines. Thus, in a system in which the receptor number contributes substantially to the reserve, a loss in number would be reflected in a gradual loss in response. If post-receptor mechanisms are primarily responsible for the reserve, loss of receptor number would lead to an abrupt drop in responsiveness.

One of the major weaknesses of an opiate receptor model of tolerance is that many studies *in vivo* have reported that tolerance occurs in the absence of changes in receptors. For this reason, several alternative mechanisms have been postulated to explain tolerance, some of which can be classified as homeostatic. In these models, non-opiate processes are activated by opiates and counteract the actions of the opiates. These mechanisms include "antiopiate" or "opiate modulating" endogenous peptides such as NeuropeptideFF (NPFF)[46–48] and brain peptides related to Tyr-MIF-1 (Tyr-Pro-Leu-Gly-NH_2). NPFF provides a model of a peptide that does not bind to opiate receptors but is able to antagonize opiate effects, presumably through actions at its own receptor. The Tyr-MIF-1–like peptides, by contrast, bind to both the mu opiate receptor[49–51] and to non-opiate Tyr-MIF-1 sites.[52–54]

Homeostatic mechanisms are thought to be activated by opiates acting at their receptors to release the modulating peptide. Although this model can account for aspects of tolerance and withdrawal that other models cannot, a theoretical limitation is that the level of activation of the modulating peptide will be maximal with saturation of the opiate receptor; this would occur relatively early in the development of tolerance, and further tolerance through this mechanism would be limited.[44] Tyr-MIF-1, Tyr-W-MIF-1 (Tyr-Pro-Trp-Gly-NH_2)[50] and related peptides, by contrast, can bind to the opiate site and, particularly during tolerance or in conditions of reduced opiate receptor reserve, can antagonize the actions of morphine and DAMGO.[55] This effect may result from its action as a partial agonist at the opiate site, from an "antiopiate" action at its own site, or both. These peptides therefore provide candidate molecules involved in a model of tolerance that combines aspects of opiate receptor and homeostatic mechanisms.[56] Such a combined model could account for aspects of tolerance that other models have difficulty explaining, such as large, continuous shifts in the dose-response curve.

In summary, several mechanisms and characteristics of agonist- and antagonist-induced changes in opiate receptors and their responsiveness to opiates have been characterized in cell lines, including the mu receptor-containing SH-SY5Y cells. These studies have implications for changes observed *in vivo* and ultimately for our understanding of the dynamic processes underlying changing responsiveness to opiates and neurotransmitters in general.

REFERENCES

1. CAMPBELL, P. T., M. HNATOWICH, B. F. O'DOWD, M. G. CARON, R. J. LEFKOWITZ & W. P. HAUSDORFF. 1991. Mutations of the human β_2-adrenergic receptor that impair coupling to G_s interfere with receptor down-regulation but not sequestration. Mol. Pharmacol. **39:** 192–198.

2. CHANG, K.-J., R. W. ECKEL & S. G. BLANCHARD. 1982. Opioid peptides induce reduction of enkephalin receptors in cultured neuroblastoma cells. Nature **296:** 446–448.

3. LAW, P.-Y., D. S. HOM & H. H. LOH. 1983. Opiate receptor down-regulation and desensitization in neuroblastoma × glioma NG108-15 hybrid cells are two separate cellular adaptation processes. Mol. Pharmacol. **24:** 413–424.

4. PUTTFARCKEN, P. S., L. L. WERLING & B. M. COX. 1988. Effects of chronic morphine exposure on opioid inhibition of adenylyl cyclase in 7315c cell membranes: A useful model for the study of tolerance and μ opiate receptors. Mol. Pharmacol. **33:** 520–527.

5. PUTTFARCKEN, P. S. & B. M. COX. 1989. Morphine-induced desensitization and down-regulation at mu-receptors in 7315c pituitary tumor cells. Life Sci. **45:** 1937–1942.

6. ZADINA, J. E., S. L. CHANG, L.-J. GE, A. J. KASTIN & R. E. HARLAN. 1990. Down-regulation of mu opiate receptors by morphine and presence of Tyr-MIF-1 binding sites in SH-SY5Y human neuroblastoma cells. *In* New Leads in Opioid Research. J. M. Van Ree, A. H. Mulder, V. M. Wiegant & T. B. Van Wimersma Greidanus, Eds: Excerpta Med. Int. Congr. Ser. no. 914. Elsevier. Amsterdam/New York.

7. ZADINA, J. E., S. L. CHANG, L.-J. GE & A. J. KASTIN. 1993. Mu opiate receptor down-regulation by morphine and up-regulation by naloxone in SH-SY5Y human neuroblastoma cells. J. Pharmacol. Exp. Ther. **265:** 254–262.

8. TOLL, L. 1990. μ-Opioid receptor binding in intact SH-SY5Y neuroblastoma cells. Eur. J. Pharmacol. **176:** 213–217.

9. YU, V. C., G. HOCHHAUS, F.-H. CHANG, M. L. RICHARDS, H. R. BOURNE & W. SADEE. 1988. Differentiation of human neuroblastoma cells: Marked potentiation of prostaglandin E-stimulated accumulation of cyclic AMP by retinoic acid. J. Neurochem. **51:** 1892–1899.

10. KAZMI, S. M. I. & R. K. MISHRA. 1987. Comparative pharmacological properties and functional coupling of μ and δ opioid receptor sites in human neuroblastoma SH-SY5Y cells. Mol. Pharmacol. **32:** 109–118.

11. YU, V. C., M. L. RICHARDS & W. SADEE. 1986. A human neuroblastoma cell line expresses μ and δ opioid receptor sites. J. Biol. Chem. **261:** 1065–1070.

12. SADEE, W., V. C. YU, M. L. RICHARDS, P. N. PREIS, M. R. SCHWAB, F. M. BRODSKY & J. L. BIEDLER. 1987. Expression of neurotransmitter receptors and myc protooncogenes in subclones of a human neuroblastoma cell line. Cancer Res. **47:** 5207–5212.

13. ZADINA, J. E., L. M. HARRISON, L.-J. GE, A. J. KASTIN & S. L. CHANG. 1994. Differential regulation of mu and delta opiate receptors by morphine, selective agonists and antagonists, and differentiating agents in SH-SY5Y human neuroblastoma cells. J. Pharmacol. Exp. Ther. **270:** 1086–1096.

14. BAUMHAKER, Y., M. GAFNI, O. KEREN & Y. SARNE. 1993. Selective and interactive down-regulation of μ- and δ-opioid receptors in human neuroblastoma SK-N-SH cells. Mol. Pharmacol. **44:** 461–467.

15. HITZEMANN, R., B. HITZEMANN & H. H. LOH. 1974. Binding of ^{3}H-naloxone in the mouse brain: Effect of ions and tolerance development. Life Sci. **14:** 2393–2404.

16. ZUKIN, R. S., J. R. SUGARMAN, M. L. FITZ-SYAGE, E. L. GARDNER, S. R. ZUKIN & A. R. GINTZLER. 1982. Naltrexone-induced opiate receptor supersensitivity. Brain Res. **245:** 285–292.

17. TEMPEL, A., E. L. GARDNER & R. S. ZUKIN. 1984. Visualization of opiate receptor upregulation by light microscopy autoradiography. Proc. Natl. Acad. Sci. USA **81:** 3893–3897.
18. YOBURN, B. C., A. DUTTAROY, S. SHAH & T. DAVIS. 1994. Opioid antagonist-induced receptor upregulation: Effects of concurrent agonist administration. Brain Res. Bull. **33:** 237–240.
19. YOBURN, B. C., F. A. NUNES, B. ADLER, G. W. PASTERNAK & C. E. INTURRISI. 1986. Pharmacodynamic supersensitivity and opioid receptor upregulation in the mouse. J. Pharmacol. Exp. Ther. **239:** 132–135.
20. BARG, J., R. LEVY & R. SIMANTOV. 1984. Up-regulation of opiate receptors by opiate antagonists in neuroblastoma-glioma cell culture: The possibility of interaction with guanosine triphosphate-binding proteins. Neurosci. Lett. **50:** 133–137.
21. FOLKESSON, R., H.-J. MONSTEIN, T. GEIJER, S. PAHLMAN, K. NILSSON & L. TERENIUS. 1988. Expression of the proenkephalin gene in human neuroblastoma cell lines. Mol. Brain Res. **3:** 147–154.
22. CHANG, S. L., J. E. ZADINA, L. SPRIGGS & S. SQUINTO. 1993. Prolonged activation of c-fos and optimal activation of pro-opiomelanocortin mRNA after repeated morphine exposure in SH-SY5Y cells. Mol. Cell. Neurosci. **4:** 25–29.
23. COTTON, R., M. G. GILES, L. MILLER, J. S. SHAW & D. TIMMS. 1984. ICI 174864: A highly selective antagonist for the opioid δ-receptor. Eur. J. Pharmacol. **97:** 331–332.
24. COSTA, T. & A. HERZ. 1989. Antagonists with negative intrinsic activity at δ opioid receptors coupled to GTP-binding proteins. Proc. Natl. Acad. Sci. USA **86:** 7321–7325.
25. DINGLEDINE, R., R. J. VALENTINO, E. BOSTOCK, M. E. KING & K.-J. CHANG. 1983. Down-regulation of δ but not μ opioid receptors in the hippocampal slice associated with loss of physiological response. Life Sci. **33:** 333–336.
26. TAO, P.-L., P.-Y. LAW & H. H. LOH. 1987. Decrease in delta and mu opioid receptor binding capacity in rat brain after chronic etorphine treatment. J. Pharmacol. Exp. Ther. **240:** 809–816.
27. HOLADAY, J. W., R. J. HITZEMANN, J. CURELL, F. C. TORTELLA & G. L. BELENKY. 1982. Repeated electroconvulsive shock or chronic morphine treatment increases the number of ^{3}H-D-Ala^2,D-Leu^5-enkephalin binding sites in rat brain membranes. Life Sci. **31:** 2359–2362.
28. ROTHMAN, R. B., J. A. DANKS, A. E. JACOBSON, T. R. BURKE, JR., K. C. RICE, F. C. TORTELLA & J. W. HOLADAY. 1986. Morphine tolerance increases μ-noncompetitive δ binding sites. Eur. J. Pharmacol. **124:** 113–119.
29. PERT, C. B. & S. H. SNYDER. 1976. Opiate receptor binding-enhancement by opiate administration *in vivo*. Biochem. Pharmacol. **25:** 847–853.
30. ZADINA, J. E., A. J. KASTIN, L.-J. GE, H. GULDEN & K. J. BUNGART. 1989. Chronic, but not acute, administration of morphine alters antiopiate (Tyr-MIF-1) binding sites in rat brain. Life Sci. **44:** 555–561.
31. ROTHMAN, R. B., V. BYKOV, J. B. LONG, L. S. BRADY, A. E. JACOBSON, K. C. RICE & J. W. HOLADAY. 1989. Chronic administration of morphine and naltrexone up-regulates μ-opioid binding sites labeled by [^{3}H][D-Ala^2,MePhe^4,Gly-ol ^{5}]enkephalin: Further evidence for two μ-binding sites. Eur. J. Pharmacol. **160:** 71–82.
32. KLEE, W. A. & R. A. STREATY. 1974. Narcotic receptor sites in morphine-dependent rats. Nature **248:** 61–63.
33. HOLLT, V., J. DUM, J. BLASIG, P. SCHUBERT & A. HERZ. 1975. Comparison of *in vivo* and *in vitro* parameters of opiate receptor binding in naive and tolerant/dependent rodents. Life Sci. **16:** 1823–1828.
34. BHARGAVA, H. N. & A. GULATI. 1990. Down-regulation of brain and spinal cord μ-opiate receptors in morphine tolerant-dependent rats. Eur. J. Pharmacol. **190:** 305–311.
35. YOBURN, B. C., B. BILLINGS & A. DUTTAROY. 1993. Opioid receptor regulation in mice. J. Pharmacol. Exp. Ther. **265:** 314–320.
36. TAO, P.-L., H.-Y. LEE, L.-R. CHANG & H. H. LOH. 1990. Decrease in μ-opioid receptor binding capacity in rat brain after chronic PL017 treatment. Brain Res. **526:** 270–275.
37. WERLING, L. L., P. N. MCMAHON & B. M. COX. 1989. Selective changes in μ opioid

receptor properties induced by chronic morphine exposure. Proc. Natl. Acad. Sci. USA **86:** 6393–6397.

38. CARTER, B. & F. MEDZIHRADSKY. 1993. Receptor mechanisms of opioid tolerance in SH-SY5Y human neural cells. Mol. Pharmacol. **43:** 465–473.
39. TSANG, D. & S. C. NG. 1980. Effect of antenatal exposure to opiates on the development of opiate receptors in rat brain. Brain Res. **188:** 199–206.
40. TEMPEL, A., J. HABAS, W. PAREDES & G. A. BARR. 1988. Morphine-induced downregulation of μ-opioid receptors in neonatal rat brain. Dev. Brain Res. **41:** 129–133.
41. HAMMER, R. P., JR., J. V. SEATRIZ & A. R. RICALDE. 1991. Regional dependence of morphine-induced μ-opiate receptor down-regulation in perinatal rat brain. Eur. J. Pharmacol. **209:** 253–256.
42. ZADINA, J. E. & A. J. KASTIN. 1986. Neonatal peptides affect developing rats: β-endorphin alters nociception and opiate receptors, corticotropin-releasing factor alters corticosterone. Dev. Brain Res. **29:** 21–29.
43. KASTIN, A. J., K. HAHN, W. A. BANKS & J. ZADINA. 1994. Delayed degradation of Tyr-MIF-1 in neonatal rat plasma. Peptides **15:** 1561–1563.
44. SMITH, A. P., P.-Y. LAW & H. H. LOH. 1988. Role of opioid receptors in narcotic tolerance/dependence. *In* The Opiate Receptors. G. W. Pasternak, Ed. Humana Press. Clifton, NJ.
45. GOLDSTEIN, A. 1974. Drug tolerance and physical dependence. *In* Principles of Drug Action. A. Goldstein, L. Aronow & S. M. Kalman, Eds. Harper & Row. New York.
46. YANG, H.-Y.T., W. FRATTA, E. A. MAJANE & E. COSTA. 1985. Isolation, sequencing, synthesis, and pharmacological characterization of two brain neuropeptides that modulate the action of morphine. Proc. Natl. Acad. Sci. USA **82:** 7757–7761.
47. MALIN, D. H., J. R. LAKE, M. V. HAMMOND, D. E. FOWLER, J. E. LEYVA, R. B. ROGILLIO, J. B. SLOAN, T. M. DOUGHERTY & K. LUDGATE. 1990. FMRF-NH2 like mammalian octapeptide in opiate dependence and withdrawal. NIDA Res. Monogr. **105:** 271–277.
48. ROTHMAN, R. B. 1992. A review of the role of anti-opioid peptides in morphine tolerance and dependence. Synapse **12:** 129–138.
49. ZADINA, J. E. & A. J. KASTIN. 1986. Interactions of Tyr-MIF-1 at opiate receptor sites. Pharmacol. Biochem. Behav. **25:** 1303–1305.
50. ERCHEGYI, J., A. J. KASTIN & J. E. ZADINA. 1992. Isolation of a novel tetrapeptide with opiate and antiopiate activity from human cortex: Tyr-Pro-Trp-Gly-NH2 (Tyr-W-MIF-1). Peptides **13:** 623–631.
51. ERCHEGYI, J., J. E. ZADINA, X.-D. QIU, D. C. KERSH, L.-J. GE, M. BROWN & A. J. KASTIN. 1993. Structure-activity relationships of analogs of the endogenous brain peptides Tyr-MIF-1 and Tyr-W-MIF-1. Peptide Res. **6:** 31–38.
52. ZADINA, J. E., A. J. KASTIN, E. F. KRIEG, JR. & D. H. COY. 1982. Characterization of binding sites for N-Tyr-MIF-1 (Tyr-Pro-Leu-Gly-NH2) in rat brain. Pharmacol. Biochem. Behav. **17:** 1193–1198.
53. ZADINA, J. E. & A. J. KASTIN. 1985. Interactions between the antiopiate Tyr-MIF-1 and the mu opiate morphiceptin at their respective binding sites in brain. Peptides **6:** 965–970.
54. ZADINA, J. E., A. J. KASTIN, L.-J. GE & V. BRANTL. 1990. Hemorphins, cytochrophins, and human beta-casomorphins bind to antiopiate (Tyr-MIF-1) as well as opiate binding sites in rat brain. Life Sci. **47:** PL-25–PL-30.
55. ZADINA, J. E., A. J. KASTIN, D. KERSH & A. WYATT. 1992. Tyr-MIF-1 and hemorphin can act as opiate agonists as well as antagonists in the guinea pig ileum. Life Sci. **51:** 869–885.
56. ZADINA, J. E., A. J. KASTIN, L.-J. GE & S. L. CHANG. 1994. Novel peptides and mechanisms of opiate tolerance. Regul. Pept. Suppl. **1:** S195–S196.

In Vivo Interactions between Opioid and Adrenergic Receptors for Peripheral Sympathetic Functional Regulation

HERMAN M. RHEE[a,b] AND DONG H. PARK[c]

[a]Food and Drug Administration
Center for Drug Evaluation and Research
Rockville, Maryland 20857

[c]Cornell University Medical College
Laboratory of Molecular Neurobiology
Burke Rehabiliation Center
White Plains, New York 10605

Methionine enkephalin (ME) produced a variety of pharmacological and/or physiological actions.[1] It decreased blood pressure (BP) although its cardiovascular actions are variable depending on routes of its administration, state of anesthesia, and animal species tested.[2] Its hypotensive action was blocked by naloxone, but not by naloxone methobromide, a quaternary analog. Hypotension was also antagonized by adrenergic α-receptor antagonists such as phentolamine.[3] ME suppressed sympathetic nerve activity before the reduction in pressure.[4] ME also affected blood flow and vascular resistance in selected organ beds of rabbits.[5] Systemic infusion of ME increased norepinephrine in microdialysis dialysate in ventrolateral medulla in rats.[6] The primary objective of this work was to elucidate the nature of opioid and adrenergic receptor interactions, which will eventually modify peripheral cardiac function.

MATERIALS AND METHODS

Sprague-Dawley (SD) or spontaneously hypertensive rats (SHR) were anesthetized with intraperitoneal Inactin (120 mg/kg). The right femoral artery and vein were cannulated for monitoring BP and drug infusion. Body temperature of the animals was maintained and their heads were fixed on stereotaxic frame. Cardiovascular parameters were recorded as described previously.[4] For intracisternal and ventrolateral medulla (C_1 area) microinjection the atlanto-occipital membrane was punctured by a 26-gauge needle that was mounted to a stereotaxic micromanipulator as reported.[6] ME (0, 1, 3, and 10 μg/kg) was injected into the cisterna magna in a volume of 5 μL. Obex (stereotaxic zero) was visualized, and C_1 coordinates were 2.0 mm anterior, ± 1.9 mm lateral, and 3.0 mm below the floor of the fourth ventricle as reported.[6] Catecholamines were determined by high-performance liquid chromatography as previously reported.[6] For the assay of thyrosine hydroxylase (TH), dopamine β-hydroxylase (DβH) and phenylethyl N-methyl transferase (PNMT), adrenal glands and C_1 area (punch biopsy) were homogenized in 1 mL and 0.25 mL of 5 mM

[b]Address correspondence to Herman M. Rhee, Ph.D., Food and Drug Administration, HFD-510 Parklawn Building, Room 14B 45, 5600 Fishers Lane, Rockville, MD 20857.

potassium phosphate buffer (pH 7.0), each containing 0.2% Triton X-100. Homogenates were centrifuged at $10,000 \times g$ for 10 min. The resultant supernatants were used for enzymes assay, based on protein concentration. For the DβH assay [^{14}C]SAM (sulfur adenosyl methionine) and 7.5×10^{-5} M Ca^{2+} were used with an appropriate volume of the supernatant. For the PNMT assay 25 µL of C_1-area supernatant, and 2 µL of adrenal supernatant were used at 37°C for 15 min in the presence of [^{3}H]SAM.

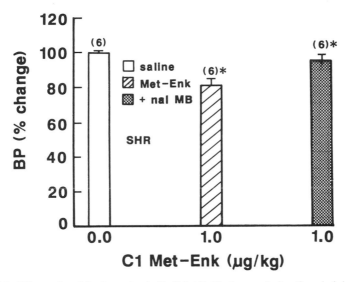

FIGURE 1. Effects of methionine enkephalin (Met-Enk) alone and after the administration of naloxone methobromide (nal MB) in spontaneously hypertensive rats (SHR). Young adult SHR rats were anesthetized with intraperitoneal Inactin (120 mg/kg). The ventrolateral medulla (C_1 area) was exposed as reported[6] on stereotaxic instrument. Met-Enk was injected into the area at a dose of 0 (control) or 1 µg/kg in a volume of 1 µL. In a different animal the procedure was repeated 10 min after an intravenous administration of nal MB (1.3 mg/kg). Systolic blood pressure was recorded as reported previously.[5] Control pressure was 116 ± 8 mmHg, which was considered 100%. The number in parentheses indicates the number of rats used and the vertical bars represent SEM. Asterisk indicates significance ($p \leq 0.05$).

RESULTS AND DISCUSSION

Intravenous (i.v.) and intra C_1 administration of ME reduced blood pressure, which was blocked by naloxone pretreatment. Intracisternal injection of ME had little effect at tested dose (30 µg/kg). Pretreatment of the animal with an intravenous naloxone methobromide did not protect the animal. As little as 1 µg/kg ME administration at C_1 produced hypotension in SD rats and in SHR (FIG. 1). Pretreatment of the animal with naloxone increased adrenal TH activity (6.4 versus 9.4 nm/mg protein per 20 min at 30 °C) and adrenal PNMT activity significantly (data not shown), suggesting that the enzyme activities might be suppressed in normal conditions. However, the PNMT and TH activities at C_1 area were not affected after

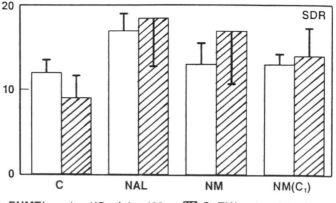

☐ C₁ PNMT(pnm/mg/15 min) x 100   ▨ C₁ TH(pm/mg/20 min) x 100

FIGURE 2. Phenylethyl N-methyltransferase (PNMT) and thyrosine hydroxylase (TH) activities at the ventrolateral medulla (C_1 area) in Sprague-Dawley rats. C_1 areas were obtained by punch biopsy, and the enzyme activity of PNMT and TH was assayed in the homogenates of the biopsy samples as described in MATERIALS AND METHODS. The PNMT and TH activities in the samples obtained from the animals that were treated with 1 mg/kg naloxone (NAL) were slightly higher than the saline treatment as control (C); however, the differences were not significant. Naloxone methobromide (NM) treatment intravenously or directly into the C_1 area [NM(C_1)] did not change the enzyme activities. Each point represents mean of at least four to seven tissues collected from three to five animals.

naloxone (1 mg/kg, i.v.), naloxone methobromide (1.3 mg/kg, i.v.), which was applied directly into C_1 area in both SHR and SD rats (FIG. 2). Furthermore, DβH activity at C_1 area was not affected by treatment of the animal. It is concluded that specific interactions between opioid and adrenergic receptors for catecholamine synthesis, and its metabolism and release in both central and peripheral systems may dictate cardiac function such as blood pressure regulation.

REFERENCES

1. OLSON, G. A., R. D. OLSON & A. J. KASTIN. 1987. Endogenous opiates: Review. Peptides 8: 1135–1164.
2. FEUERSTEIN, G. & A. L. SIREN. 1987. The opioid peptides. A role in hypertension? Hypertension 9: 561–565.
3. EULIE, P. & H. M. RHEE. 1984. Reduction by phentolamine of the hypotensive effect of methionine enkephalin in anesthetized rabbits. Br. J. Pharmacol. 83: 783–790.
4. RHEE, H. M., P. J. EULIE & D. F. PETERSON. 1985. Suppression of renal nerve activity by methionine enkephalin in anesthetized rabbits. J. Pharmacol. Exp. Ther. 234: 534–537.
5. EULIE, P. J., H. M. RHEE & M. H. LAUGHLIN. 1987. Effects of met-enkephalin on regional blood flow and vascular resistance in rabbits. Eur. J. Pharmacol. 137(1): 25–31.
6. RHEE, H. M., J. A. STRICKLAND & P. A. MASON. 1992. Microdialysis of noradrenaline in rostral ventrolateral medulla after intravenous methionine enkephalin administration in anesthetized rats. Can. J. Cardiol. 8(5): 527–535.

Receptor-Receptor Interactions and Their Relevance for Receptor Diversity

Focus on Neuropeptide/Dopamine Interactions[a]

K. FUXE,[b,c] X.-M. LI,[b] S. TANGANELLI,[d] P. HEDLUND,[b]
W. T. O'CONNOR,[e] L. FERRARO,[d] U. UNGERSTEDT,[e]
AND L. F. AGNATI[f]

[b]Department of Neuroscience
[e]Department of Physiology and Pharmacology
Karolinska Institutet
Stockholm, Sweden

[d]Department of Pharmacology
University of Ferrara
Ferrara, Italy

[f]Department of Human Physiology
University of Modena
Modena, Italy

Receptor diversity appears to be a general phenomenon occurring for both G-protein–coupled and ion channel–coupled receptors involved in neuronal communication. In our own work we analyzed the role of receptor diversity in the dopamine (DA) receptor systems of the basal ganglia.[1-5] Five subtypes of DA receptors exist, namely, D_1 to D_5. The two major DA receptor subtypes are the D_1 receptors, which by activation of the G_s-proteins increase adenylate cyclase activity and phospholipase C activity, and the D_2 receptors which, via G_i-proteins, are coupled to multiple transduction pathways involving inhibition of adenylate cyclase and phospholipase C activity, regulation of calcium influx, opening of potassium channels, and increases of arachidonic acid release.[6,7]

This paper introduces the hypothesis that one meaning of receptor diversity is that it allows the development of discrete interactions between receptor subtypes of the same transmitter as well as for different transmitters, leading to the development of a new type of plasticity in synaptic (wiring) transmission (WT) and volume transmission (VT).[8]

POSSIBLE FUNCTIONAL MEANING OF RECEPTOR DIVERSITY

Multiple reasons probably exist for the development of a high degree of receptor diversity in large numbers of receptor systems for the transmitters of the nervous system. In the case of G-protein–coupled receptors, it seems likely that the receptor

[a]This work was supported by grant 04X-715 from the Swedish Medical Research Council and by a grant from the Knut and Alice Wallenberg Foundation.

[c]Address correspondence to Prof. Kjell Fuxe, Department of Neuroscience, Division of Cellular and Molecular Neurochemistry, Karolinska Institutet, 171 77 Stockholm, Sweden.

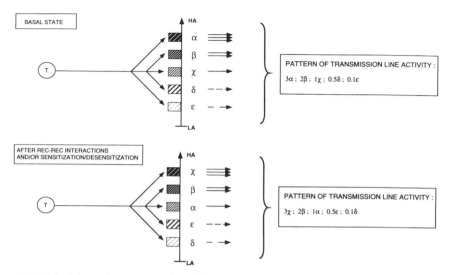

FIGURE 1. Schematic representation of the possible role of receptor-receptor interactions in the pattern of subtype receptor (α, β, γ, ...) activation. The scheme is based on the following conditions: (1) equal amount of transmitter (T) released in the basal state and after receptor-receptor interactions or sensitization/desensitization; (2) equal receptor subtype densities; and (3) equal access of T to the various receptor subtypes. Note that basal state can probably be observed only *in vitro*.

diversity allows the receptors to couple to different types of G-proteins, which leads to the activation or inhibition of multiple transduction mechanisms as is the case, for example, for the DA receptor family. Because of such multiple transduction mechanisms it becomes possible for the transmitter, DA, for example, to increase the flow of information over the synapses. Instead, in the case where the receptor subtypes have the same transduction mechanism, the existence of the receptor subtypes allows the redundancy of the transmission so that the safety of the transmission processes can be insured.

The existence of receptor subtypes with a relatively high and a relatively low affinity for the transmitter—for example DA—makes it possible to have a transmission process with very low release of the transmitter, and also to recruit, with increasing impulse flow, new receptor subtypes having a higher affinity for the transmitter, but located further away from the site of release.

By recruiting in the transmission process different receptor subtypes such as D_1 and D_2 receptors, it also becomes possible to develop positive or negative cooperation between the receptor subtypes.[9] As an example, the DAergic inhibition of the sodium potassium ATPase requires the recruitment of both D_1 and D_2 receptors.[10] It may also be surmised that whereas low-affinity receptors can be involved in the WT, high-affinity receptors may also be involved in the VT.

The existence of receptor diversity makes possible the development of receptor subtype-specific interactions with other transmitter receptors so that the plasticity of transmission can be substantially increased (FIG. 1). Thus, by selectively antagonizing the transduction over the D_2 receptor subtype, transmission over the D_1 receptor subtype will be favored,[11] as illustrated in the selective heteroregulation of D_2 receptors by neurotensin (NT) receptors. Other selective regulators of D_2 receptors are the cholecystokinin (CCK) A and B receptors.[12-15] A powerful antagonistic regulator specifically involved in the inhibitory control of D_2 receptor transduction is adenosine operating via A_{2a} receptors.[16] It should be noted that, unlike the neuropeptides, the A_{2a} receptors not only can control the modulation within an affinity state but can also trigger a switching in the proportion of the two affinity states (FIG. 2).

Another functional meaning of receptor diversity is probably also the possibility to desensitize or sensitize the various receptor subtypes for one receptor in a differential manner. This may be brought about either via the second messenger mechanisms involving protein phosphorylation of the receptor subtype proteins, but could also involve the receptor-receptor interactions within the membrane. Thus, it seems possible that neuropeptide receptors, such as the NT and the CCK_A and CCK_B receptors can modulate the desensitization process via the regulation and modulation of receptor affinities. Such phenomena will be capable of regulating the duration of the postsynaptic responses induced by the transmission process.

Thus, it seems clear that the receptor diversity in combination with receptor-receptor subtype specific interactions, which can be antagonistic or synergistic in character, markedly increase plasticity in WT and VT in the nervous system. In this way, switching among transmission lines for the various DA receptor subtypes will become possible. Thus, both the peak and the duration of the transmission process

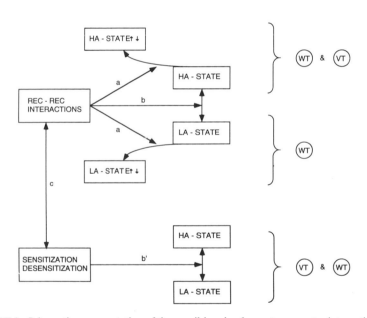

FIGURE 2. Schematic representation of the possible role of receptor-receptor interactions and sensitization/desensitization in the control of receptor affinity. It should be observed that receptor-receptor interactions can modulate both receptor affinity within an affinity state (**a**), as well as trigger the switch in the proportion between the two different affinity states (**b**). Also indicated in the scheme is that sensitization/desensitization processes can affect receptor-receptor interactions (**c**), as well as switch the receptor affinity from the high-affinity (HA) to the low-affinity (LA) state and vice versa (**b'**). The preferential role of the affinity state of a receptor for volume transmission (VT) and wiring transmission (WT) is also indicated. For further details, see text.

can become substantially modulated in relation to the state of activity of the target neuron and its other afferent inputs.

Some of these aspects are supported by our work on selective modulation of D_2 receptors by CCK and NT. In this paper we will illustrate how such receptor subtype-specific interactions among receptors can represent a substrate of neuronal plasticity.

SELECTIVE REGULATION OF D_2 RECEPTORS VIA CCK RECEPTOR SUBTYPES AND NEUROTENSIN RECEPTORS MAY UNDERLIE CCK/DOPAMINE AND NEUROTENSIN/DOPAMINE INTERACTIONS IN THE BASAL GANGLIA

Studies at the Membrane Level

It was early demonstrated that an antagonistic intramembrane regulation of postsynaptic striatal D_2 receptors by CCK receptors may exist that may underlie the neuroleptic-like actions found after central CCK-8 administration.[12,13,15,17-19] The results obtained in striatal membranes, both from the dorsal and ventral part, demonstrated that CCK-8 could produce a selective reduction of the D_2 agonist affinity without any change in the B_{max} value. Recently, it has been shown by Li *et al.*[15] that 0.1 nM of CCK-8 increases the K_d value of the D_2 agonist [^{3}H]N-propylnor-apomorphine (NPA) binding sites by 42%, an action blocked by the CCK_B antagonist PD134308. This increase in the K_d value by CCK-8 was probably related to a reduction of the association rate constant of [^{3}H]NPA by 45% induced by CCK-8. In contrast, NT, which also has been found to increase the K_d value of the D_2 agonist binding sites, did so instead by increasing the dissociation rate constant.[20] The fact that CCK-8 reduces the association rate constant[15] whereas NT increases the dissociation rate constant in their modulation of the K_d value of the D_2 agonist binding sites may explain the recent observation of synergistic interactions between NT and CCK-8 in their inhibitory control of D_2 receptors.[21] A synergistic interaction was not obtained when a high concentration of the neuropeptides was used, suggesting that the two types of neuropeptide receptors can interact with a common regulatory mechanism in the D_2 receptor transduction.

It should be underlined that only the CCK_B receptors are involved in the reduction of the K_d value of the D_2 agonist binding sites in rat striatal membranes, inasmuch as the CCK_A antagonist L364718 was ineffective in counteracting the increase of the K_d value by 1 nM of CCK-8.[15] These results are also in line with the early studies of Agnati, Fuxe, and colleagues that also CCK-4, a selective CCK_B agonist, reduces the affinity of the D_2 agonist binding sites in striatal membranes.[12,13] Thus, it seems clear that a receptor subtype of the CCK receptor family, the CCK_B receptor subtype, can selectively interact in an inhibitory way with the D_2 receptor subtype of the DA receptor family, illustrating the receptor subtype selectivity involving both the interacting receptors. However, it must be emphasized that the D_3 and D_4 subtypes of DA receptors have not yet been tested for their interaction with the CCK and NT receptors, so that the absolute specificity of these interactions still remains to be clarified. It is of substantial interest that the C-terminal NT-(8–13) fragment potently and antagonistically modulates rat neostriatal D_2 receptors[22] and that neuromedin N (NN) also is a potent modulator of D_2 receptor agonist binding in rat neostriatal membranes.[23] In view of the higher potency of NN versus NT to regulate neostriatal D_2 receptors—in contrast to the higher potency of NT versus NN to bind to the cloned NT receptors—the NN-activated neostriatal NT, receptors involved in the regulation of the D_2 receptors, may represent a distinct subtype of NT receptors.[23]

However, in competition experiments a different type of modulation of D_2 receptors by CCK-8 has been observed.[15] CCK-8 (1 nM) was found to reduce the K_H and K_L values of DA for the D_2 antagonist [³H]raclopride binding sites in the order of 50% (FIG. 3). These increases in affinity were found to be blocked by both CCK_A and CCK_B antagonists. Of substantial interest was the demonstration that the D_1 antagonist SCH23390 counteracted the CCK-8–induced reductions in the K_H and K_L values of DA. Thus, it seems clear that upon a joint activation of D_1 and D_2 receptors, CCK-8 via activation of both CCK_A and CCK_B receptors, will increase and not reduce the affinity of D_2 receptors for DA. The pattern of DA receptor subtype activation will determine whether CCK-8 will antagonize or enhance D_2 receptor

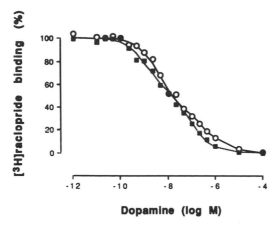

FIGURE 3. Representative competition curves illustrating the effect of 1 nM of cholecystokinin octapeptide (CCK-8) on dopamine (DA)-induced inhibition of D_2 antagonist [³H]raclopride (2 nM) binding in rat neostriatal membranes. Competition experiments with 20 concentrations of DA (1 pM–0.1 mM) were performed by incubating the neostriatal membranes for 30 min at 25 °C in the presence of 1 nM of CCK-8. Using iterative nonlinear regression fitting procedure, the K_H and the K_L values were 4.10 nM and 284 nM, respectively, for the control curve (○), and 1.17 nM and 92 nM, respectively, for the 1 nM of CCK-8 curve (■). The R_H values were 51 and 46%, respectively.

transduction within the brain. It appears that especially the D_1 receptor exerts a switching action in D_2 regulation, so that upon D_1 receptor activation CCK-8 then enhances the affinity of the D_2 receptors instead of reducing it. Thus, the pattern of DA receptor subtype activation will determine the type of modulation of the D_2 receptors induced by activation of CCK receptor subtypes. This view has recently been strengthened by observations that CCK-8 *in vitro* and *in vivo* can strongly regulate striatal D_2 receptors in sections of rat forebrain in the same way as observed in the striatal membrane preparations.[24] Of particular interest in this analysis was the demonstration by Li *et al.*[24] that a stronger CCK/DA interaction was found in sections versus that found in membrane preparations, indicating that either cytosolic factors and/or intact membranes are necessary for the full development of this type of receptor interaction. Another interesting finding was the demonstration that within the CCK/DA costoring region of the nucleus accumbens a stronger modulation of D_2 receptor binding characteristics was found to take place after the *in vitro* and *in vivo* treatment with CCK-8. Therefore, these types of intramembrane receptor-receptor interactions may have a special role in cotransmission. Such a strong modulation of D_2 receptors has also been found in rat striatal sections when using NT/NN peptides,[25] again emphasizing the importance of intracellular factors and/or of the intact membrane structure. Taken together these studies underline the importance of receptor-receptor interactions exerted at the membrane level between neuropeptide receptors and D_2 receptors, which are determined at least in part by the ongoing activity at D_1 receptors.

It is of interest to note that the switching role of D_1 receptors is abolished when NT and CCK-8 are added jointly to regulate the D_2 receptor binding characteristics in rat neostriatal membranes.[21] Thus, in these experiments threshold concentrations of CCK-8 and NT significantly increased the K_d value of the high-affinity D_2 receptors as studied in competition experiments with the [³H]raclopride versus DA and in saturation experiments involving a D_2 agonist radioligand [³H]NPA. In this case the activation of NT receptors will not allow the activated D_1 receptor to convert the CCK receptor regulation of the D_2 receptors into one of enhancement of the affinity. Instead, the results demonstrate that the reduction of affinity will dominate and that the two neuropeptides synergize in the inhibitory regulation of the D_2 receptor affinity. It becomes increasingly clear that the modulation of D_2 receptor subtype is not dependent upon a single interaction but is determined by a set of directly and indirectly interacting receptors activated by several neurotransmitters impinging on the same striatal cells. The results obtained in the membrane binding studies certainly imply that the same striatal nerve cells must contain both D_1, D_2, NT, and CCK receptors. The cellular colocalization of these receptors, however, still remains to be directly demonstrated.

In conclusion, it seems possible that every neuron operates with a preferred constellation of receptor-receptor interactions, possibly involving the formation of receptor mosaics.[26]

Studies at the Network Level

When using intracerebral microdialysis in combination with studies on DA release, *in vivo* evidence has been obtained that the CCK_B/D_2 antagonistic receptor interaction exists at the presynaptic level in the striatal DA nerve terminal networks.[27] Thus, CCK-8 perfused by the microdialysis probe in the halothane anesthetized rat was able, in a concentration-related way (1 nM to 1 μM), to counteract the inhibitory actions of systemically given apomorphine (0.05 mg/kg, s.c.) on the DA release (FIG. 4), an action blocked by a CCK antagonist. These results seem to give a functional correlate to the antagonistic interaction between CCK_B and D_2 receptors demonstrated in the striatal membrane preparations. Thus, activation of presumable CCK_B receptors may reduce the D_2 autoreceptor affinity leading to a reduction of the apomorphine-induced inhibition of DA release. Studies on GABA release within the nucleus accumbens support the existence of an antagonistic CCK_B/D_2 interaction also within the postsynaptic cells by the demonstration that CCK-8 (1 μM) increased both GABA and DA release by 35 and 43%, respectively.[28] It is also possible to obtain a functional correlate to the synergistic interaction demonstrated between NT and CCK-8 in the control of D_2 receptors.[21] Thus, the two neuropeptides were found to synergistically antagonize the apomorphine-induced inhibition of DA release as evaluated by means of intrastriatal microdialysis. In the presence of 1 nM of CCK-8,

NT in subthreshold concentrations (0.01–1 nM) counteracts the apomorphine (0.05 mg/kg, s.c.) induced inhibition of DA release by 70%, but in the absence of CCK-8, NT in the low concentrations has no action (FIG. 5).

Also in the case of NT/D_2 receptor interactions, it has been possible—by means of intrastriatal and intraaccumbens microdialysis—to obtain a functional correlate to the receptor interactions found in the membrane preparations from the striatum. Thus, the presynaptic NT receptors located on the striatal DA terminals will antagonize the inhibitory actions exerted by apomorphine on DA release mediated by activation of D_2 autoreceptors (FIG. 6).[29] Furthermore, in the awake and unrestrained male rat NT can also antagonize the inhibitory effects of D_2 agonists on extracellular levels of DA, DOPAC, and HVA. These results give evidence that the

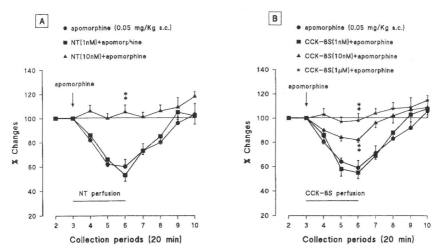

FIGURE 4. Effects of local perfusion with neurotensin (NT) (*panel A*) and CCK-8 (*panel B*) on dopamine (DA) extracellular levels in the dorsal striatal dialysates from the apomorphine-treated halothane anesthetized rat. The results were expressed as a percentage of the mean of three basal values. Mean ± SEM are shown. The absolute value of basal DA outflow was 126 ± 5 fmol/20 min. The statistical analysis was carried out according to one-way ANOVA followed by Newman-Keuls test for multiple comparisons. **p < 0.01 versus apomorphine alone and plus 1 nM NT (*panel A*) or 1 nM CCK-8 (*panel B*). Significances are shown only for the peak effect.

presynaptic NT receptors located on DA terminals also can counteract transduction occurring at D_2 autoreceptors leading to inhibition of DA release. Recent studies on D_2-regulated GABA release indicate that D_2 receptors located in the nerve cell membranes on the GABA/enkephalin (ENK) striopallidal neurons projecting to the external globus pallidus are antagonistically regulated by the NT receptors.[11] Thus, in the awake unrestrained male rat perfusion with NT by microdialysis is capable of counteracting the inhibitory actions of the D_2 agonist pergolide on the extracellular levels of GABA. These results provided a functional correlate to the binding experiments postulating the existence of antagonistic NT/D_2 receptor interactions in neostriatal membranes. Thus, the neuroleptic activity of NT peptides[11] and NN peptides[23] may be produced via antagonistic actions on D_2 receptor transduction.

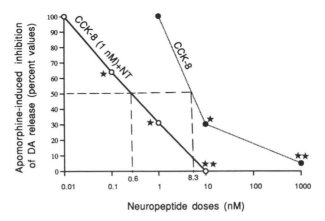

Neuropeptide doses (nM)

FIGURE 5. Effects of CCK-8 alone or in the presence of increasing concentrations of neurotensin (NT) (0.1, 1, and 10 nM). The area below curves during the perfusion period has been considered (see FIG. 4, curve B) and expressed as percent values of the mean area under the apomorphine curve in the corresponding perfusion period. Concentrations NT and CCK-8 up to 1 nM were by themselves ineffective. Percentage values for NT + CCK-8 and CCK-8 alone have been plotted. From these two dose-response curves an approximate evaluation of the respective ED_{50} values can be obtained as indicated. $n = 5$–7 rats. *$p < 0.05$; **$p < 0.01$ versus the apomorphine alone group according to one-way ANOVA followed by Neuman-Keuls test for multiple comparisons.

Instead, combined treatment with the D_1 agonist SKF38393 and NT, but not with the D_1 agonist alone, leads to significant increases in the extracellular striatal levels of GABA. In this way the striatal NT receptor can selectively reduce the transmission over the D_2 receptors, leading to a switching of DA transmission towards D_1

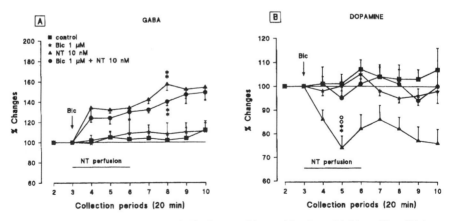

FIGURE 6. Effects of neurotensin (NT) alone and in combination with bicuculline (Bic) on GABA (*panel A*) and dopamine (DA) (*panel B*) extracellular levels from the posteromedial nucleus accumbens of the halothane anesthetized rat. The results are expressed as a percentage of the mean of three basal values. Mean ± SEM are shown. The basal absolute values were 99 ± 6 fmol/20 min for DA and 608 ± 40 fmol/20 min for GABA. $n = 5$–7. **$p < 0.01$ versus control as well as Bic alone. ${}^x p < 0.01$ versus NT plus Bic.

receptor-mediated neurotransmission, increasing the extracellular GABA levels. Following activation of striatal NT receptors DA transmission will therefore mainly operate via D_1 receptor-mediated excitation of the strionigral GABAergic system.

In this way plastic responses based on receptor diversity and receptor-receptor interactions are made possible for both WT and VT. The data summarized in this paper illustrate these plastic responses for DA neurotransmission, where DA receptor diversity and receptor-receptor interactions among different types of receptors selective for a certain subtype—in this case the D_2 receptor subtype—substantially increase the DA transmission plasticity.

Recently, we focused our attention on the nucleus accumbens, a brain area where the interaction of the two neuropeptides with the DAergic and GABAergic systems also seems to take place. Our results demonstrated that the local perfusion with NT (10 nM) induced a long-lasting increase of basal GABA outflow (+45%), but, surprisingly, produced a prolonged inhibition of DA release (−25%). Pretreatment with the GABA$_A$ antagonist bicuculline (1 μM, a concentration by itself ineffective) abolished the NT-induced reduction of dopamine outflow without affecting the associated increase of GABA release (FIG. 6A and B).[28] These findings provided strong evidence that NT-induced inhibition of DA release is mediated by a local increase in the GABA outflow. The activation of GABA release induced by the peptide could be related to selective postsynaptic antagonistic NT/D_2 receptor-receptor interaction on the GABA neurons innervating the nucleus accumbens.

Taken together, the results obtained with the perfusion of NT in the dorsal striatum and in the nucleus accumbens demonstrate that the neuropeptide differentially influences DA and GABA transmission in these brain areas. Furthermore, they suggest that, in contrast to the dorsal striatum, the presynaptic but not the postsynaptic NT/D_2 receptor-receptor interaction is missing in the nucleus accumbens.[30-32] Thus, it seems possible that the NT-induced activation of GABA release could either directly inhibit the DA release or, indirectly, reduce the activity of a tonic excitatory input on DAergic terminals in the nucleus accumbens. On the contrary, in the dorsal striatum, the lack of DA inhibition could be due to the activation of NT receptors present on the DAergic terminals,[33] which via the presynaptic NT/D_2 receptor interaction increase DA release.

In conclusion, these functional microdialysis data suggest that the modulation by NT and CCK-8 of the striatal and accumbens DAergic and GABAergic systems may be relevant for the postulated antipsychotic actions of these neuropeptides.[34,35]

CCK$_B$/D_2 AND NT/D_2 RECEPTOR INTERACTIONS AND THEIR RELEVANCE FOR SCHIZOPHRENIA

The present studies open up the possibility that the neuroleptic-like actions seen following central NT and CCK-8 administration is the result of an antagonistic NT and CCK$_B$ receptor modulation of the postsynaptic D_2 receptors in the neostriatum and the nucleus accumbens.[18,36] These antagonistic intramembrane interactions involving the postsynaptic D_2 receptors probably take place in the striopallidal GABAergic neurons involving both the dorsal and ventral components of this pathway, the ventral component being of particular interest in relation to schizophrenia in view of its role in controlling the output from the limbic system. Thus, it may be surmised that schizophrenia can be the final outcome of different neurochemical lesions such as alterations in CCK-8 and/or NT release, as well as in CCK and NT receptor interactions with the D_2 receptors. These alterations could lead to a pathological pattern in DA communication (WT and VT). However, abnormal spatial/temporal patterns in the DA communication may also depend on a miswiring of the system inasmuch as structural abnormalities have been described in the brain of some patients with schizophrenia.[11,37,38]

This new way of looking at schizophrenia may represent a more precise picture of the pathogenetic mechanisms and it may also suggest novel therapeutic approaches.

SUMMARY

Receptor diversity in combination with receptor-receptor subtype specific interactions, which can be antagonistic or synergistic in character, markedly increase plasticity in WT and VT in the nervous system. In this way switching among transmission lines for the various DA receptor subtypes becomes possible. Some of these aspects are supported by our work on selective modulation of D_2 receptors by CCK and NT. Selective regulation of D_2 receptors via CCK-8 receptor subtypes and NT receptors may underlie CCK/DA interactions and NT/DA interactions in the basal ganglia. These studies underline the importance of receptor-receptor interactions exerted at the membrane level between neuropeptide receptors and D_2 receptors, which are determined at least in part by the ongoing activity at D_1 receptors. In the case of both CCK/D_2 and NT/D_2 receptor interactions, it has been possible, by means of intrastriatal and intraaccumbens microdialysis, to obtain a functional correlate to the receptor interactions found in the membrane preparations from the striatum.

Schizophrenia may be in part related to reduced release of CCK and/or NT peptides or to alterations in their receptor interactions with the D_2 receptor. This view may lead to new therapeutic approaches.

ACKNOWLEDGMENTS

We are grateful to Anne Edgren for expert secretarial assistance.

REFERENCES

1. BUNZOW, J. R., H. H. M. VAN TOL, D. K. GRANDY, P. ALBERT, J. SALÓN, M. C. CHRISTIE, C. A. MACHIDA, K. A. NEVE & O. CIVELLI. 1988. Cloning and expression of a rat D_2 dopamine receptor cDNA. Nature **336:** 783–787.
2. FUXE, K. & L. AGNATI. 1985. Receptor-receptor interactions in the central nervous system. A new integrative mechanism in synapses. Med. Res. Rev. **5:** 441–482.
3. FUXE, K. & L. AGNATI. 1987. Receptor-Receptor Interactions. A New Intramembrane Integrative Mechanism. Macmillan Press. London.
4. VAN TOL, H. H. M., J. B. BUNZOW, H. C. GUAN, R. K. SUNAHARA, P. SEEMAN, H. B. NIZNIK & O. CIVELLI. 1991. Cloning of the gene for a human D_4 receptor with high affinity for the antipsychotic clozapine. Nature **350:** 610–614.
5. ZOLI, M., L. F. AGNATI, P. B. HEDLUND, X.-M. LI, S. FERRE & K. FUXE. 1993. Receptor-receptor interactions as an inegrative mechanism in nerve cells. Mol. Neurobiol. 293–334.
6. STRANGE, P. G. 1990. Aspects of the structure of the D_2 dopamine receptor. Trends Neurosci. **13:** 373–378.
7. VALLAR, L. & J. MELDOLESI. 1989. Mechanisms of signal transduction at DA D_2 receptors. Trends Pharmacol. Sci. **10:** 74–77.
8. AGNATI, L. F., B. BJELKE & K. FUXE. 1992. Volume transmission in the brain. Am. Sci. **80:** 362–373.

9. CLARK, D. & F. J. WHITE. 1987. D_1 dopamine receptor—the search for a function: A critical evaluation of the D_1/D_2 dopamine receptor classification and functional implications. Synapse **1:** 347–388.

10. BERTORELLO, A. M., J. F. HOPFIELD, A. APERIA & P. GREENGARD. 1990. Inhibition of dopamine of (Na^+K^+) ATPase activity in neostriatal neurons through D_1 and D_2 dopamine synergism. Nature **347:** 386–388.

11. FUXE, K., G. VON EULER, L. F. AGNATI, E. MERLO PICH, W. T. O'CONNOR, S. TANGANELLI, X.-M. LI, B. TINNER, A. CINTRA, C. CARANI & F. BENFENATI. 1992. Intramembrane interactions between neurotensin receptors and dopamine D_2 receptors as a major mechanism for the neuroleptic-like action of neurotensin. Ann. N. Y. Acad. Sci. **668:** 186–204.

12. AGNATI, L., K. FUXE, F. BENFENATI, V. MUTT & T. HÖKFELT. 1983. Differential modulation by CCK-8 and CCK-4 of $[^3H]$spiperone binding sites linked to dopamine and 5-HT receptors in the brain of the rat. Neurosci. Lett. **35:** 179–183.

13. AGNATI, L. F., M. F. CELANI & K. FUXE. 1983. Cholecystokinin peptides in vitro modulate the characteristics of the striatal $[^3H]$N-propylnor-apomorphine sites. Acta Physiol. Scand. **118:** 79–81.

14. FUXE, K., L. F. AGNATI, G. VON EULER, S. TANGANELLI, W. T. O'CONNOR, S. FERRÉ, P. HEDLUND & M. ZOLI. 1992. Neuropeptide, excitatory amino acid and adenosine A_2 receptors regulate D_2 receptors via intramembrane receptor-receptor interactions. Relevance for Parkinson's disease and schizophrenia. Neurochem. Int. **20(Suppl.):** 215S–224S.

15. LI, X.-M., P. B. HEDLUND, L. F. AGNATI & K. FUXE. 1994. Dopamine D_1 receptors are involved in the modulation of D_2 receptors induced by cholecystokinin receptor subtypes in rat neostriatal membranes. Brain Res. **650:** 289–298.

16. FERRÉ, S., A. RUBIO & K. FUXE. 1991. Stimulation of adenosine A_2 receptors induces catalepsy. Neurosci. Lett. **130:** 162–164.

17. FUXE, K., L. F. AGNATI, F. BENFENATI, M. CIMMINO, S. ALGERI & T. HÖKFELT. 1981. Modulation by cholecystokinins of $[^3H]$spiroperidol binding in rat striatum: Evidence for increased affinity and reduction in the number of binding sites. Acta Physiol. Scand. **113:** 567–569.

18. FUXE, K., L. AGNATI, M. ZOLI, E. M. PICH, R. GRIMALDI, B. BJELKE, S. TANGANELLI, G. V. EULER & A. JANSON. 1991. Neuroplasticity of mesotelencephalic dopamine neurons at the network and receptor level: New aspects of the role of dopamine in schizophrenia and possible pharmacological treatments. In Advances in Neuropsychiatry and Psychopharmacology, Vol. 1. Schizophrenia Research. C. Tamminga & S. Schultz, Eds.: 51–76. Raven Press. New York.

19. ZETLER, G. 1985. Neuropharmacological profile of cholecystokinin-like peptides. Ann. N.Y. Acad. Sci. **448:** 448–469.

20. VON EULER, G. 1991. Biochemical characterization of the intramembrane interaction between neurotensin and dopamine D_2 receptors in the rat brain. Brain Res. **561:** 93–98.

21. TANGANELLI, S., X.-M. LI, L. FERRARO, G. VON EULER, W. T. O'CONNOR, C. BIANCHI, L. BEANI & K. FUXE. 1993. Neurotensin and cholecystokinin octapeptide control synergistically dopamine release and dopamine D_2 receptor affinity in rat neostriatum. Eur. J. Pharmacol. **230:** 159–166.

22. LI, X.-M., G. VON EULER, P. B. HEDLUND, U. B. FINNMAN & K. FUXE. 1993. The C-terminal neurotensin-(8-13) fragment potently modulates rat neostriatal dopamine D_2 receptors. Eur. J. Pharmacol. **234:** 125–128.

23. LI, X.-M., U.-B. FINNMAN, G. VON EULER, P. B. HEDLUND & K. FUXE. 1993. Neuromedin N is a potent modulator of dopamine D_2 receptor agonist binding in rat neostriatal membranes. Neurosci. Lett. **155:** 121–124.

24. LI, X.-M., P. B. HEDLUND & K. FUXE. 1994. Cholecystokinin octapeptide in vitro and in vivo strongly regulates striatal dopamine D_2 receptors in sections of rat forebrain. Eur. J. Neurosci. In press.

25. LI, X.-M., P. B. HEDLUND & K. FUXE. 1994. Strong effects of NT/NN peptides on DA D_2 receptors in rat neostriatal sections. NeuroReport **5:** 1621–1624.

26. AGNATI, L. F., K. FUXE, M. ZOLI, C. RONDANINI & S.-O. ÖGREN. 1982. New vistas on synaptic plasticity: Mosaic hypothesis on the engram. Med. Biol. **60:** 183–190.

27. TANGANELLI, S., K. FUXE, G. VON EULER, L. F. AGNATI, L. FERRARO & U. UNGERSTEDT. 1990. Involvement of cholecystokinin receptors in the control of striatal dopamine autoreceptors. Naunyn Schmiedebergs Arch. Pharmakol. **342:** 300–304.

28. TANGANELLI, S., W. T. O'CONNOR, J. FERRARO, C. BIANCHI, L. BEANI, U. UNGERSTEDT & K. FUXE. 1994. Facilitation of GABA release by neurotensin is associated with a reduction of dopamine release in rat nucleus accumbens. Neuroscience **63:** 649–657.

29. TANGANELLI, S., G. VON EULER, K. FUXE, L. F. AGNATI & U. UNGERSTEDT. 1989. Neurotensin counteracts apomorphine-induced inhibition of dopamine release as studied by microdialysis in rat neostriatum. Brain Res. **502:** 319–325.

30. DUNCAN, C. C. & V. G. ERWIN. 1992. Neurotensin modulates K^+-stimulated dopamine release from the caudate-putamen but not the nucleus accumbens of mice with differential sensitivity to ethanol. Alcohol **9:** 23–29.

31. QUIRION, R., C. C. CHIUEH, H. D. HEVERIST & A. PERT. 1985. Comparative localization of neurotensin receptors on nigrostriatal and mesolimbic dopaminergic terminals. Brain Res. **327:** 385–389.

32. REYNEKE, L., V. A. RUSSELL & J. J. TALJAARD. 1990. Evidence that the stimulatory effect of neurotensin on dopamine release in rat nucleus accumbens slices is independent of dopamine D_2 receptor activation. Brain Res. **534:** 188–194.

33. AZZI, M., D. GULLY, M. HEAULME, A. BEROD, D. PELARAT, P. KITABGI, R. BOIGEGRAIN, J. P. MAFFRAND, G. LEFUR & W. ROSTENE. 1994. Neurotensin receptor interaction with dopaminergic systems in the guinea-pig brain shown by neurotensin receptor antagonist. Eur. J. Pharmacol. **255:** 167–174.

34. NEMEROFF, C. B. 1986. The interaction of neurotensin with dopaminergic pathways in the central nervous system: Basic neurobiology and implications for the pathogenesis and treatment of schizophrenia. Psychoneuroendocrinology **11:** 15–37.

35. VAN REE, J. M., O. GADFORI & D. DEWIED. 1983. In rats, the behavioral profile of CCK-8 related peptides resembles that of antipsychotic agents. Eur. J. Pharmacol. **93:** 63–78.

36. FUXE, K., L. AGNATI, G. VON EULER, F. BENFENATI & S. TANGANELLI. 1990. Modulation of dopamine D_1 and D_2 transmission lines in the central nervous system. *In* Current Aspects in the Neurosciences, Vol. 1. N. Osborne, Ed.: 203–243. MacMillan Press. Basingstoke, UK.

37. KELSOE, J. R., J. L. CADET, D. PICKAR & D. R. WEINBERGER. 1988. Quantitative neuroanatomy in schizophrenia. Arch. Gen. Psychiatry **45:** 533–541.

38. STEVENS, C. D., L. L. ALTSHULER, B. BOGERTS & P. FALKAI. 1988. Quantitative study of gliosis in schizophrenia and Huntington's chorea. Biol. Psychiatry **24:** 687–700.

Anatomy and Mechanisms
of Neurotensin-Dopamine Interactions
in the Central Nervous System

PHILIP D. LAMBERT,[a] ROBIN GROSS,
CHARLES B. NEMEROFF, AND CLINTON D. KILTS

Department of Psychiatry and Behavioral Sciences
Emory University School of Medicine
P.O. Drawer AF
Atlanta, Georgia 30322

Neurotensin (NT), a tridecapeptide,[1] is widely distributed in the brain of various species and is recognized as a putative peptide neurotransmitter in the mammalian brain. There now exists a growing body of anatomical, neurochemical, electrophysiological, behavioral, and pharmacological evidence that NT interacts with brain dopamine (DA) systems.

NEUROCHEMISTRY

The modulatory influence of NT on brain DA neuronal function is widespread with effects reported for retinal to hypothalamic[2] DA neurons. Light microscopic autoradiographic analyses have shown a dense localization of NT receptors on DA-containing neurons in the substantia nigra zona compacta and the ventral tegmental area (VTA).[3] A facilitatory effect of NT on [³H]-DA release from rat mesencephalic cells in primary culture represents a functional corollary of this receptor distribution.[4] Several studies using 6-hydroxydopamine-induced lesions of the nigrostriatal or mesolimbic DA system in the rat have also demonstrated the presence of NT receptors on presynaptic dopaminergic innervation.[5–8] Along with the dopaminergic terminals, a large percentage of NT receptors is lost in the caudate putamen, nucleus accumbens, and olfactory tubercle following 6-hydroxydopamine lesion of the median forebrain bundle.[8] As well as receptors for both NT and DA existing in the same anatomical location, the neurotransmitters themselves are colocalized within neurons originating from the ventral tegmental area and projecting to either the nucleus accumbens[9] or the prefrontal cortex[10] in the rat. Whether an interaction would have the same physiological relevance in man is unclear because, in contrast to the rat, NT and DA in man are not colocalized in neurons projecting to the prefrontal cortex. The presence of NT receptors on presynaptic dopaminergic terminals in the caudate-putamen correlates well with demonstrated biochemical effects of NT on this pathway. It has been shown that NT modifies DA metabolism in the striatum[11] and facilitates the *in vitro* release of DA from rat striatal slices.[12]

A modulatory interaction between NT and DA receptors on receptor function is also supported by accumulating evidence. *In vitro* findings demonstrate that NT reduces the affinity of [³H]-DA binding sites in membranes from the subcortical

[a] Corresponding author.

limbic forebrain and striatum.[13] It has been shown, both *in vivo* and *in vitro*, that exogenous NT decreases the affinity of D2 and D3 DA receptor agonist binding.[14–16] Both D2 and D3 DA receptors are thought to be coupled to G-proteins; however, the effect of NT on the affinity of DA receptors would seem to be via an as-yet-unknown, G-protein-independent mechanism.[14,15] Data indicates that DA receptor activation produces a significant reduction in affinity (increased K_d), and a significant increase in the number of [³H]-NT binding sites (increased B_{max}) in subcortical limbic membranes.[17] Furthermore, rat striatal membranes from the 6-hydroxydopamine-lesioned hemisphere exhibit a significantly enhanced effect of DA (10 nM) on the K_d for [³H]-NT binding sites compared to membranes from the intact side.[6] These data support an inhibitory effect of DA on NT neurotransmission, perhaps mediated by an intramembrane feedback loop of receptor-receptor interactions. Chronic treatment with the neuroleptic haloperidol enhances NT receptor binding in the substantia nigra of the human (postmortem tissue from patients with schizophrenia) and rat brain.[18] Chronic DA receptor blockade leads to a significant increase in NT receptor density in both species. The predicted increase in the number of DA receptors (particularly the D2 subtype) was also seen. Interestingly, electron microscopic analysis revealed an increased number of DA terminals in animals treated with neuroleptics that positively correlated with the cellular increase in the number of NT receptors.[19] These results give further support to the idea that heteroregulatory NT receptor–DA receptor interactions exist within dopaminergic neurons. The increased density of NT receptors following treatment with neuroleptics could lead to increased excitability of DA cells as a compensatory response to the effects of DA receptor blockade.

BEHAVIOR

NT produces a large number of distinct physiological and behavioral effects after systemic and central administration. As examples, the systemic administration of NT produces hypotension, hyperglycemia, smooth muscle contraction, and inhibition of gastric acid secretion.[20] In contrast, after intracerebroventricular (i.c.v.) administration, NT potentiates barbiturate-induced sedation[21] and produces hypothermia,[22] muscle relaxation,[23] antinociception,[24] a reduced food consumption,[25] catalepsy,[26] alterations in locomotor activity[27] and increased serum corticosterone.[2] None of these effects is seen after intravenous NT, indicating that they are centrally mediated. Many of the central effects of NT mimic those seen after administration of neuroleptics, and considerable data now exists to support the hypothesis that NT possesses a pharmacological profile similar to that of neuroleptics. It has been proposed that NT may be an endogenous neuroleptic-like peptide. Differences in the central effects of NT and neuroleptics do exist. For instance, an increase in serum corticosterone levels was not seen following treatment with neuroleptics and although NT does have a stimulatory action on tuberoinfundibular DA neurons, the increase in serum corticosterone does not seem to be mediated via a dopaminergic mechanism.[2] The dose of NT needed to activate DA neurons was much greater than the lowest dose producing a significant increase in serum corticosterone levels, an effect that was not blocked with haloperidol.[2]

Like systemically administered DA antagonists, i.c.v. administration of NT attenuates rat locomotor activity induced by indirect-acting DA agonists.[28] This behavioral interaction of NT with brain DA systems appears to be selective for mesolimbic DA projections. Discrete injection of NT into the nucleus accumbens, like haloperidol, blocked d-amphetamine-induced locomotor behavior;[29] injection of

NT into the caudate-putamen did not antagonize d-amphetamine-induced stereo-typic behavior, although haloperidol was potent in this regard. NT has also been shown to antagonize directly the effects of DA in the mesolimbic DA system. NT injection either i.c.v. or directly into the nucleus accumbens produced a dose-dependent reduction in both locomotion and rearing produced by intraaccumbal DA injection.[30] Neurotensin (i.c.v.) has also been shown to block hyperactivity induced by other direct-acting DA receptor agonists such as n-propylnorapomorphine and ADTN.[31,27] These data support the view that NT does not act primarily on presynap-tic, but rather on postsynaptic sites of DA neurotransmission in the nucleus accumbens. Further evidence exists to support the multilevel ability of NT to modulate selective brain DA systems. Bilateral injection of NT into the ventral tegmental area produces a dose-dependent increase in locomotion in the rat.[32-34] This effect is consistent with an antagonism by NT of the autoinhibitory effects of DA released from dendrites. Thus, although the behavioral effects of NT, after discrete injection into different brain regions, would seem to be quite different, they can be explained by a common mechanism of interaction with, and antagonism of, DA systems.

ELECTROPHYSIOLOGY

NT has been shown to produce an increase in the *in vitro* firing rate of DA neurons in the zona compacta of the substantia nigra and VTA,[35] while exerting negligible effects on nondopaminergic neurons.[36] *In vitro* studies have demonstrated a linear dose-response curve for the effect of NT on DA cell firing rate,[35,37] although doses exceeding approximately 300 nM produced a decreased firing rate followed by cessation of measurable activity.[35,38] The C-terminal peptide fragment, NT 8–13, possesses similar activity to NT, whereas direct application of the N-terminal fragments, NT 1–8 and NT 1–11, produced no measurable alteration in the firing rate of dopaminergic neurons *in vitro*. Experiments conducted in high-magnesium and low-calcium media suggest that the NT-induced increase in neuronal firing occurs via a calcium-dependent postsynaptic mechanism.[35] Initial data, both *in vitro* and *in vivo*, showed that NT attenuated the DA-induced inhibition of firing in dopaminergic neurons.[39,40] The mechanism of action of this effect remains unknown.

The effects of NT on neuronal firing rate in brain areas outside the mesencepha-lon are largely unknown. However, a small number of studies show that NT increased the firing rate of individual neurons in various cortical regions.[41,42] The ability of NT to depolarize neurons in slice preparations of medial prefrontal cortex was not affected by pretreatment with phentolamine, propranolol, sulpiride or fluphenazine. *In vitro* electrophysiological studies have found variable effects of NT on hypotha-lamic neurons. The direct application of NT to preoptic-anterior hypothalamic neurons produced a significant increase in firing rate that persisted in a low-calcium medium;[43,44] however, little or no response was seen when NT was applied to arcuate nucleus neurons.[45] An *in vivo* study has shown a significant inhibitory effect of NT on the firing rate of neurons within the rat thalamus and a variable effect on neurons within the hippocampus.[41]

The electrophysiological effects of acute and chronic treatment with antipsy-chotic drugs on midbrain dopaminergic neurons have been extensively studied.[46] Most clinically efficacious antipsychotic drugs tested increased the firing rate of dopaminergic neurons in the VTA and zona compacta;[47] repeated administration induced a depolarization inactivation of dopaminergic neurons in the VTA and substantia nigra.[48,49] Bath application of haloperidol produces an attenuation of the

DA-induced inhibition of firing rate in the substantia nigra,[50] whereas no effect was seen after treatment with clozapine.[51]

The available electrophysiological data are in good agreement with neurochemical and behavioral studies regarding an interaction between NT and dopaminergic systems. The actions of NT appear to be independent of other neurotransmitter receptors and the C-terminus of the NT molecule is essential for activity. The electrophysiological data for NT are remarkably similar to those produced by antipsychotic drugs and further support behavioral and electrophysiological data suggesting that NT may be an endogenous antipsychotic.

STRUCTURE AND FUNCTION OF NEUROTENSIN AND DOPAMINE RECEPTORS—INTERACTIONS AT THE LEVEL OF RECEPTOR SIGNAL TRANSDUCTION MECHANISMS

Both the rat and human NT receptor have been molecularly cloned. The cloned rat cDNA encodes a 424-amino-acid peptide,[52] whereas the human cDNA encodes a putative peptide of 418 amino acids with 84% homology to the rat NT receptor.[53] The inferred membrane topology of both the rat and human NT receptor contains seven transmembrane segments, suggesting that the NT receptor represents a member of the superfamily of G-protein-coupled receptors that includes the family of DA receptor subtypes.[54] Recent investigations of the properties of the native NT receptor, as well as NT receptors expressed by both clonal cell lines and in transfected cell systems, have generated a greatly improved understanding of the functional significance and variants of NT receptors. Both autoradiographic and *in situ* hybridization histochemical analyses indicated that brain NT receptors exhibit multiple patterns of ontogenic development that vary greatly between brain regions.[55,56] NT receptor binding sites and mRNA in the rat neocortex demonstrated a transient expression that peaked in the first postnatal week and declined thereafter to adult levels. A second pattern of binding site and mRNA expression indicated an initial expression in late prenatal periods with a gradual development to adult levels within the second week of postnatal life. This pattern was observed in the DA cell body groups of the ventral mesencephalon, the diagonal band, substantia innominata, suprachiasmatic nucleus, and medial habenular nucleus.[56] A third pattern of ontogenic expression was represented by postnatal development of NT receptor binding sites and mRNA with a gradual small decline thereafter. This pattern is apparent in the developing allocortex, tenia tecta, and hippocampus. The fact that the transient expression of neocortical NT receptors occurs long before a neural network is established suggests that NT may play an important role in the development of the neocortical brain structures that precedes its neurotransmitter-like role in the adult brain. A potentially important relevant experiment may involve the use of NT receptor antagonists or immunoinactivation strategies to explore the trophic role of NT in the development of the dopaminergic innervation of the rat neocortex.

The desensitization of the NT receptor response to agonist occupancy by the prolonged exposure to NT receptor agonists has been documented in multiple cell systems and demonstrated for the native receptor. The decrease in receptor number thought to underlie NT receptor desensitization is mediated by a putative mechanism of NT receptor sequestration in response to prolonged agonist exposure. An agonist-induced decrease in cell surface NT receptors has been reported in clonal cell lines[57,58] and in primary cultures of rodent neurons.[59,60] An internalization of rat neostriatal NT receptors and NT followed by their retrograde axonal transport in nigrostriatal DA neurons has been reported in the intact rat brain.[61] These results support the possibility that NT may be involved in modulating the function of DA neurons at intracellular sites (e.g., gene expression) with long-term actions. Finally, pharmacological studies using either novel NT receptor antagonists (see below) or peptide and pseudo-peptide NT analogs[62] support the concept of a multiplicity of NT receptors. Collectively, the results of these studies support the dissociation of the effects of NT receptor agonists and antagonists on the hypothermic or analgesic effects of NT and the modulatory influences of NT on DA neurons. A subtype of NT receptors apparently mediates the hypothermic and analgesic responses to the peptide, whereas the DA-releasing effects of NT appear to be mediated by a distinct NT receptor subtype that is similar to the cloned high-affinity rat brain NT receptor. An additional level of functional diversity for brain NT systems is provided by the demonstration that the posttranslational processing of the NT/neuromedin N precursor, a 169-residue polypeptide containing one copy each of NT and neuromedin N, is differentially processed in different brain regions.[63]

The regulation by both NT and DA receptors of G-protein-regulated signal transduction systems, as inferred from structural and functional receptor studies, supports the possibility that NT-DA receptor interactions may be expressed in a brain region-specific pattern at the level of post-receptor signaling pathways. Indeed, it was demonstrated in clonal cell lines and transfected cell systems that NT receptor occupancy is associated with alterations in the intracellular content of inositol phosphates,[64] cyclic AMP,[65] cyclic GMP,[66] and calcium.[67] All of these effector systems were shown to be regulated by activation of members of the DA receptor family. Although the NT receptor in these cell lines and systems appears to be pharmacologically similar to the native brain NT receptor, it remains unestablished as to whether the signal transduction mechanisms for the NT receptor in these model cell systems accurately model mechanisms of transmembrane signaling for brain NT receptors. Relatively few studies have investigated these signaling pathways associated with brain NT receptors,[68,69] with no systematic studies having been performed.

Our group has investigated the coupling of rat brain NT receptors with the adenylate cyclase second messenger system and the possible interaction between brain NT and DA receptors at the level of this transduction cascade (TABLE 1). The literature regarding the coupling of the NT receptor with cyclic AMP formation in model cell systems is at the same time both consistent and conflicting. The NT receptor expressed by neuroblastoma N1E115 cells mediates an inhibition of cyclic AMP formation,[70,71] whereas the cloned NT receptor expressed in mammalian cells[72] or Chinese hamster ovary (CHO) cells[73] mediates a stimulation of cAMP formation. The present attempt to characterize the coupling between rat brain NT receptors and cyclic AMP formation examined the effects of NT on cyclic AMP efflux from superfused slice preparations of the rat neostriatum or amygdaloid complex. The selection of a brain slice preparation (i.e., $1.0 \times 0.4 \times 0.4$ mm dimensions) for these studies was based on the observation that neuropeptide receptor-effector coupling has been demonstrated to be dependent on or best demonstrated using intact cell versus homogenate preparations. Also, this preparation, in contrast to cell-free brain homogenates, represents an attempt to replicate for the intact rat brain NT receptor the results of investigations of NT receptor-adenylate cyclase coupling in clonal cell lines and transfected cell systems expressing the NT receptor. The amygdaloid complex was selected for the study of NT receptor coupling with the adenylate cyclase second messenger system because it contains a high concentration of NT as well as NT-immunopositive cells and fibers, with both markers relatively high or dense in the central amygdaloid nucleus.[74] The rat amygdaloid complex also contains a moderate-to-high density of [^{125}I] Tyr$_3$-NT binding sites.[75]

The effect of NT on the efflux of cyclic AMP from the rat amygdaloid complex

TABLE 1. Effect of Neurotensin on Cyclic AMP Efflux from the Rat Amygdaloid Complex and Caudate-Putamen

Treatment	Cyclic AMP Efflux (% of basal)	
	Amygdaloid Complex	Caudate-Putamen
NT (0.3)	95 ± 6	103 ± 5
NT (3)	84 ± 4	109 ± 7
Forskolin (10)	283 ± 28	—
+ NT (0.3)	221 ± 14[a]	—
Forskolin (10)	322 ± 51	561 ± 66
+ NT (3)	385 ± 40	663 ± 79
PGE$_1$ (30)	285 ± 15	—
+ NT (3)	331 ± 18	—
Isoproterenol (3)	226 ± 15	—
+ NT (3)	208 ± 18	—
DA (50) + Sulp (30)	—	189 ± 18
+ NT (0.3)	—	173 ± 12
+ NT (3)	—	138 ± 11[a]

NOTE: Values represent determinations (n = 5–8) for 5-, 7.5- or 15-min periods of superfusion. Numbers in parentheses represent drug concentrations (μM). Sulp, sulpiride.
[a]$p < 0.05$.

and caudate-putamen was assessed under both basal conditions as well as receptor and non-receptor-mediated increases in cyclic AMP efflux. Superfusion of minced slice preparations of the amygdaloid complex or caudate nucleus with NT-containing buffer did not affect significantly the efflux of cyclic AMP from either region compared to efflux assessed in the absence of NT (TABLE 1). In contrast to the effects of NT on cyclic AMP efflux from neuroblastoma cells expressing NT receptors,[70,71] the forskolin-induced increase in cyclic AMP efflux from the amygdaloid complex or caudate-putamen was not consistently altered by NT. The inhibitory effect of 0.3 but not 3 μM NT on forskolin-stimulated amygdaloid cyclic AMP efflux suggests a coupling of the high-affinity NT receptor with amygdaloid adenylate cyclase. The possible dependence of NT–adenylate cyclase coupling on the interaction with specific receptor types was assessed for prostaglandin E$_1$ (PGE$_1$), β-adrenergic receptor, or D1 DA receptor-stimulated cyclic AMP efflux. NT did not affect the PGE$_1$- or isoproterenol-induced increase in amygdaloid cyclic AMP efflux. NT did, however, produce a concentration-dependent decrease in the D$_1$ receptor-induced increase in cyclic AMP efflux from the caudate-putamen. NT-DA receptor interactions on cyclic AMP formation were not assessed in the amygdaloid complex as amygdaloid D$_1$ DA receptors are not coupled to adenylate cyclase. The results obtained in rat brain slice preparations are consistent with a role for effectors other than adenylate cyclase in the transduction of amygdaloid NT receptor signals in the rat brain. Results obtained using experimental controls indicate that these negative findings are not attributable to the degradation of NT during the course of superfusion, an agonist-induced desensitization of NT-receptor function (see above), or the interaction of NT with levocabastine-sensitive binding sites. These results do, however, support an inhibitory influence of the rat striatal NT receptor on transmembrane signaling for the striatal D$_1$ DA receptor. These latter findings are consistent with a G-protein-mediated effect of NT on the agonist affinity state of the striatal D$_1$ DA receptor.[76] These results also highlight the conclusion that the generalization of receptor effects from clonal cell lines or transfected cell systems to receptor physiology in the intact central nervous system must await the generation of direct supporting evidence. Experiments using primary neuronal cultures may be of value in bridging results from model cell systems and brain preparations.

EFFECT OF ANTIPSYCHOTICS ON BRAIN NEUROTENSIN CONTENT AND GENE EXPRESSION

It is increasingly apparent that alterations in DA mechanisms represent insufficient explanations of the pathophysiology or pharmacotherapy of schizophrenia. The similarities between the pharmacological effects of antipsychotics and the central effects of NT (see above) suggest that NT may represent an important element of the final common pathway of the mechanisms of action of antipsychotic drugs. Govoni and co-workers[77] initially reported an increase in NT-like immunoreactivity (NT-LI) in selective areas of the rat brain after treatment with neuroleptics. An increase in NT-LI in the rat nucleus accumbens and striatum was observed following chronic or acute treatment with haloperidol, chlorpromazine, trifluoperazine or pimozide. No drug-induced changes in the concentration of NT-LI in areas of the brain such as the hypothalamus and amygdala were observed. A subsequent preclinical study expanded the investigation of the effects of antipsychotic drugs on brain NT systems by an enhanced anatomical resolution of drug effects and the determination of the comparative effects on NT-LI of a neuroleptic and an atypical antipsychotic agent.[74] The prolonged (14 days) intraperitoneal administration of haloperidol or clozapine did not affect the content of NT-LI in the great majority of the 38 rat brain nuclei or areas examined. However, there was a differential effect of the two antipsychotics in the caudate-putamen, with haloperidol, but not clozapine, increasing the concentration of NT-LI drugs (FIG. 1). Both compounds increased NT-LI in the nucleus

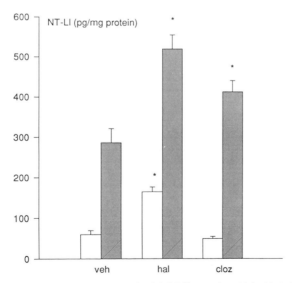

FIGURE 1. Effect of treatment (14 days, i.p.) with 0.3% tartaric acid (veh), haloperidol (hal, 1 mg/kg) or clozapine (cloz, 20 mg/kg) on neurotensin-like immunoreactivity (NT-LI) in the dorsolateral caudate (*open bars*) or the nucleus accumbens (*closed bars*). Results represent the mean (± SEM) of 8 animals per group. *$p < 0.05$ compared to vehicle-treated rats.

accumbens (FIG. 1). The administration of both drugs also produced significant decreases in NT-LI in the medial prefrontal cortex and bed nucleus of the stria terminalis. The brain region- and drug-specific effects of haloperidol and clozapine implicate distinct NT systems in the therapeutic as well as motor side effects of antipsychotic drugs. The findings of a study by Radke and co-workers,[78] consistent with the nontolerating therapeutic effects of haloperidol, showed that long-term (8 months) oral administration of haloperidol produced an enduring increase in NT-LI in the rat nucleus accumbens similar to that observed following 3 weeks of drug administration. Data also indicated that drug-induced alterations in NT concentrations in the rat brain represent a specific response to antipsychotic drugs. Acute (single injection) or repeated treatment (3 weeks) with drugs from other major neuroactive groups, such as tricyclic antidepressants (desipramine), anxiolytics (chlordiazepoxide), and H1 histamine receptor antagonists (diphenhydramine), did not alter rat striatal NT-LI.[79] The drug-induced changes in brain regional NT concentrations also vary in response to different atypical antipsychotics. For example, the atypical antipsychotic drugs sulpiride and rimcazole, in contrast to clozapine, do produce significant increases of NT-LI in the caudate nucleus, but not in the nucleus accumbens.[80] These data further suggest that brain region-specific changes in NT may be related to the distinct therapeutic and side effect profiles of different antipsychotics. Although a direct effect of treatment with antipsychotics on the central NT system has not been demonstrated in man, a normalization of the subnormal cerebrospinal fluid concentrations of NT in a subgroup of drug-free schizophrenics was noted after a course of treatment with antipsychotics.[81] Subsequent clinical studies have largely confirmed and extended these findings.[82]

Although a large amount of accumulated evidence supports brain region–specific and drug dose–dependent changes in NT content in response to antipsychotic drug treatment, the mechanisms underlying these drug effects are not known. Over the past five years, Merchant and co-workers have extensively examined the effects of the typical neuroleptic haloperidol and the atypical antipsychotic clozapine on striatal NT gene expression in the rat brain. A single dose of haloperidol rapidly (within 30 min) increased the expression of NT/N mRNA in the dorsolateral striatum, whereas clozapine produced virtually no effect.[83] The maximal effect of haloperidol was seen at 7 h postinjection, at which time the levels of NT/N mRNA were an order of magnitude higher than those of controls. A small but significant increase in NT/N mRNA was seen in the shell of the nucleus accumbens after treatment with either haloperidol or clozapine.[83] These data support further the possibility that NT neurons in the nucleus accumbens may in part mediate the therapeutic effects common to diverse antipsychotics, whereas NT neurons in the caudate-putamen may be involved in mediating the motor side effects of neuroleptics. Consistent with this theme, Merchant and Dorsa[84] also showed that drug-induced increases in the level of NT/N mRNA within the caudate nucleus were specific to neuroleptic antipsychotics (haloperidol, fluphenazine), whereas increases in the nucleus accumbens occurred after treatment with either neuroleptic or atypical antipsychotics (haloperidol, fluphenazine, clozapine, remoxipride, and thioridazine). Interestingly, they also showed that drug-induced increases in NT/N mRNA expression in the dorsolateral striatum in response to acute treatment with antipsychotics were preceded by a rapid and transient increase in c-fos mRNA.[84] None of the antipsychotics affected c-fos mRNA in the nucleus accumbens. A c-fos antisense oligonucleotide specifically attenuated the haloperidol-induced increase in NT/N mRNA expression, suggesting that an activation of c-fos expression is essential for the effect of haloperidol on NT/N mRNA.[85] These data suggest mechanistic differences in the effects of antipsychotics on discrete NT systems.

Maximal therapeutic benefit of antipsychotics is derived following prolonged drug administration, so an effect of chronic, rather than acute, antipsychotic treatment on NT/N mRNA would be of greater clinical relevance. An initial repeated dosing study showed an increase in NT receptor mRNA in the rat substantia nigra after treatment with haloperidol (2 weeks), but showed no effect after treatment with clozapine.[86] An increase in NT/N mRNA in the rat caudate nucleus has been shown following 28 days of continuous haloperidol administration, but not following chronic clozapine treatment.[87] However, this drug response was only half that caused by acute haloperidol administration. Increases in NT/N mRNA, of similar magnitude to the effect of acute drug treatments, were seen in the nucleus accumbens after chronic treatment with haloperidol or clozapine.[87]

All of the data collected to date investigating the effect of antipsychotics on NT/N mRNA demonstrate a region-specific pattern of effects on NT/N mRNA which discriminates between neuroleptic and atypical antipsychotics. Effects in the limbic sector of the neostriatum would seem more likely to be involved with therapeutic effects of antipsychotics, and NT/N gene induction in the motor striatal regions may indicate cellular events associated with motor side effects of neuroleptics. However, a direct functional connection has not yet been shown between regional changes in NT/N mRNA expression and antipsychotic drug efficacy or side effect profile.

ROLE FOR ENDOGENOUS CENTRAL NEUROTENSIN IN ATTENTIVE FUNCTIONS SENSITIVE TO ANTIPSYCHOTIC DRUGS AND SCHIZOPHRENIA

The lack of selective, potent nonpeptide NT receptor antagonists has delayed the clarification of the neurophysiology and pharmacology of NT. SR48692 (2-[(1-(7-chloro-4-quinolinyl)-5-(2,6-dimethoxyphenyl) pyrazol-3-yl) carbinylamino] tricyclo (3.3.1.13.7) decan-2-carboxylic acid) was recently proposed as the first selective, high-affinity nonpeptide NT receptor antagonist. Initial *in vitro* data showed SR48692 had a high affinity and selectivity for a wide range of mammalian brain NT receptors.[88] SR48692 also had affinity for peripheral NT receptors, inhibiting the increase in blood pressure induced by exogenous NT in the anesthetized guinea-pig and potently antagonizing the tachycardia and inotropic responses induced by NT in the isolated guinea-pig atria.[89] Few experiments have been carried out with SR48692 *in vivo,* in part due to the limited solubility of the compound in physiological solutions. However, SR48692 has been shown to be active *in vivo,* inhibiting the turning behavior in the mouse induced by intrastriatal NT after both i.v. and oral administration.[88] These initial data suggested that SR48692 may be a useful tool for investigating the interactions between endogenous NT and central dopaminergic pathways. In striatal slices the increase in potassium-evoked [³H]-DA release induced by NT was antagonized by SR48692. However, Steinberg and co-workers[90] showed that the increase in DA metabolism in the VTA following intra-VTA microinjection of NT was not antagonized by SR48692. Using *in vivo* voltammetric techniques, they also found no effect of SR48692 on the increased release of DA in the nucleus accumbens evoked by NT injected into the VTA. The lack of effect of SR48692 in these two studies highlights a complexity of NT interactions with brain DA systems and suggests a NT receptor heterogeneity. Further evidence for NT receptor subtypes was provided by the demonstration that SR48692 failed to antagonize NT-induced hypothermia in the rat or mouse.[90]

Our group has an ongoing research interest in latent inhibition (LI) paradigms as

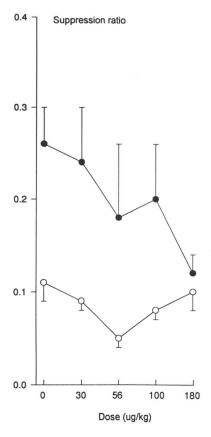

FIGURE 2. Effect of the NT receptor antagonist SR48692 (dose µg/kg) on the latent inhibition (LI) of condition response suppression by stimulus preexposure. Groups of animals receiving 20 preexposures to the to-be-conditioned stimulus (*filled circles*) were compared to non-preexposed rats (*open circles*). Results represent the mean (± SEM) of 6–12 animals per group.

animal behavioral models of stimulus filtering/attentional brain functions sensitive to schizophrenia and targets of the therapeutic actions of antipsychotic drugs.[91] LI represents the negative impact of preexposure to a neutral stimulus on the subsequent ability of that stimulus to serve as a conditioned stimulus. In order to evaluate the antipsychotic potential of endogenous NT, we examined the effects of SR48692 on LI of a conditioned response by stimulus preexposures. We hypothesized that SR48692 would produce an opposite pharmacology to that observed for haloperidol and other antipsychotics. Our initial results (FIG. 2) confirmed this hypothesis by demonstrating a dose-related inhibition of LI by SR48692, in contrast to the increase in suppression ratio seen with haloperidol. These data suggest that NT is recruited in the expression of LI and provide further support for NT as an endogenous antipsychotic.

REFERENCES

1. CARRAWAY, R. & S. E. LEEMAN. 1973. J. Biol. Chem. **248:** 6854–6861.
2. GUDELSKY, G. A., S. A. BERRY & H. Y. MELTZER. 1989. Neuroendocrinology **49:** 604–609.
3. SZIGETHY, E. & A. BEAUDET. 1989. J. Comp. Neurol. **279:** 128–137.

4. BROUARD, A., M. HEAULME & R. LEYRIS *et al.* 1994. Eur. J. Pharmacol. **253:** 289–291.
5. PALACIOS, J. M. & M. J. KUHAR. 1981. Nature **295:** 587–589.
6. FUXE, K., L. F. AGNATI, M. MARTIRE *et al.* 1986. Acta Physiol. Scand. **126**(1): 147–149.
7. QUIRION, R., C. C. CHIUEH, H. D. EVERIST & A. PERT. 1985. Brain Res. **327:** 385–389.
8. SCHOTTE, A. & J. E. LEYSEN. 1989. J. Chem. Neuroanat. **2:** 253–257.
9. KALIVAS, P. W. & J. S. MILLER. 1984. Brain Res. **300:** 157–160.
10. BEAN, A. J., M. J. DURING & R. H. ROTH. 1989. J. Neurochem. **53:** 655–657.
11. DRUMHELLER, A. D., M. A. GAGNE, S. ST-PIERRE & F. B. JOLICOEUR. 1990. Neuropeptides **15:** 169–178.
12. OKUMA, Y., Y. FUKUDA & Y. OSUMI. 1983. Eur. J. Pharmacol. **93:** 27.
13. AGNATI, L. F., K. FUXE, F. BENEFANTI & N. BATTISTINI. 1983. Acta Physiol. Scand. **119:** 459–461.
14. LIU, Y., M. HILLEFORS-BERGLUND & G. VON EULER. 1994. Brain Res. **643:** 343–348.
15. VON EULER, G., I. VAN DER PLOEG, B. B. FREDHOLM & J. FUXE. 1991. J. Neurochem. **56:** 178–183.
16. VON EULER G., G. MEISTER, T. HOKFELT, P. ENEROTH & K. FUXE. 1990. Brain Res. **531:** 253–262.
17. AGANTI, L. F., K. FUXE, N. BATTISTINI *et al.* 1985. Am. Physiol. Scand. **124:** 125–128.
18. UHL, G. R. & M. J. KUHUR. 1984. Nature **309:** 350–352.
19. BENES, F. M., P. A. PASKEVICH & V. B. DOMESICK. 1983. Science **221:** 969–971.
20. LEEMAN, S. E., N. ARONIN & C. FERRIS. 1982. Horm. Res. **38:** 93–132.
21. NEMEROFF, C. B., G. BISSETTE, A. J. PRANGE, JR. *et al.* 1977. Brain Res. **128:** 485–498.
22. BISSETTE, G., C. B. NEMEROFF, P. T. LOOSEN *et al.* 1976. Nature **262:** 607–609.
23. OSBAHR, A. J., III., C. B. NEMEROFF, P. J. MANBERG & A. J. PRANGE, JR. 1979. Eur. J. Pharmacol. **54:** 229–302.
24. OSBAHR, A. J., III., C. B. NEMEROFF, D. LUTTINGER *et al.* 1981. J. Pharmacol. Exp. Ther. **217:** 645–651.
25. LUTTINGER, D., R. A. KING, D. SHEPPARD *et al.* 1982. Eur. J. Pharmacol. **81:** 499–504.
26. SNIDJERS, R., N. R. KRAMARCY, R. W. HURD *et al.* 1982. Neuropharmacology **21:** 465–468.
27. JOLICOEUR, F. B., G. DE MICHELE & A. BARBEAU. 1983. Neurosci. Biobehav. Rev. **7:** 385–390.
28. NEMEROFF, C. B., D. LUTTINGER, D. E. HERNANDEZ *et al.* 1983. J. Pharmacol. Exp. Ther. **225:** 337–345.
29. ERVIN, G. M., L. S. BIRKEMO, C. B. NEMEROFF & A. J. PRANGE. 1981. Nature **291:** 73–76.
30. KALIVAS, P. W., C. B. NEMEROFF & A. J. PRANGE, JR. 1984. Neuroscience **11:** 919–930.
31. JOLICOEUR, F. B. & A. BARBEAU. 1982. Ann. N. Y. Acad. Sci. **400:** 440–441.
32. KALIVAS, P. W., C. B. NEMEROFF & A. J. PRANGE, JR. 1981. Brain Res. **229:** 525–529.
33. KALIVAS, P. W., C. B. NEMEROFF & A. J. PRANGE, JR. 1982. Eur. J. Pharmacol. **78:** 471–474.
34. KALIVAS, P. W., S. K. BURGESS, C. B. NEMEROFF & A. J. PRANGE, JR. 1983. Neuroscience **8:** 495–505.
35. SEUTIN, V., L. MASSOTTE & A. DRESSE. 1989. Neuropharmacology **9:** 949–954.
36. INNIS, R. B., R. ANDRADE & G. K. AGHAJANIAN. 1984. Soc. Neurosci. Abstr. **10:** 241.9.
37. PINNOCK, R. D. 1985. Brain Res. **338:** 151–154.
38. POZZA, M. F., E. KUNG, S. BISCHOFF & H.-R. OLPE. 1988. Eur. J. Pharmacol. **145:** 341–343.
39. SHI, W.-X. & B. S. BUNNEY. 1987. Soc. Neurosci. Abstr. **13:** 259.1.
40. SHI, W.-X. & B. S. BUNNEY. Brain Res. In press.
41. DAO, W. P. C., H. YAJIMA, J. KITAGAWA & R. J. WALKER. 1980. Adv. Physiol. Sci. **14:** 249–254.
42. AUDINANT, E., J.-M. HERMEL & F. CREPEL. 1989. Exp. Brain Res. **78:** 358–368.
43. BALDINO, F., JR., G. A. HIGGINS, M. T. MOKE & B. WOLFSON. 1985. Peptides **6:** 249–256.
44. BALDINO, F., JR. & B. WOLFSON. 1985. Brain Res. **342:** 266–272.
45. HERBISON, A. E., J. I. HUBBARD & N. E. SIRET. 1986. Brain Res. **364:** 391–395.
46. BUNNEY, B. S. 1988. Ann. N. Y. Acad. Sci. **537:** 77–85.
47. BUNNEY, B. S., J. R. WALTERS, R. H. ROTH *et al.* 1973. J. Pharmacol. Exp. Ther. **185:** 560–571.

48. CHIODO, L. A. & B. S. BUNNEY. 1983. J. Neurosci. **3:** 1607–1619.
49. WHITE, F. J. & R. Y. WANG. 1983. Science **221:** 1054–1057.
50. PINNOCK, R. D. 1984. Br. J. Pharmacol. **81:** 631–635.
51. SUPPES, T. & R. J. PINNOCK. 1987. Neuropharmacology **26:** 331–337.
52. TANAKA, K., M. MASU & S. NAKANISHI. 1990. Neuron **4:** 847–854.
53. VITA, N., P. LAURENT, S. LEFORT et al. 1993. FEBS Lett. **317(1,2):** 139–142.
54. DOHLMAN, H. G., J. THORNER, M. G. CARON & R. J. LEFKOWITZ. 1991. Annu. Rev. Biochem. **60:** 653–688.
55. PALACIOS, J. M., A. PAZOS, M. M. DIETL et al. 1988. Neuroscience **25(1):** 307–317.
56. SATO, M., H. KIYAMA & M. TOHYAMA. 1992. Neuroscience **48(1):** 137–149.
57. DI PAOLA, E. D., B. CUSACK, M. YAMADA & E. RICHELSON. 1993. J. Pharmacol. Exp. Ther. **264(1):** 1–5.
58. YAMADA, M., M. YAMAGA & E. RICHELSON. 1993. Biochem. Pharmacol. **45(10):** 2149–2154.
59. VANISBERG, M. A., J. M. MALOTEAUX, J. N. OCTAVE & P. M. LADURON. 1991. Biochem. Pharmacol. **42:** 2265–2274.
60. MAZELLA, J., K. LEONARD, J. CHABRY et al. 1991. Brain Res. **564:** 249–255.
61. CASTEL, M. N., D. FAUCHER, F. CUINE et al. 1991. J. Neurochem. **56:** 1816–1818.
62. LABBE-JULLIE, C., I. DUBUC, A. BROUARD et al. 1994. J. Pharmacol. Exp. Ther. **268(1):** 328–336.
63. KITABGI, P., Y. MASUO, A. NICOT et al. 1991. Neurosci. Lett. **124:** 9–12.
64. AMAR, S., P. KITABGI & J.-P. VINCENT. 1987. J. Neurochem. **49:** 999–1006.
65. BOZOU, J.-C., S. AMAR, J.-P. VINCENT & P. KITABGI. 1986. Mol. Pharmacol. **29:** 489–496.
66. GILBERT, J. A., C. J. MOSES, M. A. PFENNING et al. 1986. Biochem. Pharmacol. **35(3):** 391–397.
67. SNIDER, R. M., C. FORRAY, M. PFENNING & E. RICHELSON. 1986. J. Neurochem. **47:** 1214–1218.
68. GOEDERT, M., R. D. PINNOCK, C. P. DOWNES et al. 1984. Brain Res. **323:** 193–197.
69. SHI, W. X. & B. S. BUNNEY. 1992. J. Neurosci. **12:** 2433–2438.
70. BOZOU, J.-C., S. AMAR, J.-P. VINCENT & P. KITABGI. 1986. Mol. Pharmacol. **29:** 489–496.
71. BOZOU, J.-C., N. ROCHET, I. MAGNALDO, J. P. VINCENT & P. KITABGI. 1989. Biochem. J. **264:** 871–878.
72. STAUCH-SLUSHER, B., A. E. ZACCO, J. A. MASLANSKI et al. 1994. Mol. Pharmacol. **46:** 115–121.
73. YAMADA, M., Y. MITSUHIKO, M. A. WATSON & E. RICHELSON. 1993. Eur. J. Pharmacol. **244:** 99–101.
74. KILTS, C. D., C. M. ANDERSON, G. BISSETTE et al. 1988. Biochem. Pharmacol. **37(8):** 1547–1554.
75. MOYSE, E., W. ROSTENE, M. VIAL et al. 1987. Neuroscience **22:** 525–536.
76. MIYOSHI, R., S. KITO, H. ISHIDA & M. NAKASHIMA. 1989. Neurochem. Int. **15(4):** 493–496.
77. GOVONI, S., J. S. HONG, H.-Y. T. YANG & E. COSTA. 1980. J. Pharmacol. Exp. Ther. **215(2):** 413–417.
78. RADKE, J. M., A. J. MACLENNAN, M. C. BEINFELD et al. 1989. Brain Res. **480:** 178–183.
79. MYERS, B., B. LEVANT, G. BISSETTE & C. B. NEMEROFF. 1992. Brain Res. **575:** 325–328.
80. LEVANT, B., G. BISSETTE, E. WINDERLOV & C. B. NEMEROFF. 1991. Regul. Peptides **32:** 193–201.
81. WIDERLOV, E., L. H. LINDSTROM, G. BESEY et al. 1982. Am. J. Psychiatry **139(9):** 1122–1126.
82. LINDSTROM, L. H., E. WIDERLOV, G. BISSETTE & C. NEMEROFF. 1988. Schizophr. Res. **1:** 55–59.
83. MERCHANT, K. M., P. R. DOBNER & D. M. DORSA. 1992. J. Neurosci. **12(2):** 652–663.
84. MERCHANT, K. M. & D. M. DORSA. 1993. Proc. Natl. Acad. Sci. USA **90:** 3447–3451.
85. MERCHANT, K. M. 1994. Mol. Cell. Neurosci. In press.
86. BOLDEN-WATSON, C., M. A. WATSON, K. D. MURRAY et al. 1993. J. Neurochem. **61:** 1141–1143.

87. MERCHANT, K. M., D. J. DOBIE, F. M. FILLOUX *et al.* 1994. J. Pharmacol. Exp. Ther. **271:** 460–471.
88. GULLY, D., M. CANTON, R. BOIGEGRAIN *et al.* 1993. Proc. Natl. Acad. Sci. USA **90:** 65–69.
89. NISATO, D., P. GUIRAUDOU & G. BARTHELEMY. 1994. Life Sci. **54(7):** 95–100.
90. STEINBERG R., P. BRUN, M. FOURNIER *et al.* 1994. Neuroscience **59(4):** 921–929.
91. DUNN, L. A., G. E. ATWATER & C. D. KILTS. 1993. Psychopharmacology **112:** 315–323.

Distribution of Neuropeptide Receptors

New Views of Peptidergic Neurotransmission Made Possible by Antibodies to Opioid Receptors[a]

ROBERT ELDE,[b,c] ULF ARVIDSSON,[b] MAUREEN RIEDL,[b]
LUCY VULCHANOVA,[b] JANG-HERN LEE,[b]
ROBERT DADO,[b] ALBERT NAKANO,[b]
SUMITA CHAKRABARTI,[d] XU ZHANG,[e]
HORACE H. LOH,[d] PING Y. LAW,[d] TOMAS HÖKFELT,[e]
AND MARTIN WESSENDORF[b]

[b]Department of Cell Biology and Neuroanatomy
[d]Department of Pharmacology
University of Minnesota Medical School
Minneapolis, Minnesota 55455

[e]Department of Neuroscience
Karolinska Institute
S-17177 Stockholm, Sweden

Chemical neurotransmission can be categorized by its major features. Temporally, one can compare fast versus slow neurotransmission; the former occurs over a time course of milliseconds whereas the latter ranges from seconds to minutes in time course. Chemical neurotransmission can also be divided by the structural arrangement of its neurotransmitter delivery system, that is, synaptic versus nonsynaptic arrangements. Synaptic transmission is a highly compartmentalized form of intercellular communication in that the site of release of a transmitter is separated by only 30 nm from receptors on the postsynaptic membrane. Nonsynaptic neurotransmission, in contrast, is less confined in that a transmitter is released and must diffuse through a significantly greater volume of extracellular space than the synaptic cleft before encountering its receptor.[1,2]

Nonsynaptic neurotransmission is now thought to be an important phenomenon because histochemical studies of the distribution of neurotransmitters and their receptors have repeatedly discovered widespread "mismatches" in the spatial relationships of transmitters and receptors.[3,4] Because of the limitations of histochemical methods, these assessments of the mismatch problem have been indirect and inferential. Developments outlined below have, for the first time, enabled the direct examination of the spatial relationship between a neuropeptide and its receptor.[5–9]

Peptidergic neurotransmission appears to be accomplished by slow, nonsynaptic mechanisms.[10,11] The cloning of receptors for neuropeptides has contributed impor-

[a]This work was supported by U.S. Public Health Service grants DA 06299, DA 05466, DA 07339, DA 08131, and DA 00564; the Swedish Medical Research Council; Stiftelsen Wenner-Gren Center; the Minnesota Medical Foundation; the Bank of Sweden Tercentenary Foundation; Marianne and Marcus Wallenberg's Stiftelse, and Gustav V's and Drottning Victoria's Stiftelse.

[c]Address correspondence to Dr. Robert Elde, Department of Cell Biology and Neuroanatomy, University of Minnesota, 321 Church Street SE, Minneapolis, MN 55455.

tantly to our understanding of signal transduction for this class of neurotransmitters. The first of the neuropeptide receptors to be cloned were the tachykinin receptors[12,13] and the receptor for neurotensin,[14] but several other families have been subsequently cloned. Cloning of these receptors has been satisfying in several regards, not the least of which was to establish that they are indeed members of the seven-transmembrane, G-protein-coupled family of receptors.

In addition, the cloning of neuropeptide receptors has allowed significant advances in our understanding of the deployment of these receptors by neurons to mediate the actions of neuropeptides. First, *in situ* hybridization studies have permitted the identification of the cells that express a given, cloned receptor.[15] Second, the ability to produce antibodies to portions of the predicted primary structure of cloned receptors[5–8,16] and to use these antibodies in immunocytochemical studies has allowed a determination of the targeting of the receptor to pre- or postsynaptic sites and also of the spatial relationship between the receptor and nerve terminals that contain its ligand.[5–9] Such findings have helped to advance our understanding of the nature of peptidergic neurotransmission; this point serves as the focus of this review. Because of the relative infancy of this field of inquiry, most of the data reviewed herein will be from the family of opioid receptors, which have been the focus of studies in the authors' laboratories.

TECHNICAL CONSIDERATIONS

Prior to the cloning of receptors for neuropeptides, the various methods of ligand-binding autoradiography provided the only histochemical method capable of revealing the distribution of these receptors.[17–20] Much was learned concerning the regional occurrence of neuropeptide receptors by autoradiography, including the conclusion that in many instances a spatial mismatch occurs between a given receptor and nerve terminals containing a ligand for this receptor. Although some investigators resorted to heroic measures,[21,22] in general, the autoradiographic methods were limited in several aspects: First, it was difficult to draw conclusions directly concerning the pre- versus postsynaptic nature of a receptor. Second, limitations in resolution precluded determination of whether a localized receptor was in the plasma membrane or in an intracellular compartment. Third, ligand-binding autoradiographic methods were not capable of revealing the distribution of receptors per se, but rather the distribution of binding sites. Although it was possible to control incubation conditions so as to maximize the likelihood that binding sites were indeed associated with receptors, it was unclear whether the correlation was perfect.

The cloning of neuropeptide receptors made possible the application of techniques that could complement and at least partially overcome the limitations of ligand-binding autoradiography. That is, knowledge of the sequence of mRNA encoding a receptor makes possible the localization of transcripts for that receptor by *in situ* hybridization, while the predicted primary structure of the receptor protein allows the development of antibodies capable of recognizing the receptor protein using immunocytochemical methods.

In Situ *Hybridization*

In situ hybridization methods for localization of neuropeptide receptor mRNA are not different than those for other transcripts, except that the abundance of most neuropeptide receptor transcripts is low. Thus, successful localizations of neuropeptide receptor transcripts were initially reported using a cocktail of several radiolabeled oligonucleotides.[15] Since then riboprobe-based *in situ* hybridization has been found to be generally useful for detection of transcripts that encode neuropeptide receptors.[23–28] These studies have been important in that they have identified the neurons whose activity might be modulated by a given neuropeptide.

Immunocytochemistry

The deduced amino acid sequence of neuropeptide receptors that resulted from cloning made possible the construction of immunogens (either fusion proteins or the synthesis of peptide fragments) for the purpose of creating antibodies useful for the localization of the receptor protein.[5,6,8,9] We used standard algorithms for predicting regions of the protein sequence that might be particularly hydrophilic and antigenically prominent (MacVector 4.1.4, Kodak). Peptides representing these regions (TABLE 1), ranging in length from 15 to 18 amino acids, were synthesized by solid-phase methods, coupled to carrier proteins, and injected into rats and rabbits with only minor modifications from those used originally for the opioid peptides.[29] After boosting, sera were screened on sections of brain and spinal cord, and on COS-7 cells electroporated with native, as well as epitope-tagged constructs[30,31] of the cloned δ, μ, and κ opioid receptors (DOR1,[32,33] MOR1,[34–36] and KOR1,[27,37–40] respectively). Sera were selected for further study if the staining of these preparations was blocked by treatment of the sera with the peptide that was used for immunization. Shorter, overlapping peptides from within the sequences used for immunization were used in additional blocking studies to determine the epitope recognized by the antisera (TABLE 1). Antisera were also tested for their ability to identify opioid receptors as determined using immunoblotting and immunoaffinity purification.[6,7,41] In some cases it was also possible to compare the ability of antisera raised to peptides from different portions of a given receptor to recognize their cognate receptor in immunoblotting and immunofluorescence procedures.[5] Finally, it has been possible to compare immunofluorescent localizations of the opioid receptors with corresponding localizations of transcripts for those receptors (i.e., using *in situ* hybridization)[6,41] and with ligand-binding autoradiographic images from the literature. Taken together, the results of these characterization studies strongly suggest that the antisera (TABLE 1) are highly selective for each receptor. In most cases, similarities of amino acid sequences within the regions selected for immunization were apparent across species (TABLE 2), a feature that has allowed the use of these antisera in brains of rodents and primates.

RESULTS AND DISCUSSION

A comprehensive account of the distribution of the cloned opioid receptors is beyond the scope of this review. However, localizations from several regions serve to represent the concepts that have emerged from observations throughout the central nervous system (CNS).

Primary Afferent Neurons and Spinal Cord

The superficial portion of the dorsal horn of the spinal cord contains terminals of fine-caliber myelinated and unmyelinated axons of primary afferent neurons, some

TABLE 1. Antipeptide Antisera against Opioid Receptors and Opioid Peptides

	Species Immunized	Amino Acid Numbers (species)	Amino Acid Sequences[a]	Reference
Receptors				
δ-Opioid receptor (DOR1)	Rabbit/rat	3–17 (mouse)	LVPSARAELQSSPLV[b]	Dado et al.[7]
		30–46 (mouse)	AGANASGSPGARSASSL[b]	Arvidsson et al.[5]
		103–120 (mouse)	PFQSAKYLMETWPFGELL[b]	
μ-Opioid receptor (MOR1)	Rabbit	384–398 (rat)	NHQLENLEAETAPLP[b]	Arvidsson et al.[6]
κ-Opioid receptor (KOR1)	Rabbit	366–380 (rat)	DPASMRDVGGMNKPV[b]	Riedl et al.[41]
Peptides				
Leu-enkephalin	Mouse	1–5 (mammals)	YGGFL	Cuello et al.[51]
Dynorphin A1–8	Rabbit	1–8 (porcine)	YGGFLRRI	Weber et al.[52]
Preprodynorphin C-peptide	Guinea pig	235–248 (rat)[c]	SQENPNTYSEDLDV[b]	Arvidsson et al.[41]

[a] Underscoring indicates the epitope recognized by the antisera.
[b] Cross-species alignment of these are shown in TABLE 2.
[c] Sequence as reported by Douglass et al.[53]

TABLE 2. Cross-species Alignment of Sequences Used for Generation of Antipeptide Antisera[a]

δ-Opioid receptor (DOR1)

Mouse	3 - LVPSARAELQSSPLV - 17	30 - AGANASGSPGARSASSL - 46	103 - PFQSAKYLMETWPFGELL - 120
Rat	PVPSARAELQFSLLA	ASANASGSPGARSASSL	PFQSAKYLMETWPFGELL
Human	PAPSAGAELQPPLFA	AGANASGPPGPGSASSL	PFQSAKYLMETWPFGELL

μ-Opioid receptor (MOR1)

Rat	384 - NHQLENLEAETAPLP - 398
Mouse	-xENLEAETAPLP
Human	NHQLENLEAETAPLP

κ-Opioid receptor (KOR1)

Rat	366 - DPASMRDVGGMNKPV - 380
Mouse	DPASMRDVGGMNKPV
Guinea pig	DPAYMRNVDGVNKPV
Human	DPAYLRDIDGMNKPV

Preprodynorphin C-peptide (ppDYN)

Rat	235 - SQENPNTYSEDLDV - 248
Porcine	SQEDPNAYYEELFDV
Human	SQEDPNAYSGELF

[a]Underscoring indicates differences from the first sequence in each series.

of which mediate nociception. Early studies on the localization of opioid receptors by autoradiographic methods determined this to be a region highly enriched in opioid receptors[18,19,42] and that a significant fraction of these receptors were presynaptic in nature and resided on terminals of primary afferent neurons.[42] Localization of DOR1 by immunocytochemical means confirmed this idea (FIG. 1C and D)[5,7] and established that this receptor is exclusively presynaptic in this and many other regions of the CNS. Localization of MOR1 by immunocytochemistry (FIG. 1A and B) has demonstrated that this receptor is frequently presynaptic in the superficial dorsal horn, is expressed by primary afferent neurons, and coexists in some of these neurons with DOR1-ir.[6] However, MOR1-ir is also found in membranes of neuronal peri-karya and dendrites within the superficial dorsal horn (FIG. 1B). The cloned KOR1 is similar to the DOR1 in that it is prominent in a subset of small- and intermediate-diameter perikarya in dorsal root ganglia (not shown) and in nerve terminals in the superficial dorsal horn (FIG. 1E and F).[41] It also resembles MOR1 in that it is found as a postsynaptic receptor on some neurons in the superficial dorsal horn (FIG. 1F). Interestingly, both enkephalin (ENK)- and preprodynorphin (ppDYN)-ir are found in axons and terminals in close proximity to DOR1, MOR1, and KOR1 (FIG. 1B, B′; D, D′; and F, F′). Although these represent close appositions, it does not imply that these relations are truly synaptic, because, in the case of the presynaptic opioid receptors on CGRP-positive primary afferent terminals, little evidence exists that such terminals receive axo-axonic synapses.[43] Furthermore, these opioid receptors are unlikely to be autoreceptors, because coexistence of any of these pairs of opioid ligands and receptors is very rare.

Locus Coeruleus

Neurons of the locus coeruleus express MOR1 mRNA and target the resultant protein to the plasma membrane of their cell bodies and dendrites (FIG. 2A and B[6]). Although the density of MOR1-ir is great in the neuropil of locus coeruleus, ENK-ir is present in nerve fibers and terminals in substantially smaller quantities than MOR1-ir (FIG. 2B and B′). DOR1-ir is also present in locus coeruleus, but, as in the spinal cord, is restricted to nerve fibers and terminals (FIG. 2C and D) where it presumably acts in a presynaptic manner. Again, as in spinal cord, ENK-ir is present in axons and terminals in close proximity to DOR1-ir (FIG. 2D and D′), but DOR1 is unlikely to be an autoreceptor because DOR1- and ENK-ir were not found to coexist.

Interpeduncular Nucleus

Nerve fibers and teriminals in the interpeduncular nuclei display complementary patterns of ENK- and DOR1-ir (FIG. 3A and B; Lee *et al.,* in preparation). The lateral portion of the nucleus contains a dense network of axons with DOR1-ir and almost no ENK-ir (FIG. 3A), whereas the converse relationship exists in the neighboring region (FIG. 3B). Similar, complementary patterns of staining have been observed in many brain regions, and they suggest that this orderly, spatial mismatch of opioid peptides and their receptors reflects a form of internuclear, nonsynaptic neurotransmission.

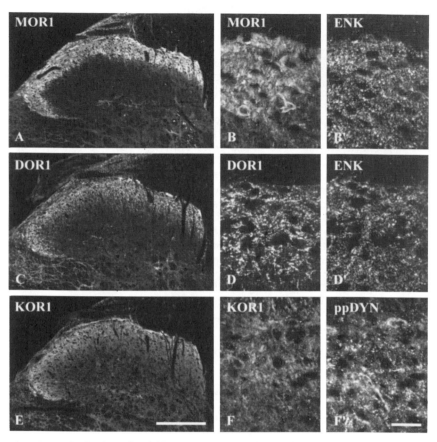

FIGURE 1. Distribution of opioid receptors and their putative endogenous ligands in the superficial dorsal horn as viewed with laser scanning confocal microscopy. Immunofluorescence images of coronal sections of the adult rat dorsal horn single stained with rabbit anti-μ-opioid receptor (MOR1; **A**) or rabbit anti-δ-opioid receptor (DOR1; **C**) or rabbit anti-κ-opioid receptor (KOR1; **E**) antisera. The opioid receptor immunoreactivity was visualized with cyanine 3.18-conjugated secondary antibodies. Note dense staining in the superficial laminae of the dorsal horn for all three opioid receptors. Two-color immunofluorescence experiments were done in order to examine the spatial distribution of opioid receptors in relation to endogenous ligands. Sections were simultaneously incubated with a mixture of rabbit anti-MOR1 (**B**) and mouse anti-enkephalin (ENK, **B′**) or rabbit anti-DOR1 (**D**) and mouse anti-ENK (**D′**) or rabbit anti-KOR1 (**F**) and guinea-pig anti-preprodynorphin (ppDYN, a marker for dynorphin; **F′**) antisera. After incubation the sections were stained with lissamine rhodamine- and fluorescein-conjugated secondary antibodies. Complementary but not overlapping distributions are seen for all three opioid receptors in relation to their putative endogenous ligands. Note also at these higher magnifications MOR1, DOR1, and KOR1 each has a distinct pattern of labeling in the dorsal horn (**B, D, F**): MOR1 labeling is preferentially seen in postsynaptic membranes, DOR1 staining is only seen in axon terminals, and KOR1 seems to label both pre- and postsynaptic structures. Scale bars = 250 μm (**A, C, E**); 25 μm (**B, B′, D, D′, F, F′**). Images **B, B′, D, D′, F, F′** are the results of projecting six optical sections taken at 0.4 μm intervals.

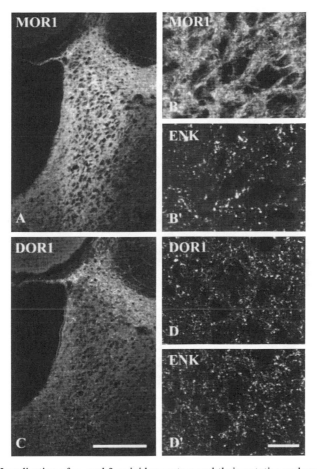

FIGURE 2. Localization of μ- and δ-opioid receptors and their putative endogenous ligand, enkephalin (ENK), in locus coeruleus as viewed with laser scanning confocal microscopy. Immunofluorescence images of coronal sections of the adult rat locus coeruleus single stained with rabbit anti-μ-opioid receptor (MOR1; **A**) or rabbit anti-δ-opioid receptor (DOR1; **C**). The opioid receptor immunoreactivity was visualized with cyanine 3.18-conjugated secondary antibodies. Note the different staining pattern for MOR1 and DOR1. To determine the relationship between ENK and MOR1 and DOR1, double-labeling immunofluorescence experiments using a mixture of either rabbit anti-MOR1 (**B**) and mouse anti-ENK (**B'**), or rabbit anti-DOR1 (**D**) and mouse anti-ENK (**D'**) were carried out. After incubation the sections were stained with lissamine rhodamine- and fluorescein-conjugated secondary antibodies. (**B–B'**) Dense staining of MOR1 was seen in the cell membrane of locus coeruleus neurons as well as in the neuropil. ENK-immunoreactive varicosities were present in close vicinity to MOR1-labeled structures; however, a mismatch in the relative abundance of ENK and MOR1 was observed. (**D–D'**) As in the case of the dorsal horn, DOR1-immunoreactivity was confined to varicose axonal profiles in locus coeruleus, suggesting a role as a presynaptic receptor. No unambiguous coexistence between DOR1 and ENK was seen, suggesting that DOR1 is not an autoreceptor for ENK in locus coeruleus. Scale bars = 200 μm (**A, C**); 25 μm (**B, B', D, D'**). Images **B, B', D, D'** are the results of projecting six optical sections taken at 0.4 μm intervals.

Ventral Forebrain

Distinctive staining patterns for each of the three cloned opioid receptors are seen in the forebrain. Patches of MOR1-ir are scattered, especially in the dorsolateral aspect of the striatum (FIG. 4A), in a pattern that strongly resembles both μ-opioid binding and the location of MOR1 as determined by *in situ* hybridization.[6] DOR1-ir is not so prominent in the striatum, but is concentrated in irregular zones in the islands of Calleja (FIG. 4B), the substantia inominata, and the septal nuclei. KOR1-ir is also prominent in the ventral forebrain, especially in the vicinity of the islands of Calleja and the shell region of nucleus accumbens (FIG. 4C).

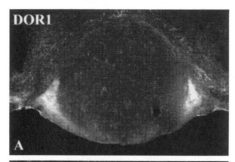
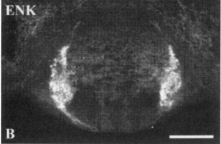

FIGURE 3. Spatial relations between δ-opioid receptor and its putative endogenous ligand, enkephalin, in the interpeduncular nuclei. Immunofluorescence images of a coronal section of the adult mouse interpeduncular nuclei as viewed with laser scanning confocal microscopy. The section was simultaneously incubated with a mixture of rabbit anti-δ-opioid receptor (DOR1, **A**) and mouse anti-enkephalin (ENK, **B**). The staining was visualized with lissamine rhodamine- and fluorescein-conjugated secondary antibodies. Note that the distribution of DOR1-immunoreactivity is found in a nucleus adjacent to the distribution of ENK-positive fibers, suggesting in this case that neurotransmission could possibly occur between nuclei. Scale bar = 250 μm (**A, B**).

Subcellular Localization of Opioid Receptors

DOR1 seems largely localized in axons and in vesicles undergoing axonal transport (FIG. 5). Surprisingly, neither DOR1-ir (in terminals in the superficial dorsal horn of the spinal cord) nor KOR1-ir (in terminals in the neural lobe of the pituitary) is prominent in the plasma membrane of nerve terminals. Rather, preliminary electron microscopic studies have demonstrated staining for these receptors is prominent in a population of large, clear vesicles (Arvidsson *et al.,* in preparation).

IMPLICATIONS OF FINDINGS

Some aspects of the neurotransmission accomplished by opioid peptides and their receptors have been thoroughly studied and thus serve as models of the more general phenomenon of peptidergic neurotransmission. In spite of these advances,

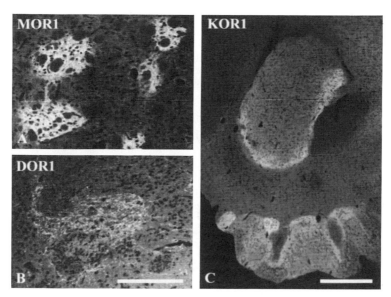

FIGURE 4. Localization of μ-, δ-, and κ-opioid receptors in the forebrain as viewed with laser scanning confocal microscopy. Immunofluorescence images of coronal sections of the adult rat forebrain single stained with rabbit anti-μ-opioid receptor (MOR1; **A**), rabbit anti-δ-opioid receptor (DOR1; **B**) or rabbit anti-κ-opioid receptor (KOR1; **C**). The opioid receptor immunoreactivity was visualized with cyanine 3.18-conjugated secondary antibodies. (**A**) Intense MOR1 staining is seen in patches in the striatum. Note also weak labeling of the neuropil between the patches. (**B**) A relatively dense network of DOR1 positive fibers is seen in the area of the islands of Calleja. (**C**) Low-power micrograph of the ventral forebrain showing dense labeling of KOR1 in the shell of nucleus accumbens and in the olfactory tubercle including the islands of Calleja. Scale bars = 250 μm (**A, B**); 500 μm (**C**).

large gaps are apparent in our understanding of opioid neurotransmission. One of these gaps has been the lack of knowledge of the spatial relationships between nerve terminals that contain opioid peptides and the membranes that possess opioid receptors. A second, and related gap, has been the lack of identification of the endogenous opioid peptides that are likely to occupy a given opioid receptor subtype under physiological or pathophysiological circumstances.

FIGURE 5. Intraaxonal localization of δ-opioid receptor (DOR1) as viewed with laser scanning confocal microscopy. High-magnification immunofluorescence image of a coronal section of the adult rat brainstem single stained with rabbit anti-δ-opioid receptor (DOR1) and visualized with cyanine 3.18-conjugated secondary antibodies. The DOR1 staining is seen in small, round structures with a size of approximately 200–300 nm. These labeled structures possibly represent vesicles transporting the receptor to its target (anterograde transport) or back to the cell (retrograde transport).

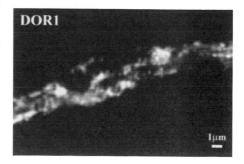

By staining for both opioid peptides and opioid receptors we have been able to determine with cellular and subcellular precision the relationship between nerve fibers and terminals that contain opioid peptides and the receptors with which they are associated. From previous studies by others as well as our own studies, we believe that opioidergic neurotransmission is generally nonsynaptic.[10,11] We have found that the arrangement of the elements involved in opioidergic transmission falls between two extremes. First, as exemplified in the dorsal horn of the spinal cord and locus coeruleus, we observed a nonsynaptic interdigitation of nerve terminals containing opioid peptides with neuronal membranes bearing opioid receptors (with the interdigitations being separated by distances on the order of microns). Second, in some cases such as the subnuclei of the interpeduncular nucleus, ENK- and DOR-ir occur in entirely different subnuclei. Although this is a striking mismatch in the spatial distribution of this transmitter/receptor combination, the complementary and orderly relationship between ligand and receptor suggests the possibility that neurotransmission in this case may occur between nuclei (over distances on the order of hundreds of microns).

A further problem in opioid neurotransmission has been the confusion in assigning endogenous ligands to subtypes of opioid receptors.[44] Previous investigations of this problem used bioassays and radioligand binding to determine the rank order of potency of endogenous ligands for each subtype of opioid receptor. This approach has pharmacological validity, but depends greatly on the stability of the ligands tested. It is known that many neurotransmitters are very unstable in such preparations because of mechanisms that terminate the process of neurotransmission. An alternative, and more physiological approach to this problem is to determine the spatial relationship and relative abundance of each of the endogenous ligands to each of the receptor subtypes. The use of multicolor immunofluorescence studies in which both receptors and opioid peptides are localized simultaneously has greatly aided this task. Based on these studies, it appears that enkephalins are likely to be physiologically relevant ligands for both DOR1 and MOR1 in the superficial dorsal horn of the spinal cord and that dynorphin-related peptides are positioned to gain access to KOR1.

Little is known concerning the mechanisms that deliver newly synthesized receptor protein to the region of plasma membrane where it might be exposed to ligands. Autoradiographic studies suggest that some opioid receptors undergo axonal transport[42,45–47] and are likely to function at presynaptic sites.[48] One of the major surprises upon our initial immunocytochemical localizations of DOR1 was that it seems to be targeted almost exclusively to the axon of neurons, and thus it is most likely that it can act in a "presynaptic" manner. However, high magnification confocal and preliminary electron microscopic studies have suggested that a great fraction of DOR1 is not on the plasma membrane of axons or their terminals, but rather is found intra-axonally associated with vesicles. This suggests the interesting possibility that presynaptic opioid receptors may be stored, awaiting exteriorization, perhaps in response to stimulation (FIG. 6). If this proves to be the case, it suggests additional regulatory steps important in receptor function. Indeed, a recent report suggests that chronic morphine treatment reduces the rate of a slow phase of axonal transport, at least in mesolimbic dopamine neurons.[49] Together, these findings suggest that certain stimulus parameters and kinetics of receptor targeting might be important mechanisms underlying tolerance.

A further surprising aspect of the targeting of neuropeptide receptors was observed in dorsal root ganglion neurons, where it was observed that DOR1 and KOR1 are transported from the perikarya into the central and peripheral processes, whereas a portion of MOR1 remains in the perikarya and appears to decorate the

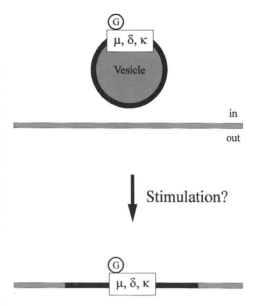

FIGURE 6. Schematic illustration of the hypothesis that presynaptic opioid receptors are inserted into the plasma membrane of the nerve terminal in an activity-dependent manner. Under normal conditions (*upper panel*) a major portion of opioid receptor immunoreactivity in nerve terminals is found associated with a distinct population of vesicles. Under conditions of increased activity (*lower panel*) the resultant fusion of the vesicle membrane with the plasma membrane of the nerve terminal results in the externalization of the receptor. G, heterotrimeric G-proteins.

plasma membrane of the perikarya.[6,50] Even more extreme, we recently determined that the Y1 receptor for neuropeptide Y is produced by a subset of dorsal root ganglion neurons, but it is almost completely retained by the perikaryon where it decorates the plasma membrane.[50] Although it has not been thought that signaling occurs within dorsal root ganglia, the position of MOR1 and the Y1 receptor suggest that blood-borne neuropeptides might gain access to ganglionic receptors, and thereby contribute to the regulation of activity of subsets of primary afferent neurons.

SUMMARY

The cloning of receptors for neuropeptides made possible studies that identified the neurons that utilize these receptors. *In situ* hybridization can detect transcripts that encode receptors and thereby identify the cells responsible for their expression, whereas immunocytochemistry enables one to determine the region of the plasma membrane where the receptor is located. We produced antibodies to portions of the predicted amino acid sequences of δ, μ, and κ opioid receptors and used them in combination with antibodies to a variety of neurotransmitters in multicolor immunofluorescence studies visualized by confocal microscopy. Several findings are notable: First, the cloned δ opioid receptor appears to be distributed primarily in axons, and therefore most likely functions in a presynaptic manner. Second, the cloned μ and κ opioid receptors are found associated with neuronal plasma membranes of dendrites and cell bodies and therefore most likely function in a postsynaptic manner. However, in certain, discrete populations of neurons, μ and κ opioid receptors appear to be distributed in axons. Third, enkephalin-containing terminals are often found in close proximity (although not necessarily synaptically linked) to membranes containing either the δ or μ opioid receptors, whereas dynorphin-containing terminals are often found in proximity to κ opioid receptors. Finally, a substantial mismatch between opioid receptors and their endogenous ligands was observed in some brain regions. However, this mismatch was characterized by complementary zones of receptor and ligand, suggesting underlying principles of organization that underlie long-distance, nonsynaptic neurotransmission.

REFERENCES

1. FUXE, K. & L. F. AGNATI. 1991. *In* Volume Transmission in the Brain: Novel Mechanisms for Neuronal Transmission. Advances in Neuroscience. K. Fuxe & L. F. Agnati, Eds. Raven Press. New York.
2. SCHELLER, R. H. & Z. W. HALL. 1992. Chemical messengers at synapses. *In* An Introduction to Molecular Neurobiology. Z. W. Hall, Ed.: 119–147. Sinauer Associates. Sunderland, MA.
3. KUHAR, M. J. 1985. The mismatch problem in receptor mapping studies. Trends Neurosci. **8:** 190–191.
4. HERKENHAM, M. 1987. Mismatches between neurotransmitter and receptor localizations in brain: Observations and implications. Neuroscience **23:** 1–38.
5. ARVIDSSON, U., R. J. DADO, M. RIEDL, J.-H. LEE, P.-Y. LAW, H. H. LOH, R. ELDE & M. W. WESSENDORF. 1995. Delta (δ)-opioid receptor immunoreactivity: Distribution in brain stem and spinal cord and relationship to biogenic amines and enkephalin. J. Neurosci. **15:** 1215–1235.
6. ARVIDSSON, U., M. RIEDL, S. CHAKRABARTI, J.-H. LEE, A. H. NAKANO, R. J. DADO, H. H. LOH, P.-Y. LAW, M. W. WESSENDORF & R. ELDE. 1995. Distribution and targeting of a μ-opioid receptor (MOR1) in brain and spinal cord. J. Neurosci. In press.
7. DADO, R. J., P. Y. LAW, H. H. LOH & R. ELDE. 1993. Immunofluorescent identification of a delta (delta)-opioid receptor on primary afferent nerve terminals. NeuroReport **5:** 341–344.
8. NAKAYA, Y., T. KANEKO, R. SHIGEMOTO, S. NAKANISHI & N. MIZUNO. 1994. Immunohistochemical localization of substance P receptor in the central nervous system of the adult rat. J. Comp. Neurol. **347:** 249–274.
9. LIU, H., J. L. BROWN, L. JASMIN, J. E. MAGGIO, S. R. VIGNA, P. W. MANTYH & A. I. BASBAUM. 1994. Synaptic relationship between substance P and the substance P receptor: Light and electron microscopic characterization of the mismatch between neuropeptides and their receptors. Proc. Natl. Acad. Sci. USA **91:** 1009–1013.
10. ELDE, R. & T. HÖKFELT. 1993. Coexistence of opioid peptides with other neurotransmitters. *In* Handbook of Experimental Pharmacology, Opioids I. A. Herz, Ed.: 585–624. Springer-Verlag. Berlin Heidelberg.
11. BEAN, A. J., X. ZHANG & T. HÖKFELT. 1994. Peptide secretion: What do we know? FASEB J. **8:** 630–638.
12. HERSHEY, A. D. & J. E. KRAUSE. 1990. Molecular characterization of a functional cDNA encoding the rat substance P receptor. Science **247:** 958–962.
13. YOKOTA, Y., Y. SASAI, K. TANAKA, T. FUJIWARA, K. TSUCHIDA, R. SHIGEMOTO, A. KAKIZUKA, H. OHKUBO & S. NAKANISHI. 1989. Molecular characterization of a functional cDNA for rat substance P receptor. J. Biol. Chem. **264:** 17649–17652.
14. TANAKA, K., M. MASU & S. NAKANISHI. 1990. Structure and functional expression of the cloned rat neurotensin receptor. Neuron **4:** 847–854.
15. ELDE, R., M. SCHALLING, S. CECCATELLI, S. NAKANISHI & T. HÖKFELT. 1990. Localization of neuropeptide receptor mRNA in rat brain: Initial observations using probes for neurotensin and substance P receptors. Neurosci. Lett. **120:** 134–138.
16. VIGNA, S. R., J. J. BOWDEN, D. M. McDONALD, J. FISHER, A. OKAMOTO, D. C. McVEY, D. G. PAYAN & N. W. BUNNETT. 1994. Characterization of antibodies to the rat substance P (NK-1) receptor and to a chimeric substance P receptor expressed in mammalian cells. J. Neurosci. **14:** 834–845.
17. ATWEH, S. F. & M. J. KUHAR. 1977. Autoradiographic localization of opiate receptors in rat brain. II. The brain stem. Brain Res. **129:** 1–12.

18. ATWEH, S. F. & M. J. KUHAR. 1977. Autoradiographic localization of opiate receptors in rat brain. I. Spinal cord and lower medulla. Brain Res. **124:** 53–67.
19. HERKENHAM, M. & C. B. PERT. 1982. Light microscopic localization of brain opiate receptors: A general autoradiographic method which preserves tissue quality. J. Neurosci. **2:** 1129–1149.
20. YOUNG, I., W. S. & M. J. KUHAR. 1979. A new method for receptor autoradiography: [³H]-opioid receptors in rat brain. Brain Res. **179:** 255–270.
21. PASQUINI, F., P. BOCHET, J. C. GARBAY, B. P. ROQUES, J. ROSSIER & A. BEAUDET. 1992. Electron microscopic localization of photoaffinity-labelled delta opioid receptors in the neostriatum of the rat. J. Comp. Neurol. **326:** 229–244.
22. GOUARDERES, C., A. BEAUDET, J. M. ZAJAC, J. CROS & R. QUIRION. 1991. High resolution radioautographic localization of [125I]FK-33-824-labelled mu opioid receptors in the spinal cord of normal and deafferented rats. Neuroscience **43:** 197–209.
23. DELFS, J. M., H. KONG, A. MESTEK, Y. CHEN, L. YU, T. REISINE & M. F. CHESSELET. 1994. Expression of mu opioid receptor mRNA in rat brain: An in situ hybridization study at the single cell level. J. Comp. Neurol. **345:** 46–68.
24. MANSOUR, A., R. C. THOMPSON, H. AKIL & S. J. WATSON. 1993. Delta opioid receptor mRNA distribution in the brain: Comparison to delta receptor binding and proenkephalin mRNA. J. Chem. Neuroanat. **6:** 351–362.
25. MANSOUR, A., C. A. FOX, F. MENG, H. AKIL & S. J. WATSON. 1994. Kappa 1 receptor mRNA distribution in the rat CNS: Comparison to kappa receptor binding and prodynorphin mRNA. Mol. Cell. Neurosci. **5:** 124–144.
26. MANSOUR, A., C. A. FOX, R. C. THOMPSON, H. AKIL & S. J. WATSON. 1994. mu-Opioid receptor mRNA expression in the rat CNS: Comparison to mu-receptor binding. Brain Res. **643:** 245–265.
27. MENG, F., G. X. XIE, R. C. THOMPSON, A. MANSOUR, A. GOLDSTEIN, S. J. WATSON & H. AKIL. 1993. Cloning and pharmacological characterization of a rat kappa opioid receptor. Proc. Natl. Acad. Sci. USA **90:** 9954–9958.
28. XIE, G. X., F. MENG, A. MANSOUR, R. C. THOMPSON, M. T. HOVERSTEN, A. GOLDSTEIN, S. J. WATSON & H. AKIL. 1994. Primary structure and functional expression of a guinea pig kappa opioid (dynorphin) receptor. Proc. Natl. Acad. Sci. USA **91:** 3779–3783.
29. ELDE, R., T. HÖKFELT, O. JOHANSSON & L. TERENIUS. 1976. Immunohistochemical studies using antibodies to leucine-enkephalin: Initial observation on the nervous system of the rat. Neuroscience **1:** 349–351.
30. VON ZASTROW, M. & B. K. KOBILKA. 1992. Ligand-regulated internalization and recycling of human β2-adrenergic receptors between the plasma membrane and endosomes containing transferrin receptors. J. Biol. Chem. **267:** 3530–3538.
31. VON ZASTROW, M., R. LINK, D. DAUNT, G. BARSH & B. KOBILKA. 1993. Subtype-specific differences in the intracellular sorting of G protein-coupled receptors. J. Biol. Chem. **268:** 763–766.
32. EVANS, C. J., D. J. KEITH, JR., H. MORRISON, K. MAGENDZO & R. H. EDWARDS. 1992. Cloning of a delta opioid receptor by functional expression. Science **258:** 1952–1955.
33. KIEFFER, B. L., K. BEFORT, R. C. GAVERIAUX & C. G. HIRTH. 1992. The delta-opioid receptor: Isolation of a cDNA by expression cloning and pharmacological characterization. Proc. Natl. Acad. Sci. USA **89:** 12048–12052.
34. CHEN, Y., A. MESTEK, J. LIU, J. A. HURLEY & L. YU. 1993. Molecular cloning and functional expression of a mu-opioid receptor from rat brain. Mol. Pharmacol. **44:** 8–12.
35. FUKUDA, K., S. KATO, K. MORI, M. NISHI & H. TAKESHIMA. 1993. Primary structures and expression from cDNAs of rat opioid receptor delta- and mu-subtypes. FEBS Lett. **327:** 311–314.
36. THOMPSON, R. C., A. MANSOUR, H. AKIL & S. J. WATSON. 1993. Cloning and pharmacological characterization of a rat mu opioid receptor. Neuron **11:** 903–913.
37. LI, S., J. ZHU, C. CHEN, Y. W. CHEN, J. K. DERIEL, B. ASHBY & C. L. LIU. 1993. Molecular cloning and expression of a rat kappa opioid receptor. Biochem. J. **295:** 629–633.
38. MINAMI, M., T. TOYA, Y. KATAO, K. MAEKAWA, S. NAKAMURA, T. ONOGI, S. KANEKO & M. SATOH. 1993. Cloning and expression of a cDNA for the rat kappa-opioid receptor. FEBS Lett. **329:** 291–295.

39. NISHI, M., H. TAKESHIMA, K. FUKUDA, S. KATO & K. MORI. 1993. cDNA cloning and pharmacological characterization of an opioid receptor with high affinities for kappa-subtype-selective ligands. FEBS Lett. **330:** 77–80.
40. YASUDA, K., K. RAYNOR, H. KONG, C. D. BREDER, J. TAKEDA, T. REISINE & G. I. BELL. 1993. Cloning and functional comparison of kappa and delta opioid receptors from mouse brain. Proc. Natl. Acad. Sci. USA **90:** 6736–6740.
41. ARVIDSSON, U., M. RIEDL, S. CHAKRABARTI, L. VULCHANOVA, J.-H. LEE, A. H. NAKANO, X. LIN, H. H. LOH, T. HÖKFELT, P.-Y. LAW, M. W. WESSENDORF & R. ELDE. 1995. The κ-opioid receptor (KOR1) is primarily postsynaptic: Combined immunohistochemical localization of the receptor and endogenous opioids. Proc. Natl. Acad. Sci. USA. In press.
42. LAMOTTE, C. C., C. B. PERT & S. H. SNYDER. 1976. Opiate receptor binding in primate spinal cord: Distribution and changes after dorsal root section. Brain Res. **112:** 407–412.
43. ALVAREZ, F. J., A. M. KAVOOKJIAN & A. R. LIGHT. 1993. Ultrastructural morphology, synaptic relationships, and CGRP immunoreactivity of physiologically identified C-fiber terminals in the monkey spinal cord. J. Comp. Neurol. **329:** 472–490.
44. CORBETT, A. D., S. J. PATERSON & H. W. KOSTERLITZ. 1993. Selectivity of ligands for opioid receptors. *In* Handbook of Experimental Pharmacology, Opioids I. A. Herz, Ed.: 645–679. Springer-Verlag. Berlin Heidelberg.
45. YOUNG, W. S. I., J. K. WAMSLEY, M. A. ZARBIN & M. J. KUHAR. 1980. Opioid receptors undergo axonal flow. Science **210:** 76–78.
46. FIELDS, H. L., P. C. EMSON, B. K. LEIGH, R. F. GILBERT & L. L. IVERSEN. 1980. Multiple opiate receptor sites on primary afferent fibres. Nature **284:** 351–353.
47. NINKOVIC, M., S. P. HUNT & J. S. KELLY. 1981. Effect of dorsal rhizotomy on the autoradiographic distribution of opiate and neurotensin receptors and neurotensin-like immunoreactivity within the rat spinal cord. Brain Res. **230:** 111–119.
48. LADURON, P. M. & M. N. CASTEL. 1990. Axonal transport of receptors. A major criterion for presynaptic localization. Ann. N.Y. Acad. Sci. **604:** 462–469.
49. BEITNER, J. D. & E. J. NESTLER. 1993. Chronic morphine impairs axoplasmic transport in the rat mesolimbic dopamine system. NeuroReport **5:** 57–60.
50. ZHANG, X., L. BAO, Z.-Q. XU, J. ROPP, U. ARVIDSSON, R. ELDE & T. HÖKFELT. 1994. Localization of neuropeptide Y Y1 receptors in the rat nervous system with special reference to somatic receptors on small dorsal root ganglion neurons. Proc. Natl. Acad. Sci. USA **91:** 11738–11742.
51. CUELLO, A. C., C. MILSTEIN, R. COUTURE, B. WRIGHT, J. V. PRIESTLEY & J. JARVIS. 1984. Characterization and immunocytochemical application of monoclonal antibodies against enkephalins. J. Histochem. Cytochem. **32:** 947–957.
52. WEBER, E., C. J. EVANS & J. D. BARCHAS. 1982. Predominance of the amino-terminal octapeptide fragment of dynorphin in rat brain regions. Nature **299:** 77–79.
53. DOUGLASS, J., C. T. MCMURRAY, J. E. GARRETT, J. P. ADELMAN & L. CALAVETTA. 1989. Characterization of the rat prodynorphin gene. Mol. Endocrinol. **3:** 2070–2078.

Localization of the Peptide Binding Domain of the NK-1 Tachykinin Receptor Using Photoreactive Analogues of Substance P

N. D. BOYD,[a] R. KAGE,[a] J. J. DUMAS,[a] S. C. SILBERMAN,[a]
J. E. KRAUSE,[b] AND S. E. LEEMAN[a]

[a]Department of Pharmacology and Experimental Therapeutics
Boston University School of Medicine
Boston, Massachusetts 02118

[b]Department of Anatomy and Neurobiology
Washington University School of Medicine
St. Louis, Missouri 63110

Substance P is a member of a family of structurally related neuropeptides known collectively as the tachykinins. The tachykinin peptides are involved in diverse biological functions both in the central nervous system and the periphery. In contrast to the smaller classical transmitters such as acetylcholine where functional diversity results from the interaction of a single transmitter with multiple receptor subtypes, a different strategy has evolved for the tachykinin peptide family. For each of the tachykinin peptides a distinct but homologous receptor exists;[1] to date there is no definitive evidence for multiple receptor subtypes. The specific receptor for substance P (SP) has been termed the NK-1 receptor. This receptor binds substance P with an affinity that is several orders of magnitude greater than for all other members of the tachykinin peptide family.[2] To determine the basis for this peptide selectivity, we are attempting to map the residues of the NK-1 receptor that are involved directly in substance P recognition and subsequent transmembrane signaling by this receptor.

For these studies we developed a new approach to photoaffinity labeling that relies on incorporating the photoactivatable benzophenone group at different locations within the 11-amino acid sequence of substance P.[3,4] The use of the benzophenone group as a photolabel has several advantages over more conventional photolabels such as the azide group. These advantages include a high efficiency of photoincorporation and the chemical stability of the benzophenone group under peptide synthesis conditions.[4,5] When photoactivated by exposure to 350 nm light, the carbonyl group of the benzophenone moiety undergoes a n-π^* transition, forming a triplet biradical oxygen. The first and rate determining step involves hydrogen abstraction from a C-H bond followed by formation of a stable C-C bond between the peptide probe and an amino acid residue present at the peptide binding site of the receptor[6,7] (FIG. 1).

p-Benzoyl-L-phenylalanine (Bpa) is an unnatural amino acid that contains a benzophenone side chain. Two photoaffinity peptides were used in this study: (Bpa[8])-SP and (Bpa[4])-SP (FIG. 2). Substitution in position 8 was chosen because studies in our own laboratory and in several others have shown that this position can accommodate a range of aliphatic and aromatic amino acids.[8] In addition, substitu-

405

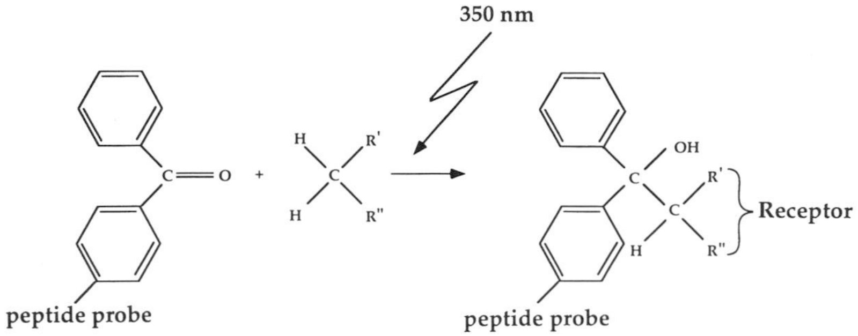

FIGURE 1. Photochemistry of benzophenone.

tion in position 8 places the photoreactive benzophenone group in the conserved C-terminal sequence that defines the tachykinin peptide family and is necessary for interaction and activation of the receptors of these peptides. Substitution in position 4 was selected because it places the photoreactive group in the middle of the divergent N-terminal sequence of substance P. Although the N-terminal sequence is not essential for binding, interactions between residues of this region and the NK-1 receptor are necessary for binding with high subnanomolar affinity and thus contribute to the selectivity of the NK-1 receptor for substance P.[2] Radioiodinated deriva-

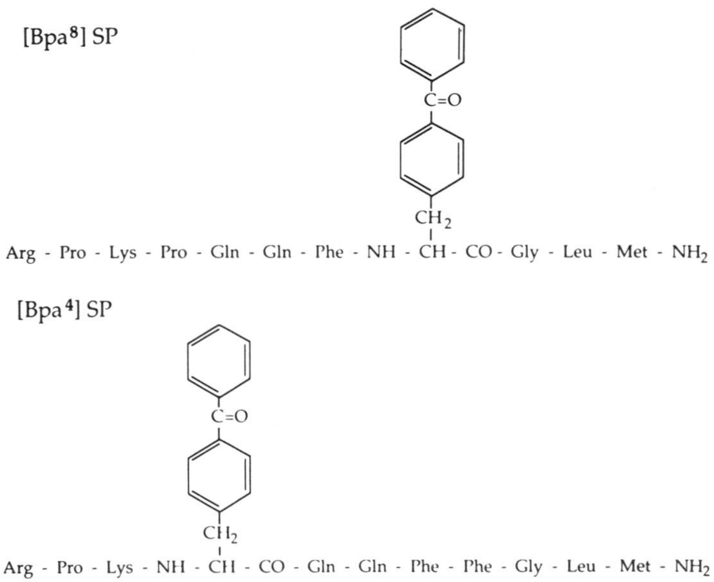

FIGURE 2. Chemical structure of substance P (SP) analogues in which the photoreactive amino acid, *p*-benzoyl-L-phenylalanine (Bpa) has been incorporated in position 4, (Bpa[4])-SP (*lower panel*), and position 8, (Bpa[8])-SP (*upper panel*).

tives of each peptide were prepared using [^{125}I]labeled Bolton Hunter reagent which acylates the ε-amino group of the lysine residue in position 3.[4]

Preliminary binding and functional studies established that both these substance P analogues are agonists that bind to NK-1 receptors with an affinity similar to that of the parent peptide, substance P.[4] However, unlike substance P, both of the Bpa-containing substance P analogues become covalently attached to the receptor upon irradiation with ultraviolet (UV) light. The photoaffinity probes have been used to detect and characterize NK-1 receptors in a variety of tissues.[2] However, for the mapping studies reviewed here, we used as a source of NK-1 receptors a stably transfected Chinese hamster ovary cell line that expresses about 5×10^6 rat NK-1 receptors per cell. In the photolabeling experiment shown in FIGURE 3, NK-1 receptors expressed in CHO cells were photolabeled with [^{125}I]-(Bpa^8)-SP or [^{125}I]-(Bpa^4)-SP. The photolabeled receptors were resolved by SDS-PAGE, and the

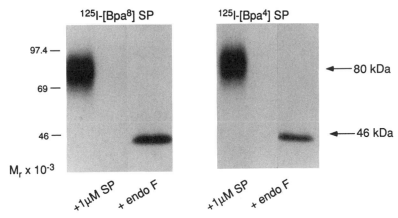

FIGURE 3. Photoaffinity labeling of NK-1 receptors expressed in Chinese hamster ovary (CHO) cells by [^{125}I]-(Bpa^4)-SP and [^{125}I]-(Bpa^8)-SP. Transfected CHO cells expressing the NK-1 receptors were equilibrated with [^{125}I]-(Bpa^4)-SP and with [^{125}I]-(Bpa^8)-SP in the absence (−) and presence (+) of 1 μM of substance P and photolyzed at 4 °C for 10 min with 350 nm light. Membranes prepared from the photolabeled cells were subjected to SDS-PAGE, and the labeled bands were visualized by autoradiography. Photolabeled membranes were also solubilized and treated with endoglycosidase F prior to analysis by SDS-PAGE/autoradiography.

position of the labeled receptor detected by autoradiography. The photolabeled receptors are visualized as a broad radioactive band centered at $M_r = 80,000$. Photolabeling by both probes is highly specific because no radiolabeled bands are detectable when photolabeling is conducted in the presence of an excess of unlabeled substance P. Photolabeling in each case is remarkably efficient; 70–80% of the bound (Bpa^8)-SP derivative and 40–50% of the bound (Bpa^4)-SP derivative undergo photoincorporation upon exposure to UV light. Treatment of the photolabeled receptors with endoglycosidase F to remove asparagine-linked carbohydrates increased the mobility of the photolabeled receptors, which are visualized as a discrete radiolabeled band of $M_r = 46,000$.[9] The M_r value for the deglycosylated receptors is in excellent agreement with the molecular mass calculated for the primary sequence of the rat NK-1 receptor.[9]

MAPPING THE PEPTIDE BINDING POCKET

Determination of the site of incorporation of photoaffinity probes such as [^{125}I]-(Bpa^8)-SP and [^{125}I]-(Bpa^4)-SP identifies directly the region of the receptor that is in contact with a specified position of the peptide. The most straightforward approach to determining the site of insertion of a photoaffinity probe is by fragmentation and sequencing. Prior to preparative scale isolation and sequencing of receptor binding domains, analyses of enzymatic and/or chemical fragmentation procedures conducted on an analytical scale are important for optimization of fragment yield and evaluation of purification steps. Inasmuch as the photoaffinity probes are themselves peptides, a key consideration in the evaluation of potential

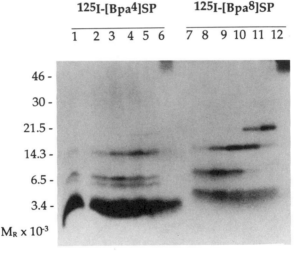

FIGURE 4. Autoradiograph of tryptic fragments of the NK-1 receptor photoaffinity labeled with [^{125}I]-(Bpa^4)-SP and [^{125}I]-(Bpa^8)-SP. Chinese hamster ovary cells expressing the NK-1 receptor were photoaffinity labeled with either [^{125}I]-(Bpa^4)-SP or [^{125}I]-(Bpa^8)-SP. Membranes containing labeled NK-1 receptor were prepared from the cells and were incubated for 60 min at 22 °C with trypsin at the following concentrations (mg/mL): 2, 0.6, 0.2, 0.06, 0.02, and 0; lanes 1–6: [^{125}I]-(Bpa^4)-SP labeled; lanes 7–12: [^{125}I]-(Bpa^8)-SP labeled. The tryptic fragments were resolved by Tricine-SDS-PAGE and visualized by autoradiography.

fragmentation schemes is the requirement that the covalently attached probe remain intact under receptor fragmentation conditions. The results of this type of experiment are shown in FIGURE 4 in which NK-1 receptors expressed in CHO cells were photolabeled with either [^{125}I]-(Bpa^8)-SP or [^{125}I]-(Bpa^4)-SP and subjected to digestion with increasing concentrations of trypsin and the resulting radiolabeled fragments analyzed by Tricine-SDS-PAGE.

Tryptic digestion of receptor photolabeled with [^{125}I]-(Bpa^8)-SP presented a different fragmentation pattern than tryptic digestion of receptor photolabeled with [^{125}I]-(Bpa^4)-SP. These results suggest that different amino acid residues serve as the site of covalent attachment for each photoaffinity probe. The experiments further suggest that by this approach it should be possible to generate sufficient quantities of labeled fragments so that amino acid attachment sites can be identified by purification and microsequencing. Extending this approach to additional substance P derivatives should provide detailed information on the amino acid residues that comprise the peptide binding pocket and thus provide an understanding at the molecular level of peptide-receptor interaction.

REFERENCES

1. NAKANISHI, S. 1991. Mammalian tachykinin receptor. Annu. Rev. Neurosci. **14**: 123–136.
2. INGI, T., Y. KITAJIMA, Y. MINAMITAKE & S. NAKANISHI. 1991. Characterization of ligand-binding properties and selectivities of three rat tachykinin receptors by transfection and functional expression of their cloned cDNAs in mammalian cells. J. Pharmacol. Exp. Ther. **259**: 969–975.
3. BOYD, N. D., C. F. WHITE, R. CERPA, E. T. KAISER & S. E. LEEMAN. 1991. Photoaffinity labeling the substance P receptor using a derivative of substance P containing *p*-benzoylphenylalanine. Biochemistry **30**: 336–342.
4. BOYD, N. D., S. G. MACDONALD, R. KAGE, J. LUBER-NAROD & S. E. LEEMAN. 1991. Substance P receptor, biochemical characterization and interactions with G proteins. Ann. N.Y. Acad. Sci. **632**: 79–93.
5. KAUER, J. C., S. ERICKSON-VIITANEN, H. R. WOLFE, JR. & W. R. DEGRADO. 1986. *p*-Benzoyl-L-phenylalanine, a new photoreactive amino acid. J. Biol. Chem. **261**: 10695–10700.
6. BRESLOW, R. 1980. Acc. Chem. Res. **13**: 170–177.
7. TURRO, N. 1978. Modern Molecular Photochemistry.: 372–377. Benjamin/Cummings. Menlo Park, CA.
8. CASCIERI, M. A., R.-R. C. HUANG, T. M. FONG, A. H. CHEUNG, S. SADOWSKI, E. BER & C. D. STRADER. 1992. Determination of the amino acid residues in substance P conferring selectivity and specificity for the rat neurokinin receptors. Mol. Pharmacol. **41**: 1096–1099.
9. KAGE, R., A. D. HERSHEY, J. E. KRAUSE, N. D. BOYD & S. E. LEEMAN. 1995. Characterization of the substance P (NK-1) receptor in tunicamycin-treated transfected cells using a photoaffinity analog of substance P. J. Neurochem. **64**: 316–321.

Effects of Peptide YY on CCK/CCK Antagonist Interactions in Cerulein-induced Pancreatic Injury

JOSEPH M. TITO, MAREK RUDNICKI,[a]
DIANE C. ROBINSON, WILLIAM B. GUINEY,
THOMAS E. ADRIAN,[b] AND MICHAEL S. GOLD

[a] The Mary Imogene Bassett Hospital
Cooperstown, New York 13326

[b] Department of Biomedical Sciences
Creighton University
Omaha, Nebraska 68178

Cholecystokinin (CCK) receptors have been shown to mediate CCK-induced pancreatic injury.[1] The beneficial effect of the CCK receptor antagonist L-364,718 on acute pancreatitis is well established.[2] We previously demonstrated an amelioration of CCK-induced pancreatic parenchymal changes in rats treated with peptide YY (PYY).[3] PYY is an ileocolonic inhibitor of pancreatic secretion that inhibits CCK-mediated effects on the pancreas.[4]

The aim of this study was to investigate the combined effects and possible interactions of PYY and L-364,718 on cerulein-induced pancreatitis.

METHODS

Twenty-four Sprague-Dawley rats underwent chronic cannulations of the jugular vein and carotid artery for drug infusion and blood sampling. After an overnight recovery from surgery they were randomly assigned to groups receiving: CCK amphibian analog cerulein (10 μg/kg/h); cerulein + L-364,718 (20 μg/kg/h); or cerulein + L-364,718 + PYY (400 pmol/kg/h). Rats in the control group received saline (1.6 cc/h). Test solutions were infused over the first 6 h of the study. Blood samples for plasma amylases were taken prior to any intervention and at 1, 3, 6, 9, and 24 h after beginning of treatment. At the conclusion of the study, the animals were sacrificed and the pancreata taken for hematoxylin and eosin staining.

RESULTS

Pancreatic specimens taken from rats infused with cerulein alone revealed a significant degree of inflammation, vacuolization, and necrosis. No differences were found among morphological pictures of the pancreatic specimen taken from rats receiving other treatments. Cerulein infusion increased plasma amylase levels (FIG. 1).

[a] Address correspondence to Marek Rudnicki, M.D., Department of Surgery, Mary Imogene Bassett Hospital, Cooperstown, NY 13326-1394.

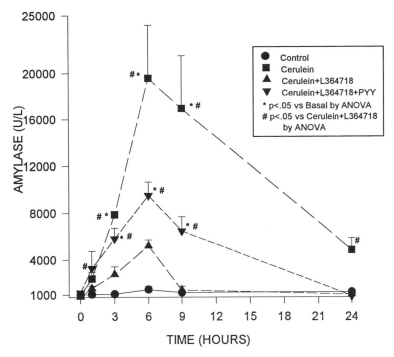

FIGURE 1. Plasma amylase levels versus time.

The CCK antagonist, L-364,718, almost completely blocked amylase release in rats treated with cerulein. PYY, when given simultaneously with cerulein and L-364,718, reduced this effect and increased plasma amylase activities at 1, 3, 6, and 9 h.

DISCUSSION

In the presence of exogenous PYY, the beneficial effect of L-364,718 on cerulein-induced hyperamylasemia was attenuated. PYY was demonstrated to decrease CCK-induced hyperamylasemia in rats.[3] Inasmuch as PYY does not alter unstimulated amylase release, it is suggested that this peptide interacts with the action of L-364,718 on intrapancreatic acinar CCK receptors overstimulated with cerulein. The demonstration of a diversity in the affinity status of CCK receptors mediating pancreatic response to cerulein implies that PYY can interfere with L-364,718 binding to these receptors.[5] Therefore, it might be concluded that exogenous PYY may diversify the functional activity of CCK receptors during cerulein-induced pancreatitis.

REFERENCES

1. SALUJA, A. K., M. SALUJA, H. PRINTZ, A. ZAVERTNIK, A. SENGUPTA & M. L. STEER. 1989. Proc. Natl. Acad. Sci. USA **86:** 8969–8971.

2. MURAYAMA, K. M., J. B. DREW, D. L. NAHRWOLD & R. J. JOEHL. 1990. Pancreas 5: 439–444.
3. TITO, J. M., M. RUDNICKI, D. H. JONES, H. D. ALPERN & M. S. GOLD. 1993. Am. J. Surg. 165: 690–696.
4. LLUIS, F., G. GOMEZ, K. FUJIMURA, G. H. GREELEY & J. C. THOMPSON. 1988. Gastroenterology 94: 137–144.
5. GOMEZ, G., R. D. BEAUCHAMP, S. RAJAMARAN, L. PADILLA, P. SINGH, G. H. GREELEY & J. C. THOMPSON. 1993. Gastroenterology 105: A305.

Aglyco Pathology of Viral Receptors in Dementias

S. BOGOCH AND E. S. BOGOCH

Center for Neurochemistry
The Nathan S. Kline Institute for Psychiatric Research
Orangeburg, New York 10962

Foundation for Research on the Nervous System
and Brain Research Inc.
Boston, Massachusetts 02215

Six previously isolated sets of studies have been brought together to define a condition we have named *aglyco pathology*[a]: (1) Glycoconjugate receptor structures in brain are known sites for viral attachment, and influenza viral neuraminidase cleaves neuraminic acid from glycoreceptors during virus entry into the cell. Thus, for example, brain gangliosides *in vitro* were shown to act as receptors for influenza virus, and *in vivo* additional gangliosides administered intracerebrally acted as a "decoy" and inhibited infection of brain neurons by influenza virus.[1-7] (2) Quantitative neurochemical studies suggested that covalently bound neuraminic acid (NA) and hexosamine (HA) in brain normally are part of a glycoconjugate intercellular recognition "sign-post" system, which forms the neural networks underlying normal brain development and behavior.[8,9] (3) A quantitative decrease in glycoconjugates, in conjugated NA and HA, occurs in schizophrenia.[10-25] That nondialyzable conjugated NA and HA in cerebrospinal fluid (CSF) are quantitatively decreased in schizophrenia is here confirmed in a double-blind study. (4) Recent epidemiological studies indicate that infection prenatally with influenza virus predisposes to schizophrenia.[26-28] (5) Recent histological and brain-scan studies in schizophrenia reveal neuronal loss.[29,30] (6) An example of aglyco pathology has been demonstrated in glioblastomas, in which the decrease of carbohydrate in the glycoconjugate brain glycoprotein 10B exposes epitopes in a constituent peptide of 10B called malignin; these epitopes in turn induce a quantified cytotoxic auto-antibody response, anti-malignin antibody (FIG. 1), with resultant destruction of the cells containing the newly exposed epitopes at picograms of antibody per cell.[31-33] This mechanism operates in other malignancies as well (FIG. 1). It is proposed that neuronal cell loss in dementia praecox (schizophrenia) and other dementias, such as those which occur in Alzheimer's disease and parkinsonism, is produced by similar aglyco pathology of viral receptors.

Aglyco pathology can be detected by quantitative determination either of the altered glycoconjugates themselves or of the cytotoxic antibody which is produced against the newly exposed epitopes. The quantitative determination of the altered glycoconjugates themselves is demonstrated in schizophrenia (FIG. 2). To be described in detail elsewhere, CSF from schizophrenic and nonschizophrenic patients at the National Institute of Mental Health, Bethesda was shipped blind in dry ice to the Foundation for Research on the Nervous System in Boston where conjugated NA and HA were determined. Each of the specimens was lyophilized, dialyzed

---

[a] Aglyco is a trademark of Aglyco, Inc.

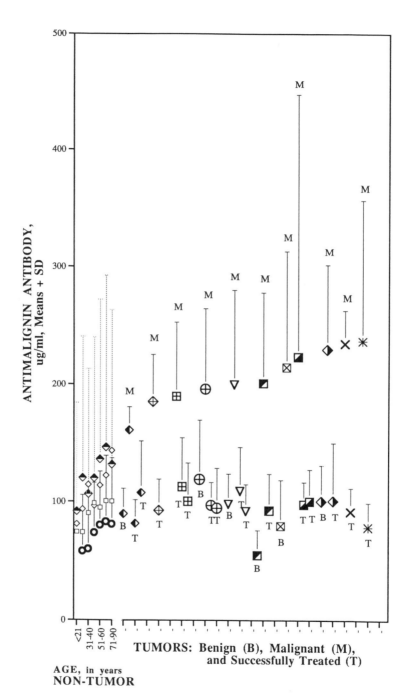

FIGURE 1. Antimalignin antibody in non-tumor and tumor populations. Number in each group appears in parentheses. **Non-tumor:** ○ Normal healthy controls (1,972); □ Screen: Unknown family history (732); ◇ Screen: +ve Family history, asymptomatic (193); ◆ Screen: +ve Family history, symptomatic (181). **Tumor:** ◇ Ovary (58); ◆ Melanoma (20); □ Colorectal (99); ○ Breast (600); ▽ Prostate (80); ◪ Genitourinary (47); ⊠ Brain (104); ◪ Lung (62); ◀ Uterus (46); ✕ Basal cell, skin (11); ✱ Lymphoma-leukemia (73). **Total number: 4,278.**

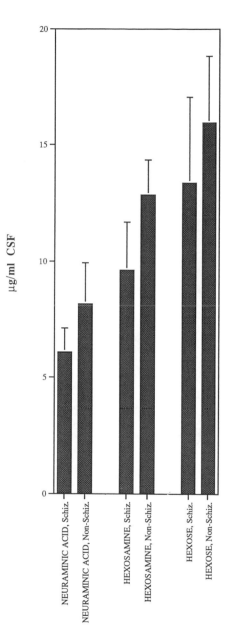

FIGURE 2. Glycoconjugate neuraminic acid, hexosamine, and hexose, in cerebrospinal fluid (CSF) of schizophrenic (Schiz.) and nonschizophrenic (Non-Schiz.) patients, in μg/mL CSF, mean ± SD.

against distilled water exhaustively (cellophane pore size approx. MW 12,000 Da) at 0–5 °C to remove free NA and free hexose, and the nondialyzable fraction was quantitatively analyzed in duplicate for conjugated NA by the Bial's orcinol method and the thiobarbituric acid method (again shown to be unreliable), and for conjugated HA and glucose as previously described.[10–25] After the neurochemical tests were completed, the code was broken. The results are shown in FIGURE 2. Conjugated NA ($p < 0.025$; 81.8% of the schizophrenic group < 7.5 μg/mL; 83.3% of the nonschizophrenic group > 7.5 μg/mL) and conjugated HA ($p < 0.006$), but not conjugated hexose, were statistically significantly lower in concentration in the schizophrenic than the nonschizophrenic group.

REFERENCES

1. BOGOCH, S. 1957. J. Am. Chem. Soc. **79:** 3286.
2. BOGOCH, S. 1958. Biochem. J. **68:** 319–324.
3. BOGOCH, S. 1957. Virology **4:** 458–462.
4. BOGOCH, S. 1970. *In* Protein Metabolism of the Nervous System. A. Lajtha, Ed.: 555–569. Plenum Press. New York.
5. BOGOCH, S., M. K. PAASONEN & U. TRENDELENBURG. 1962. Br. J. Pharmacol. **18:** 325–329.
6. BOGOCH, S., P. C. RAJAM & P. C. BELVAL. 1964. Nature **204:** 73–75.
7. BOGOCH, S., P. C. BELVAL, W. H. SWEET, W. SACKS & G. KORSH. 1968. Protides Biol. Fluids **15:** 129–135.
8. BOGOCH, S. 1968. The Biochemistry of Memory: With an Inquiry into the Function of the Brain Mucoids. Oxford University Press. New York.
9. BOGOCH, S. 1975. *In* The Nervous System. National Institute of Neurological Diseases and Communicative Disorders and Stroke. D. B. Tower, Ed. Vol. **1:** 591–600. Raven Press. New York.
10. BOGOCH, S. 1957. Am. J. Psychiatry **114:** 172.
11. BOGOCH, S. 1958. AMA Arch. Neurol. Psychiatry **80:** 221–224.
12. BOGOCH, S. 1959. Nature **184:** 1628–1629.
13. BOGOCH, S. 1960. J. Biol. Chem. **235:** 16–20.
14. BOGOCH, S. 1960. Am. J. Psychiatry **116:** 743–747.
15. BOGOCH, S., K. T. DUSSIK & P. G. LEVER. 1959. AMA Arch. Gen. Psychiatry **1:** 441–446.
16. BOGOCH, S., K. T. DUSSIK, C. FENDER & P. CONRAN. 1960. Am. J. Psychiatry **117:** 409–415.
17. BOGOCH, S., W. SACKS & G. SIMPSON. 1963. Neurology **13:** 355.
18. BOGOCH, S., K. T. DUSSIK & P. CONRAN. 1961. N. Engl. J. Med. **264:** 251–258.
19. BOGOCH, S., P. C. BELVAL, K. T. DUSSIK & P. CONRAN. 1962. Am. J. Psychiatry **119:** 128–133.
20. BOGOCH, S., P. EVANS. 1962. Nature **195:** 180.
21. CAMPBELL, R. J., S. BOGOCH, M. J. SCOLARO & P. C. BELVAL. 1967. Am. J. Psychiatry **123:** 952–962.
22. ROBINS, E., A. B. CRONINGER, M. K. SMITH & A. C. MOODY. 1962. Ann. N.Y. Acad. Sci. **96:** 390–391.
23. CHRISTONI, G. & R. ZAPPOLI. 1960. Am. J. Psychiatry **117:** 246.
24. PAPADOPOULOS, N. M., J. E. MCLANE, D. O'DOHERTY & W. C. HESS. 1959. J. Nerv. Ment. Dis. **128:** 450.
25. SIROTA, P., H. BESSLER, D. ALLALOUF, M. DALDETTI & H. LEVINSKY. 1988. Prog. Neuro-Psychopharmacol & Biol-Psychiatry **12(1):** 103–107.
26. O'CALLAGHAN, E. O., P. SHAM, N. TAKEI & R. M. MURRAY. 1991. Lancet **337:** 1248–1250.
27. MEDNICK, S. A., R. A. MACHON, M. O. HUTTENNEN & D. BONNET. 1988. Arch. Gen. Psychiatry **45:** 189–192.
28. O'CALLAGHAN, E. O., P. SHAM, N. TAKEI, G. MURRAY, G. GLOVER, E. HARE & R. M. MURRAY. 1993. Schizophr. Res. **9:** 138.
29. BLOOM, F. E. 1993. Arch. Gen. Psychiatry **50:** 224–227.
30. CONRAD, A. J. & A. B. SCHEIBEL. 1987. Schizophr. Bull. **13(4):** 577–587.

31. BOGOCH, S. & E. S. BOGOCH. 1991. Lancet **337:** 977.
32. ABRAMS, M. B., K. T. BEDNAREK, S. BOGOCH, E. S. BOGOCH, H. J. DARDICK, R. DOWDEN, S. C. FOX, E. E. GOINS, G. GOODFRIEND, R. A. HERRMAN, J. IMPERIO, W. JACKSON, S. KEUER, M. KILLACKEY, G. KIMEL, R. E. LAYTON, A. H. LIEBENTRITT, D. MARSDEN, D. MCCABE, D. MENASHA, K. ORTEN, M. PASMANTIER, T. PILLAI, V. B. PILLAI, W. PROBST, W. REIMER, S. SMITH, J. THORNTHWAITE, W. J. TURNER & R. T. WHITLOCK. 1994. Canc. Detect. Prev. **18(1):** 65–78.
33. BOGOCH, S. & E. S. BOGOCH. 1994. J. Cell. Biochem. **19:** 172–185.

Excitatory Amino Acid Metabotropic Receptor Subtypes and Calcium Regulation

MAX RÉCASENS[a] AND MICHEL VIGNES

Institut National de la Santé et de la Recherche Medicale (INSERM)
Unité 254
Hôpital Saint Charles
300 rue A. Broussonnet
34295 Montpellier Cedex 5, France

Calcium (Ca^{2+}) possesses a pivotal role in a large variety of cellular processes. In neurons, it represents an essential step in the mechanisms triggering synaptic plasticity, long-term potentiation,[1,2] long-term depression,[3–5] cytoskeletal organization,[6] exocytosis,[7,8] and delayed neurotoxicity.[9] Considering all these distinct fundamental roles of Ca^{2+}, one may logically suppose that they result from specific changes in intracellular Ca^{2+} concentration, which differ in their amplitude, duration, and location. All these subcellular Ca^{2+} variations should be tightly regulated to avoid the emergence of uncontrolled mechanisms, often leading ultimately to cell death. Many systems contribute to this regulation, namely, membrane sodium (Na^+)/Ca^{2+} exchangers, Ca^{2+} pumps (Ca^{2+}-ATPases), Ca^{2+} binding proteins, and Ca^{2+} channels. The later category could be divided into five main subtypes: voltage-dependent Ca^{2+} channels (VDCC), receptor-operated Ca^{2+} channels (ROC), G-protein-operated Ca^{2+} channel (GOC), second messenger-operated Ca^{2+} channels (SMOC), and finally Ca^{2+} release-activated channel (CRAC), as recently proposed[10] (FIG. 1). Glutamate (Glu), which is the main excitatory neurotransmitter in the brain, may directly or indirectly modulate most of the above-mentioned Ca^{2+} channels. Its action occurs via two main classes of receptors, namely, the ionotropic and the metabotropic receptors as shown in FIGURE 2. Ionotropic Glu receptors (iGluR) are composed of subunit proteins, which form an integral ligand-gated ion channel. iGluR can be subdivided into two main categories, the *N*-methyl-D-aspartate (NMDA) receptors and the non-NMDA ones. The NMDA receptor family are receptor-channels, permeable to Ca^{2+} (ROC). The non-NMDA receptors, composed of α-amino-3-hydroxy-5-methyl-4-isoxazole propionic acid (AMPA) receptors and kainic acid (KA) receptors, are also receptor-channels, generally almost impermeable to Ca^{2+}. The exception to this last statement is the existence of AMPA receptors lacking the GluR2 subunit, which are permeable to Ca^{2+}. The activation of the non-NMDA receptors produces a depolarization, which opens the VDCC. The second class of receptors, the metabotropic glutamate receptors (mGluR) are linked to G-proteins. Their stimulation generates the formation of second messengers and/or regulates ion channel function. Molecular cloning by cross-hybridization and polymerase chain reaction (PCR) has revealed the existence of at least seven subtypes of mGluR.[11–16] The mGluR can be subdivided into three main subgroups according to DNA sequence similarities, receptor-associated signal transduction, and the agonist selectivities[17] (FIG. 2). In agreement with their high sequence

---

[a] Corresponding author.

homology, both mGluR1 and mGluR5 stimulate the phosphoinositide metabolism, leading to IP3 synthesis and Ca^{2+} mobilization from the endoplasmic reticulum. The most potent agonist of these receptors is quisqualate (QA), as previously reported,[18] whereas *trans*-1-aminocyclopentane-1,3-dicarboxylic acid (*t*-ACPD) is a very weak agonist. These receptors also induce arachidonic acid release. The two other mGluR subgroups are composed of receptors coupled to the inhibition of forskolin-

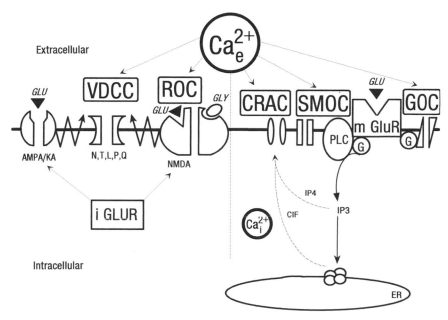

FIGURE 1. Schematic representation of possible mechanisms and pathways of Ca^{2+} influx by excitatory amino acid receptor activation. Ca^{2+} influx may occur through voltage-dependent Ca^{2+} channels (VDCC), receptor-operated Ca^{2+} channels (ROC), Ca^{2+}-release activated Ca^{2+} channels (CRAC), second messenger-operated Ca^{2+} channels (SMOC), and G-protein-operated Ca^{2+} channels (GOC). Ionotropic glutamate receptors (iGluR) are all able to activate VDCC, whereas only the *N*-methyl-D-aspartate (NMDA) is permeable to Ca^{2+} (ROC). The metabotropic glutamate receptor (mGluR) could be associated via a G-protein to GOC or could activate indirectly CRAC or SMOC. The SMOC activation may be due to the inositol-1,4,5-trisphosphate (IP3) synthesis from membrane inositol phospholipids, catalyzed by the phospholipase C enzymes (PLC), or inositol-1,3,4,5-tetrakisphosphate (IP4), or Ca^{2+}, or Ca^{2+} influx factor (CIF). CRAC may result from emptying the endoplasmic reticulum (ER) Ca^{2+} pool. Other abbreviations: Glu, glutamate; Gly, glycine; AMPA, RS-α-amino-3-hydroxy-5-methyl-4-isoxazole propionate; KA, kainate.

stimulated adenylate cyclase activity. Their discrimination arises from their agonist selectivity. mGluR2 and mGluR3 are potently activated by *t*-ACPD and Glu, whereas QA is almost ineffective. mGluR4, mGluR6, and mGluR7 are preferentially activated by 2-amino-4-phosphonobutyric acid (AP4), which is more potent than Glu by one order of magnitude. *t*-ACPD is a very weak agonist at these receptors, and QA presents almost no agonist activity.

    Altogether, these facts indicate that Ca^{2+} represents a converging step for most, if not all, excitatory amino acid (EAA) receptor types. In turn, Ca^{2+} ions possess numerous targets: calmodulin, which activates Ca^{2+} calmodulin kinase; protein kinases C (PKC), which are translocated from the cytosol to the plasma membrane; phospholipase C; phospholipase A2; early genes; and channels. As feedback, all the stimulated targets may directly or indirectly regulate the activity of the EAA receptor subtypes. Indeed, EAA receptors themselves could serve as substrates for the various kinases, which modulate their respective activity by phosphorylation. For instance, PKC activation is known to enhance NMDA currents.[19] The phosphoryla-

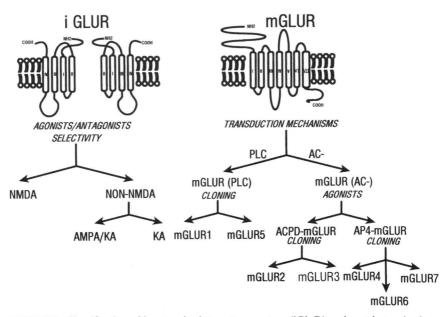

FIGURE 2. Classification of ionotropic glutamate receptors (iGluR) and metabotropic glutamate receptors (mGluR). iGluR are divided into N-methyl-D-aspartate (NMDA) and non-NMDA types according to their respective agonist and antagonist selectivity. The non-NMDA is further subdivided into AMPA (RS-α-amino-3-hydroxy-5-methyl-4-isoxazole propionate)/KA (kainate) and KA with respect to the agonist selectivity of these two substances. The mGluR are classified into two groups—those linked to the phospholipase C (PLC) enzymes, named mGluR (PLC), and those coupled to the inhibition of adenylate cyclase (AC-), named mGluR (AC-). This latter group is made up of two types, defined by the agonist selectivity of 2-amino-4-phosphonobutyrate (AP4) and *trans*-1-aminocyclopentane-1,3-dicarboxylate (ACPD). Finally, a last subdivision is obtained from receptor cloning, suggesting that up to seven receptor types exist.

tion of several distinct sites on the NR1 subunit of the NMDA receptor may be catalyzed by PKC.[20,21] Consequently, Ca^{2+} represents one of the main centers for the cross talk between most, if not all, EAA receptor subtypes. This concept could likely be extended to many other neurotransmitter receptors.

This paper, however, is restricted to the role of EAA metabotropic receptors linked to phospholipase C (PLC) in the regulation of intracellular Ca^{2+} concentration ($[Ca^{2+}]_i$) during development and, reciprocally, on the modulation by Ca^{2+} of the EAA-stimulated inositol phosphate (IP) metabolite formation. The activation of

PLC-linked receptors is known to induce the formation of two second messengers, namely, inositol 1,4,5-trisphosphate (IP3) and diacylglycerol (DAG) from the cleavage of membrane inositol lipid, phosphatidylinositol 4,5-bisphosphate (PIP2).[22,23] Whereas DAG activates protein kinase C (for a review, see ref. 24), IP3 releases Ca^{2+} from intracellular organelles possessing IP3 receptors, such as the endoplasmic reticulum.[25] However, it became apparent that IP3-sensitive Ca^{2+} stores have a limited capacity and could not completely account for the substantial $[Ca^{2+}]_i$ increase, subsequent to EAA receptor activation. Moreover, it was reported that, in many cell types, receptor-mediated $[Ca^{2+}]_i$ augmentation invariably results from both a Ca^{2+} mobilization from intracellular stores and a Ca^{2+} influx from the extracellular medium.[26-32] The mechanism(s) as well as the molecular nature of the channel(s) that mediate IP3-induced Ca^{2+} influx remain unknown. Two main hypotheses have been proposed.[33] First, IP3 and/or inositol 1,3,4,5-tetrakisphosphate (IP4) directly gate a Ca^{2+} channel on the plasma membrane.[34] The other alternative, termed the capacitive Ca^{2+} entry hypothesis, is that emptying the intracellular Ca^{2+} stores by IP3 represents the triggering stimulus for Ca^{2+} entry (FIG. 1). This later hypothesis necessitates a communication between the endoplasmic reticulum, which represents the store to be emptied, and the plasma membrane, where the Ca^{2+} channel is located. The three possibilities proposed[33] for this communication are a diffusible messenger, a signal given by a low Ca^{2+} concentration near the plasma membrane, and protein-protein interaction (a direct interaction between IP3 receptors and Ca^{2+} channel, for instance). The first possibility is supported by the discovery of a soluble small molecule, named Ca^{2+} influx factor (CIF)[35] and by electrophysiological patch-clamp experiments.[36] The second possibility is far less substantiated although it has been demonstrated that decreasing cytosolic Ca^{2+} could increase Ca^{2+} entry.[37] The third proposal could be evidenced by the fact that intracellular Ca^{2+} increases the sensitivity of IP3 receptors, likely directly associated with the plasma membrane Ca^{2+} channel and to IP3, which may represent a Ca^{2+} sensor for evaluating the Ca^{2+} level of the store.[38,39] The involvement of G-proteins has also been proposed on the basis of whole-cell patch-clamp experiments showing that GTPγS, but not ALF_4^-, prevents the Ca^{2+} influx.

We report evidence indicating that, in nerve terminals, the intracellular Ca^{2+} release from IP3-sensitive stores following EAA metabotropic receptor activation directly or indirectly produces a depolarization, likely by opening a Ca^{2+}-activated nonspecific cation channel (CAN). Subsequently, the activation of this CAN channel leads to an extracellular Ca^{2+} influx through a new voltage-dependent, Mn^{2+} impermeable Ca^{2+} channel. This influx activates Ca^{2+}-sensitive PLC, which results in an increased IP3 production.

CALCIUM INCREASE MAINLY RESULTS FROM EXCITATORY AMINO ACID METABOTROPIC RECEPTOR ACTIVATION IN SYNAPTONEUROSOMES

We previously showed that all the classical antagonists of the iGluR, namely, D-2-amino-5-phosphonovalerate (APV), 6-nitro-7-cyanoquinoxaline-2,3 dion (CNQX), glutamate-aminomethyl-sulfonate, and γ-D-glutamylglycine neither affected the QA-, Glu- nor the t-ACPD-induced IP formation in 8-day-old rat forebrain synaptoneurosomes.[18,40] The agonists QA, t-ACPD, and Glu produce a dose-dependent increase in intracellular Ca^{2+} (reaching up to 70 nM) as measured by Fura-2 fluorescence. The pharmacology of this mGluR agonist-elicited Ca^{2+} increase is highly correlated with that reported for the stimulation of IP metabolism[41]

(FIG. 3). Moreover, the iGluR antagonists did not block the QA-, Glu- or t-ACPD-induced Ca^{2+} increase.

Taken together these data strongly suggest that the intracellular Ca^{2+} rise in synaptoneurosomes essentially originates from mGluR activation.

ORIGIN OF Ca^{2+} PRODUCING THE INTRACELLULAR Ca^{2+} RISE

QA (10 μM) induces a long-lasting increase in intracellular Ca^{2+} of about 70 nM in the presence of an extracellular Ca^{2+} concentration ($[Ca^{2+}]_e$) of 150 μM. In the presence of 150 μM of the Ca^{2+} chelator 1,2-bis (2-aminophenoxy)ethane-N,N,N',N'-tetraacetic acid (BAPTA), which reduces the $[Ca^{2+}]_e$ to 0.3 μM, QA only elicits a transient intracellular Ca^{2+} rise of 16 nM, which is no longer measurable after a 5-min delay. This indicates that the major long-lasting $[Ca^{2+}]_i$ increase is due to a Ca^{2+} influx from the extracellular medium (TABLE 1).

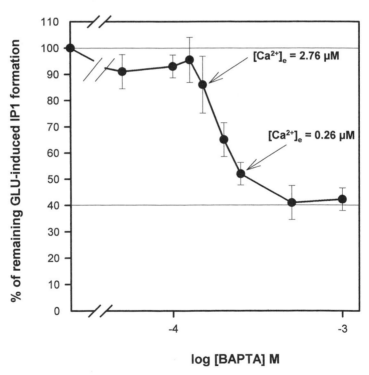

log [BAPTA] M

FIGURE 3. Effect of increasing concentrations of the Ca^{2+} chelator (BAPTA) on the IP1 formation elicited by Glu (1 mM). The IP1 formation elicited by Glu in the presence of 150 μM extracellular Ca^{2+} was taken as 100%. The values are the mean ± SD of three to four separate experiments. The method was previously described.[18] Briefly, 8-day-old rat forebrain synapto-neurosomes, preloaded with [³H]-myo-inositol for 1 h at 37 °C (about 1 mg protein/mL) were incubated at 37 °C for 13 min in the presence of Li+ (10 mM). The stimulation by QA lasts for a further 20 min, then the inositol phosphates (IPs) were extracted and separated by ion-exchange chromatography, and the radioactivity contained in the IP1 eluate was counted.

TABLE 1. Intracellular Ca^{2+} Concentration Variations Elicited by Quisqualate in the Presence of Different Extracellular Ca^{2+} Concentrations

mGluR Agonist	$\Delta[Ca^{2+}]_i$ (nM) (1 min after QA application)	$\Delta[Ca^{2+}]_i$ (nM) (5 min after QA application)	$[Ca^{2+}]_e$ (μM)
QA (10 μM)	65 ± 7	71 ± 6	150
QA (10 μM)	16 ± 5	0 ± 2	0.3

Ca^{2+} INFLUX AND ITS CHARACTERISTICS

It has been extensively demonstrated that an influx of Ca^{2+}, either due to the action of depolarizing agents or to Ca^{2+} ionophores, stimulates the IP production.[42-47] Using increasing concentrations of BAPTA to reduce the $[Ca^{2+}]_e$, we found that about 60% of the Glu-induced inositol monophosphate (IP1) accumulation is due to the Ca^{2+} influx from the extracellular medium (FIG. 3). The remaining response (40%) corresponds to the direct IP1 accumulation solely resulting from the mGluR activation.

The VDCC antagonists, including verapamil, nifedipine, ω-conotoxin, flunarizine, ω-agatoxin IVA, and SC 38249 (a gift of Prof. J. Meldolesi) neither affect the Ca^{2+} influx nor the EAA-elicited IP production. This clearly indicates that the voltage-dependent N, L, T, and P type channels are not involved in these two effects.

At 100 μM, Zn^{2+}, Ni^{2+}, Co^{2+}, Ba^{2+}, and La^{3+} are without effect on the Glu-induced IP metabolism, whereas Mn^{2+} only slightly affects this response. Cd^{2+} and Hg^{2+} totally inhibit the Glu-elicited IP response.[47,48] However, this effect does not appear to be due to competitive blockade of the Ca^{2+} channel by these two ions. Cd^{2+} and Hg^{2+} most probably bind to free -SH groups of proteins involved in the transduction system.[48]

Mn^{2+}, which usually crosses the plasma membrane by the same pathways as Ca^{2+}—in particular by the VDCC—and quenches the Ca^{2+}-Fura-2 fluorescences,[49] is often used as a surrogate ion to monitor Ca^{2+} entry. However, in synaptoneurosomes, Mn^{2+} does not decrease the Ca^{2+}-Fura-2 fluorescence stimulated by QA. This further reinforces the conclusion that the Ca^{2+} influx does not occur via a classical VDCC.

Using stopped flow spectrofluorimetry, we demonstrated that the Ca^{2+} influx triggered by QA (10 μM) is delayed by about 300 ms as compared to that elicited by K^+ (30 mM) ions or the Ca^{2+} ionophore A23187 (2 μM).[41] This specific kinetic likely indicates that the Ca^{2+} influx is not directly mediated by the metabotropic receptor associated with a Ca^{2+} channel via a G-protein (GOC). Rather this delayed influx suggests the involvement of a multistep mechanism in which IP3 participates. It has previously been reported that IP3 synthesis requires about 150 ms after odorant stimulation of the olfactory cilia[50] and about 200 ms after thrombin stimulation of the human platelets.[51]

It does not appear that CRAC channels are involved in this response because La^{3+} ions, known to inhibit the current carried through CRAC channels, have no effect on the IP formation elicited by mGluR agonists[47] (TABLE 2).

Quinine and quinidine, able to inhibit CAN channels,[52] block in a dose-dependent manner the QA-elicited IP formation (data not shown). This result suggests the involvement of such CAN channels in the calcium entry, which follows

TABLE 2. Comparison of Ca^{2+} Permeable Channels[a]

Channel	Trigger Direct or Indirect	Conductance	Permeation	Inhibitors
CRAC	EGTA BAPTA IP3 Ionomycin Thapsigargin	20 fS	$Ca^{2+} > Ba^{2+}, Sr^{2+}$ $> Mn^{2+}$	Cd^{2+} La^{3+} Imidazole derivatives Cytochrome P450 Inhibitors (econazole)
SMOC	Ca^{2+} IP3 IP4	2–25 pS	$Ba^{2+} \sim Ca^{2+}$ $\sim Mn^{2+}$	?
CAN	IP3 Thapsigargin Ca^{2+}	10–50 pS	Ca^{2+}, Na^+, K^+	Quinine Quinidine Replacement of Na^+ for NMDG

[a]CRAC, Ca^{2+} release-activated Ca^{2+} channel; SMOC, second messenger-operated Ca^{2+} channel; CAN, Ca^{2+}-activated nonspecific cation channel.

EAA metabotropic receptor activation. Moreover, substitution of Na^+ ions for N-methyl-D-glucamine, which also blocks this type of channel, inhibits the QA-evoked IP response.[46]

FACTORS THAT TRIGGER THE Ca^{2+} INFLUX

Taking into account the results from the rapid kinetics of the Ca^{2+} influx, one could accept that the synthesis of IP3 is required to trigger the Ca^{2+} influx. The next question then is, does IP3 activate the Ca^{2+} influx directly or indirectly, via the released Ca^{2+} from the endoplasmic reticulum pool, via the IP4 synthesis, or via the release of a diffusible factor such as CIF?[35] To answer this question, we by-passed the whole transduction system associated with the EAA metabotropic receptor and the synthesis of IP3. This was performed thanks to thapsigargin, which blocks the Ca^{2+} ATPase responsible for pumping the cytosolic Ca^{2+} into the endoplasmic reticulum.[52,53] This leads to an increase in intracellular Ca^{2+}, independent of the IP3 synthesis. Thapsigargin produces a strong depolarization accompanied by a Ca^{2+} influx in synaptoneurosomes. The chronology of these two events has been elucidated by showing that (1) thapsigargin still depolarizes the synaptoneurosomal membrane when Ca^{2+} influx is eliminated by reducing free extracellular Ca^{2+} with the chelating agent BAPTA; and (2) an intracellular Ca^{2+} chelator, BAPTA-AM (1,2-bis(2-aminophenoxy)ethane-N,N,N',N'-tetraacetic acetoxymethylester), strongly decreases the thapsigargin-induced depolarization and the Ca^{2+} influx. Assuming that the first step of action of thapsigargin is to increase intracellular Ca^{2+} by blocking the Ca^{2+}-ATPase, one can deduce that the second event is the intracellular Ca^{2+}-induced depolarization, and finally the influx of Ca^{2+} from the extracellular medium. In addition, a strong correlation exists between the IP1 formation elicited by increasing concentration of thapsigargin and the $[Ca^{2+}]_i$ increase (FIG. 4). The correlation also holds true between thapsigargin-induced membrane depolarization and Ca^{2+} increase, as simultaneously measured using two fluorescent dyes, $DiSC_2(5)$ and Fura-2, respectively (FIG. 5). One could deduce that a high linkage exists between these three phenomena: IP3 synthesis, membrane depolarization, and intracellular Ca^{2+} increase.

DO EAA AND THAPSIGARGIN TRIGGER THE SAME MECHANISMS FROM THE FIRST INTRACELLULAR Ca^{2+} INCREASE?

From the above assumption, one could expect that thapsigargin and QA have no cumulative effects on the IP formation. Moreover, considering that only 60% of the IP1 formation induced by QA is dependent on the Ca^{2+} influx as previously demonstrated, one also could expect that QA-induced IP accumulation is larger than thapsigargin-elicited IP accumulation; these two assumptions are confirmed by experiments.[41] In addition, thapsigargin and QA both produce a Ca^{2+} entry, which is impermeable to Mn^{2+}. The hypothesis that EAA and thapsigargin activate the same mechanism is further reinforced by the fact that IP3 and thapsigargin are supposed to affect the same intracellular Ca^{2+} stores.[54,55] Finally, the Glu-elicited IP response

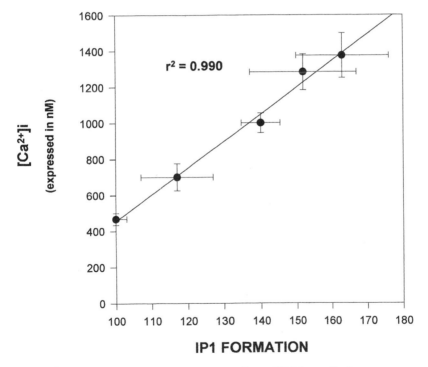

FIGURE 4. Dose-dependent effects of thapsigargin: Correlation between the intracellular Ca^{2+} increase and the IP1 formation produced by increasing concentrations of thapsigargin in 8-day-old rat forebrain synaptoneurosomes. Thapsigargin was tested at concentrations ranging from 0.1 to 10 μM. Each value is the mean ± SD of four separate experiments. Experiments were as previously described.[18,48]

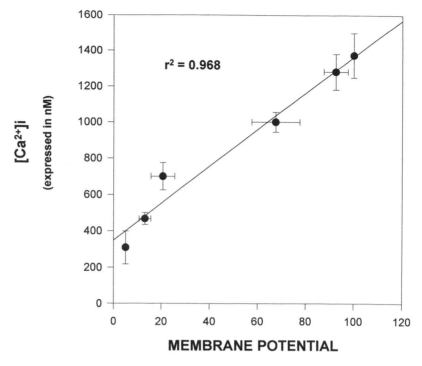

FIGURE 5. Dose-dependent effects of thapsigargin: Correlation between the intracellular Ca^{2+} increase and the membrane potential produced by increasing concentrations of thapsigargin in 8-day-old rat forebrain synaptoneurosomes as simultaneously measured using two fluorescent probes, Fura-2 for Ca^{2+} and $DiSC_2(5)$ for membrane potential. Thapsigargin was tested at concentrations ranging from 0.1 to 10 μM. Each value is the mean ± SD of four separate experiments. Methods were as previously described.[41,48]

is also partially blocked by BAPTA-AM. One could then suppose that thapsigargin and EAA stimulate the same mechanism and that Ca^{2+} is directly implicated in triggering Ca^{2+} influx from the extracellular medium rather than a releasable substance like CIF.

Because most neurotransmitter receptors linked to the phosphoinositide metabolism produces IP3 synthesis, one could speculate that $[Ca^{2+}]_i$-induced depolarization and a subsequent Ca^{2+} influx is a general phenomenon. However, the specificity of such a mechanism may arise from the fact that there is a concomitant association of the neurotransmitter receptor with the IP3-thapsigargin sensitive store, a $[Ca^{2+}]_i$-sensitive ionic channel producing the depolarization, an adequate voltage-dependent Ca^{2+} channel, sensitive to this depolarization, and the presence of Ca^{2+}-sensitive PLC. It appears that such an association is not always found, because in our experimental model, a muscarinic agonist, carbachol, does not at all trigger the same mechanism.[41,46–48]

In conclusion, we found that the stimulation of mGluR by glutamate, or related agonists, produces a self-maintained increase in $[Ca^{2+}]_i$ (see FIG. 6) via a first stimulation of IP3 formation, which in turn release Ca^{2+} from intracellular stores. The subsequent modest increase in $[Ca^{2+}]_i$ is sufficient to activate a $[Ca^{2+}]_i$-dependent ionic channels, which produces a depolarization. This depolarization opens a nonclassical voltage-dependent Ca^{2+} channel and allows a massive Ca^{2+} influx from the extracellular medium. The subsequent additional $[Ca^{2+}]_i$ increase activates Ca^{2+}-sensitive PLC, likely but not necessarily different from that directly linked to the mGluR. This induces a further IP3 production and the above cycle could begin again.

Many mechanisms could be involved in stopping this cycle, for example, a $[Ca^{2+}]_i$ threshold for activating the Na^+/Ca^{2+} exchanger to inhibit the IP3 receptor or the $[Ca^{2+}]_i$-dependent ionic channel, the exhaustion of the intracellular Ca^{2+} pool or the membrane phosphoinositide precursor pool. Nevertheless, we noticed a transient enhanced activity of this transduction mechanism, which appears to be specific for the glutamatergic system, at a period of development when synaptic contact is being formed.[56,57] Consequently, this mechanism may play a key role, when high $[Ca^{2+}]_i$ is required for inducing cytoskeletal rearrangement and gene expression likely necessary for shaping and maintaining the newly formed synaptic contacts.

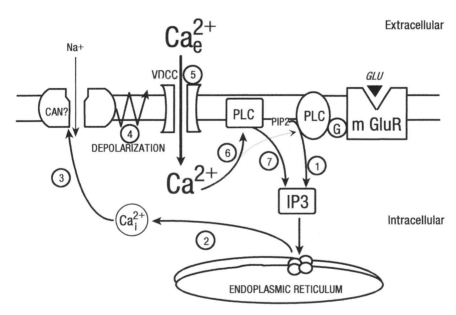

FIGURE 6. Putative mechanism triggered by the activation of metabotropic glutamate receptors (mGluR). (1) Synthesis of inositol-1,4,5-trisphosphate (IP3) from the membrane precursor inositol phospholipid: phosphatidyinositol-4,5-bisphosphate (PIP2). (2) Release of Ca^{2+} from the endoplasmic reticulum following IP3 interaction with its receptor. (3) Activation by intracellular Ca^{2+} of an ion channel, likely a Ca^{2+}-activated nonspecific cation channel (CAN). (4) Activation of CAN produces a local depolarization, which (5) opens a nonclassical voltage-dependent Ca^{2+} channel (VDCC). (6) The massive Ca^{2+} influx originating from the extracellular medium activates Ca^{2+}-sensitive phospholipase C (PLC), and possibly the PLC linked to the mGluR. (7) Activation of the Ca^{2+}-dependent PLC further increases the IP3 formation. This proposed mechanism leads to a self-sustained or even self-amplified intracellular Ca^{2+} concentration increase.

REFERENCES

1. LYNCH, G., J. LARSON, G. BARRIONUEVO & F. SCHOTTLER. 1983. Nature **305:** 719–721.
2. MALENKA, R. C., J. A. KAUER, R. S. ZUCKER & R. A. NICOLL. Science **242:** 81–84.
3. ITO, M. 1989. Annu. Rev. Neurosci. **12:** 85–102.
4. ANWYL, R. 1991. Trends Pharmacol. Sci. **12:** 324–326.
5. BRÖCHER, S., A. ARTOLA & W. SINGER. 1992. Proc. Natl. Acad. Sci. USA **89:** 123–127.
6. MATTSON, M. P., P. B. GUTHRIE & S. B. KATER. 1988. J. Neurosci. Res. **20:** 331–345.
7. MULKEY, R. M. & R. S. ZUCKER. 1991. Nature **350:** 153–155.
8. VON GERSDORFF, H. & G. MATTHEWS. 1994. Nature **367:** 735–739.
9. CHOI, D. W. 1985. Neurosci. Lett. **58:** 293–297.
10. FASOLATO, C., B. INNOCENTI & T. POZZAN. 1994. Trends Pharmacol. Sci. **15:** 77–83.
11. MASU, M., Y. TANABE, K. TSUCHIDA, R. SHIGEMOTO & S. NAKANISHI. 1991. Nature **349:** 760–765.
12. HOUAMED, K. M., J. L. KIUJPER, T. L. GILBERT, B. A. HALDEMAN, P. J. O'HARA, E. R. MULVIHILL, W. ALMERS & F. S. HAGEN. 1991. Science **252:** 1318–1321.
13. TANABE, Y., M. MASU, T. ISHII, R. SHIGEMOTO & S. NAKANISHI. 1992. Neuron **8:** 169–179.
14. ABE, T., H. SUGIHARA, H. NAWA, R. SHIGEMOTO, N. MIZUNO & S. NAKANISHI. 1992. J. Biol. Chem. **267:** 13361–13368.
15. NAKAJIMA, Y., H. IWAKABE, C. AKAZAWA, H. NAWA, R. SHIGEMOTO, N. MIZUNO & S. NAKANISHI. 1993. J. Biol. Chem. **268:** 11868–11873.
16. OKAMOTO, N., S. HORI, C. AKAZAWA, Y. HAYASHI, R. SHIGEMOTO, N. MIZUNO & S. NAKANISHI. 1994. J. Biol. Chem. **269:** 1231–1236.
17. NAKANISHI, S. 1992. Science **258:** 597–603.
18. RÉCASENS, M., J. GUIRAMAND, A. NOURIGAT, I. SASSETTI & G. DEVILLIERS. 1988. Neurochem. Int. **4:** 463–467.
19. BEN-ARI, Y., L. ANIKSZTEJN & P. BREGESTOVSKI. 1992. Trends Neurosci. **15:** 333–339.
20. WHITTEMORE, G. T., K. W. ROCHE, A. K. THOMPSON & R. L. HUGANIR. 1993. Nature **364:** 70–73.
21. MORIYOSHI, K., M. MASU, T. ISHII, R. SHIGEMOTO, N. MIZUNO & S. NAKANISCHI. 1991. Nature **354:** 31–37.
22. BERRIDGE, M. J. 1984. Biochem. J. **220:** 345–360.
23. BERRIDGE, M. J. & R. F. IRVINE. 1984. Nature **312:** 315–321.
24. NISHIZUKA, Y. 1984. Science **22:** 1365–1370.
25. BERRIDGE, M. J. & R. F. IRVINE. 1989. Nature **341:** 197–205.
26. TSIEN, R. Y., T. POZZAN & T. J. RINK. Nature **295:** 68–70.
27. POZZAN, T., P. ARSLAN, R. Y. TSIEN & T. J. RINK. J. Cell Biol. **94:** 335–345.
28. PUTNEY, J. W., JR., H. TAKEMURA, A. R. HUGHES, D. A. HORSTMAN & O. THASTRUP. 1989. FASEB J. **3:** 1899–1905.
29. MELDOLESI, J., E. CLEMENTI, C. FASOLATO, D. ZACHETTI & T. POZZAN. 1991. Trends Pharmacol. Sci. **12:** 289–292.
30. CLEMENTI, E., H. SHEER, D. ZACHETTI, C. FASOLATO, T. POZZAN & J. MELDOLESI. 1992. J. Biol. Chem. **267:** 2164–2172.
31. BERRIDGE, M. J. 1993. Nature **361:** 315–325.
32. HOTH, M. & R. PENNER. 1993. J. Physiol. **465:** 359–386.
33. IRVINE, R. F. 1992. FASEB J. **6:** 3085–3091.
34. MORRIS, A. P., D. V. GALLACHER, R. F. IRVINE & O. H. PETERSEN. 1987. Nature **330:** 653–655.
35. RANDRIAMAMPITA, C. & R. Y. TSIEN. 1993. Nature **364:** 809–814.
36. PAREKH, A. B., H. TERLAU & W. STÜHMER. 1993. Nature **364:** 814–818.
37. POGGIOLO, J., J.-P. MAUGER, F. GUESDON & M. CLARET. 1985. J. Biol. Chem. **260:** 3289–3294.
38. MISSIAEN, L., H. DESMEDT, G. DROOGSMANS & R. CASTEELS. 1992. Nature **357:** 599–602.
39. IRVINE, R. F. 1990. FEBS Lett. **263:** 5–9.
40. RÉCASENS, M. & J. GUIRAMAND. 1991. The quisqualate metabotropic receptor: Characterization and putative role. *In* Excitatory Amino Acid Antagonists. B. S. Meldrum, Ed.: 195–215. Blackwell Scientific Publications. Oxford, UK.

41. VIGNES, M. & M. RÉCASENS. 1994. Eur. J. Neurosci. In press.
42. KENDALL, D. A. & S. R. NAHORSKI. 1984. J. Neurochem. **42:** 1388–1394.
43. EBERHARD, D. A. & R. W. HOLZ. 1988. Trends Neurosci. **11:** 517–520.
44. GONZALES, R. A. & F. T. CREWS. 1988. J. Neurochem. **50:** 1522–1528.
45. GUIRAMAND, J., A. NOURIGAT, I. SASSETTI & M. RÉCASENS. 1989. Neurosci. Lett. **98:** 222–228.
46. GUIRAMAND, J., M. VIGNES, E. MAYAT, F. LEBRUN, I. SASSETTI & M. RÉCASENS. 1991. J. Neurochem. **57:** 1488–1500.
47. GUIRAMAND, J., M. VIGNES & M. RÉCASENS. 1991. J. Neurochem. **57:** 1500–1509.
48. VIGNES, M., J. GUIRAMAND, I. SASSETTI & M. RÉCASENS. 1993. Eur. J. Neurosci. **5:** 327–334.
49. MERRIT, J. E., R. JACOB & T. J. HALLAM. 1989. J. Biol. Chem. **264:** 1522–1527.
50. BREER, H., I. BOECKHOFF & E. TAREILUS. 1990. Nature **345:** 65–68.
51. SAGE, S. O. & T. R. RINK. 1987. J. Biol. Chem. **262:** 16364–16369.
52. PARTRIDGE, L. D. & D. SWANDULLA. 1988. Trends Pharmacol. Sci. **11:** 69–72.
53. TAKEMURA, H., O. THASTRUP & J. W. PUTNEY, JR. 1990. Cell Calcium **11:** 11–17.
54. THASTRUP, O., P. J. CULLEN, B. K. DROBAK, M. R. HANLEY & P. S. DANNIES. 1990. Proc. Natl. Acad. Sci. USA **87:** 2466–2470.
55. IRVING, A. J., G. L. COLLINGRIDGE & J. G. SCHOFIELD. 1992. Cell Calcium **13:** 293–301.
56. GUIRAMAND, J., I. SASSETTI & M. RÉCASENS. 1989. Int. J. Dev. Neurosci. **7:** 257–266.
57. MAYAT, E., F. LEBRUN, I. SASSETTI & M. RÉCASENS. 1994. Int. J. Dev. Neurosci. **12:** 1–17.

Polyamines as Endogenous Modulators of the *N*-Methyl-D-Aspartate Receptor

EDYTHE D. LONDON[a–d] AND ALEXEY MUKHIN[a,e]

ªNeuroimaging and Drug Action Section
Intramural Research Program
National Institute on Drug Abuse
National Institutes of Health
Baltimore, Maryland 21224

ᵇDepartment of Radiology
Johns Hopkins Medical Institutions
Baltimore, Maryland 21205

ᶜDepartment of Pharmacology and Experimental Therapeutics
School of Medicine
University of Maryland
Baltimore, Maryland 21201

Within the context of receptor interactions, ligand-gated ion channels are of particular interest because their activity is governed by signals from a variety of modulatory sites, each of which contributes to coordinated function. In this regard, the *N*-methyl-D-aspartate (NMDA) receptor is a multisubunit structure that bears binding sites for agonists and antagonists.[1] Both glutamate and glycine are required to effect receptor function, which is gating of a cationic channel that can be blocked by noncompetitive antagonists, such as phencyclidine, 1-[1-(2-thienyl)cyclohexyl]piperidine (TCP), and dizocilpine. In fact, function of the channel has been evaluated by assays of the binding of radiolabeled noncompetitive antagonists. Some of these studies revealed cooperative interactions between glutamate, glycine, and spermidine, suggesting that these modulators act at distinct, but allosterically coupled sites on the receptor.[2]

Polyamines are ubiquitous molecules that have been implicated in a variety of cellular functions, including growth and development, biosynthesis of nucleic acids and proteins, and regulation of mitochondrial Ca^{2+}.[3] They are formed via a pathway that involves the decarboxylation of ornithine to form putrescine. Subsequent steps involve biosynthesis of the charged triamine and tetramine, spermidine and spermine, respectively. Positive charges on polyamine molecules presumably could facilitate their interactions with various important biologically important sites, such as negatively charged moieties of nucleic acids and phospholipids of membranes.

Evidence for a neuromodulatory role of polyamines derives from a variety of sources.[4] Polyamines are present in brain at high concentrations (about 400 μM);[3,5,6] they influence behavioral indices of central nervous system excitability[7] and they inhibit the synaptosomal uptake of choline and dopamine.[8] Furthermore, spermidine is transported axonally in sciatic nerves,[9] and studies of cortical slices have revealed a high-affinity uptake system and Ca^{2+}-stimulated release of polyamines.[10] Release of

ᵈAddress correspondence to Edythe D. London, Ph.D., Chief, Neuroimaging and Drug Action Section, NIDA Division of Intramural Research, P.O. Box 5180, Baltimore, MD 21224.
ᵉPresent address: Departments of Neurology and Pharmacology, Georgetown University, Washington, DC 20007.

polyamines from synaptosomes by a high concentration of KCl also has been observed.[11]

Specific interactions of polyamines with the NMDA receptor have been observed both biochemically and electrophysiologically. Evidence that polyamines facilitate NMDA receptor activation is evidenced by observations that they enhance binding of [³H]dizocilpine and [³H]TCP to sites within the NMDA receptor channel.[12–16] Supporting data, obtained in physiological recordings from *Xenopus* oocytes, expressing NMDA receptor, have revealed that spermine potentiates responses to NMDA.[5,17,18]

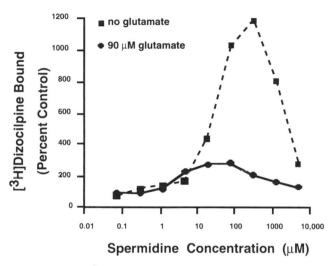

FIGURE 1. Enhancement of [³H]dizocilpine binding by spermidine. Well-washed membranes (7 washes, 15 μg protein) obtained from rat forebrain (frontal cortex + hippocampus) were incubated with 2 nM [³H]dizocilpine in 0.3 mL 10 mM HEPES · KOH buffer, pH 7.4, containing 30 μM glycine at 20 °C. Incubations were run for 6 h in the absence or presence of 30 μM added L-glutamate. Binding was terminated by filtration through GF/B filters presoaked in 0.5% polyethyleneimine. Nonspecific binding was determined in the presence of 1 μM cold, unlabeled dizocilpine. Each point represents the average specific binding, expressed as percent control, of triplicate determinations from one experiment. Control specific binding of [³H]dizocilpine was 0.13 ± 0.01 pmol/mg protein and 0.83 ± 0.05 pmol/mg protein, the absence or presence of 90 μM added L-glutamate, respectively (mean ± SEM for triplicates). Similar results were obtained in two additional experiments.

The mechanisms by which polyamines apparently facilitate opening of the NMDA receptor channel have been investigated by assessing their effects on the binding of [³H]dizocilpine under various conditions.[19] In most of these studies, spermine and spermidine stimulate [³H]dizocilpine binding.[12,21–23] Without added glutamate, the stimulation is substantial (about 12-fold in our studies), at the maximally effective concentration of spermidine (about 500 μM) (FIG. 1). This effect could at least in part be accounted for by an enhancement in the affinity of the NMDA receptor for agonist. Nonetheless, spermidine also produces stimulation in the presence of a saturating concentration of glutamate[24] (FIG. 1), although the stimulation is much smaller in magnitude. This stimulation could not reflect an increased affinity for the agonist, and suggests that spermidine exerts a positive action on the response to agonist binding, facilitating agonist-induced channel opening.

Although the putative mechanism by which polyamines may alter the functional response to agonist binding, as evidenced by increased binding of noncompetitive agonists, is not clarified, direct evidence is available regarding the effects of polyamines on NMDA recognition sites. In this regard, polyamines enhance the binding of [3H](+)-3-(2-carboxypiperazin-4-yl)-propyl-1-phosphonic acid ([3H]CPP), a competitive antagonist of NMDA or glutamate at these sites.[13] A similar action has been observed on the binding of [3H]CGP 39653, another competitive antagonist that is highly selective for NMDA recognition sites[25] (see FIG. 2).

The ability to stimulate [3H]CGP 39653 binding appears to be a property common to all polyamines, as it is observed when incubations are performed with spermine and putrescine as well as with spermidine (FIG. 2). The rank order of potencies of the amines is inversely related to the number of amino groups in the molecule, indicating that the number of positive charges determines potency.

The view that polyamines enhance binding to NMDA recognition sites by an enhancement of affinity for the radioligands is supported by a recent report in which spermine reduced the K_d for [3H]CGP 39653 binding from about 15 nM to 3 nM.[25] However, our studies of [3H]CGP 39653 binding to well-washed membranes of rat forebrain indicated behavior consistent with a two-site model (FIG. 3), and with the existence of two states of NMDA recognition sites, exhibiting K_d values of approximately 6 nM and 200 nM for high- and lower-affinity sites, respectively. Furthermore, as the high-affinity sites represented a small fraction (about 25%) of the total density

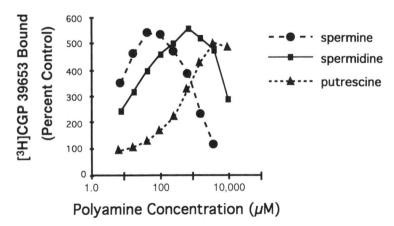

FIGURE 2. Stimulation of [3H]CGP 39653 binding by polyamines. Well-washed membranes of rat forebrain (24 μg protein) were incubated with 3.2 nM [3H]CGP 39653 in 0.3 mL of 10 mM HEPES · KOH buffer, pH 8.0, containing various concentrations of polyamines. Incubations were run for 1 h at 4 °C. Binding was terminated by filtration through GF/B filters presoaked in 0.5 M Tris HCl, pH 8.0, containing 1 M KCl and 0.1 mM glutamate. Nonspecific binding was determined in the presence of 100 μM NMDA. Each point represents the average specific binding, expressed as percent control, of triplicate determinations from one experiment. Specific binding of [3H]CGP 39653 in the absence of added polyamines (control) was 0.51 ± 0.03 pmol/mg protein (mean ± SEM for triplicates). Similar results were obtained in two to four additional experiments.

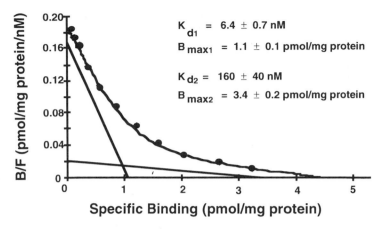

FIGURE 3. Scatchard analysis of [^{3}H]CGP 39653 binding to well-washed membranes from rat forebrain. Membranes (18 μg protein) were incubated with each of 11 concentrations of [^{3}H]CGP 39653 (0.3–300 nM) in 0.3 mL 10 mM HEPES · KOH buffer, pH 8.0, at 4 °C for 1 h. Nonspecific binding was determined in the presence of 500 μM NMDA. Binding was terminated by filtration through GF/B filters presoaked in 0.5 M Tris HCl pH 8.0, containing 1 M KCl and 0.1 mM glutamate. Each point represents the mean of 16 replicates (quadruplicate determinations from four separate assays performed on the same tissue preparation). Values of K_d and B_{max} indicate the means and standard errors obtained from a single nonlinear, least squares regression analysis of the data from this experiment. Similar results were obtained in eight additional assays.

of specific binding sites in rat forebrain, additional studies of polyamine interactions with NMDA recognition sites seemed warranted.

Saturation studies of [^{3}H]CGP 39653 binding, therefore, were performed in the absence of added polyamines, and were compared to data obtained when 500 μM spermidine was added to the incubations (FIG. 4). The concentration of spermidine was selected on the basis of its ability to induce an approximately fivefold increase in the binding of [^{3}H]CGP 39653 (FIG. 2). When spermidine was added, Scatchard analysis of the resultant data did not support a two-site model, because only one population of sites was apparent (FIG. 4B). The affinity of the sites corresponded closely to that of the high-affinity portion of binding obtained in the absence of added spermidine (FIGS. 3 and 4A). Furthermore, in the presence of spermidine, the density of sites (B_{max}) was nearly equal to the sum of the densities of high- and low-affinity sites assayed in the control condition, and close to the density of sites assayed with [^{3}H]dizocilpine (about 5 pmol/mg protein). Therefore, it appeared that the major reason for the spermidine-induced enhancement of [^{3}H]CGP 39653 binding was not to increase the affinity of the sites, but rather to convert the low-affinity sites to a high-affinity conformation. Nonetheless, the present data suggest that spermidine may also produce a slight enhancement in affinity of the high-affinity sites.

Competition assays revealed that the sites converted by spermidine from a low- to a high-affinity conformation had a pharmacological profile consistent with that of NMDA recognition sites, and similar or identical to that of high-affinity sites labeled in the absence of added spermidine. Under both conditions, ligands for NMDA recognition sites (NMDA, glutamate, CPP, and (±)2-amino-5-phosphonopentanoic

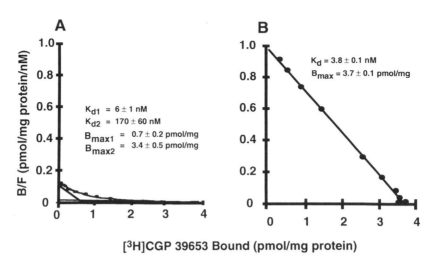

[³H]CGP 39653 Bound (pmol/mg protein)

FIGURE 4. Effect of spermidine on parameters of [³H]CGP 39653 binding in well-washed membranes of rat forebrain (see FIG. 3 legend for assay procedure). Results are shown of an experiment performed in quadruplicate using the same membrane preparation in the absence (**A**) and presence (**B**) of 500 μM spermidine. Values of K_d and B_{max} are the means ± SEM for individual Scatchard analyses. Similar results were obtained in three additional experiments.

acid) had much higher potencies than ligands for non-NMDA glutamate binding sites (quisqualic acid, kainic acid, cystine, *trans*-1-amino-cyclopentyl-1,3-decarboxylate); and the potencies of the inhibitors in the absence of spermidine were highly correlated with corresponding potencies when spermidine was added.

The results of these studies are consistent with the view that polyamines modulate activity of the NMDA receptor by at least two mechanisms. One of these, inferred from the stimulation of [³H]dizocilpine binding in the presence of saturating concentrations of glutamate, is independent of an action on affinity of NMDA recognition sites. Another mechanism of polyamine-induced activation of the NMDA receptor, as evidenced by an increase in the binding of noncompetitive antagonists to the channel, appears to involve an enhancement in the sensitivity of NMDA recognition sites. Although previous studies have indicated that this action reflects an increase in affinity of the sites, the present report demonstrates that a major aspect of the spermidine-induced increases in binding to these sites is through a conversion of NMDA recognition sites from a low-affinity conformation to one of high affinity. This action of polyamines is a function of the number of amine groups on the molecule and is shared by a variety of organic and inorganic cations.[26] Therefore, relatively nonspecific stimulatory effects of cations on the NMDA receptor could be relevant to receptor function *in vivo,* and this has implications for the interpretation of *in vitro* assays of NMDA receptors.

REFERENCES

1. NAKANISHI, S. 1992. Molecular diversity of glutamate receptors and implications for brain function. Science **258:** 597–603.
2. RANSOM, R. W. & N. L. STEC. 1988. Cooperative modulation of [³H]MK-801 binding to the *N*-methyl-D-aspartate receptor-ion channel complex by L-glutamate, glycine, and polyamines. J. Neurochem. **51:** 830–836.

3. TABOR, C. W. & H. TABOR. 1984. Polyamines. Annu. Rev. Biochem. **53:** 749–790.
4. SHAW, G. G. 1979. The polyamines in the central nervous system. Biochem. Pharmacol. **28:** 1–6.
5. SEILER, N. & F. N. BOLKENIUS. 1985. Polyamine reutilization and turnover in brain. Neurochem. Res. **10:** 529–544.
6. SLOTKIN, T. A. & J. BARTOLOME. 1986. Role of ornithine decarboxylase and the polyamines in nervous system development: A review. Brain Res. Bull. **17:** 307–320.
7. ANDERSON, D. J., J. CROSSLAND & G. G. SHAW. 1975. The actions of sperimidine and spermine on the central nervous system. Neuropharmacology **14:** 571–577.
8. LAW, C.-L., P. C. L. WONG & W.-F. FONG. 1984. Effects of polyamines on the uptake of neurotransmitters by rat brain synaptosomes. J. Neurochem. **42:** 870–872.
9. LINDQUIST, T. D., J. A. STURMAN, R. M. GOULD & N. A. INGOGLIA. 1985. Axonal transport of polyamines in intact and regenerating axons of the rat sciatic nerve. J. Neurochem. **44:** 1913–1919.
10. HARMAN, R. J. & G. G. SHAW. 1981. High-affinity uptake of spermine by slices of rat cerebral cortex. J. Neurochem. **36:** 1609–1615.
11. GILAD, G. M. & V. H. GILAD. 1991. Polyamine uptake, binding and release in rat brain. Eur. J. Pharmacol. **193:** 41–46.
12. WILLIAMS, K., C. ROMANO & P. B. MOLINOFF. 1989. Effects of polyamines on the binding of [^{3}H]MK-801 to the N-methyl-D-aspartate receptor: Pharmacological evidence for the existence of a polyamine recognition site. Mol. Pharmacol. **36:** 575–581.
13. CARTER, C. J., K. G. LLOYD, B. ZIVKOVIC & B. SCATTON. 1990. Ifenprodil and SL 82.0715 as cerebral antiischemic agents. III. Evidence for antagonistic effects at the polyamine modulatory site within the N-methyl-D-aspartate receptor complex. J. Pharmacol. Exp. Ther. **253:** 475–482.
14. ANIS, N., S. SHERBY, R. GOODNOW, JR. *et al.* 1990. Structure-activity relationships of philanthotoxin analogs and polyamines on N-methyl-D-aspartate and nicotinic acetylcholine receptors. J. Pharmacol. Exp. Ther. **254:** 764–773.
15. STEELE, J. E., D. M. BOWEN, P. T. FRANCIS, A. R. GREEN & A. J. CROSS. 1990. Spermidine enhancement of [^{3}H]MK-801 binding to the NMDA receptor complex in human cortical membranes. Eur. J. Pharmacol. **189:** 195–200.
16. ROBINSON, T. N., C. ROBERTSON, A. J. CROSS & A. R. GREEN. 1990. Modulation of [^{3}H]-dizocilpine ([^{3}H]-MK-801) binding to rat cortical N-methyl-D-aspartate receptors by polyamines. Mol. Neuropharmacol. **1:** 31–35.
17. BRACKLEY, P. R., J. R. GOODNOW, K. NAKANISHI, H. L. SUDAN & P. N. R. USHERWOOD. 1990. Spermine and philanthotoxin potentiate excitatory amino acid responses of *Xenopus* oocytes injected with rat and chick brain RNA. Neurosci. Lett. **114:** 51–56.
18. MCGURK, J. F., M. V. L. BENNETT & R. S. ZUKIN. 1990. Polyamines potentiate responses of N-methyl-D-aspartate receptors expressed in *Xenopus* oocytes. Proc. Natl. Acad. Sci. USA **87:** 9971–9974.
19. WILLIAMS, K., C. ROMANO, M. A. DICHTER & P. B. MOLINOFF. 1991. Modulation of the NMDA receptor by polyamines. Life Sci. **48:** 469–498.
20. BAKKER, M. H. M., R. M. MCKERNAN, E. H. F. WONG & A. C. FOSTER. 1991. [^{3}H]MK-801 binding to N-methyl-D -aspartate receptors solubilized from rat brain: Effects of glycine site ligands, polyamines, ifenprodil, and desipramine. J. Neurochem. **57:** 39–45.
21. JAVITT, D. C., M. J. FRUSCIANTE & S. R. ZUKIN. 1994. Activation-related and activation-independent effects of polyamines on phencyclidine receptor binding within the N-methyl-D-aspartate receptor complex. J. Pharmacol. Exp. Ther. **270:** 604–613.
22. SUBRAMANIAM, S. & P. MCGONIGLE. 1991. Quantitative autoradiographic characterization of the binding of (+)-5-methyl-10,11-dihydro-5H-dibenzo[a,d]cyclo-hepten-5,10-imine ([^{3}H]MK-801) in rat brain: Regional effects of polyamines. J. Pharmacol. Exp. Ther. **256:** 811–819.
23. ROMANO, C., K. WILLIAMS & P. B. MOLINOFF. 1991. Polyamines modulate the binding of [^{3}H]MK-801 to the solubilized N-methyl-D-aspartate receptor. J. Neurochem. **57:** 811–818.
24. ROMANO, C., K. WILLIAMS, S. DEPRIEST, *et al.* 1992. Effects of mono-, di-, and triamines on

the N-methyl-D-aspartate receptor complex: A model of the polyamine recognition site. Mol. Pharmacol. **41:** 785–792.

25. REYNOLDS, I. J. 1994. [^{3}H]CGP 39653 binding to the agonist site of the N-methyl-D-aspartate receptor is modulated by Mg^{2+} and polyamines independently of arcaine-sensitive polyamine site. J. Neurochem. **62:** 54–62.

26. MUKHIN, A., E. S. KOVALEVA & E. D. LONDON. 1994. Effects of mono- and divalent cations on binding properties of NMDA recognition sites. Submitted.

Effects of Systematically Administered Polyamines on Imipramine Immobility Action in Rats

K. SZCZAWIŃSKA

AND D. CENAJEK-MUSIAŁ

Department of Pharmacology
K. Marcinkowski University of Medical Sciences, Poznań
10 Fredry Street
61-701 Poznań, Poland

The main natural polyamines (PA)—spermidine (SPD) and spermine (SPM)—and their precursor putrescine (PUT) are organic cations of low molecular weight, ubiquitously distributed in living organisms. They are present in high concentration in the adult mammalian nervous system with wide regional variations.[1] Several and different experimental results suggest that natural PA may function as modulators of different central neurotransmitter systems, but little information concerning the possible pharmacological activity of PA or their interaction with drugs is available at present.[2,3] However, recent studies of particular significance to pharmacologists report on aminoglycoside antibiotic neomycin[4] and neuroprotective agent ifenprodil[5] interactions at a PA site on the N-methyl-D-aspartate (NMDA) receptor. We found previously that exogenously administered PA modulate the anxiolytic effects of some 1,4-benzodiazepines.[6] The present work was aimed at determining whether antidepressive activity of imipramine (IMI) is also affected by PA. Moreover, the MK-801, a noncompetitive NMDA receptor antagonist[7] was used because it was found to affect the IMI activity in rats.[8]

MATERIALS AND METHODS

The experiments were carried out on nonfasted male Wistar rats. IMI (10 mg/kg) was administered intraperitoneally twice: 24 and 1 h before the test; in a second experiment it was given nine times: once a day for nine days with the last dose given 1 h before the test. PUT (200 mg/kg), SPD (80 mg/kg), and SPM (40 mg/kg) were given 1 h prior to the test. MK-801 (0.1 mg/kg) was injected 30 min after the last dose of drugs. The antidepressant effect of IMI was assessed in the forced swimming test.

RESULTS

After a single injection of PA, only the PUT-treated group was immobile for a shorter time than control: 171.0 ± 13.6 s and 231.0 ± 6.0 s, respectively ($p < 0.05$ vs. control). PUT was as effective as IMI given twice: 157.8 ± 7.1 s ($p < 0.05$ vs. control). When PA were co-administered with IMI the reduced immobility was observed after injection of IMI + SPM: 106.7 ± 8.1 s ($p < 0.05$ vs. IMI-treated group). The combined treatment with SPD + IMI + MK-801 and SPM + IMI + MK-801

437

reduced significantly the immobility time: 92.5 ± 18.2 s and 97.0 ± 10.6 s ($p < 0.05$ vs. control and IMI-treated group), respectively.

In summary, these results suggest that different effects of PA on IMI activity could be mediated by an excitatory amino acid (EAA)-ergic system; however, for PUT + IMI, the GABA-ergic system must also be considered.

REFERENCES

1. SHAW, G. G. 1979. Biochem. Pharmacol. **28:** 1–6.
2. BAZZANI, C. & S. GENEDANI. 1989. J. Pharm. Pharmacol. **41:** 651–653.
3. FERCHMIN, P. A. & V. A. ETEROVIC. 1990. Pharmacol. Biochem. Behav. **37:** 445–449.
4. PULLAN, L. M., R. J. STUMPO, R. J. POWEL, K. A. PASCHETTO & M. BRITT. 1992. J. Neurochem. **59:** 2087–2093.
5. REYNOLDS, I. J. & R. MILLER. 1990. Mol. Pharmacol. **36:** 758–765.
6. SZCZAWINSKA, K. & D. CENAJEK-MUSIAL. 1995. Pol. J. Pharmacol. Pharm. In press.
7. WONG, E. H. F., A. R. KNIGHT & G. N. WOODRUFF. 1988. J. Neurochem. **50:** 274–281.
8. MAJ, J., Z. ROGOZ, G. SKUZA & H. SOWINSKA. 1992. Eur. Neuropsychopharmacol. **50:** 274–281.

Glutamate Receptor Modulation of [³H]GABA Release and Intracellular Calcium in Chick Retina Cells[a]

ARSÉLIO P. CARVALHO,[b] ILDETE L. FERREIRA,
ANA L. CARVALHO, AND CARLOS B. DUARTE

Center for Neurosciences of Coimbra
Department of Zoology
University of Coimbra
3049 Coimbra Codex, Portugal

The retina has served in many studies as a model for studying neuronal function because it is easily accessible and because of its laminar arrangements of a limited number of basic neuron types. The retina contains almost every neurotransmitter known, but glutamate and γ-aminobutyric acid (GABA) predominate; glutamate is present in the photoreceptors and bipolar and ganglion cells,[1,2] whereas GABA is the neurotransmitter in the horizontal cells and in many amacrine cells.[1-4] Each neurotransmitter can generate various types of responses at its receptors, and, apparently, the specific glutamate and GABA transporters involved in neurotransmitter uptake are also involved in determining the synaptic activity levels. Thus, although we normally think of vesicular release of the neurotransmitters, studies with retina cells have provided evidence that the membrane transporters are important in releasing GABA.[5-9]

The existing data on the mechanism(s) of GABA release by GABAergic neurons, including retina cells, are not consistent, and at least two alternate hypotheses for GABA release have been proposed: one favoring the classical Ca^{2+}-dependent exocytotic mechanism[10,11] and another postulating the reversal of the Na^+-dependent carrier.[5,8,9,12,13] Some recent studies carried out with embryonic chick retina cells[6] called our attention to the fact that these cells apparently do not exhibit Ca^{2+}-dependent release of GABA in response to K^+ depolarization or to glutamate, suggesting that the release would be entirely nonvesicular, although contradictory reports now are found in the literature.[5,8,9,14] The presence of only Ca^{2+}-independent release of GABA in chick retina cells due to depolarization would be unequivocal evidence that nonvesicular release may constitute a physiological means for GABA release.

Therefore, we used cultured embryonic chick retina cells to study in detail the mechanism(s) of [³H]GABA release stimulated by K^+ depolarization or by glutamate through its receptors, and by the specific glutamate receptor agonists, N-methyl-D-aspartate (NMDA), kainate (KA), quisqualate (QA), α-amino-3-hydroxy-5-methyl-4-isoxazolepropionic acid (AMPA) or (1S,3R)-1-aminocyclopentane-1,3-dicarboxylic acid (1S,3R-ACPD). We studied in particular the Ca^{2+} requirement for [³H]GABA release under conditions in which the carrier mediated [³H]GABA release was blocked by 1-(2-(((diphenylmethylene)amino)oxy)ethyl)-1,2,5,6-tetrahydro-3-pyridinecarboxylic acid (NNC-711), a specific inhibitor of the GABA carrier.[15] The NMDA receptor is permeable to Ca^{2+}, Na^+, and K^+, which will depolarize the

[a] This work was supported by JNICT (Portuguese Research Council), Portugal.
[b] Corresponding author.

439

membrane and thus activate voltage-sensitive Ca^{2+}-channels (VSCCs), resulting in a composite $[Ca^{2+}]_i$ signal due to Ca^{2+} influx through the receptor-associated channel and through VSCCs.[16,17] The KA and AMPA receptors are generally regarded as low Ca^{2+} permeability receptors and activation of these receptors is thought to increase the $[Ca^{2+}]_i$ mainly via the VSCCs, although some Ca^{2+} permeable KA and AMPA receptors have been described.[18–20] Also, the metabotropic glutamate receptors may increase the $[Ca^{2+}]_i$ through activation of phospholipase C, leading to the formation of inositol 1,4,5-triphosphate which releases Ca^{2+} from intracellular stores.[21]

Because the Ca^{2+} signals generated by either K^+ depolarization or by activation of glutamate receptors are diverse, they probably have distinct functions within the neurons. Thus, we considered it of importance to dissect the $[Ca^{2+}]_i$ signals generated by K^+ depolarization and by activation of each of the glutamate receptors in the same chick retina cell cultures enriched in amacrine-like neurons.[6]

EMBRYONIC CHICK RETINA CELLS DISPLAY CALCIUM-DEPENDENT RELEASE OF [³H]GABA

We used monolayers of embryonic retina cells previously loaded with the fluorescent indicator Indo-1 or [³H]GABA, and followed, respectively, the $[Ca^{2+}]_i$ responses and the release of [³H]GABA induced by 50 mM KCl. Depolarization rapidly increased the $[Ca^{2+}]_i$ by 748.8 ± 43.8 nM (FIG. 1A), and the initial peak was followed by a slow decrease of the $[Ca^{2+}]_i$ towards a plateau at 172 ± 16.5 nM ($n = 7$) above the resting value (FIG. 1A), suggesting that the cells nearly recovered the initial $[Ca^{2+}]_i$ shortly after depolarization.[8]

The results of parallel studies show that the KCl depolarization in the absence of Ca^{2+} released about 1% of the total [³H]GABA accumulated and that Ca^{2+} further increased the release of [³H]GABA to over 2% release (FIG. 1B). Both the changes in $[Ca^{2+}]_i$ and the release of [³H]GABA could be blocked partially by nitrendipine (0.1 μM), but ω-conotoxin GVIA (ω-CgTx; 0.5 μM), which blocked the changes in $[Ca^{2+}]_i$ by about 27%, did not influence the [³H]GABA release (TABLE 1). Thus, it appears that in retina cells in culture Ca^{2+} entry through the L-type of VSCCs is coupled to the release of [³H]GABA.[8]

These results lead us to conclude that depolarization of cultured retina cells with 50 mM KCl induces Ca^{2+}-dependent release of [³H]GABA, in addition to carrier-mediated release (FIG. 1B). Thus, at least some of the embryonic cells of the retina have the exocytotic mechanism for releasing [³H]GABA. Furthermore, the Ca^{2+}-dependent release appears to be initiated by Ca^{2+} entering through the L-type Ca^{2+} channels rather than through the N-type, which seems more common for brain tissue.

The Ca^{2+}-independent release of [³H]GABA induced by K^+ depolarization can be blocked by NNC-711 (10 μM), which has been shown to block the GABA carrier.[15] In FIGURE 1B (inset), we show that, in the absence of Ca^{2+}, NNC-711 completely abolished the release of [³H]GABA induced by 50 mM KCl, and that, in the presence of Ca^{2+} (1 mM), the GABA carrier blocker inhibited the release by about 66%, but did not appear to affect the Ca^{2+}-dependent release.

GLUTAMATE INCREASES THE $[Ca^{2+}]_i$ BUT STIMULATES Ca^{2+}-INDEPENDENT RELEASE OF [³H]GABA

Glutamate has also been found to increase the intracellular $[Ca^{2+}]$ in several cell type cultures, but there is controversy as to whether the effect of glutamate on the release of [³H]GABA is mediated through a rise in [Ca²⁺].[5–9,13,14] We found that in chick retina cell cultures, which exhibit Ca²⁺-dependent release of [³H]GABA when stimulated with 50 mM KCl (see above), Ca²⁺ was inhibitory when the release was induced by glutamate (FIG. 2A). As in the case of KCl depolarization, glutamate depolarization increased the [Ca²⁺]ᵢ (Δ[Ca²⁺]ᵢ; FIG. 2B) through activation of L-type VSCCs because nitrendipine inhibited the Δ[Ca²⁺]ᵢ by about 50%, whereas ω-CgTx was without effect.[9]

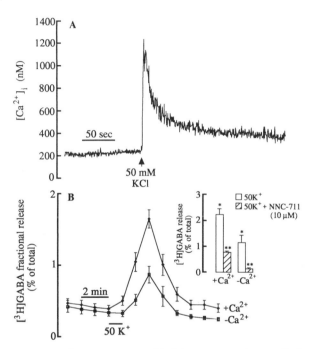

FIGURE 1. Intracellular Ca²⁺ concentration ([Ca²⁺]ᵢ) response (A) and [³H]GABA release (B) stimulated by KCl depolarization. (A) The cells were loaded with the fluorescent indicator Indo-1, and the [Ca²⁺]ᵢ was determined as previously described.[8,9] After 4 min of preincubation in Na⁺ medium (in mM: 132 NaCl, 4 KCl, 1.4 MgCl₂, 6 glucose, and 10 HEPES-Na, pH 7.4), the cells were stimulated with 50 mM KCl. (B) The cells were loaded with [³H]GABA for 1 h at 37 °C and perfused, using a superfusion system, with Na⁺ medium, with or without added Ca²⁺. Where indicated, the cells were depolarized by replacing NaCl isoosmotically by 50 mM KCl (horizontal bar). The release of [³H]GABA is expressed as previously described.[8,9] (*Inset*) Effect of NNC-711 on the release of [³H]GABA induced by K⁺ depolarization. When NNC-711 (10 μM) was tested, a preincubation of 1 min with the drug was performed. The data are presented as means ± SEM (n = 4–11). *Significantly different from control, $p < 0.005$.

Furthermore, the effect of glutamate on the [³H]GABA release is in fact mediated through glutamate receptors because 1 μM (+)-5-methyl-10,11-dihydro-5H-dibenzo[a,b]cyclohepten-5,10-imine maleate (MK-801) inhibited it by about 40%, and the combination of 1 μM MK-801 and 10 μM 6-cyano-7-nitroquinoxaline-2,3-dione (CNQX) nearly inhibited it by 100%.[9] The release of [³H]GABA induced by glutamate in the presence or in the absence of Ca²⁺ was nearly all carrier mediated, as shown in FIGURE 3 depicting the effect of 10 μM NNC-711 on [³H]GABA release. This drug blocked the uptake of [³H]GABA in chick retina cells

TABLE 1. Effect of Ca^{2+} Channel Blockers on $\Delta[Ca^{2+}]_i$ and $[^3H]GABA$ Release Induced by KCl Depolarization[a]

	Control	Nitrendipine (0.1 μM)	ω-Cg Tx (0.5 μM)
$\Delta[Ca^{2+}]_i$ (nM)	748.8 ± 43.8 ($n = 8$)	265.5 ± 14.4 ($n = 5$)	546.5 ± 27.7 ($n = 7$)
$[^3H]GABA$ release (% of total)	2.22 ± 0.2 ($n = 11$)	1.32 ± 0.11 ($n = 4$)	1.96 ± 0.3 ($n = 7$)

NOTE: The experimental conditions were similar to those described in FIGURE 1. When present, nitrendipine was applied 1 min prior to stimulation with KCl, whereas for ω-CgTx the cells were treated as described previously.[8]

[a]Adapted from Duarte et al.[8]

(not shown) and completely blocked the release of $[^3H]GABA$ evoked by 100 μM glutamate in the absence of Ca^{2+} (FIG. 3). At 1 mM, Ca^{2+} again inhibited the release in the absence of NNC-711; in the presence of the drug the release of $[^3H]GABA$ evoked by glutamate, in Ca^{2+}-containing or in Ca^{2+}-free media, was not significantly different ($p = 0.097$). The results suggest that relatively little or no exocytotic (Ca^{2+}-dependent) release was induced by glutamate. We studied further the Ca^{2+}-dependent release of $[^3H]GABA$ mediated by the glutamate receptor agonists, NMDA, KA, QA, and AMPA, under conditions of blockade of the GABA carrier by NNC-711, to avoid carrier mediated release (see below).

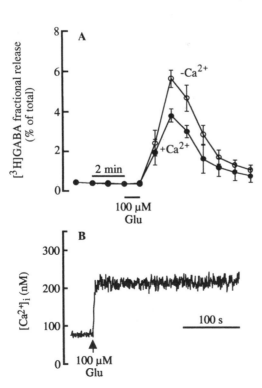

FIGURE 2. Glutamate stimulated $[^3H]GABA$ release (A) and $[Ca^{2+}]_i$ changes (B) in cultured chick retina cells. (A) The cells were stimulated with 100 μM glutamate, as indicated by the horizontal bar, in Na^+ medium with or without added Ca^{2+}. The release of $[^3H]GABA$ is expressed as previously described.[8,9] The data are presented as means ± SEM. ($n = 7–8$). (B) The cells were preincubated for 4 min in Na^+-medium and, where indicated, 100 μM glutamate was added.

INDIVIDUAL GLUTAMATE AGONISTS INDUCE BOTH Ca^{2+}-INDEPENDENT AND Ca^{2+}-DEPENDENT RELEASE OF [³H]GABA

The results described above suggest that nearly 100% of the [³H]GABA release evoked by glutamate was Ca^{2+}-independent and occurs through the GABA carrier. In this section we show that stimulation of the individual ionotropic glutamate receptors, namely the NMDA and AMPA receptors, induced a significant Ca^{2+}-dependent release of [³H]GABA, although most release is Ca^{2+} independent, as in the case of glutamate presented in the previous section.

Stimulation of the cells with 200 μM NMDA (under 50 mM K^+ depolarization) in the presence of glycine (3 μM) and in the absence of Ca^{2+}, produced a release of [³H]GABA of about 7% per minute of the total [³H]GABA accumulated, as

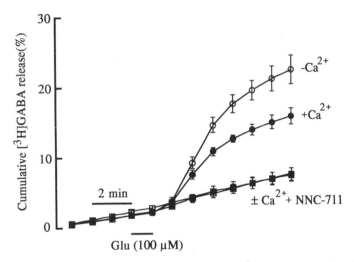

FIGURE 3. Effect of NNC-711 on the cumulative release of [³H]GABA induced by glutamate (100 μM) in Na^+ medium, with or without added Ca^{2+}. When the effect of NNC-711 (10 μM) was tested, the cells were perfused with NNC-711 1 min prior to stimulation (horizontal bar) with glutamate. The release of [³H]GABA is expressed as previously described.[8,9] The data are presented as means ± SEM (*n* = 3–8).

compared to about 1% per minute of the release produced by 50 mM KCl (FIG. 4). This effect was completely blocked by 1 μM MK-801.[9] Glutamate alone had an effect similar to that of NMDA + K^+ + glycine. Thus, either glutamate or NMDA causes massive release of [³H]GABA which is Ca^{2+} independent. Similar results were obtained for KA (100 μM) and QA (10 μM), and this effect was nearly all reversed by 10 μM CNQX (FIG. 5). Thus, chick retina cells stimulated with KA released about 6.44% of the total [³H]GABA/min in a Na^+ medium without Ca^{2+}, and QA released about 1.68% per minute under the same conditions. These results suggest that both agents are acting on non-NMDA receptors and that the effect is Ca^{2+} independent.

To evaluate the effect of Ca^{2+} on the release of [³H]GABA evoked by the glutamate receptor agonists, we stimulated the cells with glutamate receptor agonists for short periods (1 min), in the presence and in the absence of Ca^{2+}. Short periods

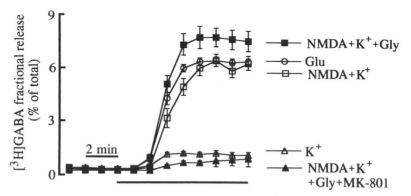

FIGURE 4. Effect of continuous stimulation with glutamate, K^+ depolarization or NMDA under K^+ depolarization on the release of [^{3}H]GABA, in the absence of added Ca^{2+}. Where indicated (horizontal bar), the superfusion solution was replaced by Na^+ medium containing 100 μM glutamate or K^+ medium prepared by isoosmotic replacement of NaCl by 50 mM KCl. When glycine (3 μM) or MK-801 (1 μM) was tested, the cells were perfused with these agents 1 min prior to stimulation with the stimulating agents. The release of [^{3}H]GABA is expressed as previously described.[8,9] The data are presented as means ± SEM (n = 3–10).

of stimulation were used in the presence of Ca^{2+} to avoid toxicity phenomena due to long exposures of the cells to glutamate.[22] Either NMDA, KA, QA, or AMPA increased the [Ca^{2+}]$_i$ in chick retina cells only if extracellular Ca^{2+} were present (not shown). The presence of Ca^{2+} significantly increased the release of [^{3}H]GABA due to stimulation by KA, QA or AMPA, but inhibited the release induced by NMDA from 7.36 ± 0.87, in the absence of Ca^{2+}, to 4.05 ± 0.39% in the presence of Ca^{2+} (TABLE 2). All agonists increased the [Ca^{2+}]$_i$ (see below), but apparently only in the case of NMDA (and glutamate) does the increase in [Ca^{2+}]$_i$ produce inhibition of [^{3}H]GABA release (TABLE 2).

The effect of Ca^{2+} on the release of [^{3}H]GABA can best be studied if we inhibit the contribution of the GABA carrier to the release. We did this by inhibiting the

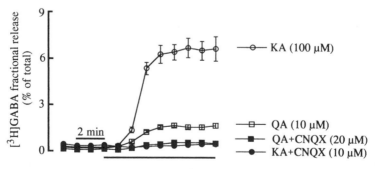

FIGURE 5. Effect of CNQX on the release of [^{3}H]GABA, evoked by kainate (KA) (100 μM) or quisqualate (10 μM). Experiments were performed as described in FIGURE 4, in the absence of added Ca^{2+}. When CNQX (10 or 20 μM) was tested, the cells were perfused with the antagonist 1 min prior to stimulation with KA or QA. The release of [^{3}H]GABA is expressed as previously described.[8,9] Results are presented as means ± SEM (n = 3–5).

carrier with NNC-711,[15] which totally inhibits glutamate-induced release in the absence of external Ca^{2+} (FIG. 3). In the presence of Ca^{2+}, we found that NNC-711 (10 μM) did not block all [^3H]GABA release induced by the glutamate agonists, and that apparently NMDA, QA, and AMPA, but not KA, caused Ca^{2+}-dependent release of [^3H]GABA in the presence of NNC-711 (FIG. 6).

It is of interest to note that the overall effect of Ca^{2+} on the release evoked by NMDA is a reduction in the release of [^3H]GABA (TABLE 2), but the exocytotic component (Ca^{2+}-dependent) is actually stimulated by Ca^{2+} (FIG. 6). This may mean that the effect of Ca^{2+} on the release is complex and is exerted not only on the exocytotic component. Evidence for this is also obtained from the influence of Ca^{2+} on the KA effect; thus, FIGURE 6 and TABLE 2 show clearly that Ca^{2+} increases the total release of [^3H]GABA induced by KA from a value of 13.23 $\pm$ 0.80% in the absence of Ca^{2+} to a value of 18.03 $\pm$ 0.63% in the presence of Ca^{2+}. Nevertheless, when we blocked the carrier with NNC-711, KA increased [^3H]GABA release to about the same extent in the presence or in the absence of Ca^{2+} (FIG. 6). Thus, it would appear that KA, under conditions in which the carrier is blocked, causes

TABLE 2. Effect of Glutamate Receptor Agonists on the $[Ca^{2+}]_i$ and [^3H]GABA Release[a]

Agonist	$\Delta[Ca^{2+}]_i$ (nM)	[^3H]GABA Release (%)	
		Ca^{2+} Absent	Ca^{2+} Present
NMDA	156.60 $\pm$ 8.25	7.36 $\pm$ 0.87	4.05 $\pm$ 0.39
	($n = 12$)	($n = 7$)	($n = 4$)
KA	276.51 $\pm$ 6.78	13.23 $\pm$ 0.80	18.03 $\pm$ 0.63
	($n = 9$)	($n = 8$)	($n = 7$)
QA	93.94 $\pm$ 5.76	3.57 $\pm$ 0.33	4.88 $\pm$ 0.52
	($n = 8$)	($n = 13$)	($n = 10$)
AMPA	120.36 $\pm$ 6.00	4.41 $\pm$ 0.75	7.07 $\pm$ 0.88
	($n = 7$)	($n = 4$)	($n = 5$)

[a]The experiments were performed in Na^+ medium as described in the legends to FIGURES 6 and 8. The release of [^3H]GABA was obtained in the absence or presence of added Ca^{2+}, without NNC-711, and for 1 min of stimulation with the various agonists.

release through a mechanism which involves neither the carrier nor the exocytotic (Ca^{2+}-dependent) mechanism.

Calcium increases the release of [^3H]GABA stimulated by either QA or AMPA, in the presence (FIG. 6) or in the absence of NNC-711 (TABLE 2), suggesting that a Ca^{2+}-dependent release occurs in the absence of carrier-mediated release. The effect on the exocytotic release is more clearly seen in the case of AMPA (50 μM), a more specific agonist for the ionotropic QA receptor. AMPA increased the $[Ca^{2+}]_i$ by 120 nM (TABLE 2) and evoked a release of 7.07 $\pm$ 0.88% of the total [^3H]GABA in the presence of Ca^{2+}, whereas only 4.41 $\pm$ 0.75% was released in the absence of Ca^{2+} (TABLE 2), and a substantial release remained in the presence of NNC-711, which was dependent on Ca^{2+} (FIG. 6).

Although QA stimulates both the ionotropic and the metabotropic receptors, (1S,3R)-ACPD, the agonist of metabotropic ACPD receptor, did not influence the release of [^3H]GABA, in the presence or in the absence of Ca^{2+}, suggesting that this receptor was not involved (results not shown).

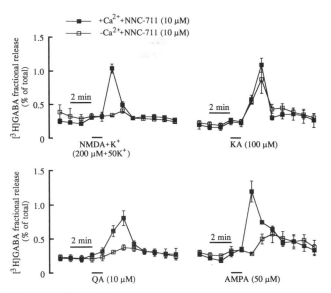

FIGURE 6. Effect of Ca^{2+} on [^{3}H]GABA release evoked by glutamate receptor agonists (NMDA, KA, QA, and AMPA), in the presence of NNC-711. The cells were stimulated with NMDA (200 μM) under K^+ depolarization, KA (100 μM), QA (10 μM) or AMPA (50 μM), and a preincubation with NNC-711 (10 μM) was made prior to stimulation with the agonists. The release of [^{3}H]GABA is expressed as previously described.[8,9] Results represent means ± SEM (n = 3–5).

MECHANISMS OF $[Ca^{2+}]_i$ REGULATION BY GLUTAMATE RECEPTOR AGONISTS

The NMDA receptor is permeable to Ca^{2+} and is expected to generate composite $[Ca^{2+}]_i$ signals due to the influx of Ca^{2+} through the receptor-associated channel and through VSCCs.[16] The KA and AMPA receptors are regarded as low-Ca^{2+} permeability receptors,[16,17] and activation of these receptors is thought to increase the $[Ca^{2+}]_i$ mainly via depolarization-induced opening of the VSCCs, although some Ca^{2+}-permeable KA and AMPA receptors have been described.[18–20] The $[Ca^{2+}]_i$ may also rise due to mobilization of intracellular Ca^{2+} stores because of activation of phospholipase C through the metabotropic glutamate receptors.[21]

The different effects of Ca^{2+} on the release of [^{3}H]GABA reported in the previous section may be related to the fact that the Ca^{2+} signals provided by activation of the glutamate receptors are diverse. Thus, it is important to separate the $[Ca^{2+}]_i$ signals generated by each glutamate receptor into its component parts. Our results show in fact that the agonists of the ionotropic glutamate receptors, NMDA, KA and AMPA, increased the $[Ca^{2+}]_i$ because of composite responses comprising Ca^{2+} entering through the receptor channels and through VSCCs.

Calcium Entry through Glutamate Receptors

To evaluate the contribution of the receptor-associated channels to the $[Ca^{2+}]_i$ response to glutamate, NMDA, KA, and AMPA, the cells were stimulated with each agonist in a Na^+-free N-methyl-D-glucamine (NMG) medium. When glutamate was added under these conditions, the effect on the rise in $[Ca^{2+}]_i$ was decreased to about 38% the value obtained in a Na^+ medium (FIG. 7A and B). Also nitrendipine was able to decrease the $[Ca^{2+}]_i$ response induced by glutamate in a Na^+ medium by about 53% (FIG. 7A), which suggests that both VSCCs and glutamate receptor-associated channels allow Ca^{2+} entry upon glutamate receptor stimulation.

In Mg^{2+}-free Na^+ medium, 100 μM NMDA evoked a sustained increase in $[Ca^{2+}]_i$ of 156.6 ± 8.2 nM that could be reversed by MK-801 (FIGS. 8A and 9A). Even in the presence of 1.4 mM Mg^{2+}, NMDA increased the $[Ca^{2+}]_i$ to a peak of about 77 nM which then declined to a plateau of about 32 nM above the resting concentration. Furthermore, when the extracellular Na^+ was replaced by NMG, the response to NMDA in the absence of Mg^{2+} was about 30 nM (FIG. 8B), which corresponds to the contribution of the NMDA receptor-associated channels to the $[Ca^{2+}]_i$ response.

Addition of 100 μM KA to the medium increased the $[Ca^{2+}]_i$ by 276 nM and then decreased to a plateau at about 74% of the value of the peak (FIG. 8C); the effect could be partially blocked by 20 μM CNQX to a value of 66 nM (FIG. 9B). When the extracellular Na^+ was replaced by NMG, the effect of KA on the $[Ca^{2+}]_i$ rise decreased to about 60% of the value obtained in the Na^+ medium, probably because of the absence of entry of Ca^{2+} through the VSCCs in the NMG medium (FIG. 8C). In the case of AMPA the receptor-associated channel seems to have a much smaller contribution to the $[Ca^{2+}]_i$ rise than that observed for the KA receptor (FIG. 8D). Activation of the AMPA receptors increased the $[Ca^{2+}]_i$ by 120 nM, but this effect was only slightly affected by CNQX (FIG. 9C).

The effect of (1S,3R)-ACPD on the $[Ca^{2+}]_i$ was evaluated in the retina cells, and we found that 200 μM (1S,3R)-ACPD did not affect the $[Ca^{2+}]_i$ in the presence (not shown) or in the absence of Ca^{2+} (FIG. 10). The presence of (1S,3R)-ACPD receptors in chick retina cells was further evaluated by measuring the accumulation of myo-[³H]inositol phosphates evoked by two minutes of stimulation with 200 μM (1S,3R)-ACPD. The agonist increased the accumulation of [³H]InsP₁, [³H]InsP₂, and

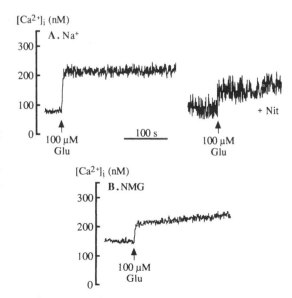

FIGURE 7. Intracellular Ca^{2+} ($[Ca^{2+}]_i$) responses to glutamate in Na^+- or N-methyl-D-glucamine (NMG) media. The cells were preincubated for 4 min in Na^+ (**A**) or NMG (**B**; containing 132 mM NMG, but otherwise identical to the Na^+ medium) media, and, where indicated, 100 μM glutamate was added. In **A** the control response to glutamate (*left*) is compared to the effect after 4 min of preincubation with 1 μM nitrendipine.

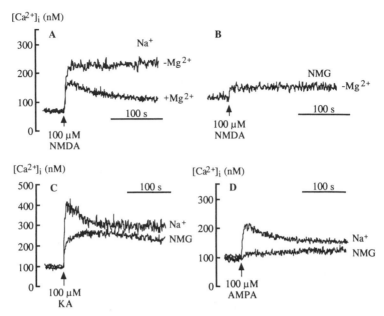

FIGURE 8. Effect of glutamate receptor agonists on the $[Ca^{2+}]_i$, in Na^+ and NMG media. The cells were stimulated with 100 μM NMDA (**A** and **B**), 100 μM KA (**C**) or 100 μM AMPA (**D**), after 4 min of preincubation in the indicated media. The response to NMDA was tested in Na^+ medium (**A**), with ($+Mg^{2+}$) or without ($-Mg^{2+}$) 1.4 mM $MgCl_2$, or in Mg^{2+}-free NMG medium (**B**).

$[^3H]InsP_3$, but this effect was not antagonized by L-AP3 (100–500 μM), a selective antagonist of some metabotropic receptors[21] (not shown).

Ca^{2+} Entry through Voltage-Sensitive Ca^{2+} Channels

The results reported in FIGURE 9 show that a large fraction of the Ca^{2+} entering the retina cells, as a result of glutamate agonists stimulation, passes through channels sensitive to nitrendipine. Indeed, nitrendipine (1 μM) decreased by 76 nM, 51 nM, and 55 nM the $[Ca^{2+}]_i$ increase evoked by NMDA, KA, and AMPA, respectively (FIG. 9A–C), suggesting that L-type channels are present in these cells. The P-type Ca^{2+} channels, identified by their sensitivity to ω-Aga IVA, also contribute significantly to the response to KA, whereas ω-Aga IVA was without effect on the $[Ca^{2+}]_i$ transients evoked by NMDA and AMPA (FIG. 9A–C). It is of interest to note that the glutamate agonists do not activate N-type VSCCs[9] although we showed previously that these channels are present in the retina cell cultures used in the experiments.[8]

The differential activation of VSCCs by NMDA, KA, and AMPA may reflect (1) that the receptors are localized in different cells which may or not contain P-type Ca^{2+} channels and that the L-type channels are always present or (2) that the receptors and the P- and N-type Ca^{2+} channels may have a heterogenous distribution within the cells.

Potentiation by TPA of the Ca²⁺ Entry through the KA Receptor-associated Channel

We took advantage of the high Ca^{2+} permeability of the KA receptor in our preparation to study the effect of protein kinase C (PKC) on the activity of the receptor. The PKC activator 12-*O*-tetradecanoylphorbol 13-acetate (TPA; 200 nM)

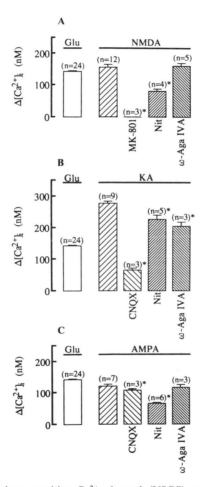

FIGURE 9. Effect of voltage-sensitive Ca^{2+} channel (VSCC) antagonists and glutamate receptor antagonists on the $[Ca^{2+}]_i$ responses to NMDA, KA or AMPA. The average initial $[Ca^{2+}]_i$ responses ($\Delta[Ca^{2+}]_i$) to 100 μM NMDA (**A**) in Mg^{2+}-free Na^+ medium, 100 μM KA (**B**) or 100 μM AMPA (**C**) in Na^+ medium, are compared with the effect of 100 μM glutamate (Glu), determined as indicated in FIGURE 2B. The effects of nitrendipine (Nit; 1.5 μM), MK-801 (7.5 μM), CNQX (20 μM), and ω-Aga IVA (200 nM) were determined after 4 min of preincubation with the antagonists. When ω-Aga IVA was tested, the cells were preincubated with the toxin for 1 h, during the loading with Indo-1. The effect of NMDA was always tested in Mg^{2+}-free Na^+ medium, whereas the $[Ca^{2+}]_i$ response to glutamate, kainate, and AMPA were determined in the presence of Mg^{2+} (Na^+ medium). Results are expressed as means ± SEM of the indicated number of experiments, carried out in different preparations. *Significantly different from control, $p < 0.05$.

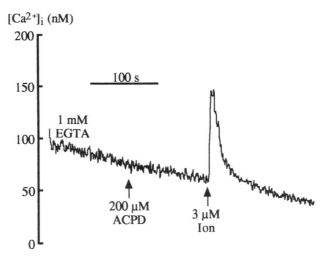

FIGURE 10. Lack of effect of (1S,3R)-ACPD on the $[Ca^{2+}]_i$. The cells were preincubated for 2 min in Ca^{2+}-free (1 mM EGTA) Na^+ medium, and, where indicated, 200 µM ACPD and 3 µM ionomycin were added. The trace is representative of experiments performed in three independent preparations.

increased the $[Ca^{2+}]_i$ response to KA in NMG medium to about 130% of the control (FIG. 11). Conversely, TPA decreased the effect of KA in Na^+ medium to about 80% of the control (FIG. 11). To test whether this inhibition was due to a decreased Ca^{2+} entry through VSCCs, we studied the effect of TPA on the $[Ca^{2+}]_i$ response to KCl depolarization, which is expected to specifically activate VSCCs. After a 4-min pre-incubation with TPA, the effect of 40 mM KCl depolarization was inhibited by about 25% (FIG. 12). Thus, although the activity of the Ca^{2+} permeable KA receptors is increased by TPA, the overall $[Ca^{2+}]_i$ response in Na^+ medium is inhibited because of the effect of the phorbol esters on the VSCCs.

DISCUSSION AND CONCLUSIONS

[³H]GABA Release by Potassium Depolarization

In the present work, we clearly show that K^+ depolarization activates L-type Ca^{2+} channels coupled to Ca^{2+}-dependent [³H]GABA release in embryonic chick retina cells. Inasmuch as [³H]GABA seems to be accumulated in horizontal and amacrine cells,[1,3,6] these cells must be the source of the [³H]GABA release. However, toad (*B. marinus*) and goldfish horizontal cells have only Ca^{2+}-independent [³H]GABA release,[5,23] and these cells do not have synaptic vesicles.[3] Therefore, the Ca^{2+}-dependent release by our preparation may represent release by amacrine cells, although Hofmann and Möckel[6] were unable to show Ca^{2+}-dependent release in a similar chick retina culture. These authors did not use a superfusion system to determine release, and it is possible that other substances released, which would accumulate in the medium, could be interfering with the Ca^{2+}-dependent (exocytotic) release.

Although most studies have shown only Ca^{2+}-independent release of $[^3H]GABA$ by various preparations of retina,[6,10,23,24] at least one report shows that goldfish retina have a "small but significant" Ca^{2+}-dependent release of $[^3H]GABA$.[5] Our studies also show that, in accordance with results from other retina cell preparations, a large fraction of the $[^3H]GABA$ release induced by K^+ depolarization is Ca^{2+}-independent (FIGS. 1 and 3) and represents the release of cytoplasmic $[^3H]GABA$ through the GABA carrier because it can be blocked by NNC-711 (FIGS. 1 and 3).

Release of [³H]GABA by Glutamate and Glutamate Agonists

One of the first observations in our studies was that Ca^{2+} decreases the release of $[^3H]GABA$ induced by glutamate (FIGS. 2 and 3). It is of interest to note that about 50% of the Ca^{2+} entry induced by glutamate uses the same L-type channels which become active because of K^+ depolarization and, in this case, increases $[^3H]GABA$ release (TABLE I). There is no clear explanation for this observation especially

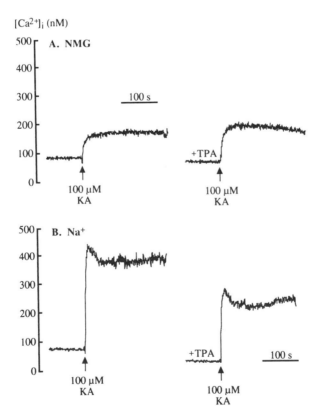

FIGURE 11. Effect of TPA on the $[Ca^{2+}]_i$ response to kainate, in NMG (**A**) or Na^+ (**B**) media. The cells were preincubated for 4 min in the indicated media, with (+TPA) or without 200 nM TPA, and, where indicated, 100 μM KA was added. The traces are representative of experiments carried out in duplicate in 6–8 different preparations.

because in both cases activation of L-type VSCCs accounts for about 50% of the Ca^{2+} entry.

The coupling of the glutamate receptors to the Ca^{2+} channels requires Na^+ in the external medium, and the $[Na^+]_i$ actually increases due to glutamate stimulation (results not shown). Our results are in accordance with earlier reports by Sucher *et al.*[25] that Ca^{2+} antagonists attenuate the rise in $[Ca^{2+}]_i$ due to glutamate in rat retinal ganglion cells, and other reports exist showing that activation of the glutamate receptors (NMDA, KA, QA) elevates the $[Ca^{2+}]_i$ indirectly by depolarizing the membrane, which then activates the Ca^{2+} influx through VSCCs.[16,18,19]

The [^{3}H]GABA release stimulated by glutamate or NMDA was more pronounced in the absence of Ca^{2+} and could be nearly totally blocked by NNC-711 (FIGS. 3 and 6A and TABLE 2). Similar Ca^{2+}-independent release of [^{3}H]GABA has been reported in other cell systems,[6,13,23,26] and in all cases Na^+ is required, which

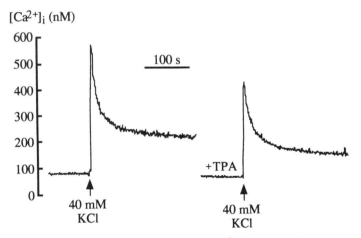

FIGURE 12. Inhibition by TPA of the K^+-stimulated $[Ca^{2+}]_i$ increase. The experiments were performed as indicated in FIGURE 1A, and where indicated (+TPA) the cells were preincubated for 4 min with 200 nM TPA. The traces are representative of duplicate experiments carried out in eight different preparations.

suggests that an increase in $[Na^+]_i$ coupled to membrane depolarization reduces the electrochemical Na^+ gradient causing reversal of the GABA carrier.

In conclusion, glutamate increases the $[Ca^{2+}]_i$ of chick retina cells and induces Na^+-dependent and Ca^{2+}-independent release of [^{3}H]GABA, which is carrier mediated. The presence of Ca^{2+} actually inhibits the total release of [^{3}H]GABA induced by glutamate, although the mechanism for this effect of Ca^{2+} is not clear. In the absence of carrier-mediated release of [^{3}H]GABA, Ca^{2+} does not affect significantly the release of [^{3}H]GABA induced by glutamate. It should be noted that most glutamate receptors may be located far from the nerve terminals, as has been suggested for chick motoneurons,[27] and in these cases the [^{3}H]GABA release may be mediated by the GABA carrier located on the soma or dendrites of the GABAergic retina cells. This localization of GABA has already been reported.[7]

The results of the studies in which we stimulated each of the glutamate receptors with specific agonists (NMDA, KA, QA, and AMPA) are in accordance with previous reports that retinal cells in culture express NMDA and non-NMDA receptors, both of which can mediate Ca^{2+}-independent release of [^{3}H]GABA.[6] We further showed that these cells also exhibit Ca^{2+}-dependent release when stimulated by the glutamate receptor agonists NMDA, QA, and AMPA (FIG. 6), when the studies are carried out in the presence of NNC-711, to inhibit the carrier-mediated release.

When we studied the total release of [^{3}H]GABA, which includes carrier-mediated release and Ca^{2+}-dependent release, Ca^{2+} had an overall inhibitory effect (FIG. 3 and TABLE 2) when either glutamate or NMDA were the stimulating agents; however, this effect was not observed for the other agonists (TABLE 2). Furthermore, in the case of KA, QA or AMPA, Ca^{2+} increased the total release (TABLE 2), and QA and AMPA, as well as NMDA, induced Ca^{2+}-dependent release of [^{3}H]GABA when the carrier was blocked (FIG. 6). The Ca^{2+} inhibition of the [^{3}H]GABA release induced by glutamate or NMDA may be related to the effect of Ca^{2+} on the binding of glutamate or NMDA to the receptors.

Mechanisms of $[Ca^{2+}]_i$ Regulation by Glutamate Receptor Agonists

Glutamate probably is the main excitatory neurotransmitter activating amacrine cells,[1,2] and the results reported here show that the agonists of the ionotropic glutamate receptors, NMDA, KA, QA, and AMPA, increase the $[Ca^{2+}]_i$ in cultures enriched in amacrine cells. The increase in $[Ca^{2+}]_i$ was found to be due to Ca^{2+} entry through the receptor-associated channels and the VSCCs, as has been reported for other cell types.[16]

The $[Ca^{2+}]_i$ increases due to Ca^{2+} entry through the NMDA receptor-associated channel, as determined in Mg^{2+}-free NMG medium, was about 25% of the total response to the agonist in Na^+ medium, which included the VSCCs contribution. Permeation of Ca^{2+} through the NMDA receptor-associated channel has been reported for several cell types,[28] but the non-NMDA ionotropic glutamate receptors generally exhibit low Ca^{2+} permeability.[18–20]

Our results show clearly that, in retina cells, the influx of Ca^{2+} through the KA receptor-associated channel was higher than that observed for the NMDA receptor-associated channel (FIG. 9); the influx of Ca^{2+} through the KA receptor-associated channel corresponds to about 60% of the initial $[Ca^{2+}]_i$ response in a Na^+ medium, and it is potentiated by activation of PKC with TPA (FIG. 11A). However, the overall effect of PKC activation in a normal Na^+ medium is to inhibit the increase in $[Ca^{2+}]_i$ induced by KA (FIG. 11B), because PKC activation has an inhibitory effect on the VSCCs triggered by KA depolarization (FIG. 12). Activation of PKC has previously been shown to modulate the activity of glutamate receptors[29,30] and of VSCCs.[31] The influx of Ca^{2+} through the KA receptor-associated channel is a Ca^{2+} pathway that may be of great interest in mediating excitatory amino acid induced phenomena of various types, particularly when NMDA receptors are not present.[32,33] Moreover, the influx of Ca^{2+} through the KA receptor is assumed to contribute to the KA-induced neurotoxicity in some neurons *in vitro*.[18] We found that domoic acid activates the KA receptor and triggers exocytotic release of [^{3}H]GABA.[34]

Cultured retina cells have at least three pharmacologically distinct classes of VSCCs: L-type, sensitive to dihydropyridines;[8,35] N-type, sensitive to ω-CgTx;[8,9] and, as shown in this work, also the presence of P-type VSCCs sensitive to ω-Aga IVA,[36] when the cells are stimulated with KA (FIG. 9), but not detected when the stimulation is with NMDA or AMPA. The P-type channels have also been reported for horizontal cells of the bass retina.[37] It should be noted that although the N-type VSCCs are present in our preparation, as detected under K^+ depolarization,[8] they do not contribute to the $[Ca^{2+}]_i$ responses to glutamate.[9]

The differential activation of the VSCCs by NMDA, KA, and AMPA (FIG. 9) may reflect that (1) the receptors may be localized in different cells, all of which would contain the L-type VSCCs, but only some cell types would contain the P-type of channels or (2) the receptors and the P- and N-type channels are distributed heterogeneously in the cells.

Finally, (1S,3R)-ACPD increased intracellular accumulation of myo-[^{3}H]inositol phosphates (not shown), but had no effect on $[Ca^{2+}]_i$ (FIG. 10). Similar results have been reported for other cell preparations.[29,38,39] The increase in the accumulation of inositol phosphates without affecting the $[Ca^{2+}]_i$ suggests that in these cells the important signal may be diacylglycerol. The activation of PKC by diacylglycerol may modulate the activity of other glutamate receptors and the influx of Ca^{2+} through VSCCs.

REFERENCES

1. BARNSTABLE, C. J. 1993. Glutamate and GABA in retinal circuitry. Curr. Opin. Neurobiol. **3:** 520–525.
2. MASSEY, S. C. 1991. Cell types using glutamate as neurotransmitter in the vertebrate retina. *In* Progress in Retinal Research. N. Osborne & G. Chandler, Eds. Vol. **9:** 399–425. Oxford Press. Oxford.
3. IUVONE, P. M. 1986. Neurotransmitters and neuromodulators in the retina: Regulations, interactions and cellular effects. *In* The Retina: A Model for Cell Biology Studies. R. Adler & D. Barber, Eds. Part 3: 1–72. Academic Press. Orlando, FL.
4. MOSINGER, J. L., S. YAZULLA & K. M. STUDHOLME. 1986. GABA-like immunoreactivity in the vertebrate retina: A species comparison. Exp. Eye Res. **42:** 631–644.
5. AYOUB, G. S. & D. K. LAM. 1987. The release of GABA from horizontal cells of goldfish (*Carassius auratus*) retina. J. Physiol. (Lond.) **355:** 191–214.
6. HOFMANN, H. D. & V. MÖCKEL. 1991. Release of γ-amino[^{3}H]butyric acid from cultured amacrine-like neurons mediated by different excitatory amino acid receptors. J. Neurochem. **56:** 923–932.
7. JAFFÉ, E. H., N. HERNANDEZ & L. HOLDER. 1984. Study of the mechanism of release of GABA from teleost retina *in vitro*. J. Neurochem. **43:** 1226–1235.
8. DUARTE, C., I. FERREIRA, P. SANTOS, C. OLIVEIRA & A. CARVALHO. 1992. Ca^{2+}-dependent release of [^{3}H]GABA in cultured chick retina cells. Brain Res. **591:** 27–32.
9. DUARTE, C., I. FERREIRA, P. SANTOS, C. OLIVEIRA & A. CARVALHO. 1993. Glutamate increases the $[Ca^{2+}]_i$ but stimulates Ca^{2+}-independent release of [^{3}H]GABA in cultured chick retina cells. Brain Res. **611:** 130–138.
10. NEAL, M. J., J. CUNNINGHAM, P. HUTSON & J. SEMARK. 1992. Calcium dependent release of acetylcholine and γ-aminobutyric acid from the rabbit retina. Neurochem. Int. **20:** 43–53.
11. NICHOLLS, D. 1989. Release of glutamate, aspartate, and γ-aminobutyric acid from isolated nerve terminals. J. Neurochem. **52:** 331–341.
12. BERNATH, S. 1991. Calcium-independent release of amino acid neurotransmitters: Fact or artifact? Prog. Neurobiol. **38:** 57–91.
13. PIN, J. P. & J. BOCKAERT. 1989. Two distinct mechanisms, differentially affected by excitatory amino acids, trigger GABA release from fetal mouse striatal neurons in primary culture. J. Neurosci. **9:** 648–656.
14. NASCIMENTO, J. L. & F. G. MELLO. 1985. Induced release of γ-aminobutyric acid by a carrier mediated, high affinity uptake of L-glutamate in cultured chick retina cells. J. Neurochem. **45:** 1820–1827.
15. SUZDAK, P. D., K. FREDERIKSEN, K. E. ANDERSEN, P. O. SØRENSEN, L. J. S. KNUTTSE &

E. B. NIELSEN. 1992. Pharmacological characterization of NNC-711, a novel potent selective GABA uptake inhibitor. Eur. J. Pharmacol. **223:** 189–198.

16. MAYER, M. L. & R. J. MILLER. 1990. Excitatory amino acid receptors, second messengers and regulation of intracellular Ca^{2+} in mammalian neurons. Trends Pharmacol. Sci. **11:** 254–260.

17. MAYER, M. L. & G. L. WESTBROOK. 1987. Permeabilization and block of N-methyl-D-aspartic acid receptor channels by divalent cations in mouse cultured neurons. J. Physiol. (Lond.) **34:** 501–527.

18. MURPHY, S. & R. MILLER. 1989. Regulation of Ca^{2+} influx into striatal neurons by kainic acid. J. Pharmacol. Exp. Ther. **249:** 184–193.

19. HALOPAINEN, I., M. LOUVE & K. E. ÅKERMAN. 1991. Interactions of glutamate receptor agonists coupled to changes in intracellular Ca^{2+} in rat cerebellar granule cells in primary culture. J. Neurochem. **57:** 1729–1734.

20. OGURA, A., M. NAKAZAWA & Y. KUDO. 1992. Further evidence for calcium permeability of non-NMDA receptor channels in hippocampal neurons. Neurosci. Res. **12:** 606–616.

21. SCHOEPP, D. D. & P. J. CONN. 1993. Metabotropic glutamate receptors in brain function and pathology. Trends Neurosci. **14:** 13–20.

22. CHOI, D. W. 1987. Ionic dependence of glutamate neurotoxicity. J. Neurosci. **7:** 369–379.

23. SCHWARTZ, E. A. 1987. Depolarization without calcium can release γ-aminobutyric acid from a retinal neuron. Science **238:** 350–355.

24. TAPIA, R. & C. ARIAS. 1979. Selective stimulation of neurotransmitter release from chick retina by kainic and glutamic acids. J. Neurochem. **39:** 1169–1178.

25. SUCHER, N. J., S. Z. LEI & S. A. LIPTON. 1991. Calcium channel antagonists attenuate NMDA receptor mediated neurotoxicity of retinal ganglion cells in culture. Brain Res. **297:** 297–302.

26. YAZULLA, S. & J. KLEINSCHMIDT. 1983. Carrier mediated release of GABA from horizontal cells. Brain Res. **243:** 63–75.

27. O'BRIEN, R. J. & G. D. FISCHBACH. 1986. Characterization of excitatory amino acid receptors expressed by embryonic chick motor neurons *in vitro*. J. Neurosci. **6:** 3275–3283.

28. ASCHER, P. & L. NOWAK. 1983. The role of divalent cations in the N-methyl-D-aspartate responses of mouse central neurons in culture. J. Physiol. (Lond.) **399:** 247–366.

29. DILDY-MAYFIELD, J. E. & R. A. HARRIS. 1994. Activation of protein kinase C inhibits kainate-induced currents in oocytes expressing glutamate receptor subunits. J. Neurochem. **62:** 1639–1642.

30. TAN, S.-E., R. J. WENTHOLD & T. R. SODERLING. 1994. Phosphorylation of AMPA-type glutamate receptors by calcium/calmodulin-dependent protein kinase II and protein kinase C in cultured hippocampal neurons. J. Neurosci. **14:** 1123–1129.

31. RANE, S. G. & K. DUNLAP. 1986. Kinase C activator 1,2-oleoylacetylglycerol attenuates voltage-dependent calcium current in sensory neurons. Proc. Natl. Acad. Sci. USA **83:** 184–188.

32. DIXON, D. & D. COPENHAGEN. 1992. Two types of glutamate receptors differentially excite amacrine cells in the tiger salamander retina. J. Physiol. (Lond.) **449:** 589–606.

33. SORIMACHI, M. 1993. Calcium permeability of non-N-methyl-D-aspartate receptor channels in immature cerebellar Purkinje cells: Studies using *Fura*-2 microfluorometry. J. Neurochem. **60:** 1236–1243.

34. ALFONSO, M., R. DURAN, C. DUARTE, I. FERREIRA & A. CARVALHO. 1994. Domoic acid induced release of [³H]GABA in cultured chick retina cells. Neurochem. Int. **24:** 267–274.

35. WEI, X. Y., A. RUTLEDGE, Q. ZHONG, J. FERRANTI & D. J. TRIGGLE. 1989. Ca^{2+} channels in chick neural retina cells characterized by 1,4-dihydropiridine antagonists and activators. Can. J. Pharmacol. **67:** 506–514.

36. MINTZ, I., V. VENEMA, K. SWIDEREK, T. LEE, B. BEAN & M. ADAMS. 1992. P-type calcium channels blocked by spider toxin ω-Aga IVA. Nature **355:** 827–829.

37. SULLIVAN, J. M. & E. M. LASATER. 1992. Sustained and transient calcium currents in horizontal cells of white bass retina. J. Gen. Physiol. **99:** 85–107.

38. BRORSON, J. R., D. BLEAKMAN, S. J. GIBBSONS & R. J. MILLER. 1991. The properties of intracellular calcium stores in cultured rat cerebral neurons. J. Neurosci. **11:** 4024–4043.
39. LINN, C. P. & B. N. CHRISTENSEN. 1992. Excitatory amino acid regulation of intracellular Ca^{2+} in isolated catfish cone horizontal cells measured under voltage- and concentration-clamp conditions. J. Neurosci. **12:** 2156–2164.

Characterization of Voltage-Sensitive Ca^{2+} Channels Activated by Presynaptic Glutamate Receptor Stimulation in Hippocampus[a]

C. M. CARVALHO, J. O. MALVA,[b] C. B. DUARTE,
AND A. P. CARVALHO

Center for Neurosciences of Coimbra
Department of Zoology
University of Coimbra
3049 Coimbra Codex, Portugal

[b]*Department of Biology*
University of Minho
Braga, Portugal

Glutamate receptor activation modulates the release of several neurotransmitters and neuromodulators. In the hippocampus, the presence of presynaptic glutamate receptors modulating the release of noradrenaline was shown for both NMDA and non-NMDA receptors.[1,2] However, the link between receptor activation and the release of the neurotransmitter has not been clearly established.

In the present study we determined the effects of stimulation by glutamate, AMPA, or NMDA on the intracellular free Ca^{2+} concentration ($[Ca^{2+}]_i$) and on the release of [^{3}H]dopamine ([^{3}H]-DA) in hippocampal synaptosomes, and we found that: (1) Ca^{2+} entry due to glutamate receptor stimulation is necessary for [^{3}H]-DA release; (2) AMPA receptor stimulation activates voltage-sensitive Ca^{2+} channels (VSCCs) and a large fraction of Ca^{2+} entry occurs through these VSCCs, because blockade by specific Ca^{2+} channel blockers caused inhibition of [^{3}H]-DA release; (3) the VSCCs involved in the release of [^{3}H]-DA due to AMPA are of the N type (51% inhibition by ω-conotoxin GVIA [ω-CgTx]) and P or Q type (54% inhibition by ω-agatoxin IVA [ω-Aga IVA]); (4) modulation of [^{3}H]-DA release due to NMDA stimulation does not seem to involve activation of VSCCs.

CHANGES IN $[Ca^{2+}]_i$ AND [^{3}H]-DA RELEASE DUE TO AMPA OR NMDA STIMULATION

The results reported in FIGURE 1[3,4] show that stimulation of hippocampal synaptosomes with AMPA (100 μM) increased the $[Ca^{2+}]_i$ by 22.4 ± 1.1 nM and that this effect was inhibited by about 59% in the presence of 10 μM CNQX. NMDA (200 μM) increased the $[Ca^{2+}]_i$ by about 10 nM (FIG. 1A). These changes in $[Ca^{2+}]_i$ are coupled to the Ca^{2+}-dependent [^{3}H]-DA release. Thus, AMPA (100 μM) caused the release of 2.1 ± 0.1% of the total [^{3}H]-DA accumulated, and this effect was reduced to about 60% by 10 μM CNQX (FIG. 1B). NMDA (200 μM) caused the release of

[a]This work was supported by JNICT (Portuguese Research Council), Portugal.

3.6 ± 0.2% of the total [³H]-DA, and this effect was reduced by 81% by MK-801 (FIG. 1B). Thus, our results show that an apparent exocytotic [³H]-DA release can be triggered by stimulating AMPA or NMDA receptors in hippocampal nerve terminals.

INVOLVEMENT OF VSCCs ON AMPA BUT NOT ON NMDA RECEPTOR STIMULATION

Our results further show that Ca^{2+} entry due to AMPA stimulation occurs mainly through VSCCs, because both the $[Ca^{2+}]_i$ increase and the coupled [³H]-DA release were inhibited by ω-CgTx and ω-Aga IVA (TABLE 1).[3,4] Thus, both ω-CgTx (0.5 μM) and ω-Aga IVA (100 nM) inhibited the change in $[Ca^{2+}]_i$ ($\Delta[Ca^{2+}]_i$) induced by AMPA to 64.8 ± 9.0% and 77.5 ± 4.1% of control, respectively, and this effect was accompanied by an inhibition in [³H]-DA release to 49 ± 3.8% or 46.1 ± 10.5% of control, respectively. No significant inhibitory effects of Ca^{2+} channel blockers were observed on the release of [³H]-DA evoked by NMDA (200 μM), presumably in this case because Ca^{2+} enters mostly through NMDA receptor channels (TABLE 1). Thus,

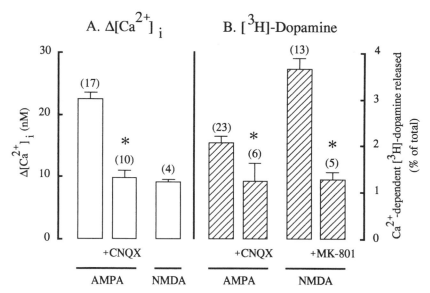

FIGURE 1. Effects of AMPA (100 μM) or NMDA (200 μM) on changes in $[Ca^{2+}]_i$ ($\Delta[Ca^{2+}]_i$) (A) and on the release of [³H]dopamine ([³H-DA]) (B) in hippocampal synaptosomes. The $[Ca^{2+}]_i$ was determined by a fluorimetric assay using Indo-1 as a probe for Ca^{2+}, and the release of preaccumulated [³H]-DA was studied by using a superfusion system, as described before.[4] Stimulation with AMPA was performed in normal Na^+ medium, whereas NMDA was tested in Mg^{2+}-free Na^+ medium in the presence of glycine (5 μM).[4] The antagonists CNQX (10 μM) and MK-801 (1 μM) were introduced 3 min before stimulation. The [³H]-DA released over basal by AMPA or NMDA, applied for 5 min, was determined in percentage of total [³H]-DA accumulated. High-performance liquid chromatography analysis showed that the released tritium was mainly [³H]-DA (75%), but some [³H]noradrenaline was also present (25%).[4] Data are presented as the mean value ± SEM of the number of ex periments indicated over the bars, performed in different synaptosomal preparations. *Significantly different from the respective control ($p < 0.05$), as determined by the Student t test.

TABLE 1. Relation of $[Ca^{2+}]_i$ Changes to $[^3H]$-Dopamine Release for Various Stimulating Agents and the Effect of Ca^{2+} Channel Blockers in Hippocampal Synaptosomes[a]

Stimulating Agent	Control	Percentage of Control		
		Nitrendipine	ω-Conotoxin GVIA	ω-Agatoxin IVA
KCl				
$\Delta[Ca^{2+}]_i$ (nM)	85.8 ± 2.6 (8)	93.9 ± 4.5 (5)	78.0 ± 7.0 (5)*	57.1 ± 2.4 (4)*
$[^3H]$-DA (%)	4.5 ± 0.4 (9)	—	—	—
Glutamate				
$\Delta[Ca^{2+}]_i$ (nM)	27.7 ± 1.2 (15)	94.2 ± 5.9 (4)	91.7 ± 7.1 (5)	72.5 ± 3.9 (4)*
$[^3H]$-DA (%)	2.1 ± 0.2 (19)	—	—	—
AMPA				
$\Delta[Ca^{2+}]_i$ (nM)	22.4 ± 1.1 (17)	105.2 ± 5.3 (4)	64.8 ± 9.0 (5)*	77.5 ± 4.1 (4)*
$[^3H]$-DA (%)	2.1 ± 0.1 (23)	100.4 ± 5.2 (5)	49.0 ± 3.8 (7)*	46.1 ± 10.5 (5)*
NMDA				
$\Delta[Ca^{2+}]_i$ (nM)	9.1 ± 0.4 (4)	—	—	—
$[^3H]$-DA (%)	3.6 ± 0.2 (13)	91.8 ± 6.5 (6)	94.9 ± 12.1 (5)	98.5 ± 9.0 (5)

NOTE: Experiments were performed as described in FIGURE 1. The Ca^{2+} channel blockers nitrendipine (1 µM), ω-CgTx (0.5 µM), or ω-Aga IVA (100 nM) were added 3 min before stimulation with various agents (5 mM KCl; 100 µM glutamate or AMPA; 200 µM NMDA). Data are presented as mean value ± SEM of changes in $[Ca^{2+}]_i$ (nM) or in evoked $[^3H]$-DA release (%). The effect of various Ca^{2+} channel blockers is expressed as % of the respective control. Asterisk indicates significantly different from control, $p < 0.05$.
[a]Data adapted partially from Malva *et al.*[3,4]

depolarization due to AMPA stimulation triggers the opening of N and P or Q type VSCCs in hippocampal synaptosomes, but not of L type, because nitrendipine had no effect on either $[Ca^{2+}]_i$ or $[^3H]$-DA release induced by AMPA (TABLE 1). The results in TABLE 1 also show that the same types of VSCCs were activated by KCl depolarization as by AMPA stimulation, although the contribution of N type VSCCs is greater with AMPA stimulation than with KCl depolarization.

In the case of glutamate (100 µM) stimulation, we observed that in addition to the effects of glutamate mediated through its receptors, part of the effects on both $[Ca^{2+}]_i$ and $[^3H]$-DA release were due to the interaction of glutamate with its carrier, because the effect could be mimicked by D-aspartate or by t-PDC,[3] a competitive inhibitor of the glutamate carrier. Glutamate transport is electrogenic and depolarizes the membrane of synaptosomes, and VSCCs are also triggered allowing Ca^{2+} entry, which could partially be blocked by ω-Aga IVA, suggesting the involvement of P or Q type VSCCs (TABLE 1).

We conclude that hippocampal synaptosomes are endowed with presynaptic glutamate receptors which modulate the release of dopamine by allowing influx of Ca^{2+} which triggers exocytosis of $[^3H]$-DA. Both AMPA and glutamate induce activation of VSCCs, whereas NMDA appears to modulate $[^3H]$-DA release due to Ca^{2+} entry probably through its receptor channel.

REFERENCES

1. PITTALUGA, A. & M. RAITERI. 1992. J. Pharmacol. Exp. Ther. **260:** 232–237.
2. WANG, J. K. T., H. ANDREWS & V. THUKRAL. 1992. J. Neurochem. **58:** 204–211.
3. MALVA, J. O., C. B. DUARTE, A. P. CARVALHO & C. M. CARVALHO. 1994. Submitted.
4. MALVA, J. O., A. P. CARVALHO & C. M. CARVALHO. 1994. Br. J. Pharmacol. **113:** 1439–1447.

Roles of Metabotropic Glutamate Receptors in Brain Plasticity and Pathology

STEPHAN MILLER,[a-c] J. PATRICK KESSLAK,[c,d]
CARMELO ROMANO,[e] AND CARL W. COTMAN[a,c,d]

*Departments of [a]Psychobiology and [d]Neurology, and
[c]Irvine Research Unit in Brain Aging
University of California, Irvine
Irvine, California 92717*

*[e]Departments of Ophthalmology and Neurobiology
Washington University Medical School
Box 8096, 660 South Euclid Avenue
St. Louis, Missouri 63110*

Receptors for the excitatory amino acid L-glutamate are widely expressed in both neurons and glia throughout the central nervous system. The glutamate transmitter system is the major system for the mediation of fast excitatory synaptic transmission and is also involved in neuroplasticity and higher cognitive functions (for review see ref. 1). Several subtypes of ionotropic glutamate receptors exist which are named for their selective agonists: *N*-methyl-D-aspartate (NMDA), kainate, and α-amino-3-hydroxy-5-methyl-4-isoxazole propionate (AMPA). These ionotropic receptors are composed of multiple subunits comprising an integral cation channel. In contrast, the metabotropic glutamate receptors (mGluRs), which are the focus of this paper, are a family of large monomeric receptors that exert their effects either on second messengers or ion channels via activation of GTP-binding proteins (G-proteins). The widespread distribution of both glutamate and the mGluRs suggests that glutamate may be the primary modulator of G-protein-coupled signal transduction in the central nervous system.

The mGluRs illustrate many of the emerging principles of receptor functional diversity represented in this volume. This paper provides a brief review of the diversity of the mGluRs including their structure, distribution, and the biochemical and electrophysiological consequences of their activation, but primarily focuses on their roles in brain plasticity and pathology and on the regulation of one receptor subtype in astrocytes. Studies from our laboratory and others demonstrated that agonists of the mGluRs can have either neuroprotective or neuropathological effects. These studies are reviewed, and possible explanations are discussed for the diversity of outcomes following mGluR agonist application. Finally, we present recent findings from our laboratory concerning an astrocyte model that we developed for the study of receptor regulation and function.

Structure and signal transduction. The members of this receptor family have a large extracellular domain (500–600 amino acids) and seven predicted transmembrane domains characteristic of G-protein-coupled receptors (FIG. 1), but have little sequence homology to other neurotransmitter receptors. The mGluRs can activate a

---

[b]Corresponding author.

variety of signal transduction mechanisms, of which the most extensively studied are the phospholipase C-mediated stimulation of phosphoinositide (PI) hydrolysis and the inhibition of adenylate cyclase activity. Hydrolysis of membrane phosphatidylinositol-4,5-bisphosphate by phospholipase C produces two second messengers: diacylglycerol, which activates protein kinase C, and inositol 1,4,5-trisphosphate (IP_3), which elicits the release of calcium from intracellular stores. Inhibition of adenylate cyclase reduces the accumulation of the second messenger cyclic AMP. The ability of glutamate and other mGluR agonists to stimulate PI hydrolysis and inhibit cyclic AMP accumulation has been characterized in a variety of preparations including tissue slices, cultured neurons, cultured astrocytes, and transfected cell lines (see ref. 2 for review). Ample evidence also exists that these receptors can couple to G-proteins that directly gate cation channels, stimulate phospholipase D activity, and stimulate adenylate cyclase activity. To date, seven subtypes (mGluR1–7)[3–9] have been identified, some of which are present in more than one splice variant

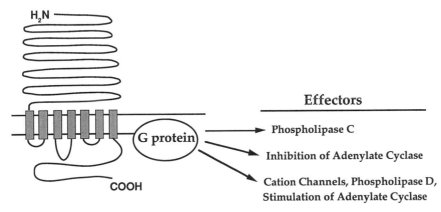

FIGURE 1. Schematic illustration demonstrating the salient features of the mGluRs which include a large extracellular domain and seven predicted membrane spanning regions. The receptors are coupled via G-proteins to a variety of effectors including phospholipase C, adenylate cyclase, and cation channels.

form.[5,10,11] These subtypes can be organized into three subfamilies[12] based upon their sequence homology, the effector system to which they couple when artificially expressed in CHO or BHK cells, and the agonist selectivity for activation of the receptor (TABLE 1).

Electrophysiological effects. The electrophysiological consequences of mGluR activation have been extensively studied and both pre- and postsynaptic effects have been identified. These studies were greatly facilitated with the identification of the selective agonist 1-aminocyclopentane-*trans*-1,3-dicarboxylic acid (*trans*-ACPD),* a conformationally restricted analog of glutamate with selectivity for metabotropic

---

*Most early studies with *trans*-ACPD used the racemic *trans*-(±)-ACPD composed of a mixture of the 1S,3R-ACPD and 1R,3S-ACPD enantiomers. More recently the separate enantiomers have also become commercially available and the more active 1S,3R-ACPD is frequently used.

receptors.[16] Activation of mGluRs results in both excitatory and inhibitory actions (for reviews see refs. 2 and 17). For example, in the hippocampus, *trans*-ACPD can produce depolarization and reduction of the after-hyperpolarization via a blockade of a calcium-activated potassium current. It has also been shown to block accommodation of cell firing, increase the amplitude of population spikes, induce generation of multiple spikes, decrease paired-pulse inhibition, and decrease evoked inhibitory postsynaptic potentials. Inhibitory actions of *trans*-ACPD in the hippocampus include reduction of the field excitatory postsynaptic potential via a presynaptic action.

Distribution. The mGluRs are expressed in neuronal and glial populations throughout the brain (see ref. 1 for review). These receptor subtypes show distinct patterns of distribution with differential expression both regionally and between cell types within a region. Each subtype has a unique pattern of expression although these patterns are sometimes overlapping. Some subtypes are widely expressed; mGluR3, for example, is prominently expressed in neurons in the cerebral cortex, thalamus, caudate putamen, and dentate gyrus and in glial cells throughout the brain.[12] The expression of other subtypes is quite restricted. For example, mGluR6

TABLE 1. Subgroups of the Metabotropic Glutamate Receptors[a]

Sub-group	Gene	Effector System	Agonist Selectivity	Ref.
I	mGluR1 mGluR5	Stimulate phospholipase C	QA > Glu ≥ Ibo > *t*-ACPD ≫ AP4	3, 4, 13, 14 6
II	mGluR2 mGluR3	Inhibit adenylate cyclase	Glu ≥ *t*-ACPD > Ibo ≫ QA ≫ AP4	5 5, 12
III	mGluR4 mGluR6 mGluR7	Inhibit adenylate cyclase	AP4 > Glu > SOP > ACPD > QA > Ibo AP4 > SOP > Glu ≫ ACPD > QA/Ibo AP4 = SOP > Glu > QA/ACPD	5, 12, 15 7 8, 9

[a]The seven metabotropic glutamate receptors can be classified into three subgroups based upon similarities in their sequence, effector coupling systems, and agonist selectivity.[12]

displays the most restricted expression of all the mGluRs, showing appreciable expression only in the inner nuclear layer of the retina, the region containing the ON-bipolar cells.[7] In addition to the regional variation in receptor expression it is interesting to note the degree of differential expression within a region. In the cerebellum for example, there is prominent expression of three separate subtypes: mGluR1 in Purkinje cells, mGluR2 in Golgi cells, and mGluR4 in granule cells. Such precise segregation of receptors implies an important role for the subtypes in functional specialization.

INVOLVEMENT OF METABOTROPIC GLUTAMATE RECEPTORS IN PLASTICITY

A substantial body of evidence now exists indicating that mGluRs have important roles in development and plasticity. For example, the developmental peak of excitatory amino acid-stimulated PI hydrolysis occurs between 6 and 12 days of age in neonatal rats and exhibits a high correlation with periods of intense synaptogenesis.[18,19] More direct evidence for a role of mGluRs in plasticity is provided by studies of long-term potentiation (LTP) and long-term depression (LTD). In the CA1 region of hippocampal slices, bath application of *trans*-ACPD can produce a form of LTP even without concomitant tetanic stimulation[20] and application of *trans*-ACPD in conjunction with tetanic stimulation potentiates the amount of LTP produced.[21] Most recently, it has been shown that the newly characterized metabotropic receptor antagonist α-methyl-4-carboxyphenylglycine (MCPG) can block the induction of LTP without affecting baseline synaptic transmission or previously established LTP.[22,23] Similar studies have indicated a role for mGluRs in LTD in both CA1 and in the parallel fiber-Purkinje cell synapse in the cerebellum.[24]

NEUROPROTECTIVE ACTIONS OF mGluR AGONISTS

Involvement of mGluRs in pathology was initially suggested when enhancements of excitatory amino acid-stimulated PI hydrolysis were demonstrated in *ex vivo* brain slice preparations following kindling,[25] hippocampal lesions,[26] and transient global ischemia.[27] These studies provided intriguing correlations between pathological conditions and mGluR activity, but could not address whether this increased receptor activity was part of the pathological cascade or part of a compensatory, possibly protective response to the injury. Additionally, such brain slice studies could not address the relative contributions of neurons and glia to these enhancements in PI hydrolysis. In the case of the ionotropic glutamate receptors, it has been well documented that overactivation of these receptors can produce pathological excitotoxicity through a process that involves calcium influx.[28] Several lines of evidence indicate that excitotoxicity is involved in the pathogenesis of trauma and ischemia as well as some neurodegenerative diseases.[29,30]

The involvement of mGluRs in excitotoxicity was first directly addressed in our laboratory when Koh *et al.*[31,32] quantified the effects of *trans*-ACPD in a mixed neuron-glial murine cortical culture model of excitotoxicity. As with the ionotropic receptors, activation of mGluRs coupled to PI hydrolysis also increases the cytoplasmic calcium concentration, leading to the initial hypothesis that mGluR agonists would also contribute to excitotoxicity. However, application of *trans*-ACPD did not produce neurotoxicity even when applied at concentrations as high as 1 mM for 24 h.[32] In contrast, extensive neuronal degeneration was produced following a 5-min exposure to 500 μM of the ionotropic receptor agonist NMDA (FIG. 2), as had been previously demonstrated. Modulation of NMDA toxicity by *trans*-ACPD was next examined with simultaneous exposure to the two agonists. Surprisingly, the presence of 100 μM *trans*-ACPD during the exposure of the cultures to NMDA markedly attenuated the NMDA-induced excitotoxicity (FIG. 2). This protective effect was not limited to glutamate receptor activation because the cholinergic receptor agonist carbachol, another activator of PI hydrolysis, had a similar but more modest protective effect, suggesting that production of excitotoxicity can be influenced by interaction between multiple receptor subtypes and transmitter systems.

The demonstration that mGluR agonists can have neuroprotective effects has now been replicated in both cortical and cerebellar cultures,[33,34] and other neuroprotective actions have been demonstrated both *in vitro* and *in vivo*. In rat hippocampal slices subjected to *in vitro* hypoxia, the presence of 1S,3R-ACPD during hypoxia was neuroprotective as measured by an enhancement in the posthypoxic recovery of field excitatory postsynaptic potentials.[35] Similarly, in an *in vivo* murine ischemia model, *trans*-ACPD administered intraperitoneally immediately after middle cerebral artery occlusion reduced the size of the infarct volume.[36] Finally, preinjection of 1S,3R-ACPD into the retina of adult rats has been shown to reduce the neurotoxicity produced by a subsequent injection of NMDA.[37]

Inasmuch as glutamate is necessary for normal physiological functioning, but

overexcitation by this transmitter can lead to pathological conditions of excitotoxicity, it seems likely that specific mechanisms must exist to control this balance between
normal functioning and pathology. Based on the results reviewed above, it is
intriguing to speculate that activation of mGluRs may be one of these mechanisms.
However, the scenario appears to be quite complex as several *in vivo* studies have
also shown that activation of these receptors can have neurotoxic consequences
rather than neuroprotective effects.

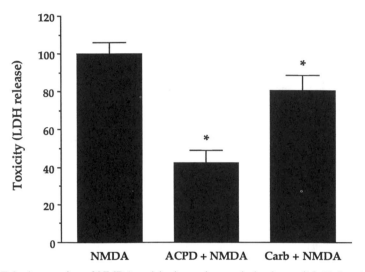

FIGURE 2. Attenuation of NMDA toxicity in murine cortical cultures (15–16 days *in vitro*).
Release into the medium of an intracellular enzyme, lactate dehydrogenase (LDH), was
quantified as an indicator of cell death. LDH release was measured 24 h after a 5-min exposure
to NMDA alone or in the presence of 100 μM *trans*-(±)-ACPD (ACPD) or 500 μM carbachol
(Carb). Data shown are mean ± SEM of three experiments performed in quadruplicate. LDH
values were scaled to the mean value in the NMDA-treated cultures after subtraction of the
mean value in sister cultures exposed to sham washes. Asterisk denotes differences from the
control ($p < 0.05$, two-tail t test with Bonferroni correction for two comparisons).

NEUROPATHOLOGICAL ACTIONS OF mGluR ACTIVATION

Several previous studies demonstrated both behavioral disturbances and neuroanatomical pathology following administration of *trans*-ACPD. Intrahippocampal
injection of 1S,3R-ACPD into adult rats produced seizure activity characterized by
akinesia, wet-dog shakes, rearing, limbic seizures, and hyperactivity.[38,39] In addition
to seizures, 1S,3R-ACPD also produced neuropathology including loss of CA1 and
CA4 pyramidal neurons and dentate gyrus granule neurons 6–7 days after injection.[39]
Striatal injections in neonatal rats produced little pathology but did potentiate
NMDA-induced injury as measured by reduction in brain weight 5 days after
injection.[40] However, injection of higher doses of *trans*-ACPD alone into the hippocampus or striatum of neonates produced loss of brain weight and signs of neuropathology including swollen perinuclear cytoplasm, extracellular debris, and shrunken
nuclei that were observed at 4 h but not 5 days following injection.[41]

We recently began studies to evaluate the effects of mGluR agonists in the hippocampus *in vivo*, focusing on the effects of lower concentrations of agonists, interaction with NMDA toxicity, and assessment of long-term anatomical outcome.[42] Intrahippocampal administration of 1S,3R-ACPD was found to produce dose-dependent damage when brains of young adult rats were examined 14 days after injection. Interaction between the ionotropic and mGluR receptor pathways was then examined using combinations of minimally toxic doses of 1S,3R-ACPD and NMDA to provide an *in vivo* parallel to the previous *in vitro* studies from our laboratory. Intrahippocampal injection of 1 μL of 10 mM 1S,3R-ACPD produced only slight neural damage to the dorsal blade of the dentate gyrus (FIG. 3B). Administration of 1 μL of 100 mM NMDA also produced limited degeneration in the dentate gyrus as well as in CA1. However, the combination of 1S,3R-ACPD and NMDA produced a much larger lesion, involving the lateral dentate gyrus and CA1 to CA3. (FIG. 3D). Although in contrast to the protective effects that have been demonstrated in other preparations, this enhancement of toxicity is consistent with the initial hypothesis that activation of mGluRs should contribute to excitotoxicity by elevating levels of intracellular calcium. Activation of an mGluR coupled to PI hydrolysis could increase calcium levels either by IP_3-mediated release of calcium from intracellular stores, or by diacylglycerol activation of protein kinase C and a subsequent facilitation of calcium current through NMDA receptors.[43]

PLASTICITY VERSUS PATHOLOGY

What are the reasons why both neuroprotective and neurotoxic effects have been reported following application of metabotropic agonists? One important consideration involves the differing methodological limitations of various preparations. Results from *in vitro* preparations must be interpreted with caution because these are artificially simplified systems, often lacking the full complement of cell types and the proper connectivity found *in vivo*. Although *in vivo* preparations are obviously desirable to examine the roles of mGluRs in intact systems, they also have inherent limitations. For example, the use of local injections often results in the introduction of extremely high local concentrations of agonists. Concentrations of 1S,3R-ACPD in the hundreds of millimolar have been used for intrahippocampal injections,[38,39] concentrations at which the selectivity of the compound for mGluRs is markedly reduced.[16] In fact, recent work from our laboratory suggests that there is pharmacological overlap between inhibitors of glutamate transport and mGluR agonists,[44] and high concentrations of *trans*-(±)-ACPD have previously been reported to inhibit glutamate uptake.[45] The resulting increase in extracellular glutamate would produce activation of both ionotropic receptors and mGluRs. In contrast, most of the methods of application that have yielded protective results with *trans*-ACPD (application of the agonist in culture medium,[31,33] in hippocampal slice bathing medium,[35] with intraperitoneal injection,[36] or intraocular injection[37]) allow greater diffusion of the agonist than what is probably achieved following intracerebral injection. Thus, crossover between uptake inhibition and receptor activation may at least partially account for the toxicity demonstrated in some preparations. Clarification of this issue will require development of increasingly selective pharmacological tools with less cross-reactivity between glutamate transport and mGluRs.

Another aspect which may contribute to the variation in the effects observed with mGluR agonists is differential expression of the mGluR subtypes in the various preparations that have been used. The ability of *trans*-ACPD to produce neurotoxic effects appears to be dependent upon a number of variables including age, brain

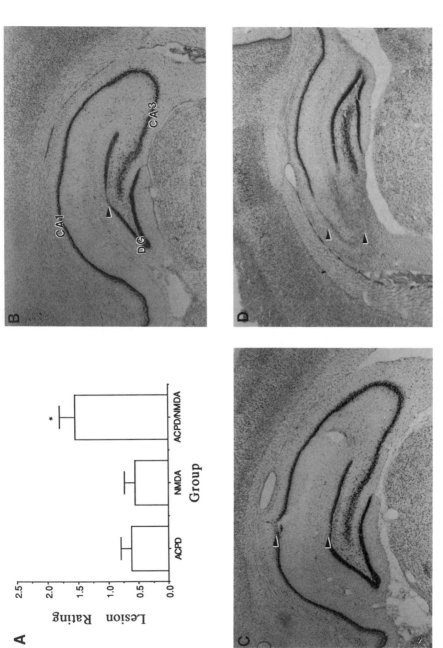

FIGURE 3. Excitotoxic interaction between 1S,3R-ACPD and NMDA following intrahippocampal injection. (**A**) Animals received unilateral intrahippocampal injections of either 1S,3R-ACPD (1 μL of 10 mM; $n = 9$) or NMDA (1 μL of 100 mM; $n = 13$). Both agonists together were administered to the contralateral hippocampus of each animal. Lesion size 14 days following injections was rated on a 3-point scale: 0 = no damage, 1 = mild, 2 = moderate, 3 = severe damage, and is presented as mean ± SEM. Asterisk denotes difference from NMDA alone ($p < 0.05$). (**B**) Injection of 1 μL of 10 mM of 1S,3R-ACPD by itself produced only minimal localized damage to the dorsal blade of the dentate gyrus (DG, *arrow*). (**C**) Injection of 1 μL of 100 mM NMDA produced minor damage to the CA1 and DG (*arrows*) surrounding the cannula tract and the site of injection. (**D**)

region, and strain of animal used. These types of differences could be due to differences in the amounts and ratios of receptor subtype expression. The mGluR subtypes can couple to a variety of signal transduction mechanisms, producing either excitatory or inhibitory effects. Thus, whether the net effect of an mGluR agonist is neuroprotective or neuropathological might depend upon the relative expression of specific metabotropic receptor subtypes coupled either to excitatory or inhibitory transduction mechanisms. As more selective pharmacological tools become available, progress can be made in evaluating the neurotoxic and neuroprotective effects of specific receptor subtypes. Further, development of experimental means with which to up-regulate or down-regulate specific mGluR subtypes may aid in examining the functional roles of these receptors.

REGULATION OF METABOTROPIC RECEPTORS—AN ASTROCYTE MODEL SYSTEM

The mGluRs of astrocytes may play a role in determining the outcome following administration of mGluR agonists *in vivo*. Astrocytes are a major component of glutamatergic pathways; their processes envelop glutamatergic synapses, and they have important roles in maintaining the balance between normal excitatory transmission and excitotoxicity, for example, through sodium-dependent high-affinity transport of glutamate.[46,47] Because activation of the mGluRs can have either neuroprotective or neurotoxic effects, we became interested in exploring the factors that regulate mGluR expression and in identifying more selective pharmacological tools with which to dissociate mGluR activation and transport inhibition. Astrocyte culture provides a system readily accessible to the study of transport pharmacology as well as mGluR signal transduction and receptor expression. Although some subtypes of mGluRs were known to be expressed in astrocytes, their function had not been previously determined. To approach an integrated functional understanding of these receptors the first step was to define a suitable *in vitro* model for their study in astrocytes.

Use of Serum-free Defined Medium

We began studies of metabotropic signal transduction and mGluR expression by analyzing *trans*-ACPD stimulation of PI hydrolysis using astrocytes cultured with conventional techniques. Primary glial cultures were prepared from neocortices of neonatal rat pups and purified by shaking as previously described.[48] After 7–8 days, secondary astrocyte cultures were established by trypsinizing and subplating into Dulbecco's Modified Eagle's Medium (DMEM) supplemented with 10% fetal calf serum (FCS).

Although widely used, this method of culture was unsatisfactory for our studies for several reasons. First, the morphology of astrocytes grown under these conditions is not representative of the morphology of the majority of astrocytes *in vivo*. Astrocytes grown in the FCS-supplemented medium were flat and polygonal in shape, having few processes (FIG. 4A), unlike the branching and stellate morphology of astrocytes *in vivo*. Second, the use of serum as a media supplement is a poor physiological model because exposure of brain cells to serum is normally limited *in vivo* by the blood-brain barrier. Third, culturing in serum-containing media has the disadvantage that serum contains a complex, undefined, and variable mixture of

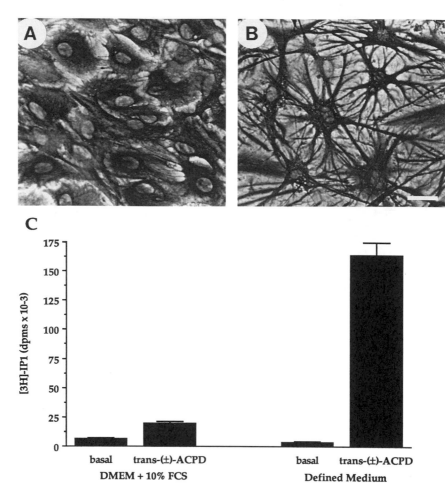

FIGURE 4. Glial fibrillary acid protein (GFAP) immunostaining of secondary astrocyte cultures maintained in either DMEM + 10% FCS (**A**) or in the serum-free defined medium (**B**). GFAP immunostaining was performed using a polyclonal rabbit primary antibody (1:2000 dilution; Dako, Carpinteria, CA) and avidin-biotin conjugated secondary antibody (ABC kit, Vector, Burlingame, CA) followed by visualization with diaminobenzidine. (**C**) Demonstration of the maximal stimulation of PI hydrolysis by *trans*-ACPD in these two types of cultures. Following a 24-h labeling period with [^{3}H]-*myo*-inositol (2 μCi/well), culture medium was aspirated and the cultures washed and preincubated for 20 min in buffer containing 10 mM LiCl (to inhibit *myo*-inositol-1-phosphatase), 116 mM NaCl, 26.2 mM NaHCO$_3$, 1 mM NaH$_2$PO$_4$, 2.5 mM KCl, 1.5 mM MgSO$_4$, 2.5 mM CaCl$_2$, and 20 mM glucose. Cultures were then stimulated with 500 μM *trans*-($\pm$)-ACPD for 60 min, followed by isolation of [^{3}H]-IP$_1$ with anion exchange chromatography and liquid scintillation counting essentially as described.[55] Values are mean $\pm$ SEM from five experiments performed in triplicate.

growth-promoting and -inhibiting components; this was a particular problem for these studies because we were interested in studying signal transduction and receptor expression in the context of neuroglial interaction, and thus required a model in which we had precise control over the concentrations of individual growth factors or cytokines. For these reasons we modified our culture techniques to the use of a serum-free chemically defined medium.[49] This method employed the same basal medium, DMEM, but rather than supplementing with FCS it was supplemented with eight defined components: transferrin (50 μg/mL), D-Biotin (10 ng/mL), selenium (5.2 ng/mL), fibronectin (1.5 μg/mL), heparin sulfate (0.5 μg/mL), epidermal growth factor (10 ng/mL), fibroblast growth factor (5 ng/mL), and insulin (5 μg/mL); a modification of the G-5 medium of Michler-Stuke *et al.*[50] When cultured under these conditions astrocytes had a highly branched, stellate shape (FIG. 4B), more similar to the morphology of astrocytes *in vivo*.

Robust Stimulation of Phosphoinositide Hydrolysis

In addition to the morphological changes, this manipulation of the growth conditions produced a radical alteration in the signal transduction properties of these cells.[49] When astrocytes were grown in secondary culture with FCS-supplemented DMEM, application of a maximally effective concentration of *trans*-ACPD (500 μM, 60 min) produced an accumulation of inositol monophosphate (IP_1), which was approximately threefold that of unstimulated cultures (3.2 ± 0.5 fold of basal, $n = 5$; FIG. 4C). However, in sister cultures maintained in secondary culture in the defined medium, *trans*-ACPD stimulated IP_1 accumulation more than 40-fold (43.2 ± 3.6 fold of basal, $n = 5$; FIG. 4C). The ability of two other mGluR agonists, quisqualate and glutamate itself, to stimulate PI hydrolysis was similarly enhanced (not shown). This exciting finding indicates a dynamic capability for astrocytes to respond to glutamate and suggests that previous studies using astrocytes cultured with serum-containing media had underestimated this potential.

Subsequently we examined the mechanism for this dramatic change in the PI response to determine whether this alteration represented a nonspecific sensitization of PI hydrolysis or whether the alteration was more selective to some component of an mGluR signaling pathway. When stimulation of PI hydrolysis by agonists of other transmitter systems was evaluated, a similar marked enhancement in PI hydrolysis was not observed. The accumulation of IP_1 stimulated by 500 μM norepinephrine was the same in DMEM + 10% FCS cultures (12.3 ± 1.3 fold of basal, $n = 3$, FIG. 5) as in defined medium cultures (12.6 ± 1.8 fold of basal, $n = 3$), whereas stimulation by the muscarinic receptor agonist carbachol (500 μM) was only moderately increased in defined medium cultures (3.1 ± 0.3 fold of basal, $n = 3$, FIG. 5) compared to DMEM + 10% FCS cultures (1.5 ± 0.2 fold of basal, $n = 3$). Because the dramatic enhancement in the stimulation of PI hydrolysis appeared to be selective for glutamate agonists, we hypothesized that exposure to the defined medium components had altered some component of a glutamatergic signal transduction system.

Regulation of mGluR Expression

We next evaluated the expression of the mGluRs themselves. Of the seven mGluR subtypes that have been characterized, two of these—mGluR1 and mGluR5— have been shown to be able to couple to PI hydrolysis.[6,14] Astrocytes were maintained

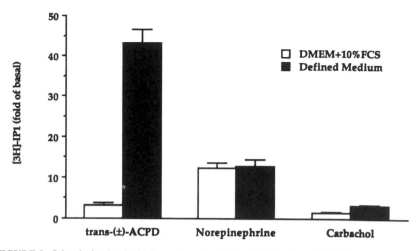

FIGURE 5. Stimulation by the indicated agonists (500 μM, 60 min) of [^{3}H]-IP$_1$ accumulation in astrocytes prelabeled with [^{3}H]-*myo*-inositol. Values are mean ± SEM from five experiments with *trans*-(±)-ACPD performed in triplicate and three experiments performed in triplicate with norepinephrine and carbachol. Basal (unstimulated) levels of [^{3}H]-IP$_1$ accumulation were 6,830 ± 820 dpm/well in DMEM + 10% FCS cultures and 4,010 ± 300 dpm/well in defined medium cultures.

for 4–5 days in secondary culture in DMEM supplemented with either 10% FCS or the defined medium supplements. Cultures were then harvested and the membranes prepared for Western immunoblotting against antibodies specific for either mGluR1α or mGluR5. Using an antibody prepared and characterized by Martin *et al.,*[51] we found no mGluR1α signal in membranes prepared from astrocytes cultured under either condition. This was not surprising as *in situ* hybridization and immunocytochemistry of brain slices have shown that mGluR1 is restricted to neuronal elements.[51,52]

However, mGluR5 does appear to be expressed in astrocytes. An affinity-purified antibody selective for mGluR5 was generated as described[53] using a synthetic peptide corresponding to the C-terminal 13 amino acids of the predicted polypeptide sequence of the mGluR5 cDNA.[6] The peptide was conjugated to thyroglobulin using glutaraldehyde, antisera prepared, and antibodies affinity-purified using standard techniques.[54] The anti-mGluR5 antibody should recognize both known splice variants mGluR5α and mGluR5β,[11] because the variants have identical C-terminal sequences, but the antibody does not recognize mGluR1α (not shown). Although there was little or no mGluR5 present in membranes prepared from DMEM + 10% FCS cultures, the antibody recognized a strong band at approximately 145 kDa in membranes from defined medium cultures (FIG. 6). Thus, a large induction in the expression of mGluR5 corresponds with the increased ability of mGluR agonists to stimulate PI hydrolysis. The results obtained with this relatively simple manipulation of the culture medium indicate that expression of the receptor subtypes can be plastic and that differential plasticity exists even between subtypes coupled to the same transduction mechanism.

SUMMARY

In summary, the mGluRs are a large family of receptor subtypes with diverse properties in terms of transduction coupling, pharmacology, and anatomical distribution. Many divergent studies have demonstrated that activation of these receptors can result in either neuroprotection or neuropathology. We hypothesized that the mGluRs of astrocytes may have a role in determining the response following administration of mGluR agonists *in vivo,* and we have defined a suitable *in vitro* model for the study of these receptors. The experimental plasticity demonstrated in the astrocyte culture model may represent a more general principle that conditions in the microenvironment may differentially alter mGluR subtype expression as part of development, functional specialization, or pathology. This astrocyte model of receptor regulation provides a system suitable for studying the effects of specific growth factors, neurotrophins, cytokines, and other substances released by neurons and glia that may act in both autocrine and paracrine fashions. Alteration in the ratios of receptors by such variables could then modify future signaling properties and neuroglial interactions, a form of conditioning of the astrocytic response that would alter the physiological output following glutamate release.

One measure of the value of this model will be its usefulness in stimulating the generation of hypotheses that can be tested *in vivo.* For example, the morphology of the astrocytes when cultured in the defined medium has similarities to the morphology of astrocytes undergoing reactive gliosis in pathological states. It is also interesting to note that treatments that have been reported to increase excitatory amino

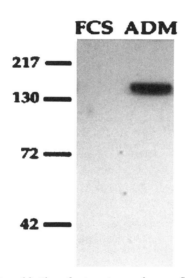

FIGURE 6. mGluR5 Western blotting of astrocyte membranes. Standard fractionation techniques were used to prepare crude membrane fractions from astrocytes cultured in either DMEM supplemented with 10% fetal calf serum (FCS) or cultured in the astrocyte defined medium (ADM). For electrophoresis, membranes were dissolved in sample buffer containing 20 mM dithiotreitol and subjected to SDS-PAGE in 7.5% gels. Separated proteins were transferred to Immobilon P membranes (Millipore, Bedford, MA) and incubated in TTBS (50 mM Tris HCl, 154 mM NaCl, 0.1% Tween-20, pH 7.5) containing 2.5% nonfat dry milk for 15 min, then overnight in the same buffer together with antibody (1:2500) and 0.1% sodium azide. After several washes in TTBS, the membranes were incubated in TTBS/2.5% milk containing goat anti-rabbit coupled to horseradish peroxidase (1:2000, GAR-HRP; Fisher, Pittsburgh, PA) for 2 h. After several washes in TTBS immunoactive bands were visualized using enhanced chemiluminescence (ECL reagent, Amersham, Arlington Heights, IL). The comparison between the two culture types was repeated four times using different astrocyte preparations. A representative blot is shown.

acid-stimulated PI hydrolysis in *ex vivo* brain slices (lesions,[26] ischemia,[27] and kindling[25]) are accompanied by reactive gliosis. Those findings combined with the present *in vitro* results lead us to speculate that mGluR5 expression may also be altered *in vivo* during reactive gliosis. If so, it will be important to examine the functional consequences of such a change with regard to the astrocytic response to injury and maintaining the balance between excitatory transmission and excitotoxicity.

ACKNOWLEDGMENTS

We thank Dr. Lee Martin for graciously providing the mGluR1α antibody, and Elisabeth Walcott, Dr. Jolanta Ulas, and Dr. Jennifer Kahle for critical readings of the manuscript.

REFERENCES

1. COTMAN, C. W., J. S. KAHLE, S. MILLER, R. J. ULAS & R. J. BRIDGES. 1995. Excitatory amino acid neurotransmission. *In* Psychopharmacology: The Fourth Generation of Progress. F. E. Bloom & D. J. Kufler, Eds.: 423. Raven Press. New York. In press.
2. SCHOEPP, D. D. & P. J. CONN. 1993. Metabotropic glutamate receptors in brain function and pathology. Trends Pharmacol. Sci. **14:** 13–20.
3. MASU, M., Y. TANABE, K. TSUCHIDA, R. SHIGEMOTO & S. NAKANISHI. 1991. Sequence and expression of a metabotropic glutamate receptor. Nature **349:** 760–765.
4. HOUAMED, K. M., J. L. KUIJPER, T. L. GILBERT, B. A. HALDEMAN, P. J. O'HARA, E. R. MULVIHILL, W. ALMERS & F. S. HAGEN. 1991. Cloning, expression, and gene structure of a G protein-coupled glutamate receptor from rat brain. Science **252:** 1318–1321.
5. TANABE, Y., M. MASU, T. ISHII, R. SHIGEMOTO & S. NAKANISHI. 1992. A family of metabotropic glutamate receptors. Neuron **8:** 169–179.
6. ABE, T., H. SUGIHARA, H. NAWA, R. SHIGEMOTO, N. MIZUNO & S. NAKANISHI. 1992. Molecular characterization of a novel metabotropic glutamate receptor mGluR5 coupled to inositol phosphate/Ca2+ signal transduction. J. Biol. Chem. **267:** 13361–13368.
7. NAKAJIMA, Y., H. IWAKABE, C. AKAZAWA, H. NAWA, R. SHIGEMOTO, N. MIZUNO & S. NAKANISHI. 1993. Molecular characterization of a novel retinal metabotropic glutamate receptor mGluR6 with a high agonist selectivity for L-2-amino-4-phosphonobutyrate. J. Biol. Chem. **268:** 11868–11873.
8. OKAMOTO, N., S. HORI, C. AKAZAWA, Y. HAYASHI, R. SHIGEMOTO, N. MIZUNO & S. NAKANISHI. 1994. Molecular characterization of a new metabotropic glutamate receptor mGluR7 coupled to inhibitory cyclic AMP signal transduction. J. Biol. Chem. **269:** 1231–1236.
9. SAUGSTAD, J. A., J. M. KINZIE, E. R. MULVIHILL, T. P. SEGERSON & G. L. WESTBROOK. 1994. Cloning and expression of a new member of the L-2-amino-4-phosphonobutyric acid-sensitive class of metabotropic glutamate receptors. Mol. Pharmacol. **45:** 367–372.
10. PIN, J. P., C. WAEBER, L. PREZEAU, J. BOCKAERT & S. F. HEINEMANN. 1992. Alternative splicing generates metabotropic glutamate receptor isoforms inducing different patterns of calcium release in Xenopus oocytes. Proc. Natl. Acad. Sci. USA **89:** 10331–10335.
11. MINAKAMI, R., F. KATSUKI & H. SUGIYAMA. 1993. A variant of metabotropic glutamate receptor subtype 5: An evolutionarily conserved insertion with no termination codon. Biochem. Biophys. Res. Commun. **194:** 622–627.
12. TANABE, Y., A. NOMURA, M. MASU, R. SHIGEMOTO, N. MIZUNO & S. NAKANISHI. 1993. Signal transduction, pharmacological properties, and expression patterns of two rat metabotropic glutamate receptors, mGluR3 and mGluR4. J. Neurosci. **13:** 1372–1378.
13. THOMSEN, C., E. R. MULVIHILL, B. HALDEMAN, D. S. PICKERING, D. R. HAMPSON & P. D.

SUZDAK. 1993. A pharmacological characterization of the mGluR1 alpha subtype of the metabotropic glutamate receptor expressed in a cloned baby hamster kidney cell line. Brain Res. **619:** 22–28.

14. ARAMORI, I. & S. NAKANISHI. 1992. Signal transduction and pharmacological characteristics of a metabotropic glutamate receptor, mGluR1, in transfected CHO cells. Neuron **8:** 757–765.

15. KRISTENSEN, P., P. D. SUZDAK & C. THOMSEN. 1993. Expression pattern and pharmacology of the rat type IV metabotropic glutamate receptor. Neurosci. Lett. **155:** 159–162.

16. PALMER, E., D. T. MONAGHAN & C. W. COTMAN. 1989. *Trans*-ACPD, a selective agonist of the phosphoinositide-coupled excitatory amino acid receptor. Eur. J. Pharmacol. **166:** 585–587.

17. MILLER, R. J. 1994. G-protein linked glutamate receptors. Semin. Neurosci. **6:** 105–115.

18. NICOLETTI, F., M. J. IADAROLA, J. T. WROBLEWSKI & E. COSTA. 1986. Excitatory amino acid recognition sites coupled with inositol phospholipid metabolism: Developmental changes and interaction with α_1-adrenoreceptors. Proc. Natl. Acad. Sci. USA **83:** 1931–1935.

19. PALMER, E., T. K. NANGEL, J. D. KRAUSE, A. ROXAS & C. W. COTMAN. 1990. Changes in excitatory amino acid modulation of phosphoinositide metabolism during development. Dev. Brain Res. **51:** 132–134.

20. BORTOLOTTO, Z. A. & G. L. COLLINGRIDGE. 1992. Activation of glutamate metabotropic receptors induces long-term potentiation. Eur. J. Pharmacol. **214:** 297–298.

21. MCGUINNESS, N., R. ANWYL & M. ROWAN. 1991. The effects of *trans*-ACPD on long-term potentiation in the rat hippocampal slice. NeuroReport **2:** 688–690.

22. BASHIR, Z. I., Z. A. BORTOLOTTO, C. H. DAVIES, N. BERRETTA, A. J. IRVING, A. J. SEAL, J. M. HENLEY, D. E. JANE, J. C. WATKINS & G. L. COLLINGRIDGE. 1993. Induction of LTP in the hippocampus needs synaptic activation of glutamate metabotropic receptors. Nature **363:** 347–350.

23. BORTOLOTTO, Z. A., Z. I. BASHIR, C. H. DAVIES & G. L. COLLINGRIDGE. 1994. A molecular switch activated by metabotropic glutamate receptors regulates induction of long-term potentiation. Nature **368:** 740–743.

24. LINDEN, D. J. 1994. Long-term synaptic depression in the mammalian brain. Neuron **12:** 457–472.

25. IADAROLA, M. J., F. NICOLETTI, J. R. NARANJO, F. PUTNAM & E. COSTA. 1986. Kindling enhances the stimulation of inositol phospholipid hydrolysis elicited by ibotenic acid in rat hippocampal slices. Brain Res. **374:** 174–178.

26. NICOLETTI, F., J. T. WROBLEWSKI, H. ALHO, C. EVA, E. FADDA & E. COSTA. 1987. Lesions of putative glutaminergic pathways potentiate the increase of inositol phospholipid hydrolysis elicited by excitatory amino acids. Brain Res. **436:** 103–112.

27. SEREN, M. S., C. ALDINIO, R. ZANONI, A. LEON & F. NICOLETTI. 1989. Stimulation of inositol phospholipid hydrolysis by excitatory amino acids is enhanced in brain slices from vulnerable regions after transient global ischemia. J. Neurochem. **53:** 1700–1705.

28. CHOI, D. W. 1987. Ionic dependence of glutamate neurotoxicity. J. Neurosci. **7:** 369–379.

29. MELDRUM, B. & J. GARTHWAITE. 1990. Excitatory amino acid neurotoxicity and neurodegenerative disease. Trends Pharmacol. **11:** 379–387.

30. CHOI, D. W. 1988. Glutamate neurotoxicity and diseases of the nervous system. Neuron **1:** 623–634.

31. KOH, J. Y., E. PALMER & C. W. COTMAN. 1991. Activation of the metabotropic glutamate receptor attenuates N-methyl-D-aspartate neurotoxicity in cortical cultures. Proc. Natl. Acad. Sci. USA **88:** 9431–9435.

32. KOH, J. Y., E. PALMER, A. LIN & C. W. COTMAN. 1991. A metabotropic glutamate receptor agonist does not mediate neuronal degeneration in cortical culture. Brain Res. **561:** 338–343.

33. PIZZI, M., C. FALLACARA, V. ARRIGHI, M. MEMO & P. F. SPANO. 1993. Attenuation of excitatory amino acid toxicity by metabotropic glutamate receptor agonists and aniracetam in primary cultures of cerebellar granule cells. J. Neurochem. **61:** 683–689.

34. BRUNO, V., A. COPANI, G. BATTAGLIA, R. RAFFAELE, H. SHINOZAKI & F. NICOLETTI. 1994.

Protective effect of the metabotropic glutamate receptor agonist, DCG-IV, against excitotoxic cell death. Eur. J. Pharmacol. **256:** 109–112.

35. OPITZ, T. & K. G. REYMANN. 1993. (1S,3R)-ACPD protects synaptic transmission from hypoxia in hippocampal slices. Neuropharmacology **32:** 103–104.

36. CHIAMULERA, C., P. ALBERTINI, E. VALERIO & A. REGGIANI. 1992. Activation of metabotropic receptors has a neuroprotective effect in a rodent model of focal ischaemia. Eur. J. Pharmacol. **216:** 335–336.

37. SILIPRANDI, R., M. LIPARTITI, E. FADDA, J. SAUTTER & H. MANEV. 1992. Activation of the glutamate metabotropic receptor protects retina against N-methyl-D-aspartate toxicity. Eur. J. Pharmacol. **219:** 173–174.

38. LIPARTITI, M., E. FADDA, G. SAVOINI, R. SILIPRANDI, J. SAUTTER, R. ARBAN & H. MANEV. 1993. In rats, the metabotropic glutamate receptor-triggered hippocampal neuronal damage is strain-dependent. Life Sci. **52:** PL85–90.

39. SACAAN, A. I. & D. D. SCHOEPP. 1992. Activation of hippocampal metabotropic excitatory amino acid receptors leads to seizures and neuronal damage. Neurosci. Lett. **139:** 77–82.

40. MCDONALD, J. W. & D. D. SCHOEPP. 1992. The metabotropic excitatory amino acid receptor agonist 1S,3R-ACPD selectively potentiates N-methyl-D-aspartate–induced brain injury. Eur. J. Pharmacol. **215:** 353–354.

41. MCDONALD, J. W., A. S. FIX, J. P. TIZZANO & D. D. SCHOEPP. 1993. Seizures and brain injury in neonatal rats induced by 1S,3R-ACPD, a metabotropic glutamate receptor agonist. J. Neurosci. **13:** 4445–4455.

42. KESSLAK, J. P., A. THEIS & C. W. COTMAN. 1994. Excitotoxic interaction between ACPD and NMDA after intrahippocampal administration. Soc. Neurosci. Abstr. **24:** 628.6.

43. ANIKSZTEJN, L., S. OTANI & Y. BEN-ARI. 1992. Quisqualate metabotropic receptors modulate NMDA currents and facilitate induction of long-term potentiation through protein kinase C. Eur. J. Neurosci. **4:** 500–505.

44. MILLER, S., R. J. BRIDGES, A. R. CHAMBERLIN & C. W. COTMAN. 1994. Pharmacological dissociation of metabotropic signal transduction pathways in cortical astrocytes. Eur. J. Pharmacol. **269:** 235–241.

45. ROBINSON, M. B., J. D. SINOR, L. A. DOWD & J. J. KERWIN. 1993. Subtypes of sodium-dependent high-affinity L-[³H]glutamate transport activity: Pharmacologic specificity and regulation by sodium and potassium. J. Neurochem. **60:** 167–179.

46. ROSENBERG, P. A., S. AMIN & M. LEITNER. 1992. Glutamate uptake disguises neurotoxic potency of glutamate agonists in cerebral cortex in dissociated cell culture. J. Neurosci. **12:** 56–61.

47. SUGIYAMA, K., A. BRUNORI & M. L. MAYER. 1989. Glial uptake of excitatory amino acids influences neuronal survival in cultures of mouse hippocampus. Neuroscience **32:** 779–791.

48. MCCARTHY, K. D. & J. DE VELLIS. 1980. Preparation of separate astroglial and oligodendroglial cell cultures from rat cerebral cultures. J. Cell Biol. **85:** 890–902.

49. MILLER, S., R. J. BRIDGES & C. W. COTMAN. 1993. Stimulation of phosphoinositide hydrolysis by trans-(±)-ACPD is greatly enhanced when astrocytes are cultured in a serum-free defined medium. Brain Res. **618:** 175–178.

50. MICHLER-STUKE, A., J. R. WOLFF & J. E. BOTTENSTEIN. 1984. Factors influencing astrocyte growth and development in defined media. Int. J. Dev. Neurosci. **2:** 575–584.

51. MARTIN, L. J., C. D. BLACKSTONE, R. L. HUGANIR & D. L. PRICE. 1992. Cellular localization of a metabotropic glutamate receptor in rat brain. Neuron **9:** 259–270.

52. GORCS, T. J., B. PENKE, Z. BOTI, Z. KATAROVA & J. HAMORI. 1993. Immunohistochemical visualization of a metabotropic glutamate receptor. NeuroReport **4:** 283–286.

53. ROMANO, C., M. A. SESMA, C. MACDONALD, K. O'MALLEY, A. N. VAN DEN POL & J. W. OLNEY. 1995. Distribution of metabotropic glutamate receptor mGluR5 immunoreactivity in the rat brain. J. Comp. Neurol. In press.

54. HARLOW, E. & D. LANE. 1988. Antibodies. A Laboratory Manual. Cold Spring Harbor Laboratory. Cold Spring Harbor, NY.

55. MILANI, D., L. FACCI, D. GUIDOLIN, A. LEON & S. D. SKAPER. 1989. Activation of polyphosphoinositide metabolism as a signal-transducing system coupled to excitatory amino acid receptors in astroglial cells. Glia **2:** 161–169.

Role of Glutamate and Glutamate Receptors in Memory Function and Alzheimer's Disease[a]

FRODE FONNUM,[b] TROND MYHRER,
RAGNHILD E. PAULSEN, KATRINE WANGEN,
AND ANNE RITA ØKSENGÅRD[c]

Norwegian Defence Research Establishment
Division for Environmental Toxicology
N-2007 Kjeller, Norway

[c] *Ullevål Hospital*
Oslo, Norway

Several investigations stress the involvement of glutamate and its receptors in learning and memory. The hippocampal region attracted early attention for being involved in such processes in man.[1,2] Later, long-term potentiation (LTP) was first ascribed to the hippocampus and several studies have developed the concept as a synaptic model for memory.[3,4] LTP is blocked by N-methyl-D-aspartate (NMDA) receptor antagonists and is localized to several different glutamergic pathways.[5] LTP is, however, a widespread phenomenon and can be produced throughout the limbic forebrain including the entorhinal area, but marked differences exist between different pathways.[6] Further, animal studies involving selective hippocampal lesion have not been able to confirm a global mnemonic role of the hippocampus.[7] According to the literature on the rat, the hippocampus appears to be more involved in working memory than in reference memory.[8,9]

Human investigations on memory function focused on the neurochemical and histological changes in Alzheimer's disease. Alzheimer's disease is a form of dementia characterized histologically by the presence of neurofibrillary tangles and senile plaque and a loss of pyramidal cells in the brain. The severity of the dementia is related to the number of neurofibrillary tangles and less to the senile plaques. Previously, the cholinergic dysfunction in Alzheimer's disease had been the dominating concept,[10] but in the last five years several reviews have pointed out a possible central role of glutamergic neurons in the development of the pathology.[11–14] The ventromedial temporal lobe including the entorhinal cortex, the hippocampus, and the amygdala are the most markedly atrophied regions in Alzheimer's disease. The association areas of the parietotemporal and prefrontal cortex are involved to an intermediate degree.[15,16]

It now seems accepted that the first cells to be affected with neurofibrillary tangles are located in the transentorhinal cortex, followed by the perirhinal cortex, the entorhinal cortex, and the pyramidal cells in CA1 and subiculum. This hierarchy of pathology occurs both in normal aging and more predominantly in Alzheimer's disease.[17,18] The affected areas in the brain seemed to be closely linked with the entorhinal cortex. An area separated by none or one synapse is more affected than an

[a] This work was supported in part by the Ring Foundation, Oslo.
[b] Corresponding author.

area separated from the entorhinal cortex by several synapses.[19] The entorhinal cortex is linked with strong projections from the association and limbic cortices.[20]

In view of the emphasis placed on entorhinal and perirhinal cortices, we investigated some glutamergic parameters after surgical transections involving these regions. Further, the effects of rearing animals under different conditions have also been investigated both neurochemically and in terms of cognitive functions. Most importantly, however, we investigated the effects of surgical lesions on learning and particularly on retention of memory. Finally, we showed how glutamatergic agonists can ameliorate the effect of lesion on learning and retention of memory and discussed their possible therapeutic potential in treatment of Alzheimer's disease.

METHODS

Rats were examined in a visual discrimination test in which the task was to differentiate one aluminium cylinder, either grey or black, from two other aluminium cylinders, which were in the other shade.[21] Testing was carried out in a Plexiglas cage (56 × 34 × 20 cm). A Plexiglas wall with an opening (10 × 10 cm) in the middle divided the apparatus into two equal compartments: the start compartment and the goal compartment. In the goal compartment there were three interchangeable aluminium cylinders with a round well in the top. The cylinders were placed with equal distance between them along the wall opposite to the partition wall. Half of the animals were trained with the black cylinder as positive and the other half with the grey cylinder as positive. The position of the positive cylinder (left, middle, right) was changed in a prearranged order and by a different set of randomized positions on days 2 and 3 and on retention. The positive cylinder was the same during all the testing for each animal.

During acquisition and retention testing the rats were deprived of water for 23½ hours a day. On the first day, each rat was allowed to explore the empty test apparatus for 15 min. On the second day, the rats were trained to run from the start to the goal compartment and allowed to explore the cylinders until they hit the positive one with water. The rats were given 10 trials and the intertrial interval was 20 s. On the third day, the animals were tested until five correct responses in succession were obtained. Both the number of errors and the number of trials until the learning criterion had been reached were noted. In order to drink or to investigate whether the well contained water, the rat had to stand on its hind legs with at least one forepaw on top of the cylinder. Error was scored when the animal mounted a negative cylinder to drink and not when they only investigated the cylinders apart from the well. The animals were operated on day 4 as described below. Intraperitoneal injections of glutamatergic agonists were given immediately after the operation and at two and ten days after surgery. The animals were tested for retention 12 days after surgery, that is, on day 16.

The animals were first operated on for the learning paradigm. Eight days later they were trained with the same procedure as described above.

Surgery

The rats were anesthetized intraperitoneally with diazepam (10 mg/kg) and fentanyl fluanisone (2 mg/kg) and placed in a stereotaxic head holder. The lesions were made mechanically by means of the sharp edge of cannula (diameter 0.5 mm) as described.[21]

Neurochemical Assays

High-affinity D-aspartate uptake was taken as a marker for glutamergic nerve terminals and assayed as described by Lund Karlsen and Fonnum.[22] Choline acetyltransferase and glutamate decarboxylase were used as markers for cholinergic and GABAergic nerve terminals, respectively. Choline acetyltransferase and glutamate decarboxylase were assayed by previously described methods.[23,24] Tissue samples were taken from the temporal cortex (TC), the lateral entorhinal cortex (LEC) or the frontal cortex.

RESULTS AND DISCUSSION

Neurochemical Studies of Entorhinal and Temporal Cortices

The importance of the entorhinal and temporal cortices in cognitive and mnemonic processes prompted us to investigate some neurochemical parameters in these regions. We found a 20% higher level of high-affinity D-aspartate uptake in a homogenate from the left LEC than from the right side in normal rats ($p < 0.05$). Such a lateralization was not found in the TC or for the other parameters such as choline acetyltransferase and glutamate decarboxylase.[25] This lateralization is also consistent with the effect of unilateral lesion on retention of memory (TABLE 1).

TABLE 1. Changes in Retention of a Visual Discrimination Task following Transections of Temporoentorhinal Connections

Lesions	Prior Learning Trials (n)	Retention (12 days)	
		Trials (n)	Errors (n)
No lesion	18	6.5	1.4
Bilateral TC/LEC	18	22.7[a]	8.6[a]
Left TC/LEC	16	12[a]	3[a]
Right TC/LEC	17	5	0
Bilateral medial perforant path	18	6.3	1.1
Bilateral lateral perforant path	19	7.4	1.3
Bilateral dorsal hippocampus	18	6.9	1.4

Abbreviations: TC, temporal cortex; LEC, lateral entorhinal cortex.
[a] $p < 0.05$.

Further support for the lateralization of glutamergic activity in LEC is the consistent enlargement of the lesions in the left LEC after systemic administration of glutamergic agonists.[26] When the TC/LEC connections were transected at the level of the perirhinal cortex, a 40% reduction in high-affinity D-aspartate uptake in both denervated cortices, demonstrating a reciprocal glutamergic connection ($p < 0.01$). No corresponding effects on glutamate decarboxylase or choline acetyltransferase were found. The remaining glutamergic activities in LEC and TC indicate other glutamergic inputs, for example, from associative cortical connections.[27]

It is well known than environmental factors can modify the development of the central nervous system. Rosenzweig and co-workers showed that rats reared in an enriched environment had increased brain weight, higher acetylcholinesterase activity, and an increased number of dendritic spines compared to rats housed in an isolated environment.[28] We focused on the effect of different rearing conditions on the glutamergic and cholinergic systems in entorhinal and temporal cortices[29] (TABLE 2). Three groups of rats (25 days old) were reared under isolated, social or enriched conditions for two months. The isolated rats were reared in a Plexiglas cage alone. The social rats were reared in groups of five in a large Plexiglas cage. The enriched group was reared in groups of five in a similar large cage containing three objects that were changed three times a week.

A significant difference in D-aspartate uptake was present in LEC, in correlation with the more enriched rearing conditions (TABLE 2). The enriched group showed a significant increase compared to the isolated group ($p < 0.01$). In the enriched group there was slightly higher uptake activity in LEC than in both temporal and frontal cortices. No differences in choline acetyltransferase were found between the different groups or between the three cortical areas. The enriched group had a slightly lower body weight than the other two groups.[29]

TABLE 2. Biochemical Parameters and Cognitive Function in Lateral Entorhinal Cortex of Rats Reared under Different Conditions

A. Biochemical Parameters			Isolated (10)[a]	Social (10)	Enriched (10)
HA DAsp uptake			1261	1436	1579
ChAT			527	528	519

B. Cognitive Function	Day 1	Acquisition (Day 2)		Retention (Day 14)	
	Errors[b]	Errors[b]	Trials	Errors	Trial[b]
Isolated (7)	4 (2–5)	1 (0–2)	17 (15–21)	1 (0–2)	8 (5–11)
Enriched (8)	2 (1–3)	0 (0–1)	15 (15–16)	0 (0)	5 (5–7)

[a]Number of samples indicated in parentheses.
[b]$p < 0.05$.

The level of D-aspartate uptake correlated significantly with both the acquisition and the retention of the brightness discrimination task (TABLE 2). These findings need not mean that uptake activity or glutamergic terminals in LEC are casually related to behavior. Together with other findings in this paper, however, they strengthen the case for an involvement of LEC in learning and memory.

Studies on Learning and Retention of Memory after Brain Lesions

We studied the acquisition and retention of memory in a series of surgical lesions including entorhinal cortex, temporal cortex, perirhinal cortex, and hippocampus.[21,30] Transection of the white matter in the rhinal sulcus disrupts the reciprocal connections between TC, perirhinal cortex, and LEC. Bilateral transections resulted in a dramatic impairment in retention of the visual discrimination task (TABLE 1). When unilateral transection was carried out involving the left side, the one with the highest D-aspartate uptake activity, a marked reduction also occurred in retention of the task. Transection involving the right-hand side showed no effect or even a slight improvement in the retention of this task.

A corresponding retention deficit could not be observed with a bilateral transection of the medial or lateral perforant path or by a bilateral removal of two-thirds of the hippocampus (TABLE 1). The lesions involving the perforant path, however, affect the animal's exploratory behavior and reaction to novelty.[31]

Role of Glutamergic Receptors in Cognitive Function and Alzheimer's Disease

The glutamergic receptors are separated into NMDA receptors, AMPA/kainate receptors, and metabotrophic receptors.[32] The NMDA receptor consists of several subunits called NMDAR1, NMDAR2A, NMDAR2B, NMDAR2C, and NMDAR2D. Heteromeric assemblies of these subunits give receptors that react very differently with glycine.[33] The AMPA/kainate receptor consists of several subunits, some with high affinity for AMPA and some with high affinity for kainate.[34] So far, six or seven metabotrophic receptors have been identified.[35]

Several examples in the literature show that NMDA receptor antagonists will block learning. The most prominent are probably the works of Morris and colleagues[36,37] who showed that intraventricular infusion of the antagonist DLAP-5 caused an impairment of spatial learning, but not the retention of previously acquired spatial information. It also blocked LTP in vivo without blocking normal synaptic transmission. Ingram et al.[38] showed a dose-dependent impairment of maze performance after giving dizocilpine (MK801). Watanabe et al.[39] showed that intraventricular injection of 7-chlorokynurenic acid, a selective antagonist at the glycine site of the NMDA receptor, inhibited LTP and impaired spatial learning in rats.

Several studies exist on changes in the sodium-independent glutamate binding, that is, receptor binding, in brains of patients with Alzheimer's disease. Earlier studies, using less specific assays, showed a dramatic fall in NMDA receptors in the hippocampus and parahippocampal region.[40] The issue was reexamined with 10 Alzheimer brains, nine controls, and six demented brains. Overall, there was a 50% loss of NMDA receptors in the pyramidal cell layer of CA1 and a 40% loss in CA3 compared to control and demented brains.[41] The decrease in NMDA receptor binding was confirmed in an investigation of a number of very old, nondemented or demented women and younger control brains[42] (27–64 years). A reduction in NMDA and kainic acid receptor binding was found in very old, nondemented women compared to controls, and a further significant reduction was found in Alzheimer brains. There was also some correlation with mental status, the number of neurofibrillary tangles, and the loss of receptor binding in CA1 of hippocampus. Jansen et al.[43] found a decrease of 50% in NMDA receptor binding, binding to the ion channel and to the glycine site in the CA1 region of hippocampus. Other studies have not been able to confirm a loss of NMDA receptors in hippocampus in Alzheimer's disease.[44–46]

Procter et al.[47] studied ^{3}H[MK801] binding to the ion channel of the NMDA receptor and reported reduced glycine stimulation in samples from frontal, parietal, and temporal cerebral cortex of Alzheimer's disease brains. This finding was later challenged.[43,48] Glycine binding can easily escape detection in membranes that are not well washed or contain an endogenous agonist. In the three cases above, a reduction of 30–50% occurred in the fully stimulated NMDA receptor binding. There was no significant loss in the parietal and temporal cortices.[47]

In general, there is a reduction in NMDA receptor binding in the hippocampus and parahippocampal regions. Little evidence supports the view that the properties of the receptor have been altered. One should keep in mind that there are several forms of the NMDA receptor, which will differ slightly in their properties. Although this has not yet been demonstrated, it may be that some forms of NMDA receptors are more resistant to loss than others during Alzheimer's disease.

Both Jansen et al.[43] and Dewar et al.,[49] but not Penney et al.,[41] found a loss of AMPA binding in CA1 and the subicular regions of the hippocampal and the parahippocampal regions. It is in this context interesting to note that aniracetam, a so-called nootropic drug, improved synaptic neurotransmission mediated by AMPA receptors.[50] Because the AMPA receptor is often colocalized with the NMDA receptor, a loss of both receptors is not surprising.

Dewar et al.[49] found a reduction in glutamate metabotrophic receptor binding in CA1, and the loss was larger than could be accounted for by the loss of neurons. The metabotrophic receptors may play a role in the release mechanism of neurotransmitters.

The memory dysfunction in rats with TC/LEC lesions could be effectively ameliorated by intraperitoneal injection of NMDA or AMPA/kainate receptor agonists (TABLE 3). The agonists were given one day after acquisition of the task and

TABLE 3. Effect of Glutamergic Agonist on Retention of Visual Discrimination Task in TC-LEC Transected Rats

	Prior Learning		Retention (12 days)	
	Dose (mg/kg)	Trials (n)	Trials (n)	Errors
Untransected + saline		18.8	6.8^a	0.5
Transected +				
Saline		17.3	22.5	9.5
NMDA	50	18.0	7.0^a	0.8
Glycine	750	17.5	5.7^a	0.3
AMPA	2.5	19.3	6.8^a	0.7
Kainate	5	17.6	14.7	4.2
Adrenaline	0.1	16.8	19.7	6.0
Bretylium + glycine	5 + 750	17.6	6.0	0.5
HA966 + glycine	30 + 750	17.8	18.5	8.2

TC, temporal cortex; LEC, lateral entorhinal cortex.
$^a p < 0.01$.

thus immediately after the surgical transection. The chemical compounds were also given two days after and sometimes ten days after the lesion. They were also given in high enough doses to pass through the blood-brain barrier. The glycine concentration used was high enough to cause temporary paralysis of the hind legs for a few minutes, and kainic acid produced "wet dog shakes." Both NMDA and glycine restored the memory retention.[51] When glycine was used in half the concentration (375 mg/kg), it had a weaker effect.[26] When HA-966, an antagonist of the glycine site in the NMDA receptor, was given 30 min before glycine, no restoration of memory function was found (Myhrer, unpublished data).

The agonists did not act by way of a peripheral adrenergic mechanism because blockade of adrenergic release with bretylium did not reduce the positive effect of glycine.[51] AMPA had a positive effect, whereas kainic acid, which had an additional toxic effect, was less effective. The results showed that both NMDA receptor agonists affecting the glycine site of the NMDA receptor and AMPA/kainate receptor agonists restored the memory function after surgical lesion of TC/LEC. Full improvement was also seen in proactive memory when glycine was given prior to training.[26]

While this work has been in progress, several other reports have been published that support a positive effect of glutamate agonists in learning and memory in animal experiments. In mice that were trained on a shock avoidance learning paradigm, it was shown that intraventricular injection of L-glutamate, L-aspartate, (±)β-*p*-chlorophenylglutamate, kainate, and quisqualate improved retention after one week in a dose-dependent manner, whereas D-glutamate did not.[52] Similarly, milacemide, a glycine prodrug, enhanced performance in a learning task in normal and amnestic rodents.[53]

Experiments with Cycloserine, a Partial NMDA Agonist

D-Cycloserine is a potent partial agonist which acts on the glycine site of the NMDA receptor. It possesses 60% of the maximal response of glycine and passes easily through the blood-brain barrier.[54] Cycloserine stimulates MK-801 binding to the NMDA receptor equally efficiently in membranes from the inferior parietal cortex of both control and Alzheimer brain.[55] Cycloserine has been used as an antibiotic against tuberculosis and has been given to humans in rather high concentration (0.5–1.0 g). When cycloserine was given in a single dose to TC/LEC transected rats, it effectively restored memory provided it was given within a few days after the operation or just prior to the retrieval. It is suggested that the action profile of cycloserine may reflect effects of both functional (LTP) and pharmacological mechanisms.

Monahan *et al.*[54] was the first to show that the partial agonist cycloserine (3 mg/kg) could improve learning and memory in a one-trial passive avoidance test and in reversal of T-maze learning. Sirvio *et al.*[56] and Fishkin *et al.*[57] extended Monahan's findings by using rats pretreated with scopolamine (1 mg/kg) in a water-maze test and found improvement in performance with varying doses of cycloserine. Flood *et al.*[58] obtained improvement in performance of a footshock test with cycloserine for both young and senescent rats treated with different doses of scopolamine. Schuster and Schmidt[59] obtained positive results with cycloserine (12 mg/kg) given to rats with hippocampal lesions 30 min before testing of working memory in an allocentric reversal test.

In contrast, Rupniak *et al.*[60] examined the effect of cycloserine on primates treated with scopolamine and did not find any improvement in a visual-spatial memory test. The concentration of cycloserine used ranged from 3 to 14 mg/kg. When we compare this dose to that given to humans (see next section), this may be too high a dose. An overview of studies involving cycloserine in cognitive function is presented in TABLE 4.

Experiments of Cycloserine in Humans

Twenty-four healthy, young male and 24 healthy, elderly male and female (age 63–75) volunteers participated in two identical studies using a double-blind placebo-controlled Latin square design.[61] The participants received scopolamine 0.5 mg (young) and 0.2 mg (old) and later 0, 5, 15 or 50 mg cycloserine. For young subjects, 15 mg D-cycloserine antagonized the scopolamine-induced decrement in three memory tasks. Less or no significant effects were seen with 5 or 50 mg doses. In older subjects, 5 mg cycloserine significantly reduced the decrement in a word recognition test. A similar but smaller effect was seen with 15 mg, whereas 50 mg was inefficient.

TABLE 4. Experiments Involving Cycloserine in Cognitive Function

Authors	Species	Status		Cycloserine (mg/kg)	Results
Monahan et al.[54]	Rat	Normal	Passive avoidance Reversal of T-maze	3	+
Sirvio et al.[56]	Rat	Scopolamine	Water maze		+
Fishkin et al.[57]	Rat	Scopolamine	Water maze	3, 10, 30	+
Flood et al.[58]	Rat	Senescent rats	Foot shock	20	+
Schuster & Schmidt[59]	Rat	Hippocampal lesion	8-Arm maze	12	+
Rupniak et al.[60]	Primates	Scopolamine	Visual-spatial memory	3–14	–
Jones et al.[61]	Human	Scopolamine	Several tests	5, 15 mg[a]	+

[a]Dose per individual.

We (Wangen et al., unpublished data) studied the effects of D-cycloserine (0, 5, and 15 mg) on six persons with Alzheimer's disease (TABLE 5). The patients were not on medication, and the symptoms ranged from those of a recently acquired illness to severe symptoms of Alzheimer's disease. The patients were nursed at home by their families. The patients received in three consecutive weeks placebo tablets, 5 mg tablets or 15 mg tablets, in a prearranged order. The studies were carried out double blind. The patients were tested before the study started to obtain the baseline level and during the last day on receiving each dose. The results show that five of six patients responded better with both doses of D-cycloserine than with placebo in the trial-making test. In the object-learning test three were better with the low dose, one was better with the high dose and the remaining two were equal to the placebo. In one instance the family reported a significant improvement in behavior, and also in another case there was also some improvement during cycloserine administration.

TABLE 5. Clinical Testing of Low or High Dose of D-Cycloserine in Patients with Alzheimer's Disease[a]

Patient	Object Learning Test[b]				Trial Making Test[c]			
			Cycloserine				Cycloserine	
	Base	Placebo	Low	High	Base	Placebo	Low	High
AA	36	33	38	32	50	42	27	32
BB	22	21	34	19	65	66	51	46
CC	25	27	28	23	119	58	83	64
DD	14	23	23	17	510	241	127	198
EE	19	23	20	23	144	217	139	184
FF	16	18	18	21	66	61	58	49

[a]Low dose of D-cycloserine = 5 mg; high dose = 15 mg.
[b]Expressed in points.
[c]Expressed in seconds.

CONCLUDING REMARKS

In this paper we focused on the glutamatergic connections between entorhinal and temporal cortices and their importance for learning and, more significantly, for retention of memory. The glutamergic transmission in the lateral entorhinal cortex is subjected to a slight degree of lateralization and also is affected by the conditions of rearing the animals. In both cases increased glutamergic activity correspond to an improved ability of learning and retention of memory. Administration of glutamergic agonists, both of the NMDA and AMPA types, restores the mnemonic function following a transection of the connection between entorhinal and temporal cortex. Several studies show cycloserine, a partial NMDA agonist, is effective in improving mnemonic processes in man and animals with brain lesions or manipulated with scopolamine, a muscarinic antagonist. The study shows that cycloserine or similar compounds may have a positive effect on patients with Alzheimer's disease and provides a basis for investigating further such a therapy.

ACKNOWLEDGMENTS

We thank Professors Ivar Reinvang and Knut Laake for permission to publish the preliminary results from our studies on cycloserine in patients with Alzheimer's disease.

REFERENCES

1. VICTOR, M., J. B. ANGEVINE, E. L. MANCALL & C. M. FISHER. 1961. Memory loss with lesions of hippocampal formation. Arch. Neurol. **5:** 244.
2. MILNER, B. 1972. Disorders of learning and memory after temporal lobe lesions in man. Clin. Neurosurg. **19:** 421–466.
3. BLISS, T. V. P. & T. LØMO. 1973. Long-lasting potentiation of synaptic transmission in the dentate area of the anaesthetized rabbit following stimulation of the perforant path. J. Physiol. **232:** 331–356.
4. BLISS, T. V. P. & G. L. COLLINGRIDGE. 1993. A synaptic model of memory: Long-term potentiation in the hippocampus. Nature **361:** 31–39.
5. COLLINGRIDGE, G. L., S. J. KEHL & H. J. MCLENNAN. 1983. Excitatory amino acids in synaptic transmission in the Schaffer collateral-commisural pathway of the rat hippocampus. J. Physiol. **334:** 33–46.
6. RACINE, R. J., N. W. MILGRAM & S. HAFNER. 1983. Long-term potentiation phenomena in the rat limbic forebrain. Brain Res. **260:** 217–231.
7. JARRARD, L. E. 1993. On the role of the hippocampus in learning and memory in the rat. Behav. Neural Biol. **60:** 9–26.
8. OLTON, D. S. 1983. Memory functions and the hippocampus. *In* Neurobiology of the Hippocampus. W. Seifert, Ed.: 335–373. Academic Press. New York.
9. RAWLINS, J. N. P. 1985. Associations across time: The hippocampus as a temporary memory store. Behav. Brain Sci. **8:** 479–497.
10. PERRY, E. K. 1986. The cholinergic hypothesis—Ten years on. Br. Med. Bull. **42(1):** 63–69.
11. PALMER, A. M. & S. GERSHON. 1990. Is the neuronal basis of Alzheimer's disease cholinergic or glutamatergic? FASEB J. **4:** 2745–2752.
12. FRANCIS, P. T., N. R. SIMS, A. W. PROCTER & D. M. BOWEN. 1993. Cortical pyramidal neurone loss may cause glutamatergic hypoactivity and cognitive impairment in Alzheimer's disease: Investigative and therapeutic perspectives. J. Neurochem. **60:** 1589–1604.
13. GREENAMYRE, J. T. & A. B. YOUNG. 1989. Excitatory amino acids and Alzheimer's disease. Neurobiol. Aging **10:** 593–602.
14. MYHRER, T. 1993. Animal models of Alzheimer's disease: Glutamatergic denervation as an alternative approach to cholinergic denervation. Neurosci. Biobehav. Rev. **17:** 195–202.

15. PEARSON, R. C. A. & T. P. S. POWELL. 1989. The neuroanatomy of Alzheimer's disease. Rev. Neurosci. 2: 101–123.
16. ESIRI, M. M., R. C. A. PEARSON, J. E. STEELE, D. M. BOWEN & T. P. S. POWELL. 1990. A quantitative study of the neurofibrillary tangles and the choline acetyltransferase activity in the cerebral cortex and the amygdala in Alzheimer's disease. J. Neurol. Neurosurg. Psychiatry 53: 161–165.
17. ARRIAGADA, P. V., K. MARZLOFF & B. T. HYMAN. 1992. Distribution of Alzheimer-type pathologic changes in nondemented elderly individuals matches the pattern in Alzheimer's disease. Neurology 42(9): 1681–1688.
18. BRAAK, H. & E. BRAAK. 1993. Entorhinal-hippocampal interaction in mnestic disorders. Hippocampus 3: 239–246.
19. ESIRI, M. 1991. Neuropathology. In Psychiatry in the Elderly. R. Jacoby & C. Oppenheimer, Eds.: 113–147. Oxford University Press. Oxford.
20. ROOM, R. & H. J. GROENEWEGEN. 1986. Connections of the parahippocampal cortex. I. Cortical afferents. J. Comp. Neurol. 251: 415–450.
21. MYHRER, T. & G. A. NÆVDAL. 1989. The temporal-hippocampal region and retention: The role of temporo-entorhinal connections in rats. Scand. J. Psychol. 30: 72–80.
22. LUND KARLSEN, R. & F. FONNUM. 1978. Evidence for glutamate as neurotransmitter in the corticofugal fibres to the dorsal lateral geniculate body and the superior culliculus in rats. Brain Res. 151: 457–467.
23. FONNUM, F. 1975. A rapid radiochemical method for the determination of choline acetyltransferase. J. Neurochem. 24: 407–409.
24. FONNUM, F., J. STORM-MATHISEN & F. WALBERG. 1970. Glutamate decarboxylase in inhibitory neurons: A study of the enzyme in Purkinje cell axons and boutons in the cat. Brain Res. 20: 259–275.
25. MYHRER, T., E. G. IVERSEN & F. FONNUM. 1989. Impaired reference memory and reduced glutamergic activity in rats with temporoentorhinal connections disrupted. Exp. Brain Res. 77: 499–506.
26. MYHRER, T., T. S. JOHANNESEN & E. SPIKKERUD. 1993. Restoration of mnemonic function in rats with glutamergic temporal systems disrupted: Dose and time of glycine injections. Pharmacol. Biochem. Behav. 45: 519–525.
27. FONNUM, F. 1984. Glutamate: A neurotransmitter in mammalian brain. J. Neurochem. 42: 1–11.
28. ROSENZWEIG, M. R. 1984. Experience, memory, and the brain. Am. Psychol. 39: 365–376.
29. MYHRER, T., L. UTSIKT, J. FJELLAND, E. G. IVERSEN & F. FONNUM. 1992. Differential rearing conditions in rats: Effects on neurochemistry in neocortical areas and cognitive behaviors. Brain Res. Bull. 28: 427–434.
30. MYHRER, T. & E. G. IVERSEN. 1990. Changes in retention of a visual discrimination task following unilateral and bilateral transections of temporo-entorhinal connections in rats. Brain Res. Bull. 25: 293–298.
31. MYHRER, T. 1988. Exploratory behaviour and reaction to novelty in rats with hippocampal perforant path systems disrupted. Behav. Neurosci. 102: 356–362.
32. NAKANISHI, S. 1992. Molecular diversity of glutamate receptors and implications for brain function. Science 258: 597–603.
33. MONYER, H., R. SPRENGEL, R. SCHOEPFER, A. HERB, M. HIGUCHI, H. LOMELI, N. BURNASHEV, B. SAKMANN & P. H. SEEBURG. 1992. Heteromeric NMDA receptors: Molecular and functional distinction of subtypes. Science 256: 1217–1221.
34. SOMMER, B. & P. H. SEEBURG. 1992. Glutamate receptor channels: Novel properties and new clones. Trends Pharmacol. Sci. 13: 291–296.
35. SCHOEPP, D. D. & P. J. CONN. 1993. Metabotropic glutamate receptors in brain function and pathology. Trends Pharmacol. Sci. 14: 13–20.
36. MORRIS, R. G. M., E. ANDERSON, G. S. LYNCH & M. BAUDRY. 1986. Selective impairment of learning and blockade of long-term potentiation by an N-methyl-D-aspartate receptor antagonist, AP5. Nature 319: 774–776.
37. MORRIS, R. G. M. 1989. Synaptic plasticity and learning: Selective impairment of learning in rats and blockade of long-term potentiation in vivo by the N-methyl-D-aspartate receptor antagonist AP5. J. Neurosci. 9(9): 3040–3057.

38. INGRAM, D. K., P. GAROFALO, E. L. SPANGLER, C. R. MANTIONE, I. ODANO & E. D. LONDON. 1992. Reduced density of NMDA receptors and increased sensitivity to dizocilpine-induced learning impairment in aged rats. Brain Res. **580:** 273–280.

39. WATANABE, Y., T. HIMI, H. SAITO & K. ABE. 1992. Involvement of glycine site associated with the NMDA receptor in hippocampal long-term potentiation and acquisition of spatial memory in rats. Brain Res. **582:** 58–64.

40. GREENAMYRE, J. T., J. B. PENNEY, C. J. D'AMATO & A. B. YOUNG. 1987. Dementia of the Alzheimer's type, changes in hippocampal L-[³H]glutamate binding. J. Neurochem. **48:** 543–551.

41. PENNEY, J. B., W. F. MARAGOS & J. T. GREENAMYRE. 1990. Excitatory amino acid binding sites in the hippocampal region of Alzheimer's disease and other dementias. J. Neurol. Neurosurg. Psychiatry **53:** 314–320.

42. REPRESSA, A., C. DUYCKAERTS, E. TREMBLAY, J. J. HAUW & Y. BEN-ARI. 1988. Is senile dementia of the Alzheimer type associated with hippocampal plasticity? Brain Res. **457:** 355–359.

43. JANSEN, K. L. R., R. L. M. FAULL, M. DRAGUNOW & B. L. SYNEK. 1990. Alzheimer's disease: Changes in hippocampal N-methyl-D-aspartate, quisqualate, neurotensin, adenosine, benzodiazepine, serotonin and opioid receptors—An autoradiographic study. Neuroscience **39:** 613–627.

44. COWBURN, R., J. HARDY, P. ROBERTS & R. BRIGGS. 1988. Regional distribution of pre- and postsynaptic glutamatergic function in Alzheimer's disease. Brain Res. **452:** 403–407.

45. PORTER, R. H. P., R. F. COWBURN, I. ALASUZOFF, R. S. J. BRIGGS & P. J. ROBERTS. 1992. Heterogeneity of NMDA receptors labelled with [³H]-((+)-2-carboxypiperazin-4-yl)propyl-1-phophonic acid ([³H]CPP): Receptor status in Alzheimer's disease brains. Eur. J. Pharmacol. Mol. Pharmacol. Sect. **225:** 195–201.

46. MOURADIAN, M. M., P. C. CONTRERAS, J. B. MONAHAN & T. N. CHASE. 1988. [³H]MK-801 binding in Alzheimer's disease. Neurosci. Lett. **93:** 225–230.

47. PROCTER, A. W., E. H. F. WONG, G. C. STRATMANN, S. L. LOWE & D. M. BOWEN. 1989. Reduced glycine stimulation of [³H]MK-801 binding in Alzheimer's disease. J. Neurochem. **53:** 698–704.

48. NINOMIYA, H., R. FUKUNAGA, T. TANIGUCHI, M. FUJIWARA, S. SHIMOHAMA & M. KAMEYAMA. 1990. [³H]N-[1-(2-thienyl)cyclohexyl]-3,4-piperidine ([³H]TCP) binding in human frontal cortex: Decreases in Alzheimer-type dementia. J. Neurochem. **54:** 526–532.

49. DEWAR, D., D. T. CHALMERS, D. I. GRAMAH & J. MCGULLOCH. 1991. Glutamate metabotropic and AMPA binding sites are reduced in Alzheimer's disease: An autoradiographic study of the hippocampus. Brain Res. **553:** 58–64.

50. TANG, C. M., Q. W. Y. SHI, A. KATCHMAN & G. LYNCH. 1991. Modulation of the time course of fast EPSCs and glutamate channel kinetics by aniracetam. Science **254:** 288–290.

51. MYHRER, T. & R. E. PAULSEN. 1992. Memory dysfunction following disruption of glutamergic systems in the temporal region of the rat: Effects of agonistic amino acid. Brain Res. **599:** 345–352.

52. FLOOD, J. F., M. L. BAKER & J. L. DAVIS. 1990. Modulation of memory processing by glutamic acid receptor agonists and antagonists. Brain Res. **521:** 197–202.

53. HANDELMANN, G. E., M. E. NEVINS, L. L. MUELLER, S. M. ARNOLDE & A. A. DORDI. 1989. Milacemide, a glycine prodrug, enhances performance of learning tasks in normal and amnestic rodents. Pharmacol. Biochem. Behav. **34:** 823–828.

54. MONAHAN, J. B., G. E. HANDELMANN, W. F. HOOD & A. A. CORDI. 1989. D-Cycloserine, a positive modulator of the N-methyl-D-aspartate receptor, enhances performance of learning task in rats. Pharmacol. Biochem. Behav. **34:** 649–653.

55. CHESSELL, I. P., A. W. PROCTER, P. T. FRANCIS & D. M. BOWEN. 1991. D-Cycloserine, a putative cognitive enhancer, facilitates activation of the N-methyl-D-aspartate receptor-ionophore complex in Alzheimer brain. Brain Res. **565:** 345–348.

56. SIRVIO, J., T. EKONSALO, P. RIEKKINEN, JR., H. LAHTINEN & P. RIEKKINEN, SR. 1992.

D-Cycloserine, a modulator of the *N*-methyl-D-aspartate receptor, improves spatial learning in rats treated with muscarinic antagonists. Neurosci. Lett. **146(2):** 215–218.

57. FISHKIN, R. J., E. S. INCE, W. A. CARLEZON, JR. & R. W. DUNN. 1993. D-Cycloserine attenuates scopolamine-induced learning and memory deficits in rats. Behav. Neural Biol. **59(2):** 150–157.

58. FLOOD, J. F., J. E. MORLEY & T. H. LANTHORN. 1992. Effect on memory processing by D-cycloserine, an agonist of the NMDA/glycine receptor. Eur. J. Pharmacol. **221(2–3):** 249–254.

59. SCHUSTER, G. M. & W. J. SCHMIDT. 1992. D-Cycloserine reverses the working memory impairment of hippocampal-lesioned rats in a spatial learning task. Eur. J. Pharmacol. **224(1):** 97–98.

60. RUPNIAK, N. M. J., M. DUCHNOWSKI, S. J. TYE, G. COOK & S. D. IVERSEN. 1992. Failure of D-cycloserine to reverse cognitive disruption induced by scopolamine or phencyclidine in primates. Life Sci. **50:** 1959–1962.

61. JONES, R. W., K. A. WESNES & J. KIRBY. 1991. Effects of NMDA modulation in scopolamine dementia. Ann. N. Y. Acad. Sci. **640:** 241–244.

Pharmacological Augmentation of NMDA Receptor Function for Treatment of Schizophrenia[a]

ILANA ZYLBERMAN,[b,d] DANIEL C. JAVITT,[b-d]
AND STEPHEN R. ZUKIN[b-e]

Departments of [b]Psychiatry and [c]Neuroscience
Albert Einstein College of Medicine of Yeshiva University
Bronx, New York 10461

[d]Bronx Psychiatric Center
Bronx, New York

Acute low-dose administration of phencyclidine (1,1-phenylcyclohexylpiperidine; PCP) induces a schizophrenia-like psychotic state in nonpsychotic subjects. By contrast to amphetamine administration, PCP psychosis incorporates negative symptoms (withdrawal, negativism, autism) and cognitive dysfunction (impairment of abstract thinking, symbolic thinking, attention, and perception), as well as positive symptoms closely resembling those in schizophrenia.[1-8] Neuropsychological tests requiring sustained attention and paired-associate learning are most affected by low-dose PCP administration,[1] suggesting that, as in schizophrenia,[9,10] prefrontal and temporohippocampal processing may be most severely disturbed by PCP. Furthermore, subjects with schizophrenia are uniquely sensitive to the psychotomimetic effects of PCP. A single dose of PCP administered to recompensated schizophrenic subjects recreated the presenting symptomatology for days[3] to weeks[2,5] without inducing symptoms and signs atypical of schizophrenia; by contrast, psychotomimetic PCP effects in normal subjects typically resolve within four to six hours, although a minority of subjects—approximately 20%—may develop symptoms for a longer period of time. Subjects predisposed to schizophrenia based upon premorbid symptoms or family history appear to be at increased risk for developing prolonged symptoms.[11] Nonschizophrenic patients with PCP psychosis are difficult to differentiate from acute schizophrenics on the basis of presenting symptomatology alone.[11,12] Administration of PCP to monkeys can simulate the abnormalities in event-related potentials seen in schizophrenic subjects.[13,14] Finally, patterns of brain metabolism determined by positron emission tomography in nonschizophrenic chronic PCP users have been reported to be abnormal and to resemble patterns seen in chronic schizophrenics.[15]

At serum concentrations (10–100 nM) calculated[16] to result from the selectively psychotomimetic human dose of 0.1 mg/kg of PCP, the only significant central nervous system (CNS) target site is the PCP receptor, which has a nanomolar affinity for PCP and binds PCP-like drugs with affinities paralleling their potencies in

[a]This work was supported in part by U.S. Public Health Service grants K11 MH00631 (DCJ) and R01 DA03383 (SRZ), a NARSAD Young Investigator Award (DCJ), the APA Dorothy C. Kempf Award (DCJ, SRZ), a grant from the Ritter Foundation (SRZ), and the Department of Psychiatry, Albert Einstein College of Medicine.

[e]Address correspondence to Stephen R. Zukin, M.D., Department of Psychiatry, Albert Einstein College of Medicine, 1300 Morris Park Avenue, F111, Bronx, NY 10461.

evoking PCP-specific behaviors (reviewed in ref. 17). The location of the PCP receptor within the interior of the ion channel gated by the N-methyl-D-aspartate (NMDA) receptor dictates that binding of PCP receptor ligands blocks NMDA receptor-mediated ion flux in a fashion that cannot be surmounted by increasing agonist (L-glutamate) concentration.

The association of PCP-induced blockade of NMDA receptor channels with induction of a schizophrenia-like psychosis in normal subjects and with long-lasting exacerbation of illness in previously stabilized schizophrenic patients suggests that an endogenous deficiency of NMDA receptor-mediated neurotransmission may play a role in schizophrenia (reviewed in ref. 17).

A particularly intriguing question is whether neuroleptic-resistant schizophrenia with prominent negative and cognitive symptoms may benefit from pharmacological augmentation of NMDA receptor-mediated neurotransmission. Of the multiple regulatory sites of the NMDA receptor (reviewed in ref. 17), the glycine site is the most logical candidate for pharmacological augmentation of NMDA receptor activation. The NMDA receptor is activated according to a multistate system in which sequential binding of two molecules of the agonist L-glutamate is required to attain a conformation from which activation can take place.[18] Once this prerequisite has occurred, the equilibrium between the activated and the resting conformations is regulated by the local concentration of glycine. Glycine acts at a specific strychnine-insensitive binding site on the NMDA receptor complex.[19] If glycine is absent, NMDA receptor activation is not observed.[20]

No high-affinity transport of glycine occurs across the blood-brain barrier, so that under normal circumstances peripheral glycine contributes little to CNS glycine levels.[21] However, pharmacological doses of glycine have been shown to increase CNS glycine levels in rodents.[22,23] Glycine is without significant side effects at doses up to 3 g/kg,[24] and has been employed in doses as high as 60 g per day without significant toxicity.[25]

In addition to increasing brain glycine levels, orally administered glycine has been shown to reverse PCP-induced behaviors in rodents,[23] suggesting that the increased brain glycine level leads to potentiation of NMDA receptor-mediated neurotransmission. Furthermore, a high-affinity glycine reuptake mechanism has recently been described, suggesting a physiological mechanism for regulation of synaptic glycine concentration at the NMDA receptor.[26]

Previous studies of glycine therapy in schizophrenia have yielded conflicting results, possibly because of variations in dosage, patient population, and rating instruments.[27-30] In a study conducted at Bronx Psychiatric Center,[31] we used a dose of glycine higher than that used in any previous trial, and employed a rating instrument designed to discriminate negative symptoms from positive symptoms and general psychopathology. Fourteen male subjects meeting DSM-III-R[32] criteria for schizophrenia were maintained on clinically determined optimal doses of neuroleptic drug and randomly assigned to receive either glycine or placebo. Glycine therapy was titrated upward to a maximum dose of 0.4 g/kg body weight (approximately 30 g per day) during the first two weeks of an 8-week treatment period. After completion of the double-blind phase, all patients were offered open-label glycine continuation for an additional 8-week period. One additional subject was treated with open-label glycine only. Subjects were rated with the Positive and Negative Syndrome Scale (PANSS),[33] the Extrapyramidal Rating Scale,[34] and the Abnormal Involuntary Movement Scale (AIMS).[35] Statistical analyses (two-tailed) were accomplished using the SPSS/PC + computer program. Values represent mean ± standard deviation.

There were no between-group differences in age (glycine, 36.0 ± 9.7 yr; placebo, 38.1 ± 7.2 yr), duration of illness (glycine, 15.5 ± 8.1; placebo, 20.0 ± 6.6 yr) or

dosage of neuroleptic drug (glycine, 1450 ± 826; placebo, 1194 ± 658 chlorpromazine equivalents per day). No between-group differences were found in baseline
ratings of negative symptoms (glycine, 21.5 ± 8.6; placebo, 24.6 ± 5.3), positive
symptoms (glycine, 23.5 ± 5.6; placebo, 22.2 ± 4.7) or general psychopathology
(glycine, 40.7 ± 7.9; placebo, 45.0 ± 8.3). A decrease in negative symptoms was
observed in all subjects who received glycine during the double-blind phase, but in
only 2 of 7 subjects who were given placebo (Fisher exact test $p < 0.025$). Subjects in
the glycine group showed a mean percentage improvement of 15.3 ± 7.7% as
compared to 1.4 ± 7.7% in the placebo-treated group (t = 5.28, $p < 0.002$). Repeated measures ANOVA with baseline as covariate across the 8-week study period
demonstrated significant effects of glycine ($F_{1,11} = 11.74$, $p = 0.006$) and time
($F_{1,8} = 8.11$, $p = 0.022$) and also a significant glycine-by-time interaction ($F_{1,8} = 8.34$,
$p = 0.02$). Subjects who received glycine during the double-blind phase did not show
further significant change in negative symptom ratings during the subsequent open-
label extension. By contrast, subjects who received placebo during the double-blind
phase showed a 19.6 ± 10.0% (t = 5.19, $p < 0.002$) improvement in negative
symptoms. Pooling of results across double-blind and open-label phases indicated a
highly significant 17.1 ± 8.6% improvement in negative symptom ratings at week 8 of
glycine treatment (t = 6.84, $p < 0.0001$). Glycine treatment had no significant
effects on ratings of positive symptoms (change scores 0.9 ± 5.2 and 0.8 ± 6.0 for the
glycine- and placebo-treatment groups, respectively), general psychopathology
(change scores 3.3 ± 5.1 and 2.9 ± 6.9, respectively), or on EPS or tardive dyskinesia
parameters. Pre- minus posttreatment EPS change scores for the glycine and placebo
groups, 1.0 ± 3.1 and 4.0 ± 8.7, did not differ from each other or from zero. AIMS
ratings were zero to minimal in both glycine and placebo groups before and after
treatment. Glycine treatment was not associated with changes in serum chemistry or
hematological values. One subject, while receiving double-blind glycine, reported
lower-extremity weakness that ceased after temporary dose reduction and did not
recur following resumption of the full dose.

The relatively modest magnitude of the glycine effect observed in this study may
stem from the fact that peripherally administered glycine gains entry to the CNS
only via passive diffusion across the blood-brain barrier. It is possible that higher
doses of glycine may be required to produce more robust effects. Other agents that
have been evaluated for treatment of neuroleptic-resistant negative symptoms,
including anticholinergics[36] and dopamine agonists,[37] may also exacerbate positive
symptoms. It is therefore noteworthy that the glycine-induced reduction in negative
symptoms was not accompanied by an increase in positive symptoms. The failure of
glycine to improve positive symptoms in this study may result from the fact that all
subjects were receiving high doses of neuroleptic drugs. We showed that in mice,
peripheral administration of haloperidol (0.5 mg/kg) 30 min after peripheral pretreatment with glycine (0.4 g/kg) or saline resulted in statistically indistinguishable brain
haloperidol levels in the glycine- and saline-pretreated groups,[31] suggesting that the
glycine-induced decrease in negative symptoms observed in our clinical study was not
due to a glycine-induced change in the effective dosage of the neuroleptic drug. The
lack of glycine-induced change in EPS scores indicates that the glycine-induced
reduction in negative symptoms did not result from a reduction in neuroleptic-
induced extrapyramidal symptoms.

These results provide additional support for the hypothesis that impaired NMDA
receptor-mediated neurotransmission may underlie symptoms of schizophrenia that
respond incompletely or not at all to treatment with neuroleptic agents, and suggest
that emerging knowledge of NMDA receptor regulation should be exploited to

develop increasingly effective pharmacological strategies for augmentation of NMDA receptor function.

REFERENCES

1. BAKKER, C. B. & F. B. AMINI. 1961. Observations on the psychotomimetic effects of Sernyl. Compr. Psychiatry **2:** 269–280.
2. LUBY, E. D., B. D. COHEN, F. ROSENBAUM, J. GOTTLIEB & R. KELLY. 1959. Study of a new schizophrenomimetic drug—Sernyl. AMA Arch. Neurol. Psychiatry **81:** 363–369.
3. BAN, T. A., J. J. LOHRENZ & H. E. LEHMANN. 1961. Observations on the action of Sernyl—a new psychotropic drug. Can. Psychiatr. Assoc. J. **6:** 150–156.
4. DAVIES, B. M. & H. R. BEECH. 1960. The effect of L-arylcyclohexylamine (Sernyl) on twelve normal volunteers. J. Ment. Sci. **106:** 912–924.
5. DOMINO, E. F. & E. LUBY. 1981. Abnormal mental states induced by phencyclidine as a model of schizophrenia. *In* PCP (Phencyclidine): Historical and Current Perspectives. E. F. Domino, Ed. NPP Books. Ann Arbor, MI.
6. RODIN, E. A., E. D. LUBY & J. S. MEYER. 1959. Electroencephalographic findings associated with Sernyl infusion. EEG Clin. Neurophysiol. **11:** 796–798.
7. ROSENBAUM, G., B. D. COHEN, E. D. LUBY, J. S. GOTTLIEB & D. YELEN. 1959. Comparisons of Sernyl with other drugs. Arch. Gen. Psychiatry **1:** 651–656.
8. COHEN, B. D., G. ROSENBAUM, E. D. LUBY & J. S. GOTTLIEB. 1961. Comparison of phencyclidine hydrochloride (Sernyl) with other drugs. Arch. Gen. Psychiatry **6:** 79–85.
9. BERMAN, K. F., B. P. ILLOWSKY & D. R. WEINBERGER. 1988. Physiological dysfunction of dorsolateral prefrontal cortex in schizophrenia. Arch. Gen. Psychiatry **45:** 616–622.
10. KOLB, B. & I. Q. WHISHAW. 1983. Performance of schizophrenic patients on tests sensitive to left or right frontal, temporal, or parietal function in neurological patients. J. Nerv. Ment. Dis. **171(7):** 435–443.
11. ERARD, R., P. V. LUISADA & R. PEELE. 1980. The PCP psychosis: Prolonged intoxication or drug-precipitated functional illness? J. Psychedelic Drugs **12:** 235–245.
12. YESAVAGE, J. A. & A. M. FREEMAN III. 1978. Acute phencyclidine (PCP) intoxication: Psychopathology and prognosis. J. Clin. Psychiatry **44:** 664–665.
13. JAVITT, D. C., P. DONESHKA, I. ZYLBERMAN, W. RITTER & H. G. VAUGHAN. 1993. Impairment of early cortical processing in schizophrenia: An event-related potential confirmation study. Biol. Psychiatry **33:** 513–519.
14. JAVITT, D. C., C. E. SCHROEDER, M. STEINSCHNEIDER, J. C. AREZZO, W. RITTER & H. G. VAUGHAN. 1994. Cognitive event-related potentials in human and non-human primates: Implications for the PCP/NMDA model of schizophrenia. Electroencephalogr. Clin. Neurophysiol. Suppl. **44:** 161–175.
15. WU, J. C., M. S. BUCHSBAUM, S. G. POTKIN, M. J. WOLF & M. E. BUNNEY, JR. 1991. Schizophr. Res. **4:** 415.
16. COOK, C. E., M. PEREZ-REYES, A. R. JEFFCOAT & D. R. DRIN. 1983. Phencyclidine disposition in humans after small doses of radiolabelled drug. Fed. Proc. **42:** 2566–2569.
17. JAVITT, D. C. & S. R. ZUKIN. 1991. Recent advances in the phencyclidine model of schizophrenia. Am. J. Psychiatry **148:** 1301–1308.
18. JAVITT, D. C., M. J. FRUSCIANTE & S. R. ZUKIN. 1990. Rat brain *N*-methyl-D-aspartate receptors require multiple molecules of agonist for activation. Mol. Pharmacol. **37:** 603–607.
19. JOHNSON, J. W. & P. ASCHER. 1987. Glycine potentiates the NMDA response in cultured mouse brain neurons. Nature **325:** 529–531.
20. KLECKNER, N. W. & R. DINGLEDINE. 1988. Requirement for glycine in activation of NMDA-receptors expressed in *Xenopus* oocytes. Science **241:** 835–837.
21. BANOS, G., P. M. DANIEL, S. R. MOORHOUSE & O. E. PRATT. 1975. The requirements of the brain for some amino acids. J. Physiol. **46:** 539–548.
22. TOTH, E. & A. LAJTHA. 1981. Elevation of cerebral levels of non-essential amino acids in vivo by administration of large doses. Neurochem. Res. **6:** 1309–1317.

23. TOTH, E. & A. LAJTHA. 1986. Antagonism of phencyclidine-induced hyperactivity by glycine in mice. Neurochem. Res. **11:** 393–400.
24. RIKER, W. F. & H. GOLD. 1942. The pharmacology of sodium hydroxyacetate with observations on the toxicity of glycine. J. Am. Pharm. Assoc. **31:** 306–312.
25. DALY, E. C. & M. H. APRISON. 1983. Glycine. *In* Handbook of Neurochemistry. A. Lajtha, Ed. Vol. 3. Plenum Press. New York.
26. LIU, Q. R., B. LOPEZ-CORCUERA, S. MANDIYAN, H. NELSON & N. NELSON. 1993. Cloning and expression of a spinal cord- and brain-specific glycine transporter with novel structural features. J. Biol. Chem. **268:** 22802–22808.
27. WAZIRI, R. 1988. Glycine therapy of schizophrenia (letter). Biol. Psychiatry **23:** 210–211.
28. ROSSE, R. B., S. K. THEUT, M. BANAY-SCHWARTZ, M. LEIGHTON, E. SCARCELLA, C. G. COHEN & S. I. DEUTSCH. 1989. Glycine adjuvant therapy to conventional neuroleptic treatment in schizophrenia: An open-label, pilot study. Clin. Neuropharmacol. **12:** 416–424.
29. COSTA, J., E. KHALED, J. SRAMEK, W. BUNNEY, JR. & S. G. POTKIN. 1990. An open trial of glycine as an adjunct to neuroleptics in chronic treatment-refractory schizophrenics. J. Clin. Psychopharmacol. **10:** 71–72.
30. POTKIN, S. G., J. COSTA, S. ROY, J. SRAMEK, Y. JIN & B. GULASEKARAM. 1992. Glycine in the treatment of schizophrenia—Theory and preliminary results. *In* Novel Antipsychotic Drugs. H. Y. Meltzer, Ed. Raven Press, New York.
31. JAVITT, D. C., I. ZYLBERMAN, S. R. ZUKIN, U. HERESCO-LEVY & J.-P. LINDENMAYER. 1994. Amelioration of negative symptoms in schizophrenia by glycine. Am. J. Psychiatry **151:** 1234–1236.
32. AMERICAN PSYCHIATRIC ASSOCIATION. 1987. Diagnostic and Statistical Manual of Mental Disorders (Third edit. rev.) DSM-III-R. American Psychiatric Association. Washington, DC.
33. KAY, S. R., A. FISZBEIN & L. A. OPLER. 1987. The positive and negative syndrome scale (PANSS) for schizophrenia. Schizophr. Bull. **13:** 261–276.
34. ALPERT, M. & M. RUSH. 1983. Comparison of affect in Parkinson's disease and schizophrenia. Psychopharmacol. Bull. **19:** 118–120.
35. GUY, W. 1976. ECDEU assessment manual for psychopharmacology. 534–537. U.S. Department of Health, Education and Welfare. Washington, DC.
36. TANDON, R., J. R. DEQUARDO, J. GOODSON, N. A. MANN & J. F. GREDEN. 1992. Effects of anticholinergics on positive and negative symptoms in schizophrenia. Psychopharmacol. Bull. **28:** 297–302.
37. LIEBERMAN, J. A., J. M. KANE & J. ALVIR. 1987. Provocative tests with psychostimulant drugs in schizophrenia. Psychopharmacology **91:** 415–433.

Excitatory Amino Acid Receptors
and Their Role in Epilepsy
and Cerebral Ischemia

BRIAN S. MELDRUM

Department of Neurology
Institute of Psychiatry
London SE5 8AF, United Kingdom

Glutamate is the principal fast excitatory neurotransmitter in the mammalian brain. Aspartate, various sulfinic or sulfonic analogues of glutamate and aspartate, and some dipeptides such as *N*-acetylaspartylglutamate may also be neurotransmitters. About 15 years ago it was proposed, on the basis of pharmacological and electrophysiological data, that glutamate acted on three subtypes of receptors directly linked to channels permeable to cations.[1,2] These are named after their preferred agonists, the α-amino-3-hydroxy-5-methyl-4-isoxazole propionate (AMPA), kainate (KA), and *N*-methyl-D-aspartate (NMDA) receptors. Glutamate also acts on a family of G-protein-coupled or "metabotropic" receptors that either activate phospholipase C or decrease the activity of adenylate cyclase.[3,4]

Glial cells possess AMPA, KA, and metabotropic glutamate receptors (but not NMDA receptors).[5] Astrocytes show a calcium-dependent release of glutamate that can be triggered by bradykinin and other agents increasing intracellular calcium concentration in astrocytes.[6] Thus glia can both respond to increases in extracellular glutamate and induce neuronal responses (via NMDA receptors[6]) to glutamate.

MOLECULAR BIOLOGY OF IONOTROPIC GLUTAMATE RECEPTORS

In 1989 the application of expression cloning led to the isolation of the first functional glutamate receptor, GluR1.[7] Since that time, 22 genes belonging to the GluR ionotropic family and 7 genes belonging to the metabotropic family have been identified.[4,8,9] These studies have confirmed that the mammalian glutamate receptors controlling ligand-gated ion channels can be grouped into three families or subtypes (FIG. 1). The four subunits that constitute the AMPA receptor family (known as GluR1-4 or GluRA-D) are of similar size (approximately 900 amino acids) and are approximately 70% homologous.[10] Each subunit exists in two forms (flip and flop) created by alternative splicing. These variants are not known to differ pharmacologically, but the flip form is more efficient than the flop form. Receptor ion-channel complexes are thought (on the basis of analogy with GABAergic receptors) to be pentameric. GluR1-4 can be expressed homomerically as functional channels; current/voltage relations are linear for GluR2, but inwardly rectifying for the other receptors. GluR2 expressed homomerically or when coexpressed with GluR1 or GluR3 does not flux Ca^{2+}.

Among the kainate receptor family three receptors, GluR5, GluR6, and GluR7, with highly (75–80%) homologous amino acid sequences have been cloned.[8,10] GluR5 has several splice variants and also shows RNA editing of the glutamine/arginine (Q/R) site on the transmembrane domain (TMD) II (as does GluR6).

GluR5 and GluR6 can be functionally expressed in oocytes and show agonist activation with an order of potency domoate > kainate > glutamate > AMPA. KA1 and KA2 show slightly greater homology with GluR5–7 than with GluR1–4. They appear to correspond to high-affinity kainate binding sites. KA2 forms functional receptors when coexpressed with GluR5 or GluR6.

The NMDAR1 subunit was isolated by expression cloning and shown to form functional homomeric units.[11,12] Three different sites for alternative splicing yield, in combination, eight splice variants that differ significantly in their pharmacology for agonists and antagonists.[13] They also show marked regional differences in their distribution.[14] These two observations support the earlier description of variations in the pharmacological properties of NMDA receptors according to the brain region.[15] The NMDA receptor ion channel is highly permeable to Ca^{2+}. Phosphorylation by protein kinase C (PKC) has a marked potentiating effect on the response to glutamate.

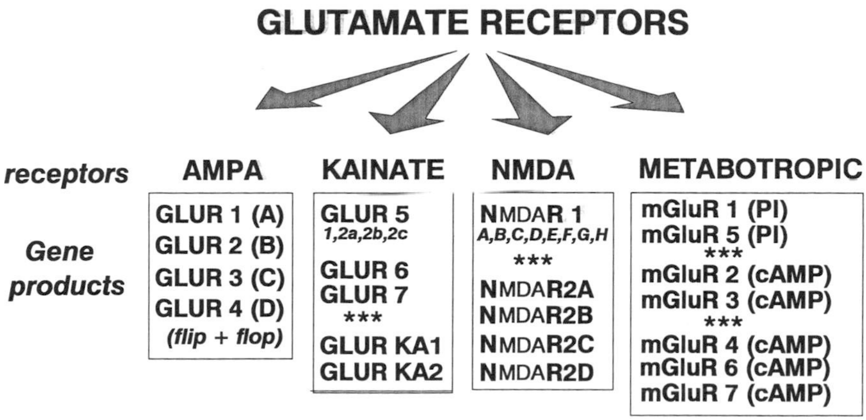

GLUTAMATE RECEPTORS

receptors	AMPA	KAINATE	NMDA	METABOTROPIC
	GLUR 1 (A)	GLUR 5	NMDAR 1	mGluR 1 (PI)
Gene	GLUR 2 (B)	1,2a,2b,2c	A,B,C,D,E,F,G,H	mGluR 5 (PI)
products	GLUR 3 (C)	GLUR 6	***	***
	GLUR 4 (D)	GLUR 7	NMDAR2A	mGluR 2 (cAMP)
	(flip + flop)	***	NMDAR2B	mGluR 3 (cAMP)
		GLUR KA1	NMDAR2C	***
		GLUR KA2	NMDAR2D	mGluR 4 (cAMP)
				mGluR 6 (cAMP)
				mGluR 7 (cAMP)

FIGURE 1. Ionotropic glutamate receptors and their constituent gene products.

NMDAR2A-D receptor subunits are functional only when coexpressed with NMDAR1. Their coexpression with NMDAR1 leads to a marked enhancement of the current amplitudes of activation compared with homomeric NMDAR1; they also show altered pharmacology. The Mg^{2+} block differs markedly between the heteromers, being strongest in NMDAR1/2A and NMDAR1/2B. The offset time is fastest in NMDAR1/2A and slowest in NMDAR1/2D. Glycine alone (that is, without glutamate) can activate NMDAR1/NMDAR2A, B or C when expressed in oocytes (but this phenomenon has not been described for any mammalian neuron). Like NMDAR1 the NMDAR2 subtypes have an asparagine at the Q/R site on the TMD II. The variety of regulatory sites identified on the NMDA receptor is indicated in TABLE 1. The properties of the redox site, at which reducing agents potentiate the action of glutamate, vary markedly according to subtype composition in recombinant NMDA receptors.[16]

Antagonists show preferential actions among heterodimeric NMDA receptors; competitive NMDA antagonists of the 2-amino-5-phosphonovalerate type act preferentially on NMDAR1/NMDAR2A;[13] ifenprodil and eliprodil (polyamine site antago-

TABLE 1. Regulatory Sites on the NMDA Receptor

Site	Agonist	Antagonist
Glutamate recognition	Glutamate	
	Aspartate	AP5, AP7
	Homocysteate	CGS19755
	NMDA	D-CPPene
	Quinolinate	
Glycine recognition	Glycine	5,7 Cl$_2$kynurenate
	D-Serine	HA-966, L-689, 560
		ACEA 1011, 1021
		MDL 105,572
Open channel		Mg^{2+}, Co^{2+}
Open channel		n-Alkyl diamines
Open channel PCP site		Phencyclidine
		Ketamine
		MK-801 (dizocilpine)
		CNS 1102 (cerestat)
Polyamine site	Spermine	Ifenprodil
	Spermidine	Eliprodil
Redox site	Dithiothreitol	DTNB
	Glutathione	
Proton site		H+
Zn^{2+} site	Zn^{2+}	Zn^{2+}, Cd^{2+}
Phosphorylation sites	Tyrosine kinases	Protein phosphatases

nists) act preferentially on NMDAR1/NMDAR2B,[17–19] and glycine site antagonists of the 7-Cl-kynurenate type act preferentially on NMDAR1/NMDAR2C.

METABOTROPIC RECEPTORS

The glutamate metabotropic receptors resemble other G-protein-coupled receptors in having seven transmembrane domains but they show no sequence homology with other G-protein-coupled receptors.[3]

mGluR1 and mGluR5 have significant sequence homology and are coupled to phospholipase C, causing hydrolysis of phosphoinositides to diacylglycerol (which activates protein kinase C) and inositol 1,4,5-triphosphate (IP3), which releases calcium from storage in the endoplasmic reticulum. Activation of this type of mGluR in hippocampal pyramidal cells causes a slow outward current through activation of a Ca^{2+} dependent K$^+$ current.[20]

mGluR2 and mGluR3 provide a family of metabotropic receptors that are negatively linked to adenylate cyclase activity. In the olfactory bulb, mGluR2 function as presynaptic inhibitors of glutamate release. In contrast to mGluR1 and mGluR5, glutamate and trans-ACPD [(±)-1-aminocydopentane-trans-1,3-dicarboxylic acid] are more potent agonists than quisqualate at these receptors.

A third family of metabotropic receptors, negatively coupled to adenylate cyclase, is provided by mGluR4, mGluR6, and mGluR7. At these receptors 2-amino-

4-phosphonobutyrate is the most potent agonist. In the retina mGluR6 in bipolar cells is coupled to cGMP formation.

SELECTIVE ANTAGONISTS

For the AMPA receptors two classes of antagonist are known, competitive antagonists such as the quinoxalinediones (NBQX and YM 900) and LY 293558[21,22] and allosteric antagonists as exemplified by the 2,3 benzodiazepines, GYKI 52466 and GYKI 53655[23,24] (see FIG. 2). NBQX has been the preferred experimental tool for *in vivo* studies because it lacks action at the glycine site of the NMDA receptor, although it suffers from poor solubility and rapid clearance. GYKI 52466 also inhibits activation of KA receptors (GluR6); NS 102 appears to be a relatively selective antagonist at the KA receptor.[25]

NMDA receptors have two agonist recognition sites (for glutamate/aspartate and for glycine/D-serine), which partially overlap with the sites at which competitive antagonists for glutamate and glycine bind. Competitive antagonists at the NMDA recognition site have been largely structural analogues of D-2 amino-5-phosphonopropionic acid (AP5) or 2-amino-7-phosphono-heptanoic acid. Glycine site antagonists show a wide variety of molecular structures but are all analogues of glycine. They have tended to have poor bioavailability and limited central nervous system (CNS) penetration, but more recently described compounds such as L-687,414 and ACEA 1021[26] have good bioavailability (see FIG. 3).

In general, potent selective antagonists for the metabotropic receptors remain to be discovered. L-2,Amino 3-phosphono-propionate appears to be an antagonist in some functional test systems for metabotropic receptor activation. (+)-α-Methyl-4-

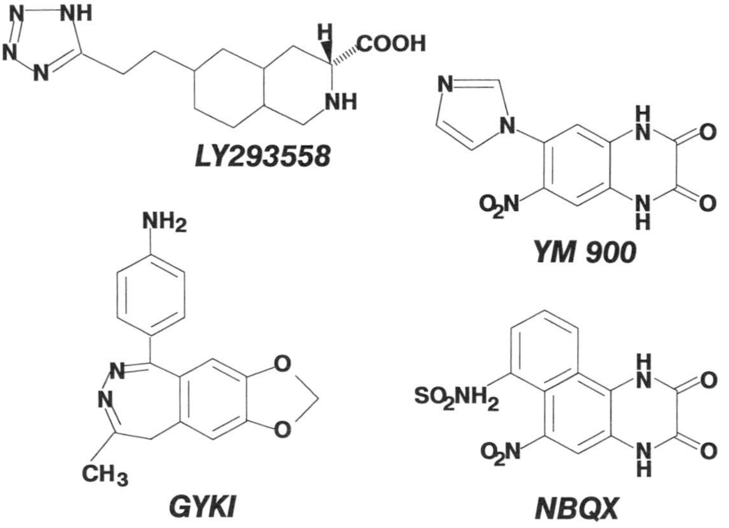

FIGURE 2. Molecular structures of AMPA/kainate antagonists.

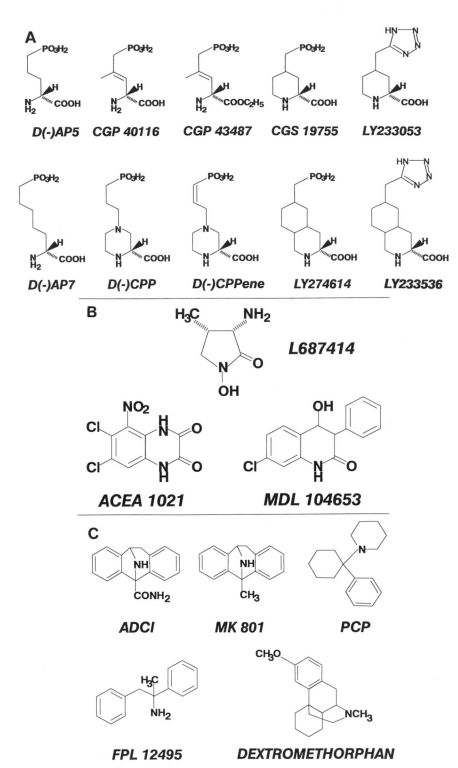

FIGURE 3. Molecular structure of (**A**) competitive NMDA antagonists, (**B**) glycine site antagonists, and (**C**) uncompetitive antagonists.

carboxyphenyl-glycine (αM4CPG) is a fairly weak antagonist for mGluR1 (pA2 4.38) and mGluR2 (pA2 4.29) and lacks effect at mGluR4.[27]

FUNCTIONAL ROLES OF GLUTAMATE RECEPTORS

Glutamate receptors, being the principal excitatory receptors in the brain, are naturally involved in all integrative functions of the CNS. Drugs interfering with glutamatergic transmission produce alterations in sensory and motor function. Particular attention has been paid to their role in learning and memory. Studies of long-term potentiation (LTP) *in vitro* have defined a critical role for NMDA receptor activation in the establishment of LTP and of AMPA-mediated excitation in its maintenance.[28] Several brain stem sites at which glutamate controls respiration and cardiovascular function have been identified.[29,30]

GENETIC DISORDERS

The question arises whether there are clinical disorders that result from mutations in the genes controlling glutamate receptor expression. Determination of the chromosomal location of GluR1-4[31] focused attention on these genes as candidate genes for various neurological disorders with similar chromosomal locations; for example, GluR3 at Xq25-26 is in the same region of the X chromosome as the oculocerebralrenal syndrome of Lowe. The suggestion that in man the location of GluR4 to 11 q22-23 made it a candidate gene for the principal gene for seizures in the epilepsy-like mouse (El-1) (on the basis of synteny with mouse chromosome 9) appears improbable if it is confirmed that a mutation in the ceruloplasmin gene at this locus explains the mouse phenotype.[32]

EPILEPSY

Epilepsy is an intermittent dynamic disorder of neural networks. The molecular and cellular basis of epilepsy is poorly understood. Genetic and acquired forms of epilepsy appear to be associated with a multiplicity of changes that may contribute to the network instability including alterations in neuronal membrane ionic conductances, alterations in GABAergic inhibitory transmission and in glutamatergic excitatory neurotransmission, alterations in neuromodulatory systems (monoamines, peptides, etc.), and changes in neuronal connectivity. We review here (1) recent findings concerning alterations in glutamatergic systems in genetic and acquired forms of epilepsy and (2) anticonvulsant effects of glutamate receptor antagonists.

It is possible that abnormalities of glutamate metabolism or transport contribute to the occurrence of seizures. Increases in plasma glutamate levels have been reported in families of patients with primary generalized epilepsy associated with 2–3 Hz cortical discharges.[33] Increases in plasma glutamate are also reported in a rodent model of limbic epilepsy.[34] Evidence of several kinds suggests that glutamate uptake into neurons or glia may be impaired in rats with kindled limbic seizures. Thus, the basal extracellular concentration of glutamate in hippocampus or cortex appears to be elevated in the fully kindled rat and uptake of [^{3}H]-D-aspartate is impaired.[35] We recently found that the levels of mRNA for the glial glutamate transporter GLT1 are decreased in the fully kindled rat brain.[36] An impairment of glutamate uptake could be a factor contributing to the increase in extracellular glutamate concentration seen in the epileptogenic hippocampus just prior to onset of complex partial seizures.[37]

CHANGES IN GLUTAMATE RECEPTORS IN ACQUIRED EPILEPSY

Autoradiographic studies visualizing the density of AMPA, kainate, and NMDA receptors in the temporal lobe in patients with complex partial seizures have shown some regional increases in the density of all three receptor subtypes.[38] Slice preparations of the hippocampus in electrically kindled rats and of the human cortex in patients with focal epilepsy have been used to assess functional changes in NMDA receptors. In the rat hippocampus enhanced sensitivity to NMDA has been shown both in the dentate gyrus and in the CA3 region.[39]

Electrophysiological study of dissociated dentate granule cells including cell-attached single-channel recordings reveals that the increased efficacy of NMDA receptors is related to a decrease in the blocking effect of Mg^{2+} and an increase in mean open times and burst lengths.[40] The efficacy of NMDA receptors is also increased in the cortical tissue surrounding an epileptic focus in man.[41] Glutamate metabotropic responses are also potentiated in the kindled rat brain (as assessed by phosphoinositide hydrolysis).[42] It is not known whether similar changes occur in the human epileptic brain.

It is also possible that gene defects might directly or indirectly modify glutamate receptor expression and function. No evidence of this is available as yet. The intrahippocampal injection of a herpes virus vector producing local overexpression of the GluR6 subunit of the kainate receptor leads within a few hours to spontaneous limbic seizures which occur recurrently over the following weeks.[43] Thus an imbalance in the expression of glutamate receptors could play a role in genetic or acquired forms of epilepsy.

FOCAL INJECTIONS OF GLUTAMATE ANTAGONISTS

Glutamate receptor antagonists injected focally into different brain structures can suppress the development and expression of seizures. It is particularly notable that limbic seizures (such as those induced by the systemic administration of high doses of pilocarpine) can be suppressed by the focal bilateral injection of NMDA receptor antagonists into (1) output pathways of the basal ganglia (substantia nigra and entopeduncular nucleus),[44–46] (2) links between the limbic and basal ganglia systems (lateral habenula, dorsomedial thalamus),[47,48] and (3) critical inputs to the amygdala and entorhinal cortex such as the deep prepyriform cortex.[49,50] These structures clearly play a crucial role in the evolution of discharges within the limbic system.

SYSTEMIC ADMINISTRATION OF GLUTAMATE ANTAGONISTS

NMDA Antagonists

Shortly after the description of potent competitive NMDA antagonists, we showed that these compounds were highly effective in rodent models of epilepsy when given intracerebroventricularly and moderately effective when given intraperitoneally or orally.[51-53] These compounds are most effective in reflex models of epilepsy including photically-induced seizures in baboons, *Papio papio*.[54] In these models of primary generalized epilepsy, NMDA antagonists have a therapeutic index (in terms of ataxia) comparable to that of established anticonvulsant drugs. They are moderately effective in maximal electroshock and chemically-induced seizures in rodents. They are least effective against electrically-induced seizures in the fully kindled rat.[55-57] Following evaluation in volunteers, one of the most potent of these competitive antagonists, D-CPPene, was administered as adjunctive therapy to a small group of patients with drug-refractory complex partial seizures.[58] No evidence of efficacy was obtained, but significant exacerbation of cognitive and neurological side effects was observed. A variety of agents appear to act as lower-affinity uncompetitive antagonists of the NMDA receptor. These include the desglycinated metabolite of remacemide,[59] an analogue of MK 801 (ADCI),[60,61] and memantine.[62] These compounds appear to dissociate more rapidly from the MK-801 binding site and to produce less severe motor and cognitive side effects.[63]

Glycine site NMDA antagonists (e.g., L-687,414 and L-689,560) are also effective anticonvulsants in reflex models of epilepsy.[26,64] Glycine site antagonists may benefit from a different selectivity for NMDA receptor subtypes and a lesser tendency to disrupt cognitive processes at anticonvulsant doses. Tests in man have not yet been reported.

AMPA Antagonists

AMPA antagonists tend to show general anesthetic-like properties; nevertheless, both competitive (quinoxalinedione) and noncompetitive (2,3 benzodiazepine) AMPA/KA antagonists are anticonvulsant in reflex models of epilepsy at doses that produce transient or negligible ataxia.[65,66] Anticonvulsant activity has also been shown in electroshock and chemoconvulsant models.[67] Synergism between the anticonvulsant action of NMDA and AMPA antagonists[68] and between AMPA antagonists and classical anticonvulsants[69] has been described. The possibility that non-NMDA antagonists could contribute to the therapy of epilepsy requires further investigation.

ACUTE NEURONAL DEGENERATION IN ISCHEMIA, HEAD INJURY, AND STATUS EPILEPTICUS

Activation of glutamate receptors is thought to play a significant role in the development of selective neuronal loss following global ischemia, in the pathogenesis of infarction in some forms of focal ischemia, in the progression of pathological processes following cranial or spinal trauma, and in selective neuronal loss after status epilepticus.[70-72] The evidence for this is partly the similar acute morphological changes observed in excitotoxic brain injury where glutamate, kainate or NMDA are the primary cause compared with acute changes in vulnerable neurons after status epilepticus and ischemia.[73-76]

GLUTAMATE RELEASE

In vivo microdialysis studies show a marked increase in the extracellular levels of glutamate and aspartate occurring 5–20 min after the onset of complete global ischemia, and with a more variable time course in relation to focal ischemia.[77–80] At the same time increases are commonly found not only in GABA and glycine concentration, but also in non-neurotransmitter amino acids such as taurine and alanine. Although under some circumstances calcium-dependent release from synaptic vesicles contributes to the extracellular changes, most of the increase appears to be associated with anoxic depolarization of neurons and to be linked with release from neuronal cell bodies and dendrites.[81,82] It is not clear what role glutamate could play during anoxic depolarization.

GLUTAMATE RECEPTOR CHANGES SECONDARY TO ISCHEMIA

In situ hybridization studies have suggested that there is a relative decrease in the mRNA for GluR2 compared with GluR1 and GluR3 in the CA1 region of the hippocampus after global ischemia.[83] It appears, however, that under a variety of circumstances all the AMPA receptor isoforms are down-regulated to an approximately equal degree.[84] Among metabotropic receptors mRNA levels are increased for mGlur2 and mGluR4, but decreased for mGluR5.[85]

PROTECTIVE EFFECTS OF GLUTAMATE ANTAGONISTS

Global Ischemia

AMPA receptor antagonists (i.e., NBQX and GYKI 52466) provide partial protection against hippocampal, striatal or neocortical damage induced by global ischemia when given in the postischemic period.[86–88] The effect is seen only when the ischemic stress is just above threshold for damaging a particular class of neurons; thus they can decrease the loss of CA1 pyramidal cells following 10, but not 20, min of global ischemia.[87]

NMDA receptor antagonists are relatively ineffective at protecting against a single episode of more-or-less complete forebrain ischemia in the rodent with normothermia maintained.[89,90] Some protection may be seen when the ischemia is less severe (as with two-vessel occlusion) and pretreatment or early posttreatment is employed.[91,92] NMDA antagonists also become more effective in the presence of hypothermia,[92,93] or when brief episodes of global ischemia are repeated at relatively close intervals.[94–96] This is consistent with the view that activation of NMDA receptors during anoxic depolarization will not significantly influence ionic distribution. In contrast, impairment of energy metabolism and partial depolarization can greatly enhance the excitotoxic effect of NMDA receptor activation.[97–99] This phenomenon has been demonstrated in *in vitro* preparations and *in vivo* using metabolic poisons.

Focal Ischemia

A reduction in the volume of cortical infarction seen at 1–7 days after transient or permanent occlusion of the middle cerebral artery in the mouse or rat can be produced by administering glutamate receptor antagonists directly after middle cerebral artery occlusion.

NMDA receptor antagonists of all types appear to be equally effective (i.e., competitive antagonists such as CPPene, noncompetitive antagonists such as dizocilpine, glycine site antagonists such as L687,414 and L689,560, and polyamine site antagonists such as ifenprodil and eliprodil).[100–103] AMPA antagonists of both types also diminish the volume of cortex showing infarction.[104–109] All the glutamate antagonists appear to have comparable therapeutic time windows except that a slow penetration into the brain (as with D-CPPene) naturally shortens the time window.

Clinical trials have been initiated for a competitive NMDA antagonist (CGS 19755), for a noncompetitive antagonist (CNS 1102, Cerestat), and for a polyamine site antagonist (Eliprodil) which shows a remarkably low incidence of cognitive side effects and may possess mechanisms of action in addition to NMDA receptor antagonism such as calcium-channel blockade. [110]

Neonatal Hypoxia/Ischemia

Protective effects of NMDA receptor antagonists have been demonstrated in rat models of neonatal hypoxia/ischemia involving unilateral carotid occlusion and reduction in the partial pressure of inspired oxygen.[111–113]

Status Epilepticus

In several animal models of limbic status epilepticus, it has been shown that the hippocampal pathology can be reduced by the administration of NMDA receptor antagonists.[114–116] This occurs even when the total duration of seizure activity is unmodified or even prolonged. The presumed mechanism of protection is that the NMDA antagonists modify the individual burst discharges, reducing the late components associated with a high entry of Ca^{2+}, and thereby reducing the excitotoxic effects of the seizure.

Cranial and Spinal Trauma

A transient increase in extracellular glutamate concentration occurs early after cranial or spinal trauma; this increase is of unknown pathological significance.[117,118] Many late events following CNS trauma may involve glutamate receptor activation.[119] These include delayed ischemia associated with cranial edema. They also include generation of excitotoxic molecules by microglia and macrophages.[119] Protective effects of glutamate antagonists have been observed in rodent models of cranial injury (fluid percussion)[120] and spinal injury.[121,122]

CHRONIC NEURODEGENERATIVE DISORDERS

A role for glutamate receptor activation has been postulated in several chronic neurodegenerative disorders. The case for activation of NMDA receptors in Huntington's disease is based on the extraordinary similarity in the selective loss of neuronal markers in the striatum in patients with Huntington's disease compared with the changes induced in the rat striatum by NMDA or quinolinic acid.[123] It is postulated that some failure in neuronal energy metabolism renders the cells vulnerable to what would otherwise be an innocuous level of activation of the NMDA receptor. In motoneuron disease (amyotrophic lateral sclerosis), it is postulated that activation of AMPA/kainate receptors on motoneurons leads to their degeneration.[124,125] Abnormalities of glutamate metabolism or transport have been repeatedly described in some forms of olivopontocerebellar atrophy.[126,127] Their role in the induction of neurodegeneration has not been defined.

GLUTAMATE ANTAGONISTS: THERAPEUTIC PROSPECTS IN EPILEPSY AND CEREBRAL ISCHEMIA

Disappointing preliminary results with the competitive NMDA antagonist D-CPPene as add-on therapy in complex partial seizures[58] may discourage further trials of NMDA antagonists as chronic therapy in epilepsy. It remains possible that more satisfactory results could be obtained with glycine site antagonists. Also the most appropriate target is probably primary generalized epilepsy. Children show less cognitive disturbance than adults following ketamine administration, suggesting that side effects of NMDA antagonists may be less in the young. On the other hand, effects on learning or synaptic remodeling might be of greater significance.

Clinical trials have been initiated with several types of NMDA antagonists in stroke, and their outcome is eagerly awaited. It may be some time before we know the optimal time for initiating and stopping treatment with glutamate antagonists. Furthermore it may not be easy to decide which antagonist is optimal in terms of efficacy and side-effect profile. Cerebral trauma may provide a more responsive clinical target in that the therapeutic time window is undoubtedly longer and patients are admitted to emergency care after a shorter time period.

REFERENCES

1. WATKINS, J. C. & R. H. EVANS. 1981. Annu. Rev. Pharmacol. Toxicol. **21:** 165–204.
2. MAYER, M. L. & G. L. WESTBROOK. 1987. Neurobiology **28:** 197–276.
3. TANABE, Y., M. MASU, T. ISHII, R. SHIGEMOTO & S. NAKANISHI. 1992. Neuron **8:** 169–179.
4. TANABE, Y., A. NOMURA, M. MASU, R. SHIGEMOTO, N. MIZUNO & S. NAKANISHI. 1993. J. Neurosci. **13:** 1372–1378.
5. MATUTE, C., K. GUTIÉRREZ-IGARZA, C. RIO & R. MILEDI. 1994. NeuroReport **5:** 1205–1208.
6. PARPURA, V., T. A. BASARSKY, F. LIU, K. JEFTINIJA, S. JEFTINIJA & P. G. HAYDON. 1994. Nature **369:** 744–747.
7. HOLLMANN, M., A. O'SHEA-GREENFIELD, S. W. ROGERS & S. HEINEMANN. 1989. Nature **342:** 643–648.
8. HOLLMANN, M. & S. HEINEMANN. 1994. Annu. Rev. Neurosci. **17:** 31–108.
9. NAKANISHI, S. 1992. Science **258:** 597–603.
10. SEEBURG, P. H. 1993. Trends Neurosci. **16:** 359–365.
11. MORIYOSHI, K., M. MASU, T. ISHII, R. SHIGEMOTO, N. MIZUNO & S. NAKANISHI. 1991. Nature **354:** 31–37.
12. ISHII, T., K. MORIYOSHI, H. SUGIHARA, K. SAKURADA, H. KADOTANI, M. YOKOI, C. AKAZAWA, R. SHIGEMOTO, N. MIZUNO, M. MASU & S. NAKANISHI. 1993. J. Biol. Chem. **268:** 2836–2843.
13. LAURIE, D. J. & P. H. SEEBURG. 1994. Eur. J. Pharmacol. **268:** 335–345.
14. MONYER, H., N. BURNASHEV, D. J. LAURIE, B. SAKMANN & P. H. SEEBURG. 1994. Neuron **12:** 529–540.
15. MONAGHAN, D. T. & J. A. BEATON. 1991. Eur. J. Pharmacol. **194:** 123–125.

16. KÖHR, G., S. ECKARDT, H. LÜDDENS, H. MONYER & P. H. SEEBURG. 1994. Neuron **12:** 1031–1040.
17. WILLIAMS, K. 1993. Mol. Pharmacol. **44:** 851–859.
18. WILLIAMS, K., A. M. ZAPPIA, D. B. PRITCHETT, Y. M. SHEN & P. B. MOLINOFF. 1994. Mol. Pharmacol. **45:** 803–809.
19. NICOLAS, C., D. FAGE & C. CARTER. 1994. J. Neurochem. **62:** 1835–1839.
20. SHIRASAKI, T., N. HARATA & N. AKAIKE. 1994. J. Physiol. (Lond.) **475:** 439–453.
21. SHEARDOWN, M. J., E. O. NIELSEN, A. J. HANSEN, P. JACOBSEN & T. HONORÉ. 1990. Science **247:** 571–574.
22. ORNSTEIN, P. L., M. B. ARNOLD, N. K. AUGENSTEIN, D. LODGE, J. D. LEANDER & D. D. SCHOEPP. 1993. J. Med. Chem. **36:** 2046–2048.
23. TARNAWA, I., I. ENGBERG & J. A. FLATMAN. 1990. *In* Chemistry, Biology, and Medicine. G. Lubec & G. A. Rosenthal, Eds.: 538–546. Escom. Leiden.
24. PALMER, A. J. & D. LODGE. 1993. Eur. J. Pharmacol. Mol. Pharmacol. **244:** 193–194.
25. NIELSEN, E. O., T. H. JOHANSEN, R. A. R. TASKER, S. M. STRAIN, L. H. JENSEN, F. WATJEN & J. DREYER. 1992. Soc. Neurosci. Abstr. **18:** 86 (Abstr.).
26. KULAGOWSKI, J. J., R. BAKER, N. R. CURTIS, P. D. LEESON, I. M. MAWER, A. M. MOSELEY *et al.* 1994. J. Med. Chem. **37:** 1402–1405.
27. HAYASHI, Y., N. SEKIYAMA, S. NAKANISHI, D. E. JANE, D. C. SUNTER, E. F. BIRSE, P. M. UDVARHELYI & J. C. WATKINS. 1994. J. Neurosci. **14:** 3370–3377.
28. BLISS, T. V. P. & G. L. COLLINGRIDGE. 1993. Nature **361:** 31–39.
29. ABRAHAMS, T. P., A. M. TAVEIRA DASILVA, P. HAMOSH, J. E. MCMANIGLE & R. A. GILLIS. 1993. Eur. J. Pharmacol. **238:** 223–233.
30. GREER, J. J., J. C. SMITH & J. L. FELDMAN. 1992. Brain Res. **576:** 355–357.
31. MCNAMARA, J. O., J. H. EUBANKS, J. D. MCPHERSON, J. J. WASMUTH, G. A. EVANS & S. F. HEINEMANN. 1992. J. Neurosci. **12:** 2555–2562.
32. GAREY, C. E., A. L. SCHWARZMAN, M. L. RISE & T. N. SEYFRIED. 1994. Nature Genet. **6:** 426–431.
33. JANJUA, N. A., T. ITANO, T. KUGOH, K. HOSOKAWA, M. NAKANO, H. MATSUI & O. HATASE. 1992. Epilepsy Res. **11:** 37–44.
34. JANJUA, N. A., H. KABUTO & A. MORI. 1992. Neurochem. Res. **17:** 293–296.
35. LEACH, M. J., R. A. O'DONNELL, K. J. COLLINS, C. M. MARDEN & A. A. MILLER. 1987. Epilepsy Res. **1:** 145–148.
36. AKBAR, M. T., R. TORP, N. C. DANBOLT, J. STORM-MATHISEN, B. S. MELDRUM & O. P. OTTERSEN. 1994. Soc. Neurosci. **20:** 927 (Abstr.).
37. DURING, M. J. & D. D. SPENCER. 1993. Lancet **341:** 1607–1610.
38. HOSFORD, D. A., B. J. CRAIN, Z. CAO, D. W. BONHAUS, A. H. FRIEDMAN, M. M. OKAZAKI, J. V. NADLER & J. O. MCNAMARA. 1991. J. Neurosci. **11:** 428–434.
39. MODY, I., P. K. STANTON & U. HEINEMANN. 1988. J. Neurophysiol. **59:** 1033–1053.
40. KÖHR, G., Y. DE KONINCK & I. MODY. 1993. J. Neurosci. **13:** 3612–3627.
41. LOUVEL, J. & R. PUMAIN. 1992. *In* Neurotransmitters, Seizures and Epilepsy. IV. J. Engel, Ed. Elsevier. Amsterdam.
42. AKIYAMA, K., A. DAIGEN, N. YAMADA, T. ITOH, I. KOHIRA, H. UJIKE & S. OTSUKI. 1992. Brain Res. **569:** 71–77.
43. DURING, M. J., G. R. MIRCHANDANI, P. LEONE, A. WILLIAMSON, N. C. DE LANEROLLE, M. D. GESCHWIND, P. J. BERGOLD & H. J. FEDEROFF. 1993. Soc. Neurosci. Abstr. **19:** 21. (Abstr.)
44. TURSKI, L., E. A. CAVALHEIRO, W. A. TURSKI & B. S. MELDRUM. 1986. Neuroscience **18:** 61–77.
45. MELDRUM, B., M. MILLAN, S. PATEL & G. DE SARRO. 1988. J. Neural Transm. **72:** 191–200.
46. PATEL, S., G. B. DE SARRO & B. S. MELDRUM. 1988. Neuroscience **22:** 837–850.
47. DE SARRO, G., B. S. MELDRUM, A. DE SARRO & S. PATEL. 1992. Brain Res. **591:** 209–222.
48. PATEL, S., M. H. MILLAN & B. S. MELDRUM. 1988. Exp. Neurol. **101:** 63–75.
49. MILLAN, M. H., S. PATEL & B. S. MELDRUM. 1988. Exp. Brain Res. **72:** 517–522.
50. PIREDDA, S. & K. GALE. 1986. Eur. J. Pharmacol. **120:** 115–118.
51. CROUCHER, M. J., J. F. COLLINS & B. S. MELDRUM. 1982. Science **216:** 899–901.

52. CHAPMAN, A. G., J. GRAHAM & B. S. MELDRUM. 1990. Eur. J. Pharmacol. **178:** 97–99.
53. CHAPMAN, A. G. & B. S. MELDRUM. 1989. Eur. J. Pharmacol. **166:** 201–211.
54. MELDRUM, B. S., M. J. CROUCHER, G. BADMAN & J. F. COLLINS. 1983. Neurosci. Lett. **39:** 101–104.
55. LÖSCHER, W. & D. HÖNACK. 1991. J. Pharmacol. Exp. Ther. **256:** 432–440.
56. LÖSCHER, W. & D. HONACK. 1993. Eur. J. Pharmacol. **238:** 191–200.
57. DÜRMÜLLER, N., M. CRAGGS & B. MELDRUM. 1994. Epilepsy Res. **17:** 167–174.
58. SVEINBJORNSDOTTIR, S., J. W. A. S. SANDER, D. UPTON, P. J. THOMPSON, P. N. PATSALOS, D. HIRT, M. EMRE, D. LOWE & J. S. DUNCAN. 1993. Epilepsy Res. **16:** 165–174.
59. PALMER, G. C., R. J. MURRAY, T. C. M. WILSON, M. S. EISMAN, R. K. RAY, R. C. GRIFFITH, J. J. NAPIER, M. FEDORCHUK, M. L. STAGNITTO & G. E. GARSKE. 1992. Epilepsy Res. **12:** 9–20.
60. GRANT, K. A., L. D. SNELL, M. A. ROGAWSKI, A. THURKAUF & B. TABAKOFF. 1992. J. Pharmacol. Exp. Ther. **260:** 1017–1022.
61. ROGAWSKI, M. A., S.-I., YAMAGUCHI, S. M. JONES, K. C. RICE, A. THURKAUF & J. A. MONN. 1991. J. Pharmacol. Exp. Ther. **259:** 30–37.
62. PARSONS, C. G., R. GRUNER, J. ROZENTAL, J. MILLAR & D. LODGE. 1993. Neuropharmacology **32:** 1337–1350.
63. ROGAWSKI, M. A. 1992. Drugs **44:** 279–292.
64. SMITH, S. E. & B. S. MELDRUM. 1992. Eur. J. Pharmacol. **211:** 109–111.
65. SMITH, S. E., N. DÜRMÜLLER & B. S. MELDRUM. 1991. Eur. J. Pharmacol. **201:** 179–183.
66. CHAPMAN, A. G., S. E. SMITH & B. S. MELDRUM. 1991. Epilepsy Res. **9:** 92–96.
67. YAMAGUCHI, S., S. D. DONEVAN & M. A. ROGAWSKI. 1993. Epilepsy Res. **15:** 179–184.
68. LÖSCHER, W., C. RUNDFELDT & D. HONACK. 1993. Eur. J. Neurosci. **5:** 1545–1550.
69. ZARNOWSKI, T., Z. KLEINROK, W. A. TURSKI & S. J. CZUCZWAR. 1993. Neuropharmacology **32:** 895–900.
70. MELDRUM, B. S. 1990. Cereb. Brain Metab. Rev. **2:** 27–57.
71. MELDRUM, B. S. 1992. *In* Emerging Strategies in Neuroprotection. P. J. Marangos & H. Lal, Eds.: 106–128. Birkhauser Boston Inc. Cambridge, MA.
72. MELDRUM, B. S. & J. GARTHWAITE. 1990. Trends Pharmacol. Sci. **11:** 379–387.
73. GRIFFITHS, T., M. C. EVANS & B. S. MELDRUM. 1983. Neuroscience **10:** 385–395.
74. SIMON, R. P., T. GRIFFITHS, M. C. EVANS, J. H. SWAN & B. S. MELDRUM. 1984. J. Cereb. Blood Flow Metab. **4:** 350–361.
75. EVANS, M., T. GRIFFITHS & B. S. MELDRUM. 1983. Neuropathol. Appl. Neurobiol. **9:** 39–52.
76. IKONOMIDOU, C., M. T. PRICE, J. L. MOSINGER, G. FRIERDICH, J. LABRUYERE, K. SHAHID SALLES & J. W. OLNEY. 1989. J. Neurosci. **9:** 1693–1700.
77. BENVENISTE, H., J. DREJER, A. SCHOUSBOE & N. H. DIEMER. 1994. J. Neurochem. **43:** 1369–1374.
78. ANDINE, P., O. ORWAR, I. JACOBSON, M. SANDBERG & H. HAGBERG. 1991. J. Neurochem. **57:** 222–229.
79. MELDRUM, B. S., J. H. SWAN, M. J. LEACH, M. H. MILLAN, R. GWINN, K. KADOTA, S. H. GRAHAM, J. CHEN & R. P. SIMON. 1992. Brain Res. **593:** 1–6.
80. MATSUMOTO, K., R. GRAF, G. ROSNER, J. TAGUCHI & W.-D. HEISS. 1993. J. Cereb. Blood Flow Metab. **13:** 586–594.
81. DREJER, J., H. BENVENISTE, N. H. DIEMER & A. SCHOUSBOE. 1985. J. Neurochem. **45:** 145–151.
82. TORP, R., B. ARVIN, E. LE PEILLET, A. G. CHAPMAN, O. P. OTTERSEN & B. S. MELDRUM. 1993. Exp. Brain Res. **96:** 365–376.
83. PELLEGRINI-GIAMPIETRO, D. E., R. S. ZUKIN, M. V. L. BENNETT, S. CHO & W. A. PULSINELLI. 1992. Proc. Natl. Acad. Sci. USA **89:** 10499–10503.
84. DIEMER, N. H. 1995. *In* Pharmcology of Cerebral Ischemia. J. Krieglstein, Ed. Wissenschaftliche Verlagsgesellschaft. Stuttgart.
85. ROSDAHL, D., D. A. SEITZBERG, T. CHRISTENSEN, T. BALCHEN & N. H. DIEMER. 1994. NeuroReport **5:** 593–596.
86. BUCHAN, A. M., H. LI, S. CHO & W. A. PULSINELLI. 1991a. Neurosci. Lett. **132:** 255–258.
87. LE PEILLET, E., B. ARVIN, C. MONCADA & B. S. MELDRUM. 1992. Brain Res. **571:** 115–120.
88. LI, H. & A. M. BUCHAN. 1993. J. Cereb. Blood Flow Metab. **13:** 933–939.

89. BUCHAN, A. & W. A. PULSINELLI. 1990. J. Neurosci. **10:** 311–316.
90. BUCHAN, A. M., H. LI, W. A. PULSINELLI. 1991. J. Neurosci. **11:** 1049–1056.
91. SWAN, J. H. & B. S. MELDRUM. 1990. J. Cereb. Blood Flow Metab. **10:** 343–351.
92. ROD, M. R. & R. N. AUER. 1989. Can. J. Neurol. Sci. **16:** 340–344.
93. IKONOMIDOU, C., J. L. MOSINGER & J. W. OLNEY. 1989. Brain Res. **487:** 184–187.
94. SHUAIB, A., S. IJAZ, R. MAZAGRI, A. SENTHILSEVLVAN. 1993. Neuroscience **56:** 915–920.
95. KATO, H., T. ARAKI & K. KOGURE. 1990. Brain Res. **516:** 175–179.
96. LIN, B., W. D. DIETRICH, M. D. GINSBERG, M. Y.-T. GLOBUS & R. BUSTO. 1993. J. Cereb. Blood Flow Metab. **13:** 925–932.
97. BEAL, M. F. 1992. Ann. Neurol. **31:** 119–130.
98. HENSHAW, R., B. G. JENKINS, J. B. SCHULZ, R. J. FERRANIE, N. W. KOWALL, B. R. ROSEN & M. F. BEAZ. 1994. Brain Res. **647:** 161–166.
99. ZEEVALK, G. D. & W. J. NICKLAS. 1992. J. Neurochem. **59:** 1211–1220.
100. PARK, C. K., D. G. NEHLE, D. I. GRAHAM, G. M. TEASDALE & J. MCCULLOCH. 1988. Ann. Neurol. **24:** 543–551.
101. PARK, C. K., J. MCCULLOCH, J. K. KANG & C. R. CHOI. 1992. Neurosci. Lett. **147:** 41–44.
102. GILL, R., C. BRAZELL, G. N. WOODRUFF & J. A. KEMP. 1991. Br. J. Pharmacol. **103:** 2030–2036.
103. GOTTI, B., D. DUVERGER, J. BERTIN, C. CARTER, R. DUPONT, J. FROST, B. GAUDILLIERE, E. T. MACKENZIE, J. ROUSSEAU, B. SCATTON & A. WICK. 1988. J. Pharmacol. Exp. Ther. **247:** 1211–1221.
104. GILL, R., L. NORDHOLM & D. LODGE. 1992. Brain Res. **580:** 35–43.
105. BUCHAN, A. M., D. XUE, Z. HUANG, K. H. SMITH & H. LESIUK. 1991. NeuroReport **2:** 473–476.
106. SMITH, S. E. & B. S. MELDRUM. 1993. Funct. Neurol. **8:** 43–48.
107. SMITH, S. E. & B. S. MELDRUM. 1992. Stroke **23:** 861–864.
108. GILL, R., L. NORDHOLM & D. LODGE. 1992. Brain Res. **580:** 35–43.
109. DEGRABA, T. J., P. OSTROW, S. HANSON & J. C. GROTTA. 1994. J. Cereb. Blood Flow Metab. **14:** 262–268.
110. BITON, B., P. GRANGER, A. CARREAU, H. DEPOORTERE, B. SCATTON & P. AVENET. 1994. Eur. J. Pharmacol. **257:** 297–301.
111. MCDONALD, J. W., F. S. SILVERSTEIN & M. V. JOHNSTON. 1987. Eur. J. Pharmacol. **140:** 359–361.
112. OLNEY, J. W., C. IKONOMIDOU, J. L. MOSINGER & G. FRIEDRICH. 1989. J. Neurosci. **9:** 1701–1704.
113. ANDINE, P., A. LEHMANN, K. ELLREN, E. WENNBERG, I. KJELLMER, T. NIELSEN & H. HAGBERG. 1988. Neurosci. Lett. **90:** 208–212.
114. FARIELLO, R. G., G. T. GOLDEN, G. G. SMITH & P. F. REYES. 1989. Epilepsy Res. **3:** 206–213.
115. CLIFFORD, D. B., J. W. OLNEY, A. M. BENZ, T. A. FULLER & C. F. ZORUMSKI. 1990. Epilepsia **31:** 382–390.
116. FUJIKAWA, D. G., A. H. DANIELS & J. S. KIM. 1994. Epilepsy Res. **17:** 207–219.
117. FADEN, A. I., P. DEMEDIUK, S. S. PANTER & R. VINK. 1989. Science **244:** 798–800.
118. GIULIAN, D. & K. VACA. 1993. Stroke (Suppl. 24): 184–192.
119. PANTER, S. S. & A. I. FADEN. 1992. Neurosci. Lett. **136:** 165–168.
120. SMITH, D. H., K. OKIYAMA, M. J. THOMAS & T. K. MCINTOSH. 1993. J. Neurosci. **13:** 5383–5392.
121. FADEN, A. I., J. A. ELLISON & L. J. NOBLE. 1990. Eur. J. Pharmacol. **175:** 165–174.
122. WRATHALL, J. R., Y. D. TENG, D. CHOINIERE & D. J. MUNDT. 1992. Brain Res. **586:** 140–143.
123. BEAL, M. F., N. W. KOWALL, D. W. ELLISON, M. F. MAZUREK, K. J. SWARTZ & J. B. MARTIN. 1996. Nature **321:** 169–171.
124. ZEMAN, S., C. LLOYD, B. MELDRUM & P. N. LEIGH. 1994. Neuropathol. Appl. Neurobiol. **20:** 219–231.
125. ROTHSTEIN, J. D., L. JIN, M. DYKES-HOBERG & R. W. KUNCL. 1993. Proc. Natl. Acad. Sci. USA **90:** 6591–6595.
126. KISH, S. J., L.-J. CHANG, L. M. DIXON, Y. ROBITAILLE & L. DISTEFANO. 1994. Metab. Brain Dis. **9:** 97–103.
127. PLAITAKIS A., S. BERL & M. D. YAHR. 1984. Ann. Neurol. **15:** 144–153.

Structure and Distribution of Multiple GABA$_A$ Receptor Subunits with Special Reference to the Cerebellum[a]

W. WISDEN

MRC Laboratory of Molecular Biology
Neurobiology Division
MRC Centre, Hills Road
Cambridge, CB2 2QH, England

γ-Aminobutyric acid (GABA) evokes membrane hyperpolarization by opening multisubunit anion channels (GABA$_A$ receptors) permeable to chloride and bicarbonate.[1-3] Like the canonical muscle nicotinic receptor, GABA$_A$ receptor subunits most likely form a "ring of five" with a central ion conducting pathway in the center.[4,5] Quantal analysis of a typical central synapse (inhibitory postsynaptic currents of hippocampal dentate granule cells) suggests the existence of GABA$_A$ receptor clusters containing 20 to 30 channels.[6] A rich variety of allosteric modulators such as 1,4-benzodiazepines (Bzs), barbiturates, divalent metal cations (especially zinc), steroids, and substituted pyrazinones[3,7-9] all modify, by differing mechanisms, the GABA-evoked channel activity.

GABA$_A$ receptor heterogeneity can be inferred from both ligand binding/ autoradiography,[9-11] and electrophysiology.[9-14] Not all GABA-activated channels exhibit all modulatory effector sites.[10,15] For example, some GABA$_A$ receptors are insensitive to benzodiazepine agonists,[15-17] and steroid and zinc modulation of GABA receptors varies in a brain region–dependent manner.[9,18] This pharmacological heterogeneity arises because different GABA$_A$ receptor subtypes are derived combinatorially from members of different sequence-related subunit classes, the subunit combinations varying with brain region and cell type. In mammals, these comprise six alpha, three beta, three gamma, one delta, and two rho subunits.[1-3] Based on predictions from primary sequence, the subunits all share the same architectural design,[4] and pairwise comparisons show approximately 30 to 40% sequence similarity between members of different subunit classes and 60 to 80% sequence identity between variants within the same class.[1] The stoichiometry of subunits within the pentamer (for example, $[\alpha 1]_2[\beta 2]_2\gamma 2$ or $\alpha 1[\beta 2]_3\gamma 2$) of any given receptor assembly is unknown. *In vivo* the subunit composition varies depending on both cell type and probably on subcellular location (e.g., soma versus dendrites).[12,19]

The frustrations of investigating GABA$_A$ receptors are universal ones faced by researchers working on any other ligand-gated ion channel that is put together from different subunit components. The arithmetical number of potential subunit combinations is very high, and it is likely that many neuronal cell types probably express more than one type of receptor;[20] it must be determined which combinations are found *in vivo* and in which brain areas. The general research program is to define receptor subtypes on any given neuronal class with a combination of techniques: *in situ* hybridization,[21-24] immunocytochemistry,[20,25-27] immunoprecipitation and immunopurification,[28-30] and patch-clamping of identified neurons in brain slices combined

---

[a] Support for the writing of this review was provided by the Medical Research Council, U.K.

with single-cell PCR.[6,31] Having defined receptor combinations, what is their functional significance? So far, only broad statements and suggestions can be made. For example, different parts of a cell are innervated by different classes of GABAergic interneurons.[32] It might be that different GABA$_A$ receptors on different parts of the cell correlate with these different GABAergic inputs, for example, different GABA$_A$ receptor subtypes on proximal dendrites, distal dendrites, and the cell body.[12] Differential subcellular distribution of subtypes is supported by immunocytochemistry using subunit specific antibodies.[20,25,27] Gene targeting and natural mutations also provide a way of assessing subunit function.[33-35] Beyond general statements, there are no clear answers to questions concerning the exact composition, distribution, and function of receptor subtypes. Subunit combination $\alpha1\beta2\gamma2$ is the most plausible one known with any certainty,[22,29] but this is complicated by having to take into account whether $\gamma2S$, $\gamma2L$ or both $\gamma2$ variants are present in the complex.[36]

FUNCTIONS OF THE SUBUNIT CLASSES

Although we do not understand the significance of receptor diversity, research has progressed in defining the mechanics and features of a generalized GABA$_A$ receptor.[1-3] The α subunits determine benzodiazepine pharmacology when coexpressed with any β subunit and a $\gamma2$ subunit.[37] Structural variations in the α subunits confer different sensitivities to Bz ligands, and the α subunits are responsible for much of the pharmacological heterogeneity reported for GABA$_A$ receptors.[16-18,33,38] The Bz pharmacology can be largely traced to a single amino acid residue in the predicted N-terminal extracellular domain of the receptor.[17] Recombinant $\alpha x\beta x\gamma2$ complexes fit nicely with the Bz pharmacological heterogeneity data established prior to the cloning of the receptors,[10] and this is further supported by the differential distribution of the α subunits combined with the more universal expression of $\gamma2$, largely fitting with the Bz1/Bz2 ligand autoradiography data.[10,22,23] The $\alpha1\beta x\gamma2$ combination is Bz1, whereas those of $\alpha2\beta x\gamma2$, $\alpha3\beta x\gamma2$, and $\alpha5\beta x\gamma2$ are variations of Bz2 subtypes.[38] Mice with specific $\alpha5$ subunit gene deletions are apparently healthy and viable.[35]

The $\alpha6\beta x\gamma2$ combination constitutes a Bz agonist insensitive subtype found on cerebellar granule cells (non Ro 15-4513 displaceable binding, Bz3 subtype).[16,33] When combined with the $\gamma2$ subunit in cell lines, the $\alpha4$ subunit has the same pharmacological features (i.e., in not binding Bz agonists) as the $\alpha6$ subunit. However, there is no Bz-agonist-insensitive binding in the brain (other than in the cerebellar granule cells), suggesting that the $\alpha4$ subunit is not found with the $\gamma2$ subunit *in vivo*, although its mRNA is abundant in forebrain.[22] The $\alpha4$ and $\alpha6$ subunits share the same pharmacology because of the absence of a conserved histidine residue that is present in all the other α subunits.[17] The $\alpha6$ subunit (or at least the absence of the critical histidine that prevents Bz agonist binding) is clearly needed for some specific functioning of the granule cells. A natural point mutation that confers Bz agonist binding onto the $\alpha6$ subunit correlates with impaired motor control when the animals are given diazepam, and even in the absence of pharmacological treatment, these animals are reported to have a diminished behavioral reaction to stress.[33] This latter observation might be explainable by an endogenous agonist acting at the Bz site.

The β subunits are important structural components of GABA$_A$ receptors, but are broadly interchangeable within the context of an $\alpha x\beta x\gamma2$ complex when used in *in vitro* assays.[39] *In vivo*, based on co-distribution patterns, $\beta2$ probably pairs throughout the central nervous system with $\alpha1$, and $\beta3$ colocalizes in many areas with $\alpha2$.[22] The distribution of β1 mRNA bears a superficial resemblance to that of α5, with both mRNAs being particularly abundant in the hippocampus.[22] However, some cell types, such as mitral cells of the olfactory bulb express all three β-subunit genes, so there are many exceptions to a simple pairing scheme. Native (but not recombinant) β subunits may differ in sensitivity to steroids and muscimol.[3,11,39] The anticonvulsant loreclezole is selective for recombinant receptors containing β2 and β3 subunits over those containing the β1 subunit.[40] Mice with β3 gene deletions have neurological deficits characterized by tremor and jerky gait.[34,35] This is not surprising considering that β3 is the dominant β subunit in the hippocampus, caudate-putamen, and spinal cord.[21,22]

All three γ subunits confer Bz responsiveness to αβ complexes,[37,41] but the most studied γ subunit has been γ2. The existence of the γ1 and γ3 subunits forces revisions on the Bz1 and Bz2 model. Receptors containing these subunits (e.g., αxβxγ1 or αxβxγ3) probably went previously undetected because of much lower affinities and efficacies for benzodiazepines.[41,42] However, these receptors are likely to be specialized. The γ1 subunit mRNA is particularly abundant in certain limbic and hypothalamic areas (medial amygdala and septum) where it is probably present in an α2β3γ1 complex.[22] The γ3 subunit mRNA is generally rare, being most abundant in the medial geniculate thalamic nucleus.[22,42] Both γ1 and particularly γ3 subunit gene expression are much more prominent in younger animals.[24] Mice with specific γ3 subunit gene deletions are apparently normal.[35]

The function of the δ subunit[43] is unknown (this is discussed more fully in the section on the cerebellum). Receptors assembled from ρ subunits are probably retinal-specific and can be activated by the agonist *cis*-4-aminocrotonic acid but are insensitive to the classical GABA$_A$ drugs, bicuculline, barbiturates, and benzodiazepines, which has lead to the proposal that ρ subunits be designated GABA$_C$.[15]

GABA$_A$ RECEPTOR EXPRESSION IN THE CEREBELLUM

The cerebellum can be used as a case study to work out and dissect GABA$_A$ receptor heterogeneity. The cerebellum is an evolutionarily conserved structure, consisting of a small number of neuronal and glial cell subtypes identifiable by their position and size with a relatively well defined synaptic circuitry (FIG. 1).[44] All neuronal cell types with the possible exception of the Golgi cells receive an inhibitory GABAergic input. The GABAergic Golgi cells synapse onto granule cells. GABAergic stellate/basket interneurons in the molecular layer synapse onto Purkinje cells. These send reciprocal inhibitory processes back onto the stellate/basket neurons as well as project to the deep cerebellar nuclei. Bluntly stated, if GABA$_A$ subunit combinations cannot be *completely* worked out for the cerebellum, it is unlikely, with perhaps the exception of the olfactory bulb and retina, that they will be worked out anywhere else in the brain, where synaptic circuitry is so much more complicated and many cell types or connections remain only partially characterized. However, even for the cerebellum, it is amazing to learn that new cell types are still being defined, for example, the unipolar brush cell that is found in certain granule cell folia.[45]

Autoradiographic studies complement our molecular biology knowledge of GABA$_A$ receptor heterogeneity in the cerebellum, but as discussed below, inferring subunit composition from this data is not necessarily simple. Additionally, although electrophysiological characteristics of GABA$_A$ receptor-channel complexes expressed *in vitro* depend upon subunit composition,[7,46] physiological distinctions between GABA$_A$ synaptic responses corresponding to the demonstrated molecular heterogeneity have not been clearly identified (see below).

MOLECULAR LAYER

In the rodent, Bz agonists (e.g., [^{3}H]flunitrazepam) bind predominantly to the molecular layer (which consists of Purkinje cell dendrites, stellate/basket cells, Bergmann glial processes, and the axon terminals of granule cells) (Fig. 1), whereas GABA_A agonists that highlight high-affinity sites ([^{3}H]GABA, [^{3}H]muscimol) bind mainly to the granule cell layer consisting of granule cell bodies and Golgi cells.

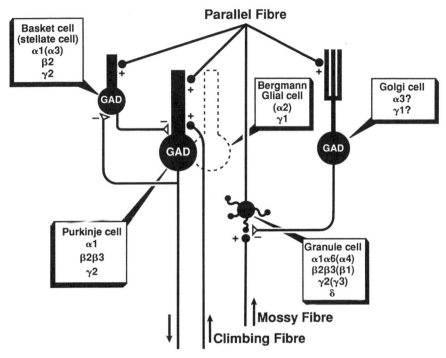

FIGURE 1. Schematic summary of the circuit diagram of the cerebellum indicating which cell types express the various subunits. Inhibitory (GABA releasing cells) are marked GAD (glutamic acid decarboxylase). Excitatory terminals are *filled circles* marked "+". Inhibitory terminals are *open triangles* marked "−". Subunits in parentheses are much more weakly expressed. The figure is modified from Farrant and Cull-Candy,[44] and the data taken from *in situ* hybridization[23] and immunocytochemistry studies.[20,25–27,29] The Golgi cell expression data are not well characterized. The tentative α3 assignment comes from bovine brain,[49] the γ1 from chicken,[54] but there is no evidence to support this combination in the rat.

However, even though high-affinity GABA_A agonist sites are scarce in the molecular layer, Bz binding is still enhanced (via allosteric mechanisms) by GABA in this sector, suggesting that the molecular Bz sites are functional GABA_A receptors.[10] GABA_A sites are more abundant in the molecular layer when assayed with the antagonist [^{3}H]bicuculline methochloride (BMC), which binds low-affinity sites.[10] The majority of Bz binding sites in the molecular layer (greater than 90%) have a selectively high affinity for β-carbolines and are thus designated Bz1.[10] The GABA_A antagonist [³H]SR-95531 binds relatively poorly to the molecular layer.[10,47] Compounds like picrotoxin and cage convulsants block GABA chloride channel openings, and their receptor sites can be radiolabeled with *t*-butyl bicyclophosphorothionate (TBPS) or the analogue *t*-butyl bicyclo-orthobenzoate (TBOB).[10] As for the other drug binding sites, the allosteric nature of the GABA$_A$ receptor complex means that the binding of TBPS and TBOP is affected by the presence of GABA, zinc, steroids, benzodiazepines, and barbiturates.[9,18] For example, [³H]TBOB binding (which in the absence of GABA$_A$ antagonists is found mainly over the molecular layer of the cerebellum) is markedly reduced by 100 μM zinc, whereas this concentration of zinc has no effect on TBOP binding in many other brain regions, for example, the thalamus, cortical laminae V-VI, and the colliculi.[9]

The major receptor subunit combination accounting for the molecular layer pharmacology is likely to be an α1β2β3γ2 combination on Purkinje cell dendrites and an α1β2γ2 combination on stellate/basket cells.[22] This is supported by immunocytochemistry showing relatively little subunit immunoreactivity on Purkinje cell soma, but large amounts on dendrites.[29,48] On the Purkinje cell dendrites it is not clear whether the β2 and β3 subunits are assembled into a single or separate receptor complex. An α1βxγ2 complex would display Bz1 pharmacology.[37] It is not known which γ2 splice forms (γ2S or γ2L) are present in the Purkinje and stellate/basket combinations, but PCR analysis on cDNA extracted from whole rat cerebellum reveals that both γ2S and γ2L are equally abundant.[36] In some systems, the presence of the γ2 subunit confers zinc insensitivity to αxβxγ2 receptors,[2,3] but because γ2 is abundant on the Purkinje cell dendrites as part of a probable α1β2/β3/γ2 complex and because the TBOP binding over the molecular layer is completely reduced by zinc, the relative abundance of the γ2 subunit cannot be the sole determinant of regional differences in the zinc sensitivity of GABA$_A$ receptors.[9]

The α3 subunit protein and mRNA is additionally present in the molecular layer, probably in stellate/basket cells.[23,25] The α2 and γ1 subunits are probably on the Bergmann glial fibers whose cell bodies originate in the Purkinje cell layer.[21,22,49] The function of such glial GABA$_A$ receptors is not known.

CEREBELLAR GRANULE CELLS

In the rodent, the cerebellar granule cell layer has far fewer high-affinity Bz agonist sites, but a much higher density of high-affinity muscimol sites than the molecular layer.[10] However, a high density of agonist-insensitive Bz binding is present in the cerebellar granule cells,[33] that is, a high proportion of the partial inverse agonist [3H]Ro 15-4513 binding in granule cell membranes is not displaceable by an excess of unlabeled flunitrazepam or diazepam—"diazepam insensitive binding." This pharmacology is explained by the unique presence of the α6 subunit (see below).[16] In contrast to the molecular layer, [³⁵S]TBPS binding over the granule cell layer is only found when GABA agonist sites have been blocked by the GABA$_A$ antagonists SR 95531 and bicuculline,[18] that is, [³⁵S]TBPS binding over the granule cell layer is more sensitive to GABA than that in the molecular layer. However, the internal rim of the granule cell layer has a small amount of [³⁵S]TBPS binding that is insensitive to 50 μM GABA[18,50] suggestive of granule cell heterogeneity.

Rat cerebellar granule cells express six subunit genes (α1, α6, β2, β3, γ2, and δ) abundantly and several others (α4, β1, γ3) weakly at the mRNA level.[23] Granule cells also contain large quantities of α1, α6, β2/β3, and δ subunits as assessed by immunocytochemistry,[20,26,27,48] and moderate amounts of γ2 immunoreactivity.[29] Even if one only takes into account the six most abundant subunits and assumes a pentameric complex, this must mean at least two abundant receptor subtypes on granule cells with a possible low background of other subtypes containing $\alpha4$, $\beta1$, and $\gamma3$ subunits. Alternatively, subsets of granule cells might express just one receptor subtype. However, in the rat, the synapses established by Golgi cell terminals on the dendrites of granule cells are immunoreactive for the $\alpha1$, $\alpha6$, and $\beta2/3$ subunits in virtually all glomeruli, indicating that the $\alpha1$ and $\alpha6$ subunits are likely to be colocalized at the same synapse.[20] The cell bodies of the granule cells (which do not receive synapses) arc negative for the $\alpha6$ subunit, although they stain with the $\alpha1$ and $\beta2/3$ subunit specific antibodies.[20] This suggests synaptically located $\alpha1$ and $\alpha6$ receptors on granule cell dendrites, and extrasynaptic $\alpha1$-containing receptors on the granule cell body.[20] It is an interesting question as to whether the $\alpha1$ and $\alpha6$ subunits in the granule cell glomeruli are assembled into a single complex or are present in two parallel receptors in the same synapse. On the one hand, immunoprecipitation studies (cited as unpublished data in ref. 20) and cotransfection experiments (transfecting $\alpha1$, $\alpha6$, $\beta2$, and $\gamma2$ encoding expression plasmids) in an embryonic kidney cell line imply that $\alpha1$ and $\alpha6$ subunits do not co-assemble.[18] However, other immunoprecipitation studies report that $\alpha6$ frequently associates with $\alpha1$, $\beta2/\beta3$, and $\gamma2$[51] or at least that there is a partial coexistence (10% of receptors) of $\alpha6$ with the $\alpha1$ subunit.[52] Beyond the specific case of the cerebellum, there is evidence that GABA$_A$ receptors can contain different α subunits within the same complex. The $\alpha1$ and $\alpha3$ subunits can co-assemble as an $\alpha1\alpha3\beta2\gamma2$ complex in embryonic kidney cells,[46] and immunopurification from neocortex suggests that a minority of GABA$_A$ receptors do contain heterologous α-subunit pairs (e.g., $\alpha2\alpha3$).[52]

The δ subunit remains a dark horse.[43] An initial observation noted that the δ-subunit gene was most prominently expressed in neurons whose dendritic processes exist in complex synaptic arrangements with other neurons—glomeruli or rosettes—and it was therefore suggested that it is involved in synapses, mediating surround inhibition.[43] Its mRNA and protein are particularly abundant in cerebellar granule cells.[24,26,43] No pharmacology data from *in vitro* studies are available. It has been suggested from immunopurification experiments that a proportion of δ subunits are associated with Bz agonist-responsive complexes containing $\alpha1$, $\alpha3$, $\beta2/\beta3$, and $\gamma2$ subunits in various forebrain regions.[30] However, the distribution of δ subunit mRNA[21,22,43] and protein[26] does not correlate with distribution of the majority of Bz agonist sites.[10] This is especially true for the granule cell layer of the cerebellum. Based on mRNA distribution, the δ subunit might be preferentially associated with non-BZ agonist binding α subunits (i.e., $\alpha4$ and $\alpha6$) in the forebrain and cerebellar granule cell layer, respectively, to perhaps form muscimol-sensitive, BZ-insensitive receptors.[10,22] Indeed, other immunoprecipitation results reveal that both $\alpha6$ and δ subunits coexist in a receptor complex that binds [^{3}H]muscimol, but not benzodiazepines (cited as unpublished data in ref. 20). Immunoreactivity for the δ subunit seems to be concentrated in the glomeruli, suggesting colocalization with the $\alpha6$ subunit.[20,26]

Do the different arrays of subunits expressed between Purkinje cells and granule cells translate into functional differences between the GABA$_A$ receptors of these two cell types? Whole-cell recordings of spontaneous inhibitory postsynaptic currents (sIPSCs) in cerebellar Purkinje cells decay with a single fast exponential, whereas in granule cells sIPCs decay with the sum of a fast and slow exponential curve.[13] This difference might arise as a product of a mixed activation of two GABA$_A$ receptor populations,[13] perhaps because of the $\alpha6$ and δ in granule cells, but not in Purkinje cells (FIG. 1). The α subunit that is functionally related to $\alpha6$ (the $\alpha4$ subunit) is also present with the δ subunit in hippocampal dentate granule cells which similarly have a double exponential sIPSC decay. However, Puia *et al.*[13] caution that slow GABA-activated Cl^- currents could be independent from the subunit assembly of the receptor and might instead depend on reversible posttranslational modifications, for example, phosphorylation, because double exponential decays are occasionally observed on Purkinje cells. This caveat is especially pertinent because these studies were made using patches of granule cell soma which may not contain the $\alpha6$ subunit, but do contain the $\alpha1$ and $\beta2/\beta3$ subunits.[20] Responses from patches of granule and Purkinje cell bodies to step applications of GABA are also very similar over a wide range of concentration,[53] but again, when looking for differences, the electrophysiology of *dendritic* granule cell receptors might be more relevant, because, as mentioned above, the soma of both granule and Purkinje cells might have an identical compliment of subunits ($\alpha1$ and $\beta2/\beta3$ and maybe $\gamma2$).

In conclusion, a combination of *in situ* hybridization and immunocytochemistry has given a good idea of the different subtypes of $GABA_A$ receptor found on the different cell types (summarized in FIG. 1). Purkinje cells probably express an $\alpha1\beta2\beta3\gamma2$ combination. Specific receptor combinations in granule cells are more problematic. Any of the recombinant $\alpha6\beta2\gamma2$, $\alpha6\beta3\gamma2$ or $\alpha6\beta2\beta3\gamma2$ combinations are likely to fit the granule-cell pharmacological print of non-Bz agonist binding. Such a combination also fits with the [^{35}S]TBPS binding modulation by GABA on granule cells[18] because recombinant $\alpha6\beta2\gamma2$ receptors produce [^{35}S]TBPS binding sites about 10-fold more sensitive to inhibition by GABA than $\alpha1\beta2\gamma2$ complexes.[18] An $\alpha6\beta x\gamma2$ complex would also have high-affinity muscimol binding which matches that seen over granule cells.[16] However, if the $\alpha1$ subunit is not in an $\alpha6$-containing complex as seems likely for the majority or receptors, but is clearly abundant in granule cells,[20,25] then why isn't $\alpha1\beta2\gamma2$ or $\alpha1\beta3\gamma2$ or $\alpha1\beta2\beta3\gamma2$ Bz1 type pharmacology abundant on the granule cells as it is on Purkinje cells? Also arguing against the abundant formation of an $\alpha1\beta x\gamma2$ complex is the finding that [^{35}S]TBPS is very sensitive to GABA in the granule cell layer.[18] Perhaps the $\gamma2$ subunit is present in limiting amounts.[29] Under these limiting conditions, $\gamma2$ may preferentially assemble with $\alpha6$. Where does the δ subunit fit in? Perhaps an $\alpha6\beta x\gamma2\delta$ or $\alpha6\beta x\gamma2$ combination is found on granule cell dendrites colocalized with an $\alpha1\beta2\beta3\delta$ complex in the same synapse. More detail on the properties of these subunit combinations will be needed before we can make any further conclusions.

[NOTE ADDED IN PROOF: Based on immunoprecipitation experiments, a model describing the composition of all $GABA_A$ receptors in the cerebellum was constructed that defined the following α (alpha) and γ/δ (gamma/delta) combinations (percentage of cerebellar $GABA_A$ receptors): $\alpha6\gamma2$ (36%), $\alpha6\delta$ (23%), $\alpha1\gamma2$ (28%), $\alpha2\gamma1$ (8%), and $\alpha3\gamma2$ (5%). An $\alpha1\delta$ combination was not found (see ref. 55)].

ACKNOWLEDGMENTS

I thank Mary-Ann Starkey for her help in the preparation of the manuscript.

REFERENCES

1. BARNARD, E. A. 1992. Receptor classes and the transmitter gated ion channels. Trends Biochem. Sci. **17:** 368–374.
2. WISDEN, W. & P. H. SEEBURG. 1992. $GABA_A$ receptor channels: From subunits to functional entities. Curr. Opin. Neurobiol. **2:** 263–269.
3. MACDONALD, R. L. & R. W. OLSEN. 1994. $GABA_A$ receptor channels. Annu. Rev. Neurosci. **17:** 569–602.

4. UNWIN, N. 1993. Nicotinic acetylcholine receptor at 9Å resolution. J. Mol. Biol. **229**: 1101–1124.

5. NAYEEM, N., T. P. GREEN, I. L. MARTIN & E. A. BARNARD. 1994. Quaternary structure of the native GABA$_A$ receptor determined by electron microscopic image analysis. J. Neurochem. **62**: 815–818.

6. EDWARDS, F. A., A. KONNERTH & B. SAKMANN. 1990. Quantal analysis of inhibitory transmission in the denate gyrus of rat hippocampal slices: A patch-clamp study. J. Physiol. **430**: 213–249.

7. VERDOORN, T. A., A. DRAGUHN, S. YMER, P. H. SEEBURG & B. SAKMANN. 1990. Functional properties of recombinant rat GABA$_A$ receptors depend upon subunit composition. Neuron **4**: 919–928.

8. IM, H. K., W. B. IM, T. M. JUDGE, R. B. GAMMILL, B. J. HAMILTON, D. B. CARTER & J. F. PREGENZER. 1993. Substituted pyrazinones, a new class of allosteric modulators for gamma-aminobutyric acid$_A$ receptors. Mol. Pharmacol. **44**: 468–472.

9. KUME, A., S. Y. SAKURAI & R. L. ALBIN. 1994. Zinc inhibition of t-[^{3}H]butylbicycloortobenzoate binding to the GABA$_A$ receptor complex. J. Neurochem. **61**: 602–607.

10. OLSEN, R. W., R. T. MCCABE & J. K. WAMSLEY. 1990. GABA$_A$ receptor subtypes: Autoradiographic comparison of GABA, benzodiazepine, and convulsant binding sites in the rat central nervous system. J. Chem. Neuroanat. **3**: 59–76.

11. BUREAU, M. H. & R. W. OLSEN. 1993. GABA$_A$ receptor subtypes: Ligand binding heterogeneity demonstrated by photoaffinity labelling and autoradiography. J. Neurochem. **61**: 1479–1491.

12. PEARCE, R. A. 1993. Physiological evidence for two distinct GABA$_A$ responses in rat hippocampus. Neuron **10**: 189–200.

13. PUIA, G., E. COSTA & S. VICINI. 1994. Functional diversity of GABA-activated Cl$^-$ currents in Purkinje versus granule neurons in rat cerebellar slices. Neuron **12**: 117–126.

14. SOLTEZ, I. & I. MODY. 1994. Patch-clamp recordings reveal powerful GABAergic inhibition in dentate hilar neurons. J. Neurosci. **14**: 2365–2376.

15. SHIMADA, S., G. CUTTING & G. R. UHL. 1992. γ-Aminobutyric acid A or C receptor? γ-Aminobutyric acid ρ1 receptor RNA induces bicuculline-, barbiturate-, and benzodiazepine-insensitive γ-aminobutyric acid responses in *Xenopus* oocytes. Mol. Pharmacol. **41**: 683–687.

16. LÜDDENS, H., D. B. PRITCHETT, M. KÖHLER, I. KILLISCH, K. KEINÄNEN, H. MONYER, R. SPRENGEL & P. H. SEEBURG. 1990. Cerebellar GABA$_A$ receptor selective for a behavioural antagonist. Nature **346**: 648–651.

17. WIELAND, H. A., H. LÜDDENS & P. H. SEEBURG. 1992. A single histidine in GABA$_A$ receptors is essential for benzodiazepine agonist binding. J. Biol. Chem. **267**: 1426–1429.

18. KORPI, E. R. & H. LÜDDENS. 1993. Regional γ-aminobutyric acid sensitivity of t-butylbicyclophosphoro[^{35}S]thionate binding depends on γ-aminobutyric acid$_A$ receptor α subunit. Mol. Pharmacol. **44**: 87–92.

19. PEREZ-VELAZQUEZ, J. L. & K. J. ANGELIDES. 1993. Assembly of GABA$_A$ receptor subunits determines sorting and localization in polarized cells. Nature **361**: 457–460.

20. BAUDE, A., J. M. SEQUIER, R. M. MCKERNAN, K. R. OLIVIER & P. SOMOGYI. 1992. Differential subcellular distribution of the α6 subunit versus the α1 and β2/β3 subunits of the GABA$_A$/benzodiazepine receptor complex in granule cells of the cerebellar cortex. Neuroscience **51**: 739–748.

21. PERSOHN, E., P. MALHERBE & J. G. RICHARDS. 1992. Comparative molecular neuroanatomy of cloned GABA$_A$ receptor subunits in the rat CNS. J. Comp. Neurol. **326**: 193–216.

22. WISDEN, W., D. J. LAURIE, H. MONYER & P. H. SEEBURG. 1992. The distribution of 13 GABA$_A$ receptor subunit mRNAs in the rat brain. I. Telencephalon, diencephalon, mesencephalon. J. Neurosci. **12**: 1040–1062.

23. LAURIE, D. J., P. H. SEEBURG & W. WISDEN. 1992. The distribution of 13 GABA$_A$ receptor

subunit mRNAs in the rat brain. II. Olfactory bulb and cerebellum. J. Neurosci. **12:** 1063–1076.

24. LAURIE, D. J., W. WISDEN & P. H. SEEBURG. 1992. The distribution of thirteen GABA$_A$ receptor subunit mRNAs in the rat brain. III. Embryonic and postnatal development. J. Neurosci. **12:** 4151–4172.

25. ZIMPRICH, F., J. ZEZULA, W. SIEGHART & H. LASSMANN. 1991. Immunohistochemical localization of the α1, α2 and α3 subunit of the GABA$_A$ receptor in the rat brain. Neurosci. Lett. **127:** 125–128.

26. BENKE, D., S. MERTENS, A. TRZECIAK, D. GILLESSEN & H. MÖHLER. 1991. Identification and immunohistochemical mapping of GABA$_A$ receptor subtypes containing the δ-subunit. FEBS Lett. **283:** 145–149.

27. TURNER, J. D., G. BODEWITZ, C. L. THOMPSON & F. A. STEPHENSON. 1993. Immunohistochemical mapping of gamma-aminobutyric acid type-A receptor alpha subunits in rat central nervous system. *In* Anxiolytic β-Carbolines: From Molecular Biology to the Clinic. Psychopharmacology series II. D. N. Stephens, Ed.: 29–49. Springer Verlag. Berlin.

28. MCKERNAN, R. M., K. QUIRK, R. PRINCE, P. A. COX, N. P. GILLARD, C. I. RAGAN & P. WHITING. 1991. GABA$_A$ receptor subtypes immunopurified from rat brain with α-subunit-specific antibodies have unique pharmacological properties. Neuron **7:** 667–676.

29. BENKE, D., S. MERTENS, A. TRZECIAK, D. GILLESSEN & H. MÖHLER. 1991. GABA$_A$ receptors display association of γ2 subunit with α1- and β2/β3 subunits. J. Biol. Chem. **266:** 4478–4483.

30. MERTENS, S., D. BENKE & H. MÖHLER. 1993. GABA$_A$ receptor populations with novel subunit combinations and drug binding profiles identified in brain by α5- and δ-subunit specific immunopurification. J. Biol. Chem. **268:** 5965–5973.

31. LAMBOLEZ, B., E. AUDINAT, P. BOCHET, F. CREPEL & J. ROSSIER. 1992. AMPA receptor subunits expressed by single Purkinje cells. Neuron **9:** 247–258.

32. BUHL, E. H., K. HALASY & P. SOMOGYI. 1994. Diverse sources of hippocampal unitary inhibitory postsynaptic potentials and the number of synaptic release sites. Nature **368:** 823–828.

33. KORPI, E. R., C. KLEINGOOR, H. KETTENMANN & P. H. SEEBURG. 1993. Benzodiazepine-induced motor impairment linked to point mutation in cerebellar GABA$_A$ receptor. Nature **361:** 356–359.

34. NAKATSU, Y., R. F. TYNDALE, T. M. DELOREY, D. DURHAM-PIERRE, J. M. GARDNER, H. J. MCDANEL, Q. NGUYEN, J. WAGSTAFF, M. LALANDE, J. M. SIKELA, R. W. OLSEN, A. J. TOBIN & M. H. BRILLIANT. 1993. A cluster of three GABA$_A$ receptor subunit genes is deleted in a neurological mutant of the mouse ρ locus. Nature **364:** 448–450.

35. CULIAT, C. T., L. J. STUBBS, C. S. MONTGOMERY, L. B. RUSSELL & E. M. RINCHIK. 1994. Phenotypic consequences of deletion of the γ3, α5 or β3 subunit of the type A γ-aminobutyric acid receptor in mice. Proc. Natl. Acad. Sci. USA **91:** 2815–2818.

36. WHITING, P., R. M. MCKERNAN & L. L. IVERSEN. 1990. Another mechanism for creating diversity in γ-aminobutyrate type A receptors: RNA splicing directs expression of two forms of γ2 subunit, one of which contains a protein kinase C phosphorylation site. Proc. Natl. Acad. Sci. USA **87:** 9966–9970.

37. PRITCHETT, D. B., H. SONTHEIMER, B. D. SHIVERS, S. YMER, H. KETTENMANN, P. R. SCHOFIELD & P. H. SEEBURG. 1989. Importance of a novel GABA$_A$ receptor subunit for benzodiazepine pharmacology. Nature **338:** 582–585.

38. PRITCHETT, D. B. & P. H. SEEBURG. 1990. γ-Aminobutyric acid$_A$ receptor α5 subunit creates novel type II benzodiazepine receptor pharmacology. J. Neurochem. **54:** 1802–1804.

39. HADINGHAM, K. L., P. B. WINGROVE, K. A. WAFFORD, C. BAIN, J. A. KEMP, K. J. PALMER, A. W. WILSON, A. S. WILCOX, J. M. SIKELA, C. I. RAGAN & P. J. WHITING. 1993. Role of the β-subunit in determining the pharmacology of human γ-aminobutyric acid type A receptors. Mol. Pharmacol. **44:** 1211–1218.

40. WAFFORD, K. A., C. J. BAIN, K. QUIRK, R. M. MCKERNAN, P. B. WINGROVE, P. J. WHITING & J. A. KEMP. 1994. A novel allosteric modulatory site on the GABA$_A$ receptor β subunit. Neuron **12:** 775–782.

41. WAFFORD, K. A., C. J. BAIN, P. J. WHITING & J. A. KEMP. 1993. Functional comparison of the role of γ subunits in recombinant human γ-aminobutyric acid$_A$/benzodiazepine receptors. Mol. Pharmacol. **44:** 437–442.

42. HERB, A., W. WISDEN, H. LÜDDENS, G. PUIA, S. VICINI & P. H. SEEBURG. 1992. The third γ subunit of the γ-aminobutyric acid type A receptor family. Proc. Natl. Acad. Sci. USA **89:** 1433–1437.

43. SHIVERS, B. D., I. KILLISCH, R. SPRENGEL, H. SONTHEIMER, M. KÖHLER, P. R. SCHOFIELD & P. H. SEEBURG. 1989. Two novel GABA$_A$ receptor subunits exist in distinct neuronal subpopulations. Neuron **3:** 327–337.

44. FARRANT, M. & S. CULL-CANDY. 1993. GABA receptors, granule cells and genes. Nature **361:** 302–303.

45. MUGNAINI, E. & A. FLORIS. 1994. The unipolar brush cell: A neglected neuron of the mammalian cerebellar cortex. J. Comp. Neurol. **339:** 174–180.

46. VERDOORN, T. A. 1994. Formation of heteromeric γ-aminobutyric acid type A receptors containing two different α subunits. Mol. Pharmacol. **45:** 475–480.

47. BRISTOW, D. R. & I. L. MARTIN. 1988. Light microscopic autoradiographic localization in rat brain of the binding sites for the GABA$_A$ receptor antagonist [^{3}H]SR 95531: Comparison with [^{3}H]GABA$_A$ distribution. Eur. J. Pharmacol. **148:** 283–288.

48. SOMOGYI, P., H. TAKAGI, J. G. RICHARDS & H. MÖHLER. 1989. Subcellular localization of benzodiazepine/GABA$_A$ receptors in the cerebellum of rat, cat and monkey using monoclonal antibodies. J. Neurosci. **9:** 2197–2209.

49. WISDEN, W., L. A. MCNAUGHTON, M. G. DARLISON, S. P. HUNT & E. A. BARNARD. 1989. Differential distribution of GABA$_A$ receptor mRNAs in bovine cerebellum— Localization of α2 mRNA in Bergmann glia layer. Neurosci. Lett. **106:** 7–12.

50. EDGAR, P. P. & R. D. SCHWARTZ. 1990. Localization and characterization of ^{35}S-t-butylbicylophosphorothionate binding in rat brain: An autoradiographic study. J. Neurosci. **10:** 603–612.

51. KHAN, Z. U., L. P. FERNANDO, P. ESCRIBA, X. BUSQUETS, J. MALLET, C. P. MIRALLES, M. FILLA & A. L. DE-BLAS. 1993. Antibodies to the human gamma 2 subunit of the gamma-aminobutyric acid A/benzodiazepine receptor. J. Neurochem. **60:** 961–971.

52. POLLARD, S., M. J. DUGGAN & F. A. STEPHENSON. 1993. Further evidence of the existence of α subunit heterogeneity within discrete γ-aminobutyric acid$_A$ receptor subpopulations. J. Biol. Chem. **268:** 3753–3757.

53. MACONOCHIE, D. J., J. M. ZEMPEL & J. H. STEINBACH. 1994. How quickly can GABA$_A$ receptors open? Neuron **12:** 61–71.

54. GLENCORSE, T. A., M. G. DARLISON, E. A. BARNARD & A. N. BATESON. 1993. Sequence and novel distribution of the chicken homologue of the mammalian γ-aminobutyric acid$_A$ receptor γ1 subunit. J. Neurochem. **61:** 2294–2302.

55. QUIRK, K., N. P. GILLARD, I. RAGAN, P. J. WHITING & R. M. MCKERNAN. 1994. Model of subunit composition of γ-aminobutyric acid A receptor subtypes expressed in rat cerebellum with respect to THGIR α (alpha) and γ/δ (gamma/delta) subunits. J. Biol. Chem. **269:** 16020–16028.

Cerebral GABA$_A$ and GABA$_B$ Receptors

Structure and Function

HIROSHI NAKAYASU,[a] HIROSHI KIMURA,[b]
AND KINYA KURIYAMA[a,c]

[a]*Department of Pharmacology*
Kyoto Prefectural University of Medicine
Kawaramachi-Hirokoji
Kamikyo-ku
Kyoto 602, Japan

[b]*Institute of Molecular Neurobiology*
Shiga University of Medical Science
Seta Otsu 52021, Japan

γ-Aminobutyric acid (GABA) is widely known as one of the major inhibitory neurotransmitters in the mammalian central nervous system (CNS).[1] In general, it has been considered that neuronal activity in the brain is inhibited by the activation of GABAergic neurons. This inhibitory action of GABA is mediated by GABA receptors, which are divided into two types—GABA$_A$ and GABA$_B$.[2] It was initially thought that GABA$_A$ receptor, a bicuculline-sensitive type, forms the GABA-gated Cl$^-$ channel, and that the activation of GABA$_A$ receptor induces the fast inhibitory postsynaptic potential (IPSP). Subsequently, the presence of GABA$_B$ receptor, which is insensitive to bicuculline and sensitive to baclofen, was revealed, and it was found that this receptor is one of the metabotropic types that couple with GTP-binding protein. The activation of the GABA$_B$ receptor is known to induce slow IPSP. Therefore, it is suggested that both GABA$_A$ and GABA$_B$ receptors have inhibitory roles in the CNS, although these actions are mediated by different molecular mechanisms.

In this paper, current concepts on GABA$_A$ and GABA$_B$ receptors in the mammalian CNS are described, with special reference to their pharmacological, neurochemical, and molecular biological characteristics.

GABA$_A$ RECEPTOR

Pharmacology of GABA$_A$ Receptor

The application of GABA induces the increase of Cl$^-$ conductance at GABAergic synapses. This increase is antagonized by bicuculline, a GABA$_A$ receptor antagonist. Picrotoxin, which induces tonic-clonic convulsions, and convulsant compounds such as t-butylbicyclophosphorothionate (TBPS) also bind to the Cl$^-$ channel in the GABA$_A$ receptor complex and depress Cl$^-$ conductance, although these compounds have no action on the GABA binding (recognition) site. Barbiturate derivatives, which are classified as one of the sedative hypnotic drugs, are also known to enhance the action of GABA by affecting the Cl$^-$ channel site of the GABA$_A$

---

[c]Corresponding author.

receptor complex. On the other hand, anxiolytic and/or hypnotic drugs such as benzodiazepine derivatives are known to activate the GABA binding site in the GABA$_A$ receptor complex. It is noteworthy that the binding site for benzodiazepines, which is coupled with the GABA$_A$ receptor and the Cl$^-$ channel, has been termed benzodiazepine receptor. Benzodiazepine receptor agonists such as diazepam and flunitrazepam, as well as benzodiazepine antagonists such as flumazenil, bind to the benzodiazepine receptor. In contrast, β-carboline derivatives are known to induce anxiety by an inhibitory modulation of GABA$_A$ receptor function through the benzodiazepine receptor. Therefore, these compounds are often termed as inverse agonist of benzodiazepine receptor. It has been also noted that various drugs and endogenous products in the brain affect the function of GABA$_A$ receptor/ benzodiazepine receptor/Cl$^-$ channel complex.[3] For example, it has been reported that neurosteroids such as 3α-hydroxy-5α-dihydroprogesterone also bind to the Cl$^-$ channel site in the GABA$_A$ receptor complex and act as putative endogenous modulators for stress and anxiety.[4,5] These characteristics of the GABA$_A$ receptor have been recognized using various pharmacological and electrophysiological approaches. Furthermore, purification and reconstitution studies on the GABA$_A$ receptor complex have provided direct evidence for these pharmacological characteristics.

Purification and Characterization of GABA$_A$ Receptor Complex

Functional as well as structural coupling of the GABA$_A$ receptor with the benzodiazepine receptor was found to be useful for the purification of the GABA$_A$ receptor complex in the CNS. We previously purified the GABA$_A$ receptor complex by means of affinity column chromatography using a benzodiazepine, 1012-S, as an immobilized ligand.[6] The purified protein of the GABA$_A$ receptor complex from rat brain had the molecular weight of approximately 300,000 and consisted of α (48,500) and β (54,500) subunits. Similarly, GABA$_A$ receptor from the bovine cerebral cortex was found to consist of α (53,000) and β (57,000) subunits.[7] These purified receptor preparations were shown to retain functional coupling between GABA$_A$ and benzodiazepine receptors even after the purification. Namely, [^{3}H]muscimol binding to the GABA$_A$ receptor was enhanced by benzodiazepine receptor agonists, whereas the enhancement was inhibited by flumazenil, an antagonist of the benzodiazepine receptor. Similarly, GABA$_A$ receptor agonists enhanced [^{3}H]flunitrazepam binding to the benzodiazepine receptor, whereas the enhancement was inhibited by bicuculline, a GABA$_A$ receptor antagonist. It was also shown that when the purified GABA$_A$ receptor complex is reconstituted into the phospholipid vesicles, the application of GABA to the reconstituted vesicles induces an increase in the ^{36}Cl$^-$ influx, whereas flunitrazepam, a benzodiazepine receptor agonist, potentiates the GABA-stimulated ^{36}Cl$^-$ influx.[8] These experimental results have provided evidence that the GABA$_A$ receptor couples with the benzodiazepine receptor and forms the GABA-gated Cl$^-$ channel.

Distribution of GABA$_A$ Receptor Complex in the Central Nervous System

The specific antibody against purified GABA$_A$ receptor complex was prepared by the immunization of albino rabbits with the purified protein.[9] The distribution of the GABA$_A$ receptor complex in rat brain was examined immunohistochemically using this specific antibody. Immunoreaction in rat brain slices was recognized in various regions of the CNS such as the ventromedial nucleus of the hypothalamus, the red nucleus, the globus pallidus, the zona compacta and zona reticulata of the substantia nigra, the layers of Purkinje cells and granular cells in the cerebellum, layers III–V of cerebral cortex, and the stratum radiatum of the hippocampus. These immunoreactive regions coincided well with the results of immunohistochemical studies using antibody against L-glutamic acid decarboxylase, the enzyme for the synthesis of GABA.

Structure of GABA_A Receptor Complex

The structural analysis of the $GABA_A$ receptor complex was advanced rapidly by the application of cDNA cloning. The cDNAs of $GABA_A$ receptor α and β subunits from bovine brain were cloned.[10] Subsequently, cDNA cloning of many subunits of the $GABA_A$ receptor complex was achieved by the application of homologous screening. The presence of α, β, γ, δ, and ρ subunits in the brain was reported, respectively.[10–15] Moreover, the heterogeneity of $GABA_A$ receptor subunits was found as shown in TABLE 1. In brief, the $GABA_A$ receptor complex consists of a combination of four or five subunits, where various combinations of these subunits result in the heterogeneity of $GABA_A$ receptor molecules.[16,17] In fact, the pharmacological properties of recombinant $GABA_A$/benzodiazepine receptor complex were

TABLE 1. Multiplicity of $GABA_A$ Receptor Subunits

α_1, α_2, α_3, α_4, α_5, α_6
β_1, β_2, β_3, β_4, $\beta_{4'}$
γ_1, γ_{2S}, γ_{2L}, γ_3, γ_4
δ
ρ_1, ρ_2

NOTE: β_4 and $\beta_{4'}$, γ_{2S} and γ_{2L} are derived from the alternative splicing, respectively.

found to be different by the use of different combinations of various subunits.[12,13,18,19] The results from cDNA cloning of the $GABA_A$ receptor complex indicated the existence of a high homology among the various subunits examined. Notably, four regions of high hydrophobicity in the structure were conserved and suggested the presence of four transmembrane (TM) domains. The structure of $GABA_A$ receptor subunits was also found to be homologous among different species. These characteristics were similar to those in other ionotropic receptors. Therefore, it was assumed that the ligand-gated ion channels constituting the $GABA_A$ receptor complex are members of a superfamily.[20] Furthermore, it was suggested that the TM II domain has an important role in the formation of the ion channel pore. It has been found that in anion channels, the common sequence, TTVLTMTTXS, in the TM II domain of both $GABA_A$ and glycine receptor subunits is hydroxy-rich, and that this sequence is not conserved as in the case of cation channels such as the nicotinic acetylcholine receptor. Consequently, it has been presumed that this region is important for the recognition of the Cl^- ion. Furthermore, it has been reported that the mutations in the channel domain of a neuronal nicotinic receptor convert ion selectively from cationic to anionic.[21] These results indicate that the structure of membrane domains is important for maintaining the selectivity of ions.

In the intracellular loop of TM III–IV in each subunit, the presence of putative sites of phosphorylation by cAMP-dependent protein kinase and/or protein kinase C was noted. In fact, it was reported that these protein kinases phosphorylate the purified $GABA_A$ receptor protein.[22,23] Furthermore, it was demonstrated that the phosphorylation of the $GABA_A$ receptor modulates the function of the $GABA_A$ receptor complex.[24]

Expression of mRNA for GABA_A Receptor Complex

It is well known that the function of the $GABA_A$ receptor is modulated by treatment with various drugs. A good example for this is down-regulation of the receptor following continuous agonist stimulation. In fact, treatment of primary cultured cerebral cortical neurons with muscimol, a selective $GABA_A$ receptor agonist, induced the reduction of expression of $GABA_A$ receptor α_1-subunit mRNA.[25] This reduction was counteracted by simultaneous treatment with bicuculline, a $GABA_A$ receptor antagonist. Moreover, the treatment with flunitrazepam, an agonist of the benzodiazepine receptor, which coupled with the $GABA_A$ receptor, also caused a decline of the expression of mRNA for $GABA_A$ receptor $\alpha 1$-subunit, and this change was also counteracted by simultaneous treatment with flumazenil, an antagonist of the central type of benzodiazepine receptor. These results suggest that continuous stimulation of the $GABA_A$ receptor complex by its agonists may induce the reduced expression of mRNA for the $GABA_A$ receptor subunits. By contrast, the treatment with inverse agonist for the benzodiazepine receptor such as β-carboline derivatives induced an increase of $GABA_A$ receptor subunit mRNAs.[26] Therefore, it is suggested that the expression of mRNA for $GABA_A$ receptor subunits may be regulated by positive and/or negative signal transductions through the $GABA_A$ receptor complex.

Recently, it has also been reported that the expression of mRNA for $GABA_A$ receptor subunits is modulated by activation of the *N*-methyl-D-aspartate-sensitive glutamate receptor.[27] Therefore, it is likely that the transcriptional mechanism for $GABA_A$ receptor subunit genes may be regulated by not only the direct stimulation or inhibition of the $GABA_A$ receptor complex, but also the functional state of neuronal networks connected to GABAergic neurons.

GABA_B RECEPTOR

Pharmacology of GABA_B Receptor

It has been reported that ($-$)baclofen decreases neurotransmitter releases in the CNS.[28] In 1981, it was established that this effect of ($-$)baclofen is exerted by the activation of a bicuculline-insensitive and non-$GABA_A$ receptor,[29] and this receptor is termed the $GABA_B$ receptor. Although GABA and ($-$)baclofen have been used as $GABA_B$ receptor agonists, studies on the physiological and pharmacological roles of the $GABA_B$ receptor in the CNS have been delayed, mainly due to lack of selective and potent antagonists for this receptor. Recently, however, selective and potent agonists and antagonists for the $GABA_B$ receptor have been introduced. For example, 3-aminopropylphosphonous acid (APPA) has been developed as a potent agonist at peripheral and central presynaptic $GABA_B$ receptors.[30] On the other hand, phaclofen and 2-hydroxy saclofen have been introduced as selective antagonists for the $GABA_B$ receptor.[31,32] More recently, it has been reported that CGP35348, having high affinity and high penetrability to the brain, is a potent blocker of the

GABA$_B$ receptor.[33] Inasmuch as the development of various GABA$_B$ receptor antagonists is under way with high hopes of their clinical applications,[34] it may be possible to obtain more potent and selective GABA$_B$ receptor antagonists in the near future.

The binding of GABA$_B$ receptor was found to be inhibited by the presence of GTP analogues, which induced a decrease in the affinity of the GABA$_B$ receptor.[35,36] In addition, it was reported that stimulation of the GABA$_B$ receptor accentuated the GTPase activity in the crude synaptic membrane.[37] This increase of GTPase activity was hindered by the pretreatment with the islet-activating protein (IAP). These results suggest that GABA$_B$ receptor may be coupled with a GTP-binding protein, such as G$_i$ or G$_o$, which is sensitive to the IAP treatment.

Activation of the cerebral GABA$_B$ receptor induced the inhibition of cAMP formation.[38,39] The increased formation of cAMP induced by the stimulation of other neurotransmitter receptors and by forskolin was also inhibited by activation of the GABA$_B$ receptor, whereas the pretreatment with IAP abolished the inhibitory action of GABA$_B$ receptor agonists.[40] Therefore, it is thought that stimulation of the GABA$_B$ receptor inhibits the adenylyl cyclase activity via IAP-sensitive GTP-binding protein.

On the other hand, it was reported that activation of the GABA$_B$ receptor resulted in the potentiation of β-adrenergic receptor-mediated cAMP formation.[41] This potentiation was also inhibited by the treatment with IAP.[42] Because the stimulation of the GABA$_B$ receptor induced an increase of binding affinity at the β-adrenergic receptor,[43] a cross-regulation of intracellular signal transduction systems, which were coupled with the β-adrenergic receptor and the GABA$_B$ receptor, was suggested. Detailed mechanisms underlying such cross talk between these two receptors, however, are presently unknown.

It is well known that several neurotransmitter receptors are positively coupled with phosphatidylinositol (PI) turnover, which generates both 1,2-diacylglycerol and inositol-1,4,5-triphosphate. Inositol phosphates formation was also inhibited by the activation of the GABA$_B$ receptor.[44,45] This GABA$_B$ receptor-mediated inhibition of PI turnover was eliminated again by the treatment with IAP.[46] Therefore, it has been suggested that the GABA$_B$ receptor is also negatively coupled with PI turnover, mediated by an IAP-sensitive GTP-binding protein. It is unclear, however, whether the same GABA$_B$ receptor inhibits both adenylyl cyclase and PI turnover systems or different types of the GABA$_B$ receptor (subclasses) are responsible for the inhibition of these intracellular signal transduction systems.

In electrophysiological studies, it was also demonstrated that the activation of the GABA$_B$ receptor resulted in the modulation of ion channels. Namely, a voltage-dependent Ca^{2+} channel was inhibited by the activation of the GABA$_B$ receptor.[47] In addition, the activation of the GABA$_B$ receptor induced an increase in K$^+$ conductance, which resulted in the occurrence of slow IPSP.[48,49] Modulation of these ion channels by the GABA$_B$ receptor was also known to be mediated by the GTP-binding protein, which is sensitive to IAP treatment. Therefore, it seems likely that the suppression of transmitter release by the activation of the presynaptic GABA$_B$ receptor is caused by a decrease in Ca^{2+} conductance and/or an increase in K$^+$ conductance.

Neurochemical Aspects of GABA$_B$ Receptor

In neurochemical approaches, the isolation of the GABA$_B$ receptor protein is important to clarify its biochemical properties. For these purposes, solubilization of the $GABA_B$ receptor from crude synaptic membranes was attempted, at first using various detergents such as CHAPS.[50] Ligand-affinity column chromatography for the purification of the $GABA_B$ receptor was then attempted using baclofen, a selective agonist for $GABA_B$ receptor, as the immobilized ligand. We developed the baclofen-epoxy-activated Sepharose 6B for this purpose.[50] Using this affinity column, we achieved approximately 11,000-fold purification of the $GABA_B$ receptor. Due to a low affinity of this baclofen-affinity gel, we faced technical difficulties for further

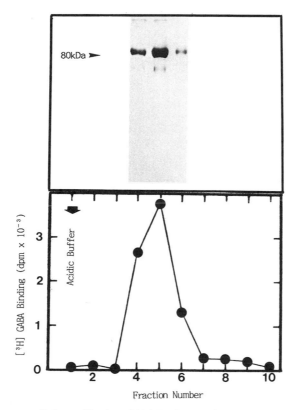

FIGURE 1. Immunoaffinity purification of 80-kDa GABA-binding protein. Solubilized synaptic membrane was incubated with monoclonal antibody-conjugated immunoaffinity beads. Elution was performed by the addition of the acidic buffer (50 mM citrate, pH 2.5) containing 5 mM CHAPS and the protease inhibitors. Eluted fractions were analyzed by SDS-PAGE (*upper panel*) and by [³H]GABA-binding assay (*lower panel*). The SDS-PAGE analysis shows that the major protein band is 80 kDa. Occasionally, one or two faintly stained bands are visible having 65 or 61 kDa (an example of 65 kDa is seen in fraction 5). For the [³H]GABA-binding assay, [³H]GABA (12.5 nM) and purified 80-kDa GABA-binding protein were incubated for 1 h in ice-cold 50 mM Tris-HCl buffer (pH 7.4) containing 2 mM $MgCl_2$, 2 mM $CaCl_2$ in the presence of a soybean phospholipid (final 4 mM). The soybean phospholipid was added to minimize an inhibitory effect of CHAPS (which was contained in the immunoaffinity elute, final 0.5 mM) on the [³H]GABA binding assay. A clear mixture (0.35 mL) became turbid with the addition of the phospholipid. Nonspecific immunoglobulin (2 mg/mL) was also added for complete sedimentation of GABA-binding proteins. Radioactivity bound to these proteins was precipitated by the addition of 0.15 mL of 50% polyethylene glycol and separated from the unbound radioactivity by filtration.

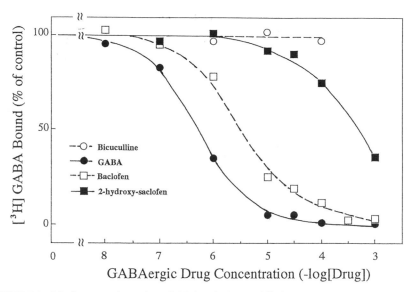

FIGURE 2. Displacement by various GABAergic drugs of [³H]GABA binding to the purified 80-kDa GABA-binding protein. GABAergic drugs at various concentrations were added to the mixture of [³H]GABA-binding assay. Both GABA_B agonist and antagonist apparently bind to GABA-binding protein, whereas GABA_A antagonist bicuculline does not.

purification using this affinity column chromatographic procedure.[51] Therefore, we decided to develop immunoaffinity chromatographic procedures using monoclonal antibody specific to the receptor.

For the production of a monoclonal antibody specific to the GABA_B receptor, the receptor protein from the bovine brain was partially purified using the above-mentioned baclofen-affinity column, and this preparation was injected into mice as an antigen. Spleen cells from mice having high titers of antisera were used for cell fusion. Positive clones were selected by Western blot analysis using the partially purified GABA_B receptor preparation, and a monoclonal antibody (GB-1) for the GABA_B receptor was obtained.[52] In Western blot analysis using the crude synaptic membrane preparation from the bovine brain, the antibody recognized only one protein band among over one hundred protein moieties. The molecular weight of the immunopositive protein was about 80 kDa. It was found that the baclofen-sensitive GABA binding activity to crude synaptic membrane was significantly reduced by the addition of this monoclonal antibody.[52] Furthermore, GABA_B receptor binding activity in the solubilized fraction obtained from bovine synaptic membrane was completely eliminated by immunoabsorbent agarose beads that were conjugated with the antibody. It was also noted that the 80 k-Da protein was immunoprecipitated without visible changes in electrophoresed profiles of total proteins on Western blot analysis.

In order to determine whether or not this GABA-binding protein is truly the GABA_B receptor, the purification of the antigen by an immunoaffinity method was attempted.[53] The 80-kDa GABA binding protein in the solubilized fraction was adsorbed on the immunobeads and eluted with an acidic buffer (pH 2.5). The eluted fraction having the GABA binding activity showed only one protein band (80 kDa) in

SDS-PAGE (FIG. 1). This binding activity to the purified protein was selectively displaced by GABA, baclofen, and 2-hydroxy saclofen, but not by bicuculline (FIG. 2).

Furthermore, an attempt was made to examine whether or not this 80-kDa GABA binding protein could function as $GABA_B$ receptor using a reconstituted system. Namely, the 80-kDa protein was reconstituted into phospholipid vesicles with partially purified GTP-binding protein (which had G_i and/or G_o, but not G_s) and adenylyl cyclase. The forskolin-stimulated adenylyl cyclase activity was inhibited by the addition of GABA or baclofen to the complete system, whereas such an effect of the $GABA_B$ receptor agonist was not detected in the system that had no 80-kDa protein, adenylyl cyclase or GTP-binding protein. This inhibition by GABA or baclofen was abolished by the simultaneous addition of 2-hydroxy saclofen (FIG. 3). Therefore, it was suggested that the stimulation of 80-kDa GABA binding protein is transduced to adenylyl cyclase via inhibitory GTP-binding protein such as G_i or G_o. These results are also strong evidence supporting the notion that the immunoaffinity-purified 80-kDa protein is indeed the $GABA_B$ receptor.

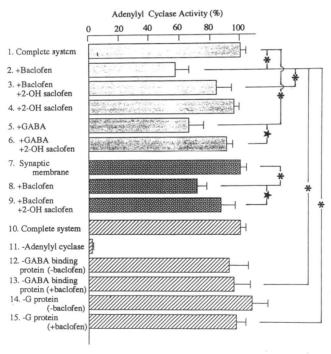

FIGURE 3. Adenylyl cyclase activity on reconstituted membrane (lanes 1–6), synaptic membrane (lanes 7–9), and incomplete reconstituted membrane (lanes 10–15). The incomplete reconstituted membrane preparations are devoid of both adenylyl cyclase (lane 11) and 80-kDa GABA-binding protein (lanes 14 and 15). Forskolin (10^{-5} M) stimulated adenylyl cyclase activity on these membranes was assayed after preincubation of the membrane with or without $GABA_B$ receptor agonist (2×10^{-5} M GABA or 2×10^{-4} M baclofen) or $GABA_B$ receptor antagonist (10^{-3} M 2-hydroxy saclofen).

CONCLUSION AND FUTURE PROSPECTS

There are two different subtypes of GABA receptors: ionotropic $GABA_A$ and metabotropic $GABA_B$. The $GABA_A$ receptor consists of a receptor complex with the Cl^- channel and the benzodiazepine receptor, whereas the $GABA_B$ receptor negatively couples with adenylyl cyclase and inositol phosphate generating system and also several calcium channels via IAP-sensitive G_i- or G_o-type GTP binding protein.[54-58] The $GABA_A$ receptor complex consists of various subunits and each subunit possesses four transmembrane domains which possibly constitute Cl^- channel. In the case of the $GABA_B$ receptor, it is highly likely that the 80-kDa protein, which we purified using an immunoaffinity column chromatographic procedure, is indeed the $GABA_B$ receptor protein. The primary structure of the $GABA_B$ receptor, however, has not been determined in detail. Molecular cloning of the $GABA_B$ receptor is under way in our laboratory. The locations of $GABA_A$ and $GABA_B$ receptors at synapses also remain to be clarified, as well as the nature of the subtype acting as presynaptic autoreceptor or heteroreceptor in the neuronal network. For these purposes, electron microscopic immunocytochemistry using monoclonal antibodies for $GABA_A$ and $GABA_B$ receptors, mentioned in this study, may be useful in future studies.

REFERENCES

1. ROBERTS, E. 1986. GABA: The road to neurotransmitter status. *In* Benzodiazepine/ GABA Receptors and Chloride: Structural and Functional Properties. R. W. Olson & J. C. Venter, Eds.: 1–39. Alan R. Liss. New York.
2. BORMANN, J. 1988. Electrophysiology of $GABA_A$ and $GABA_B$ receptor subtypes. Trends Neurosci. **11:** 112–116.
3. SIEGHART, W. 1992. $GABA_A$ receptors: Ligand-gated Cl^- channels modulated by multiple drug-binding sites. Trends Pharmacol. Sci. **13:** 446–450.
4. PUIA, G., M. SANTI, S. VICINI, D. B. PRITCHETT, R. H. PURDY, S. M. PAUL, P. H. SEEBURG & E. COSTA. 1990. Neurosteroids act on recombinant human $GABA_A$ receptors. Neuron **4:** 759–765.
5. MAJEWSKA, M. D., S. DEMIRGOREN, C. E. SPIVAK & E. D. LONDON. 1990. The neurosteroid dehydroepiandrosterone sulfate is an allosteric antagonist of the $GABA_A$ receptor. Brain Res. **526:** 143–146.
6. TAGUCHI, J. & K. KURIYAMA. 1984. Purification of γ-aminobutyric acid (GABA) receptor from rat brain by affinity column chromatography using a new benzodiazepine, 1012-S, as an immobilized ligand. Brain Res. **323:** 219–226.
7. KURIYAMA, K. & J. TAGUCHI. 1987. Purification of γ-aminobutyric acid receptor, benzodiazepine receptor and Cl channel from bovine cerebral cortex by benzodiazepine affinity gel column chromatography. Neurochem. Int. **10:** 253–263.
8. HIROUCHI, M., J. TAGUCHI, T. UEHA & K. KURIYAMA. 1987. GABA-stimulated $^{36}Cl^-$ influx into reconstituted vesicles with purified $GABA_A$/benzodiazepine receptor complex. Biochem. Biophys. Res. Commun. **146:** 1471–1477.
9. TAGUCHI, J., T. KURIYAMA, Y. OHMORI & K. KURIYAMA. 1989. Immunohistochemical studies on distribution of $GABA_A$ receptor complex in the rat brain using antibody against purified $GABA_A$ receptor complex. Brain Res. **483:** 395–401.
10. SCHOFIELD, P. R., M. G. DARLISON, N. FUJITA, D. R. BURT, F. A. STEPHENSON, H. RODRIGUEZ, L. M. RHEE, J. RAMACHANDRAN, V. REALE, T. A. GLENCORSE, P. H. SEEBURG & E. A. BARNARD. 1987. Sequence and functional expression of the $GABA_A$ receptor shows a ligand-gated receptor super-family. Nature **328:** 221–227.
11. HIROUCHI, M., R. KUWANO, T. KATAGIRI, Y. TAKAHASHI & K. KURIYAMA. 1989. Nucleotide and deduced amino acid sequences of the $GABA_A$ receptor α-subunit from human brain. Neurochem. Int. **15:** 33–38.

12. PRITCHETT, D. B., H. SONTHEIMER, B. D. SHIVERS, S. YMER, H. KETTENMANN, P. R. SCHOFIELD & P. H. SEEBURG. 1989. Importance of a novel GABA$_A$ receptor subunit for benzodiazepine pharmacology. Nature **338:** 582–585.
13. PRITCHETT, D. B., H. LUDDENS & P. H. SEEBURG. 1989. Type I and type II GABA$_A$-benzodiazepine receptors produced in transfected cells. Science **245:** 1389–1392.
14. SHIVERS, B. D., I. KILLISCH, R. SPRENGEL, H. SONTHEIMER, M. KOHLER, P. R. SCHOFIELD & P. H. SEEBURG. 1989. Two novel GABA$_A$ receptor subunits exist in distinct neuronal subpopulations. Neuron **3:** 327–337.
15. CUTTING, G. R., L. LU, B. F. O'HARA, L. M. KASCH, C. MONTROSE-RAFIZADER, D. M. DONOVAN, S. SHIMADA, S. E. ANTONARAKIS, W. B. GUGGINO, G. R. UHL & H. H. KAZAZIAN. 1991. Cloning of the γ-aminobutyric acid (GABA) ρ1 cDNA: A GABA receptor subunit highly expressed in the retina. Proc. Natl. Acad. Sci. USA **88:** 2673–2677.
16. OLSEN, R. W., M. H. BEREAU, S. ENDO & G. SMITH. 1991. The GABA$_A$ receptor family in the mammalian brain. Neurochem. Res. **16:** 317–325.
17. BURT, D. R. & G. L. KAMATCHI. 1991. GABA$_A$ receptor subtypes: From pharmacology to molecular biology. FASEB J. **5:** 2916–2923.
18. LEVITAN, E. S., P. R. SCHOFIELD, D. R. BURT, L. M. RHEE, W. WISDEN, M. KOHLER, N. FUJITA, H. F. RODRIGUEZ, A. STEPHENSON, M. G. DARLISON, E. A. BARNARD & P. H. SEEBURG. 1988. Structural and functional basis for GABA$_A$ receptor heterogeneity. Nature **335:** 76–79.
19. YMER, S., P. R. SCHOFIELD, A. DRAGUHN, P. WERNER, M. KOHLER & P. H. SEEBURG. 1989. GABA$_A$ receptor β subunit heterogeneity: Functional expression of cloned cDNAs. EMBO J. **8:** 1665–1670.
20. BARNARD, E. A., M. G. DARLISON & P. SEEBURG. 1987. Molecular biology of the GABA$_A$ receptor: The receptor/channel superfamily. Trends Neurosci. **10:** 502–509.
21. GALZI, J. L., A. DEVILLERSTHIERY, N. HUSSY, S. BERTRAND, J. P. CHANGEUX & D. BERTRAND. 1992. Mutations in the channel domain of a neuronal nicotinic receptor convert ion selectivity from cationic to anionic. Nature **359:** 500–505.
22. KIRKNESS, E. F., C. F. BOVENKERK, T. UEDA & A. J. TURNER. 1989. Phosphorylation of γ-aminobutyrate (GABA)/benzodiazepine receptors by cyclic AMP-dependent protein kinase. Biochem. J. **259:** 613–616.
23. BROWING, M. D., M. BEREAU, E. M. DUDEK & R. W. OLSEN. 1990. Protein kinase C and cAMP-dependent protein kinase phosphorylate the β-subunit of the purified γ-aminobutyric acid A receptor. Proc. Natl. Acad. Sci. USA **87:** 1315–1318.
24. HEUSCHNEIDER, G. & R. D. SCHWARTZ. 1989. cAMP and forskolin decrease γ-aminobutyric acid-gated chloride flux in rat brain synaptoneurosomes. Proc. Natl. Acad. Sci. USA **86:** 2938–2942.
25. HIROUCHI, M., S. OHKUMA & K. KURIYAMA. 1992. Muscimol-induced reduction of GABA$_A$ receptor α1-subunit mRNA in primary cultured cerebral cortical neurons. Mol. Brain Res. **15:** 327–331.
26. PRIMUS, R. J. & D. W. GALLAGER. 1992. GABA$_A$ receptor subunit mRNA levels are differentially influenced by chronic FG 7142 and diazepam exposure. Eur. J. Pharmacol. **226:** 21–28.
27. MEMO, M., P. BOVOLIN, E. COSTA & D. R. GRAYSON. 1991. Regulation of γ-aminobutyric acid$_A$ receptor subunit expression by activation of N-methyl-D-aspartate-sensitive glutamate receptors. Mol. Pharmacol. **39:** 599–603.
28. BOWERY, N. G., D. R. HILL, A. L. HUDSON, A. DOBLE, D. N. MIDDLEMISS, J. SHAW & M. TURNBULL. 1980. (−)Baclofen decreases neurotransmitter release in the mammalian CNS by an action at a novel GABA receptor. Nature **283:** 92–94.
29. HILL, D. R. & N. G. BOWERY. 1981. ^{3}H-Baclofen and ^{3}H-GABA bind to bicuculline-insensitive GABA$_B$ sites in rat brain. Nature **290:** 149–152.
30. ONG, J., N. L. HARRISON, R. G. HALL, J. L. BARKER, G. A. R. JOHNSTON & D. I. B. KERR. 1990. 3-Aminopropanephosphinic acid is a potent agonist at peripheral and central presynaptic GABA$_B$ receptors. Brain Res. **526:** 138–142.
31. KERR, D. I. B., J. ONG, R. H. PRAGER, B. D. GYNTHER & D. R. CURTIS. 1987. Phaclofen: A peripheral and entral baclofen antagonist. Brain Res. **405:** 150–154.

32. Curtis, D. R., B. D. Gynther, D. T. Beattie, D. I. B. Kerr & R. H. Prager. 1988. Baclofen antagonism by 2-hydroxy-saclofen in the cat spinal cord. Neurosci. Lett. **92:** 97–101.

33. Olpe, H. R., G. Karlsson, M. F. Pozza, F. Brugger, M. Steinmann, H. V. Riezen, G. Fagg, R. G. Hall, W. Froestl & H. Bittiger. 1990. CGP 35348: A centrally active blocker of $GABA_B$ receptors. Eur. J. Pharmacol. **187:** 27–38.

34. Waldmeier, P. C., P. Wicki, H. Bittiger & P. A. Baumann. 1992. The regulation of GABA release by $GABA_B$ autoreceptors: Some intriguing findings and their possible implications. Pharmacol. Res. Commun. **2:** 1–2.

35. Hill, D. R., N. G. Bowery & A. L. Hudson. 1984. Inhibition of $GABA_B$ receptor binding by guanyl nucleotides. J. Neurochem. **42:** 652–657.

36. Asano, T., M. Ui & N. Ogasawara. 1985. Prevention of the agonist binding to γ-aminobutyric acid$_B$ receptors by guanine nucleotides and islet-activating protein, pertussis toxin, in bovine cerebral cortex. J. Biol. Chem. **260:** 12653–12658.

37. Ohmori, Y., M. Hirouchi, J. Taguchi & K. Kuriyama. 1990. Functional coupling of the γ-aminobutyric acid$_B$ receptor with calcium ion channel and GTP-binding protein and its alteration following solubilization of the γ-aminobutyric acid$_B$ receptor. J. Neurochem. **54:** 80–85.

38. Wojcik, W. J. & N. H. Neff. 1984. γ-Aminobutyric acid$_B$ receptors are negatively coupled to adenylate cyclase in brain, and in the cerebellum these receptors may be associated with granule cells. Mol. Pharmacol. **25:** 24–28.

39. Nishikawa, M. & K. Kuriyama. 1989. Functional coupling of cerebral γ-aminobutyric acid $(GABA)_B$ receptor with adenylate cyclase: Effect of phaclofen. Neurochem. Int. **14:** 85–90.

40. Xu, J. & W. J. Wojcik. 1986. Gamma aminobutyric acid$_B$ receptor-mediated inhibition of adenylate cyclase in cultured cerebellar granule cells: Blockade by islet-activating protein. J. Pharmacol. Exp. Ther. **239:** 568–573.

41. Karbon, E. W. & S. J. Enna. 1985. Characterization of the relationship between γ-aminobutyric acid$_B$ agonist and transmitter-coupled cyclic nucleotide-generating systems in rat brain. Mol. Pharmacol. **27:** 53–59.

42. Wojcik, W. J., M. Ulivi, X. Paez & E. Costa. 1989. Islet-activating protein inhibits the β-adrenergic receptor facilitation elicited by γ-aminobutyric acid$_B$ receptors. J. Neurochem. **53:** 753–758.

43. Scherer, R. W., J. W. Ferkany, E. W. Karbon & S. J. Enna. 1989. γ-Aminobutyric acid$_B$ receptor activation modifies agonist binding to β-adrenergic receptors in rat brain cerebral cortex. J. Neurochem. **53:** 989–991.

44. Crawford, M. L. A. & J. M. Young. 1988. $GABA_B$ receptor-mediated inhibition of histamine H1-receptor-induced inositol phosphate formation in slices of rat cerebral cortex. J. Neurochem. **51:** 1441–1447.

45. Godfrey, P. P., D. G. Grahame-Smith & J. A. Gray. 1988. $GABA_B$ receptor activation inhibits 5-hydroxytryptamine-stimulated inositol phospholipid turnover in mouse cerebral cortex. Eur. J. Pharmacol. **152:** 185–188.

46. Ohmori, Y. & K. Kuriyama. 1989. Negative coupling of γ-aminobutyric acid $(GABA)_B$ receptor with phosphatidylinositol turnover in the brain. Neurochem. Int. **15:** 359–363.

47. Holz, G. G., S. G. Rane & K. Dunlap. 1986. GTP-binding proteins mediate transmitter inhibition of voltage-dependent calcium channels. Nature **319:** 670–672.

48. Newberry, N. R. & R. A. Nicoll. 1984. Direct hyperpolarizing action of baclofen on hippocampal pyramidal cells. Nature **308:** 450–452.

49. Dutar, P. & R. A. Nicoll. 1988. A physiological role for $GABA_B$ receptors in the central nervous system. Nature **332:** 156–158.

50. Kerr, D. I. B., J. Ong, G. A. R. Johnston, J. Abbenante & R. H. Prager. 1988. 2-Hydroxy-saclofen: An improved antagonist at central and peripheral $GABA_B$ receptors. Neurosci. Lett. **92:** 92–96.

51. Kuriyama, K., H. Mizutani & H. Nakayasu. 1992. Purification and identification of 61 kilodalton GABA (γ-aminobutyric acid)$_B$ receptor from bovine brain. Mol. Neuropharmacol. **2:** 155–157.

52. Nakayasu, H., H. Mizutani, K. Hanai, H. Kimura & K. Kuriyama. 1992. Monoclonal

antibody to GABA binding protein, a possible $GABA_B$ receptor. Biochem. Biophys. Res. Commun. **182:** 722–726.

53. NAKAYASU, H., M. NISHIKAWA, H. MIZUTANI, H. KIMURA & K. KURIYAMA. 1993. Immunoaffinity purification and characterization of γ-aminobutyric acid $(GABA)_B$ receptor from bovine cerebral cortex. J. Biol. Chem. **268:** 8658–8664.

54. BOWERY, N. G. 1993. $GABA_B$ receptor pharmacology. Annu. Rev. Pharmacol. Toxicol. **33:** 109–147.

55. MINTZ, I. M. & B. P. BEAN. 1993. $GABA_B$ receptor inhibition of P-type Ca^{2+} channels in central neurons. Neuron **10:** 889–898.

56. SCANZIANI, M., M. CAPOGNA, B. H. GAHWILER & S. M. THOMPSON. 1992. Presynaptic inhibition of miniature exitatory synaptic currents by baclofen and adenosine in the hypocampus. Neuron **9:** 919–927.

57. THOMPSON, S. M., M. CAPOGNA & M. SCANZIANI. 1993. Presynaptic inhibition in the hippocampus. Trends Neurosci. **16:** 222–227.

58. PFRIEGER, F. W., K. GOTTMANN & H. LUX. 1994. Kinetics of $GABA_B$ receptor-mediated inhibition of calcium currents and exitatory synaptic transmission in hippocampal neurons in vitro. Neuron **12:** 97–107.

Subject Index

Index of Contributors